Graduate Texts in Physics

Graduate Texts in Physics

Graduate Texts in Physics publishes core learning/teaching material for graduate- and advanced-level undergraduate courses on topics of current and emerging fields within physics, both pure and applied. These textbooks serve students at the MS- or PhD-level and their instructors as comprehensive sources of principles, definitions, derivations, experiments and applications (as relevant) for their mastery and teaching, respectively. International in scope and relevance, the textbooks correspond to course syllabi sufficiently to serve as required reading. Their didactic style, comprehensiveness and coverage of fundamental material also make them suitable as introductions or references for scientists entering, or requiring timely knowledge of, a research field.

More information about this series at http://www.springer.com/series/8431

Rainer Dick

Advanced Quantum Mechanics

Materials and Photons

Second Edition

Springer

Rainer Dick
Department of Physics and Engineering Physics
University of Saskatchewan
Saskatoon, Saskatchewan
Canada

ISSN 1868-4513 ISSN 1868-4521 (electronic)
Graduate Texts in Physics
ISBN 978-3-319-79826-4 ISBN 978-3-319-25675-7 (eBook)
DOI 10.1007/978-3-319-25675-7

Softcover reprint of the hardcover 2nd edition 2016

Printed on acid-free paper

This Springer imprint is published by Springer Nature
The registered company is Springer International Publishing AG Switzerland

Preface to the Second Edition

The second edition features 62 additional end of chapter problems and many sections were edited for clarity and improvement of presentation. Furthermore, the chapter on Klein-Gordon and Dirac fields has been expanded and split into Chapter 21 on relativistic quantum fields and Chapter 22 on applications of quantum electrodynamics. This was motivated by the renewed interest in the notions and techniques of relativistic quantum theory due to their increasing relevance for materials research. Of course, relativistic quantum theory has always been an important tool in subatomic physics and in quantum optics since the dynamics of photons or high energy particles is expressed in terms of relativistic quantum fields. Furthermore, relativistic quantum mechanics has also always been important for chemistry and condensed matter physics through the impact of relativistic corrections to the Schrödinger equation, primarily through the Pauli term and through spin-orbit couplings. These terms usually dominate couplings to magnetic fields and relativistic corrections to energy levels in materials, and spin-orbit couplings became even more prominent due to their role in manipulating spins in materials through electric fields. Relativistic quantum mechanics has therefore always played an important foundational role throughout the physical sciences and engineering.

However, we have even seen discussions of fully quasirelativistic wave equations in materials research in recent years. This development is driven by discoveries of materials like Graphene or Dirac semimetals, which exhibit low energy effective Lorentz symmetries in sectors of momentum space. In these cases c and m become effective low energy parameters which parametrize quasirelativistic cones or hyperboloids in regions of $(E, \boldsymbol{k})$ space. As a consequence, materials researchers now do not only deal with Pauli and spin-orbit terms, but with representations of γ matrices and solutions of Dirac equations in various dimensions.

To prepare graduate students in the physical sciences and engineering better for the increasing number of applications of (quasi-)relativistic quantum physics, Section 21.5 on the non-relativistic limit of the Dirac equation now also contains a detailed discussion of the Foldy-Wouthuysen transformation including a derivation of the general spin-orbit coupling term and a discussion of the origin of Rashba terms, and the Section 21.6 on quantization of the Maxwell field in Lorentz gauge has been added. The discussion of applications of quantum electrodynamics now also includes the new Section 22.2 on electron-nucleus scattering. Finally, the new Appendix I discusses the transformation properties of scalars, spinors and gauge fields under parity or time reversal.

Saskatoon, SK, Canada Rainer Dick

Preface to the First Edition

Quantum mechanics was invented in an era of intense and seminal scientific research between 1900 and 1928 (and in many regards continues to be developed and expanded) because neither the properties of atoms and electrons, nor the spectrum of radiation from heat sources could be explained by the classical theories of mechanics, electrodynamics and thermodynamics. It was a major intellectual achievement and a breakthrough of curiosity driven fundamental research which formed quantum theory into one of the pillars of our present understanding of the fundamental laws of nature. The properties and behavior of every elementary particle is governed by the laws of quantum theory. However, the rule of quantum mechanics is not limited to atomic and subatomic scales, but also affects macroscopic systems in a direct and profound manner. The electric and thermal conductivity properties of materials are determined by quantum effects, and the electromagnetic spectrum emitted by a star is primarily determined by the quantum properties of photons. It is therefore not surprising that quantum mechanics permeates all areas of research in advanced modern physics and materials science, and training in quantum mechanics plays a prominent role in the curriculum of every major physics or chemistry department.

The ubiquity of quantum effects in materials implies that quantum mechanics also evolved into a major tool for advanced technological research. The construction of the first nuclear reactor in Chicago in 1942 and the development of nuclear technology could not have happened without a proper understanding of the quantum properties of particles and nuclei. However, the real breakthrough for a wide recognition of the relevance of quantum effects in technology occurred with the invention of the transistor in 1948 and the ensuing rapid development of semiconductor electronics. This proved once and for all the importance of quantum mechanics for the applied sciences and engineering, only 22 years after publication of the Schrödinger equation! Electronic devices like transistors rely heavily on the quantum mechanical emergence of energy bands in materials, which can be considered as a consequence of combination of many atomic orbitals or as a consequence of delocalized electron states probing a lattice structure. Today the rapid developments of spintronics, photonics and nanotechnology provide continuing testimony to the technological relevance of quantum mechanics.

As a consequence, every physicist, chemist and electrical engineer nowadays has to learn aspects of quantum mechanics, and we are witnessing a time when also mechanical and aerospace engineers are advised to take at least a 2nd year course, due to the importance of quantum mechanics for elasticity and stability properties of materials. Furthermore, quantum information appears to become increasingly relevant for computer science and information technology, and a whole new area of quantum technology will likely follow in the wake of this development. Therefore it seems safe to posit that within the next two generations, 2nd and 3rd year quantum mechanics courses will become as abundant and important in the curricula of science and engineering colleges as first and second year calculus courses.

Quantum mechanics continues to play a dominant role in particle physics and atomic physics – after all, the Standard Model of particle physics is a quantum theory, and the spectra and stability of atoms cannot be explained without quantum mechanics. However, most scientists and engineers use quantum mechanics in advanced materials research. Furthermore, the dominant interaction mechanisms in materials (beyond the nuclear level) are electromagnetic, and many experimental techniques in materials science are based on photon probes. The introduction to quantum mechanics in the present book takes this into account by including aspects of condensed matter theory and the theory of photons at earlier stages and to a larger extent than other quantum mechanics texts. Quantum properties of materials provide neat and very interesting illustrations of time-independent and time-dependent perturbation theory, and many students are better motivated to master the concepts of quantum mechanics when they are aware of the direct relevance for modern technology. A focus on the quantum mechanics of photons and materials is also perfectly suited to prepare students for future developments in quantum information technology, where entanglement of photons or spins, decoherence, and time evolution operators will be key concepts.

Other novel features of the discussion of quantum mechanics in this book concern attention to relevant mathematical aspects which otherwise can only be found in journal articles or mathematical monographs. Special appendices include a mathematically rigorous discussion of the completeness of Sturm-Liouville eigenfunctions in one spatial dimension, an evaluation of the Baker-Campbell-Hausdorff formula to higher orders, and a discussion of logarithms of matrices. Quantum mechanics has an extremely rich and beautiful mathematical structure. The growing prominence of quantum mechanics in the applied sciences and engineering has already reinvigorated increased research efforts on its mathematical aspects. Both students who study quantum mechanics for the sake of its numerous applications, as well as mathematically inclined students with a primary interest in the formal structure of the theory should therefore find this book interesting.

This book emerged from a quantum mechanics course which I had introduced at the University of Saskatchewan in 2001. It should be suitable both for advanced undergraduate and introductory graduate courses on the subject. To make advanced quantum mechanics accessible to wider audiences which might not have been exposed to standard second and third year courses on atomic physics, analytical mechanics, and electrodynamics, important aspects of these topics are briefly, but concisely introduced in special chapters and appendices. The success and relevance of quantum mechanics has reached far beyond the realms of physics research, and physicists have a duty to disseminate the knowledge of quantum mechanics as widely as possible.

Saskatoon, SK, Canada Rainer Dick

To the Students

Congratulations! You have reached a stage in your studies where the topics of your inquiry become ever more interesting and more relevant for modern research in basic science and technology.

Together with your professors, I will have the privilege to accompany you along the exciting road of your own discovery of the bizarre and beautiful world of quantum mechanics. I will aspire to share my own excitement that I continue to feel for the subject and for science in general.

You will be introduced to many analytical and technical skills that are used in everyday applications of quantum mechanics. These skills are essential in virtually every aspect of modern research. A proper understanding of a materials science measurement at a synchrotron requires a proper understanding of photons and quantum mechanical scattering, just like manipulation of qubits in quantum information research requires a proper understanding of spin and photons and entangled quantum states. Quantum mechanics is ubiquitous in modern research. It governs the formation of microfractures in materials, the conversion of light into chemical energy in chlorophyll or into electric impulses in our eyes, and the creation of particles at the Large Hadron Collider.

Technical mastery of the subject is of utmost importance for understanding quantum mechanics. Trying to decipher or apply quantum mechanics without knowing how it really works in the calculation of wave functions, energy levels, and cross sections is just idle talk, and always prone for misconceptions. Therefore we will go through a great many technicalities and calculations, because you and I (and your professor!) have a common goal: You should become an expert in quantum mechanics.

However, there is also another message in this book. The apparently exotic world of quantum mechanics is *our* world. Our bodies and all the world around us is built on quantum effects and ruled by quantum mechanics. It is not apparent and only visible to the *cognoscenti*. Therefore we have developed a mode of thought and explanation of the world that is based on classical pictures – mostly waves and particles in mechanical interaction. This mode of thought was amended by the notions of gravitational and electromagnetic forces, thus culminating in a powerful tool called classical physics. However, by 1900 those who were paying attention had caught enough glimpses of the underlying non-classical world to embark on the exciting journey of discovering quantum mechanics. Indeed, every single atom in your body is ruled by the laws of quantum mechanics, and could not even exist as a classical particle. The electrons that provide the light for your long nights of studying generate this light in stochastic quantum leaps from a state of a single electron to a state of an electron and a photon. And maybe the most striking example of all: There is *absolutely nothing classical* in the sunlight that provides the energy for all life on Earth.

Quantum theory is not a young theory any more. The scientific foundations of the subject were developed over half a century between 1900 and 1949, and many of the mathematical foundations were even developed in the 19th century. The steepest ascent in the development of quantum theory appeared between 1924 and 1928, when matrix mechanics, Schrödinger's equation, the Dirac equation and field quantization were invented. I have included numerous references to original papers from this period, not to ask you to read all those papers – after all, the primary purpose of a textbook is to put major achievements into context, provide an introductory overview at an appropriate level, and replace often indirect and circuitous original derivations with simpler explanations – but to honour the people who brought the then nascent theory to maturity. Quantum theory is an extremely well established and developed theory now, which has proven itself on numerous occasions. However, we still continue to improve our collective understanding of the theory and its wide ranging applications, and we test its predictions and its probabilistic interpretation with ever increasing accuracy. The implications and applications of quantum mechanics are limitless, and we are witnessing a time when many technologies have reached their "quantum limit", which is a misnomer for the fact that any methods of classical physics are just useless in trying to describe or predict the behavior of atomic scale devices. It is a "limit" for those who do not want to learn quantum physics. For you, it holds the promise of excitement and opportunity if you are prepared to work hard and if you can understand the calculations.

Quantum mechanics combines power and beauty in a way that even supersedes advanced analytical mechanics and electrodynamics. Quantum mechanics is universal and therefore incredibly versatile, and if you have a sense for mathematical beauty: The structure of quantum mechanics is breathtaking, indeed.

I sincerely hope that reading this book will be an enjoyable and exciting experience for you.

To the Instructor

Dear Colleague,

As professors of quantum mechanics courses, we enjoy the privilege of teaching one of the most exciting subjects in the world. However, we often have to do this with fewer lecture hours than were available for the subject in the past, when at the same time we should include more material to prepare students for research or modern applications of quantum mechanics. Furthermore, students have become more mobile between universities (which is good) and between academic programs (which can have positive and negative implications). Therefore we are facing the task to teach an advanced subject to an increasingly heterogeneous student body with very different levels of preparation. Nowadays the audience in a fourth year undergraduate or beginning graduate course often includes students who have not gone through a course on Lagrangian mechanics, or have not seen the covariant formulation of electrodynamics in their electromagnetism courses. I deal with this problem by including one special lecture on each topic in my quantum mechanics course, and this is what Appendices A and B are for. I have also tried to be as inclusive as possible without sacrificing content or level of understanding by starting at a level that would correspond to an advanced second year Modern Physics or Quantum Chemistry course and then follow a steeply ascending route that takes the students all the way from Planck's law to the photon scattering tensor.

The selection and arrangement of topics in this book is determined by the desire to develop an advanced undergraduate and introductory graduate level course that is useful to as many students as possible, in the sense of giving them a head start into major current research areas or modern applications of quantum mechanics without neglecting the necessary foundational training.

There is a core of knowledge that every student is expected to know by heart after having taken a course in quantum mechanics. Students must know the Schrödinger equation. They must know how to solve the harmonic oscillator and the Coulomb problem, and they must know how to extract information from the wave function. They should also be able to apply basic perturbation theory, and they should understand that a wave function $\langle \boldsymbol{x}|\psi(t)\rangle$ is only one particular representation of a quantum state $|\psi(t)\rangle$.

In a North American physics program, students would traditionally learn all these subjects in a 300-level Quantum Mechanics course. Here these subjects are discussed in Chapters 1–7 and 9. This allows the instructor to use this book also in 300-level courses or introduce those chapters in a 400-level or graduate course if needed. Depending on their specialization, there will be an increasing number of students from many different science and engineering programs who will have to learn these subjects at M.Sc. or beginning Ph.D. level before they can learn about photon scattering or quantum effects in materials, and catering to these students will also become an increasingly important part of the mandate of physics departments. Including chapters 1–7 and 9 with the book is part of the philosophy of being as inclusive as possible to disseminate knowledge in advanced quantum mechanics as widely as possible.

Additional training in quantum mechanics in the past traditionally focused on atomic and nuclear physics applications, and these are still very important topics in fundamental and applied science. However, a vast number of our current students in quantum mechanics will apply the subject in materials science in a broad sense encompassing condensed matter physics, chemistry and engineering. For these students it is beneficial to see Bloch's theorem, Wannier states, and basics of the theory of covalent bonding embedded with their quantum mechanics course. Another important topic for these students is quantization of the Schrödinger field. Indeed, it is also useful for students in nuclear and particle physics to learn quantization of the Schrödinger field because it makes quantization of gauge fields and relativistic matter fields so much easier if they know quantum field theory in the non-relativistic setting.

Furthermore, many of our current students will use or manipulate photon probes in their future graduate and professional work. A proper discussion of photon-matter interactions is therefore also important for a modern quantum mechanics course.

This should include minimal coupling, quantization of the Maxwell field, and applications of time-dependent perturbation theory for photon absorption, emission and scattering.

Students should also know the Klein-Gordon and Dirac equations after completion of their course, not only to understand that Schrödinger's equation is not the final answer in terms of wave equations for matter particles, but to understand the nature of relativistic corrections like the Pauli term or spin-orbit coupling.

The scattering matrix is introduced as early as possible in terms of matrix elements of the time evolution operator on states in the interaction picture, $S_{fi}(t,t') = \langle f|U_D(t,t')|i\rangle$, cf. equation (13.26). This representation of the scattering matrix appears so naturally in ordinary time-dependent perturbation theory that it makes no sense to defer the notion of an S-matrix to the discussion of scattering in quantum field theory with two or more particles in the initial state. It actually mystifies the scattering matrix to defer its discussion until field quantization has been introduced. On the other hand, introducing the scattering matrix even earlier in the framework of scattering off static potentials is counterproductive, because its natural and useful definition as matrix elements of a time evolution operator cannot properly be introduced at that level, and the notion of the scattering matrix does not really help with the calculation of cross sections for scattering off static potentials.

I have also emphasized the discussion of the various roles of transition matrix elements depending on whether the initial or final states are discrete or continuous. It helps students to understand transition probabilities, decay rates, absorption cross sections and scattering cross sections if the discussion of these concepts is integrated in one chapter, cf. Chapter 13. Furthermore, I have put an emphasis on canonical field quantization. Path integrals provide a very elegant description for free-free scattering, but bound states and energy levels, and basic many-particle quantum phenomena like exchange holes are very efficiently described in the canonical formalism. Feynman rules also appear more intuitive in the canonical formalism of explicit particle creation and annihilation.

The core advanced topics in quantum mechanics that an instructor might want to cover in a traditional 400-level or introductory graduate course are included with Chapters 8, 11–13, 15–18, and 21. However, instructors of a more inclusive course for general science and engineering students should include materials from Chapters 1–7 and 9, as appropriate.

The direct integration of training in quantum mechanics with the foundations of condensed matter physics, field quantization, and quantum optics is very important for the advancement of science and technology. I hope that this book will help to achieve that goal. I would greatly appreciate your comments and criticism. Please send them to rainer.dick@usask.ca.

Contents

Chapter 1
The Need for Quantum Mechanics

1.1 Electromagnetic spectra and evidence for discrete energy levels

Quantum mechanics was initially invented because classical mechanics, thermodynamics and electrodynamics provided no means to explain the properties of atoms, electrons, and electromagnetic radiation. Furthermore, it became clear after the introduction of Schrödinger's equation and the quantization of Maxwell's equations that we cannot explain *any* physical property of matter and radiation without the use of quantum theory. We will see a lot of evidence for this in the following chapters. However, in the present chapter we will briefly and selectively review the early experimental observations and discoveries which led to the development of quantum mechanics over a period of intense research between 1900 and 1928.

The first evidence that classical physics was incomplete appeared in unexpected properties of electromagnetic spectra. Thin gases of atoms or molecules emit line spectra which contradict the fact that a classical system of electric charges can oscillate at any frequency, and therefore can emit radiation of any frequency. This was a major scientific puzzle from the 1850s until the inception of the Schrödinger equation in 1926.

Contrary to a thin gas, a hot body does emit a continuous spectrum, but even those spectra were still puzzling because the shape of heat radiation spectra could not be explained by classical thermodynamics and electrodynamics. In fact, classical physics provided no means at all to predict any sensible shape for the spectrum of a heat source! But at last, hot bodies do emit a continuous spectrum and therefore, from a classical point of view, their spectra are not quite as strange and unexpected as line spectra. It is therefore not surprising that the first real clues for a solution to the puzzles of electromagnetic spectra emerged when Max Planck figured out a way to calculate the spectra of heat sources under the simple, but classically

R. Dick, *Advanced Quantum Mechanics*, Graduate Texts in Physics,
DOI 10.1007/978-3-319-25675-7_1

extremely counterintuitive assumption that the energy in heat radiation of frequency f is *quantized* in integer multiples of a minimal energy quantum hf,

$$E = nhf, \quad n \in \mathbb{N}. \tag{1.1}$$

The constant h that Planck had introduced to formulate this equation became known as Planck's constant and it could be measured from the shape of heat radiation spectra. A modern value is $h = 6.626 \times 10^{-34}\,\mathrm{J \cdot s} = 4.136 \times 10^{-15}\,\mathrm{eV \cdot s}$.

We will review the puzzle of heat radiation and Planck's solution in the next section, because Planck's calculation is instructive and important for the understanding of incandescent light sources and it illustrates in a simple way how quantization of energy levels yields results which are radically different from predictions of classical physics.

Albert Einstein then pointed out that equation (1.1) also explains the photoelectric effect. He also proposed that Planck's quantization condition is not a property of any particular mechanism for generation of electromagnetic waves, but an intrinsic property of electromagnetic waves. However, once equation (1.1) is accepted as an intrinsic property of electromagnetic waves, it is a small step to make the connection with line spectra of atoms and molecules and conclude that these line spectra imply existence of discrete energy levels in atoms and molecules. Somehow atoms and molecules seem to be able to emit radiation only by jumping from one discrete energy state into a lower discrete energy state. This line of reasoning, combined with classical dynamics between electrons and nuclei in atoms then naturally leads to the Bohr-Sommerfeld theory of atomic structure. This became known as *old quantum theory*.

Apparently, the property which underlies both the heat radiation puzzle and the puzzle of line spectra is discreteness of energy levels in atoms, molecules, and electromagnetic radiation. Therefore, *one major motivation for the development of quantum mechanics was to explain discrete energy levels in atoms, molecules, and electromagnetic radiation.*

It was Schrödinger's merit to find an explanation for the discreteness of energy levels in atoms and molecules through his wave equation[1] ($\hbar \equiv h/2\pi$)

$$\mathrm{i}\hbar\frac{\partial}{\partial t}\psi(\boldsymbol{x},t) = -\frac{\hbar^2}{2m}\Delta\psi(\boldsymbol{x},t) + V(\boldsymbol{x})\psi(\boldsymbol{x},t). \tag{1.2}$$

A large part of this book will be dedicated to the discussion of Schrödinger's equation. An intuitive motivation for this equation will be given in Section 1.6.

Ironically, the fundamental energy quantization condition (1.1) for electromagnetic waves, which precedes the realization of discrete energy levels in atoms and molecules, cannot be derived by solving a wave equation, but emerges from the quantization of Maxwell's equations. This is at the heart of understanding photons

[1] E. Schrödinger, Annalen Phys. 386, 109 (1926).

and the quantum theory of electromagnetic waves. We will revisit this issue in Chapter 18. However, we can and will discuss already now the early quantum theory of the photon and what it means for the interpretation of spectra from incandescent sources.

1.2 Blackbody radiation and Planck's law

Historically, Planck's deciphering of the spectra of incandescent heat and light sources played a key role for the development of quantum mechanics, because it included the first proposal of energy quanta, and it implied that line spectra are a manifestation of energy quantization in atoms and molecules. Planck's radiation law is also extremely important in astrophysics and in the technology of heat and light sources.

Generically, the heat radiation from an incandescent source is contaminated with radiation reflected from the source. Pure heat radiation can therefore only be observed from a non-reflecting, i.e. perfectly black body. Hence the name blackbody radiation for pure heat radiation. Physicists in the late 19th century recognized that the best experimental realization of a black body is a hole in a cavity wall. If the cavity is kept at temperature T, the hole will emit perfect heat radiation without contamination from any reflected radiation.

Suppose we have a heat radiation source (or thermal emitter) at temperature T. The power per area radiated from a thermal emitter at temperature T is denoted as its *exitance* (or *emittance*) $e(T)$. In the blackbody experiments $e(T) \cdot A$ is the energy per time leaking through a hole of area A in a cavity wall.

To calculate $e(T)$ as a function of the temperature T, as a first step we need to find out how it is related to the density $u(T)$ of energy stored in the heat radiation. One half of the radiation will have a velocity component towards the hole, because all the radiation which moves under an angle $\vartheta \leq \pi/2$ relative to the axis going through the hole will have a velocity component $v(\vartheta) = c\cos\vartheta$ in the direction of the hole. To find out the average speed v of the radiation in the direction of the hole, we have to average $c\cos\vartheta$ over the solid angle $\Omega = 2\pi$ sr of the forward direction $0 \leq \varphi \leq 2\pi, 0 \leq \vartheta \leq \pi/2$:

$$v = \frac{c}{2\pi}\int_0^{2\pi} d\varphi \int_0^{\pi/2} d\vartheta\, \sin\vartheta \cos\vartheta = \frac{c}{2}.$$

The effective energy current density towards the hole is energy density moving in forward direction $\times$ average speed in forward direction:

$$\frac{u(T)}{2}\frac{c}{2} = u(T)\frac{c}{4},$$

and during the time t an amount of energy

$$E = u(T)\frac{c}{4}tA$$

will escape through the hole. Therefore the emitted power per area $E/(tA) = e(T)$ is

$$e(T) = u(T)\frac{c}{4}. \tag{1.3}$$

However, Planck's radiation law is concerned with the *spectral exitance* $e(f,T)$, which is defined in such a way that

$$e_{[f_1,f_2]}(T) = \int_{f_1}^{f_2} df\, e(f,T)$$

is the power per area emitted in radiation with frequencies $f_1 \leq f \leq f_2$. In particular, the total exitance is

$$e(T) = e_{[0,\infty]}(T) = \int_0^\infty df\, e(f,T).$$

Operationally, the spectral exitance is the power per area emitted with frequencies $f \leq f' \leq f + \Delta f$, and normalized by the width Δf of the frequency interval,

$$e(f,T) = \lim_{\Delta f\to 0}\frac{e_{[f,f+\Delta f]}(T)}{\Delta f} = \lim_{\Delta f\to 0}\frac{e_{[0,f+\Delta f]} - e_{[0,f]}(T)}{\Delta f} = \frac{\partial}{\partial f}e_{[0,f]}(T).$$

The spectral exitance $e(f,T)$ can also be denoted as the *emitted power per area and per unit of frequency* or as the *spectral exitance in the frequency scale*.

The spectral energy density $u(f,T)$ is defined in the same way. If we measure the energy density $u_{[f,f+\Delta f]}(T)$ in radiation with frequency between f and $f + \Delta f$, then the energy per volume and per unit of frequency (i.e. the spectral energy density in the frequency scale) is

$$u(f,T) = \lim_{\Delta f\to 0}\frac{u_{[f,f+\Delta f]}(T)}{\Delta f} = \frac{\partial}{\partial f}u_{[0,f]}(T), \tag{1.4}$$

and the total energy density in radiation is

$$u(T) = \int_0^\infty df\, u(f,T).$$

The equation $e(T) = u(T)c/4$ also applies separately in each frequency interval $[f, f + \Delta f]$, and therefore must also hold for the corresponding spectral densities,

$$e(f,T) = u(f,T)\frac{c}{4}. \tag{1.5}$$

The following facts were known before Planck's work in 1900.

- The prediction from classical thermodynamics for the spectral exitance $e(f, T)$ (Rayleigh-Jeans law) was wrong, and actually non-sensible!
- The exitance $e(T)$ satisfies Stefan's law (Stefan, 1879; Boltzmann, 1884)

$$e(T) = \sigma T^4,$$

with the Stefan-Boltzmann constant

$$\sigma = 5.6704 \times 10^{-8} \, \frac{\text{W}}{\text{m}^2\text{K}^4}.$$

- The spectral exitance $e(\lambda, T) = e(f, T)\Big|_{f=c/\lambda} \cdot c/\lambda^2$ per unit of wavelength (i.e. *the spectral exitance in the wavelength scale*) has a maximum at a wavelength

$$\lambda_{max} \cdot T = 2.898 \times 10^{-3} \, \text{m} \cdot \text{K} = 2898 \, \mu\text{m} \cdot \text{K}.$$

This is Wien's displacement law (Wien, 1893).

The puzzle was to explain the observed curves $e(f, T)$ and to explain why classical thermodynamics had failed. We will explore these questions through a calculation of the spectral energy density $u(f, T)$. Equation (1.5) then also yields $e(f, T)$.

The key observation for the calculation of $u(f, T)$ is to realize that $u(f, T)$ can be split into two factors. If we want to know the radiation energy density $u_{[f,f+df]} = u(f, T)df$ in the small frequency interval $[f, f + df]$, then we can first ask ourselves how many different electromagnetic oscillation modes per volume, $\varrho(f)df$, exist in that frequency interval. Each oscillation mode will then contribute an energy $\langle E \rangle(f, T)$ to the radiation energy density, where $\langle E \rangle(f, T)$ is the expectation value of energy in an electromagnetic oscillation mode of frequency f at temperature T,

$$u(f, T)df = \varrho(f)df \langle E \rangle(f, T).$$

The spectral energy density $u(f, T)$ can therefore be calculated in two steps:

1. Calculate the number $\varrho(f)$ of oscillation modes per volume and per unit of frequency ("counting of oscillation modes").
2. Calculate the mean energy $\langle E \rangle(f, T)$ in an oscillation of frequency f at temperature T.

The results can then be combined to yield the spectral energy density $u(f, T) = \varrho(f)\langle E \rangle(f, T)$.

The number of electromagnetic oscillation modes per volume and per unit of frequency is an important quantity in quantum mechanics and will be calculated explicitly in Chapter 12, with the result

$$\varrho(f) = \frac{8\pi f^2}{c^3}. \tag{1.6}$$

The corresponding density of oscillation modes in the wavelength scale is

$$\varrho(\lambda) = \varrho(f)\Big|_{f=c/\lambda} \cdot \frac{c}{\lambda^2} = \frac{8\pi}{\lambda^4}.$$

Statistical physics predicts that the probability $P_T(E)$ to find an oscillation of energy E in a system at temperature T should be exponentially suppressed,

$$P_T(E) = \frac{1}{k_B T}\exp\!\left(-\frac{E}{k_B T}\right). \tag{1.7}$$

The possible values of E are not restricted in classical physics, but can vary continuously between $0 \leq E < \infty$. For example, for any classical oscillation with fixed frequency f, continually increasing the amplitude yields a continuous increase in energy. The mean energy of an oscillation at temperature T according to classical thermodynamics is therefore

$$\langle E\rangle\Big|_{classical} = \int_0^\infty dE\, E P_T(E) = \int_0^\infty dE\,\frac{E}{k_B T}\exp\!\left(-\frac{E}{k_B T}\right) = k_B T. \tag{1.8}$$

Therefore the spectral energy density in blackbody radiation and the corresponding spectral exitance according to classical thermodynamics should be

$$u(f,t) = \varrho(f)k_B T = \frac{8\pi f^2}{c^3}k_B T, \quad e(f,T) = u(f,T)\frac{c}{4} = \frac{2\pi f^2}{c^2}k_B T,$$

but this is obviously nonsensical: it would predict that every heat source should emit a diverging amount of energy at high frequencies/short wavelengths! This is the *ultraviolet catastrophe* of the Rayleigh-Jeans law.

Max Planck observed in 1900 that he could derive an equation which matches the spectra of heat sources perfectly if he assumes that the energy in electromagnetic waves of frequency f is quantized in multiples of the frequency,

$$E = nhf = n\frac{hc}{\lambda}, \quad n \in \mathbb{N}.$$

The exponential suppression of high energy oscillations then reads

$$P_T(E) = P_T(n) \propto \exp\!\left(-\frac{nhf}{k_B T}\right),$$

but due to the discreteness of the *energy quanta hf*, the normalized probabilities are now

$$\begin{aligned} P_T(E) = P_T(n) &= \left[1 - \exp\!\left(-\frac{hf}{k_B T}\right)\right]\exp\!\left(-\frac{nhf}{k_B T}\right) \\ &= \exp\!\left(-n\frac{hf}{k_B T}\right) - \exp\!\left(-(n+1)\frac{hf}{k_B T}\right), \end{aligned}$$

such that $\sum_{n=0}^{\infty} P_T(n) = 1$.

The resulting mean energy per oscillation mode is

$$\begin{aligned}
\langle E\rangle &= \sum_{n=0}^{\infty} nhfP_T(n) \\
&= \sum_{n=0}^{\infty} nhf \exp\left(-n\frac{hf}{k_BT}\right) - \sum_{n=0}^{\infty} nhf \exp\left(-(n+1)\frac{hf}{k_BT}\right) \\
&= \sum_{n=0}^{\infty} nhf \exp\left(-n\frac{hf}{k_BT}\right) - \sum_{n=0}^{\infty} (n+1)hf \exp\left(-(n+1)\frac{hf}{k_BT}\right) \\
&\quad + hf\sum_{n=0}^{\infty} \exp\left(-(n+1)\frac{hf}{k_BT}\right)
\end{aligned}$$

The first two sums cancel, and the last term yields the mean energy in an electromagnetic wave of frequency f at temperature T as

$$\langle E\rangle(f,T) = hf\frac{\exp\left(-\frac{hf}{k_BT}\right)}{1-\exp\left(-\frac{hf}{k_BT}\right)} = \frac{hf}{\exp\left(\frac{hf}{k_BT}\right)-1}. \tag{1.9}$$

Combination with $\varrho(f)$ from equation (1.6) yields Planck's formulas for the spectral energy density and spectral exitance in heat radiation,

$$u(f,T) = \frac{8\pi hf^3}{c^3}\frac{1}{\exp\left(\frac{hf}{k_BT}\right)-1}, \quad e(f,T) = \frac{2\pi hf^3}{c^2}\frac{1}{\exp\left(\frac{hf}{k_BT}\right)-1}. \tag{1.10}$$

These functions fitted the observed spectra perfectly! The spectrum $e(f,T)$ and the emitted power $e_{[0,f]}(T)$ with maximal frequency f are displayed for $T = 5780$ K in Figures 1.1 and 1.2.

1.3 Blackbody spectra and photon fluxes

Their technical relevance for the quantitative analysis of incandescent light sources makes it worthwhile to take a closer look at blackbody spectra. Blackbody spectra are also helpful to elucidate the notion of spectra more closely, and to explain that a maximum in a spectrum strongly depends on the choice of independent variable (e.g. wavelength or frequency) and dependent variable (e.g. energy flux or photon flux). In particular, it is sometimes claimed that our sun has maximal radiation output at a wavelength $\lambda_{max} \simeq 500$ nm. This statement is actually very misleading if the notion of "radiation output" is not clearly defined, and if no explanation

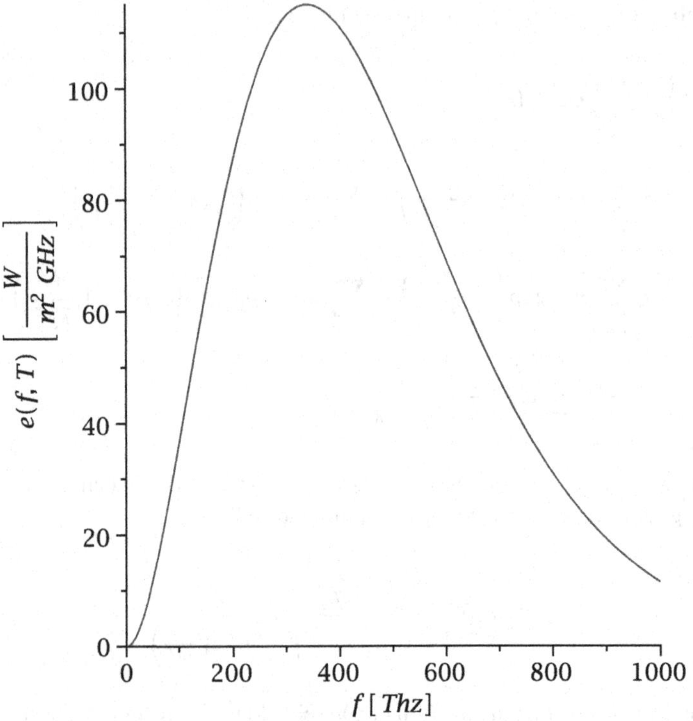

Fig. 1.1 The spectral emittance $e(f, T)$ for a heat source of temperature $T = 5780$ K

is included that different perfectly suitable notions of radiation output yield very different wavelengths or frequencies of maximal emission. We will see below that the statement above only applies to *maximal power output per unit of wavelength*, i.e. if we use a monochromator which slices the *wavelength* axis into intervals of equal length $d\lambda = c|df|/f^2$, then we find maximal power output in an interval around $\lambda_{max} \simeq 500$ nm. However, we will also see that if we use a monochromator which slices the *frequency* axis into intervals of equal length $df = c|d\lambda|/\lambda^2$, then we find maximal power output in an interval around $f_{max} \simeq 340$ THz, corresponding to a wavelength $c/f_{\max} \simeq 880$ nm. If we ask for maximal photon counts instead of maximal power output, we find yet other values for peaks in the spectra.

Since Planck's radiation law (1.10) yielded perfect matches to observed black-body spectra, it must also imply Stefan's law and Wien's law. Stefan's law is readily derived in the following way. The emitted power per area is

$$e(T) = \int_0^\infty df\, e(f,T) = \int_0^\infty d\lambda\, e(\lambda,T) = 2\pi \frac{k_B^4 T^4}{h^3 c^2} \int_0^\infty dx\, \frac{x^3}{\exp(x) - 1}.$$

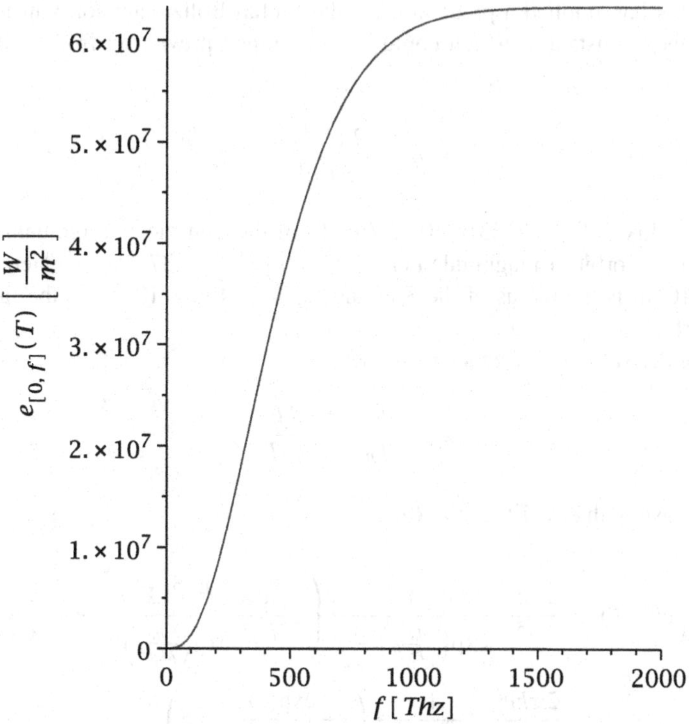

Fig. 1.2 The emittance $e_{[0,f]}(T) = \int_0^f df'\, e(f', T)$ (i.e. emitted power per area in radiation with maximal frequency f) for a heat source of temperature $T = 5780$ K. The asymptote for $f \to \infty$ is $e_{[0,\infty]}(T) \equiv e(T) = \sigma T^4 = 6.33 \times 10^7\ \mathrm{W/m^2}$ for the temperature $T = 5780$ K

Evaluation of the integral

$$\begin{aligned}
\int_0^\infty dx \frac{x^3}{\exp(x) - 1} &= \int_0^\infty dx\, x^3 \sum_{n=0}^{\infty} \exp[-(n+1)x] \\
&= -\sum_{n=1}^{\infty} \frac{d^3}{dn^3} \int_0^\infty dx\, \exp(-nx) = -\sum_{n=1}^{\infty} \frac{d^3}{dn^3} \frac{1}{n} \\
&= \sum_{n=1}^{\infty} \frac{6}{n^4} = 6\zeta(4) = \frac{\pi^4}{15}
\end{aligned}$$

implies

$$e(T) = \frac{2\pi^5 k_B^4}{15 h^3 c^2} T^4,$$

i.e. Planck's law implied a prediction for the Stefan-Boltzmann constant in terms of the Planck constant h, which could be determined previously from a fit to the spectra,

$$\sigma = \frac{2\pi^5 k_B^4}{15h^3c^2}.$$

An energy flux $e(T) = 6.33 \times 10^7\,\mathrm{W/m^2}$ from the Sun yields a remnant energy flux at Earth's orbit of magnitude $e(T) \times (R_\odot/r_\oplus)^2 = 1.37\,\mathrm{kW/m^2}$. Here $R_\odot = 6.955 \times 10^8$ m is the radius of the Sun and $r_\oplus = 1.496 \times 10^{11}$ m is the radius of Earth's orbit.

For the derivation of Wien's law, we set

$$x = \frac{hc}{\lambda k_B T} = \frac{hf}{k_B T}.$$

Then we have with $e(\lambda, T) = e(f, T)|_{f=c/\lambda} c/\lambda^2$,

$$\begin{aligned}\frac{\partial}{\partial\lambda}e(\lambda, T) &= \frac{2\pi hc^2}{\lambda^5}\frac{1}{\exp\left(\frac{hc}{\lambda k_B T}\right) - 1}\left(\frac{hc}{\lambda^2 k_B T}\frac{\exp\left(\frac{hc}{\lambda k_B T}\right)}{\exp\left(\frac{hc}{\lambda k_B T}\right) - 1} - \frac{5}{\lambda}\right)\\ &= \frac{2\pi hc^2}{\lambda^6}\frac{1}{\exp(x) - 1}\left(x\frac{\exp(x)}{\exp(x) - 1} - 5\right),\end{aligned}$$

which implies that $\partial e(\lambda, T)/\partial\lambda = 0$ is satisfied if and only if

$$\exp(x) = \frac{5}{5 - x}.$$

This condition yields $x \simeq 4.965$. The wavelength of maximal spectral emittance $e(\lambda, T)$ therefore satisfies

$$\lambda_{\max} \cdot T \simeq \frac{hc}{4.965 k_B} = 2898\,\mu\mathrm{m} \cdot \mathrm{K}.$$

For a heat source of temperature $T = 5780$ K, like the surface of our sun, this yields

$$\lambda_{\max} = 501\,\mathrm{nm}, \quad \frac{c}{\lambda_{\max}} = 598\,\mathrm{THz},$$

see Figure 1.3.

One can also derive an analogue of Wien's law for the frequency $f_{\max}$ of maximal spectral emittance $e(f, T)$. We have

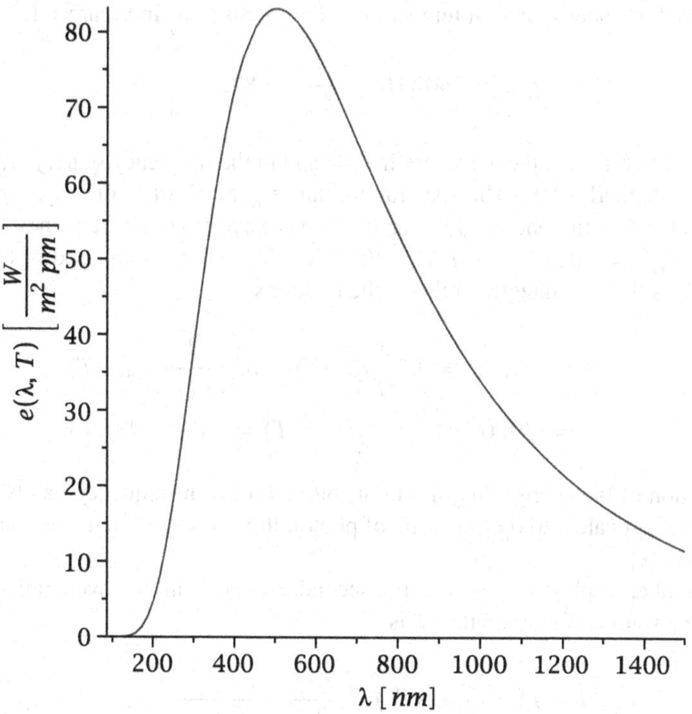

Fig. 1.3 The spectral emittance $e(\lambda, T)$ for a heat source of temperature $T = 5780$ K

$$\frac{\partial}{\partial f}e(f,T) = \frac{2\pi hf^2}{c^2}\frac{1}{\exp\left(\frac{hf}{k_BT}\right)-1}\left(3-\frac{hf}{k_BT}\frac{\exp\left(\frac{hc}{\lambda k_BT}\right)}{\exp\left(\frac{hc}{\lambda k_BT}\right)-1}\right)$$

$$= \frac{2\pi hf^2}{c^2}\frac{1}{\exp(x)-1}\left(3-x\frac{\exp(x)}{\exp(x)-1}\right),$$

which implies that $\partial e(f,T)/\partial f = 0$ is satisfied if and only if

$$\exp(x) = \frac{3}{3-x},$$

with solution $x \simeq 2.821$. The frequency of maximal spectral emittance $e(f,T)$ therefore satisfies

$$\frac{f_{\max}}{T} \simeq 2.821\frac{k_B}{h} = 58.79\,\frac{\text{GHz}}{\text{K}}.$$

This yields for a heat source of temperature $T = 5780$ K, as in Figure 1.1,

$$f_{\max} = 340\,\text{THz}, \qquad \frac{c}{f_{\max}} = 882\,\text{nm}.$$

The photon fluxes in the wavelength scale and in the frequency scale, $j(\lambda, T)$ and $j(f, T)$, are defined below. The spectral emittance per unit of frequency, $e(f, T)$, is directly related to the photon flux per fractional wavelength or frequency interval $d\ln f = df/f = -d\ln\lambda = -d\lambda/\lambda$. We have with the notations used in (1.4) for spectral densities and integrated fluxes the relations

$$\begin{aligned} e(f, T) &= hfj(f, T) = hf\frac{\partial}{\partial f}j_{[0,f]}(T) = h\frac{\partial}{\partial \ln(f/f_0)}j_{[0,f]}(T) \\ &= hj(\ln(f/f_0), T) = h\lambda j(\lambda, T) = hj(\ln(\lambda/\lambda_0), T). \end{aligned}$$

Optimization of the energy flux of a light source for given frequency bandwidth df is therefore equivalent to optimization of photon flux for fixed fractional bandwidth $df/f = |d\lambda/\lambda|$.

The number of photons per area, per second, and per unit of wavelength emitted from a heat source of temperature T is

$$j(\lambda, T) = \frac{\lambda}{hc}e(\lambda, T) = \frac{2\pi c}{\lambda^4}\frac{1}{\exp\left(\frac{hc}{\lambda k_B T}\right) - 1}.$$

This satisfies

$$\frac{\partial}{\partial\lambda}j(\lambda, T) = \frac{j(\lambda, T)}{\lambda}\left(x\frac{\exp(x)}{\exp(x) - 1} - 4\right) = 0$$

if

$$\exp(x) = \frac{4}{4 - x}.$$

This has the solution $x \simeq 3.921$. The wavelength of maximal spectral photon flux $j(\lambda, T)$ therefore satisfies

$$\lambda_{\max} \cdot T \simeq \frac{hc}{3.921 k_B} = 3670\,\mu\text{m} \cdot \text{K}.$$

This yields for a heat source of temperature $T = 5780$ K

$$\lambda_{\max} = 635\,\text{nm}, \qquad \frac{c}{\lambda_{\max}} = 472\,\text{THz},$$

see Figure 1.4.

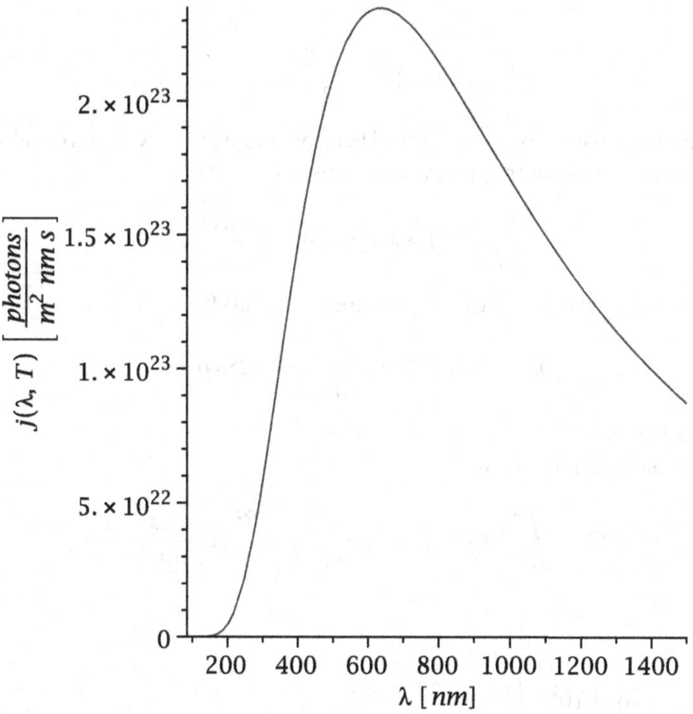

Fig. 1.4 The spectral photon flux $j(\lambda, T)$ for a heat source of temperature $T = 5780$ K

The photon flux in the wavelength scale, $j(\lambda, T)$, is also related to the energy fluxes per fractional wavelength or frequency interval $d\ln\lambda = d\lambda/\lambda = -d\ln f = -df/f$,

$$j(\lambda, T) = \frac{\lambda}{hc} e(\lambda, T) = \frac{1}{hc} e(\ln(\lambda/\lambda_0), T) = \frac{f}{hc} e(f, T) = \frac{1}{hc} e(\ln(f/f_0), T).$$

Therefore optimization of photon flux for fixed wavelength bandwidth $d\lambda$ is equivalent to optimization of energy flux for fixed fractional bandwidth $d\lambda/\lambda = |df/f|$.

Finally, the number of photons per area, per second, and per unit of frequency emitted from a heat source of temperature T is

$$j(f, T) = \frac{e(f, T)}{hf} = \frac{2\pi f^2}{c^2} \frac{1}{\exp\left(\frac{hf}{k_B T}\right) - 1}.$$

This satisfies

$$\frac{\partial}{\partial f} j(f, T) = \frac{j(f, T)}{f} \left(2 - x \frac{\exp(x)}{\exp(x) - 1}\right) = 0$$

if

$$\exp(x) = \frac{2}{2-x}.$$

This condition is solved by $x \simeq 1.594$. Therefore the frequency of maximal spectral photon flux $j(f, T)$ in the frequency scale satisfies

$$\frac{f_{\max}}{T} \simeq 1.594\frac{k_B}{h} = 33.21\,\frac{\text{GHz}}{\text{K}}.$$

This yields for a heat source of temperature $T = 5780$ K

$$f_{\max} = 192\,\text{THz}, \qquad \frac{c}{f_{\max}} = 1.56\,\mu\text{m},$$

see Figure 1.5.

The flux of emitted photons is

$$j(T) = \int_0^\infty df\, j(f, T) = 2\pi\frac{k_B^3 T^3}{h^3 c^2}\int_0^\infty dx\,\frac{x^2}{\exp(x) - 1}.$$

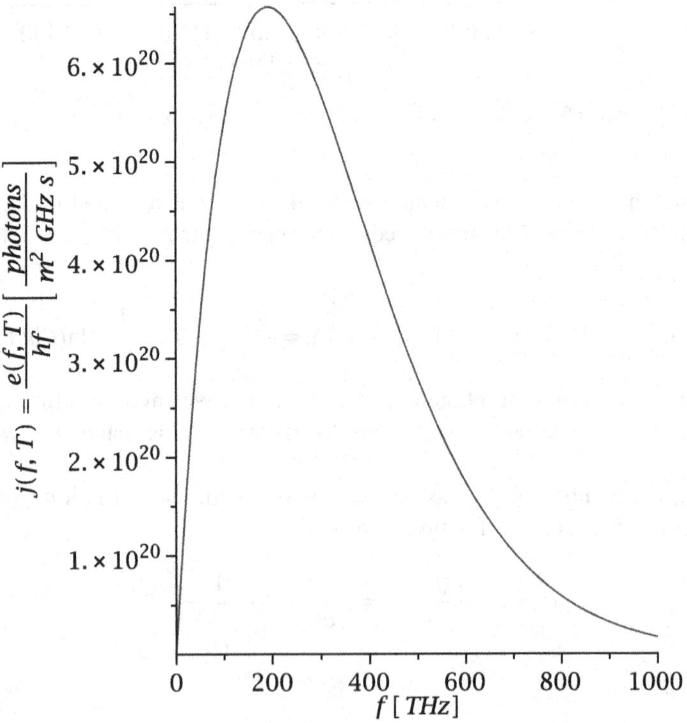

Fig. 1.5 The spectral photon flux $j(f, T)$ for a heat source of temperature $T = 5780$ K

Evaluation of the integral

$$\int_0^\infty dx\,\frac{x^2}{\exp(x)-1} = \int_0^\infty dx\,x^2\sum_{n=0}^\infty \exp[-(n+1)x]$$
$$= \sum_{n=1}^\infty \frac{d^2}{dn^2}\int_0^\infty dx\,\exp(-nx) = \sum_{n=1}^\infty \frac{d^2}{dn^2}\frac{1}{n}$$
$$= \sum_{n=1}^\infty \frac{2}{n^3} = 2\zeta(3)$$

yields

$$j(T) = \frac{4\pi\zeta(3)k_B^3}{h^3c^2}T^3 = 1.5205\times 10^{15}\frac{T^3}{\mathrm{m}^2\cdot\mathrm{s}\cdot\mathrm{K}^3}.$$

A surface temperature $T = 5780$ K for our sun yields a photon flux at the solar surface $2.94\times 10^{26}\,\mathrm{m}^{-2}\,\mathrm{s}^{-1}$ and a resulting photon flux at Earth's orbit of $6.35\times 10^{21}\,\mathrm{m}^{-2}\,\mathrm{s}^{-1}$. The average photon energy $e(T)/j(T) = 1.35\,\mathrm{eV}$ is in the infrared.

1.4 The photoelectric effect

The notion of energy quanta in radiation was so revolutionary in 1900 that Planck himself speculated that this must somehow be related to the emission mechanism of radiation from the material of the source. In 1905 Albert Einstein pointed out that hitherto unexplained properties of the photoelectric effect can also be explained through energy quanta hf in ultraviolet light, and proposed that this energy quantization is likely an intrinsic property of electromagnetic waves irrespective of how they are generated. In short, the photoelectric effect observations by J.J. Thomson and Lenard revealed the following key properties:

- An ultraviolet light source of frequency f will generate photoelectrons of maximal kinetic energy $hf - hf_0$ if $f > f_0$, where $hf_0 = \phi$ is the minimal energy to liberate photoelectrons from the photocathode.
- Increasing the intensity of the incident ultraviolet light without changing its frequency will increase the photocurrent, but not change the maximal kinetic energy of the photoelectrons. Increasing the intensity must therefore liberate more photoelectrons from the photocathode, but does not impart more energy on single electrons.

Einstein realized that this behavior can be explained if the incident ultraviolet light of frequency f comes in energy parcels of magnitude hf, and if the electrons in the metal can (predominantly) only absorb a single of these energy parcels.

1.5 Wave-particle duality

When X-rays of wavelength λ_0 are scattered off atoms, one observes scattered X-rays of the same wavelength λ_0 in all directions. However, in the years 1921–1923 Arthur H. Compton observed that under every scattering angle ϑ against the direction of incidence, there is also a component of scattered X-rays with a longer wavelength

$$\lambda = \lambda_0 + \lambda_C(1 - \cos\vartheta) .$$

The constant $\lambda_C = 2.426$ pm has the same value for every atom. Compton (and also Debye) recognized that this longer wavelength component in the scattered radiation can be explained as a consequence of particle like collisions of Planck's and Einstein's energy parcels hf with weakly bound electrons *if the energy parcels also carry momentum* h/λ. Energy conservation during the collision of the electromagnetics energy parcels (meanwhile called photons) with weakly bound electrons ($\boldsymbol{p}'_e$ is the momentum of the recoiling electron),

$$m_e c + \frac{h}{\lambda_0} = \sqrt{p_e'^2 + m_e^2 c^2} + \frac{h}{\lambda},$$

yields

$$p_e'^2 = \frac{h^2}{\lambda_0^2} + \frac{h^2}{\lambda^2} - 2\frac{h^2}{\lambda\lambda_0} + 2m_e hc\left(\frac{1}{\lambda_0} - \frac{1}{\lambda}\right),$$

while momentum conservation implies

$$p_e'^2 = \frac{h^2}{\lambda_0^2} + \frac{h^2}{\lambda^2} - 2\frac{h^2}{\lambda\lambda_0}\cos\vartheta.$$

This yields for the wavelength of the scattered photon

$$\lambda = \lambda_0 + \frac{h}{m_e c}(1 - \cos\vartheta) , \tag{1.11}$$

with excellent numerical agreement between $h/m_e c$ and the measured value of λ_C.

From the experimental findings on blackbody radiation, the photoelectric effect, and Compton scattering, and the ideas of Planck, Einstein, and Compton, an electromagnetic wave of frequency $f = c/\lambda$ appears like a current of particles with energy hf and momentum h/λ. However, electromagnetic waves also show wavelike properties like diffraction and interference. The findings of Planck, Einstein, and Compton combined with the wavelike properties of electromagnetic waves (observed for the first time by Heinrich Hertz) constitute the first observation of *wave-particle duality*. Depending on the experimental setup, a physical system can sometimes behave like a wave and sometimes behave like a particle.

However, the puzzle did not end there. Louis de Broglie recognized in 1923 that the orbits of the old Bohr model could be explained through closed circular electron waves if the electrons are assigned a wavelength $\lambda = h/p$, like photons. Soon thereafter, wavelike behavior of electrons was observed by Clinton Davisson and Lester Germer in 1927, when they observed interference of non-relativistic electrons scattered off the surface of Nickel crystals. At the same time, George Thomson was sending high energy electron beams (with kinetic energies between 20 keV and 60 keV) through thin metal foils and observed interference of the transmitted electrons, thus also confirming the wave nature of electrons. We can therefore also conclude that *another major motivation for the development of quantum mechanics was to explain wave-particle duality*.

1.6 Why Schrödinger's equation?

The foundations of quantum mechanics were developed between 1900 and 1950 by some of the greatest minds of the 20th century, from Max Planck and Albert Einstein to Richard Feynman and Freeman Dyson. The inner circle of geniuses who brought the nascent theory to maturity were Heisenberg, Born, Jordan, Schrödinger, Pauli, Dirac, and Wigner. Among all the outstanding contributions of these scientists, Schrödinger's invention of his wave equation (1.2) was likely the most important *single step* in the development of quantum mechanics. Understanding this step, albeit in a simplified pedagogical way, is important for learning and understanding quantum mechanics.

Ultimately, basic equations in physics have to prove themselves in comparison with experiments, and the Schrödinger equation was extremely successful in that regard. However, this does not explain how to come up with such an equation. Basic equations in physics cannot be derived from any rigorous theoretical or mathematical framework. There is no algorithm which could have told Newton to come up with Newton's equation, or would have told Schrödinger how to come up with his equation (or could tell us how to come up with a fundamental theory of quantum gravity). Basic equations in physics have to be invented in an act of creative ingenuity, which certainly requires a lot of brainstorming and diligent review of pertinent experimental facts and solutions of related problems (where known).

It is much easier to accept an equation and start to explore its consequences if the equation makes intuitive sense – if we can start our discussion of Schrödinger's equation with the premise "yes, the hypothesis that Schrödinger's equation solves the problems of energy quantization and wave-particle duality seems intuitively promising and is worth pursuing".

Therefore I will point out how Schrödinger *could* have invented the Schrödinger equation (although his actual thought process was much more involved and was motivated by the connection of the quantization rules of old quantum mechanics with the Hamilton-Jacobi equation of classical mechanics [39]).

The problem is to come up with an equation for the motion of particles, which explains both quantization of energy levels and wave-particle duality.

As a starting point, we recall that the motion of a non-relativistic particle under the influence of a conservative force $\boldsymbol{F}(\boldsymbol{x}) = -\nabla V(\boldsymbol{x})$ is classically described by Newton's equation

$$m\frac{d^2\boldsymbol{x}(t)}{dt^2} = -\nabla V(\boldsymbol{x}(t)),$$

and this equation also implies energy conservation,

$$E = \frac{\boldsymbol{p}^2}{2m} + V(\boldsymbol{x}). \tag{1.12}$$

However, this cannot be the whole story, because Davisson and Germer, and G.P. Thomson had shown that at least electrons sometimes also behave like waves with wavelength $\lambda = h/p$, as predicted by de Broglie. Furthermore, Compton has demonstrated that photons of energy $E = hf$ satisfy the relation $\lambda = h/p$ between wavelength and momentum. This motivates the hypothesis that a non-relativistic particle might also satisfy the relation $E = hf$. A monochromatic plane wave of frequency f, wavelength λ, and direction of motion $\hat{\boldsymbol{k}}$ can be described by a wave function

$$\psi(\boldsymbol{x},t) = A\exp\left[2\pi\mathrm{i}\left(\frac{\hat{\boldsymbol{k}}\cdot\boldsymbol{x}}{\lambda} - ft\right)\right].$$

Substitution of the relations

$$\lambda = \frac{h}{p}, \quad E = hf = \frac{\boldsymbol{p}^2}{2m}$$

yields with $\hbar \equiv h/2\pi$

$$\psi(\boldsymbol{x},t) = A\exp\left[\mathrm{i}\left(\frac{\boldsymbol{p}\cdot\boldsymbol{x}}{\hbar} - \frac{\boldsymbol{p}^2}{2m\hbar}t\right)\right].$$

Under the supposition of wave-particle duality, we have to assume that this wave function must somehow be related to the wave properties of free particles as observed in the electron diffraction experiments. However, this wave function satisfies a differential equation

$$\mathrm{i}\hbar\frac{\partial}{\partial t}\psi(\boldsymbol{x},t) = E\psi(\boldsymbol{x},t) = \frac{\boldsymbol{p}^2}{2m}\psi(\boldsymbol{x},t) = -\frac{\hbar^2}{2m}\Delta\psi(\boldsymbol{x},t), \tag{1.13}$$

because under the assumption of wave-particle duality we had to replace f with E/h in the exponent, and we used $E = \boldsymbol{p}^2/2m$ for a free particle.

This does not yet tell us how to calculate the wave function which would describe motion of particles in a potential $V(\boldsymbol{x})$. However, comparison of the differential equation (1.13) with the classical energy equation (1.12) can give us the idea to try

$$\mathrm{i}\hbar\frac{\partial}{\partial t}\psi(\boldsymbol{x},t) = -\frac{\hbar^2}{2m}\Delta\psi(\boldsymbol{x},t) + V(\boldsymbol{x})\psi(\boldsymbol{x},t) \tag{1.14}$$

as a starting point for the calculation of wave functions for particles moving in a potential $V(\boldsymbol{x})$. Schrödinger actually found this equation after he had found the time-independent Schrödinger equation (3.3) below, and he had demonstrated that these equations yield the correct spectrum for hydrogen atoms, where

$$V(\boldsymbol{x}) = -\frac{e^2}{4\pi\epsilon_0|\boldsymbol{x}|}.$$

Schrödinger's solution of the hydrogen atom will be discussed in Chapter 7.

1.7 Interpretation of Schrödinger's wave function

The Schrödinger equation was a spectacular success right from the start, but it was not immediately clear what the physical meaning of the complex wave function $\psi(\boldsymbol{x},t)$ is. A natural first guess would be to assume that $|\psi(\boldsymbol{x},t)|^2$ corresponds to a physical density of the particle described by the wave function $\psi(\boldsymbol{x},t)$. In this interpretation, an electron in a quantum state $\psi(\boldsymbol{x},t)$ would have a spatial mass density $m\,|\psi(\boldsymbol{x},t)|^2$ and a charge density $-e\,|\psi(\boldsymbol{x},t)|^2$. This interpretation would imply that waves would have prevailed over particles in wave-particle duality.

However, quantum leaps are difficult to reconcile with a physical density interpretation for $|\psi(\boldsymbol{x},t)|^2$, and Schrödinger, Bohr, Born and Heisenberg developed a statistical interpretation of the wave function which is still the leading paradigm for quantum mechanics. Already in June 1926, the view began to emerge that the wave function $\psi(\boldsymbol{x},t)$ should be interpreted as a *probability density amplitude*[2] in

[2]E. Schrödinger, Annalen Phys. 386, 109 (1926), paragraph on pp. 134–135, sentences 2–4: "$\psi\overline{\psi}$ is a kind of *weight function* in the configuration space of the system. The *wave mechanical* configuration of the system is a *superposition* of many, strictly speaking of *all*, kinematically possible point mechanical configurations. Thereby each point mechanical configuration contributes with a certain *weight* to the true wave mechanical configuration, where the weight is just given by $\psi\overline{\psi}$." Of course, a weakness of this early hint at the probability interpretation is the vague reference to a "true wave mechanical configuration". A clearer formulation of this point was offered by Born essentially simultaneously, see the following reference. While there was (and always has been) agreement on the importance of a probabilistic interpretation, the question of the concept which underlies those probabilities was a contentious point between Schrödinger, who at that time may have preferred to advance a de Broglie type pilot wave interpretation, and Bohr and Born and their particle-wave complementarity interpretation. In the end the complementarity picture prevailed:

the sense that

$$P_V(t) = \int_V d^3\boldsymbol{x}\, |\psi(\boldsymbol{x},t)|^2 \tag{1.15}$$

is the probability to find a particle (or rather, an excitation of the vacuum with minimal energy mc^2 and certain other quantum numbers) in the volume V at time t. This equation implies that $|\psi(\boldsymbol{x},t)|^2$ is the *probability density* to find the particle in the location $\boldsymbol{x}$ at time t. The expectation value for the location of the particle at time t is then

$$\langle \boldsymbol{x} \rangle(t) = \int d^3\boldsymbol{x}\, \boldsymbol{x}\, |\psi(\boldsymbol{x},t)|^2, \tag{1.16}$$

where integrals without explicit limits are taken over the full range of the integration variable, i.e. here over all of $\mathbb{R}^3$. Many individual particle measurements will yield the location $\boldsymbol{x}$ with a frequency proportionally to $|\psi(\boldsymbol{x},t)|^2$, and averaging over the observations will yield the expectation value (1.16) with a variance e.g. for the x coordinate

$$\begin{aligned}\Delta x^2(t) &= \langle (x - \langle x\rangle)^2\rangle(t) = \langle x^2\rangle(t) - \langle x\rangle^2(t) \\ &= \int d^3\boldsymbol{x}\, x^2\, |\psi(\boldsymbol{x},t)|^2 - \left(\int d^3\boldsymbol{x}\, x\, |\psi(\boldsymbol{x},t)|^2\right)^2.\end{aligned}$$

This interpretation of the relation between the wave function and particle properties was essentially proposed by Max Born in an early paper on quantum mechanical scattering[3].

The Schrödinger equation (1.2) implies a local conservation law for probability

$$\frac{\partial}{\partial t}|\psi(\boldsymbol{x},t)|^2 + \nabla\cdot\boldsymbol{j}(\boldsymbol{x},t) = 0 \tag{1.17}$$

with the probability current density

$$\boldsymbol{j}(\boldsymbol{x},t) = \frac{\hbar}{2im}\Big(\psi^+(\boldsymbol{x},t)\cdot\nabla\psi(\boldsymbol{x},t) - \nabla\psi^+(\boldsymbol{x},t)\cdot\psi(\boldsymbol{x},t)\Big). \tag{1.18}$$

There are fundamental degrees of freedom with certain quantum numbers. These degrees of freedom are quantal excitations of the vacuum, and mathematically they are described by quantum fields. Depending on the way they are probed, they exhibit wavelike or corpuscular properties. Whether or not to denote these degrees of freedom as particles is a matter of convenience and tradition.

[3] M. Born, Z. Phys. 38, 803 (1926).

The conservation law (1.17) is important for consistency of the probability interpretation of Schrödinger theory. We assume that the integral

$$P(t) = \int d^3x \, |\psi(x,t)|^2$$

over $\mathbb{R}^3$ converges. *A priori* this should yield a time-dependent function $P(t)$. However, equation (1.17) implies

$$\frac{d}{dt}P(t) = 0, \tag{1.19}$$

whence $P(t) \equiv P$ is a positive constant. This allows for rescaling $\psi(x,t) \to \psi(x,t)/\sqrt{P}$ such that the new wave function still satisfies equation (1.2) and yields a normalized integral

$$\int d^3x \, |\psi(x,t)|^2 = 1. \tag{1.20}$$

This means that the probability to find the particle anywhere at time t is 1, as it should be. The equations (1.15) and (1.16) make sense only in conjunction with the normalization condition (1.20)

We can also substitute the Schrödinger equation or the local conservation law (1.17) into

$$\langle p\rangle(t) = m\frac{d}{dt}\langle x\rangle(t) = m\int d^3x\, x\frac{\partial}{\partial t}\,|\psi(x,t)|^2 \tag{1.21}$$

to find

$$\langle p\rangle(t) = \int d^3x\, \psi^+(x,t)\frac{\hbar}{i}\nabla\psi(x,t). \tag{1.22}$$

Equations (1.16) and (1.22) tell us how to extract particle like properties from the wave function $\psi(x,t)$. At first sight, equation (1.22) does not seem to make a lot of intuitive sense. Why should the momentum of a particle be related to the gradient of its wave function? However, recall the Compton-de Broglie relation $p = h/\lambda$. Wave packets which are composed of shorter wavelength components oscillate more rapidly as a function of x, and therefore have a larger average gradient. Equation (1.22) is therefore in agreement with a basic relation of wave-particle duality.

A related argument in favor of equation (1.22) arises from substitution of the Fourier transforms[4]

[4]Fourier transformation is reviewed in Section 2.1.

$$\psi(\boldsymbol{x},t) = \frac{1}{\sqrt{2\pi}^3}\int d^3\boldsymbol{k}\,\exp(\mathrm{i}\boldsymbol{k}\cdot\boldsymbol{x})\psi(\boldsymbol{k},t),$$

$$\psi^+(\boldsymbol{x},t) = \frac{1}{\sqrt{2\pi}^3}\int d^3\boldsymbol{k}\,\exp(-\mathrm{i}\boldsymbol{k}\cdot\boldsymbol{x})\psi^+(\boldsymbol{k},t)$$

in equations (1.20) and (1.22). This yields

$$\int d^3\boldsymbol{k}\,|\psi(\boldsymbol{k},t)|^2 = 1$$

and

$$\langle \boldsymbol{p}\rangle(t) = \int d^3\boldsymbol{k}\,\hbar\boldsymbol{k}\,|\psi(\boldsymbol{k},t)|^2,$$

in perfect agreement with the Compton-de Broglie relation $\boldsymbol{p} = \hbar\boldsymbol{k}$. Apparently $|\psi(\boldsymbol{k},t)|^2$ is a probability density in $\boldsymbol{k}$ space in the sense that

$$P_{\tilde{V}}(t) = \int_{\tilde{V}} d^3\boldsymbol{k}\,|\psi(\boldsymbol{k},t)|^2$$

is the probability to find the particle with a wave vector $\boldsymbol{k}$ contained in a volume $\tilde{V}$ in $\boldsymbol{k}$ space.

We can also identify an expression for the energy of a particle which is described by a wave function $\psi(\boldsymbol{x},t)$. The Schrödinger equation (1.2) implies the conservation law

$$\frac{d}{dt}\int d^3\boldsymbol{x}\,\psi^+(\boldsymbol{x},t)\left(-\frac{\hbar^2}{2m}\Delta + V(\boldsymbol{x})\right)\psi(\boldsymbol{x},t) = 0. \tag{1.23}$$

Here it plays a role that we assumed time-independent potential[5]. In classical mechanics, the conservation law which appears for motion in a time-independent potential is energy conservation. Therefore, we expect that the expectation value for energy is given by

$$\langle E\rangle = \int d^3\boldsymbol{x}\,\psi^+(\boldsymbol{x},t)\left(-\frac{\hbar^2}{2m}\Delta + V(\boldsymbol{x})\right)\psi(\boldsymbol{x},t). \tag{1.24}$$

We will also rederive this at a more advanced level in Chapter 17. From the classical relation (1.12) between energy and momentum of a particle, we should also have

$$\langle E\rangle = \frac{\langle \boldsymbol{p}^2\rangle}{2m} + \langle V(\boldsymbol{x})\rangle. \tag{1.25}$$

[5]Examples of the Schrödinger equation with time-dependent potentials will be discussed in Chapter 13 and following chapters.

Comparison of equations (1.22) and (1.24) yields

$$\langle p^2\rangle(t) = \int d^3x\, \psi^+(x,t)(-i\hbar\nabla)^2\psi(x,t),$$

such that calculation of expectation values of powers of momentum apparently amounts to corresponding powers of the differential operator $-i\hbar\nabla$ acting on the wave function $\psi(x,t)$.

Maybe one of the most direct observational confirmations of the statistical interpretation of the wave function was the observation of single particle interference by Tonomura, Endo, Matsuda and Kawasaski[6] in 1988. Electrons are passing through a double slit with a time difference that makes it extremely unlikely that two electrons interfere during their passages through the slit. Behind the slit the electrons are observed with a scintillation screen or a camera. Each individual electron is observed to generate only a single dot on the screen. This is the behavior expected from a pointlike particle which is not spread over a physical density distribution. The first few electrons seem to generate a random pattern of dots. However, when more and more electrons hit the screen, their dots generate a collective pattern which exactly corresponds to a distribution $|\psi(x,t)|^2$ for double slit interference. This implies that $|\psi(x,y,z_0,t)|^2$ is indeed the probability density for an electron to hit the point $\{x,y\}$ on the screen which is located at z_0, but it is not the physical density of a spatially extended electron[7].

A recent three-slit experiment also confirmed the statistical interpretation of the wave function by proving that the interference patterns from many sequential single particle paths agree with the probability density interpretation of $|\psi(x,t)|^2$ for single slit diffraction, double-slit interference, and triple-slit interference[8].

1.8 Problems

1.1. Plot the emittance $e_{[0,\lambda]}(T)$ of our sun.

1.2. Suppose that the resolution of a particular monochromator scales with $1/f$, i.e. if the monochromator is set to a particular frequency f the product $fdf = df^2/2$ of frequency and bandwidth is constant. Furthermore, assume that the monochromator is coupled to a device which produces a signal proportional to the energy of the incident radiation. In the limit $df \to 0$, is the signal curve from this apparatus proportional to $e(f,T)$, $e(\lambda,T)$, $j(f,T)$ or $j(\lambda,T)$?

---

[6] A. Tonomura, J. Endo, T. Matsuda, T. Kawasaki, Amer. J. Phys. 57, 117 (1989).

[7] It has been argued that Bohmian mechanics can also explain the Tonomura experiment through a pilot wave interpretation of the wave function. However, Bohmian mechanics has other problems. We will briefly return to Bohmian mechanics in Problem 7.17.

[8] U. Sinha, C. Couteau, T. Jennewein, R. Laflamme, G. Weihs, Science 329, 418 (2010).

1.3. Suppose that the resolution of a particular monochromator scales with f, i.e. if the monochromator is set to a particular frequency f the fractional bandwidth df/f is constant. The monochromator is coupled to a device which produces a signal proportional to the energy of the incident radiation. The device is used for observation of a Planck spectrum. For which relation between frequency and temperature does this device yield maximal signal?

1.4. Derive the probability conservation law (1.17) from the Schrödinger equation. Hint: Multiply the Schrödinger equation with $\psi^+(\boldsymbol{x}, t)$ and use also the complex conjugate equation.

1.5. We will often deal with quantum mechanics in d spatial dimensions. There are many motivations to go beyond the standard case $d = 3$. E.g. $d = 0$ is the number of spatial dimensions for an idealized quantum dot, $d = 1$ is often used for pedagogical purposes and also for idealized quantum wires or nanowires, and $d = 2$ is used for physics on surfaces and interfaces.

We consider a normalized wave function $\psi(\boldsymbol{x}, t)$ in d dimensions. What are the SI units of the wave function? What are the SI units of the d-dimensional current density $\boldsymbol{j}$ for the wave function $\psi(\boldsymbol{x}, t)$?

1.6. Derive equation (1.22) from (1.21).

1.7. Show that the Schrödinger equation (1.14) implies the conservation laws

$$\frac{d}{dt}\int d^3x\,\psi^+(\boldsymbol{x}, t)\left(-\frac{\hbar^2}{2m}\Delta + V(\boldsymbol{x})\right)^n \psi(\boldsymbol{x}, t) = 0, \quad n \in \mathbb{N}_0. \tag{1.26}$$

Two particular cases of this equation appeared in Section 1.7. Which are those cases and what are the related conserved quantities?

Why is there usually not much interest in the infinitely many higher order conservation laws (1.26) for $n > 1$? Hint: Think about the classical interpretation of these conservation laws.

Why do the higher order conservation laws nevertheless matter in quantum mechanics? Hint: Equation (1.26) is generically different from the "similar" conservation law $d(\langle E\rangle^n)/dt = 0$. Is there an interesting implication of the two conservation laws for $n = 2$?

1.8. Equation (1.21) implies that the equation $\boldsymbol{p}(t) = m d\boldsymbol{x}(t)/dt$ from non-relativistic classical mechanics is realized as an equation between expectation values in non-relativistic quantum mechanics. Show that Newton's law holds in the following sense in non-relativistic quantum mechanics (Ehrenfest's theorem),

$$\frac{d}{dt}\langle\boldsymbol{p}\rangle(t) = -\langle\nabla V(\boldsymbol{x})\rangle(t). \tag{1.27}$$

Chapter 2
Self-adjoint Operators and Eigenfunction Expansions

The relevance of waves in quantum mechanics naturally implies that the decomposition of arbitrary wave packets in terms of monochromatic waves, commonly known as Fourier decomposition after Jean-Baptiste Fourier's *Théorie analytique de la Chaleur* (1822), plays an important role in applications of the theory. Dirac's δ function, on the other hand, gained prominence primarily through its use in quantum mechanics, although today it is also commonly used in mechanics and electrodynamics to describe sudden impulses, mass points, or point charges. Both concepts are intimately connected to the completeness of eigenfunctions of self-adjoint operators. From the quantum mechanics perspective, the problem of completeness of sets of functions concerns the problem of enumeration of all possible states of a quantum system.

2.1 The δ function and Fourier transforms

Let $f(x)$ be a smooth function in the interval $[a, b]$. Dirichlet's equation [7]

$$\lim_{\kappa\to\infty}\int_a^b dx'\,\frac{\sin(\kappa(x-x'))}{\pi(x-x')}f(x') = \begin{cases} 0, & x\notin[a,b],\\ f(x), & x\in(a,b),\end{cases} \tag{2.1}$$

motivates the formal definition

$$\begin{aligned}\delta(x) &= \lim_{\kappa\to\infty}\frac{\sin(\kappa x)}{\pi x} = \lim_{\kappa\to\infty}\frac{1}{2\pi}\int_{-\kappa}^{\kappa}dk\,\exp(ikx)\\ &= \frac{1}{2\pi}\int_{-\infty}^{\infty}dk\,\exp(ikx),\end{aligned} \tag{2.2}$$

R. Dick, *Advanced Quantum Mechanics*, Graduate Texts in Physics,
DOI 10.1007/978-3-319-25675-7_2

such that equation (2.1) can (in)formally be written as

$$\int_a^b dx'\,\delta(x-x')f(x') = \begin{cases} 0, & x \notin [a,b], \\ f(x), & x \in (a,b). \end{cases}$$

A justification for Dirichlet's equation is given below in the derivation of equation (2.8).

The generalization to three dimensions follows immediately from Dirichlet's formula in a three-dimensional cube, and exhaustion of an arbitrary three-dimensional volume V by increasingly finer cubes. This yields

$$\delta(\boldsymbol{x}) = \prod_{i=1}^{3} \lim_{\kappa_i\to\infty} \frac{\sin(\kappa_i x_i)}{\pi x_i} = \frac{1}{(2\pi)^3}\int d^3k\,\exp(\mathrm{i}\boldsymbol{k}\cdot\boldsymbol{x}), \tag{2.3}$$

$$\int_V d^3x'\,\delta(\boldsymbol{x}-\boldsymbol{x}')f(\boldsymbol{x}') = \begin{cases} 0, & \boldsymbol{x} \notin V, \\ f(\boldsymbol{x}), & \boldsymbol{x} \text{ inside } V. \end{cases}$$

The case $\boldsymbol{x} \in \partial V$ ($\boldsymbol{x}$ on the boundary of V) must be analyzed on a case-by-case basis.

Equation (2.3) implies

$$\begin{aligned} \psi(\boldsymbol{x},t) &= \int d^3x'\,\delta(\boldsymbol{x}-\boldsymbol{x}')\psi(\boldsymbol{x}',t) \\ &= \frac{1}{(2\pi)^3}\int d^3k\,\exp(\mathrm{i}\boldsymbol{k}\cdot\boldsymbol{x})\int d^3x'\,\exp(-\mathrm{i}\boldsymbol{k}\cdot\boldsymbol{x}')\psi(\boldsymbol{x}',t). \end{aligned}$$

This can be used to introduce Fourier transforms by splitting the previous equation into two equations,

$$\psi(\boldsymbol{x},t) = \frac{1}{\sqrt{2\pi}^3}\int d^3k\,\exp(\mathrm{i}\boldsymbol{k}\cdot\boldsymbol{x})\psi(\boldsymbol{k},t), \tag{2.4}$$

with

$$\psi(\boldsymbol{k},t) = \frac{1}{\sqrt{2\pi}^3}\int d^3x\,\exp(-\mathrm{i}\boldsymbol{k}\cdot\boldsymbol{x})\psi(\boldsymbol{x},t). \tag{2.5}$$

Use of $\psi(\boldsymbol{x},t)$ corresponds to the ***x*-*representation*** of quantum mechanics. Use of $\psi(\boldsymbol{k},t)$ corresponds to the ***k*-*representation*** or *momentum-representation* of quantum mechanics.

The notation above for Fourier transforms is a little sloppy, but convenient and common in quantum mechanics. From a mathematical perspective, the Fourier transformed function $\psi(\boldsymbol{k},t)$ should actually be denoted by $\tilde{\psi}(\boldsymbol{k},t)$ to make it clear that it is *not* the same function as $\psi(\boldsymbol{x},t)$ with different symbols for the first three variables. The physics notation is motivated by the observation that $\psi(\boldsymbol{x},t)$ and $\psi(\boldsymbol{k},t)$ are just different representations of the same quantum mechanical state ψ.

Another often used convention for Fourier transforms is to split the factor $(2\pi)^{-3}$ asymmetrically, or equivalently replace it with a factor 2π in the exponents,

$$\psi(\boldsymbol{x},t) = \frac{1}{(2\pi)^3}\int d^3k\, \exp(\mathrm{i}\boldsymbol{k}\cdot\boldsymbol{x})\psi(\boldsymbol{k},t),$$

$$\psi(\boldsymbol{k},t) = \int d^3x\, \exp(-\mathrm{i}\boldsymbol{k}\cdot\boldsymbol{x})\psi(\boldsymbol{x},t),$$

or equivalently

$$\psi(\boldsymbol{x},t) = \int d^3\tilde{\boldsymbol{\nu}}\, \exp(2\pi\mathrm{i}\tilde{\boldsymbol{\nu}}\cdot\boldsymbol{x})\psi(\tilde{\boldsymbol{\nu}},t),$$

$$\psi(\tilde{\boldsymbol{\nu}},t) = \int d^3x\, \exp(-2\pi\mathrm{i}\tilde{\boldsymbol{\nu}}\cdot\boldsymbol{x})\psi(\boldsymbol{x},t),$$

with the vector of wave numbers

$$\tilde{\boldsymbol{\nu}} = \frac{\boldsymbol{k}}{2\pi}.$$

The conventions (2.4, 2.5) are used throughout this book.

The following is an argument for equation (2.1) and its generalizations to other representations of the δ function. The idea is to first construct a limit for the Heaviside step function or Θ function

$$\Theta(x) = \begin{cases} 1, & x > 0, \\ 0, & x < 0, \end{cases}$$

and go from there. The value of $\Theta(0)$ is often chosen to suite the needs of the problem at hands. The choice $\Theta(0) = 1/2$ seems intuitive and is also mathematically natural in the sense that any decomposition of a discontinuous functions in a complete set of functions (e.g. Fourier decomposition) will approximate the mean value between the left and right limit for a finite discontinuity, but in many applications other values of $\Theta(0)$ are preferred.

The Θ function helps us to explain Dirichlet's equation (2.1) through the following construction. Suppose $d(x)$ is a normalized function,

$$\int_{-\infty}^{\infty} dx\, d(x) = 1. \tag{2.6}$$

The integral

$$D(x) = \int_{-\infty}^{x} d\xi\, d(\xi)$$

satisfies

$$\lim_{\kappa\to\infty} D(\kappa\cdot x) = \Theta(x), \tag{2.7}$$

where we apparently defined $\Theta(0)$ as $\Theta(0) = \int_{-\infty}^{0} d\xi\, d(\xi)$, but this plays no role for the following reasoning.

Equation (2.7) yields for $f(x)$ differentiable in $[a, b]$

$$\int_a^b dx\, \kappa\, d(\kappa\cdot x) f(x) = D(\kappa\cdot x) f(x)\Big|_a^b - \int_a^b dx\, D(\kappa\cdot x) f'(x),$$

$$\begin{aligned}
\lim_{\kappa\to\infty} \int_a^b dx\, \kappa\, d(\kappa\cdot x) f(x) &= \Theta(b)f(b) - \Theta(a)f(a) - \int_a^b dx\, \Theta(x) f'(x) \\
&= \Theta(b)f(b) - \Theta(a)f(a) - \Theta(b)[f(b) - f(0)] + \Theta(a)[f(a) - f(0)] \\
&= [\Theta(b) - \Theta(a)] f(0),
\end{aligned} \tag{2.8}$$

where we simply split

$$\int_a^b dx\, \Theta(x) f'(x) = \int_0^b dx\, \Theta(x) f'(x) - \int_0^a dx\, \Theta(x) f'(x)$$

to arrive at the final result. Equation (2.8) confirms

$$\lim_{\kappa\to\infty} \kappa\, d(\kappa x) = \delta(x), \tag{2.9}$$

or after shifting the argument,

$$\lim_{\kappa\to\infty} \kappa\, d[\kappa(x - x_0)] = \delta(x - x_0).$$

From a mathematical perspective, equations like (2.9) mean that the action of the δ distribution on a smooth function corresponds to integration with a kernel $\kappa\, d(\kappa x)$ and then taking the limit $\kappa \to \infty$.

Equation (2.2) is an important particular realization of equation (2.9) with the normalized sinc function $d(x) = \mathrm{sinc}(x)/\pi = \sin(x)/\pi x$. Another important realization uses the function $d(x) = (\pi + \pi x^2)^{-1}$,

$$\begin{aligned}
\delta(x) &= \lim_{\kappa\to\infty} \frac{1}{\pi} \frac{\kappa}{1 + \kappa^2 x^2} = \lim_{a\to 0} \frac{1}{\pi} \frac{a}{a^2 + x^2} \\
&= \lim_{a\to 0} \frac{1}{2\pi} \int_{-\infty}^{\infty} dk\, \exp(ikx - a|k|).
\end{aligned} \tag{2.10}$$

Note that we did not require $d(x)$ to have a maximum at $x = 0$ to derive (2.9), and indeed we do not need this requirement. Consider the following example,

$$d(x) = \frac{1}{2}\sqrt{\frac{\alpha}{\pi}}\exp[-\alpha(x-a)^2] + \frac{1}{2}\sqrt{\frac{\beta}{\pi}}\exp[-\beta(x-b)^2].$$

This function has two maxima if $\alpha \cdot \beta \neq 0$ and if a and b are sufficiently far apart, and it even has a minimum at $x = 0$ if $\alpha = \beta$ and $a = -b$. Yet we still have

$$\lim_{\kappa\to\infty} \kappa\, d(\kappa \cdot x) = \lim_{\kappa\to\infty} \left(\frac{\kappa}{2}\sqrt{\frac{\alpha}{\pi}}\exp[-\alpha(\kappa x - a)^2] + \frac{\kappa}{2}\sqrt{\frac{\beta}{\pi}}\exp[-\beta(\kappa x - b)^2]\right) = \delta(x),$$

because the scaling with κ scales the initial maxima near a and b to $a/\kappa \to 0$ and $b/\kappa \to 0$.

Sokhotsky-Plemelj relations

The Sokhotsky-Plemelj relations are very useful relations involving a δ distribution[1],

$$\frac{1}{x - \mathrm{i}\epsilon} = \mathcal{P}\frac{1}{x} + \mathrm{i}\pi\delta(x), \qquad \frac{1}{x + \mathrm{i}\epsilon} = \mathcal{P}\frac{1}{x} - \mathrm{i}\pi\delta(x). \tag{2.11}$$

Indeed, for the practical evaluation of integrals involving singular denominators, we virtually never use these relations but evaluate the integrals with the left hand sides directly using the Cauchy and residue theorems. The primary use of the Sokhotsky-Plemelj relations in physics and technology is to establish relations between different physical quantities. The relation between retarded Green's functions and local densities of states is an example for this and will be derived in Section 20.1.

I will give a brief justification for the Sokhotsky-Plemelj relations. The relations

$$\frac{1}{x + \mathrm{i}\epsilon} = \frac{1}{\mathrm{i}}\int_0^\infty dk\, \exp[\mathrm{i}k(x + \mathrm{i}\epsilon)] = \frac{1}{\mathrm{i}}\int_{-\infty}^0 dk\, \exp[-\mathrm{i}k(x + \mathrm{i}\epsilon)]$$

[1] Yu.V. Sokhotsky, Ph.D. thesis, University of St. Petersburg, 1873; J. Plemelj, Monatshefte Math. Phys. 19, 205 (1908). The "physics" version (2.11) of the Sokhotsky-Plemelj relations is of course more recent than the original references because the δ distribution was only introduced much later.

imply

$$\Im\frac{1}{x+\mathrm{i}\epsilon}=-\frac{1}{2}\int_{-\infty}^{\infty}dk\,\cos(kx)=-\pi\delta(x).$$

On the other hand, the real part is

$$\Re\frac{1}{x+\mathrm{i}\epsilon}=\frac{1}{2(x+\mathrm{i}\epsilon)}+\frac{1}{2(x-\mathrm{i}\epsilon)}=\frac{x}{x^2+\epsilon^2}.$$

This implies for integration with a bounded function $f(x)$ in $[a,b]$

$$\int_a^b dx\,\frac{f(x)}{x+\mathrm{i}\epsilon}=\int_a^b dx\,\frac{xf(x)}{x^2+\epsilon^2}-\mathrm{i}\pi[\Theta(b)-\Theta(a)]f(0).$$

However, the weight factor

$$K_\epsilon(x)=\frac{x}{x^2+\epsilon^2}$$

essentially cuts the region $-3\epsilon < x < 3\epsilon$ symmetrically from the integral $\int_a^b dx f(x)/x$ (the value 3ϵ is chosen because $xK_\epsilon(x) = 0.9$ for $x = \pm 3\epsilon$), see Figure 2.1. Therefore we can use this factor as one possible definition of a principal value integral,

$$\mathcal{P}\int_a^b dx\frac{f(x)}{x}=\lim_{\epsilon\to 0}\int_a^b dx\,K_\epsilon(x)f(x).$$

2.2 Self-adjoint operators and completeness of eigenstates

The statistical interpretation of the wave function $\psi(\boldsymbol{x},t)$ implies that the wave functions of single stable particles should be normalized,

$$\int d^3\boldsymbol{x}\,|\psi(\boldsymbol{x},t)|^2=1. \tag{2.12}$$

Time-dependence plays no role and will be suppressed in the following investigations.

Indeed, we have to require a little more than just normalizability of the wave function $\psi(\boldsymbol{x})$ itself, because the functions $\nabla\psi(\boldsymbol{x})$, $\Delta\psi(\boldsymbol{x})$, and $V(\boldsymbol{x})\psi(\boldsymbol{x})$ for admissible potentials $V(\boldsymbol{x})$ should also be square integrable. We will therefore also encounter functions $f(\boldsymbol{x})$ which may not be normalized, although they are square integrable,

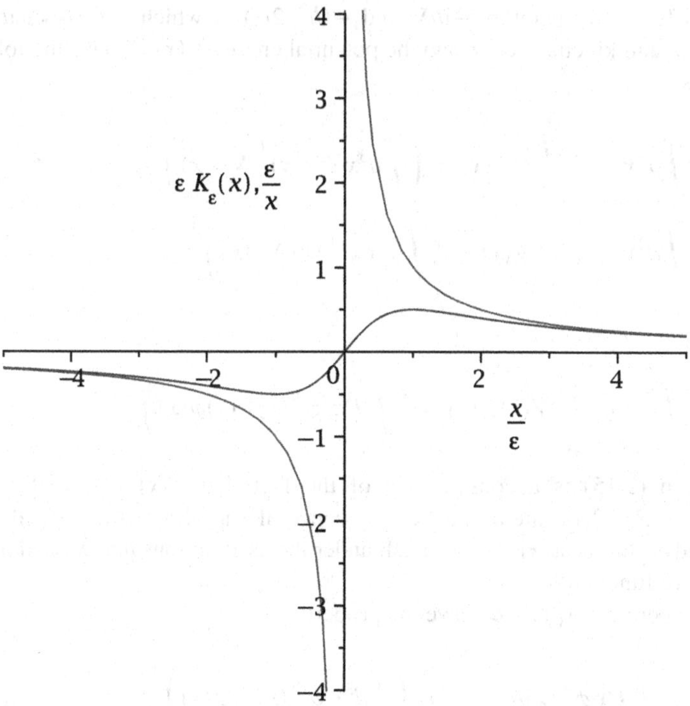

Fig. 2.1 Comparison of $1/x$ with the weight factor $K_\epsilon(x)$

$$\int d^3x\, |f(x)|^2 < \infty.$$

Let $\psi(x)$ and $\phi(x)$ be two square integrable functions. The identity

$$\int d^3x\, |\psi(x) - \lambda\phi(x)|^2 \geq 0$$

yields with the choice

$$\lambda = \frac{\int d^3x\, \phi^+(x)\psi(x)}{\int d^3x\, |\phi(x)|^2}$$

the Schwarz inequality

$$\left|\int d^3x\, \phi^+(x)\psi(x)\right|^2 \leq \int d^3x\, |\psi(x)|^2 \int d^3x'\, |\phi(x')|^2 .$$

The differential operators $-i\hbar\nabla$ and $-(\hbar^2/2m)\Delta$, which we associated with momentum and kinetic energy, and the potential energy $V(\boldsymbol{x})$ all have the following properties,

$$\int d^3\boldsymbol{x}\,\phi^+(\boldsymbol{x})\frac{\hbar}{\mathrm{i}}\nabla\psi(\boldsymbol{x}) = \left(\int d^3\boldsymbol{x}\,\psi^+(\boldsymbol{x})\frac{\hbar}{\mathrm{i}}\nabla\phi(\boldsymbol{x})\right)^+, \tag{2.13}$$

$$\int d^3\boldsymbol{x}\,\phi^+(\boldsymbol{x})\Delta\psi(\boldsymbol{x}) = \left(\int d^3\boldsymbol{x}\,\psi^+(\boldsymbol{x})\Delta\phi(\boldsymbol{x})\right)^+, \tag{2.14}$$

and

$$\int d^3\boldsymbol{x}\,\phi^+(\boldsymbol{x})V(\boldsymbol{x})\psi(\boldsymbol{x}) = \left(\int d^3\boldsymbol{x}\,\psi^+(\boldsymbol{x})V(\boldsymbol{x})\phi(\boldsymbol{x})\right)^+. \tag{2.15}$$

Equation (2.15) is a consequence of the fact that $V(\boldsymbol{x})$ is a real function. Equations (2.13, 2.14) are a direct consequence of partial integrations and the fact that boundary terms at $|\boldsymbol{x}| \to \infty$ vanish under the assumptions that we had imposed on the wave functions.

If two operators A_x and B_x have the property

$$\int d^3\boldsymbol{x}\,\phi^+(\boldsymbol{x})A_x\psi(\boldsymbol{x}) = \left(\int d^3\boldsymbol{x}\,\psi^+(\boldsymbol{x})B_x\phi(\boldsymbol{x})\right)^+, \tag{2.16}$$

for *all* wave functions of interest, then B_x is denoted as *adjoint* to the operator A_x. The mathematical notation for the adjoint operator to A_x is A_x^+,

$$B_x = A_x^+.$$

Complex conjugation of (2.16) then immediately tells us $B_x^+ = A_x$.

An operator with the property $A_x^+ = A_x$ is denoted as a *self-adjoint* or *hermitian* operator[2]. Self-adjoint operators are important in quantum mechanics because they yield real expectation values,

[2]We are not addressing matters of definition of domains of operators in function spaces, see e.g. [21] or Problem 2.6. If the operators A_x^+ and A_x can be defined on different classes of functions, and $A_x^+ = A_x$ holds on the intersections of their domains, then A_x is usually denoted as a *symmetric operator*. The notion of self-adjoint operator requires identical domains for both A_x and A_x^+ such that the domain of neither operator can be extended. If the conditions on the domains are violated, we can e.g. have a situation where A_x has no eigenfunctions at all, or where the eigenvalues of A_x are complex and the set of eigenfunctions is overcomplete. Hermiticity is sometimes defined as equivalent to symmetry or as equivalent to the more restrictive notion of self-adjointness of operators. We define Hermiticity as self-adjointness.

$$
\begin{aligned}
(\langle A\rangle_\psi)^+ &= \left(\int d^3x\,\psi^+(x)A_x\psi(x)\right)^+ = \int d^3x\,\psi^+(x)A_x^+\psi(x)\\
&= \int d^3x\,\psi^+(x)A_x\psi(x) = \langle A\rangle_\psi.
\end{aligned}
$$

Observable quantities like energy or momentum or location of a particle are therefore implemented through self-adjoint operators, e.g. momentum $\boldsymbol{p}$ is implemented through the self-adjoint differential operator $-i\hbar\nabla$. We have seen one method to figure this out in equation (1.21). We will see another method in equations (4.26, 4.27).

Self-adjoint operators have the further important property that their eigenfunctions yield *complete sets* of functions. Schematically this means the following: Suppose we can enumerate all constants a_n and functions $\psi_n(x)$ which satisfy the equation

$$
A_x\psi_n(x) = a_n\psi_n(x) \tag{2.17}
$$

with the set of discrete indices $\boldsymbol{n}$. The constants a_n are *eigenvalues* and the functions $\psi_n(x)$ are *eigenfunctions* of the operator A_x. Hermiticity of the operator A_x implies orthogonality of eigenfunctions for different eigenvalues,

$$
\begin{aligned}
a_n\int d^3x\,\psi_m^+(x)\psi_n(x) &= \int d^3x\,\psi_m^+(x)A_x\psi_n(x)\\
&= \left(\int d^3x\,\psi_n^+(x)A_x\psi_m(x)\right)^+\\
&= a_m\int d^3x\,\psi_m^+(x)\psi_n(x)
\end{aligned}
$$

and therefore

$$
\int d^3x\,\psi_m^+(x)\psi_n(x) = 0 \text{ if } a_n \neq a_m.
$$

However, even if $a_n = a_m$ for different indices $\boldsymbol{n} \neq \boldsymbol{m}$ (i.e. if the eigenvalue a_n is *degenerate* because there exist at least two eigenfunctions with the same eigenvalue), one can always chose orthonormal sets of eigenfunctions for a degenerate eigenvalue. We therefore require

$$
\int d^3x\,\psi_m^+(x)\psi_n(x) = \delta_{m,n}. \tag{2.18}
$$

Completeness of the set of functions $\psi_n(x)$ means that an "arbitrary" function $f(x)$ can be expanded in terms of the eigenfunctions of the self-adjoint operator A_x in the form

$$
f(x) = \sum_n c_n\psi_n(x) \tag{2.19}
$$

with expansion coefficients

$$c_n = \int d^3x\, \psi_n^+(x) f(x). \tag{2.20}$$

If we substitute equation (2.20) into (2.19) and (in)formally exchange integration and summation, we can express the completeness property of the set of functions $\psi_n(\boldsymbol{x})$ in the *completeness relation*

$$\sum_n \psi_n(x)\psi_n^+(x') = \delta(x - x'). \tag{2.21}$$

Both the existence and the meaning of the series expansions (2.19, 2.20) depends on what large a class of "arbitrary" functions $f(\boldsymbol{x})$ one considers. Minimal constraints require boundedness of $f(\boldsymbol{x})$, and continuity if the series (2.19) is supposed to converge pointwise. The default constraints in non-relativistic quantum mechanics are continuity of wave functions $\psi(\boldsymbol{x})$ to ensure validity of the Schrödinger equation with at most finite discontinuities in potentials $V(\boldsymbol{x})$, and normalizability. Under these circumstances the expansion (2.19, 2.20) for a wave function $f(\boldsymbol{x}) \equiv \psi(\boldsymbol{x})$ will converge pointwise to $\psi(\boldsymbol{x})$. However, it is convenient for many applications of quantum mechanics to use limiting forms of wave functions which are not normalizable in the sense of equation (2.12) any more, e.g. plane wave states $\psi_{\boldsymbol{k}}(\boldsymbol{x}) \propto \exp(i\boldsymbol{k}\cdot\boldsymbol{x})$, and we will frequently also have to expand non-continuous functions, e.g. functions of the form $f(\boldsymbol{x}) = V(\boldsymbol{x})\psi(\boldsymbol{x})$ with a discontinuous potential $V(\boldsymbol{x})$. However, finally we only have to use expansions of the form (2.19, 2.20) in the evaluation of integrals of the form $\int d^3x g^+(x) f(x)$, and here the concept of *convergence in the mean* comes to our rescue in the sense that substitution of the series expansion (2.19, 2.20) in the integral will converge to the same value of the integral, even if the expansion (2.19, 2.20) does not converge pointwise to the function $f(\boldsymbol{x})$.

A more thorough discussion of completeness of sets of eigenfunctions of self-adjoint operators in the relatively simple setting of wave functions confined to a finite one-dimensional interval is presented in Appendix C. However, for a first reading I would recommend to accept the series expansions (2.19, 2.20) with the assurance that substitutions of these series expansions is permissible in the calculation of observables in quantum mechanics.

2.3 Problems

2.1. Suppose the function $f(x)$ has only first order zeros, i.e. we have non-vanishing slope at all nodes x_i of the function,

$$f(x_i) = 0 \Rightarrow f'(x_i) \equiv \left.\frac{df(x)}{dx}\right|_{x=x_i} \neq 0.$$

Prove the following property of the δ function:

$$\delta(f(x)) = \sum_i \frac{1}{|f'(x_i)|}\delta(x - x_i).$$

2.2. Calculate the Fourier transforms of the following functions, where in all cases $-\infty < x < \infty$. Do not use any electronic integration program.

2.2a. $\psi_1(x) = \exp(-ax^2)$, $\Re a > 0$,

2.2b. $\psi_2(x) = 1/(a^2 + x^2)$, $a > 0 \in \mathbb{R}$,

2.2c. $\psi_3(x) = x^n \exp(-a|x|)$, $a > 0 \in \mathbb{R}$, where n is a natural number.

2.3. The functions $f_1(x) = \exp(-x^2)$ and $f_2(x) = \exp(-|x|)$ are normalizable to functions $d(x)$ in the sense of equation (2.6). Use this to find other derivations of the Fourier representation of the δ function similar to equation (2.10).

2.4. We consider a finite interval $[a, b]$ together with the set $C^{(1,\alpha)}[a, b]$ of complex valued functions which are continuous in $[a, b]$ and differentiable in (a, b), and satisfy the pseudo-periodicity condition

$$\psi(b) = \exp(i\alpha)\psi(a), \quad \alpha \in \mathbb{R}.$$

Show that the differential operator $-id/dx$ is self-adjoint on $C^{(1,\alpha)}[a, b]$. Give a complete set of eigenstates of $-id/dx$ in $C^{(1,\alpha)}[a, b]$.

2.5. We consider the finite interval $[a, b]$ together with the set $C^{(2),0}[a, b]$ of complex valued functions which are continuous in $[a, b]$ and second order differentiable in (a, b), and satisfy the boundary conditions

$$\psi(a) = \psi(b) = 0.$$

Show that that the differential operator d^2/dx^2 is self-adjoint on $C^{(2),0}[a, b]$. Give a complete set of eigenstates of d^2/dx^2 in $C^{(2),0}[a, b]$.

2.6. We consider the finite interval $[a, b]$ together with the set $C^{(1),0}[a, b]$ of complex valued functions which are continuous in $[a, b]$ and differentiable in (a, b), and satisfy the boundary conditions

$$\psi(a) = \psi(b) = 0.$$

Show that the symmetric differential operator $h_1 = -id/dx$ with domain $C^{(1),0}[a, b]$ is not self-adjoint in the sense that h_1^+ can be defined on the larger set $L_2[a, b]$ of square integrable functions over $[a, b]$.

Show that h_1 has no eigenstates, while h_1^+ has complex eigenvalues and an overcomplete set of eigenstates.

Chapter 3
Simple Model Systems

One-dimensional models and models with piecewise constant potentials have been used as simple model systems for quantum behavior ever since the inception of Schrödinger's equation. These models vary in their levels of sophistication, but their generic strength is the clear demonstration of important general quantum effects and effects of dimensionality of a quantum system at very little expense in terms of effort or computation. Simple model systems are therefore more than just pedagogical tools for teaching quantum mechanics. They also serve as work horses for the modeling of important quantum effects in nanoscience and technology, see e.g. [4, 20].

3.1 Barriers in quantum mechanics

Widely used models for quantum behavior in solid state electronics are described by piecewise constant potentials $V(\boldsymbol{x})$. This means that $V(\boldsymbol{x})$ attains constant values in different regions of space, and the transition between those regions of constant $V(\boldsymbol{x})$ appears through discontinuous jumps in the potential. Figure 3.1 shows an example of a piecewise constant potential.

The Schrödinger equation with a piecewise constant potential is easy to solve, and the solutions provide instructive examples for the impact of quantum effects on the motion of charge carriers through semiconductors and insulating barriers. We will first discuss the case of a rectangular barrier.

Figure 3.1 shows a cross section of a non-symmetric rectangular square barrier. The piecewise constant potential has values

$$V(x) = \begin{cases} 0, & x < 0, \\ \Phi_1, & 0 \leq x \leq L, \\ \Phi_2, & x > L. \end{cases}$$

R. Dick, *Advanced Quantum Mechanics*, Graduate Texts in Physics,
DOI 10.1007/978-3-319-25675-7_3

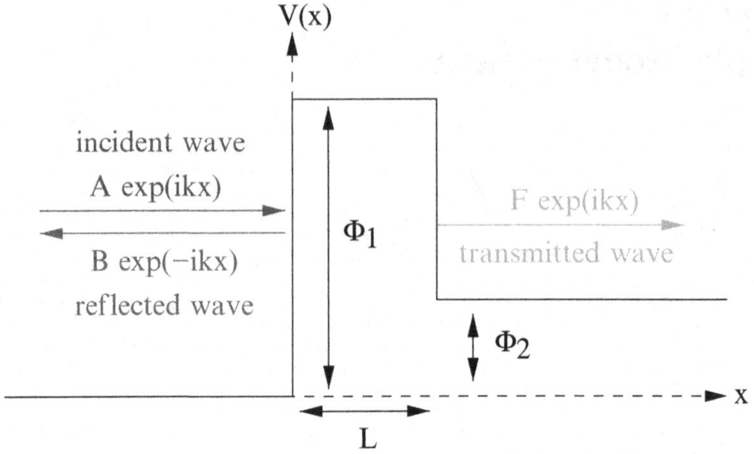

Fig. 3.1 A non-symmetric square barrier

with $\Phi_1 > \Phi_2 > 0$. This barrier impedes motion in the x direction. It can be used e.g. as a simple quantum mechanical model for a metal coated with an insulating layer. The region $x < 0$ would be inside the metal and the potential Φ_2 would be the energy which is required to liberate an electron from the metal if there would not be the insulating layer of thickness L. The energy Φ_1 is the energy which would classically be required for an electron to penetrate the layer.

Quantum problems with time-independent potentials are conveniently analyzed by using a Fourier transformation[1] from time t to energy E,

$$\psi(\boldsymbol{x},t) = \frac{1}{\sqrt{2\pi\hbar}}\int_{-\infty}^{\infty} dE\, \exp\!\left(-\frac{\mathrm{i}}{\hbar}Et\right)\psi(\boldsymbol{x},E), \tag{3.1}$$

$$\psi(\boldsymbol{x},E) = \frac{1}{\sqrt{2\pi}}\int_{-\infty}^{\infty} dt\, \exp\!\left(\frac{\mathrm{i}}{\hbar}Et\right)\psi(\boldsymbol{x},t). \tag{3.2}$$

Substitution into the time-dependent Schrödinger equation (1.2) yields the time-independent Schrödinger equation[2]

$$E\psi(\boldsymbol{x},E) = -\frac{\hbar^2}{2m}\Delta\psi(\boldsymbol{x},E) + V(\boldsymbol{x})\psi(\boldsymbol{x},E). \tag{3.3}$$

[1]The normalization condition (1.20) implies that the function $\psi(\boldsymbol{x},E)$ does not exist in the sense of classical Fourier theory. We will therefore see in Section 5.2 that $\psi(\boldsymbol{x},E)$ is rather a series of δ-functions of the energy. This difficulty is usually avoided by using an *exponential ansatz* $\psi(\boldsymbol{x},t) = \psi(\boldsymbol{x},E)\exp(-\mathrm{i}Et/\hbar)$ instead of a full Fourier transformation. However, if one accepts the δ-function and corresponding extensions of classical Fourier theory, the transition to the time-independent Schrödinger equation through a formal Fourier transformation to the energy variable is logically more satisfactory.

[2]E. Schrödinger, Annalen Phys. 384, 361 (1926). Schrödinger found the time-independent equation first and published the time-dependent equation (1.2) five months later.

The potential on Figure 3.1 depends only on x. In this case we can also eliminate the derivatives with respect to y and z through further Fourier transformations,

$$\psi(\boldsymbol{x}, E) = \frac{1}{2\pi}\int_{-\infty}^{\infty} dk_2 \int_{-\infty}^{\infty} dk_3 \exp[\mathrm{i}(k_2 y + k_3 z)]\,\psi(x, k_2, k_3, E)$$

to find the time-independent Schrödinger equation for motion in the x direction,

$$E_1\psi(x, E_1) = -\frac{\hbar^2}{2m}\frac{\partial^2}{\partial x^2}\psi(x, E_1) + V(x)\psi(x, E_1). \tag{3.4}$$

Here

$$E_1 \equiv E - \hbar^2\frac{k_2^2 + k_3^2}{2m}, \quad \psi(x, E_1) \equiv \psi(x, k_2, k_3, E).$$

E_1 is the kinetic energy for motion in the x direction in the region $x < 0$.

Within each of the three separate regions $x < 0$, $0 < x < L$, and $x > L$ the potential attains a constant value, and equation (3.4) can be solved with a final Fourier transformation from x to k_1,

$$\psi(x, E_1) = \begin{cases} A\exp(\mathrm{i}k_1 x) + B\exp(-\mathrm{i}k_1 x), \ k_1 = \sqrt{2mE_1}/\hbar, \ x < 0, \\ C\exp(\mathrm{i}k_1'' x) + D\exp(-\mathrm{i}k_1'' x), \\ \quad k_1'' = \sqrt{2m(E_1 - \Phi_1)}/\hbar, \ 0 < x < L, \\ F\exp(\mathrm{i}k_1' x) + G\exp(-\mathrm{i}k_1' x), \\ \quad k_1' = \sqrt{2m(E_1 - \Phi_2)}/\hbar, \ x > L. \end{cases} \tag{3.5}$$

We must have $E_1 > 0$ because the absolute minimum of the potential determines a lower bound for the energy of a particle moving in the potential. However, the wave numbers k_1'' and k_1' can be real or imaginary depending on the magnitude of E_1. We define

$$k_1'' = -\mathrm{i}\kappa, \quad k_1' = \mathrm{i}\kappa',$$

with the conventions $\kappa > 0$, $\kappa' > 0$, if k_1'' or k_1' are imaginary.

The wave function (3.5) is not yet the complete solution to our problem, because we have to impose junction conditions on the coefficients at the transition points $x = 0$ and $x = L$ to ensure that the Schrödinger equation is also satisfied in those points. This will be done below. However, we can already discuss the meaning of the six different exponential terms appearing in (3.5). The wave function $\psi(x, E_1)$ is multiplied by the time-dependent exponential $\exp(-\mathrm{i}E_1 t/\hbar)$ in the transition from $\psi(x, E_1)$ to the time-dependent wave function $\psi(x, t)$ for motion in x direction,

$$\psi(x, t) = \frac{1}{\sqrt{2\pi\hbar}}\int_0^{\infty} dE_1 \exp\left(-\frac{\mathrm{i}}{\hbar}E_1 t\right)\psi(x, E_1). \tag{3.6}$$

A single monochromatic component therefore corresponds to a time-dependent wave function proportional to $\psi(x, E_1)\exp(-iE_1 t/\hbar)$. The term $A\exp[i(k_1 x - E_1 t/\hbar)]$ corresponds to a right moving wave in the region $x < 0$, while the term $B\exp[-i(k_1 x + E_1 t/\hbar)]$ is a left moving wave. Similar identifications apply to the C and D components if k_1'' is real, and to the F and G components if k_1' is real. Otherwise, these components will correspond to exponentially damped or growing wave functions, which requires $G = 0$ if $\kappa' = -ik_1' > 0$ is real, to avoid divergence of the wave function for $x \to \infty$.

There is a subtle point here that needs to be emphasized because it is also relevant for potential scattering theory in three dimensions. We have just realized that the monochromatic wave function $\psi(x, E_1)$ describes a particle of energy E_1 (for the motion in x direction) simultaneously as left and right moving particles in the regions where the wave number is real. The energy dependent wave function always simultaneously describes all states of the particle with energy E_1, but does not yield a time resolved picture of what happens to a particle in the presence of the potential $V(x)$. Let us e.g. assume that we shoot a particle of energy E_1 at the potential $V(x)$ from the left. The component $A\exp[i(k_1 x - E_1 t/\hbar)]$ describes the initially incident particle, while the component $B\exp[-i(k_1 x + E_1 t/\hbar)]$ describes a particle that is reflected by the barrier. The component $F\exp[i(k_1' x - E_1 t/\hbar)]$, on the other hand, describes a particle which went across the barrier (if $E_1 > \Phi_1$), or a particle that penetrated the barrier (without damaging the barrier!) if $\Phi_1 > E_1 > \Phi_2$.

The calculation of expectation values sheds light on the property of the monochromatic wave function $\psi(x, E_1)\exp(-iE_1 t/\hbar)$ to describe all states of a particle of energy E_1 simultaneously. The expectation values both for location $\langle x\rangle$ and momentum $\langle p\rangle$ of a particle described by a monochromatic wave function are time-independent, i.e. a single monochromatic wave function can never describe the time evolution of motion of a particle in the sense of first corresponding to an incident wave from the left, and later either to a reflected wave or a transmitted wave. A time resolved picture describing sequential events really requires superposition of several monochromatic components (3.6) with contributions from many different energies. Stated differently, the wave function of a moving particle can never correspond to only one exact value for the energy of the particle. Building wave functions for moving particles will always require superposition of different energy values, which corresponds to an uncertainty in the energy of the particle. Stated in yet another way: The energy resolved picture described by the Schrödinger equation in the energy domain (3.3) describes all processes happening with energy E, whereas the time-dependent Schrödinger equation describes processes happening at time t. If the time-dependent wave function of the system is indeed monochromatic, $\psi(x, t) = \psi(x, E_1)\exp(-iE_1 t/\hbar)$, then we imply that all these processes at energy E_1 happen simultaneously, e.g. because we have a continuous particle beam of energy E_1 incident on the barrier.

The monochromatic wave function can still tell us a lot about the behavior of particles in the presence of the potential barrier $V(x)$. We choose as an initial condition a particle moving against the barrier from the left. Then we have to set $G = 0$ in the solution above irrespective of whether k_1' is real or imaginary, because in the real case this component would correspond to a particle hitting the barrier from the right, and in the imaginary case $G = 0$ was imposed anyway from the requirement that the wave function cannot diverge.

Before we can proceed, we have to discuss junction conditions for wave functions at points where the potential is discontinuous.

A finite jump in $V(x)$ translates through the time-independent Schrödinger equation into a finite jump in $d^2\psi(x)/dx^2$, which means a jump in the slope of $d\psi(x)/dx$, but not a discontinuity in $d\psi(x)/dx$. Therefore both $\psi(x)$ and $d\psi(x)/dx$ have to remain continuous across a finite jump in the potential[3]. This means that the wave function $\psi(x)$ remains smooth across a finite jump in $V(x)$. On the other hand, an infinite jump in $V(x)$ only requires continuity, but not smoothness of $\psi(x)$.

The requirement of smoothness of the wave function yields the junction conditions

$$\begin{aligned} A + B &= C + D \\ k_1(A - B) &= k_1''(C - D) \\ C\exp(ik_1''L) + D\exp(-ik_1''L) &= F\exp(ik_1'L) \\ k_1''[C\exp(ik_1''L) - D\exp(-ik_1''L)] &= k_1'F\exp(ik_1'L) \end{aligned}$$

Elimination of C and D yields

$$\begin{aligned} 2k_1k_1''A &= \left[k_1''(k_1 + k_1')\cos(k_1''L) - i(k_1k_1' + k_1''^2)\sin(k_1''L)\right]F\exp(ik_1'L), \\ 2k_1k_1''B &= \left[k_1''(k_1 - k_1')\cos(k_1''L) - i(k_1k_1' - k_1''^2)\sin(k_1''L)\right]F\exp(ik_1'L). \end{aligned}$$

Note that

$$\cos(k_1''L) = \cosh(\kappa L), \quad \sin(k_1''L) = -i\sinh(\kappa L).$$

If we decompose the wave function to the left and the right of the barrier into incoming, reflected, and transmitted components

$$\psi_{in}(x) = A\exp(ik_1x), \quad \psi_{re}(x) = B\exp(-ik_1x), \quad \psi_{tr}(x) = F\exp(ik_1'x),$$

then the probability current density (1.18) yields

$$j_{in} = \frac{\hbar k_1}{m}|A|^2, \quad j_{re} = -\frac{\hbar k_1}{m}|B|^2, \quad j_{tr} = \frac{\hbar}{m}|F|^2\Re k_1'.$$

[3]The time-dependent Schrödinger equation permits discontinuous wave functions $\psi(x,t)$ even for smooth potentials, because there can be a trade-off between the derivative terms, see e.g. Problem 3.15.

In the last equation we used that k_1' is either real or imaginary. The reflection and transmission coefficients from the barrier are then

$$R = \frac{|j_{re}|}{|j_{in}|} = \frac{|B|^2}{|A|^2}, \quad T = \frac{|j_{tr}|}{|j_{in}|} = \frac{|F|^2 \Re k_1'}{|A|^2 k_1}.$$

This yields in all cases $0 \leq T = 1 - R \leq 1$. The transmission coefficient is $T = 0$ for $0 < E_1 \leq \Phi_2$,

$$\begin{aligned} T &= 4\sqrt{E_1(E_1 - \Phi_2)}(\Phi_1 - E_1) \\ &\times \left[(\Phi_1 - E_1)\left(2E_1 - \Phi_2 + 2\sqrt{E_1(E_1 - \Phi_2)}\right) \right. \\ &\left. + \Phi_1(\Phi_1 - \Phi_2)\sinh^2\!\left(\sqrt{2m(\Phi_1 - E_1)}L/\hbar\right) \right]^{-1} \end{aligned}$$

for $\Phi_2 \leq E_1 \leq \Phi_1$, and

$$\begin{aligned} T &= 4\sqrt{E_1(E_1 - \Phi_2)}(E_1 - \Phi_1) \\ &\times \left[(E_1 - \Phi_1)\left(2E_1 - \Phi_2 + 2\sqrt{E_1(E_1 - \Phi_2)}\right) \right. \\ &\left. + \Phi_1(\Phi_1 - \Phi_2)\sin^2\!\left(\sqrt{2m(E_1 - \Phi_1)}L/\hbar\right) \right]^{-1} \end{aligned}$$

for $E_1 \geq \Phi_1$. Classical mechanics, on the other hand predicts $T = 0$ for $E_1 < \Phi_1$ and $T = 1$ for $E_1 > \Phi_1$, in stark contrast to the quantum mechanical transmission coefficient shown in Figure 3.2.

The phenomenon that particles can tunnel through regions even when they do not have the required energy is denoted as *tunnel effect*. It has been observed in many instances in nature and technology, e.g. in the α decay of radioactive nuclei (Gamow, 1928) or electron tunneling in heavily doped pn junctions (Esaki, 1958). Esaki diodes actually provide a beautiful illustration of the interplay of two quantum effects, *viz.* energy bands in solids and tunneling. Charge carriers can tunnel from one energy band into a different energy band in heavily doped pn junctions. We will discuss energy bands in Chapter 10.

Quantum mechanical tunneling is also used e.g. in scanning tunneling microscopes (Binnig & Rohrer, 1982), and in flash memory and magnetic tunnel junction devices[4].

It is easy to understand from our results for the transmission probability why quantum mechanical tunneling plays such an important role in modern memory devices. If we want to have a memory device which is electrically controlled, then

[4]Magnetic tunnel junctions provide yet another beautiful example of the interplay of two quantum effects – tunneling and exchange interactions. Exchange interactions will be discussed in Chapter 17.

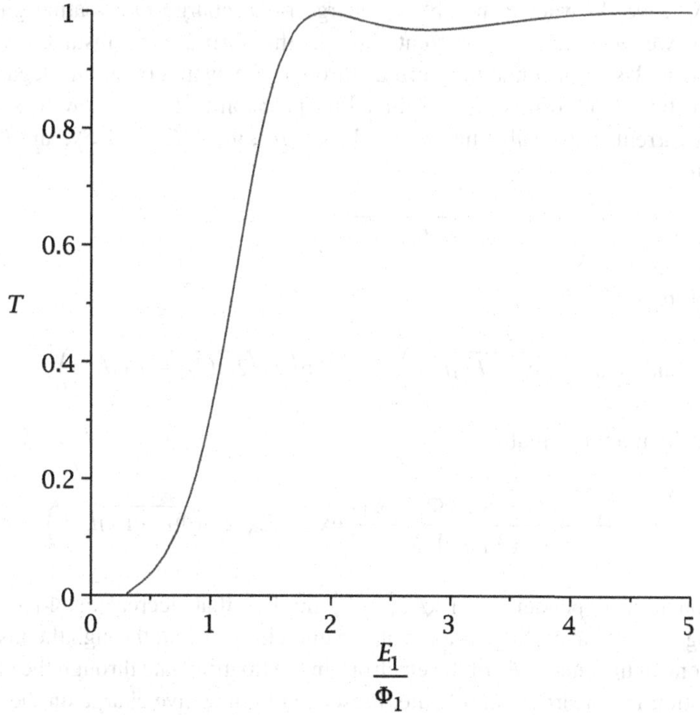

Fig. 3.2 The transmission coefficient for a non-symmetric square barrier. The curve calculated here corresponds to $m = 511\,\mathrm{keV}/c^2$, $\Phi_1 = 10$ eV, $\Phi_2 = 3$ eV, $L = 2$ Å

Fig. 3.3 A simplified schematic of a flash memory cell. The tunneling barrier is the thin section of the insulator between the floating gate and the semiconductor

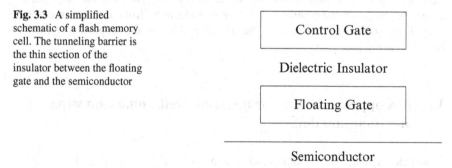

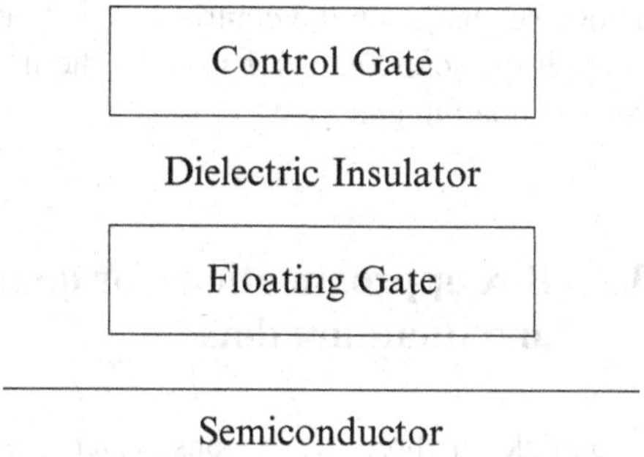

apparently the information bits 0 and 1 can be encoded through the two states of a device being electrically charged or neutral. If we also want to maintain storage of the information even when the power supply is switched off (a *non-volatile memory*), then the device should not discharge spontaneously, i.e. it should be electrically insulated. The device should therefore be a conductor which is surrounded by insulating material. Such a device is called a *floating gate* in flash memory devices, see Figure 3.3.

However, we do want to be able to charge or discharge the floating gate, i.e. eventually we want to run a current through the surrounding insulator without destroying it. Using a tunneling current through the insulator is an elegant way to achieve this. Our results for the tunneling probability tell us how to switch a tunneling current. If we substitute $m = 511\ \mathrm{keV}/c^2$, $\Phi_1 - E_1 \simeq 1$ eV, and $L \simeq 10$ nm, we find

$$\sqrt{2m(\Phi_1 - E_1)}L/\hbar \simeq 51$$

and therefore

$$\sinh^2\!\left(\sqrt{2m(\Phi_1 - E_1)}L/\hbar\right) \simeq \frac{1}{4}\exp\!\left(2\sqrt{2m(\Phi_1 - E_1)}L/\hbar\right),$$

i.e. in excellent approximation

$$T \simeq 16\frac{\sqrt{E_1(E_1 - \Phi_2)}(\Phi_1 - E_1)}{\Phi_1(\Phi_1 - \Phi_2)}\exp\!\left(-2\sqrt{2m(\Phi_1 - E_1)}L/\hbar\right).$$

The exponential dependence on $\sqrt{\Phi_1 - E_1}$ implies that decreasing $\Phi_1 - E_1$ by increasing E_1 will have a huge impact on the tunneling current through the insulator. We can control the energy E_1 of the electrons in the floating gate through the electron concentration in a nearby control gate. Presence of a negative charge on the nearby control gate will increase the energy of any electrons stored in the floating gate and allow them to tunnel into a conducting sink (usually a semiconductor) opposite to the control gate. This process will discharge the floating gate. On the other hand, a positive charge on the control gate will attract electrons from an electron current through the semiconductor towards the insulating barrier and help them to tunnel into the floating gate.

3.2 Box approximations for quantum wells, quantum wires and quantum dots

A particle in three dimensions which can move freely in two directions, but is confined in one direction, is said to be confined in a quantum well. A particle which can move freely only in one direction but is confined in two directions is confined in a quantum wire. Finally, a particle which is confined to a small region of space is confined to a quantum dot. We will discuss energy levels and wave functions of particles in all three situations in the approximation of confinement to rectangular (box-like) regions. For the quantum well this means that our particle will be confined to the region $0 < x < L_1$, but it can move freely in y and z direction. The particle in the quantum wire is confined in x and y direction to $0 < x < L_1$, $0 < y < L_2$, but it can move freely in the z direction. Finally, box approximation for a quantum dot means that the particle is confined to the box $0 < x < L_1, 0 < y < L_2, 0 < z < L_3$.

We will assume strict confinement in this section, i.e. the wave function of the particle vanishes outside of the allowed region while the wave function inside the region must continuously go to zero at the boundaries of the allowed region.

We gauge the energy axis such that in the allowed region the potential energy of the particle vanishes, $V(\boldsymbol{x}) = 0$, i.e. the time-independent three-dimensional Schrödinger equation in the allowed region takes the form

$$E\psi(\boldsymbol{x}) = -\frac{\hbar^2}{2m}\Delta\psi(\boldsymbol{x}). \tag{3.7}$$

Substitution of the Fourier decomposition

$$\psi(\boldsymbol{x}) = \frac{1}{\sqrt{2\pi}^3}\int d^3\boldsymbol{k}\,\psi(\boldsymbol{k})\exp(\mathrm{i}\boldsymbol{k}\cdot\boldsymbol{x})$$

yields $k = \sqrt{2mE}/\hbar$ and the general solution for given energy E takes the form

$$\psi(\boldsymbol{x}) = \int d^2\hat{\boldsymbol{k}}\,A(\hat{\boldsymbol{k}})\exp\!\left(\frac{\mathrm{i}}{\hbar}\sqrt{2mE}\hat{\boldsymbol{k}}\cdot\boldsymbol{x}\right), \quad \hat{\boldsymbol{k}}^2 = 1.$$

On the other hand, equation (3.7) tells us that the energy of a plane wave $\psi(\boldsymbol{x}) = \exp(\mathrm{i}\boldsymbol{k}\cdot\boldsymbol{x})/\sqrt{2\pi}^3$ of momentum $\boldsymbol{p} = \hbar\boldsymbol{k}$ is

$$E = \frac{\hbar^2\boldsymbol{k}^2}{2m}. \tag{3.8}$$

If we have no confinement condition at all, our particle is a free particle and equation (3.8) is the kinetic energy of a free non-relativistic particle of momentum $\boldsymbol{p} = \hbar\boldsymbol{k}$.

Energy levels in a quantum well

If we have a confinement condition in x-direction, e.g. $\psi(0, y, z) = 0$ and $\psi(L_1, y, z) = 0$, then we have to superimpose plane wave solutions in x direction to form a standing wave with nodes at the boundary points, and we find solutions

$$\psi_{n_1,k_2,k_3}(\boldsymbol{x}) = \frac{1}{\pi\sqrt{2L_1}}\exp[\mathrm{i}(k_2 y + k_3 z)]\sin\!\left(\frac{n_1\pi x}{L_1}\right), \tag{3.9}$$

with integer $n_1 \in \mathbb{N}$ and energy

$$E_{n_1,k_2,k_3} = \frac{\hbar^2}{2m}\left(k_2^2 + k_3^2 + \frac{n_1^2\pi^2}{L_1^2}\right).$$

The energy of the particle is therefore determined by the discrete quantum number n_1 and the continuous wave numbers k_2 and k_3.

Energy levels in a quantum wire

If the particle is confined both in the x-direction to the region $0 < x < L_1$ and in the y-direction to the region $0 < y < L_2$, the boundary conditions $\psi(0, y, z) = 0$, $\psi(L_1, y, z) = 0$, $\psi(x, 0, z) = 0$ and $\psi(x, L_2, z) = 0$ yield

$$\psi_{n_1,n_2,k_3}(\boldsymbol{x}) = \sqrt{\frac{2}{\pi L_1 L_2}} \exp(\mathrm{i}k_3 z) \sin\left(\frac{n_1\pi x}{L_1}\right) \sin\left(\frac{n_2\pi x}{L_2}\right), \tag{3.10}$$

and the energy of the particle is determined by the discrete quantum numbers n_1 and n_2 and the continuous wave number k_3 for motion in z direction,

$$E_{n_1,n_2,k_3} = \frac{\pi^2\hbar^2}{2m}\left(\frac{n_1^2}{L_1^2} + \frac{n_2^2}{L_2^2}\right) + \frac{\hbar^2 k_3^2}{2m}.$$

Energy levels in a quantum dot

If the particle is confined to the region $0 < x < L_1$, $0 < y < L_2$, $0 < z < L_3$, the conditions of vanishing wave function on the boundaries yields normalized states

$$\psi_{n_1,n_2,n_3}(\boldsymbol{x}) = \sqrt{\frac{8}{L_1 L_2 L_3}} \sin\left(\frac{n_1\pi x}{L_1}\right) \sin\left(\frac{n_2\pi y}{L_2}\right) \sin\left(\frac{n_3\pi z}{L_3}\right), \tag{3.11}$$

and the energy levels are determined in terms of three discrete quantum numbers,

$$E_{n_1,n_2,n_3} = \frac{\pi^2\hbar^2}{2m}\left(\frac{n_1^2}{L_1^2} + \frac{n_2^2}{L_2^2} + \frac{n_3^2}{L_3^2}\right). \tag{3.12}$$

Degeneracy of quantum states

If two or more different quantum states have the same energy, the quantum states are said to be degenerate, and the corresponding energy level is also denoted as degenerate. This happens e.g. for the quantum wire and the quantum dot if at least two of the length scales L_i have the same value. We will discuss the quantum dot (3.12, 3.11) with $L_1 = L_2 = L_3 \equiv L$ as an example. This cubic quantum dot has energy levels

$$E_{n_1,n_2,n_3} = \left(n_1^2 + n_2^2 + n_3^2\right) \frac{\pi^2 \hbar^2}{2mL^2}.$$

The lowest energy level

$$E_{1,1,1} = 3\frac{\pi^2 \hbar^2}{2mL^2}$$

corresponds to a unique quantum state $\psi_{1,1,1}(\boldsymbol{x})$ and is therefore not degenerate. However, the next allowed energy value

$$E_{1,1,2} = E_{1,2,1} = E_{2,1,1} = 6\frac{\pi^2 \hbar^2}{2mL^2}$$

is realized for three different wave functions $\psi_{1,1,2}(\boldsymbol{x})$, $\psi_{1,2,1}(\boldsymbol{x})$ and $\psi_{2,1,1}(\boldsymbol{x})$, and is therefore *three-fold degenerate*. Three-fold degeneracy is also realized for the next two energy levels

$$E_{1,2,2} = E_{2,1,2} = E_{2,2,1} = 9\frac{\pi^2 \hbar^2}{2mL^2}$$

and

$$E_{1,1,3} = E_{1,3,1} = E_{3,1,1} = 11\frac{\pi^2 \hbar^2}{2mL^2}.$$

The next energy level is again non-degenerate,

$$E_{2,2,2} = 12\frac{\pi^2 \hbar^2}{2mL^2}.$$

Then follows a six-fold degenerate energy level,

$$E_{1,2,3} = E_{2,3,1} = E_{3,1,2} = E_{1,3,2} = E_{3,2,1} = E_{2,1,3} = 14\frac{\pi^2 \hbar^2}{2mL^2}.$$

3.3 The attractive δ function potential

The attractive δ function potential

$$V(x) = -\mathcal{W}\delta(x), \quad \mathcal{W} > 0,$$

provides a simple model system for co-existence of free states and bound states of particles in a potential.

Positive energy solutions of the stationary Schrödinger equation for the δ function potential must have the form

$$\psi_k(x) = \sum_{\pm} \Theta(\pm x)\left[A_\pm \exp(ikx) + B_\pm \exp(-ikx)\right], \quad \hbar k = \sqrt{2mE},$$

and normalizability limits the negative energy solutions to the from

$$\psi_\kappa(x) = \sum_{\pm} \Theta(\pm x) C_\pm \exp(\mp\kappa x), \quad \hbar\kappa = \sqrt{-2mE}.$$

These solutions must be continuous in order not to generate $\delta'(x)$ terms which would violate the Schrödinger equation,

$$A_+ + B_+ = A_- + B_-, \quad C_+ = C_-. \tag{3.13}$$

On the other hand, integrating the Schrödinger equation from $x = -\epsilon$ to $x = \epsilon$ and taking the limit $\epsilon \to 0+$ yields the junction conditions

$$\lim_{\epsilon\to 0+}\left(\left.\frac{d\psi(x)}{dx}\right|_{x=\epsilon} - \left.\frac{d\psi(x)}{dx}\right|_{x=-\epsilon}\right) = -\frac{2m}{\hbar^2}\mathcal{W}\psi(0).$$

This implies

$$ik(A_+ - B_+ - A_- + B_-) = -\frac{m}{\hbar^2}\mathcal{W}(A_+ + B_+ + A_- + B_-) \tag{3.14}$$

for the free states and

$$\kappa = \frac{m}{\hbar^2}\mathcal{W} \tag{3.15}$$

for the bound states.

Equation (3.15) tells us that there exists one bound state for $\mathcal{W} > 0$ with energy

$$E_\kappa = -\frac{m}{2\hbar^2}\mathcal{W}^2.$$

The normalized bound state is

$$\psi_\kappa(x) = \sqrt{\kappa}\exp(-\kappa|x|). \tag{3.16}$$

For the free states, we first look at solutions which are right or left moving plane waves $\exp(\pm ikx)/\sqrt{2\pi}$ on the half-line $x > 0$, i.e. we solve equations (3.13) and (3.14) first under the conditions $A_+ = 1/\sqrt{2\pi}$, $B_+ = 0$, and then under the conditions $A_+ = 0$, $B_+ = 1/\sqrt{2\pi}$. This yields solutions

$$\psi_{+k}(x) = \frac{1}{\sqrt{2\pi}}\exp(ikx) + \sqrt{\frac{2}{\pi}}\Theta(-x)\frac{\kappa}{k}\sin(kx), \tag{3.17}$$

$$\psi_{-k}(x) = \frac{1}{\sqrt{2\pi}}\exp(-ikx) + \sqrt{\frac{2}{\pi}}\Theta(-x)\frac{\kappa}{k}\sin(kx).$$

The free solutions can be unified if we also allow for negative values of k (recall that until now k was defined positive from $\hbar k = \sqrt{2mE}$),

$$\psi_k(x) = \frac{1}{\sqrt{2\pi}}\exp(ikx) + \sqrt{\frac{2}{\pi}}\Theta(-x)\frac{\kappa}{k}\sin(kx). \tag{3.18}$$

These states are free right or left moving plane waves for $x > 0$, but they do not provide orthonormal bases for the scattering states in the attractive δ potential. We will construct two orthonormal bases below in (3.21, 3.22) and (3.25, 3.26), respectively.

Another useful representation for the free states is motivated by considering the outgoing waves with amplitudes A_+ and B_- as consequences of the incident waves with amplitudes A_- and B_+. The junction conditions (3.13, 3.14) yield

$$\begin{pmatrix} A_+ \\ B_- \end{pmatrix} = \frac{1}{k - i\kappa}\begin{pmatrix} k & i\kappa \\ i\kappa & k \end{pmatrix} \cdot \begin{pmatrix} A_- \\ B_+ \end{pmatrix}. \tag{3.19}$$

The unitary matrix

$$\underline{S} = \frac{1}{k - i\kappa}\begin{pmatrix} k & i\kappa \\ i\kappa & k \end{pmatrix} = \frac{1}{\sqrt{E} - i\sqrt{B}}\begin{pmatrix} \sqrt{E} & i\sqrt{B} \\ i\sqrt{B} & \sqrt{E} \end{pmatrix} \tag{3.20}$$

is also known as a *scattering matrix* because it describes scattering of incoming waves off the potential. Here $B \equiv -E_\kappa$ is the binding energy of the bound state.

The scattering matrix can be used to read off the reflection and transmission coefficients for the δ function potential,

$$R = \left|\frac{\partial B_-}{\partial A_-}\right|^2 = \left|\frac{\partial A_+}{\partial B_+}\right|^2 = \frac{\kappa^2}{k^2 + \kappa^2} = \frac{B}{E + B},$$

$$T = \left|\frac{\partial A_+}{\partial A_-}\right|^2 = \left|\frac{\partial B_-}{\partial B_+}\right|^2 = \frac{k^2}{k^2 + \kappa^2} = \frac{E}{E + B}.$$

In many situations it is also convenient to use even and odd solutions of the Schrödinger equation. Odd (or *negative parity*) solutions $\psi(x) = -\psi(-x)$ must satisfy $A_+ = -B_-$, $B_+ = -A_-$. Solving equations (3.13) and (3.14) with these conditions yields the negative parity solutions

$$\psi_{k,-}(x) = \frac{1}{\sqrt{\pi}}\sin(kx). \tag{3.21}$$

The positive energy solutions of positive parity follow from $A_+ = B_-$, $B_+ = A_-$ and equations (3.13), (3.14) in the form

$$\psi_{k,+}(x) = \frac{1}{\sqrt{\pi}} \frac{k\cos(kx) - \kappa\sin(k|x|)}{\sqrt{\kappa^2 + k^2}}. \tag{3.22}$$

The wave number k in (3.21) and (3.22) is constrained to the positive half-line $k = \sqrt{2mE}/\hbar > 0$.

The solutions (3.16), (3.21) and (3.22) satisfy the usual orthonormalization conditions for bound or free states, respectively (see Problem 3.9), and their completeness relation is

$$\psi_\kappa(x)\psi_\kappa(x') + \int_0^\infty dk\left[\psi_{k,-}(x)\psi_{k,-}(x') + \psi_{k,+}(x)\psi_{k,+}(x')\right] = \delta(x-x'). \tag{3.23}$$

The state (3.17) describes a situation in which a particle is incident on the δ potential from the left, and therefore on the right side of the potential ($x > 0$) we only have the right moving transmitted component, whereas the wave function for $x < 0$ contains both incoming and reflected components. We can find a corresponding normalized solution and construct a basis of scattering states which describe scattering of particles incident from the left or from the right by applying a unitary transformation on the basis of even and odd scattering states (3.22, 3.21):

$$\begin{pmatrix} \psi_{k,l}(x) \\ \psi_{k,r}(x) \end{pmatrix} = \frac{1}{\sqrt{2(\kappa^2 + k^2)}} \begin{pmatrix} \sqrt{\kappa^2 + k^2} & \kappa + \mathrm{i}k \\ \sqrt{\kappa^2 + k^2} & -\kappa - \mathrm{i}k \end{pmatrix} \begin{pmatrix} \psi_{k,+}(x) \\ \psi_{k,-}(x) \end{pmatrix}. \tag{3.24}$$

This yields a basis with states describing incidence of particles from the left,

$$\psi_{k,l}(x) = \frac{k\exp(\mathrm{i}kx) + 2\kappa\Theta(-x)\sin(kx)}{\sqrt{2\pi(\kappa^2 + k^2)}}, \tag{3.25}$$

and incidence of particles from the right,

$$\psi_{k,r}(x) = \frac{k\exp(-\mathrm{i}kx) - 2\kappa\Theta(x)\sin(kx)}{\sqrt{2\pi(\kappa^2 + k^2)}}. \tag{3.26}$$

The completeness relation is

$$\psi_\kappa(x)\psi_\kappa(x') + \int_0^\infty dk\left[\psi_{k,l}(x)\psi_{k,l}(x') + \psi_{k,r}(x)\psi_{k,r}(x')\right] = \delta(x-x'). \tag{3.27}$$

There is no bound state solution for a repulsive δ potential

$$V(x) = \mathcal{W}\delta(x) = \frac{\hbar^2\kappa}{m}\delta(x)$$

and the even parity energy eigenstates become

$$\phi_{k,+}(x) = \frac{1}{\sqrt{\pi}} \frac{k\cos(kx) + \kappa\sin(k|x|)}{\sqrt{\kappa^2 + k^2}}. \tag{3.28}$$

The completeness relation for the eigenfunctions of the repulsive δ potential is therefore

$$\int_0^\infty dk \left[\psi_{k,-}(x)\psi_{k,-}(x') + \phi_{k,+}(x)\phi_{k,+}(x')\right] = \delta(x - x'). \tag{3.29}$$

3.4 Evolution of free Schrödinger wave packets

Another important model system for quantum behavior is provided by free wave packets. We will discuss in particular free Gaussian wave packets because they provide a simple analytic model for dispersion of free wave packets. This example will also demonstrate that the spatial and temporal range of free particle models is constrained in quantum physics. We will see that free wave packets of subatomic particles disperse on relatively short time scales, which are however too long to interfere with lab experiments involving free electrons or nucleons. Nevertheless, the discussion of the dispersion of free wave packets makes it also clear that simple interpretations of particles in quantum mechanics as highly localized free wave packets which every now and then get disturbed through interactions with other wave packets are not feasible. Particles can exist in the form of not too small free wave packets for a little while, but atomic or nuclear size wave packets must be stabilized by interactions to avoid rapid dispersion. We will see examples of stable wave packets in Chapters 6 and 7.

The free Schrödinger propagator

Substitution of a Fourier *ansatz*

$$\psi(x,t) = \frac{1}{2\pi}\int_{-\infty}^{\infty} dk \int_{-\infty}^{\infty} d\omega\, \psi(k,\omega)\exp[\mathrm{i}(kx - \omega t)]$$

into the free Schrödinger equation shows that the general solution of that equation in one dimension is given in terms of a wave packet

$$\psi(k,\omega) = \sqrt{2\pi}\,\psi(k)\delta\!\left(\omega - \frac{\hbar k^2}{2m}\right),$$

$$\psi(x,t) = \frac{1}{\sqrt{2\pi}}\int_{-\infty}^{\infty} dk\, \psi(k)\exp\!\left[\mathrm{i}\left(kx - \frac{\hbar k^2}{2m}t\right)\right]. \tag{3.30}$$

The amplitude $\psi(k)$ is related to the initial condition $\psi(x, 0)$ through inverse Fourier transformation

$$\psi(k) = \frac{1}{\sqrt{2\pi}} \int_{-\infty}^{\infty} dx\, \psi(x, 0) \exp(-ikx),$$

and substitution of $\psi(k)$ into (3.30) leads to the expression

$$\psi(x, t) = \int_{-\infty}^{\infty} dx'\, U(x - x', t)\psi(x', 0) \tag{3.31}$$

with the free *propagator*

$$U(x, t) = \frac{1}{2\pi} \int_{-\infty}^{\infty} dk \exp\left[\mathrm{i}\left(kx - \frac{\hbar k^2}{2m}t\right)\right]. \tag{3.32}$$

This is sometimes formally integrated as[5]

$$U(x, t) = \sqrt{\frac{m}{2\pi \mathrm{i}\hbar t}} \exp\left(\mathrm{i}\frac{mx^2}{2\hbar t}\right). \tag{3.33}$$

The propagator is the particular solution of the free Schrödinger equation

$$\mathrm{i}\hbar \frac{\partial}{\partial t} U(x, t) = -\frac{\hbar^2}{2m} \frac{\partial^2}{\partial x^2} U(x, t)$$

with initial condition $U(x, 0) = \delta(x)$. It yields the corresponding *retarded Green's function*

$$\mathrm{i}\hbar \frac{\partial}{\partial t} \mathcal{G}(x, t) + \frac{\hbar^2}{2m} \frac{\partial^2}{\partial x^2} \mathcal{G}(x, t) = \delta(t)\delta(x), \tag{3.34}$$

$$\mathcal{G}(x, t)\Big|_{t<0} = 0, \tag{3.35}$$

through

$$\mathcal{G}(x, t) = \frac{\Theta(t)}{\mathrm{i}\hbar} U(x, t). \tag{3.36}$$

This can also be derived from the Fourier decomposition of equation (3.34), which yields

[5]The propagator is commonly denoted as $K(x, t)$. However, we prefer the notation $U(x, t)$ because the propagator is nothing but the x representation of the time evolution operator $U(t)$ introduced in Chapter 13.

$$\mathcal{G}(x,t) = \frac{1}{(2\pi)^2\hbar}\int_{-\infty}^{\infty}dk\int_{-\infty}^{\infty}d\omega\,\frac{\exp[\mathrm{i}(kx-\omega t)]}{\omega-(\hbar k^2/2m)+\mathrm{i}\epsilon}.$$

The negative imaginary shift of the pole $(\hbar k^2/2m) - \mathrm{i}\epsilon$, $\epsilon \to +0$, in the complex ω plane ensures that the condition (3.35) is satisfied. We will encounter time evolution operators and Green's functions in many places in this book. The designation *propagator* is often used both for the time evolution operator $U(x,t)$ and for the related Green's function $\mathcal{G}(x,t)$. $U(x,t)$ propagates initial conditions as in equation (3.31) while $\mathcal{G}(x,t)$ propagates perturbations or source terms in the Schrödinger equation.

Width of Gaussian wave packets

A wave packet $\psi(x,t)$ is denoted as a *Gaussian wave packet* if $|\psi(x,t)|^2$ is a Gaussian function of x. We will see below through direct Fourier transformation that $\psi(x,t)$ is a Gaussian wave packet in x if and only if $\psi(k,t)$ is a Gaussian wave packet in k.

Normalized Gaussian wave packets have the general form

$$\psi(x,t) = \left(\frac{2\alpha(t)}{\pi}\right)^{\frac{1}{4}}\exp\!\left(-\alpha(t)[x-x_0(t)]^2+\mathrm{i}\varphi(x,t)\right), \tag{3.37}$$

and we will verify that the real coefficient $\alpha(t)$ is related to the variance through $\Delta x^2(t) = 1/4\alpha(t)$. The expectation values of x and x^2 are readily evaluated,

$$\begin{aligned}\langle x\rangle(t) &= \sqrt{\frac{2\alpha(t)}{\pi}}\int_{-\infty}^{\infty}dx\,x\exp\!\left(-2\alpha(t)[x-x_0(t)]^2\right)\\ &= \sqrt{\frac{2\alpha(t)}{\pi}}\int_{-\infty}^{\infty}d\xi\,[\xi+x_0(t)]\exp\!\left(-2\alpha(t)\xi^2\right) = x_0(t),\end{aligned}$$

$$\begin{aligned}\langle x^2\rangle(t) &= \sqrt{\frac{2\alpha(t)}{\pi}}\int_{-\infty}^{\infty}dx\,x^2\exp\!\left(-2\alpha(t)[x-x_0(t)]^2\right)\\ &= \sqrt{\frac{2\alpha(t)}{\pi}}\int_{-\infty}^{\infty}d\xi\,[\xi+x_0(t)]^2\exp\!\left(-2\alpha(t)\xi^2\right)\\ &= \sqrt{\frac{2\alpha(t)}{\pi}}\left(x_0^2(t)-\frac{1}{2}\frac{d}{d\alpha(t)}\right)\int_{-\infty}^{\infty}d\xi\,\exp\!\left(-2\alpha(t)\xi^2\right)\\ &= x_0^2(t)+\frac{1}{4\alpha(t)},\end{aligned}$$

and therefore we find indeed

$$\Delta x^2(t) = \langle x^2\rangle(t) - \langle x\rangle^2(t) = \frac{1}{4\alpha(t)}. \tag{3.38}$$

Free Gaussian wave packets in Schrödinger theory

We assume that the wave packet of a free particle at time $t = 0$ was a Gaussian wave packet of width Δx,

$$\psi(x,0) = \frac{1}{(2\pi\Delta x^2)^{1/4}}\exp\left(-\frac{(x-x_0)^2}{4\Delta x^2} + \mathrm{i}k_0 x\right). \tag{3.39}$$

This yields a Gaussian wave packet of constant width

$$\Delta k = \frac{1}{2\Delta x}$$

in k space,

$$\begin{aligned}
\psi(k) &= \frac{1}{\sqrt{2\pi}}\int_{-\infty}^{\infty} dx\,\psi(x,0)\exp(-\mathrm{i}kx) \\
&= \frac{1}{(2\pi)^{3/4}(\Delta x^2)^{1/4}}\int_{-\infty}^{\infty} dx\,\exp\left(-\frac{(x-x_0)^2}{4\Delta x^2} + \mathrm{i}(k_0-k)x\right) \\
&= \frac{\exp[\mathrm{i}(k_0-k)x_0]}{(2\pi)^{3/4}(\Delta x^2)^{1/4}}\int_{-\infty}^{\infty} d\xi\,\exp\left[-\frac{\xi^2}{4\Delta x^2} + \mathrm{i}(k_0-k)\xi\right] \\
&= \left(\frac{2\Delta x^2}{\pi}\right)^{\frac{1}{4}}\exp\left[-\Delta x^2(k-k_0)^2 - \mathrm{i}(k-k_0)x_0\right],
\end{aligned} \tag{3.40}$$

$$\psi(k,t) = \psi(k)\exp\left(-\mathrm{i}\frac{\hbar k^2}{2m}t\right). \tag{3.41}$$

Substitution of $\psi(k)$ into equation (3.30) then yields

$$\begin{aligned}
\psi(x,t) &= \left(\frac{\Delta x^2}{2\pi^3}\right)^{\frac{1}{4}}\exp\left(-\Delta x^2 k_0^2 + \mathrm{i}k_0x_0\right) \\
&\quad\times\int_{-\infty}^{\infty} dk\,\exp\left[-\left(\Delta x^2 + \mathrm{i}\frac{\hbar t}{2m}\right)k^2 + \left(2\Delta x^2 k_0 + \mathrm{i}(x-x_0)\right)k\right] \\
&= \frac{(2\pi\Delta x^2)^{1/4}}{\left[2\pi\Delta x^2 + \mathrm{i}\pi(\hbar t/m)\right]^{1/2}}\exp\left(-\Delta x^2 k_0^2 + \mathrm{i}k_0x_0\right)
\end{aligned}$$

$$\times \exp\left[\frac{\left[2\Delta x^2 k_0 + \mathrm{i}(x-x_0)\right]^2}{4\Delta x^2 + 2\mathrm{i}(\hbar t/m)}\right]$$

$$= \frac{(2\pi\Delta x^2)^{1/4}}{\left[2\pi\Delta x^2 + \mathrm{i}\pi(\hbar t/m)\right]^{1/2}} \exp\left[-\frac{[x - x_0 - (\hbar k_0/m)t]^2}{4\Delta x^2 + (\hbar^2 t^2/m^2\Delta x^2)}\right]$$

$$\times \exp\left[\mathrm{i}\left(k_0 x - \frac{\hbar k_0^2}{2m}t + \frac{\hbar t}{8m}\frac{[x - x_0 - (\hbar k_0/m)t]^2}{(\Delta x^2)^2 + (\hbar^2 t^2/4m^2)}\right)\right]. \tag{3.42}$$

Comparison of equation (3.42) with equations (3.37, 3.38) yields

$$\Delta x^2(t) = \Delta x^2 + \frac{\hbar^2 t^2}{4m^2\Delta x^2}, \tag{3.43}$$

i.e. a strongly localized packet at time $t = 0$ will disperse very fast, because the dispersion time scale τ is proportional to Δx^2. The reason for the fast dispersion is that a strongly localized packet at $t = 0$ comprises many different wavelengths. However, each monochromatic component in a free wave packet travels with its own phase velocity

$$v(k) = \frac{\omega}{k} = \frac{\hbar k}{2m},$$

and a free strongly localized packet therefore had to emerge from rapid collapse and will disperse very fast. On the other hand, a poorly localized packet is almost monochromatic and therefore slowly changes in shape.

The relevant time scale for decay of the wave packet is

$$\tau = \frac{2m\Delta x^2}{\hbar}. \tag{3.44}$$

Electron guns often have apertures in the millimeter range. Assuming $\Delta x = 1$ mm for an electron wave packet yields $\tau \simeq 2 \times 10^{-2}$ seconds. This sounds like a short time scale for dispersion of the wave packet. However, on the time scales of a typical lab experiment involving free electrons, dispersion of electron wave packets is completely negligible, see e.g. Problem 3.17.

On the other hand, suppose we can produce a free electron wave packet with atomic scale localization, $\Delta x = 1$ Å. This wave packet would disperse with an extremely short time scale $\tau \simeq 2 \times 10^{-16}$ seconds, which means that the wave function of that electron would be smeared across the planet within a minute. See also Problem 3.18 for a corresponding discussion for neutrons.

We will see in Chapters 6 and 7 that wave packets can remain localized under the influence of forces, i.e. the notion of stable electrons in atoms makes sense, although the notion of highly localized *free* electrons governed by the *free* Schrödinger equation is limited to small distance and time scales.

We can infer from the example of the free Gaussian wave packet that the kinetic term in the Schrödinger equation drives wave packets apart. If there is no attractive potential term, the kinetic term decelerates any eventual initial contraction of a free wave packet and ultimately pushes the wave packet towards accelerated dispersion. We will see that this action of the kinetic term can be compensated by attractive potential terms in the Schrödinger equation. Balance between the collapsing force from attractive potentials and the dispersing force from the kinetic term can stabilize quantum systems.

Comparison of equation (3.41) with equations (3.37, 3.38) yields constant width of the wave packet in k space and therefore

$$\Delta p = \hbar \Delta k = \frac{\hbar}{2\Delta x},$$

i.e. there is no dispersion in momentum. The product of uncertainties of momentum and location of the particle satisfies $\Delta p \Delta x(t) \geq \hbar/2$, in agreement with Heisenberg's uncertainty relation, which will be derived for general wave packets in Section 5.1.

The energy expectation value and uncertainty of the wave packet are

$$\langle E \rangle = \frac{\hbar^2}{2m}\left(k_0^2 + \frac{1}{4\Delta x^2}\right)$$

and

$$\Delta E = \frac{\hbar^2}{2m}\sqrt{\frac{k_0^2}{\Delta x^2} + \frac{1}{8\Delta x^4}}.$$

Suppose we want to observe strong localization of a free particle. The decay time (3.44) then defines a measure for the time window Δt of observability of the particle. This satisfies

$$\Delta E \Delta t = \hbar\sqrt{\frac{1}{8} + k_0^2 \Delta x^2} \geq \frac{\hbar}{\sqrt{8}},$$

in agreement with the qualitative energy-time uncertainty relation (5.7), which we will encounter in Section 5.1.

The free Gaussian wave packet reproduces momentum eigenstates in the limit $\Delta x^2 \to \infty$ in the sense

$$\lim_{\Delta x^2 \to \infty} \left(\frac{\Delta x^2}{2\pi}\right)^{1/4} \psi(k) = \delta(k - k_0),$$

$$\lim_{\Delta x^2 \to \infty} \left(\frac{\Delta x^2}{2\pi}\right)^{1/4} \psi(x,t) = \frac{1}{\sqrt{2\pi}} \exp\left[\mathrm{i}\left(k_0 x - \frac{\hbar t}{2m} k_0^2\right)\right].$$

3.5 Problems

3.1. Show that the tunneling probability for the square barrier in Figure 3.1 always satisfies $T < 1$ if $\Phi_2 > 0$.

Remark. Don't be fooled by Figure 3.2. The first transmission maximum at $E_1 \simeq 2\Phi_1$ corresponds already to $T \simeq 0.998$ and the next transmission maximum is even closer to 1, but it only looks like the transmission probability would reach 1 in Figure 3.2.

Tunneling resonances $T = 1$ occur for $\Phi_2 = 0$. For which values of E_1 and k_1 do this tunneling resonances occur? Which geometric matching condition holds between the wavelength of the incident particles and the square barrier for the tunneling resonances?

3.2a. Calculate the reflection and transmission coefficients for a particle that falls down a potential step of height V_0.

3.2b. Calculate the reflection and transmission coefficients for a particle that runs against a potential step of height V_0 both for particle energy $0 < E < V_0$ and for $E > V_0$.

3.3. Calculate the reflection and transmission coefficients for a particle in the (x, y) plane which moves in the potential $V(x, y) = V_0\Theta(x)$. Assume that the particle initially approached the potential step from the left,

$$\psi_{in}(x, y) = A\exp[\mathrm{i}(k_x x + k_y y)], \quad x < 0, \quad k_x > 0.$$

Solution. The reflected wave function is

$$\psi_r(x, y) = B\exp[\mathrm{i}(-k_x x + k_y y)], \quad x < 0.$$

The time-independent Schrödinger equation for the transmitted component of the wave function

$$\psi_t(x, y) = C\exp[\mathrm{i}(k'_x x + k'_y y)], \quad x > 0,$$

implies

$$k_x^2 + k_y^2 = k_x'^2 + k_y'^2 + \frac{2mV_0}{\hbar^2}.$$

However, smoothness of the wave function across the step does not only have to hold for all times (this is what requires equal particle energy from equal time-dependence $\exp(-\mathrm{i}Et/\hbar)$ of the wave functions on both sides of the step), but also for all locations along the step, i.e. for all y values. The latter requirement implies equal $\exp(ik_y y)$ factors on both sides of the step, and we find with $k'_y = k_y$ exactly the same set of conditions that apply to the one-dimensional step from Problem 2,

$$k_x^2 = k_x'^2 + \frac{2mV_0}{\hbar^2}, \quad A + B = C, \quad k_x(A - B) = k'_x C,$$

i.e. we find the same results as in Problem 2, except for the substitution $E \to E_x = \hbar^2 k_x^2/2m$ if we want to express results in terms of energies rather than wave numbers.

3.4. A barrier for motion of a particle consists of a combination of two repulsive δ function potentials with separation a,

$$V(x) = \mathcal{W}\delta(x) + \mathcal{W}\delta(x - a).$$

Calculate the reflection and transmission coefficients for particles with momentum $\hbar k$.

3.5. Why is there a simple relation between momentum uncertainty and energy level in the box model for a quantum dot? What is the relation?

3.6. A very simple cubic model for a color center in an alkali halide crystal consists of an electron confined to a cube of length L. How large is the length L if the electron absorbs photons of energy 2.3 eV?

Mollwo had found the empirical relation $\nu d^2 = 5.02 \times 10^{-5}\,\mathrm{m}^2\,\mathrm{Hz}$ between absorption frequencies ν of color centers and lattice constants d in alkali halide crystals. For the simple cubic model, which relation between L and d follows from Mollwo's relation?

A spherical model is also very simple, but gives a better estimate for the ratio between size of the color center and lattice constant, see the corresponding problem in Chapter 7.

3.7. Calculate the momentum uncertainty in the bound state (3.16).

3.8. Calculate the scattering matrix for the square barrier from Section 3.1.

3.9. Show that the bound state (3.16) of the attractive δ potential is orthogonal to the free states (3.21, 3.22),

$$\int_{-\infty}^{\infty} dx\, \psi_{k,\pm}(x)\psi_\kappa(x) = 0.$$

Show also that the free eigenstates (3.21, 3.22) and (3.28) of the attractive or repulsive δ potential are normalized according to

$$\int_{-\infty}^{\infty} dx\, \psi_{k,-}(x)\psi_{k',-}(x) = \delta(k - k'), \quad \int_{-\infty}^{\infty} dx\, \psi_{k,+}(x)\psi_{k',+}(x) = \delta(k - k'),$$

$$\int_{-\infty}^{\infty} dx\, \phi_{k,+}(x)\phi_{k',+}(x) = \delta(k - k').$$

Remark. You have to use that the wave numbers k and k' are positive.

3.10. Construct a basis of scattering states for the repulsive δ potential which consists of states describing incidence of particles from the left or from the right on the potential.

3.11. Show that all the momentum expectation values $\langle p^n \rangle$ are conserved for a free particle.

If the particle is moving in a potential $V(x)$, find a necessary and sufficient condition for $V(x)$ such that $\langle p^n \rangle$ is constant.

3.12. Suppose $\psi(x,t)$ is a normalizable free wave packet in one dimension, e.g. the Gaussian wave packet from Section 3.4. Which classical quantity of the particle corresponds to the integral $\int dx\, j(x)$ of the current density? Does a similar result hold for $\int d^3\boldsymbol{x}\, \boldsymbol{j}(\boldsymbol{x})$ in three dimensions?

3.13. The wave function of a free particle at time $t = 0$ is

$$\psi(x,0) = \sqrt{\frac{2a^3}{\pi}} \frac{1}{x^2 + a^2}.$$

How large are the uncertainties $\Delta x(t)$ and Δp in location and momentum of the particle?

Remark. The wave function $\psi(x,t)$ of the particle can be expressed in terms of complex error functions, but it is easier to use the wave function $\psi(k,t)$ in k space for the calculation of the uncertainties.

3.14. The wave function of a free particle at time $t = 0$ is

$$\psi(x,0) = \sqrt{\kappa}\exp(-\kappa|x|). \tag{3.45}$$

One could produce this state as initial state of a free particle by first capturing the particle in the bound state of an attractive δ potential and then switching off the potential.

Calculate the wave function $\psi(k,t)$ of the particle.

How large are the uncertainties $\Delta x(t)$ and Δp in location and momentum of the particle?

3.15. The wave function of a free particle at time $t = 0$ is

$$\psi(x,0) = \frac{\Theta(x+a)\Theta(a-x)}{\sqrt{2a}} = \frac{\Theta(a-x) - \Theta(-x-a)}{\sqrt{2a}}$$
$$= \frac{\Theta(x+a) - \Theta(x-a)}{\sqrt{2a}}. \tag{3.46}$$

Calculate the wave function $\psi(x,t)$ of the particle.

Solution. The wave function in momentum space is

$$\psi(k) = \frac{1}{2\sqrt{\pi a}} \int_{-a}^{a} dx \, \exp(-ikx) = \frac{\sin(ka)}{\sqrt{\pi a} k},$$

$$\psi(k,t) = \psi(k) \exp\left(-\mathrm{i}\frac{\hbar k^2}{2m} t\right).$$

The wave function $\psi(x,t)$ can be evaluated numerically from the first line of the following representations,

$$\begin{aligned}\psi(x,t) &= \frac{1}{2\pi\mathrm{i}\sqrt{2a}} \int_{-\infty}^{\infty} dk \, \frac{\exp[\mathrm{i}k(x+a)] - \exp[\mathrm{i}k(x-a)]}{k} \exp\left(-\mathrm{i}\frac{\hbar t}{2m} k^2\right) \\ &= \frac{1}{2\pi\mathrm{i}\sqrt{2a}} \int_{-\infty}^{\infty} dk \, \frac{\exp[\mathrm{i}k(x+a)] - \exp[\mathrm{i}k(x-a)]}{k + \mathrm{i}\epsilon} \exp\left(-\mathrm{i}\frac{\hbar t}{2m} k^2\right) \\ &= \frac{1}{2\pi\mathrm{i}\sqrt{2a}} \int_{-\infty}^{\infty} dk \, \frac{\exp[\mathrm{i}k(x+a)] - \exp[\mathrm{i}k(x-a)]}{k - \mathrm{i}\epsilon} \exp\left(-\mathrm{i}\frac{\hbar t}{2m} k^2\right).\end{aligned}$$

However, we can also proceed with the analytical evaluation of the integrals by using the observation

$$\pm\frac{\partial}{\partial a} \int_{-\infty}^{\infty} dk \, \frac{\exp[\mathrm{i}k(x \pm a)]}{k} \exp\left(-\mathrm{i}\frac{\hbar t}{2m} k^2\right) = \mathrm{i}\sqrt{\frac{2\pi m}{\mathrm{i}\hbar t}} \exp\left(\mathrm{i}\frac{m}{2\hbar t}(x \pm a)^2\right).$$

Integration with respect to the parameter a then yields

$$\psi(x,t) = \frac{1}{2\sqrt{2a}} \left[\operatorname{erf}\left(\sqrt{\frac{m}{2\mathrm{i}\hbar t}}(x+a)\right) - \operatorname{erf}\left(\sqrt{\frac{m}{2\mathrm{i}\hbar t}}(x-a)\right)\right], \tag{3.47}$$

where the error function is defined as

$$\operatorname{erf}(z) = \frac{2}{\sqrt{\pi}} \int_0^z du \, \exp(-u^2).$$

One can easily check that the error functions $\operatorname{erf}[\sqrt{m/2\mathrm{i}\hbar t}(x - x_0)]$ satisfy the free Schrödinger equation. The wave function $\psi(x,t)$ from equation (3.47) also satisfies the initial condition (3.46) through

$$\lim_{t \to 0 - \mathrm{i}\epsilon} \operatorname{erf}\left(\sqrt{\frac{m}{2\mathrm{i}\hbar t}}(x - x_0)\right) = \Theta(x - x_0) - \Theta(x_0 - x).$$

3.16. The initial condition (3.39) yielded a Gaussian wave packet that had its minimal spread in location x exactly at the time $t = 0$. Before that particular moment, the wave packet was contracting and afterwards it was spreading.

Find an initial condition $\psi(x, 0)$ for a Gaussian wave packet that will continue to contract for some time Δt before it expands.

3.17. In their famous verification of the wave properties of electrons through diffraction off the surface of a nickel crystal, Davisson and Germer[6] used an electron gun with an aperture of about 1 mm. The electron beam that produced the most prominent diffraction pattern had a kinetic energy of 54 eV, and the distance from the electron gun to the nickel target was 7 mm. Show that dispersion of the electron wave packet on the way from the electron gun to the target is completely negligible.

3.18. Suppose we could produce a free neutron with a nuclear scale width $\Delta x =$ 1 fm. How large is the dispersion time scale (3.44) for the neutron?

Neutron beam experiments to measure e.g. the β decay of free neutrons use beams with apertures of a few centimeters[7]. How large is the dispersion time scale for neutrons with $\Delta x \simeq 3\,\text{cm}$? How does this compare to the lifetime of free neutrons?

---

[6] C. Davisson, L.H. Germer, Phys. Rev. 30, 705 (1927).

[7] See e.g. J.M. Robson, Phys. Rev. 83, 349 (1951) for one of the early lifetime measurements of free neutrons, or H.P. Mumm *et al.*, Rev. Sci. Instrum. 75, 5343 (2004), for a modern experimental setup.

Chapter 4
Notions from Linear Algebra and Bra-Ket Notation

The Schrödinger equation (1.14) is linear in the wave function $\psi(\boldsymbol{x},t)$. This implies that for any set of solutions $\psi_1(\boldsymbol{x},t)$, $\psi_2(\boldsymbol{x},t),\ldots$, any linear combination $\psi(\boldsymbol{x},t) = C_1\psi_1(\boldsymbol{x},t) + C_2\psi_2(\boldsymbol{x},t) + \ldots$ with complex coefficients C_n is also a solution. The set of solutions of equation (1.14) for fixed potential V will therefore have the structure of a complex vector space, and we can think of the wave function $\psi(\boldsymbol{x},t)$ as a particular vector in this vector space. Furthermore, we can map this vector bijectively into different, but equivalent representations where the wave function depends on different variables. An example of this is Fourier transformation (2.5) into a wave function which depends on a wave vector $\boldsymbol{k}$,

$$\psi(\boldsymbol{k},t) = \frac{1}{\sqrt{2\pi}^3}\int d^3\boldsymbol{x}\,\exp(-\mathrm{i}\boldsymbol{k}\cdot\boldsymbol{x})\,\psi(\boldsymbol{x},t).$$

We have already noticed that this is sloppy notation from the mathematical point of view. We should denote the Fourier transformed function with $\tilde{\psi}(\boldsymbol{k},t)$ to make it clear that $\tilde{\psi}(\boldsymbol{k},t)$ and $\psi(\boldsymbol{x},t)$ have different dependencies on their arguments (or stated differently, to make it clear that $\psi(\boldsymbol{k},t)$ and $\psi(\boldsymbol{x},t)$ are really different functions). However, there is a reason for the notation in equations (2.4, 2.5). We can switch back and forth between $\psi(\boldsymbol{x},t)$ and $\psi(\boldsymbol{k},t)$ using Fourier transformation. This implies that any property of a particle that can be calculated from the wave function $\psi(\boldsymbol{x},t)$ in $\boldsymbol{x}$ space can also be calculated from the wave function $\psi(\boldsymbol{k},t)$ in $\boldsymbol{k}$ space. Therefore, following Dirac, we nowadays do not think any more of $\psi(\boldsymbol{x},t)$ as a wave function of a particle, but we rather think more abstractly of $\psi(t)$ as a time-dependent *quantum state*, with particular *representations* of the quantum state $\psi(t)$ given by the wave functions $\psi(\boldsymbol{x},t)$ or $\psi(\boldsymbol{k},t)$. There are infinitely more possibilities to represent the quantum state $\psi(t)$ through functions. For example, we could perform a Fourier transformation only with respect to the y variable and represent $\psi(t)$ through the wave function $\psi(x,k_y,z,t)$, or we could perform an invertible transformation to completely different independent variables. In 1939, Paul Dirac

R. Dick, *Advanced Quantum Mechanics*, Graduate Texts in Physics,
DOI 10.1007/978-3-319-25675-7_4

introduced a notation in quantum mechanics which emphasizes the vector space and representation aspects of quantum states in a very elegant and suggestive manner. This notation is Dirac's bra-ket notation, and it is ubiquitous in advanced modern quantum mechanics. It is worthwhile to use bra-ket notation from the start, and it is most easily explained in the framework of linear algebra.

4.1 Notions from linear algebra

The mathematical structure of quantum mechanics resembles linear algebra in many respects, and many notions from linear algebra are very useful in the investigation of quantum systems. Bra-ket notation makes the linear algebra aspects of quantum mechanics particularly visible and easy to use. Therefore we will first introduce a few notions of linear algebra in standard notation, and then rewrite everything in bra-ket notation.

Tensor products

Suppose $\mathcal{V}$ is an N-dimensional real vector space with a Cartesian basis[1] $\hat{\boldsymbol{e}}_a$, $1 \leq a \leq N$, $\hat{\boldsymbol{e}}_a{}^{\mathrm{T}} \cdot \hat{\boldsymbol{e}}_b = \delta_{ab}$. Furthermore, assume that u^a, v^a are Cartesian components of the two vectors $\boldsymbol{u}$ and $\boldsymbol{v}$,

$$\boldsymbol{u} = \sum_{a=1}^{N} u^a \hat{\boldsymbol{e}}_a \equiv u^a \hat{\boldsymbol{e}}_a.$$

Here we use *summation convention*: Whenever an index appears twice in a multiplicative term, it is automatically summed over its full range of values. We will continue to use this convention throughout the remainder of the book.

The *tensor product* $\underline{\boldsymbol{M}} = \boldsymbol{u} \otimes \boldsymbol{v}^{\mathrm{T}}$ of the two vectors yields an $N \times N$ matrix with components $M^{ab} = u^a v^b$ in the Cartesian basis:

$$\underline{\boldsymbol{M}} = \boldsymbol{u} \otimes \boldsymbol{v}^{\mathrm{T}} = u^a v^b \hat{\boldsymbol{e}}_a \otimes \hat{\boldsymbol{e}}_b{}^{\mathrm{T}}. \tag{4.1}$$

Tensor products appear naturally in basic linear algebra e.g. in the following simple problem: Suppose $\boldsymbol{u} = u^a \hat{\boldsymbol{e}}_a$ and $\boldsymbol{w} = w^a \hat{\boldsymbol{e}}_a$ are two vectors in an N-dimensional vector space, and we would like to calculate the part $\boldsymbol{w}_\parallel$ of the vector $\boldsymbol{w}$ that is parallel to $\boldsymbol{u}$. The unit vector in the direction of $\boldsymbol{u}$ is $\hat{\boldsymbol{u}} = \boldsymbol{u}/|\boldsymbol{u}|$, and we have

$$\boldsymbol{w}_\parallel = \hat{\boldsymbol{u}}|\boldsymbol{w}|\cos(\boldsymbol{u}, \boldsymbol{w}), \tag{4.2}$$

[1]We write scalar products of vectors initially as $\boldsymbol{u}^{\mathrm{T}} \cdot \boldsymbol{v}$ to be consistent with proper tensor product notation used in (4.1), but we will switch soon to the shorter notations $\boldsymbol{u} \cdot \boldsymbol{v}$, $\boldsymbol{u} \otimes \boldsymbol{v}$ for scalar products and tensor products.

where $\cos(\boldsymbol{u},\boldsymbol{w}) = \hat{\boldsymbol{u}}\cdot\hat{\boldsymbol{w}}$ is the cosine of the angle between $\boldsymbol{u}$ and $\boldsymbol{w}$. Substituting the expression for $\cos(\boldsymbol{u},\boldsymbol{w})$ into (4.2) yields

$$\begin{aligned}\boldsymbol{w}_{\parallel} &= \hat{\boldsymbol{u}}(\hat{\boldsymbol{u}}\cdot\boldsymbol{w}) = \hat{u}^a\hat{u}^b w^c\hat{\boldsymbol{e}}_a(\hat{\boldsymbol{e}}_b{}^{\mathrm{T}}\cdot\hat{\boldsymbol{e}}_c) = \hat{u}^a\hat{u}^b w^c(\hat{\boldsymbol{e}}_a\otimes\hat{\boldsymbol{e}}_b{}^{\mathrm{T}})\cdot\hat{\boldsymbol{e}}_c \\ &= (\hat{\boldsymbol{u}}\otimes\hat{\boldsymbol{u}}^T)\cdot\boldsymbol{w}, \qquad (4.3)\end{aligned}$$

i.e. the tensor product $\underline{P}_{\parallel} = \hat{\boldsymbol{u}}\otimes\hat{\boldsymbol{u}}^T$ is the projector onto the direction of the vector $\boldsymbol{u}$.

The matrix $\underline{\boldsymbol{M}}$ is called a *2nd rank tensor* due to its transformation properties under linear transformations of the vectors appearing in the product.

Suppose we perform a transformation of the Cartesian basis vectors $\hat{\boldsymbol{e}}_a$ to a new set $\hat{\boldsymbol{e}}'_i$ of basis vectors,

$$\hat{\boldsymbol{e}}_a \to \hat{\boldsymbol{e}}'_i = \hat{\boldsymbol{e}}_a R^a{}_i, \qquad (4.4)$$

subject to the constraint that the new basis vectors also provide a Cartesian basis,

$$\hat{\boldsymbol{e}}'_i\cdot\hat{\boldsymbol{e}}'_j = \delta_{ab}R^a{}_i R^b{}_j = R^a{}_i R_{aj} = \delta_{ij}. \qquad (4.5)$$

Linear transformations which map Cartesian bases into Cartesian bases are denoted as rotations.

We defined $R_{aj} \equiv \delta_{ab}R^b{}_j$ in equation (4.5), i.e. numerically $R_{aj} = R^a{}_j$. Equation (4.5) is in matrix notation

$$\underline{R}^T\cdot\underline{R} = \underline{1}, \qquad (4.6)$$

i.e. $\underline{R}^T = \underline{R}^{-1}$.

However, a change of basis in our vector space does nothing to the vector $\boldsymbol{v}$, except that the vector will have different components with respect to the new basis vectors,

$$\boldsymbol{v} = \hat{\boldsymbol{e}}_a v^a = \hat{\boldsymbol{e}}'_i v'^i = \hat{\boldsymbol{e}}_a R^a{}_i v'^i. \qquad (4.7)$$

Equations (4.7) and (4.5) and the uniqueness of the decomposition of a vector with respect to a set of basis vectors imply

$$v^a = R^a{}_i v'^i, \quad v'^i = (R^{-1})^i{}_a v^a = (R^T)^i{}_a v^a = v^a R_a{}^i. \qquad (4.8)$$

This is the *passive interpretation* of transformations: The transformation changes the reference frame, but not the physical objects (here: vectors). Therefore the expansion coefficients of the physical objects change inversely (or *contravariant*) to the transformation of the reference frame. We will often use the passive interpretation for symmetry transformations of quantum systems.

The transformation laws (4.4) and (4.8) define *first rank tensors*, because the transformation laws are linear (or first order) in the transformation matrices $\underline{R}$ or $\underline{R}^{-1}$.

The tensor product $\underline{M} = \boldsymbol{u} \otimes \boldsymbol{v}^{\mathrm{T}} = u^a v^b \hat{\boldsymbol{e}}_a \otimes \hat{\boldsymbol{e}}_b{}^{\mathrm{T}}$ then defines a *second rank tensor*, because the components and the basis transform quadratically (or in second order) with the transformation matrices $\underline{R}$ or $\underline{R}^{-1}$,

$$M'^{ij} = u'^i v'^j = (R^{-1})^i{}_a (R^{-1})^j{}_b u^a v^b = (R^{-1})^i{}_a (R^{-1})^j{}_b M^{ab}, \tag{4.9}$$

$$\hat{\boldsymbol{e}}'_i \otimes \hat{\boldsymbol{e}}'^{\mathrm{T}}_j = \hat{\boldsymbol{e}}_a \otimes \hat{\boldsymbol{e}}_b{}^{\mathrm{T}} R^a{}_i R^b{}_j. \tag{4.10}$$

The concept immediately generalizes to n-th order tensors.

Writing the tensor product explicitly as $\boldsymbol{u} \otimes \boldsymbol{v}^{\mathrm{T}}$ reminds us that the a-th row of $\underline{M}$ is just the row vector $u^a \boldsymbol{v}^{\mathrm{T}}$, while the b-th column is just the column vector $\boldsymbol{u} v^b$. However, usually one simply writes $\boldsymbol{u} \otimes \boldsymbol{v}$ for the tensor product, just as one writes $\boldsymbol{u} \cdot \boldsymbol{v}$ instead of $\boldsymbol{u}^{\mathrm{T}} \cdot \boldsymbol{v}$ for the scalar product.

Dual bases

We will now complicate things a little further by generalizing to more general sets of basis vectors which may not be orthonormal. Strictly speaking this is overkill for the purposes of quantum mechanics, because the infinite-dimensional basis vectors which we will use in quantum mechanics are still mutually orthogonal, just like Euclidean basis vectors in finite-dimensional vector spaces. However, sometimes it is useful to learn things in a more general setting to acquire a proper understanding, and besides, non-orthonormal basis vectors are useful in solid state physics (as explained in an example below) and unavoidable in curved spaces.

Let $\boldsymbol{a}_i$, $1 \leq i \leq N$, be another basis of the vector space $\mathcal{V}$. Generically this basis will not be orthonormal: $\boldsymbol{a}_i \cdot \boldsymbol{a}_j \neq \delta_{ij}$. The corresponding *dual basis* with basis vectors $\boldsymbol{a}^i$ is defined through the requirements

$$\boldsymbol{a}^i \cdot \boldsymbol{a}_j = \delta^i{}_j. \tag{4.11}$$

Apparently a basis is *self-dual* ($\boldsymbol{a}^i = \boldsymbol{a}_i$) if and only if it is orthonormal (i.e. Cartesian).

For the explicit construction of the dual basis, we observe that the scalar product of the N vectors $\boldsymbol{a}_i$ defines a symmetric $N \times N$ matrix

$$g_{ij} = \boldsymbol{a}_i \cdot \boldsymbol{a}_j.$$

This matrix is not degenerate, because otherwise it would have at least one vanishing eigenvalue, i.e. there would exist N numbers X^i (not all vanishing) such that $g_{ij}X^j = 0$. This would imply existence of a non-vanishing vector $\boldsymbol{X} = X^i \boldsymbol{a}_i$ with vanishing length,

$$\boldsymbol{X}^2 = X^i X^j \boldsymbol{a}_i \cdot \boldsymbol{a}_j = X^i g_{ij} X^j = 0.$$

The matrix g_{ij} is therefore invertible, and we denote the inverse matrix with g^{ij},

$$g^{ij}g_{jk} = \delta^i{}_k.$$

The inverse matrix can be used to construct the dual basis vectors as

$$\boldsymbol{a}^i = g^{ij}\boldsymbol{a}_j. \tag{4.12}$$

The condition for dual basis vectors is readily verified,

$$\boldsymbol{a}^i \cdot \boldsymbol{a}_k = g^{ij}\boldsymbol{a}_j \cdot \boldsymbol{a}_k = g^{ij}g_{jk} = \delta^i{}_k.$$

For an example for the construction of a dual basis, consider Figure 4.1. The vectors $\boldsymbol{a}_1$ and $\boldsymbol{a}_2$ provide a basis. The angle between $\boldsymbol{a}_1$ and $\boldsymbol{a}_2$ is $\pi/4$ radian, and their lengths are $|\boldsymbol{a}_1| = 2$ and $|\boldsymbol{a}_2| = \sqrt{2}$.

The matrix g_{ij} therefore has the following components in this basis,

$$\underline{g} = \begin{pmatrix} g_{11} & g_{12} \\ g_{21} & g_{22} \end{pmatrix} = \begin{pmatrix} \boldsymbol{a}_1 \cdot \boldsymbol{a}_1 & \boldsymbol{a}_1 \cdot \boldsymbol{a}_2 \\ \boldsymbol{a}_2 \cdot \boldsymbol{a}_1 & \boldsymbol{a}_2 \cdot \boldsymbol{a}_2 \end{pmatrix} = \begin{pmatrix} 4 & 2 \\ 2 & 2 \end{pmatrix}.$$

The inverse matrix is then

$$\underline{g}^{-1} = \begin{pmatrix} g^{11} & g^{12} \\ g^{21} & g^{22} \end{pmatrix} = \frac{1}{2}\begin{pmatrix} 1 & -1 \\ -1 & 2 \end{pmatrix}.$$

This yields with (4.12) the dual basis vectors

$$\boldsymbol{a}^1 = \frac{1}{2}\boldsymbol{a}_1 - \frac{1}{2}\boldsymbol{a}_2, \quad \boldsymbol{a}^2 = -\frac{1}{2}\boldsymbol{a}_1 + \boldsymbol{a}_2.$$

These equations determined the vectors $\boldsymbol{a}^i$ in Figure 4.1.

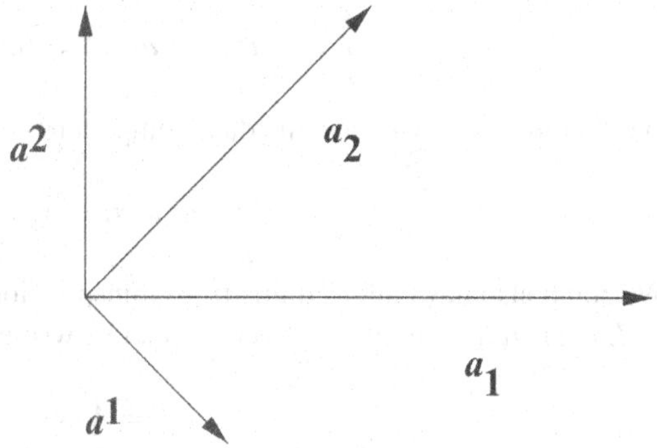

Fig. 4.1 The blue vectors are the basis vectors $\boldsymbol{a}_i$. The red vectors are the dual basis vectors $\boldsymbol{a}^i$

Decomposition of the identity

Equation (4.11) implies that the decomposition of a vector $\boldsymbol{v} \in \mathcal{V}$ with respect to the basis $\boldsymbol{a}_i$ can be written as (note summation convention)

$$\boldsymbol{v} = \boldsymbol{a}_i(\boldsymbol{a}^i \cdot \boldsymbol{v}), \tag{4.13}$$

i.e. the projection of $\boldsymbol{v}$ onto the i-th basis vector $\boldsymbol{a}_i$ (the component v^i in standard notation) is given through scalar multiplication with the dual basis vector $\boldsymbol{a}^i$:

$$v^i = \boldsymbol{a}^i \cdot \boldsymbol{v}.$$

The right hand side of equation (4.13) contains three vectors in each summand, and brackets have been employed to emphasize that the scalar product is between the two rightmost vectors in each term. Another way to make that clear is to write the combination of the two leftmost vectors in each term as a tensor product:

$$\boldsymbol{v} = \boldsymbol{a}_i \otimes \boldsymbol{a}^i \cdot \boldsymbol{v}.$$

If we first evaluate all the tensor products and sum over i, we have for every vector $\boldsymbol{v} \in \mathcal{V}$

$$\boldsymbol{v} = (\boldsymbol{a}_i \otimes \boldsymbol{a}^i) \cdot \boldsymbol{v},$$

which makes it clear that the sum of tensor products in this equation adds up to the identity matrix,

$$\boldsymbol{a}_i \otimes \boldsymbol{a}^i = \underline{1}. \tag{4.14}$$

This is the statement that every vector can be uniquely decomposed in terms of the basis $\boldsymbol{a}_i$, and therefore this is a basic example of a *completeness relation*.

Note that we can just as well expand $\boldsymbol{v}$ with respect to the dual basis:

$$\boldsymbol{v} = v_i \boldsymbol{a}^i = \boldsymbol{a}^i(\boldsymbol{a}_i \cdot \boldsymbol{v}) = (\boldsymbol{a}^i \otimes \boldsymbol{a}_i) \cdot \boldsymbol{v},$$

and therefore we also have the dual completeness relation

$$\boldsymbol{a}^i \otimes \boldsymbol{a}_i = \underline{1}. \tag{4.15}$$

We could also have inferred this from transposition of equation (4.14).

Linear transformations of vectors can be written in terms of matrices,

$$\boldsymbol{v}' = \underline{A} \cdot \boldsymbol{v}.$$

If we insert the decompositions with respect to the basis $\boldsymbol{a}_i$,

$$\boldsymbol{v}' = \boldsymbol{a}_i \otimes \boldsymbol{a}^i \cdot \boldsymbol{v}' = \boldsymbol{a}_i \otimes \boldsymbol{a}^i \cdot \underline{\mathbf{A}} \cdot \boldsymbol{a}_j \otimes \boldsymbol{a}^j \cdot \boldsymbol{v},$$

we find the equation in components $v'^i = A^i{}_j v^j$, with the matrix elements of the operator A,

$$A^i{}_j = \boldsymbol{a}^i \cdot \underline{\mathbf{A}} \cdot \boldsymbol{a}_j. \tag{4.16}$$

Using (4.14), we can also infer that

$$\underline{\mathbf{A}} = \boldsymbol{a}_i \otimes \boldsymbol{a}^i \cdot \underline{\mathbf{A}} \cdot \boldsymbol{a}_j \otimes \boldsymbol{a}^j = A^i{}_j \boldsymbol{a}_i \otimes \boldsymbol{a}^j. \tag{4.17}$$

An application of dual bases in solid state physics: The Laue conditions for elastic scattering off a crystal

Non-orthonormal bases and the corresponding dual bases play an important role in solid state physics. Assume e.g. that $\boldsymbol{a}_i$, $1 \leq i \leq 3$, are the three fundamental translation vectors of a three-dimensional lattice L. They generate the lattice according to

$$\boldsymbol{\ell} = \boldsymbol{a}_i m^i, \quad m^i \in \mathbb{Z}.$$

In three dimensions one can easily construct the dual basis vectors using cross products:

$$\boldsymbol{a}^i = \epsilon^{ijk} \frac{\boldsymbol{a}_j \times \boldsymbol{a}_k}{2\boldsymbol{a}_1 \cdot (\boldsymbol{a}_2 \times \boldsymbol{a}_3)} = \frac{1}{2V} \epsilon^{ijk} \boldsymbol{a}_j \times \boldsymbol{a}_k, \tag{4.18}$$

where $V = \boldsymbol{a}_1 \cdot (\boldsymbol{a}_2 \times \boldsymbol{a}_3)$ is the volume of the lattice cell spanned by the basis vectors $\boldsymbol{a}_i$.

The vectors $\boldsymbol{a}^i$, $1 \leq i \leq 3$, generate the *dual lattice* or *reciprocal lattice* $\tilde{L}$ according to

$$\tilde{\boldsymbol{\ell}} = n_i \boldsymbol{a}^i, \quad n_i \in \mathbb{Z},$$

and the volume of a cell in the dual lattice is

$$\tilde{V} = \boldsymbol{a}^1 \cdot (\boldsymbol{a}^2 \times \boldsymbol{a}^3) = \frac{1}{V}. \tag{4.19}$$

Max von Laue derived in 1912 the conditions for constructive interference in the coherent elastic scattering off a regular array of scattering centers. If the directions

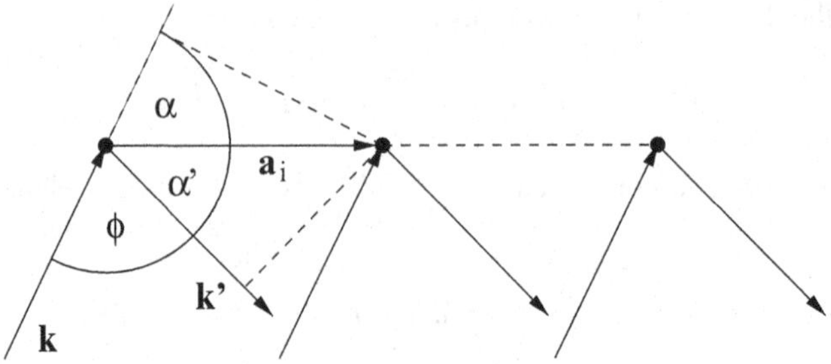

Fig. 4.2 The Laue equation (4.20) is the condition for constructive interference between scattering centers along the line generated by the primitive basis vector $\boldsymbol{a}_i$

of the incident and scattered waves of wavelength λ are $\hat{\boldsymbol{e}}_k$ and $\hat{\boldsymbol{e}}'_k$, as shown in Figure 4.2, the condition for constructive interference from all scattering centers along a line generated by $\boldsymbol{a}_i$ is

$$|\boldsymbol{a}_i|\left(\cos\alpha' - \cos\alpha\right) = \left(\hat{\boldsymbol{e}}'_k - \hat{\boldsymbol{e}}_k\right)\cdot\boldsymbol{a}_i = n_i\lambda, \tag{4.20}$$

with integer numbers n_i.

In terms of the wavevector shift

$$\Delta\boldsymbol{k} = \boldsymbol{k}' - \boldsymbol{k} = \frac{2\pi}{\lambda}\left(\hat{\boldsymbol{e}}'_k - \hat{\boldsymbol{e}}_k\right)$$

equation (4.20) can be written more neatly as

$$\Delta\boldsymbol{k}\cdot\boldsymbol{a}_i = 2\pi n_i. \tag{4.21}$$

If we want to have constructive interference from all scattering centers in the crystal this condition must hold for all three values of i. In case of surface scattering equation (4.21) must only hold for the two vectors $\boldsymbol{a}_1$ and $\boldsymbol{a}_2$ which generate the lattice structure of the scattering centers on the surface.

In 1913 W.L. Bragg observed that for scattering from a bulk crystal equations (4.21) are equivalent to constructive interference from specular reflection from sets of equidistant parallel planes in the crystal, and that the Laue conditions can be reduced to the Bragg equation in this case. However, for scattering from one or two-dimensional crystals[2] and for the Ewald construction one still has to use the Laue conditions.

[2]For scattering off two-dimensional crystals the Laue conditions can be recast in simpler forms in special cases. E.g. for orthogonal incidence a plane grating equation can be derived from the Laue conditions, or if the momentum transfer $\Delta\boldsymbol{k}$ is in the plane of the crystal a two-dimensional Bragg equation can be derived.

If we study scattering off a three-dimensional crystal, we know that the three dual basis vectors $\boldsymbol{a}^i$ span the whole three-dimensional space. Like any three-dimensional vector, the wavevector shift can then be expanded in terms of the dual basis vectors according to

$$\Delta\boldsymbol{k} = \boldsymbol{a}^i(\boldsymbol{a}_i \cdot \Delta\boldsymbol{k}),$$

and substitution of equation (4.21) yields

$$\Delta\boldsymbol{k} = 2\pi n_i \boldsymbol{a}^i,$$

i.e. the condition for constructive interference from coherent elastic scattering off a three-dimensional crystal is equivalent to the statement that $\Delta\boldsymbol{k}/(2\pi)$ is a vector in the dual lattice $\tilde{L}$. Furthermore, energy conservation in the elastic scattering implies $|\boldsymbol{p}'| = |\boldsymbol{p}|$,

$$\Delta\boldsymbol{k}^2 + 2\boldsymbol{k} \cdot \Delta\boldsymbol{k} = 0. \tag{4.22}$$

Equations (4.21) and (4.22) together lead to the Ewald construction for the momenta of elastically scattered beams (see Figure 4.3): Draw the dual lattice and multiply all distances by a factor 2π. Then draw the vector $-\boldsymbol{k}$ from one (arbitrary) point of this rescaled dual lattice. Draw a sphere of radius $|\boldsymbol{k}|$ around the endpoint of $-\boldsymbol{k}$. Any point in the rescaled dual lattice which lies on this sphere corresponds to the $\boldsymbol{k}'$ vector of an elastically scattered beam; $\boldsymbol{k}'$ points from the endpoint of $-\boldsymbol{k}$ (the center of the sphere) to the rescaled dual lattice point on the sphere.

We have already noticed that for scattering off a planar array of scattering centers, equation (4.21) must only hold for the two vectors $\boldsymbol{a}_1$ and $\boldsymbol{a}_2$ which generate the lattice structure of the scattering centers on the surface. And if we have only a

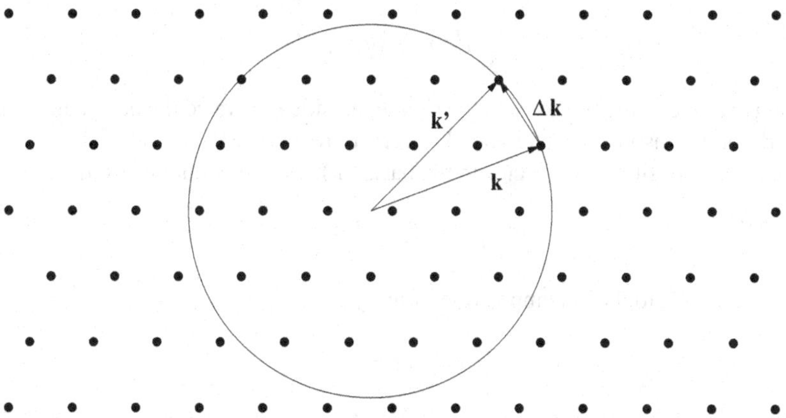

Fig. 4.3 The Ewald construction of the wave vectors of elastically scattered beams. The points correspond to the reciprocal lattice stretched with the factor 2π

linear array of scattering centers, equation (4.21) must only hold for the vector $\boldsymbol{a}_1$ which generates the linear array. In those two cases the wavevector shift can be decomposed into components orthogonal and parallel to the scattering surface or line, and the Laue conditions then imply that the parallel component is a vector in the rescaled dual lattice,

$$\Delta\boldsymbol{k} = \Delta\boldsymbol{k}_\perp + \Delta\boldsymbol{k}_\| = \Delta\boldsymbol{k}_\perp + \boldsymbol{a}^i(\boldsymbol{a}_i \cdot \Delta\boldsymbol{k}) = \Delta\boldsymbol{k}_\perp + 2\pi n_i \boldsymbol{a}^i.$$

The rescaled dual lattice is also important in the *umklapp* processes in phonon-phonon or electron-phonon scattering in crystals. Lattices can only support oscillations with wavelengths larger than certain minimal wavelengths, which are determined by the crystal structure. As a result momentum conservation in phonon-phonon or electron-phonon scattering involves the rescaled dual lattice,

$$\sum \boldsymbol{k}_{in} - \sum \boldsymbol{k}_{out} \in 2\pi \times \tilde{L},$$

see textbooks on solid state physics.

Bra-ket notation in linear algebra

The translation of the previous notions in linear algebra into bra-ket notation starts with the notion of a *ket vector* for a vector, $\boldsymbol{v} = |v\rangle$, and a *bra vector* for a transposed vector[3], $\boldsymbol{v}^{\mathrm{T}} = \langle v|$. The tensor product is

$$\boldsymbol{u} \otimes \boldsymbol{v}^{\mathrm{T}} = |u\rangle\langle v|,$$

and the scalar product is

$$\boldsymbol{u}^{\mathrm{T}} \cdot \boldsymbol{v} = \langle u|v\rangle.$$

The appearance of the brackets on the right hand side motivated the designation "bra vector" for a transposed vector and "ket vector" for a vector.

The decomposition of a vector in the basis $|a_i\rangle$, using the dual basis $|a^i\rangle$ is

$$|v\rangle = |a_i\rangle\langle a^i|v\rangle,$$

and corresponds to the decomposition of unity

$$|a_i\rangle\langle a^i| = \underline{1}.$$

[3]In the case of a complex finite-dimensional vector space, the "bra vector" would actually be the transpose complex conjugate vector, $\langle v| = \boldsymbol{v}^+ = \boldsymbol{v}^{*\mathrm{T}}$.

A linear operator maps vectors $|v\rangle$ into vectors $|v'\rangle$, $|v'\rangle = A|v\rangle$. This reads in components

$$\langle a^i|v'\rangle = \langle a^i|A|v\rangle = \langle a^i|A|a_j\rangle\langle a^j|v\rangle,$$

where

$$A^i{}_j \equiv \langle a^i|A|a_j\rangle$$

are the matrix elements of the linear operator A. There is no real advantage in using bra-ket notation in the linear algebra of finite-dimensional vector spaces, but it turns out to be very useful in quantum mechanics.

4.2 Bra-ket notation in quantum mechanics

We can represent a state as a probability amplitude in $\boldsymbol{x}$-space or in $\boldsymbol{k}$-space, and we can switch between both representations through Fourier transformation. The state itself is apparently independent from which representation we choose, just like a vector is independent from the particular basis in which we expand the vector. In Chapter 7 we will derive a wave function $\psi_{1s}(\boldsymbol{x},t)$ for the relative motion of the proton and the electron in the lowest energy state of a hydrogen atom. However, it does not matter whether we use the wave function $\psi_{1s}(\boldsymbol{x},t)$ in $\boldsymbol{x}$-space or the Fourier transformed wave function $\psi_{1s}(\boldsymbol{k},t)$ in $\boldsymbol{k}$-space to calculate observables for the ground state of the hydrogen atom. Every information on the state can be retrieved from each of the two wave functions. We can also contemplate more exotic possibilities like writing the ψ_{1s} state as a linear combination of the oscillator eigenstates that we will encounter in Chapter 6. There are infinitely many possibilities to write down wave functions for one and the same quantum state, and all possibilities are equivalent. Therefore wave functions are only particular representations of a state, just like the components $\langle a^i|v\rangle$ of a vector $|v\rangle$ in an N-dimensional vector space provide only a representation of the vector with respect to a particular basis $|a_i\rangle$, $1 \leq i \leq N$.

This motivates the following adaptation of bra-ket notation: The (generically time-dependent) state of a quantum system is $|\psi(t)\rangle$, and the $\boldsymbol{x}$-representation is just the specification of $|\psi(t)\rangle$ in terms of its projection on a particular basis,

$$\psi(\boldsymbol{x},t) = \langle \boldsymbol{x}|\psi(t)\rangle,$$

where the "basis" is given by the non-enumerable set of "**x**-eigenkets":

$$\mathbf{x}|\boldsymbol{x}\rangle = \boldsymbol{x}|\boldsymbol{x}\rangle. \tag{4.23}$$

Here $\mathbf{x}$ is the operator, or rather a vector of operators $\mathbf{x} = (\mathrm{x},\mathrm{y},\mathrm{z})$, and $\boldsymbol{x} = (x,y,z)$ is the corresponding vector of eigenvalues.

In advanced quantum mechanics, the operators for location or momentum of a particle and their eigenvalues are sometimes not explicitly distinguished in notation, but for the experienced reader it is always clear from the context whether e.g. $\boldsymbol{x}$ refers to the operator or the eigenvalue. We will denote the operators $\mathbf{x}$ and $\mathbf{p}$ for location and momentum and their Cartesian components with upright notation, $\mathbf{x} = (\mathrm{x}, \mathrm{y}, \mathrm{z})$, $\mathbf{p} = (\mathrm{p}_x, \mathrm{p}_y, \mathrm{p}_z)$, while their eigenvalue vectors and Cartesian eigenvalues are written in cursive notation, $\boldsymbol{x} = (x, y, z)$ and $\boldsymbol{p} = \hbar\boldsymbol{k} = (p_x, p_y, p_z)$. However, this becomes very clumsy for non-Cartesian components of the operators $\mathbf{x}$ and $\mathbf{p}$, but once we are at the stage where we have to use e.g. both location operators and their eigenvalues in polar coordinates, you will have so much practice with bra-ket notation that you will infer from the context whether e.g. r refers to the operator $r = \sqrt{\mathrm{x}^2 + \mathrm{y}^2 + \mathrm{z}^2}$ or to the eigenvalue $r = \sqrt{x^2 + y^2 + z^2}$. Some physical quantities have different symbols for the related operator and its eigenvalues, e.g. H for the energy operator and E for its eigenvalues,

$$H|E\rangle = E|E\rangle,$$

so that in these cases the use of standard cursive mathematical notation for the operators and the eigenvalues cannot cause confusion.

Expectation values of observables are often written in terms of the operator or the observable, e.g. $\langle x\rangle \equiv \langle \mathrm{x}\rangle$, $\langle E\rangle \equiv \langle H\rangle$ etc., but explicit matrix elements of operators should always explicitly use the operator, e.g. $\langle\psi|\mathrm{x}|\psi\rangle$, $\langle\psi|H|\psi\rangle$.

The "momentum-eigenkets" provide another basis of quantum states of a particle,

$$\mathbf{p}|\boldsymbol{k}\rangle = \hbar\boldsymbol{k}|\boldsymbol{k}\rangle, \tag{4.24}$$

and the change of basis looks like the corresponding equation in linear algebra: If we have two sets of basis vectors $|a_i\rangle$, $|b_a\rangle$, then the components of a vector $|v\rangle$ with respect to the new basis $|b_a\rangle$ are related to the $|a_i\rangle$-components via (just insert $|v\rangle = |a_i\rangle\langle a^i|v\rangle$)

$$\langle b^a|v\rangle = \langle b^a|a_i\rangle\langle a^i|v\rangle,$$

i.e. the transformation matrix $T^a{}_i = \langle b^a|a_i\rangle$ is just given by the components of the old basis vectors in the new basis.

The corresponding equation in quantum mechanics for the $|\boldsymbol{x}\rangle$ and $|\boldsymbol{k}\rangle$ bases is

$$\langle\boldsymbol{x}|\psi(t)\rangle = \int d^3\boldsymbol{k}\,\langle\boldsymbol{x}|\boldsymbol{k}\rangle\langle\boldsymbol{k}|\psi(t)\rangle = \frac{1}{\sqrt{2\pi}^3}\int d^3\boldsymbol{k}\,\exp(\mathrm{i}\boldsymbol{k}\cdot\boldsymbol{x})\langle\boldsymbol{k}|\psi(t)\rangle,$$

which tells us that the expansion coefficients of the vectors $|\boldsymbol{k}\rangle$ with respect to the $|\boldsymbol{x}\rangle$-basis are just

$$\langle\boldsymbol{x}|\boldsymbol{k}\rangle = \frac{1}{\sqrt{2\pi}^3}\exp(\mathrm{i}\boldsymbol{k}\cdot\boldsymbol{x}). \tag{4.25}$$

The Fourier decomposition of the δ-function implies that these bases are self-dual, e.g.

$$\langle \boldsymbol{x}|\boldsymbol{x}'\rangle = \int d^3k\, \langle \boldsymbol{x}|\boldsymbol{k}\rangle\langle \boldsymbol{k}|\boldsymbol{x}'\rangle = \frac{1}{(2\pi)^3}\int d^3k\, \exp[\mathrm{i}\boldsymbol{k}\cdot(\boldsymbol{x}-\boldsymbol{x}')] = \delta(\boldsymbol{x}-\boldsymbol{x}').$$

The scalar product of two states can be written in terms of $|\boldsymbol{x}\rangle$-components or $|\boldsymbol{k}\rangle$-components

$$\begin{aligned}\langle\varphi(t)|\psi(t)\rangle &= \int d^3x\, \langle\varphi(t)|\boldsymbol{x}\rangle\langle\boldsymbol{x}|\psi(t)\rangle = \int d^3x\, \varphi^+(\boldsymbol{x},t)\psi(\boldsymbol{x},t)\\ &= \int d^3x\, \langle\varphi(t)|\boldsymbol{k}\rangle\langle\boldsymbol{k}|\psi(t)\rangle = \int d^3x\, \varphi^+(\boldsymbol{k},t)\psi(\boldsymbol{k},t).\end{aligned}$$

To get some practice with bra-ket notation let us derive the $\boldsymbol{x}$-representation of the momentum operator. We know equation (4.24) and we want to find out what the $\boldsymbol{x}$-components of the state $\boldsymbol{p}|\psi(t)\rangle$ are. We can accomplish this by inserting the decomposition

$$|\psi(t)\rangle = \int d^3k\, |\boldsymbol{k}\rangle\langle\boldsymbol{k}|\psi(t)\rangle$$

into $\langle\boldsymbol{x}|\mathbf{p}|\psi(t)\rangle$,

$$\langle\boldsymbol{x}|\mathbf{p}|\psi(t)\rangle = \int d^3k\, \langle\boldsymbol{x}|\mathbf{p}|\boldsymbol{k}\rangle\langle\boldsymbol{k}|\psi(t)\rangle = \int d^3k\, \hbar\boldsymbol{k}\langle\boldsymbol{x}|\boldsymbol{k}\rangle\langle\boldsymbol{k}|\psi(t)\rangle. \tag{4.26}$$

However, equation (4.25) implies

$$\hbar\boldsymbol{k}\langle\boldsymbol{x}|\boldsymbol{k}\rangle = \frac{\hbar}{\mathrm{i}}\nabla\langle\boldsymbol{x}|\boldsymbol{k}\rangle,$$

and substitution into equation (4.26) yields

$$\langle\boldsymbol{x}|\mathbf{p}|\psi(t)\rangle = \frac{\hbar}{\mathrm{i}}\nabla\int d^3k\, \langle\boldsymbol{x}|\boldsymbol{k}\rangle\langle\boldsymbol{k}|\psi(t)\rangle = \frac{\hbar}{\mathrm{i}}\nabla\langle\boldsymbol{x}|\psi(t)\rangle. \tag{4.27}$$

This equation yields in particular the matrix elements of the momentum operator in the $|\boldsymbol{x}\rangle$-basis,

$$\langle\boldsymbol{x}|\mathbf{p}|\boldsymbol{x}'\rangle = \frac{\hbar}{\mathrm{i}}\nabla\delta(\boldsymbol{x}-\boldsymbol{x}').$$

Equation (4.27) means that the $\boldsymbol{x}$-expansion coefficients $\langle\boldsymbol{x}|\mathbf{p}|\psi(t)\rangle$ of the new state $\mathbf{p}|\psi(t)\rangle$ can be calculated from the expansion coefficients $\langle\boldsymbol{x}|\psi(t)\rangle$ of the old state

$|\psi(t)\rangle$ through application of $-i\hbar\nabla$. In sloppy terminology this is the statement "the $\boldsymbol{x}$-representation of the momentum operator is $-i\hbar\nabla$", but the proper statement is equation (4.27),

$$\langle \boldsymbol{x}|\mathbf{p}|\psi(t)\rangle = \frac{\hbar}{\mathrm{i}}\nabla\langle \boldsymbol{x}|\psi(t)\rangle.$$

The quantum operator $\mathbf{p}$ acts on the quantum state $|\psi(t)\rangle$, the differential operator $-i\hbar\nabla$ acts on the expansion coefficients $\langle \boldsymbol{x}|\psi(t)\rangle$ of the state $|\psi(t)\rangle$.

The corresponding statement in linear algebra is that a linear transformation A transforms a vector $|v\rangle$ according to

$$|v\rangle \;\to\; |v'\rangle = A|v\rangle,$$

and the transformation in a particular basis reads

$$\langle a^i|v'\rangle = \langle a^i|A|v\rangle = \langle a^i|A|a_j\rangle\langle a^j|v\rangle.$$

The operator A acts on the vector, and its representation $\langle a^i|A|a_j\rangle$ in a particular basis acts on the components of the vector in that basis.

Bra-ket notation requires a proper understanding of the distinction between quantum operators (like $\mathbf{p}$) and operators that act on expansion coefficients of quantum states in a particular basis (like $-i\hbar\nabla$). Bra-ket notation appears in virtually every equation of advanced quantum mechanics and quantum field theory. It provides in many respects the most useful notation for recognizing the elegance and power of quantum theory.

Equations equivalent to equations (4.23, 4.24, 4.27) are contained in

$$\mathbf{x} = \int d^3\boldsymbol{x}\,|\boldsymbol{x}\rangle\boldsymbol{x}\langle\boldsymbol{x}| = \int d^3\boldsymbol{k}\,|\boldsymbol{k}\rangle\mathrm{i}\frac{\partial}{\partial\boldsymbol{k}}\langle\boldsymbol{k}|, \tag{4.28}$$

$$\mathbf{p} = \int d^3\boldsymbol{k}\,|\boldsymbol{k}\rangle\hbar\boldsymbol{k}\langle\boldsymbol{k}| = \int d^3\boldsymbol{x}\,|\boldsymbol{x}\rangle\frac{\hbar}{\mathrm{i}}\frac{\partial}{\partial\boldsymbol{x}}\langle\boldsymbol{x}|. \tag{4.29}$$

Here we used the very convenient notation $\nabla \equiv \partial/\partial\boldsymbol{x}$ for the del operator in $\boldsymbol{x}$ space, and $\partial/\partial\boldsymbol{k}$ for the del operator in $\boldsymbol{k}$ space. One often encounters several copies of several vector spaces in an equation, and this notation is extremely useful to distinguish the different del operators in the different vector spaces.

Functions of operators are operators again. An important example are the operators $V(\mathbf{x})$ for the potential energy of a particle. The eigenkets of $\mathbf{x}$ are also eigenkets of $V(\mathbf{x})$,

$$V(\mathbf{x})|\boldsymbol{x}\rangle = V(\boldsymbol{x})|\boldsymbol{x}\rangle,$$

and the matrix elements in $\boldsymbol{x}$ representation are

$$\langle\boldsymbol{x}|V(\mathbf{x})|\boldsymbol{x}'\rangle = V(\boldsymbol{x}')\delta(\boldsymbol{x}-\boldsymbol{x}').$$

The single particle Schrödinger equation (1.14) is in representation free notation

$$i\hbar\frac{d}{dt}|\psi(t)\rangle = H|\psi(t)\rangle = \frac{\mathbf{p}^2}{2m}|\psi(t)\rangle + V(\mathbf{x})|\psi(t)\rangle. \tag{4.30}$$

We recover the $\boldsymbol{x}$ representation already used in (1.14) through projection on $\langle\boldsymbol{x}|$ and substitution of

$$1 = \int d^3x'\,|\boldsymbol{x}'\rangle\langle\boldsymbol{x}'|,$$

$$i\hbar\frac{\partial}{\partial t}\langle\boldsymbol{x}|\psi(t)\rangle = -\frac{\hbar^2}{2m}\Delta\langle\boldsymbol{x}|\psi(t)\rangle + V(\boldsymbol{x})\langle\boldsymbol{x}|\psi(t)\rangle.$$

The definition of adjoint operators in representation-free bra-ket notation is

$$\langle\varphi|A|\psi\rangle = \langle\psi|A^+|\varphi\rangle^+. \tag{4.31}$$

This implies in particular that the "bra vector" $\langle\Psi|$ adjoint to the "ket vector" $|\Psi\rangle = A|\psi\rangle$ satisfies

$$\langle\Psi| = \langle\psi|A^+. \tag{4.32}$$

This is an intuitive equation which can be motivated e.g. from matrix algebra of complex finite-dimensional vector spaces. However, it deserves a formal derivation. We have for any third state $|\phi\rangle$ the relation

$$\langle\Psi|\phi\rangle = (\langle\phi|\Psi\rangle)^+ = (\langle\phi|A|\psi\rangle)^+ = \langle\psi|A^+|\phi\rangle,$$

where we used the defining property of adjoint operators in the last equation. Since this equation holds for every state $|\phi\rangle$, the operator equation (4.32) follows: Projection[4] onto the state $|\Psi\rangle = A|\psi\rangle$ is equivalent to action of the operator A^+ followed by projection onto the state $|\psi\rangle$.

Self-adjoint operators (e.g. $\mathrm{p}^+ = \mathrm{p}$) have real expectation values and in particular real eigenvalues:

$$\langle\psi|\mathrm{p}|\psi\rangle = \langle\psi|\mathrm{p}^+|\psi\rangle^+ = \langle\psi|\mathrm{p}|\psi\rangle^+.$$

Observables are therefore described by self-adjoint operators in quantum mechanics.

[4]Strictly speaking, we can think of multiplication of a state $|\phi\rangle$ with $\langle\Psi|$ as projecting onto a component parallel to $|\Psi\rangle$ only if $|\Psi\rangle$ is normalized. It is convenient, though, to denote multiplication with $\langle\Psi|$ as projection, although in the general case this will only be *proportional* to the coefficient of the $|\Psi\rangle$ component in $|\phi\rangle$.

Unitary operators ($U^+ = U^{-1}$) do not change the norm of a state: Substitution of $|\psi\rangle = U|\varphi\rangle$ into $\langle\psi|\psi\rangle$ yields

$$\langle\psi|\psi\rangle = \langle\psi|U|\varphi\rangle = \langle\varphi|U^+|\psi\rangle^+ = \langle\varphi|U^+U|\varphi\rangle^+ = \langle\varphi|\varphi\rangle^+ = \langle\varphi|\varphi\rangle.$$

Time evolution and symmetry transformations of quantum systems are described by unitary operators.

4.3 The adjoint Schrödinger equation and the virial theorem

We consider a matrix element

$$\langle\psi(t)|A(t')|\phi(t')\rangle = (\langle\phi(t')|A^+(t')|\psi(t)\rangle)^+. \tag{4.33}$$

We assume that $|\psi(t)\rangle$ satisfies the Schrödinger equation

$$\mathrm{i}\hbar\frac{d}{dt}|\psi(t)\rangle = H|\psi(t)\rangle,$$

while $A(t')$ and $|\phi(t')\rangle$ are an arbitrary operator and state, respectively. We have artificially taken the state $|\Phi(t')\rangle = A(t')|\phi(t')\rangle$ at another time t', because we are particularly interested in the time-dependence of the matrix element $\langle\psi(t)|A(t')|\phi(t')\rangle$ which arises from the time-dependence of $|\psi(t)\rangle$.

Equation (4.33), the Schrödinger equation, and hermiticity of H imply

$$\frac{d}{dt}\langle\psi(t)|A(t')|\phi(t')\rangle = \left(\langle\phi(t')|A^+(t')\frac{d}{dt}|\psi(t)\rangle\right)^+$$
$$= \left(\frac{1}{\mathrm{i}\hbar}\langle\phi(t')|A^+(t')H|\psi(t)\rangle\right)^+ = \frac{\mathrm{i}}{\hbar}\langle\psi(t)|HA(t')|\phi(t')\rangle.$$

Since this holds for every operator $A(t')$ and state $|\phi(t')\rangle$, we have an operator equation

$$\left(\frac{d}{dt}\langle\psi(t)|\right) = \frac{\mathrm{i}}{\hbar}\langle\psi(t)|H. \tag{4.34}$$

With the brackets on the left hand side, this equation also holds for projection on time-dependent states of the form $A(t)|\phi(t)\rangle$: Projection of any state $|\Phi(t)\rangle$ on $(d\langle\psi(t)|/dt)$ is equivalent to action of H on $|\Phi(t)\rangle$ followed by projection of $H|\Phi(t)\rangle$ on $(\mathrm{i}/\hbar)\langle\psi(t)|$,

$$\frac{d}{dt}\langle\psi(t)|A(t)|\phi(t)\rangle = \frac{\mathrm{i}}{\hbar}\langle\psi(t)|HA(t)|\phi(t)\rangle + \langle\psi(t)|\frac{dA(t)}{dt}|\phi(t)\rangle$$
$$+ \langle\psi(t)|A(t)\frac{d}{dt}|\phi(t)\rangle.$$

In particular, if $|\phi(t)\rangle$ also satisfies the Schrödinger equation, we have

$$\frac{d}{dt}\langle\psi(t)|A(t)|\phi(t)\rangle = \frac{\mathrm{i}}{\hbar}\langle\psi(t)|[H,A(t)]|\phi(t)\rangle + \langle\psi(t)|\frac{dA(t)}{dt}|\phi(t)\rangle. \quad (4.35)$$

The operator equation (4.34) is the adjoint Schrödinger equation. In general it is an operator equation, but it reduces to the complex conjugate of the Schrödinger equation if it is projected onto $\mathbf{x}$ eigenkets,

$$\begin{aligned}\frac{d}{dt}\langle\psi(t)|\boldsymbol{x}\rangle &= \frac{\mathrm{i}}{\hbar}\int d^3x'\,\langle\psi(t)|\boldsymbol{x}'\rangle\left(-\frac{\hbar^2}{2m}\frac{\partial^2}{\partial\boldsymbol{x}'^2} + V(\boldsymbol{x}')\right)\delta(\boldsymbol{x}'-\boldsymbol{x})\\ &= \frac{\mathrm{i}}{\hbar}\left(-\frac{\hbar^2}{2m}\frac{\partial^2}{\partial\boldsymbol{x}^2} + V(\boldsymbol{x})\right)\langle\psi(t)|\boldsymbol{x}\rangle.\end{aligned}$$

The result (4.35) for the time-dependence of matrix elements appears in many different settings in quantum mechanics, but one application that we will address now concerns the particular choice of the virial operator $\mathbf{x}\cdot\mathbf{p}$ for the operator A. In classical mechanics, Newton's equation and $m\dot{\boldsymbol{x}} = \boldsymbol{p}$ imply that the time derivative of the virial $\boldsymbol{x}\cdot\boldsymbol{p}$ is

$$\frac{d}{dt}\boldsymbol{x}\cdot\boldsymbol{p} = \frac{\boldsymbol{p}^2}{m} - \boldsymbol{x}\cdot\nabla V(\boldsymbol{x}).$$

Application of the time averaging operation $\lim_{T\to\infty}\int_0^T dt\ldots$ on both sides of this equation then yields the classical virial theorem for the time average $\langle K\rangle_T$ of the kinetic energy $K = \boldsymbol{p}^2/2m$,

$$2\langle K\rangle_T = \langle\boldsymbol{x}\cdot\nabla V(\boldsymbol{x})\rangle_T. \quad (4.36)$$

The equation (4.35) applied to $A = \mathbf{x}\cdot\mathbf{p}$ implies that the same relation holds for all matrix elements of the operators $K = \mathbf{p}^2/2m$ and $\mathbf{x}\cdot\nabla V(\mathbf{x})$. We have

$$\frac{\mathrm{i}}{\hbar}[H,\mathbf{x}\cdot\mathbf{p}] = \frac{\mathbf{p}^2}{m} - \mathbf{x}\cdot\frac{\partial}{\partial\mathbf{x}}V(\mathbf{x}),$$

and therefore

$$\frac{d}{dt}\langle\psi(t)|\mathbf{x}\cdot\mathbf{p}|\phi(t)\rangle = 2\langle\psi(t)|K|\phi(t)\rangle - \langle\psi(t)|\mathbf{x}\cdot\frac{\partial}{\partial\mathbf{x}}V(\mathbf{x})|\phi(t)\rangle. \quad (4.37)$$

Time averaging then yields a quantum analog of the classical virial theorem,

$$2\langle\psi(t)|K|\phi(t)\rangle_T = \langle\psi(t)|\mathbf{x}\cdot\frac{\partial}{\partial\mathbf{x}}V(\mathbf{x})|\phi(t)\rangle_T. \quad (4.38)$$

However, if $|\psi(t)\rangle$ and $|\phi(t)\rangle$ are energy eigenstates,

$$|\psi(t)\rangle = |\psi\rangle \exp(-\mathrm{i}E_\psi t/\hbar), \quad |\phi(t)\rangle = |\phi\rangle \exp(-\mathrm{i}E_\phi t/\hbar),$$

then equation (4.37) yields

$$\frac{\mathrm{i}}{\hbar}(E_\psi - E_\phi)\langle\psi|\mathbf{x}\cdot\mathbf{p}|\phi\rangle = 2\langle\psi|K|\phi\rangle - \langle\psi|\mathbf{x}\cdot\frac{\partial}{\partial\mathbf{x}}V(\mathbf{x})|\phi\rangle. \tag{4.39}$$

In this case, the classical time averaging cannot yield anything interesting, but if we assume that our energy eigenstates are degenerate normalizable states,

$$E_\psi = E_\phi, \quad \langle\psi|\psi\rangle = \langle\phi|\phi\rangle = 1,$$

then we find the quantum virial theorem for matrix elements of degenerate normalizable energy eigenstates[5],

$$2\langle\psi|K|\phi\rangle = \langle\psi|\mathbf{x}\cdot\frac{\partial}{\partial\mathbf{x}}V(\mathbf{x})|\phi\rangle. \tag{4.40}$$

Furthermore, if $V(\boldsymbol{x})$ is homogeneous of order ν,

$$V(a\boldsymbol{x}) = a^\nu V(\boldsymbol{x}),$$

then

$$\boldsymbol{x}\cdot\nabla V(\boldsymbol{x}) = \nu V(\boldsymbol{x})$$

and

$$2\langle\psi|K|\phi\rangle = \nu\langle\psi|V|\phi\rangle. \tag{4.41}$$

The relations (4.40) and (4.41) hold in particular for the expectation values of normalizable energy eigenstates. Special cases for the appearance of physically relevant homogeneous potential functions include harmonic oscillators, $\nu = 2$, and the three-dimensional Coulomb potential, $\nu = -1$. We will discuss harmonic oscillators and the Coulomb problem in Chapters 6 and 7, respectively. Equation (4.41) has also profound implications for hypothetical physics in higher dimensions, see Problem 20.5.

[5]Normalizability is important for the correctness of equation (4.40), because for states in an energy continuum the left hand side of equation (4.39) may not vanish in the degenerate limit $E_\psi \to E_\phi$, see Problem 9.

4.4 Problems

4.1. We consider again the rotation (4.4) of a Cartesian basis,

$$\hat{\boldsymbol{e}}_a \to \hat{\boldsymbol{e}}'_i = \hat{\boldsymbol{e}}_a R^a{}_i,$$

but this time we insist on keeping the expansion coefficients v^a of the vector $\boldsymbol{v} = v^a \hat{\boldsymbol{e}}_a$. Rotation of the basis with fixed expansion coefficients $\{v^1, \ldots v^N\}$ will therefore generate a new vector

$$\boldsymbol{v} \to \boldsymbol{v}' \equiv v^i \hat{\boldsymbol{e}}'_i.$$

This is the *active interpretation* of transformations, because the change of the reference frame is accompanied by a change of the physical objects.

In the active interpretation, transformations of the expansion coefficients are defined by the condition that the transformed expansion coefficients describe the expansion of the *new* vector $\boldsymbol{v}'$ with respect to the *old* basis $\hat{\boldsymbol{e}}_a$,

$$\boldsymbol{v}' \equiv v^i \hat{\boldsymbol{e}}'_i = v'^a \hat{\boldsymbol{e}}_a. \tag{4.42}$$

How are the new expansion coefficients v'^a related to the old expansion coefficients v^i for an active transformation?

In the active interpretation, rotations are special by preserving the lengths of vectors and the angles between vectors.

Equation (4.42) implies that we can describe an active transformation either through a transformation of the basis with fixed expansion coefficients, or equivalently through a transformation of the expansion coefficients with a fixed basis. This is very different from the passive transformation, where a transformation of the basis is always accompanied by a compensating contragredient transformation of the expansion coefficients.

4.2. Two basis vectors $\boldsymbol{a}_1$ and $\boldsymbol{a}_2$ have length one and the angle between the vectors is $\pi/3$. Construct the dual basis.

4.3. Nickel atoms form a regular triangular array with an interatomic distance of 2.49 Å on the surface of a Nickel crystal. Particles with momentum $p = h/\lambda$ are incident on the crystal. Which conditions for coherent elastic scattering off the Nickel surface do we get for orthogonal incidence of the particle beam? Which conditions for coherent elastic scattering do we get for grazing incidence in the plane of the surface?

4.4. Suppose $V(\boldsymbol{x})$ is an analytic function of $\boldsymbol{x}$. Write down the $\boldsymbol{k}$-representation of the time-dependent and time-independent Schrödinger equations. Why is the $\boldsymbol{x}$-representation usually preferred for solving the Schrödinger equation?

4.5. Sometimes we seem to violate the symmetric conventions (2.4, 2.5) in the Fourier transformations of the Green's functions that we will encounter later on. We will see that the asymmetric split of powers of 2π that we will encounter in these cases is actually a consequence of the symmetric conventions (2.4, 2.5) for the Fourier transformation of wave functions.

Suppose that the operator G has translation invariant matrix elements,

$$\langle \boldsymbol{x}|G|\boldsymbol{x}'\rangle = G(\boldsymbol{x}-\boldsymbol{x}').$$

Show that the Fourier transformed matrix elements $\langle \boldsymbol{k}|G|\boldsymbol{k}'\rangle$ satisfy $\langle \boldsymbol{k}|G|\boldsymbol{k}'\rangle = G(\boldsymbol{k})\delta(\boldsymbol{k}-\boldsymbol{k}')$ with

$$G(\boldsymbol{k}) = \int d^3\boldsymbol{x}\, G(\boldsymbol{x})\exp(-\mathrm{i}\boldsymbol{k}\cdot\boldsymbol{x}),$$

$$G(\boldsymbol{x}) = \frac{1}{(2\pi)^3}\int d^3\boldsymbol{k}\, G(\boldsymbol{k})\exp(\mathrm{i}\boldsymbol{k}\cdot\boldsymbol{x}). \tag{4.43}$$

4.6. Suppose that the Hamilton operator depends on a real parameter λ, $H = H(\lambda)$. This parameter dependence will influence the energy eigenvalues and eigenstates of the Hamiltonian,

$$H(\lambda)|\psi_n(\lambda)\rangle = E_n(\lambda)|\psi_n(\lambda)\rangle.$$

Use $\langle\psi_m(\lambda)|\psi_n(\lambda)\rangle = \delta_{mn}$ (this could also be a δ function normalization), to show that[6]

$$\delta_{mn}\frac{dE_n(\lambda)}{d\lambda} = \langle\psi_m(\lambda)|\frac{dH(\lambda)}{d\lambda}|\psi_n(\lambda)\rangle + [E_m(\lambda)-E_n(\lambda)]\,\langle\psi_m(\lambda)|\frac{d}{d\lambda}|\psi_n(\lambda)\rangle. \tag{4.44}$$

For $m = n$ discrete this is known as the Hellmann-Feynman theorem[7] [15]. The theorem is important for the calculation of forces in molecules.

4.7. We consider particles of mass m which are bound in a potential $V(\boldsymbol{x})$. The potential does not depend on m. How do the energy levels of the bound states change if we increase the mass of the particles?

The eigenstates for different energies will usually have different momentum uncertainties Δp. Do the energy levels with large or small Δp change more rapidly with mass?

[6] P. Güttinger, Diplomarbeit, ETH Zürich, Z. Phys. 73, 169 (1932). Exceptionally, there is *no* summation convention used in equation (4.44).

[7] R.P. Feynman, Phys. Rev. 56, 340 (1939).

4.8. Show that the free propagator (3.32, 3.33) is the x representation of the one-dimensional free time evolution operator,

$$U(t) = \exp\left(-\mathrm{i}\frac{t-\mathrm{i}\epsilon}{2m\hbar}\mathrm{p}^2\right), \quad U(x-x',t) = \langle x|U(t)|x'\rangle.$$

Here a small negative imaginary part was added to the time variable to ensure convergence of a Gaussian integral.

Show also that the free time-evolution operator in three dimensions satisfies

$$\begin{aligned} U(\boldsymbol{x}-\boldsymbol{x}',t) &= \langle\boldsymbol{x}|\exp\left(-\mathrm{i}\frac{t-\mathrm{i}\epsilon}{2m\hbar}\mathbf{p}^2\right)|\boldsymbol{x}'\rangle \\ &= \sqrt{\frac{m}{2\pi\mathrm{i}\hbar(t-\mathrm{i}\epsilon)}}^{\,3}\exp\left(\mathrm{i}\frac{m}{2\hbar(t-\mathrm{i}\epsilon)}(\boldsymbol{x}-\boldsymbol{x}')^2\right). \end{aligned} \tag{4.45}$$

For later reference we also note that this implies the formula

$$\begin{aligned} \exp\left(\mathrm{i}\hbar\frac{t-\mathrm{i}\epsilon}{2m}\frac{\partial^2}{\partial\boldsymbol{x}^2}\right)\delta(\boldsymbol{x}-\boldsymbol{x}') &= \sqrt{\frac{m}{2\pi\mathrm{i}\hbar(t-\mathrm{i}\epsilon)}}^{\,3} \\ &\quad\times\exp\left(\mathrm{i}\frac{m}{2\hbar(t-\mathrm{i}\epsilon)}(\boldsymbol{x}-\boldsymbol{x}')^2\right). \end{aligned} \tag{4.46}$$

4.9. Apply equation (4.39) in the case $V(\boldsymbol{x}) = 0$ to plane wave states. Show that in this case the left hand side does not vanish in the limit $E(\boldsymbol{k}) \to E(\boldsymbol{k}')$. Indeed, the equation remains correct in this case only *because* the left hand side does not vanish.

4.10. Use the calculation of $\boldsymbol{p}$ or $\boldsymbol{x}$ expectation values in the wave vector representation and in the momentum representation of the state $|\psi\rangle$ to show that momentum and wave vector eigenstates in d spatial dimensions are related according to $|\boldsymbol{p}\rangle = |\boldsymbol{k}\rangle/\hbar^{d/2}$. Does this comply with proper δ function normalization of the two bases?

Chapter 5
Formal Developments

We have to go through a few more formalities before we can resume our discussion of quantum effects in physics. In particular, we need to address minimal uncertainties of observables in quantum mechanics, and we have to discuss transformation and solution properties of differential operators.

I have also included an introduction to the notion of length dimensions of states, since this is useful for understanding the meaning of matrix elements in scattering theory in Chapters 11 and 13. Furthermore, I have included a section on frequency-time Fourier transformation, although that can only be defined in a distributional sense for time-dependent wave functions. However, it is sometimes useful to represent the decompositions of states $|\psi(t)\rangle$ in terms of energy eigenmodes $|\psi_\alpha\rangle$, $H|\psi_\alpha\rangle = E_\alpha|\psi_\alpha\rangle$, in the framework of Fourier transformation to a frequency-dependent state $|\psi(\omega)\rangle$. The frequency-dependent states vanish if $\hbar\omega$ is not part of the spectrum of H, and they contain δ-functions for $\hbar\omega$ in the discrete spectrum of H.

5.1 Uncertainty relations

The statistical interpretation of the wave function naturally implies uncertainty in an observable A_o if the wave function is not an eigenstate of the hermitian operator A that corresponds to A_o. Suppose that A has eigenvalues a_n,

$$A|\phi_n\rangle = a_n|\phi_n\rangle, \quad \langle\phi_m|\phi_n\rangle = \delta_{mn}.$$

Substitution of the expansion

$$|\psi\rangle = \sum_n |\phi_n\rangle\langle\phi_n|\psi\rangle$$

R. Dick, *Advanced Quantum Mechanics*, Graduate Texts in Physics,
DOI 10.1007/978-3-319-25675-7_5

into the formula for the expectation value $\langle A\rangle = \langle\psi|A|\psi\rangle$ in the state $|\psi\rangle$ yields

$$\langle A\rangle = \sum_n a_n\,|\langle\phi_n|\psi\rangle|^2,$$

i.e. $|\langle\phi_n|\psi\rangle|^2$ is a probability that the value a_n for the observable A_o will be observed if the system is in the state $|\psi\rangle$.

If the distribution $|\langle\phi_n|\psi\rangle|^2$ is strongly concentrated around a particular index ℓ, then it is very likely that a measurement of A_o will find the value a_ℓ with very little uncertainty. However, if the probability distribution $|\langle\phi_n|\psi\rangle|^2$ covers a broad range of indices or has maxima e.g. for two separated indices, then there will be high uncertainty of the value of the observable A_o, and observation of A_o for many copies of the system in the state $|\psi\rangle$ will yield a large scatter of observed values.

If A has a continuous spectrum of eigenvalues, e.g. if $A = \mathrm{x}$ is the operator for the location x of a particle in one dimension, then $|\langle x|\psi\rangle|^2\,dx$ is the probability to find the system with a value of x in the interval $[x, x+dx]$.

Heisenberg found in 1927 an intuitive estimate for the minimal product of uncertainties Δx and Δp in location and momentum of a particle[1]. His arguments were easily made rigorous and generalized to other pairs of observables using the statistical formalism of quantum mechanics.

Suppose that two observables A_o and B_o are represented by the two hermitian operators A and B. The expectation value of the observable A_o in a state $|\psi\rangle$ is

$$\langle A\rangle = \langle\psi|A|\psi\rangle,$$

and the uncertainty ΔA of the expectation value $\langle A\rangle$ is defined through

$$\Delta A^2 = \langle(A-\langle A\rangle)^2\rangle = \langle A^2\rangle - \langle A\rangle^2.$$

Minimal values of the uncertainty ΔA with which the observable A_o can be measured are directly related to the *commutator* of the operator A with other operators. The commutator of the two operators A and B is defined through

$$[A, B]|\psi\rangle \equiv AB|\psi\rangle - BA|\psi\rangle,$$

where $AB|\psi\rangle$ is the action of the operator B on the state $|\psi\rangle$ followed by the action of the operator A on the new state $|\psi'\rangle = B|\psi\rangle$.

The commutator of two hermitian operators yields a new hermitian operator C,

$$[A, B] = \mathrm{i}C,$$

and it is easy to show that the magnitude of the expectation value $\langle C\rangle$ yields a lower bound on the product of uncertainties $\Delta A\cdot\Delta B$,

$$\Delta A\Delta B \geq \frac{1}{2}|\langle C\rangle|. \tag{5.1}$$

[1] W. Heisenberg, Z. Phys. 43, 172 (1927).

For the proof of this relation, we use a real parameter ξ. The function

$$\begin{aligned}0 \le f(\xi) &= \langle (A - \langle A\rangle - \mathrm{i}\xi B + \mathrm{i}\xi\langle B\rangle)\,(A - \langle A\rangle + \mathrm{i}\xi B - \mathrm{i}\xi\langle B\rangle)\rangle \\ &= \Delta A^2 - \xi\langle C\rangle + \xi^2\Delta B^2 \end{aligned} \tag{5.2}$$

has minimal value for

$$\xi = \frac{\langle C\rangle}{2\Delta B^2}$$

and substitution into (5.2) yields

$$0 \le \Delta A^2 - \frac{\langle C\rangle^2}{4\Delta B^2}.$$

This implies the result (5.1).

For the inequality in equation (5.2) note that

$$\begin{aligned}&\langle\psi|\,(A - \langle A\rangle - \mathrm{i}\xi B + \mathrm{i}\xi\langle B\rangle)\,(A - \langle A\rangle + \mathrm{i}\xi B - \mathrm{i}\xi\langle B\rangle)\,|\psi\rangle \\ &= |(A - \langle A\rangle + \mathrm{i}\xi B - \mathrm{i}\xi\langle B\rangle)\,|\psi\rangle|^2 .\end{aligned}$$

Equation (5.1) implies for the operators x and p for location and momentum of a particle Heisenberg's uncertainty relation

$$\Delta x\Delta p \ge \frac{\hbar}{2}, \tag{5.3}$$

or in tensorial form for the three-dimensional operators **x** and **p**

$$\Delta\boldsymbol{x} \otimes \Delta\boldsymbol{p} \ge \frac{\hbar}{2}\underline{1}$$

If the state $|\psi\rangle$ should satisfy the uncertainty relation for $\Delta A \cdot \Delta B$ with the equality sign (minimal product of uncertainties), then we must have

$$\begin{aligned}0 &= \langle\psi|\left[(A - \langle A\rangle)^2 - \frac{\langle C\rangle^2}{4\Delta B^4}\,(B - \langle B\rangle)^2\right]|\psi\rangle \\ &= \langle\psi|\left[A - \langle A\rangle - \mathrm{i}\langle C\rangle\frac{B - \langle B\rangle}{2\Delta B^2}\right]\left[A - \langle A\rangle + \mathrm{i}\langle C\rangle\frac{B - \langle B\rangle}{2\Delta B^2}\right]|\psi\rangle \\ &= \left|\left[A - \langle A\rangle + \mathrm{i}\langle C\rangle\frac{B - \langle B\rangle}{2\Delta B^2}\right]|\psi\rangle\right|^2,\end{aligned}$$

where in the second equation the commutator $[A, B] = \mathrm{i}C$ has been used. This is equivalent to

$$\left[A - \langle A\rangle + \mathrm{i}\langle C\rangle \frac{B - \langle B\rangle}{2\Delta B^2}\right]|\psi\rangle = 0.$$

In particular we have minimal $\Delta p \Delta x$ if and only if

$$\left[p - p_0 - \mathrm{i}\hbar \frac{x - x_0}{2\Delta x^2}\right]|\psi\rangle = 0. \tag{5.4}$$

This implies in the x-representation

$$\left[\frac{d}{dx} - \mathrm{i}\frac{p_0}{\hbar} + \frac{x - x_0}{2\Delta x^2}\right]\langle x|\psi\rangle = 0$$

and yields up to an arbitrary constant phase factor the Gaussian wave packet (3.39)

$$\langle x|\psi\rangle = \frac{1}{(2\pi\Delta x^2)^{1/4}} \exp\!\left(-\frac{(x - x_0)^2}{4\Delta x^2} + \mathrm{i}\frac{p_0 x}{\hbar} - \mathrm{i}\frac{p_0 x_0}{2\hbar}\right). \tag{5.5}$$

Equation (5.4) is in p-representation

$$\left[2\Delta x^2 (p - p_0) + \mathrm{i}\hbar x_0 + \hbar^2 \frac{d}{dp}\right]\langle p|\psi\rangle = 0,$$

and this yields (again up to an arbitrary constant phase factor)

$$\langle p|\psi\rangle = \left(\frac{2\Delta x^2}{\pi\hbar^2}\right)^{\frac{1}{4}} \exp\!\left(-\frac{\Delta x^2}{\hbar^2}(p - p_0)^2 - \mathrm{i}\frac{p x_0}{\hbar} + \mathrm{i}\frac{p_0 x_0}{2\hbar}\right), \tag{5.6}$$

which explicitly confirms

$$\Delta p = \frac{\hbar}{2\Delta x}$$

and corresponds to the wave packet (3.40).

For comparison of the solutions, we note that

$$\begin{aligned}
\langle p|\psi\rangle &= \frac{1}{\sqrt{2\pi\hbar}} \int_{-\infty}^{\infty} dx\, \exp\!\left(-\mathrm{i}\frac{px}{\hbar}\right)\langle x|\psi\rangle \\
&= \frac{1}{\sqrt{2\pi\hbar}} \frac{1}{(2\pi\Delta x^2)^{1/4}}
\end{aligned}$$

$$\times \int_{-\infty}^{\infty} dx \exp\left[-\frac{1}{4\Delta x^2}\left(x - x_0 + \mathrm{i}\frac{2\Delta x^2}{\hbar}(p - p_0)\right)^2\right]$$
$$\times \exp\left(-\frac{\Delta x^2}{\hbar^2}(p - p_0)^2 - \mathrm{i}\frac{(p - p_0)x_0}{\hbar}\right)$$
$$= \left(\frac{2\Delta x^2}{\pi\hbar^2}\right)^{\frac{1}{4}} \exp\left(-\frac{\Delta x^2}{\hbar^2}(p - p_0)^2 - \mathrm{i}\frac{px_0}{\hbar} + \mathrm{i}\frac{p_0 x_0}{2\hbar}\right).$$

The phase factor $\exp(-\mathrm{i}p_0x_0/2\hbar)$ was included in in (5.5) to ensure that (5.5) and (5.6) are also related through direct Fourier transformations. Otherwise there would have been a mismatch in a phase $\propto p_0x_0/2\hbar$.

We have seen in equation (3.43) that the width Δx^2 will remain minimal only for a certain moment in time if a Gaussian packet follows a free evolution $\mathrm{i}d|\psi(t)\rangle/dt \propto p^2|\psi(t)\rangle$, while the uncertainty Δp^2 in momentum remains constant. Therefore a freely evolving Gaussian packet will satisfy the minimal condition $\Delta x\Delta p = \hbar/2$ only for a moment in time.

A Gaussian wave packet following an evolution $\mathrm{i}d|\psi(t)\rangle/dt \propto x^2|\psi(t)\rangle$ would have constant Δx^2, but Δp^2 would have the minimal possible value $\hbar/(2\Delta x)$ only for a moment in time. Such a hypothetical quantum system would correspond to an oscillator without kinetic energy, and it could move uniformly along the p axis.

In Chapter 6 we will find that a harmonic oscillator evolution for Gaussian wave packets, $\mathrm{i}d|\psi(t)\rangle/dt = (\alpha p^2 + \beta x^2)|\psi(t)\rangle$, yields constant widths both in x and in p direction, and the minimal uncertainty condition $\Delta x\Delta p = \hbar/2$ will be satisfied at all times.

There is also an uncertainty relation between energy and time, which is not as strict as the relations (5.1, 5.3), and cannot be proven by the same rigorous mathematical methods. The relation involves the minimal time window Δt which is required to observe a system with energy uncertainty ΔE. Smaller energy uncertainty requires a longer observation window, or a longer time to form the system,

$$\Delta t\Delta E \gtrsim \mathcal{O}(\hbar). \tag{5.7}$$

This order of magnitude estimate is often written as $\Delta t\Delta E \gtrsim \hbar/2$ for symmetry with the Heisenberg uncertainty relation (5.3), but it should not be mistaken to indicate a strict lower bound as in equation (5.3).

Equation (5.7) cannot be derived in the same way as equation (5.3) because time is not an observable, but a parameter in quantum mechanics. Therefore there is no related expectation value, nor is there any corresponding definition of Δt as the variance of an expectation value.

There exist a few simple heuristic derivations to motivate equation (5.7) from equation (5.3), but we will find the best justification for (5.7) in the equations of time-dependent perturbation theory in Section 13.8.

5.2 Frequency representation of states

Fourier transformation from $\boldsymbol{x}$-dependent wave functions $\psi(\boldsymbol{x},t)$ to $(\boldsymbol{k}=\boldsymbol{p}/\hbar)$-dependent wave functions $\psi(\boldsymbol{k},t)$ is defined within the framework of classical analysis as long as we are only dealing with square integrable functions, which is the case for bound states. However, even for Fourier transformation between the $\boldsymbol{x}$ and $\boldsymbol{p}$ representations of wave functions we had to go beyond classical analysis and invoke the δ function to deal e.g. with plane wave states.

Fourier transformation of wave functions between frequency and time is also important in quantum mechanics, but it is clear that we always have to define it in terms of distributions when it comes to the frequency representation, because wave functions $\psi(\boldsymbol{x},t)$ are never square integrable (nor e.g. absolutely integrable) with respect to time. Therefore the standard classical criteria for existence of Fourier transforms from t to $\omega = E/\hbar$ in the sense of classical analysis will never apply for a quantum system. In spite of this verdict, we will see that time-frequency Fourier transformation automatically appears in quantum mechanics if we combine completeness of energy eigenstates with the time evolution implied by the Schrödinger equation.

Recall that the eigenstates of a stationary Hamiltonian,

$$H|\psi_\alpha\rangle = E_\alpha|\psi_\alpha\rangle$$

form a complete basis,

$$\sum\!\!\!\!\!\!\int d\alpha\,|\psi_\alpha\rangle\langle\psi_\alpha| = 1, \tag{5.8}$$

see Section 2.2 and Appendix C. The notation $\sum\!\!\!\!\!\!\int d\alpha$ stands for summation over discrete quantum numbers and integration over continuous quantum numbers, see e.g. (3.23), which we can write in bra-ket notation in the form

$$|\kappa\rangle\langle\kappa| + \int_0^\infty dk\,(|k,-\rangle\langle k,-| + |k,+\rangle\langle k,+|) = 1. \tag{5.9}$$

Another example is the completeness of the states (3.10) in a cubic quantum wire,

$$\sum_{n_1=1}^{\infty}\sum_{n_2=1}^{\infty}\int_{-\infty}^{\infty} dk\,|n_1,n_2,k\rangle\langle n_1,n_2,k| = 1.$$

On the other hand, for every time-dependent state which evolves with the Hamiltonian H, the Schrödinger equation

$$i\hbar\frac{d}{dt}|\psi(t)\rangle = H|\psi(t)\rangle$$

implies

$$|\psi(t)\rangle = \exp\left(-\frac{\mathrm{i}}{\hbar}Ht\right)|\psi(0)\rangle. \tag{5.10}$$

Substitution of the decomposition $|\psi(0)\rangle = \sum\!\!\!\!\!\int d\alpha |\psi_\alpha\rangle\langle\psi_\alpha|\psi(0)\rangle$ into (5.10) yields

$$|\psi(t)\rangle = \sum\!\!\!\!\!\!\int d\alpha\, |\psi_\alpha\rangle\langle\psi_\alpha|\psi(0)\rangle \exp\left(-\frac{\mathrm{i}}{\hbar}E_\alpha t\right).$$

We can interprete this as a frequency-time Fourier transformation[2]

$$|\psi(t)\rangle = \frac{1}{\sqrt{2\pi}}\int d\omega\, |\psi(\omega)\rangle \exp(-\mathrm{i}\omega t) \tag{5.12}$$

with inversion

$$|\psi(\omega)\rangle = \frac{1}{\sqrt{2\pi}}\int dt\, |\psi(t)\rangle \exp(\mathrm{i}\omega t) \tag{5.13}$$

if we define

$$\begin{aligned} |\psi(\omega)\rangle &= \sqrt{2\pi}\sum\!\!\!\!\!\!\int d\alpha\, |\psi_\alpha\rangle\langle\psi_\alpha|\psi(0)\rangle\delta(\omega - E_\alpha/\hbar) \\ &= \sqrt{2\pi}\delta(\omega - H/\hbar)|\psi(t=0)\rangle. \end{aligned} \tag{5.14}$$

The sum over continuous indices α will include an integration over $\delta(\omega - E_\alpha/\hbar)$ for E_α in the continuous parts of the spectrum of H, and although the Fourier transformation from $|\psi(t)\rangle$ to a frequency dependent state does not exist in the sense of classical Fourier theory, $|\psi(\omega)\rangle$ exists as a sum of δ functions over the discrete spectrum of the Hamiltonian plus a sum over continuous states. E.g. an initial state $|\psi(0)\rangle$ moving in the attractive δ potential from Section 3.3 (see also (3.23)) evolves according to

$$\begin{aligned} |\psi(t)\rangle = \exp(-\mathrm{i}Ht/\hbar)|\psi(0)\rangle &= \exp(\mathrm{i}\hbar\kappa^2 t/2m)|\kappa\rangle\langle\kappa|\psi(0)\rangle \\ &+ \sum_\pm \int_0^\infty dk\, \exp(-\mathrm{i}\hbar k^2 t/2m)|k,\pm\rangle\langle k,\pm|\psi(0)\rangle \end{aligned}$$

[2]Frequency-time Fourier transformation for Green's functions and potentials, which generically will depend on two arguments, will often appear asymmetric due to translation invariance $G(t,t') = G(t-t') \leftrightarrow G(\omega,\omega') = G(\omega)\delta(\omega-\omega')$:

$$G(t) = \frac{1}{2\pi}\int d\omega\, G(\omega)\exp(-\mathrm{i}\omega t), \quad G(\omega) = \int dt\, G(t)\exp(\mathrm{i}\omega t). \tag{5.11}$$

in the time domain, and corresponds to

$$|\psi(\omega)\rangle = \sqrt{2\pi}\delta(\omega - H/\hbar)|\psi(0)\rangle = \sqrt{2\pi}|\kappa\rangle\langle\kappa|\psi(0)\rangle\delta\left(\omega + \frac{\hbar\kappa^2}{2m}\right)$$
$$+\Theta(\omega)\sqrt{\frac{\pi m}{\hbar\omega}}\sum_{\pm}|\sqrt{2m\omega/\hbar}, \pm\rangle\langle\sqrt{2m\omega/\hbar}, \pm|\psi(0)\rangle$$

in the frequency domain[3].

In a formal sense $|\psi(\omega)\rangle$ solves the Schrödinger equation in frequency representation, $(\hbar\omega - H)|\psi(\omega)\rangle = 0$, with initial condition $|\psi(t = 0)\rangle$ by decomposing the initial state in energy eigenmodes. Nevertheless, the notion of frequency-time Fourier transformation in quantum mechanics is delicate and can easily be abused to draw incorrect conclusions if not used carefully. A major source of error is to confuse the frequency representation $|\psi(\omega)\rangle$ of a state, which only exists in the distribution sense, with energy eigenstates $|\psi_\alpha\rangle$ of H which satisfy $H|\psi_\alpha\rangle = E_\alpha|\psi_\alpha\rangle$ in the sense of classical analysis. To avoid possible confusion, we will abstain from using frequency itself as a quantum number for classifying energy eigenstates like $|\psi_\alpha\rangle \equiv |\psi(E_\alpha)\rangle$.

The energy expectation value in the state $|\psi(t)\rangle$ is

$$\langle E\rangle = \langle\psi(t)|H|\psi(t)\rangle = \sum\!\!\!\!\!\!\int d\alpha\, E_\alpha\, |\langle\psi_\alpha|\psi(0)\rangle|^2 .$$

The corresponding energy uncertainty ΔE follows from

$$\Delta E^2 = \sum\!\!\!\!\!\!\int d\alpha\, E_\alpha^2\, |\langle\psi_\alpha|\psi(0)\rangle|^2 - \left(\sum\!\!\!\!\!\!\int d\alpha\, E_\alpha\, |\langle\psi_\alpha|\psi(0)\rangle|^2\right)^2 .$$

5.3 Dimensions of states

A simple, but useful concept for checking consistency in quantum mechanical calculations is the concept of length dimension of a state. To introduce this concept, note that the completeness relation for the eigenstates of the one-dimensional attractive δ potential

$$|\kappa\rangle\langle\kappa| + \int_0^\infty dk\,(|k,-\rangle\langle k,-| + |k,+\rangle\langle k,+|) = 1$$

[3]Free states with initial conditions $\langle x|\psi\rangle \equiv \langle x|\psi(t = 0)\rangle$ yield frequency representations $\langle x|\psi(\omega)\rangle$ in terms of convolutions of $\langle x|\psi\rangle$ with Bessel functions. This is explained in Appendix J, especially equations (J.32)and (J.34–J.36). However, if you began only recently to learn quantum mechanics, don't let yourself become distracted by the technicalities of Appendix J. Save it for later.

implies that the discrete bound state $|\kappa\rangle$ is dimensionless, while the continuous unbound states $|k,\pm\rangle$ have the dimension of length$^{1/2}$.

Similarly, the completeness relations for continuous one-dimensional x or free momentum eigenstates

$$\int dx\,|x\rangle\langle x| = 1, \quad \int dk\,|k\rangle\langle k| = 1 \tag{5.15}$$

imply that $|x\rangle$ has the dimension of length$^{-1/2}$ and $|k\rangle$ has the dimension of length$^{1/2}$. The wave functions $\langle x|\kappa\rangle$ and $\langle k|\kappa\rangle$ therefore have dimension of length$^{-1/2}$ or dimension of length$^{1/2}$, respectively, while the representations or wave functions $\langle x|k\rangle$ or $\langle x|k,\pm\rangle$ are dimensionless. The momentum representations of the unbound states of the attractive δ potential,

$$\langle k'|k,\pm\rangle = \frac{1}{\sqrt{2\pi}}\int_{-\infty}^{\infty} dx\,\exp(-\mathrm{i}k'x)\psi_{k,\pm}(x)$$

have length dimension 1, e.g.

$$\langle k'|k,-\rangle = \frac{1}{\mathrm{i}\sqrt{2}}\left[\delta(k-k') - \delta(k+k')\right].$$

In three dimensions, the states $|n_1,n_2,n_3\rangle$ in a cubic quantum dot are dimensionless while the states $|n_1,n_2,k\rangle$ (3.10) in a cubic quantum wire have the dimension of length$^{1/2}$ in agreement with their completeness relation

$$\sum_{n_1=1}^{\infty}\sum_{n_2=1}^{\infty}\int_{-\infty}^{\infty} dk\,|n_1,n_2,k\rangle\langle n_1,n_2,k| = 1.$$

The continuous state $|\boldsymbol{x}\rangle$ apparently has dimension of length$^{-3/2}$ and $|\boldsymbol{k}\rangle$ has the dimension of length$^{3/2}$. The representation $\langle\boldsymbol{x}|\boldsymbol{k}\rangle$ of a plane wave state is therefore dimensionless, while the representation $\langle\boldsymbol{x}|n_1,n_2,n_3\rangle$ of a quantum dot state has the dimension length$^{-3/2}$.

A state $|\psi\rangle$ in d spatial dimensions is usually specified in terms of $N \geq d$ quantum numbers. If c_k of these quantum numbers are continuous wave-number like quantum numbers and c_x quantum numbers are continuous position like quantum numbers, the length dimension of the state is length$^{(c_k-c_x)/2}$. The representation (= transformation matrix element = wave function) $\langle\psi|\psi'\rangle$ has length dimension length$^{(c_k+c_k'-c_x-c_x')/2}$. This is as trivial as calculating with units, and as useful for checking consistency of results. Furthermore, length dimensions of initial and final states will also provide important hints for us to identify the use and the physical meaning of transition matrix elements in time-dependent perturbation theory in Chapters 13 and 22.

5.4 Gradients and Laplace operators in general coordinate systems

The transformation properties of the gradient ∇ and Laplace operator $\Delta = \nabla^2$ under coordinate transformations are basic aspects of mathematics that we have to discuss due to the importance of those operators in quantum mechanics. We use three dimensions for the discussions in this section, but the methods apply in any number of dimensions.

Suppose we wish to use coordinates ξ^α, $1 \leq \alpha \leq 3$, instead of Cartesian coordinates x^i. The coordinate maps

$$\xi^\alpha \to x^i(\boldsymbol{\xi}), \quad x^i \to \xi^\alpha(\boldsymbol{x}) \tag{5.16}$$

define Jacobi matrices

$$\partial_\alpha x^i(\boldsymbol{\xi}) \equiv \frac{\partial x^i(\boldsymbol{\xi})}{\partial \xi^\alpha} \tag{5.17}$$

and $\partial_i \xi^\alpha(\boldsymbol{x})$. The Jacobi matrix (5.17) in particular allows us to calculate tangent vectors to the new ξ^α coordinate lines. We can easily figure this out by observing that the map (5.16) can be written as (recall summation convention)

$$\xi^\alpha \to \boldsymbol{r}(\boldsymbol{\xi}) = x^i(\boldsymbol{\xi})\boldsymbol{e}_i,$$

where the vectors $\boldsymbol{e}_i$ are the Cartesian basis vectors along the x^i coordinate lines. Infinitesimal coordinate shifts generate a vector

$$d\boldsymbol{r} = dx^i\boldsymbol{e}_i = d\xi^\alpha \partial_\alpha x^i \boldsymbol{e}_i = \boldsymbol{a}_\alpha d\xi^\alpha,$$

and this tells us that the vector

$$\boldsymbol{a}_\alpha(\boldsymbol{\xi}) = \partial_\alpha \boldsymbol{r}(\boldsymbol{\xi}) = \partial_\alpha x^i(\boldsymbol{\xi})\boldsymbol{e}_i. \tag{5.18}$$

is a tangent vector along the ξ^α coordinate line in the point $\boldsymbol{r}(\boldsymbol{\xi})$.

The products of these tangent vectors define the components of the metric in the new coordinate system,

$$g_{\alpha\beta}(\boldsymbol{\xi}) = \boldsymbol{a}_\alpha(\boldsymbol{\xi}) \cdot \boldsymbol{a}_\beta(\boldsymbol{\xi}),$$

because the length squared of the shift vector $d\boldsymbol{r}$ is

$$ds^2 = d\boldsymbol{r}^2 = \boldsymbol{a}_\alpha \cdot \boldsymbol{a}_\beta d\xi^\alpha d\xi^\beta = g_{\alpha\beta} d\xi^\alpha d\xi^\beta.$$

The inverse metric $g^{\alpha\beta}$ yields the dual basis vectors according to equations (4.11, 4.12),

$$\boldsymbol{a}^{\alpha}(\boldsymbol{\xi}) = g^{\alpha\beta}(\boldsymbol{\xi})\boldsymbol{a}_{\beta}(\boldsymbol{\xi}),$$

and the Jacobian matrix $\partial_i\xi^{\alpha}(\boldsymbol{x})$ connects the dual basis vectors,

$$\boldsymbol{a}^{\alpha} = \boldsymbol{e}^i\partial_i\xi^{\alpha}. \tag{5.19}$$

Simple consequences of $\boldsymbol{a}_{\alpha}(\boldsymbol{\xi}) = \partial_{\alpha}\boldsymbol{r}(\boldsymbol{\xi})$ and $\boldsymbol{a}^{\alpha}(\boldsymbol{\xi})\cdot\boldsymbol{a}_{\beta}(\boldsymbol{\xi}) = \delta^{\alpha}{}_{\beta}$ are

$$\partial_{\alpha}\boldsymbol{a}_{\beta} = \partial_{\beta}\boldsymbol{a}_{\alpha}$$

and

$$\boldsymbol{a}^{\alpha}\cdot\partial_{\gamma}\boldsymbol{a}_{\beta} = -\boldsymbol{a}_{\beta}\cdot\partial_{\gamma}\boldsymbol{a}^{\alpha}.$$

The $\boldsymbol{\nabla}$ operator is defined as

$$\boldsymbol{\nabla} = \boldsymbol{e}^i\frac{\partial}{\partial x^i} = \boldsymbol{e}^i\frac{\partial\xi^{\alpha}}{\partial x^i}\frac{\partial}{\partial\xi^{\alpha}} = \boldsymbol{a}^{\alpha}\frac{\partial}{\partial\xi^{\alpha}}. \tag{5.20}$$

Note that once we accept the Cartesian equation $\boldsymbol{\nabla} = \boldsymbol{e}^i\partial_i$ with the recognition that the Cartesian vectors $\boldsymbol{e}^i = \boldsymbol{e}_i$ appearing in the $\boldsymbol{\nabla}$ operator are actually the dual basis vectors, the representation $\boldsymbol{\nabla} = \boldsymbol{a}^{\alpha}\partial_{\alpha}$ in the new coordinate system is a direct consequence of the chain rule of differentiation. Furthermore, we can write equation (5.19) also in the form

$$\boldsymbol{a}^{\alpha} = \boldsymbol{\nabla}\xi^{\alpha} \tag{5.21}$$

and the inverse metric is

$$g^{\alpha\beta} = (\boldsymbol{\nabla}\xi^{\alpha})\cdot(\boldsymbol{\nabla}\xi^{\beta}). \tag{5.22}$$

These equations are particularly convenient if the new coordinates are given in terms of the Cartesian coordinates x^i, $\xi^{\alpha} = \xi^{\alpha}(\boldsymbol{x})$.

The new dual basis vectors $\boldsymbol{a}^{\alpha} = \boldsymbol{a}^{\alpha}(\boldsymbol{\xi})$ generically depend on the coordinates $\boldsymbol{\xi}$, because the new coordinates will often be curvilinear. We have to take this into account when calculating the Laplace operator in the new coordinate system,

$$\Delta = \boldsymbol{\nabla}^2 = \boldsymbol{a}^{\alpha}(\boldsymbol{\xi})\partial_{\alpha}\circ\boldsymbol{a}^{\beta}(\boldsymbol{\xi})\partial_{\beta} = g^{\alpha\beta}(\boldsymbol{\xi})\partial_{\alpha}\partial_{\beta} + \boldsymbol{a}^{\alpha}(\boldsymbol{\xi})\cdot(\partial_{\alpha}\boldsymbol{a}^{\beta}(\boldsymbol{\xi}))\partial_{\beta}. \tag{5.23}$$

We can also write this as

$$\Delta = g^{\alpha\beta}(\boldsymbol{\xi})\left(\partial_{\alpha}\partial_{\beta} + \boldsymbol{a}_{\beta}(\boldsymbol{\xi})\cdot\partial_{\alpha}\boldsymbol{a}^{\gamma}(\boldsymbol{\xi})\partial_{\gamma}\right) = g^{\alpha\beta}(\boldsymbol{\xi})\left(\partial_{\alpha}\partial_{\beta} - \Gamma^{\gamma}{}_{\beta\alpha}(\boldsymbol{\xi})\partial_{\gamma}\right),$$

where the coefficients

$$\Gamma^{\gamma}{}_{\beta\alpha}(\boldsymbol{\xi}) = -\boldsymbol{a}_{\beta}(\boldsymbol{\xi}) \cdot \partial_{\alpha}\boldsymbol{a}^{\gamma}(\boldsymbol{\xi}) = \boldsymbol{a}^{\gamma}(\boldsymbol{\xi}) \cdot \partial_{\alpha}\boldsymbol{a}_{\beta}(\boldsymbol{\xi}) \tag{5.24}$$

are known as *Christoffel symbols*.

We can use equation (5.23) e.g. to calculate the Laplace operator in spherical coordinates. It is convenient to use the common column vector notation for components with respect to the Cartesian $\boldsymbol{e}_i = \boldsymbol{e}^i$ in the actual calculation. The transformation

$$\boldsymbol{r} = \begin{pmatrix} x^1 \\ x^2 \\ x^3 \end{pmatrix} = \begin{pmatrix} r \sin\vartheta \cos\varphi \\ r \sin\vartheta \sin\varphi \\ r \cos\vartheta \end{pmatrix}$$

yields tangent vectors

$$\boldsymbol{a}_r = \partial_r \boldsymbol{r} = \begin{pmatrix} \sin\vartheta \cos\varphi \\ \sin\vartheta \sin\varphi \\ \cos\vartheta \end{pmatrix}, \quad \boldsymbol{a}_{\vartheta} = \partial_{\vartheta} \boldsymbol{r} = \begin{pmatrix} r\cos\vartheta \cos\varphi \\ r\cos\vartheta \sin\varphi \\ -r\sin\vartheta \end{pmatrix},$$

and

$$\boldsymbol{a}_{\varphi} = \partial_{\varphi} \boldsymbol{r} = \begin{pmatrix} -r\sin\vartheta \sin\varphi \\ r\sin\vartheta \cos\varphi \\ 0 \end{pmatrix}.$$

The metric and the inverse metric in spherical coordinates are

$$\underline{g} = \begin{pmatrix} 1 & 0 & 0 \\ 0 & r^2 & 0 \\ 0 & 0 & r^2\sin^2\vartheta \end{pmatrix}, \quad \underline{g}^{-1} = \begin{pmatrix} 1 & 0 & 0 \\ 0 & r^{-2} & 0 \\ 0 & 0 & r^{-2}\sin^{-2}\vartheta \end{pmatrix},$$

and the dual vectors are

$$\boldsymbol{a}^r = \begin{pmatrix} \sin\vartheta \cos\varphi \\ \sin\vartheta \sin\varphi \\ \cos\vartheta \end{pmatrix}, \quad \boldsymbol{a}^{\vartheta} = \frac{1}{r}\begin{pmatrix} \cos\vartheta \cos\varphi \\ \cos\vartheta \sin\varphi \\ -\sin\vartheta \end{pmatrix}, \quad \boldsymbol{a}^{\varphi} = \frac{1}{r\sin\vartheta}\begin{pmatrix} -\sin\varphi \\ \cos\varphi \\ 0 \end{pmatrix}.$$

The non-vanishing products

$$\boldsymbol{a}^{\vartheta} \cdot \partial_{\vartheta}\boldsymbol{a}^r = \boldsymbol{a}^{\varphi} \cdot \partial_{\varphi}\boldsymbol{a}^r = \frac{1}{r}, \quad \boldsymbol{a}^{\varphi} \cdot \partial_{\varphi}\boldsymbol{a}^{\vartheta} = \frac{\cot\vartheta}{r^2}$$

yield

$$\Delta = \frac{\partial^2}{\partial r^2} + \frac{2}{r}\frac{\partial}{\partial r} + \frac{1}{r^2}\frac{\partial^2}{\partial\vartheta^2} + \frac{\cot\vartheta}{r^2}\frac{\partial}{\partial\vartheta} + \frac{1}{r^2\sin^2\vartheta}\frac{\partial^2}{\partial\varphi^2}. \tag{5.25}$$

Later we will also use the normalized spherical tangent vectors

$$\boldsymbol{e}_r \equiv \boldsymbol{a}_r, \quad \boldsymbol{e}_\vartheta = \frac{1}{r}\boldsymbol{a}_\vartheta, \quad \boldsymbol{e}_\varphi = \frac{1}{r\sin\vartheta}\boldsymbol{a}_\varphi.$$

Another useful representation of the Laplace operator which follows from equation (5.23) is

$$\Delta = \frac{1}{\sqrt{g}}\partial_\alpha\left(\sqrt{g}g^{\alpha\beta}\partial_\beta\right), \tag{5.26}$$

where $g(\boldsymbol{\xi})$ is the determinant of the metric tensor $g_{\alpha\beta}(\boldsymbol{\xi})$. To demonstrate equivalence of (5.26) with (5.23) one has to use the properties

$$\partial_\alpha g = g g^{\beta\gamma}\partial_\alpha g_{\beta\gamma} = -g g_{\beta\gamma}\partial_\alpha g^{\beta\gamma}$$

and

$$g_{\beta\gamma}\partial_\alpha g^{\beta\gamma} = 2\boldsymbol{a}_\gamma\cdot\partial_\alpha\boldsymbol{a}^\gamma = -2\boldsymbol{a}^\gamma\cdot\partial_\alpha\boldsymbol{a}_\gamma = -2\boldsymbol{a}^\gamma\cdot\partial_\gamma\boldsymbol{a}_\alpha = 2\boldsymbol{a}_\alpha\cdot\partial_\gamma\boldsymbol{a}^\gamma.$$

It is also useful to point out that $\sqrt{g}(\boldsymbol{\xi})$ is the volume measure in the new coordinates ξ^α,

$$d^3x = dx^1dx^2dx^3 = d\xi^1 d\xi^2 d\xi^3\,\sqrt{g}(\boldsymbol{\xi}). \tag{5.27}$$

This follows from the fact that the matrix relation

$$g_{\alpha\beta}(\boldsymbol{\xi}) = \partial_\alpha\boldsymbol{r}(\boldsymbol{\xi})\cdot\partial_\beta\boldsymbol{r}(\boldsymbol{\xi}) = \partial_\alpha x^i(\boldsymbol{\xi})\cdot\partial_\beta x^j(\boldsymbol{\xi})\delta_{ij}$$

implies that the determinant $g = \det(g_{\alpha\beta})$ is the square of the Jacobian determinant,

$$g(\boldsymbol{\xi}) = \left(\det(\partial_\alpha x^i(\boldsymbol{\xi}))\right)^2. \tag{5.28}$$

The familiar form for transformation of volume measures,

$$dx^1dx^2dx^3 = d\xi^1 d\xi^2 d\xi^3\,\left|\det(\partial_\alpha x^i(\boldsymbol{\xi}))\right|, \tag{5.29}$$

then yields (5.27).

5.5 Separation of differential equations

Separation of variables, where applicable, is a very powerful and useful tool for solution of partial differential equations. The purpose of this section is to point out that separation of variables for separable hermitian operators is not just a matter of convenient choice, but a must for the determination of eigenfunctions of these operators. Eigenstates of separable hermitian differential operators will automatically factorize into eigenstates of the corresponding lower-dimensional operators.

We assume a three-dimensional space with coordinates ξ^i, $1 \leq i \leq 3$, but the reader will again recognize that the arguments presented in this section do not depend on the number of dimensions.

In quantum mechanics we often encounter Hamiltonians with the following property: If we choose a suitable set of coordinates $\boldsymbol{\xi}$, then the time-independent Schrödinger equation

$$E|\psi(E)\rangle = H|\psi(E)\rangle$$

will separate in the form

$$E\langle\boldsymbol{\xi}|\psi(E)\rangle = \sum_i \langle\boldsymbol{\xi}|H_i|\psi(E)\rangle = \sum_i \mathcal{D}_i\langle\boldsymbol{\xi}|\psi(E)\rangle, \tag{5.30}$$

where each of the hermitian differential operators $\mathcal{D}_i$ has the property to contain only the coordinate ξ_i and the corresponding derivative $\partial/\partial\xi^i$. However, the results on completeness of eigenstates of one-dimensional hermitian operators from Appendix C imply that each of the operators has its own complete set of eigenfunctions,

$$\mathcal{D}_i\langle\xi_i|\psi_i(E_i)\rangle = E_i\langle\xi_i|\psi_i(E_i)\rangle, \tag{5.31}$$

where the different eigenfunctions $\langle\xi_i|\psi_i(E_i)\rangle$ are labeled by the eigenvalues E_i.

This can give us the idea to decompose the function $\langle\boldsymbol{\xi}|\psi(E)\rangle$ with respect to the eigenfunctions $\langle\xi_1|\psi_1(E_1)\rangle$,

$$\langle\boldsymbol{\xi}|\psi(E)\rangle = \sum\!\!\!\!\!\!\int dE_1\, \langle\xi_1|\psi_1(E_1)\rangle\langle E_1,\xi_2,\xi_3|\psi(E)\rangle,$$

$$\langle E_1,\xi_2,\xi_3|\psi(E)\rangle = \int d\xi_1\, \langle\psi_1(E_1)|\xi_1\rangle\langle\xi_1,\xi_2,\xi_3|\psi(E)\rangle.$$

We can then repeat the decomposition with respect to the eigenfunctions of $\mathcal{D}_2$ and $\mathcal{D}_3$. This leads finally to the decomposition

$$\begin{aligned}\langle\boldsymbol{\xi}|\psi(E)\rangle = {} & \sum\!\!\!\!\!\!\int dE_1 \sum\!\!\!\!\!\!\int dE_2 \sum\!\!\!\!\!\!\int dE_3\, \langle\xi_1|\psi_1(E_1)\rangle\langle\xi_2|\psi_2(E_2)\rangle\langle\xi_3|\psi_3(E_3)\rangle \\ & \times\langle E_1,E_2,E_3|\psi(E)\rangle,\end{aligned} \tag{5.32}$$

$$\begin{aligned}\langle E_1,E_2,E_3|\psi(E)\rangle = {} & \int d\xi_1 \int d\xi_2 \int d\xi_3\, \langle\psi_1(E_1)|\xi_1\rangle\langle\psi_2(E_2)|\xi_2\rangle \\ & \times\langle\psi_3(E_3)|\xi_3\rangle\langle\xi_1,\xi_2,\xi_3|\psi(E)\rangle.\end{aligned}$$

Substitution of the decomposition (5.32) into the Schrödinger equation (5.30) yields

$$E \sum\!\!\!\!\!\!\int dE_1 \sum\!\!\!\!\!\!\int dE_2 \sum\!\!\!\!\!\!\int dE_3\, \langle \xi_1 | \psi_1(E_1)\rangle \langle \xi_2 | \psi_2(E_2)\rangle \langle \xi_3 | \psi_3(E_3)\rangle$$
$$\times \langle E_1, E_2, E_3 | \psi(E)\rangle$$
$$= \sum\!\!\!\!\!\!\int dE_1 \sum\!\!\!\!\!\!\int dE_2 \sum\!\!\!\!\!\!\int dE_3\, (E_1 + E_2 + E_3)\, \langle \xi_1 | \psi_1(E_1)\rangle \langle \xi_2 | \psi_2(E_2)\rangle \langle \xi_3 | \psi_3(E_3)\rangle$$
$$\times \langle E_1, E_2, E_3 | \psi(E)\rangle. \tag{5.33}$$

The orthogonality properties of different eigenfunctions,

$$\int d\xi_i\, \langle \psi_i(E_i) | \xi_i\rangle \langle \xi_i | \psi_i(E_i')\rangle \Big|_{E_i \neq E_i'} = 0$$

then imply that for all combinations of eigenvalues E and E_i the condition

$$(E_1 + E_2 + E_3 - E)\, \langle \xi_1 | \psi_1(E_1)\rangle \langle \xi_2 | \psi_2(E_2)\rangle \langle \xi_3 | \psi_3(E_3)\rangle$$
$$\times \langle E_1, E_2, E_3 | \psi(E)\rangle = 0$$

must be satisfied, i.e. the eigenvalue E in equation (5.33) must equal one particular sum of eigenvalues,

$$E = E_1 + E_2 + E_3, \tag{5.34}$$

and the related eigenfunction is

$$\langle \boldsymbol{\xi} | \psi(E)\rangle = \langle \xi_1 | \psi_1(E_1)\rangle \langle \xi_2 | \psi_2(E_2)\rangle \langle \xi_3 | \psi_3(E_3)\rangle. \tag{5.35}$$

These observations tell us that we can and indeed should use a separation ansatz for Hamiltonians of the form (5.30).

The previous argument works for Hamiltonian operators which split into a sum of one-dimensional operators in suitable coordinates. However, the arguments are easily generalized to the more general case

$$E\langle \boldsymbol{\xi} | \psi(E)\rangle = f(\xi_3) g(\xi_2) \mathcal{D}_1 \langle \boldsymbol{\xi} | \psi(E)\rangle + f(\xi_3) \mathcal{D}_2 \langle \boldsymbol{\xi} | \psi(E)\rangle$$
$$+ \mathcal{D}_3 \langle \boldsymbol{\xi} | \psi(E)\rangle. \tag{5.36}$$

Since $\mathcal{D}_1$ commutes with the Hamiltonian operator on the right hand side, the space of eigenfunctions $\langle \boldsymbol{\xi} | \psi(E)\rangle$ with fixed eigenvalue E must be generated by a set of eigenstates of the hermitian differential operator $\mathcal{D}_1$ with eigenvalues $\tilde{E}_1$,

$$\langle \boldsymbol{\xi} | \psi(E)\rangle = \sum\!\!\!\!\!\!\int d\tilde{E}_1\, \langle \xi_1 | \psi_1(\tilde{E}_1)\rangle \langle \tilde{E}_1, \xi_2, \xi_3 | \psi(E)\rangle.$$

Non-degeneracy of the one-dimensional eigenvalues $\tilde{E}_1$ and linear independence of the corresponding eigenstates then implies

$$E\langle\tilde{E}_1,\xi_2,\xi_3|\psi(E)\rangle = f(\xi_3)\left[g(\xi_2)\tilde{E}_1 + \mathcal{D}_2\right]\langle\tilde{E}_1,\xi_2,\xi_3|\psi(E)\rangle + \mathcal{D}_3\langle\tilde{E}_1,\xi_2,\xi_3|\psi(E)\rangle.$$

Now the Hamiltonian on the right hand side commutes with the parameter-dependent hermitian one-dimensional operator

$$\mathcal{D}_2' = \mathcal{D}_2 + g(\xi_2)\tilde{E}_1,$$

which has parameter-dependent eigenvalues $\tilde{E}_2(\tilde{E}_1)$. The further decomposition

$$\langle\boldsymbol{\xi}|\psi(E)\rangle = \sum\!\!\!\!\!\!\int d\tilde{E}_1 \sum\!\!\!\!\!\!\int d\tilde{E}_2(\tilde{E}_1)\,\langle\xi_1|\psi_1(\tilde{E}_1)\rangle\langle\xi_2|\psi_2'(\tilde{E}_2(\tilde{E}_1))\rangle \times\langle\tilde{E}_1,\tilde{E}_2(\tilde{E}_1),\xi_3|\psi(E)\rangle$$

then yields the one-dimensional hermitian problem

$$E\langle\tilde{E}_1,\tilde{E}_2(\tilde{E}_1),\xi_3|\psi(E)\rangle = f(\xi_3)\tilde{E}_2(\tilde{E}_1)\langle\tilde{E}_1,\tilde{E}_2(\tilde{E}_1),\xi_3|\psi(E)\rangle + \mathcal{D}_3\langle\tilde{E}_1,\tilde{E}_2(\tilde{E}_1),\xi_3|\psi(E)\rangle,$$

which finally yields eigenvalues $E_n(\tilde{E}_1,\tilde{E}_2)$ and the solution

$$\langle\boldsymbol{\xi}|\psi(E_n(\tilde{E}_1,\tilde{E}_2))\rangle = \langle\xi_1|\psi_1(\tilde{E}_1)\rangle\langle\xi_2|\psi_2'(\tilde{E}_2(\tilde{E}_1))\rangle \times\langle\tilde{E}_1,\tilde{E}_2(\tilde{E}_1),\xi_3|\psi(E_n(\tilde{E}_1,\tilde{E}_2))\rangle. \tag{5.37}$$

5.6 Problems

5.1. We consider a particle of mass m in a one-dimensional infinite square well, i.e. the energy eigenstates $|n\rangle$ are labelled by natural numbers n. How large are the energy expectation value and energy uncertainty in the state

$$|\psi\rangle = \sqrt{e-1}\sum_{n=1}^{\infty}\exp(-n/2)|n\rangle?$$

5.2. Calculate the Laplace operator in spherical coordinates from equation (5.26).

5.3. Calculate the tangent vectors, the ∇ operator and the Laplace operator in parabolic coordinates

$$x = 2\sqrt{\xi\eta}\cos\varphi, \quad y = 2\sqrt{\xi\eta}\sin\varphi, \quad z = \xi - \eta,$$

$$2\xi = r + z, \quad 2\eta = r - z, \quad \varphi = \arctan\frac{y}{x}.$$

Equation (5.26) is more convenient than (5.23) for the calculation of the Laplace operator in parabolic coordinates.

5.4. Show that the Christoffel symbols (5.24) can also be expressed in terms of the metric and inverse metric components,

$$\Gamma^{\gamma}{}_{\beta\alpha} = \frac{1}{2}g^{\gamma\delta}\left(\partial_\beta g_{\delta\alpha} + \partial_\alpha g_{\delta\beta} - \partial_\delta g_{\alpha\beta}\right). \tag{5.38}$$

5.5. Prove the following statement: The Euler-Lagrange equation (see Appendix A, in particular equation (A.3))

$$\frac{d}{d\tau}\frac{\partial L}{\partial\dot{\xi}^\alpha} - \frac{\partial L}{\partial\xi^\alpha} = 0$$

for the Lagrange function

$$L(\xi,\dot{\xi}) = \frac{1}{2}g_{\alpha\beta}(\xi(\tau))\dot{\xi}^\alpha(\tau)\dot{\xi}^\beta(\tau) \tag{5.39}$$

yields the equation

$$\ddot{\xi}^\alpha(\tau) + \Gamma^\alpha{}_{\beta\gamma}(\xi(\tau))\dot{\xi}^\beta(\tau)\dot{\xi}^\gamma(\tau) = 0. \tag{5.40}$$

This is a most useful lemma for the calculation of Christoffel symbols. For given metric $g_{\alpha\beta}(\boldsymbol{\xi})$, one simply calculates the Euler-Lagrange equations for the Lagrange function (5.39) and then reads off the Christoffel symbols from the quadratic terms in the velocities $\dot{\xi}(\tau)$.

Equation (5.40) is known as the geodesic equation, because in a general space it yields lines $\xi(\tau)$ of stationary length (e.g. shortest or longest lines). In the flat spaces that we are dealing with in this book, equation (5.40) is the condition for a straight line in terms of the curvilinear coordinates ξ^α.

5.6. Find the eigenvalues and eigenfunctions of the two-dimensional differential operator $H_{x,y} = y\partial_x^2 - \mathrm{i}\partial_y$.

Chapter 6
Harmonic Oscillators and Coherent States

The harmonic oscillator is the general approximation for the dynamics of small fluctuations around a minimum of a potential. This is the reason why harmonic oscillators are very important model systems both in mechanics and in quantum mechanics. In addition there is another reason why we have to discuss the quantum harmonic oscillator in detail. For the discussion of quantum mechanical reactions between particles later on, we have to go beyond ordinary quantum mechanics and use a technique called second quantization or canonical quantum field theory. The techniques of second quantization are based on linear superpositions of infinitely many oscillators. Therefore it is important to have a very good understanding of oscillator eigenstates and of the calculational techniques involved with oscillation operators.

6.1 Basic aspects of harmonic oscillators

The classical motion of a particle in the three-dimensional isotropic potential

$$V(\boldsymbol{x}) = \frac{m}{2}\omega^2\boldsymbol{x}^2$$

without external driving forces is described by the classical solution

$$\begin{aligned}\boldsymbol{x}(t) &= \boldsymbol{X}\cos(\omega t) + \frac{\boldsymbol{P}}{m\omega}\sin(\omega t),\\ \boldsymbol{p}(t) &= \boldsymbol{P}\cos(\omega t) - m\omega\boldsymbol{X}\sin(\omega t),\end{aligned} \tag{6.1}$$

where $\boldsymbol{X} = \boldsymbol{x}(0)$, $\boldsymbol{P} = m\dot{\boldsymbol{x}}(0)$ are the values of location and momentum of the particle at time $t = 0$.

R. Dick, *Advanced Quantum Mechanics*, Graduate Texts in Physics,
DOI 10.1007/978-3-319-25675-7_6

The corresponding Schrödinger equation is

$$i\hbar\frac{d}{dt}|\psi(t)\rangle = \left(\frac{\mathbf{p}^2}{2m} + \frac{m}{2}\omega^2\mathbf{x}^2\right)|\psi(t)\rangle$$

or after substitution of energy-time Fourier transformation (5.12),

$$E|\psi(E)\rangle = \left(\frac{\mathbf{p}^2}{2m} + \frac{m}{2}\omega^2\mathbf{x}^2\right)|\psi(E)\rangle. \tag{6.2}$$

The corresponding differential equation in $\boldsymbol{x}$ representation

$$E\langle\boldsymbol{x}|\psi(E)\rangle = \left(-\frac{\hbar^2}{2m}\Delta + \frac{m}{2}\omega^2\boldsymbol{x}^2\right)\langle\boldsymbol{x}|\psi(E)\rangle \tag{6.3}$$

can be decomposed into three one-dimensional problems through separation of the spatial variables. The separation ansatz

$$\langle\boldsymbol{x}|\psi(E)\rangle = \prod_{i=1}^{3}\langle x_i|\psi_i(E_i)\rangle \tag{6.4}$$

yields

$$E = \sum_{i=1}^{3} E_i, \tag{6.5}$$

where the three energy values E_i and wave functions $\langle x_i|\psi(E_i)\rangle$ have to satisfy the one-dimensional equation

$$E\langle x|\psi(E)\rangle = -\frac{\hbar^2}{2m}\frac{d^2}{dx^2}\langle x|\psi(E)\rangle + \frac{m}{2}\omega^2x^2\langle x|\psi(E)\rangle. \tag{6.6}$$

Indeed, the results of Section 5.5 imply that the solutions of the three-dimensional equation (6.3) will always have the separated form (6.4, 6.5).

6.2 Solution of the harmonic oscillator by the operator method

The one-dimensional oscillator equation (6.6) is in representation free notation

$$E|\psi(E)\rangle = \left(\frac{\mathrm{p}^2}{2m} + \frac{m}{2}\omega^2\mathrm{x}^2\right)|\psi(E)\rangle. \tag{6.7}$$

There exists a powerful and elegant method to solve equation (6.7) through a transformation from the self-adjoint operators x and p to mutually adjoint operators a and a^+. The substitutions

$$a = \frac{1}{\sqrt{2\hbar}}\left(\sqrt{m\omega}\mathrm{x} + \mathrm{i}\frac{\mathrm{p}}{\sqrt{m\omega}}\right), \quad a^+ = \frac{1}{\sqrt{2\hbar}}\left(\sqrt{m\omega}\mathrm{x} - \mathrm{i}\frac{\mathrm{p}}{\sqrt{m\omega}}\right), \tag{6.8}$$

yield the commutation relation

$$[a, a^+] = 1, \tag{6.9}$$

the inverse transformation

$$\mathrm{x} = \sqrt{\frac{\hbar}{2m\omega}}\left(a + a^+\right), \quad \mathrm{p} = -\mathrm{i}\sqrt{\frac{m\omega\hbar}{2}}\left(a - a^+\right), \tag{6.10}$$

and the Hamiltonian in the form

$$H = \frac{1}{2}\hbar\omega(aa^+ + a^+a) = \hbar\omega\left(a^+a + \frac{1}{2}\right).$$

The equations

$$[H, a] = -\hbar\omega a, \quad [H, a^+] = \hbar\omega a^+ \tag{6.11}$$

and

$$H|\psi(E)\rangle = E|\psi(E)\rangle, \tag{6.12}$$

imply that the operator a decreases energy eigenvalues and the operator a^+ increases energy eigenvalues in units of $\hbar\omega$,

$$Ha|\psi(E)\rangle = (E - \hbar\omega)a|\psi(E)\rangle, \quad Ha^+|\psi(E)\rangle = (E + \hbar\omega)a^+|\psi(E)\rangle. \tag{6.13}$$

The operator a is therefore denoted as an *annihilation operator* or *lowering operator*, while a^+ is a *creation operator* or a *raising operator*. Together, they are also known as *ladder operators*.

Stability of the system requires existence of a lowest energy state $|\Omega\rangle$. This state must be annihilated by the operator a since otherwise $a|\Omega\rangle$ would be a state of lower energy,

$$\exists|\Omega\rangle: \ a|\Omega\rangle = 0, \ \Rightarrow \ H|\Omega\rangle = \frac{1}{2}\hbar\omega|\Omega\rangle.$$

The standard notation for this lowest energy state or vacuum state is $|\Omega\rangle = |0\rangle$.

The excited energy eigenstates are then

$$|n\rangle = \frac{(a^+)^n}{\sqrt{n!}}|0\rangle, \tag{6.14}$$

and the corresponding energy eigenvalues follow from

$$H|n\rangle = \hbar\omega\left(n + \frac{1}{2}\right)|n\rangle$$

as

$$E_n = \hbar\omega\left(n + \frac{1}{2}\right).$$

These relations are equivalent to

$$a^+a|n\rangle = n|n\rangle,$$

and therefore a^+a returns the level number of a state. The operator a^+a is denoted as an *occupation number operator*, or *number operator* for short, because this operator enumerates how many energy quanta $\hbar\omega$ are contained in an energy level.

For an explanation of the normalization of the states (6.14), we note that (4.32) implies that the adjoint of the state $(a^+)^n|0\rangle$ is $\langle 0|a^n$, and therefore the inner product of the state is $\langle 0|a^n(a^+)^n|0\rangle$. We can evaluate this product by using the property

$$[a, (a^+)^n] = n(a^+)^{n-1}, \tag{6.15}$$

which is easily proved by induction. We use (6.15) in the second step of the following calculation (and then in $n-1$ additional steps to arrive at the final result),

$$\begin{aligned}\langle 0|a^n(a^+)^n|0\rangle &= \langle 0|a^{n-1}[a, (a^+)^n]|0\rangle = n\langle 0|a^{n-1}(a^+)^{n-1}|0\rangle \\ &= n!\langle 0|0\rangle = n! \end{aligned} \tag{6.16}$$

On a more formal level, the proof of (6.16) would also involve an induction step with respect to n.

We have for arbitrary states $|\psi\rangle$

$$\langle 0|a^+|\psi\rangle = (\langle\psi|a|0\rangle)^+ = 0$$

and therefore the projector $\langle 0|a^+$ annihilates every state,

$$\langle 0|a^+ = 0.$$

6.3 Construction of the states in the x-representation

We construct the expansion coefficients $\langle x|n\rangle$ of the states with respect to the $|x\rangle$-basis. In the first step we construct the components $\langle x|0\rangle$ of the ground state $|0\rangle$. The equation $a|0\rangle = 0$ in x representation,

$$\langle x|a|0\rangle = \left(\sqrt{\frac{m\omega}{2\hbar}}x + \sqrt{\frac{\hbar}{2m\omega}}\frac{d}{dx}\right)\langle x|0\rangle = 0,$$

yields

$$\langle x|0\rangle = \left(\frac{m\omega}{\pi\hbar}\right)^{\frac{1}{4}}\exp\left(-\frac{m\omega}{2\hbar}x^2\right),$$

and from this the x-components of the higher states can be calculated in the following way:

$$\begin{aligned}\langle x|n\rangle &= \frac{1}{\sqrt{n!}}\langle x|(a^+)^n|0\rangle \\ &= \frac{1}{\sqrt{n!}}\left(\frac{m\omega}{\pi\hbar}\right)^{\frac{1}{4}}\left(\sqrt{\frac{m\omega}{2\hbar}}x - \sqrt{\frac{\hbar}{2m\omega}}\frac{d}{dx}\right)^n\exp\left(-\frac{m\omega}{2\hbar}x^2\right) \\ &= \frac{1}{\sqrt{2^n n!}}\left(\frac{m\omega}{\pi\hbar}\right)^{\frac{1}{4}}H_n\left(\sqrt{\frac{m\omega}{\hbar}}x\right)\exp\left(-\frac{m\omega}{2\hbar}x^2\right).\end{aligned} \tag{6.17}$$

The functions $H_n(x)$ are the Hermite polynomials

$$H_n(x) = \exp\left(\frac{1}{2}x^2\right)\left(x - \frac{d}{dx}\right)^n\exp\left(-\frac{1}{2}x^2\right), \tag{6.18}$$

$$H_0(x) = 1, \quad H_1(x) = 2x, \quad H_2(x) = 4x^2 - 2, \ldots$$

Properties of Hermite polynomials are discussed in Appendix D.

The general state of the one-dimensional harmonic oscillator in the x-representation is

$$\langle x|\psi(t)\rangle = \sum_{n\geq 0}\langle x|n\rangle\langle n|\psi(t)\rangle = \sum_{n\geq 0}\langle x|n\rangle\langle n|\psi(0)\rangle\exp\left(-\mathrm{i}\left(n+\frac{1}{2}\right)\omega t\right),$$

and the general normalizable state of a 1-dimensional system with Hamiltonian H (here $\dot{H} = 0$) in the x-representation can be expanded in oscillator eigenstates

$$\begin{aligned}\langle x|\psi(t)\rangle &= \sum_{n\geq 0}\langle x|n\rangle\langle n|\psi(t)\rangle = \sum_{n\geq 0}\langle x|n\rangle\langle n|\exp(-\mathrm{i}Ht/\hbar)|\psi(0)\rangle \\ &= \sum_{n\geq 0}\langle x|\exp(-\mathrm{i}Ht/\hbar)|n\rangle\langle n|\psi(0)\rangle.\end{aligned}$$

These expansions are particular examples of the completeness relations of Sturm-Liouville eigenfunctions discussed in Appendix C. They hold pointwise for every continuous square integrable function $\langle x|\psi(t)\rangle$, and they hold for the derivatives as long as the derivatives are continuous (otherwise they remain valid in the mean).

Oscillator eigenstates in k space and bilinear relations for Hermite polynomials

The k space representations of the oscillator energy eigenstates can be constructed in the same way as the x representations. The equation

$$\langle k|a|0\rangle = \mathrm{i}\left(\sqrt{\frac{m\omega}{2\hbar}}\frac{d}{dk} + \sqrt{\frac{\hbar}{2m\omega}}k\right)\langle k|0\rangle = 0$$

yields

$$\langle k|0\rangle = \left(\frac{\hbar}{\pi m\omega}\right)^{\frac{1}{4}}\exp\!\left(-\frac{\hbar}{2m\omega}k^2\right),$$

and from this the k-components of the higher states can be calculated,

$$\begin{aligned}\langle k|n\rangle &= \frac{1}{\sqrt{n!}}\langle k|(a^+)^n|0\rangle \\ &= \frac{\mathrm{i}^n}{\sqrt{n!}}\left(\frac{\hbar}{\pi m\omega}\right)^{\frac{1}{4}}\left(\sqrt{\frac{m\omega}{2\hbar}}\frac{d}{dk} - \sqrt{\frac{\hbar}{2m\omega}}k\right)^n\exp\!\left(-\frac{\hbar}{2m\omega}k^2\right) \\ &= \frac{(-\mathrm{i})^n}{\sqrt{2^n n!}}\left(\frac{\hbar}{\pi m\omega}\right)^{\frac{1}{4}} H_n\!\left(\sqrt{\frac{\hbar}{m\omega}}k\right)\exp\!\left(-\frac{\hbar}{2m\omega}k^2\right).\end{aligned}$$

From this and the previous result we find an expression for the decomposition of plane waves,

$$\begin{aligned}\langle x|k\rangle &= \frac{1}{\sqrt{2\pi}}\exp(\mathrm{i}kx) = \frac{1}{\sqrt{\pi}}\sum_{n=0}^{\infty}\frac{\mathrm{i}^n}{2^n n!}H_n\!\left(\sqrt{\frac{m\omega}{\hbar}}x\right)H_n\!\left(\sqrt{\frac{\hbar}{m\omega}}k\right) \\ &\quad\times\exp\!\left[-\frac{1}{\hbar\omega}\left(\frac{\hbar^2k^2}{2m} + \frac{m}{2}\omega^2x^2\right)\right].\end{aligned} \tag{6.19}$$

This reads in scaled variables

$$\frac{1}{\sqrt{2\pi}}\exp(\mathrm{i}KX)=\frac{1}{\sqrt{\pi}}\sum_{n=0}^{\infty}\frac{\mathrm{i}^n}{2^n n!}H_n(X)\,H_n(K)\exp\!\left(-\frac{1}{2}\left(K^2+X^2\right)\right).$$

This equation follows from the Mehler formula (D.8) in the limit $z\to\mathrm{i}$.
For comparison, we must also have

$$\langle x|y\rangle=\delta(x-y)=\sqrt{\frac{m\omega}{\pi\hbar}}\sum_{n=0}^{\infty}\frac{1}{2^n n!}H_n\!\left(\sqrt{\frac{m\omega}{\hbar}}x\right)H_n\!\left(\sqrt{\frac{m\omega}{\hbar}}y\right)\times\exp\!\left(-\frac{m\omega}{2\hbar}\left(x^2+y^2\right)\right),\tag{6.20}$$

or in scaled variables

$$\delta(X-Y)=\frac{1}{\sqrt{\pi}}\sum_{n=0}^{\infty}\frac{1}{2^n n!}H_n(X)\,H_n(Y)\exp\!\left(-\frac{1}{2}\left(X^2+Y^2\right)\right).$$

This equation follows from the Mehler formula (D.8) in the limit $z\to 1$ and using $\lim_{\kappa\to\infty}\kappa\exp\left(-\kappa^2x^2\right)=\sqrt{\pi}\delta(x)$.

6.4 Lemmata for exponentials of operators

Exponentials

$$\exp(A)\equiv\sum_{n=0}^{\infty}\frac{1}{n!}A^n\tag{6.21}$$

of operators appear in many applications of quantum mechanics, because

- exponentials of Hamiltonians generate time evolution in quantum systems (time evolution operators);
- exponentials of operators generate continuous transformations (e.g. translations or rotations or phase rotations etc.) in quantum systems; and because
- the exponential $\exp(-\lambda A/C)$ shifts the eigenvalues of the complementary operator A_c (where $[A,A_c]=C=const.$) by λ, i.e. $\exp(A)$ maps one eigenstate of A_c into another eigenstate. In this case the exponential operator is also denoted as a *shift operator*. We will see explicit examples for all these uses of exponentials of operators in later sections.

The definition (6.21) implies the property

$$\frac{d}{d\lambda}\exp(\lambda A) = A\exp(\lambda A) = \exp(\lambda A)A,$$

where λ is a complex variable. The property $\exp(A)\exp(-A) = 1$ is also easily proven, see Problem 6.5.

in addition, there are three other very useful theorems for products involving operator exponentials. The formulation of the first two of these theorems requires the notion of higher order commutators $\overset{n}{[A,B]}$, which can be recursively defined through

$$\overset{0}{[A,B]} \equiv B, \quad \overset{0}{[A,B]} \equiv A, \quad \overset{1}{[A,B]} = \overset{1}{[A,B]} = [A,B],$$

$$\overset{n+1}{[A,B]} = [A,\overset{n}{[A,B]}], \quad [A,\overset{n+1}{B]} = [\overset{n}{[A,B]},B].$$

With these definitions, we can state **Lemma 1:**

$$\exp(A)B\exp(-A) = \sum_{n\geq 0}\frac{1}{n!}\overset{n}{[A,B]}. \tag{6.22}$$

The proof simply proceeds by Taylor expansion of $\exp(\lambda A)B\exp(-\lambda A)$ with respect to λ. This uses the property

$$\frac{d^n}{d\lambda^n}\exp(\lambda A)B\exp(-\lambda A) = \exp(\lambda A)\overset{n}{[A,B]}\exp(-\lambda A),$$

which is proven by induction.

The second lemma is useful to combine certain products of three operator exponentials.

Lemma 2.

$$\exp(A)\exp(B)\exp(-A) = \exp[\exp(A)B\exp(-A)]. \tag{6.23}$$

The proof proceeds by applying the Taylor expansion $\exp(C) = \sum_{n=0}^{\infty} C^n/n!$ for $C = \exp(A)B\exp(-A)$.

The exponent on the right hand side of Lemma 2 can be evaluated with Lemma 1.

The third useful lemma concerns the combination of two exponentials of operators and requires that all higher order commutators of two operators A and B vanish:

$$\overset{2}{[A,B]} = 0, \quad \overset{2}{[B,A]} = 0. \tag{6.24}$$

Then the following equations hold[1],

[1]There is a generalization of equation (6.25) known as the Baker-Campbell-Hausdorff formula, which holds if the higher order commutators of A and B do not vanish. The recursive construction of higher order terms is outlined in Appendix E.

Lemma 3. *If equations (6.24) hold, then*

$$\exp(A)\exp(B) = \exp\left(A + B + \frac{1}{2}[A,B]\right) \tag{6.25}$$

$$= \exp(A+B)\exp\left(\frac{1}{2}[A,B]\right). \tag{6.26}$$

Proof. First we note that (6.26) is a direct consequence of (6.25) if we apply this equation to $\exp(A')\exp(B')$ with operators $A' = A+B$ and $B' = [A,B]/2$. Therefore it is enough to prove (6.25), which we will prove in the equivalent form

$$\exp(\lambda A)\exp(\lambda B) = \exp\left(\lambda A + \lambda B + \frac{\lambda^2}{2}[A,B]\right). \tag{6.27}$$

This equation is certainly correct for $\lambda = 0$. For the first order derivative of the left hand side of equation (6.27) one finds with (6.22) and (6.24)

$$\begin{aligned}\frac{d}{d\lambda}\exp(\lambda A)\exp(\lambda B) &= (A + B + \lambda[A,B])\exp(\lambda A)\exp(\lambda B)\\ &= \exp(\lambda A)\exp(\lambda B)(A + B + \lambda[A,B]),\end{aligned} \tag{6.28}$$

while the first order derivative of the right hand side of (6.27) is

$$\begin{aligned}&\frac{d}{d\lambda}\exp\left(\lambda A + \lambda B + \frac{\lambda^2}{2}[A,B]\right) = (A + B + \lambda[A,B])\\ &\times\exp\left(\lambda A + \lambda B + \frac{\lambda^2}{2}[A,B]\right) = \exp\left(\lambda A + \lambda B + \frac{\lambda^2}{2}[A,B]\right)\\ &\times(A + B + \lambda[A,B]).\end{aligned} \tag{6.29}$$

Therefore we also have

$$\begin{aligned}\left[\frac{d}{d\lambda}\exp(\lambda A)\exp(\lambda B)\right]_{\lambda=0} &= A + B\\ &= \left[\frac{d}{d\lambda}\exp\left(\lambda A + \lambda B + \frac{\lambda^2}{2}[A,B]\right)\right]_{\lambda=0}.\end{aligned} \tag{6.30}$$

Equations (6.28, 6.29) then also yield that in general

$$\begin{aligned}F_n &\equiv \left[\frac{d^n}{d\lambda^n}\exp(\lambda A)\exp(\lambda B)\right]_{\lambda=0}\\ &= \left[\frac{d^n}{d\lambda^n}\exp\left(\lambda A + \lambda B + \frac{\lambda^2}{2}[A,B]\right)\right]_{\lambda=0}\end{aligned}$$

by induction:

$$\begin{aligned}
F_{n+1} &= \left[\frac{d^{n+1}}{d\lambda^{n+1}}\exp(\lambda A)\exp(\lambda B)\right]_{\lambda=0} \\
&= \left[\frac{d^{n}}{d\lambda^{n}}(A+B+\lambda[A,B])\exp(\lambda A)\exp(\lambda B)\right]_{\lambda=0} \\
&= (A+B)\left[\frac{d^{n}}{d\lambda^{n}}\exp(\lambda A)\exp(\lambda B)\right]_{\lambda=0} \\
&\quad + n[A,B]\left[\frac{d^{n-1}}{d\lambda^{n-1}}\exp(\lambda A)\exp(\lambda B)\right]_{\lambda=0} \\
&= (A+B)\left[\frac{d^{n}}{d\lambda^{n}}\exp\!\left(\lambda A+\lambda B+\frac{\lambda^2}{2}[A,B]\right)\right]_{\lambda=0} \\
&\quad + n[A,B]\left[\frac{d^{n-1}}{d\lambda^{n-1}}\exp\!\left(\lambda A+\lambda B+\frac{\lambda^2}{2}[A,B]\right)\right]_{\lambda=0} \\
&= \left[\frac{d^{n}}{d\lambda^{n}}(A+B+\lambda[A,B])\exp\!\left(\lambda A+\lambda B+\frac{\lambda^2}{2}[A,B]\right)\right]_{\lambda=0} \\
&= \left[\frac{d^{n+1}}{d\lambda^{n+1}}\exp\!\left(\lambda A+\lambda B+\frac{\lambda^2}{2}[A,B]\right)\right]_{\lambda=0}.
\end{aligned}$$

Therefore the two operators have the same expansion in λ, and since they also agree for $\lambda = 0$ they must be the same.

We will often use these *lemmata* in quantum mechanical calculations.

6.5 Coherent states

Coherent states were introduced by Schrödinger in 1926 as quantum states which reproduce the classical oscillatory motion of a harmonic oscillator on the level of expectation values[2].

Equation (6.22) implies

$$\exp(-\lambda a^+)a\exp(\lambda a^+) = a+\lambda,$$

and therefore

$$a\exp(\lambda a^+)|0\rangle = \exp(\lambda a^+)(a+\lambda)|0\rangle = \lambda\exp(\lambda a^+)|0\rangle,$$

[2]E. Schrödinger, Naturwissenschaften 14, 664 (1926).

i.e. the state $\exp(\lambda a^+)|0\rangle$ is an eigenstate of the annihilation operator a with eigenvalue λ. It is not yet normalized, however. We can remedy this by replacing the shift operator $\exp(\lambda a^+)$ by a unitary shift operator with the same effect on a,

$$\exp(\lambda^+ a - \lambda a^+)a\exp(\lambda a^+ - \lambda^+ a) = a + \lambda,$$

and therefore the normalized eigenstate of a is

$$|\lambda\rangle = \exp(\lambda a^+ - \lambda^+ a)|0\rangle = \exp\left(-\frac{1}{2}|\lambda|^2\right)\exp(\lambda a^+)|0\rangle$$
$$= \exp\left(-\frac{1}{2}|\lambda|^2\right)\sum_{n=0}^{\infty}\frac{\lambda^n}{\sqrt{n!}}|n\rangle. \tag{6.31}$$

Here we used Lemma 3,

$$\exp(\lambda a^+)\exp(-\lambda^+ a) = \exp(\lambda a^+ - \lambda^+ a)\exp\left(\frac{1}{2}|\lambda|^2\right).$$

We also used an implicit convention that a state $|n\rangle$ labelled by an integer is an eigenstate of the Hamiltonian H of the harmonic oscillator, while a state $|\lambda\rangle$ labelled by a complex number is an eigenstate of a. Only the lowest energy state $|0\rangle$ is an eigenstate of both operators.

The states $|\lambda\rangle$ for $\lambda \neq 0$ are apparently superpositions of all energy eigenstates, and they are known as *coherent states*. They can be used to generate quantum states which move like a classical particle in the oscillator potential.

Classical motion with initial values $x(0) = X$ and $p(0) = P$ is described by

$$x_{cl}(t) = X\cos(\omega t) + \frac{P}{m\omega}\sin(\omega t),$$
$$p_{cl}(t) = P\cos(\omega t) - m\omega X\sin(\omega t), \tag{6.32}$$

and what we want to construct is a state $|\lambda(t)\rangle$ with exactly these time dependences of its expectation values,

$$\langle\lambda(t)|\mathrm{x}|\lambda(t)\rangle = \sqrt{\frac{\hbar}{2m\omega}}\langle\lambda(t)|a + a^+|\lambda(t)\rangle = \sqrt{\frac{\hbar}{2m\omega}}\left[\lambda(t) + \lambda^+(t)\right]$$
$$= x_{cl}(t) = X\cos(\omega t) + \frac{P}{m\omega}\sin(\omega t),$$
$$\langle\lambda(t)|\mathrm{p}|\lambda(t)\rangle = -\mathrm{i}\sqrt{\frac{m\omega\hbar}{2}}\langle\lambda(t)|a - a^+|\lambda(t)\rangle = -\mathrm{i}\sqrt{\frac{m\omega\hbar}{2}}\left[\lambda(t) - \lambda^+(t)\right]$$
$$= p_{cl}(t) = P\cos(\omega t) - m\omega X\sin(\omega t).$$

This yields

$$\lambda(t) = \sqrt{\frac{m\omega}{2\hbar}}x_{cl}(t) + \mathrm{i}\frac{p_{cl}(t)}{\sqrt{2m\omega\hbar}}$$
$$= \left(\sqrt{\frac{m\omega}{2\hbar}}X + \mathrm{i}\frac{P}{\sqrt{2m\omega\hbar}}\right)\exp(-\mathrm{i}\omega t). \tag{6.33}$$

We can also write the coherent state in terms of the operators x and p,

$$|\lambda(t)\rangle = \exp\!\big[\lambda(t)a^{+} - \lambda^{+}(t)a\big]|0\rangle = \exp\!\left(\frac{\mathrm{i}}{\hbar}\left[p_{cl}(t)\mathrm{x} - x_{cl}(t)\mathrm{p}\right]\right)|0\rangle. \tag{6.34}$$

We still have to show that the coherent states (6.34) satisfy the Schrödinger equation for the harmonic oscillator. We have with $|\lambda|^2(t) = |\lambda|^2$ time-independent,

$$\mathrm{i}\hbar\frac{d}{dt}|\lambda(t)\rangle = \mathrm{i}\hbar\dot{\lambda}(t)a^{+}|\lambda(t)\rangle = \hbar\omega\lambda(t)a^{+}|\lambda(t)\rangle = \hbar\omega a^{+}a|\lambda(t)\rangle$$
$$= H|\lambda(t)\rangle - \frac{1}{2}\hbar\omega|\lambda(t)\rangle.$$

Therefore the oscillating state which satisfies the Schrödinger equation of the harmonic oscillator including the zero point energy term is $|\lambda(t)\rangle\exp(-\mathrm{i}\omega t/2)$.

For the x representation of the coherent states, we notice

$$\langle x|\lambda(t)\rangle = \exp\!\left(-\frac{1}{2}|\lambda(t)|^2\right)\langle x|\exp\!\big(\lambda(t)a^{+}\big)|0\rangle$$
$$= \exp\!\left(-\frac{1}{2}|\lambda(t)|^2\right)\langle x|\exp\!\left[\frac{\lambda(t)}{\sqrt{2\hbar}}\left(\sqrt{m\omega}x - \mathrm{i}\frac{p}{\sqrt{m\omega}}\right)\right]|0\rangle$$
$$= \exp\!\left(\lambda(t)\sqrt{\frac{m\omega}{2\hbar}}x\right)\exp\!\left(-\frac{1}{2}|\lambda(t)|^2 - \frac{1}{4}\lambda^2(t)\right)$$
$$\times\langle x|\exp\!\left(-\mathrm{i}\lambda(t)\frac{p}{\sqrt{2m\omega\hbar}}\right)|0\rangle$$
$$= \exp\!\left(\lambda(t)\sqrt{\frac{m\omega}{2\hbar}}x\right)\exp\!\left(-\frac{1}{2}|\lambda(t)|^2 - \frac{1}{4}\lambda^2(t)\right)$$
$$\times\langle x - \lambda(t)\sqrt{\frac{\hbar}{2m\omega}}|0\rangle$$
$$= \left(\frac{m\omega}{\pi\hbar}\right)^{\frac{1}{4}}\exp\!\left(\lambda(t)\sqrt{\frac{m\omega}{2\hbar}}x\right)\exp\!\left(-\frac{1}{2}|\lambda(t)|^2 - \frac{1}{4}\lambda^2(t)\right)$$

$$\times \exp\left(-\frac{m\omega}{2\hbar}x^2 + \sqrt{\frac{m\omega}{2\hbar}}\lambda(t)x - \frac{1}{4}\lambda^2(t)\right)$$
$$= \left(\frac{m\omega}{\pi\hbar}\right)^{\frac{1}{4}} \exp\left(-\frac{m\omega}{2\hbar}x^2 + \sqrt{\frac{2m\omega}{\hbar}}\lambda(t)x - \frac{1}{2}|\lambda(t)|^2 - \frac{1}{2}\lambda^2(t)\right).$$

This yields with

$$\frac{1}{2}|\lambda(t)|^2 + \frac{1}{2}\lambda^2(t) = \frac{m\omega}{2\hbar}x_{cl}^2(t) + \mathrm{i}\frac{x_{cl}(t)p_{cl}(t)}{2\hbar}$$

the result

$$\langle x|\lambda(t)\rangle = \left(\frac{m\omega}{\pi\hbar}\right)^{\frac{1}{4}} \exp\left(-\frac{m\omega}{2\hbar}(x - x_{cl}(t))^2 + \mathrm{i}\frac{xp_{cl}(t)}{\hbar} - \mathrm{i}\frac{x_{cl}(t)p_{cl}(t)}{2\hbar}\right). \quad (6.35)$$

Comparison of equation (6.35) with equations (5.5) and (5.6) or direct evaluation yields the momentum representation of the coherent states,

$$\langle p|\lambda(t)\rangle = \frac{1}{(\pi m\omega\hbar)^{\frac{1}{4}}} \exp\left(-\frac{(p - p_{cl}(t))^2}{2m\omega\hbar} - \mathrm{i}\frac{x_{cl}(t)p}{\hbar} + \mathrm{i}\frac{x_{cl}(t)p_{cl}(t)}{2\hbar}\right). \quad (6.36)$$

The variances of the expectation values $\langle x\rangle(t) = x_{cl}(t)$ and $\langle p\rangle(t) = p_{cl}(t)$ of the coherent state $|\lambda(t)\rangle$ (6.33, 6.34) are

$$\Delta x^2 = \frac{\hbar}{2m\omega}, \quad \Delta p^2 = \frac{m\omega\hbar}{2}. \quad (6.37)$$

In terms of the force constant $K = m\omega^2$ of the harmonic potential the width of the coherent states is $\Delta x^2 \propto 1/\sqrt{mK}$. This equation holds with an n-dependent factor also for the energy eigenstates $|n\rangle$, see Problem 6.2. We have seen in Section 3.4 that the kinetic term $p^2/2m$ drives wave packets apart. The attractive potential $V(x) = Kx^2/2$ on the other hand tries to collapse wave packets, and the balance of these terms yields the stable wave packets (6.17) and (6.33, 6.34). This is consistent with $\Delta x^2 \propto 1/\sqrt{mK}$, because $mK \to \infty$ would correspond to domination of the attractive potential while $mK \to 0$ would imply domination of the kinetic term. In the next chapter we will see that the same basic mechanism also stabilizes the bound states of atoms. Balance between kinetic terms driving wave packets apart and the attractive Coulomb potential trying to contract the wave function generates minimal possible sizes of wave functions for given kinetic parameters $1/m$ and force constants Ze^2, thus preventing electrons from the classically inevitable core collapse.

Scalar products and overcompleteness of coherent states

We use $\overline{\lambda}$ instead of λ^+ to denote the complex conjugate of λ in this section.

The decomposition (6.31) yields for the product of two coherent state the result

$$\begin{aligned}\langle\lambda|\mu\rangle &= \exp\!\left(-\frac{1}{2}\left(|\lambda|^2+|\mu|^2\right)\right)\sum_{n=0}^{\infty}\frac{\left(\overline{\lambda}\mu\right)^n}{n!}\\ &= \exp\!\left(\overline{\lambda}\mu-\frac{1}{2}\left(|\lambda|^2+|\mu|^2\right)\right),\end{aligned} \tag{6.38}$$

and therefore we also have

$$|\langle\lambda|\mu\rangle|^2 = \exp(-|\lambda-\mu|^2),$$

i.e. coherent states are never orthogonal. Therefore any completeness relation cannot be unique, and we will find indeed that coherent states are *overcomplete*, i.e. we can decompose every state in a series of coherent states in infinitely many different ways. However, we can still identify a kind of "canonical" completeness relation in which all coherent states contribute with the same weight. We use

$$dz = \exp(\mathrm{i}\varphi)dr + \mathrm{i}zd\varphi, \quad \frac{d\overline{z}\wedge dz}{2\mathrm{i}} = d\Re z\wedge d\Im z = dr\wedge d\varphi\, r$$

to find the particular completeness relation

$$\begin{aligned}\int\frac{d\overline{z}\wedge dz}{2\pi\mathrm{i}}|z\rangle\langle z| &= \int\frac{d\overline{z}\wedge dz}{2\pi\mathrm{i}}\exp(-|z|^2)\sum_{m,n}|m\rangle\frac{z^m\overline{z}^n}{\sqrt{m!n!}}\langle n|\\ &= \frac{1}{\pi}\int_0^\infty dr\int_0^{2\pi}d\varphi\,\exp(-r^2)\sum_{m,n}|m\rangle\frac{r^{m+n+1}}{\sqrt{m!n!}}\exp[\mathrm{i}(n-m)\varphi]\,\langle n|\\ &= 2\int_0^\infty dr\,r\exp(-r^2)\sum_n|n\rangle\frac{r^{2n}}{n!}\langle n|\\ &= \sum_n\int_0^\infty du\left(-\frac{d}{d\alpha}\right)^n\exp(-\alpha u)\bigg|_{\alpha=1}\frac{1}{n!}|n\rangle\langle n|\\ &= \sum_n\left(-\frac{d}{d\alpha}\right)^n\frac{1}{\alpha}\bigg|_{\alpha=1}\frac{1}{n!}|n\rangle\langle n| = \sum_n|n\rangle\langle n| = 1.\end{aligned} \tag{6.39}$$

For example, substitution of

$$\langle z|n\rangle = \frac{\overline{z}^n}{\sqrt{n!}}\exp\!\left(-\frac{1}{2}|z|^2\right)$$

yields the following decomposition of the energy eigenstates of the harmonic oscillator in terms of coherent states,

$$|n\rangle = \int \frac{d\bar{z} \wedge dz}{2\pi i} |z\rangle \frac{\bar{z}^n}{\sqrt{n!}} \exp\left(-\frac{1}{2}|z|^2\right).$$

For another example of an expansion in terms of coherent states, substitution of equation (6.38) yields a decomposition of coherent states in terms of coherent states,

$$|\zeta\rangle = \int \frac{d\bar{z} \wedge dz}{2\pi i} |z\rangle \exp\left(\bar{z}\zeta - \frac{1}{2}\left(|z|^2 + |\zeta|^2\right)\right). \tag{6.40}$$

This lack of orthogonality of the coherent states implies that we can shift contributions of different coherent states to the expansion of a state $|\psi\rangle$, see e.g. Problem 6.10.

Note that for the coherent state parameter $\lambda(t)$ (6.33)

$$\frac{d\overline{\lambda}(t) \wedge d\lambda(t)}{2\pi i} = \frac{dx_{cl}(t) \wedge dp_{cl}(t)}{h} = \frac{dX \wedge dP}{h},$$

or if we denote the classical parameters simply with x and p,

$$\int \frac{dx \wedge dp}{h} |\lambda_{x,p}(t)\rangle\langle\lambda_{x,p}(t)| = 1.$$

Later we will encounter the measure $dx \wedge dp/h$ in phase space also in the density of states.

Squeezed states

We try to construct new oscillation operators b, b^+ from the oscillation operators (6.8). Substitution of the linear *ansatz*

$$b = Aa + Ba^+, \quad b^+ = \overline{A}a^+ + \overline{B}a$$

into the condition $[b, b^+] = 1$ yields

$$|A|^2 - |B|^2 = 1,$$

i.e. we find

$$b = \exp(i\alpha)\cosh(u)a + \exp(i\beta)\sinh(u)a^+,$$

$$b^+ = \exp(-i\alpha)\cosh(u)a^+ + \exp(-i\beta)\sinh(u)a.$$

The phase factors are irrelevant in the following. Therefore we study the ladder operators

$$a(u) = \cosh(u)a + \sinh(u)a^+ = \frac{1}{\sqrt{2\hbar}}\left(\sqrt{m\omega}\mathrm{x}\exp(u) + \mathrm{i}\frac{\mathrm{p}\exp(-u)}{\sqrt{m\omega}}\right)$$

and

$$a^+(u) = \cosh(u)a^+ + \sinh(u)a = \frac{1}{\sqrt{2\hbar}}\left(\sqrt{m\omega}\mathrm{x}\exp(u) - \mathrm{i}\frac{\mathrm{p}\exp(-u)}{\sqrt{m\omega}}\right).$$

The x and p operators of the original oscillator are

$$\mathrm{x} = \sqrt{\frac{\hbar}{2m\omega}}\left(a + a^+\right) = \sqrt{\frac{\hbar}{2m\omega}}\exp(-u)\left(a(u) + a^+(u)\right)$$

and

$$\mathrm{p} = -\mathrm{i}\sqrt{\frac{m\omega\hbar}{2}}\left(a - a^+\right) = -\mathrm{i}\sqrt{\frac{m\omega\hbar}{2}}\exp(u)\left(a(u) - a^+(u)\right).$$

We can think of the new operators $a(u)$ and $a^+(u)$ as oscillation operators for a harmonic oscillator with a u-dependent product of mass and frequency,

$$(m\omega)(u) = (m\omega)\exp(2u). \tag{6.41}$$

The coherent state for the new oscillation operators $|\lambda\rangle(u) = \exp(\lambda a^+(u) - \lambda^+ a(u))|0\rangle_u$, $a(u)|0\rangle_u = 0$, has expectation values and variances

$$\langle x\rangle(u) = \sqrt{\frac{\hbar}{2m\omega}}\exp(-u)\left(\lambda + \lambda^+\right) = \langle x\rangle\Big|_{u=0} \times \exp(-u),$$

$$\langle p\rangle(u) = -\mathrm{i}\sqrt{\frac{m\omega\hbar}{2}}\exp(u)\left(\lambda - \lambda^+\right) = \langle p\rangle\Big|_{u=0} \times \exp(u),$$

$$\Delta x^2(u) = \frac{\hbar}{2m\omega}\exp(-2u) = \Delta x^2\Big|_{u=0} \times \exp(-2u),$$

$$\Delta p^2(u) = \frac{m\omega\hbar}{2}\exp(2u) = \Delta p^2\Big|_{u=0} \times \exp(2u),$$

i.e. the uncertainty in x or p direction is squeezed at the expense of a corresponding increase of the uncertainty in the complementary direction.

We could formally write

$$\lambda a^+(u) - \lambda^+ a(u) = \lambda(u)a^+ - \lambda^+(u)a$$

with $\lambda(u) = \lambda\cosh(u) - \lambda^+\sinh(u)$. However, $|0\rangle_u \neq |0\rangle$ and therefore $|\lambda\rangle_u \neq |\lambda(u)\rangle$. Without the change in the vacuum, the variances could not change.

For the actual transformation, we note that

$$a(u) = \exp\Big(\frac{u}{2}\big[a^2 - (a^+)^2\big]\Big)\, a \exp\Big(-\frac{u}{2}\big[a^2 - (a^+)^2\big]\Big)$$

and therefore

$$|0\rangle_u = \exp\Big(\frac{u}{2}\big[a^2 - (a^+)^2\big]\Big)\, |0\rangle, \quad |\lambda\rangle_u = \exp\Big(\frac{u}{2}\big[a^2 - (a^+)^2\big]\Big)\, |\lambda\rangle.$$

6.6 Problems

6.1. Write down the p-representation of the Schrödinger equation for the one-dimensional harmonic oscillator. Which transformations between the parameters m and ω map the p-representation into the x-representation?

6.2. Calculate the widths Δx_n and Δp_n of the n-th energy eigenstate of the harmonic oscillator.

Remark. This is most conveniently done using the annihilation and creation operators.

6.3. A one-dimensional oscillator at time $t = 0$ is in a state

$$|\psi_\alpha\rangle = \cos\alpha|0\rangle + \exp(\mathrm{i}\varphi)\sin\alpha|1\rangle.$$

6.3a. Calculate the expectation values $\langle x\rangle(t)$, $\langle p\rangle(t)$ and $\langle E\rangle$ for the oscillator.

6.3b. Calculate the uncertainties $\Delta x(t)$, $\Delta p(t)$ and ΔE for the oscillator.

6.4. For the oscillator from Problem 6.3, how large is the probability density to find the oscillator in the location x at time t?

6.5. Show that

$$\exp(A)\exp(-A) = 1. \tag{6.42}$$

Hint: Define a corresponding λ-dependent operator $F(\lambda)$ by rescaling the operator A in (6.42) with the complex number λ. Show $dF(\lambda)/d\lambda = 0$. This implies $F(\lambda) = F(0)$.

6.6. Show that every coherent state $|\lambda\rangle = \exp(\lambda a^+ - \lambda^+ a)|0\rangle$ has the variances (6.37).

6.7. Calculate the energy expectation value $\langle E\rangle$ and the energy uncertainty ΔE for the coherent state $|\lambda\rangle$. Which values do you find in particular for the state $|\lambda(t)\rangle$ (6.34) which reproduces the classical trajectories (6.32)?

6.8. Show that

$$\begin{aligned}\exp\left(\frac{\mathrm{i}}{\hbar}\left[p_{cl}(t)\mathrm{x}-x_{cl}(t)\mathrm{p}\right]\right) &= \exp\left(\frac{\mathrm{i}}{\hbar}p_{cl}(t)\mathrm{x}\right)\exp\left(-\frac{\mathrm{i}}{\hbar}x_{cl}(t)\mathrm{p}\right)\\ &\quad\times\exp\left(-\frac{\mathrm{i}}{2\hbar}x_{cl}(t)p_{cl}(t)\right)\\ &= \exp\left(-\frac{\mathrm{i}}{\hbar}x_{cl}(t)\mathrm{p}\right)\exp\left(\frac{\mathrm{i}}{\hbar}p_{cl}(t)\mathrm{x}\right)\\ &\quad\times\exp\left(\frac{\mathrm{i}}{2\hbar}x_{cl}(t)p_{cl}(t)\right).\end{aligned}$$

6.9. Use the results from Problem 6.8 and equation (6.34) to re-derive the x and p representations (6.35, 6.36) of the coherent state $|\boldsymbol{\lambda}(t)\rangle$.

6.10. Suppose $K\subset\mathbb{C}$ is a subset of the complex plane. Show that the relation (6.40) implies a completeness relation

$$\begin{aligned}1 &= \int_{\mathbb{C}\setminus K}\frac{d\overline{z}\wedge dz}{2\pi\mathrm{i}}|z\rangle\left(\langle z|+\int_K\frac{d\overline{\zeta}\wedge d\zeta}{2\pi\mathrm{i}}\exp\left(\overline{z}\zeta-\frac{1}{2}\left(|z|^2+|\zeta|^2\right)\right)\langle\zeta|\right)\\ &\quad+\int_K\frac{d\overline{z}\wedge dz}{2\pi\mathrm{i}}|z\rangle\int_K\frac{d\overline{\zeta}\wedge d\zeta}{2\pi\mathrm{i}}\exp\left(\overline{z}\zeta-\frac{1}{2}\left(|z|^2+|\zeta|^2\right)\right)\langle\zeta|.\end{aligned}$$

What we have done here is to use (6.40) to redistribute contributions from coherent states with eigenvalues in K and in $\mathbb{C}\setminus K$ to the completeness relation. Since there are non-enumerably many possibilities to choose the subset K, there are non-enumerably many completeness relations for coherent states.

6.11. Construct a coherent state $|\boldsymbol{\lambda}(t)\rangle$ that follows the orbit (6.1) in terms of its expectation values $\langle\boldsymbol{\lambda}(t)|\mathbf{x}|\boldsymbol{\lambda}(t)\rangle$, $\langle\boldsymbol{\lambda}(t)|\mathbf{p}|\boldsymbol{\lambda}(t)\rangle$.

How large are the uncertainties of the coordinate and momentum expectation values?

6.12. The classical solution (6.1) of the three-dimensional isotropic oscillator describes a curve on a five-dimensional ellipsoid in six-dimensional phase space.

How long are the main axes of the ellipsoid?

How can you then think geometrically of the evolution of the coherent state from Problem 6.11?

Chapter 7
Central Forces in Quantum Mechanics

Radially symmetric problems appear if the interaction between two particles depends only on their separation r. We will first see how the dynamical problem of the motion of the two particles can be separated in terms of center of mass motion and relative motion and then write the effective Hamiltonian for the relative motion of the two particles in spherical coordinates.

7.1 Separation of center of mass motion and relative motion

The separation of center of mass motion and relative motion proceeds like in classical mechanics. The Hamiltonian of the 2-particle system is

$$H = \frac{\mathbf{p}_1^2}{2m_1} + \frac{\mathbf{p}_2^2}{2m_2} + V(|\mathbf{x}_1 - \mathbf{x}_2|) = \frac{\mathbf{P}^2}{2M} + \frac{\mathbf{p}^2}{2\mu} + V(r), \tag{7.1}$$

where

$$M = m_1 + m_2, \quad \mu = \frac{m_1 m_2}{m_1 + m_2} \tag{7.2}$$

are the total and reduced mass,

$$\mathbf{R} = \frac{m_1\mathbf{x}_1 + m_2\mathbf{x}_2}{m_1 + m_2}, \quad \mathbf{r} = \mathbf{x}_1 - \mathbf{x}_2 \tag{7.3}$$

are the operators for center of mass and relative coordinates, and

$$\mathbf{P} = M\dot{\mathbf{R}} = \mathbf{p}_1 + \mathbf{p}_2, \quad \mathbf{p} = \mu\dot{\mathbf{r}} = \frac{\mu}{m_1}\mathbf{p}_1 - \frac{\mu}{m_2}\mathbf{p}_2 = \frac{m_2\mathbf{p}_1 - m_1\mathbf{p}_2}{m_1 + m_2} \tag{7.4}$$

R. Dick, *Advanced Quantum Mechanics*, Graduate Texts in Physics,
DOI 10.1007/978-3-319-25675-7_7

are the momentum operators of center of mass motion and relative motion. The relative motion of the two original particles also comes with an angular momentum

$$\boldsymbol{l} = \mathbf{x}_1 \times \mathbf{p}_1 + \mathbf{x}_2 \times \mathbf{p}_2 - \mathbf{R} \times \mathbf{P} = \mathbf{r} \times \mathbf{p}. \tag{7.5}$$

The inverse transformations are

$$\mathbf{x}_1 = \mathbf{R} + \frac{m_2}{M}\mathbf{r}, \quad \mathbf{x}_2 = \mathbf{R} - \frac{m_1}{M}\mathbf{r}, \quad \mathbf{p}_1 = \frac{m_1}{M}\mathbf{P} + \mathbf{p}, \quad \mathbf{p}_2 = \frac{m_2}{M}\mathbf{P} - \mathbf{p}, \tag{7.6}$$

and if we assume $m_2 \geq m_1$,

$$m_2 = \frac{M + \sqrt{M(M - 4\mu)}}{2}, \quad m_1 = \frac{M - \sqrt{M(M - 4\mu)}}{2}.$$

The ∇ operators transform as

$$\frac{\partial}{\partial \boldsymbol{R}} = \frac{\partial}{\partial \boldsymbol{x}_1} + \frac{\partial}{\partial \boldsymbol{x}_2}, \quad \frac{\partial}{\partial \boldsymbol{r}} = \frac{m_2}{M}\frac{\partial}{\partial \boldsymbol{x}_1} - \frac{m_1}{M}\frac{\partial}{\partial \boldsymbol{x}_2},$$

$$\frac{\partial}{\partial \boldsymbol{x}_1} = \frac{m_1}{M}\frac{\partial}{\partial \boldsymbol{R}} + \frac{\partial}{\partial \boldsymbol{r}}, \quad \frac{\partial}{\partial \boldsymbol{x}_1} = \frac{m_2}{M}\frac{\partial}{\partial \boldsymbol{R}} - \frac{\partial}{\partial \boldsymbol{r}}.$$

These are the same transformations for operators as the corresponding transformations for classical coordinates and momenta in classical mechanics. From the quantum mechanics perspective this is not surprising, since the transformation equations for the operators are linear and therefore also hold for the expectation values of the operators, hence for the classical variables. What becomes particularly relevant for quantum mechanics is that the transformations preserve canonical commutation relations,

$$[\mathbf{x}_1, \mathbf{p}_1] = \mathrm{i}\hbar\underline{1}, \; [\mathbf{x}_2, \mathbf{p}_2] = \mathrm{i}\hbar\underline{1} \quad \Leftrightarrow \quad [\mathbf{R}, \mathbf{P}] = \mathrm{i}\hbar\underline{1}, \; [\mathbf{r}, \mathbf{p}] = \mathrm{i}\hbar\underline{1}.$$

Since the interaction does not depend on the center of mass coordinates, we can separate the center of mass motion with momentum $\boldsymbol{P} = \hbar\boldsymbol{K}$ in the wave function for the time-independent Schrödinger equation,

$$\Psi(\boldsymbol{x}_1, \boldsymbol{x}_2) = \frac{1}{\sqrt{2\pi}^3} \exp(\mathrm{i}\boldsymbol{K} \cdot \boldsymbol{R})\psi(\boldsymbol{r}), \tag{7.7}$$

and the energy eigenvalue problem $H|\Psi\rangle = E_{\mathrm{total}}|\Psi\rangle$ reduces to an eigenvalue problem for the relative motion,

$$E\psi(\boldsymbol{r}) = -\frac{\hbar^2}{2\mu}\Delta\psi(\boldsymbol{r}) + V(r)\psi(\boldsymbol{r}), \tag{7.8}$$

where

$$E_{\mathrm{total}} = E + \frac{\hbar^2 \boldsymbol{K}^2}{2M} \tag{7.9}$$

is the total energy in the center of mass motion and relative motion.

The discussion of separated solutions in Section 5.5 implies that the solutions for the Hamiltonian (7.1) should have the separated form with respect to center of mass motion and relative motion, $\Psi(\boldsymbol{x}_1,\boldsymbol{x}_2) = \Psi(\boldsymbol{R})\psi(\boldsymbol{r})$. However, the reasoning there also implies that we cannot find a solution of the Schrödinger equation which is separated in the actual coordinates of the two particles, $\Psi(\boldsymbol{x}_1,\boldsymbol{x}_2) \neq \psi_2(\boldsymbol{x}_2)\psi_1(\boldsymbol{x}_1)$. The wave functions of interacting particles are always entangled. This entanglement is easy to understand. Suppose that the interaction between two particles is attractive and strong enough to generate a bound state between the two particles. If we observe one particle at location $\boldsymbol{x}_1$, we know that the second particle has to be nearby at a location which is determined probabilistically by $|\psi(\boldsymbol{x}_1 - \boldsymbol{x}_2)|^2$. On the other hand, if the interaction is weak or repulsive, interactions with other particles will soon dominate each of the two particles, and their two-particle wave function $\Psi(\boldsymbol{x}_1,\boldsymbol{x}_2) = \Psi(\boldsymbol{R})\psi(\boldsymbol{r})$ is not a viable description any more: their mutual entanglement is destroyed by interactions with other particles.

We have based this discussion on the wave function (7.7) which appears in the $(\boldsymbol{x}_1,\boldsymbol{x}_2)$ representation of the time-independent two-particle Schrödinger equation $H|\Psi\rangle = E_{\text{total}}|\Psi\rangle$ with the Hamiltonian (7.1). If one starts from a time-dependent two-particle Schrödinger equation

$$\begin{aligned} \mathrm{i}\hbar\frac{\partial}{\partial t}\Psi(\boldsymbol{x}_1,\boldsymbol{x}_2,t) = &-\frac{\hbar^2}{2m_1}\frac{\partial^2}{\partial \boldsymbol{x}_1^2}\Psi(\boldsymbol{x}_1,\boldsymbol{x}_2,t) - \frac{\hbar^2}{2m_2}\frac{\partial^2}{\partial \boldsymbol{x}_2^2}\Psi(\boldsymbol{x}_1,\boldsymbol{x}_2,t) \\ &+ V(|\boldsymbol{x}_1 - \boldsymbol{x}_2|)\Psi(\boldsymbol{x}_1,\boldsymbol{x}_2,t), \end{aligned}$$

the separation *ansatz* for the center of mass motion

$$\begin{aligned} \Psi(\boldsymbol{x}_1,\boldsymbol{x}_2,t) &= \Psi(\boldsymbol{x}_1,\boldsymbol{x}_2)\exp\!\left(-\mathrm{i}\frac{E_{\text{total}}}{\hbar}t\right) \\ &= \frac{1}{\sqrt{2\pi}^3}\exp\!\left(\mathrm{i}\boldsymbol{K}\cdot\boldsymbol{R} - \mathrm{i}\frac{\hbar K^2}{2M}t\right)\psi(\boldsymbol{r},t) \\ &= \frac{1}{\sqrt{2\pi}^3}\exp\!\left(\mathrm{i}\boldsymbol{K}\cdot\boldsymbol{R} - \mathrm{i}\frac{\hbar K^2}{2M}t\right)\psi(\boldsymbol{r})\exp\!\left(-\mathrm{i}\frac{E}{\hbar}t\right) \end{aligned}$$

leads again to equation (7.8).

Separation of the center of mass motion in the present form works for any potential $V(\boldsymbol{r})$ which only depends on the separation vector $\boldsymbol{r}$ of the two particles. More general, if the 2-particle system moves in a potential of the form

$$V(\boldsymbol{x}_1,\boldsymbol{x}_2) = V(\boldsymbol{r}) + W(\boldsymbol{R}),$$

we can separate center of mass motion in the potential $W(\boldsymbol{R})$ from relative motion with the interaction potential $V(\boldsymbol{r})$ and find two independent effective single-particle Schrödinger equations (or Newton equations in mechanics) for the system. An example for this situation would be a hydrogen atom trapped in a potential well, e.g. in ice. If we model the potential well through a three-dimensional oscillator potential, the center of mass motion could be described by oscillator eigenstates

$\Psi_N(\boldsymbol{R},t)$ while the relative motion between the electron and the proton would be described by the wave functions $\psi_{n,\ell}(r)Y_{\ell,m}(\vartheta,\varphi)\exp(-\mathrm{i}E_n t/\hbar)$ derived in Sections 7.5 and 7.8.

7.2 The concept of symmetry groups

The effective single particle equation (7.8) for relative motion has the same form in every coordinate system which is related to the coordinates $\boldsymbol{r}$ through a rotation $r'^i = R^i{}_a r^a$, or in column vector notation,

$$\boldsymbol{r}' = \underline{R}\cdot\boldsymbol{r}, \quad \underline{R}^T\cdot\underline{R} = \underline{1},$$

cf. Section 4.1 and in particular equations (4.6) and (4.8). Contrary to Section 4.1, here we use the common left multiplication convention for linear coordinate transformations, $r^a \to r'^i = R^i{}_a r^a$, i.e. our rotation matrix $\underline{R}$ in the present section corresponds to $\underline{R}^T = \underline{R}^{-1}$ in Section 4.1.

Rotations have the following four basic properties:

1. The combination of two rotations $\underline{R}_1$ and $\underline{R}_2$ yields again a rotation $\underline{R}_2\cdot\underline{R}_1$,

$$\boldsymbol{r}'' = \underline{R}_2\cdot\boldsymbol{r}' = \underline{R}_2\cdot(\underline{R}_1\cdot\boldsymbol{r}) = (\underline{R}_2\cdot\underline{R}_1)\cdot\boldsymbol{r}.$$

2. The identity transformation $\underline{1}$ is a particular rotation.
3. For every rotation there is an inverse rotation, $\underline{R}^{-1}\cdot\underline{R} = \underline{1}$.
4. Combination of rotations is associative,

$$\underline{R}_3\cdot(\underline{R}_2\cdot\underline{R}_1) = (\underline{R}_3\cdot\underline{R}_2)\cdot\underline{R}_1.$$

These four algebraic properties are common to all sets of symmetry transformations of physical systems, and they have far reaching consequences in the sense that many other interesting properties of symmetry transformations can be derived from these four properties. Every set of mathematical objects having these four properties is therefore denoted as a *group*, and the study of groups is a subdiscipline of algebra denoted as group theory.

Groups which are particularly relevant for quantum mechanics include the following sets:

1. The group of proper rotations is the set of all rotations which does not include inversion of an odd number of axes. Matrices which generate proper rotations do not only satisfy the orthogonality condition

$$\underline{R}^T\cdot\underline{R} = \underline{1},$$

but also the special additional condition

$$\det\underline{R} = 1.$$

The group of proper rotations in three dimensions is therefore also denoted as the *special orthogonal group of rotations in three dimensions*, or SO(3) for short.

2. A group which is closely related to the group SO(3) is the group of unitary 2×2 matrices with determinant 1,

$$\underline{U}^{+} \cdot \underline{U} = \underline{1}, \quad \det\underline{U} = 1.$$

This is the *special group of unitary transformations in two dimensions* SU(2).

3. The Poincaré group and its various subgroups, including in particular the proper orthochronous Lorentz group SO(1,3) of proper rotations and Lorentz boosts in Minkowski spacetime, are important for relativistic quantum mechanics.
4. SO(1,3) is closely related to the group of complex 2×2 matrices with determinant 1. This group is often denoted as SL(2,ℂ).
5. Discrete symmetry groups involve e.g. translations along lattice vectors in a regular lattice, or inversions of axes, or rotations by fixed angles. Discrete groups are also important in many applications of quantum mechanics.
6. The known basic particle interactions (besides gravity) are related to the group U(1) of phase transformations, and also to the special unitary groups SU(2) and SU(3).

In this and the following chapter we are primarily concerned with the groups SO(3) and SU(2), and we will develop the relevant aspects of these groups and their matrix representations along the way. Students who would like to acquire a deeper understanding of groups and their representations from a physics perspective should consult the excellent texts by Barut and Raczka [2] or Cornwell [6] for groups in general, or Sexl and Urbantke [37] for emphasis on the Poincaré and Lorentz groups. However, this is not required for understanding the following chapters.

7.3 Operators for kinetic energy and angular momentum

The kinetic operator in spherical coordinates follows from (5.25) as

$$\begin{aligned}\frac{\mathbf{p}^2}{2\mu} &= -\frac{\hbar^2}{2\mu}\int d^3\boldsymbol{r}\,|\boldsymbol{r}\rangle\left(\frac{1}{r}\frac{\partial^2}{\partial r^2}r + \frac{1}{r^2}\frac{\partial^2}{\partial\vartheta^2} + \frac{\cot\vartheta}{r^2}\frac{\partial}{\partial\vartheta} + \frac{1}{r^2\sin^2\vartheta}\frac{\partial^2}{\partial\varphi^2}\right)\langle\boldsymbol{r}| \\ &= -\frac{\hbar^2}{2\mu}\int d^3\boldsymbol{r}\,|\boldsymbol{r}\rangle\frac{1}{r}\frac{\partial^2}{\partial r^2}r\langle\boldsymbol{r}| + \frac{\boldsymbol{M}^2}{2\mu r^2},\end{aligned}$$

where $\boldsymbol{M}$ is the angular momentum operator

$$\boldsymbol{M} = \mathbf{r}\times\mathbf{p} = \frac{\hbar}{\mathrm{i}}\int d^3\boldsymbol{r}\,|\boldsymbol{r}\rangle\boldsymbol{r}\times\nabla\langle\boldsymbol{r}|$$

$$= \frac{\hbar}{\mathrm{i}} \int d^3\boldsymbol{r}\, |\boldsymbol{r}\rangle \left(\boldsymbol{e}_\varphi \frac{\partial}{\partial \vartheta} - \frac{1}{\sin\vartheta} \boldsymbol{e}_\vartheta \frac{\partial}{\partial \varphi} \right) \langle \boldsymbol{r}|$$

$$= \frac{\hbar}{\mathrm{i}} \int d^3\boldsymbol{r}\, |\boldsymbol{r}\rangle \Big[\boldsymbol{e}_x \left(-\sin\varphi \frac{\partial}{\partial \vartheta} - \cot\vartheta \cdot \cos\varphi \frac{\partial}{\partial \varphi} \right)$$

$$+ \boldsymbol{e}_y \left(\cos\varphi \frac{\partial}{\partial \vartheta} - \cot\vartheta \cdot \sin\varphi \frac{\partial}{\partial \varphi} \right) + \boldsymbol{e}_z \frac{\partial}{\partial \varphi} \Big] \langle \boldsymbol{r}|. \tag{7.10}$$

The property

$$\boldsymbol{M}^2 = -\hbar^2 \int d^3\boldsymbol{r}\, |\boldsymbol{r}\rangle \left(\frac{\partial^2}{\partial \vartheta^2} + \cot\vartheta \frac{\partial}{\partial \vartheta} + \frac{1}{\sin^2\vartheta} \frac{\partial^2}{\partial \varphi^2} \right) \langle \boldsymbol{r}| \tag{7.11}$$

follows from

$$\partial_\varphi \boldsymbol{e}_\varphi = -\cos\varphi\, \boldsymbol{e}_x - \sin\varphi\, \boldsymbol{e}_y = -\sin\vartheta\, \boldsymbol{e}_r - \cos\vartheta\, \boldsymbol{e}_\vartheta.$$

The energy eigenvalue problem for the relative motion therefore reads

$$\left(-\frac{\hbar^2}{2\mu} \frac{1}{r} \frac{\partial^2}{\partial r^2} r + V(r) \right) \langle \boldsymbol{r}|\psi\rangle + \frac{1}{2\mu r^2} \langle \boldsymbol{r}|\boldsymbol{M}^2|\psi\rangle = E\langle \boldsymbol{r}|\psi\rangle, \tag{7.12}$$

and we can deal with the angular part in the equation by first solving the eigenvalue problem for the operator $\boldsymbol{M}^2$.

A useful tool for the analysis of angular momentum operators is the ϵ *tensor* or *Eddington tensor.* The ϵ tensor in an n-dimensional flat space is the completely anti-symmetric tensor of n-th order

$$\epsilon_{i_1 i_2 \ldots i_{k-1} i_k i_{k+1} \ldots i_{m-1} i_m i_{m+1} \ldots i_{n-1} i_n} = -\epsilon_{i_1 i_2 \ldots i_{k-1} i_m i_{k+1} \ldots i_{m-1} i_k i_{m+1} \ldots i_{n-1} i_n}$$

with the normalization

$$\epsilon_{123\ldots n} = 1.$$

This tensor has n^n components. Anti-symmetry and the normalization imply that $n!/2$ of the components have the value 1, $n!/2$ of the components have the value -1, and $n^n - n!$ components vanish. We often use the ϵ tensor in three dimensions,

$$\epsilon_{123} = \epsilon_{231} = \epsilon_{312} = -\epsilon_{213} = -\epsilon_{132} = -\epsilon_{321} = 1.$$

This tensor appears e.g. if we express cross products of vectors in terms of their components in a Cartesian basis $\boldsymbol{e}_i$,

$$\boldsymbol{a} = a_i \boldsymbol{e}_i, \quad \boldsymbol{b} = b_i \boldsymbol{e}_i \quad \Rightarrow \quad \boldsymbol{a} \times \boldsymbol{b} = \boldsymbol{e}_i \epsilon_{ijk} a_j b_k. \tag{7.13}$$

This relation can be verified directly from the explicit definition of cross products like $(\boldsymbol{a} \times \boldsymbol{b})_1 = a_2 b_3 - a_3 b_2$ etc., or it can be considered as a consequence of a relation for the cross product of Cartesian basis vectors,

$$\boldsymbol{e}_i \times \boldsymbol{e}_j = \epsilon_{ijk}\boldsymbol{e}_k.$$

An example of (7.13) which involves the gradient operator is the curl of a vector field,

$$\boldsymbol{B} = \nabla \times \boldsymbol{A} = \boldsymbol{e}_i \epsilon_{ijk}\partial_j A_k.$$

A useful identity is

$$\epsilon_{ijk}\epsilon_{k\ell m} = \delta_{i\ell}\delta_{jm} - \delta_{im}\delta_{j\ell}. \tag{7.14}$$

This identity yields e.g. the relations

$$\boldsymbol{a} \times (\boldsymbol{b} \times \boldsymbol{c}) = (\boldsymbol{a} \cdot \boldsymbol{c})\boldsymbol{b} - (\boldsymbol{a} \cdot \boldsymbol{b})\boldsymbol{c},$$

$$\nabla \times (\nabla \times \boldsymbol{A}) = \nabla(\nabla \cdot \boldsymbol{A}) - \Delta \boldsymbol{A}.$$

Equation (7.13) implies that the Cartesian components of the angular momentum operator are related to the Cartesian components of position and momentum operators according to

$$M_i = \epsilon_{ijk}\mathrm{x}_j\mathrm{p}_k = \frac{\hbar}{\mathrm{i}}\int d^3\boldsymbol{r}|\boldsymbol{r}\rangle\epsilon_{ijk}x_j\frac{\partial}{\partial x^k}\langle\boldsymbol{r}|.$$

The first of these relations and the canonical commutation relations $[\mathrm{x}_i, \mathrm{p}_j] = \mathrm{i}\hbar\delta_{ij}$ imply the angular momentum commutation relations

$$[M_i, M_j] = \mathrm{i}\hbar\epsilon_{ijk}M_k. \tag{7.15}$$

Determination of the eigenvalues of $\boldsymbol{M}^2$ is equivalent to the determination of all hermitian matrix representations of the Lie algebra (7.15), which in turn is equivalent to the determination of all the matrix representations of the rotation group. We will find that all those matrix representations are realized in rotationally symmetric quantum systems. Therefore our next task is the determination of all the matrix representations of (7.15).

7.4 Matrix representations of the rotation group

We will start the study of matrix representations of the rotation group by looking at the defining representation, and then derive the general matrix representation.

The defining representation of the three-dimensional rotation group

In Section 4.1 we found the condition

$$\underline{R}\cdot\underline{R}^T = \underline{1}$$

for rotation matrices. This leaves the following possibilities for the matrix[1] $\underline{X} = \ln\underline{R}$,

$$\underline{X}^T = -\underline{X} + 2\pi i n \underline{1}. \tag{7.16}$$

The equation

$$\det(\exp\underline{X}) = \exp(\mathrm{tr}\underline{X}), \tag{7.17}$$

which follows from the existence of a Jordan canonical form (F.2) for every matrix, implies then

$$\det\underline{R} = \exp\left(\mathrm{tr}\,\frac{\underline{X}+\underline{X}^T}{2}\right) = (-1)^n,$$

i.e. $\det\underline{R} = \pm 1$. Pure rotations have $\det\underline{R} = 1$, whereas additional inversion of an odd number of axes[2] yields $\det\underline{R} = -1$. We will focus on pure rotations.

The general solution of equation (7.16) in three dimensions and with $n = 0$ is

$$\underline{X} = \begin{pmatrix} 0 & \varphi_3 & -\varphi_2 \\ -\varphi_3 & 0 & \varphi_1 \\ \varphi_2 & -\varphi_1 & 0 \end{pmatrix} = \varphi_i \underline{L}_i = \boldsymbol{\varphi}\cdot\underline{\boldsymbol{L}},$$

where the basis of anti-symmetric real 3×3 matrices

$$\underline{L}_1 = \begin{pmatrix} 0 & 0 & 0 \\ 0 & 0 & 1 \\ 0 & -1 & 0 \end{pmatrix}, \quad \underline{L}_2 = \begin{pmatrix} 0 & 0 & -1 \\ 0 & 0 & 0 \\ 1 & 0 & 0 \end{pmatrix}, \quad \underline{L}_3 = \begin{pmatrix} 0 & 1 & 0 \\ -1 & 0 & 0 \\ 0 & 0 & 0 \end{pmatrix} \tag{7.18}$$

was introduced. We can write the equations above in short form $(\underline{L}_i)_{jk} = \epsilon_{ijk}$. The general orientation preserving rotation in three dimensions therefore has the form

$$\underline{R}(\boldsymbol{\varphi}) = \exp(\boldsymbol{\varphi}\cdot\underline{\boldsymbol{L}}).$$

[1]See Appendix F for the calculation of the logarithm of an invertible matrix.
[2]Inversion of three axes is equivalent to inversion of one axis combined with a rotation.

Expansion of the exponential function and ordering into even and odd powers of $\boldsymbol{\varphi} \cdot \underline{\boldsymbol{L}}$ yields the representation

$$\underline{R}(\boldsymbol{\varphi}) = \hat{\boldsymbol{\varphi}} \otimes \hat{\boldsymbol{\varphi}}^T + \left(\underline{1} - \hat{\boldsymbol{\varphi}} \otimes \hat{\boldsymbol{\varphi}}^T\right) \cos\varphi + \hat{\boldsymbol{\varphi}} \cdot \underline{\boldsymbol{L}} \sin\varphi. \tag{7.19}$$

Application of the matrix $\hat{\boldsymbol{\varphi}} \cdot \underline{\boldsymbol{L}}$ on a vector $\boldsymbol{r}$ generates a vector product,

$$(\hat{\boldsymbol{\varphi}} \cdot \underline{\boldsymbol{L}}) \cdot \boldsymbol{r} = -\hat{\boldsymbol{\varphi}} \times \boldsymbol{r},$$

i.e. for every vector $\boldsymbol{r}$, the first term in (7.19) preserves the part $\boldsymbol{r}_\parallel = \hat{\boldsymbol{\varphi}} \otimes \hat{\boldsymbol{\varphi}}^T \cdot \boldsymbol{r}$ of the vector which is parallel to the vector $\boldsymbol{\varphi}$, the second term multiplies the orthogonal part $\boldsymbol{r}_\perp = \boldsymbol{r} - \boldsymbol{r}_\parallel$ by the factor $\cos\varphi$, and the third part takes the orthogonal part, rotates it by $\pi/2$ and multiplies it by the factor $\sin\varphi$,

$$\underline{R}(\boldsymbol{\varphi}) \cdot \boldsymbol{r} = \boldsymbol{r}_\parallel + \boldsymbol{r}_\perp \cos\varphi - \hat{\boldsymbol{\varphi}} \times \boldsymbol{r} \sin\varphi.$$

This also implies that the direction $\hat{\boldsymbol{\varphi}}$ of the vector $\boldsymbol{\varphi}$ is the direction of the axis of rotation.

Exponentiation of the linear combinations $\boldsymbol{\varphi} \cdot \underline{\boldsymbol{L}}$ of the matrices (7.18) thus generates rotations in three dimensions, and therefore these matrices are also denoted as three-dimensional representations of *generators* of the rotation group. They satisfy the commutation relations

$$[\underline{L}_i, \underline{L}_j] = -\epsilon_{ijk} \underline{L}_k. \tag{7.20}$$

We will also use the hermitian matrices

$$\underline{M}_i = -\mathrm{i}\hbar \underline{L}_i, \quad [\underline{M}_i, \underline{M}_j] = \mathrm{i}\hbar \epsilon_{ijk} \underline{M}_k, \quad [\underline{M}_i, \underline{\boldsymbol{M}}^2] = 0. \tag{7.21}$$

It is no coincidence that the angular momentum operators

$$M_i = \epsilon_{ijk} \mathrm{x}_j \mathrm{p}_k$$

satisfy the same commutation relations. We will see that angular momentum operators also generate rotations, and a set of operators M_i generates rotations if and only if the operators satisfy the commutation relations (7.21). It is a consequence of the general Baker-Campbell-Hausdorff formula in Appendix E that the combination of any two rotations to a new rotation is completely determined by the commutation relations (7.21) of the generators of rotations.

The general matrix representations of the rotation group

We wish to classify all possible representations of the commutation relations (7.21) in vector spaces. To accomplish this, it is convenient to change the basis from $M_x \equiv$

M_1 and $M_y \equiv M_2$ to

$$M_\pm = M_1 \pm \mathrm{i}M_2, \quad M_z \equiv M_3.$$

The product $\boldsymbol{M}^2 \equiv M_i M_i$ is then

$$\boldsymbol{M}^2 = \frac{1}{2}(M_+M_- + M_-M_+) + M_z{}^2 = M_-M_+ + M_z{}^2 + \hbar M_z,$$

and we have the commutation relations in the new basis,

$$[M_z, M_\pm] = \pm\hbar M_\pm, \quad [M_+, M_-] = 2\hbar M_z.$$

Hermiticity[3] implies that we can use a basis where M_z is diagonal with real eigenvalues,

$$M_z|m\rangle = \hbar m|m\rangle, \quad m \in \mathbb{R}.$$

The commutation relations then imply

$$\begin{aligned} M_\pm|m\rangle &= \hbar C_\pm(m)|m \pm 1\rangle, \\ C_+(m-1)C_-(m) &= 2\hbar m + C_-(m+1)C_+(m), \end{aligned} \tag{7.22}$$

and $M_+{}^+ = M_-$ implies

$$C_-(m) = \langle m-1|M_-|m\rangle = (\langle m|M_+{}^+|m-1\rangle)^+ = C_+(m-1)^+.$$

Substitution in equation (7.22) yields

$$|C_+(m)|^2 = |C_+(m-1)|^2 - 2\hbar^2 m.$$

Since the left hand side cannot become negative, there must exist some maximal value ℓ for m such that $C_+(\ell) = 0$, $M_+|\ell\rangle = 0$, and we have

$$|C_+(\ell-1)|^2 = 2\hbar^2\ell, \quad |C_+(\ell-2)|^2 = 2\hbar^2(2\ell-1),$$

and after $n-1$ steps

$$|C_+(\ell-n)|^2 = |C_-(\ell-n+1)|^2 = \hbar^2[2n\ell - n(n-1)]. \tag{7.23}$$

Again, the left hand side cannot become negative, and therefore the expression on the right hand side must terminate for some value N of n, $C_-(\ell - N + 1) = 0$, $M_-|\ell - N + 1\rangle = 0$. This implies existence of an integer N such that $2\ell = N - 1$ and

$$C_+(\ell - N) = C_+(-(N+1)/2) = C_-((1-N)/2) = C_-(-\ell) = 0, \tag{7.24}$$

[3] We could do the following calculations in slightly more general form without using hermiticity, and then find hermiticity of the finite-dimensional representations along the way.

where an irrelevant possible phase factor was excluded. Therefore we have boundaries

$$-\ell = \frac{1-N}{2} \le m \le \frac{N-1}{2} = \ell \tag{7.25}$$

and $N = 2\ell + 1$ possible values for m both for integer ℓ and half-integer ℓ.

Equation (7.23) yields with

$$n = \frac{N-1}{2} - m$$

the equation

$$\begin{aligned} n(N-n) &= \frac{N^2}{4} - \left(m + \frac{1}{2}\right)^2 = \frac{N^2-1}{4} - m(m+1) \\ &= \ell(\ell+1) - m(m+1), \end{aligned}$$

and therefore

$$\begin{aligned} C_+(m)^2 &= \hbar^2\left(\frac{N^2-1}{4} - m(m+1)\right) = \hbar^2\left[\ell(\ell+1) - m(m+1)\right], \\ C_-(m)^2 &= C_+(m-1)^2 = \hbar^2\left[\ell(\ell+1) - m(m-1)\right]. \end{aligned}$$

We have found all the hermitian matrix representations of the commutation relations (7.21). The magnetic quantum number m can take values $-\ell \le m \le \ell$, the number of dimensions is $N = 2\ell+1 \in \mathbb{N}$, and the actions of the angular momentum operators are

$$M_z|\ell,m\rangle = \hbar m|\ell,m\rangle, \quad 2\ell \in \mathbb{N}_0, \quad m \in \{-\ell, -\ell+1, \ldots, \ell-1, \ell\}, \tag{7.26}$$

$$M_+|\ell,m\rangle = \hbar\sqrt{\ell(\ell+1) - m(m+1)}|\ell,m+1\rangle \tag{7.27}$$

$$M_-|\ell,m\rangle = \hbar\sqrt{\ell(\ell+1) - m(m-1)}|\ell,m-1\rangle \tag{7.28}$$

$$\begin{aligned} M_x|\ell,m\rangle &= \frac{\hbar}{2}\sqrt{\ell(\ell+1) - m(m+1)}|\ell,m+1\rangle \\ &\quad + \frac{\hbar}{2}\sqrt{\ell(\ell+1) - m(m-1)}|\ell,m-1\rangle, \end{aligned} \tag{7.29}$$

$$\begin{aligned} M_y|\ell,m\rangle &= \frac{\hbar}{2\mathrm{i}}\sqrt{\ell(\ell+1) - m(m+1)}|\ell,m+1\rangle \\ &\quad - \frac{\hbar}{2\mathrm{i}}\sqrt{\ell(\ell+1) - m(m-1)}|\ell,m-1\rangle, \end{aligned} \tag{7.30}$$

$$\boldsymbol{M}^2|\ell,m\rangle = \left(C_+(\ell,m)^2 + \hbar^2 m(m+1)\right)|\ell,m\rangle = \hbar^2\ell(\ell+1)|\ell,m\rangle. \tag{7.31}$$

7.5 Construction of the spherical harmonic functions

We now want to construct the $\boldsymbol{r}$ representations of the angular momentum eigenstates $|\ell, m\rangle$, i.e. we are seeking the solutions $\langle\vartheta,\varphi|\ell,m\rangle \equiv Y_{\ell,m}(\vartheta,\varphi)$ of the differential equations

$$\begin{aligned}\langle\vartheta,\varphi|\boldsymbol{M}^2|\ell,m\rangle &\equiv -\hbar^2\left(\frac{\partial^2}{\partial\vartheta^2}+\cot\vartheta\frac{\partial}{\partial\vartheta}+\frac{1}{\sin^2\vartheta}\frac{\partial^2}{\partial\varphi^2}\right)Y_{\ell,m}(\vartheta,\varphi)\\ &= \hbar^2\ell(\ell+1)Y_{\ell,m}(\vartheta,\varphi)\end{aligned}$$

and

$$\langle\vartheta,\varphi|M_z|\ell,m\rangle \equiv \frac{\hbar}{\mathrm{i}}\frac{\partial}{\partial\varphi}Y_{\ell,m}(\vartheta,\varphi) = \hbar m Y_{\ell,m}(\vartheta,\varphi).$$

Here we used that the angular momentum operators act in $\boldsymbol{r}$ space as differential operators with respect to ϑ and φ, and therefore do not determine the radial dependence of wave functions. The radial part can therefore be left out in their representation[4],

$$\begin{aligned}\boldsymbol{M}^2 &= -\hbar^2\int_0^{2\pi}d\varphi\int_0^{\pi}d\vartheta\,\sin\vartheta\,|\vartheta,\varphi\rangle\left(\frac{\partial^2}{\partial\vartheta^2}+\cot\vartheta\frac{\partial}{\partial\vartheta}+\frac{1}{\sin^2\vartheta}\frac{\partial^2}{\partial\varphi^2}\right)\langle\vartheta,\varphi|,\\ M_z &= \frac{\hbar}{\mathrm{i}}\int_0^{2\pi}d\varphi\int_0^{\pi}d\vartheta\,\sin\vartheta\,|\vartheta,\varphi\rangle\frac{\partial}{\partial\varphi}\langle\vartheta,\varphi|,\\ M_+ &= \hbar\int_0^{2\pi}d\varphi\int_0^{\pi}d\vartheta\,\sin\vartheta\,|\vartheta,\varphi\rangle\exp(\mathrm{i}\varphi)\left(\mathrm{i}\cot\vartheta\frac{\partial}{\partial\varphi}+\frac{\partial}{\partial\vartheta}\right)\langle\vartheta,\varphi|,\\ M_- &= \hbar\int_0^{2\pi}d\varphi\int_0^{\pi}d\vartheta\,\sin\vartheta\,|\vartheta,\varphi\rangle\exp(-\mathrm{i}\varphi)\left(\mathrm{i}\cot\vartheta\frac{\partial}{\partial\varphi}-\frac{\partial}{\partial\vartheta}\right)\langle\vartheta,\varphi|.\end{aligned}$$

The equation

$$\langle\vartheta,\varphi|M_z|\ell,\ell\rangle = \frac{\hbar}{\mathrm{i}}\frac{\partial}{\partial\varphi}\langle\vartheta,\varphi|\ell,\ell\rangle = \hbar\ell\langle\vartheta,\varphi|\ell,\ell\rangle \tag{7.32}$$

implies

$$\langle\vartheta,\varphi|\ell,\ell\rangle = Y_{\ell,\ell}(\vartheta,\varphi) = f_\ell(\vartheta)\exp(\mathrm{i}\ell\varphi).$$

Single valuedness of the eigenstates implies $\ell\in\mathbb{N}_0$.

The equation

$$\langle\vartheta,\varphi|M_+|\ell,\ell\rangle = \hbar\exp(\mathrm{i}\varphi)\left(\frac{\partial}{\partial\vartheta}+\mathrm{i}\cot\vartheta\frac{\partial}{\partial\varphi}\right)\langle\vartheta,\varphi|\ell,\ell\rangle = 0 \tag{7.33}$$

[4]Stated differently, we leave out a factor $1=\int_0^\infty dr r^2|r\rangle\langle r|$.

implies

$$\left(\frac{\partial}{\partial\vartheta} - \ell\cot\vartheta\right) f_\ell(\vartheta) = 0,$$

with the solution $f_\ell(\vartheta) = N_\ell^{-1/2}\sin^\ell\vartheta$. The normalization constants are chosen to ensure

$$\int_0^\pi d\vartheta\, f_\ell^2(\vartheta)\sin\vartheta = 1.$$

They can be calculated recursively if we note that

$$N_0 = \int_0^\pi d\vartheta\,\sin\vartheta = 2$$

and

$$\begin{aligned} N_{\ell\geq 1} &= \int_0^\pi d\vartheta\,\sin^{2\ell+1}\vartheta = \int_{-1}^1 d\xi\,(1-\xi^2)^\ell \\ &= \xi(1-\xi^2)^\ell\Big|_{-1}^1 + \int_{-1}^1 d\xi\,2\ell\xi^2(1-\xi^2)^{\ell-1} \\ &= -2\ell N_\ell + 2\ell N_{\ell-1}. \end{aligned}$$

This yields

$$\begin{aligned} N_\ell &= \frac{2\ell}{2\ell+1}N_{\ell-1} = \frac{2^2\ell(\ell-1)}{(2\ell+1)(2\ell-1)}N_{\ell-2} = \ldots = \frac{2^\ell\ell!}{(2\ell+1)!!}2 \\ &= 2\frac{2^{2\ell}(\ell!)^2}{(2\ell+1)!}, \end{aligned}$$

and therefore

$$\langle\vartheta,\varphi|\ell,\ell\rangle = Y_{\ell,\ell}(\vartheta,\varphi) = \frac{(-)^\ell}{2^{\ell+1}\ell!}\sqrt{\frac{(2\ell+1)!}{\pi}}\exp(i\ell\varphi)\sin^\ell\vartheta. \tag{7.34}$$

We can get the other eigenfunctions from $\langle\vartheta,\varphi|\ell,\ell\rangle$ through repeated applications of the lowering operator M_-,

$$\begin{aligned} Y_{\ell,m}(\vartheta,\varphi) &= \langle\vartheta,\varphi|\ell,m\rangle = \frac{\langle\vartheta,\varphi|M_-|\ell,m+1\rangle}{\hbar\sqrt{(\ell+m+1)(\ell-m)}} \\ &= \frac{\langle\vartheta,\varphi|(M_-)^2|\ell,m+2\rangle}{\hbar^2\sqrt{(\ell+m+1)(\ell+m+2)(\ell-m)(\ell-m-1)}} = \ldots \\ &= \frac{\langle\vartheta,\varphi|(M_-)^{\ell-m}|\ell,\ell\rangle}{\hbar^{\ell-m}}\,[(\ell+m+1)(\ell+m+2)\times\ldots\times 2\ell \end{aligned}$$

$$\times(\ell-m)(\ell-m-1)\times\ldots\times 1]^{-1/2}$$

$$=\frac{1}{\hbar^{\ell-m}}\sqrt{\frac{(\ell+m)!}{(2\ell)!(\ell-m)!}}\langle\vartheta,\varphi|(M_-)^{\ell-m}|\ell,\ell\rangle$$

$$=\sqrt{\frac{(\ell+m)!}{(2\ell)!\cdot(\ell-m)!}}$$

$$\times\left[\exp(-\mathrm{i}\varphi)\left(\mathrm{i}\cot\vartheta\frac{\partial}{\partial\varphi}-\frac{\partial}{\partial\vartheta}\right)\right]^{\ell-m}Y_{\ell,\ell}(\vartheta,\varphi). \tag{7.35}$$

If we substitute $Y_{\ell,\ell}(\vartheta,\varphi)$ from (7.34) into (7.35), we find

$$Y_{\ell,m}(\vartheta,\varphi)=\frac{(-)^m}{2^{\ell+1}\ell!}\sqrt{\frac{(2\ell+1)\cdot(\ell+m)!}{\pi\cdot(\ell-m)!}}\exp(\mathrm{i}m\varphi)$$

$$\times\left[\prod_{n=m+1}^{\ell}\left(n\cot\vartheta+\frac{d}{d\vartheta}\right)\right]\sin^\ell\vartheta$$

$$=\frac{(-)^m}{2^{\ell+1}\ell!}\sqrt{\frac{(2\ell+1)\cdot(\ell+m)!}{\pi\cdot(\ell-m)!}}\exp(\mathrm{i}m\varphi)$$

$$\times\left[\prod_{n=m+1}^{\ell}\left(\sin^{-n}\vartheta\frac{d}{d\vartheta}\sin^n\vartheta\right)\right]\sin^\ell\vartheta$$

$$=\frac{(-)^\ell}{2^{\ell+1}\ell!}\sqrt{\frac{(2\ell+1)\cdot(\ell+m)!}{\pi\cdot(\ell-m)!}}\exp(\mathrm{i}m\varphi)$$

$$\times\sin^{-m}\vartheta\frac{d^{\ell-m}}{d(\cos\vartheta)^{\ell-m}}\sin^{2\ell}\vartheta. \tag{7.36}$$

Equations (7.35) or (7.36) provide a solution to the problem to construct the spherical harmonic functions. However, it is common to make the connection to orthogonal polynomials in the interval $[-1,1]$. If we use the following equation for the associated Legendre polynomials,

$$P_\ell^m(x)=(-)^m\frac{(\ell+m)!}{2^\ell\ell!\cdot(\ell-m)!}\left(1-x^2\right)^{-m/2}\frac{d^{\ell-m}}{dx^{\ell-m}}\left(x^2-1\right)^\ell,$$

we can also write

$$Y_{\ell,m}(\vartheta,\varphi)=(-)^m\sqrt{\frac{(2\ell+1)\cdot(\ell-m)!}{4\pi\cdot(\ell+m)!}}\exp(\mathrm{i}m\varphi)P_\ell^m(\cos\vartheta). \tag{7.37}$$

The identity

$$(x^2-1)^{m/2}\sqrt{\frac{(\ell-m)!}{(\ell+m)!}}\frac{d^{\ell+m}}{dx^{\ell+m}}\left(x^2-1\right)^{\ell}$$

$$=\left(x^2-1\right)^{-m/2}\sqrt{\frac{(\ell+m)!}{(\ell-m)!}}\frac{d^{\ell-m}}{dx^{\ell-m}}\left(x^2-1\right)^{\ell}$$

implies

$$Y_{\ell,-m}(\vartheta,\varphi)=(-)^m Y^+_{\ell,m}(\vartheta,\varphi).$$

The spherical harmonic functions provide an orthonormal basis for functions on the sphere. The completeness relations are

$$\int_0^{2\pi}d\varphi\int_0^{\pi}d\vartheta\,\sin\vartheta\,Y_{\ell,m}(\vartheta,\varphi)Y^+_{\ell',m'}(\vartheta,\varphi)=\delta_{\ell,\ell'}\delta_{m,m'} \tag{7.38}$$

and

$$\sum_{\ell=0}^{\infty}\sum_{m=-\ell}^{\ell}Y_{\ell,m}(\vartheta,\varphi)Y^+_{\ell,m}(\vartheta',\varphi')=\delta(\hat{\boldsymbol{r}}-\hat{\boldsymbol{r}}')$$
$$=\delta(\cos\vartheta-\cos\vartheta')\delta(\varphi-\varphi'). \tag{7.39}$$

In bra-ket notation we can write these completeness relations for functions on the sphere as

$$\int_0^{\pi}d\vartheta\int_0^{2\pi}d\varphi\,\sin\vartheta\,|\vartheta,\varphi\rangle\langle\vartheta,\varphi|=\sum_{\ell=0}^{\infty}\sum_{m=-\ell}^{\ell}|\ell,m\rangle\langle\ell,m|=1, \tag{7.40}$$

i.e. the spherical harmonics $Y_{\ell,m}(\vartheta,\varphi)=\langle\vartheta,\varphi|\ell,m\rangle$ discretize functions on the sphere in the same sense as the Fourier monomials discretize functions on a circle or a finite interval (Sec. 10.1 contains a brief review of Fourier monomials on a finite interval). Knowledge of a continuous function $\langle\vartheta,\varphi|f\rangle$ on the sphere is equivalent to knowing the enumerably many numbers $\langle\ell,m|f\rangle$.

The lowest order spherical harmonics are

$$Y_{0,0}(\vartheta,\varphi)=\frac{1}{\sqrt{4\pi}},\quad Y_{1,0}(\vartheta,\varphi)=\sqrt{\frac{3}{4\pi}}\cos\vartheta,$$

$$Y_{1,\pm1}(\vartheta,\varphi)=\mp\sqrt{\frac{3}{8\pi}}\exp(\pm i\varphi)\sin\vartheta,\quad Y_{2,0}(\vartheta,\varphi)=\frac{1}{4}\sqrt{\frac{5}{\pi}}(3\cos^2\vartheta-1),$$

$$Y_{2,\pm1}(\vartheta,\varphi) = \mp\frac{1}{4}\sqrt{\frac{30}{\pi}}\exp(\pm i\varphi)\sin\vartheta\cos\vartheta,$$

$$Y_{2,\pm2}(\vartheta,\varphi) = \frac{1}{8}\sqrt{\frac{30}{\pi}}\exp(\pm 2i\varphi)\sin^2\vartheta.$$

7.6 Basic features of motion in central potentials

Separation of the wave function in equation (7.12)

$$\psi(\boldsymbol{r}) = \psi(r)Y_{\ell,m}(\vartheta,\varphi) \tag{7.41}$$

and use of

$$\boldsymbol{M}^2|\ell,m\rangle = \hbar^2\ell(\ell+1)|\ell,m\rangle$$

yields the radial Schrödinger equation

$$-\frac{\hbar^2}{2\mu}\frac{1}{r}\frac{d^2}{dr^2}r\psi(r) + \left(\frac{\hbar^2\ell(\ell+1)}{2\mu r^2} + V(r)\right)\psi(r) = E\psi(r). \tag{7.42}$$

The effective potential for the radial part of the relative motion of the two particles is therefore

$$V_{\text{eff}}(r) = V(r) + \frac{\hbar^2\ell(\ell+1)}{2\mu r^2},$$

with a "centrifugal barrier" term $\boldsymbol{M}^2/(2\mu r^2)$ just like in classical mechanics. The reason for this term is essentially the same consistency requirement as in classical mechanics. Classically, two particles with non-vanishing relative angular momentum $\boldsymbol{M}$ can never be in the same location, and the centrifugal barrier term simply reflects this property. Quantum mechanically, non-vanishing relative angular momentum $\boldsymbol{M}$ implies that the particular value $\psi(r=0)$ of the radial wave function must be suppressed, and it must be more strongly suppressed for larger $\boldsymbol{M}^2$.

Equation (7.42) is usually solved by the Sommerfeld method. In the first step one studies the asymptotic equations for small r and for large r, and keeps only the normalizable solutions or those solutions which approximate Fourier monomials in the asymptotic regions. In the next step one makes an *ansatz* for the full solution by multiplying the asymptotic solutions with a polynomial. Before we apply this method to the hydrogen atom, we will do something that one might find odd at first sight. The simplest case of a radially symmetric potential is $V = 0$, i.e. free motion. It is of interest both for scattering theory and for ionization or decay of rotationally symmetric systems to discuss free motion with defined angular momentum, when the wave function for a free particle has the form (7.41).

7.7 Free spherical waves: The free particle with sharp $M_z, \boldsymbol{M}^2$

The radial Schrödinger equation for a free particle with fixed angular momentum $M_z, \boldsymbol{M}^2$ and energy $E = \hbar^2 k^2/2\mu$ is

$$-\frac{\hbar^2}{2\mu}\frac{1}{r}\frac{d^2}{dr^2}r\psi(r) + \frac{\hbar^2\ell(\ell+1)}{2\mu r^2}\psi(r) = E\psi(r),$$

or

$$\left(\frac{d^2}{dr^2} - \frac{\ell(\ell+1)}{r^2} + k^2\right) r\psi(r) = 0. \tag{7.43}$$

The regular solution for $\ell = 0$ is

$$\psi_{k,0}(r) = \sqrt{\frac{2}{\pi}}\frac{\sin(kr)}{kr},$$

where the pre-factor was determined from the normalization condition

$$\int_0^\infty dr\, r^2 \psi_{k,0}(r)\psi_{k',0}(r) = \frac{1}{kk'}\delta(k-k').$$

For the study of solutions $\psi_{k,\ell}(r)$ for higher ℓ, we observe that solutions of equation (7.43) for $kr \ll \sqrt{\ell(\ell+1)}$ are $\psi(r) \propto r^\ell$ or $\psi(r) \propto r^{-\ell-1}$. We will only study solutions which are regular for $r = 0$, i.e. for $kr \ll \sqrt{\ell(\ell+1)}$ our solutions must approximate r^ℓ. Therefore we substitute $\psi_{k,\ell}(r) = r^\ell f_{k,\ell}(r)$ into equation (7.43),

$$\left(\frac{d^2}{dr^2} + \frac{2}{r}(\ell+1)\frac{d}{dr} + k^2\right) f_{k,\ell}(r) = \left(\frac{1}{r}\frac{d^2}{dr^2}r + \frac{2\ell}{r}\frac{d}{dr} + k^2\right) f_{k,\ell}(r) = 0.$$

It is useful to write this as

$$\left(A_r B_r + 2\ell A_r + k^2\right) f_{k,\ell}(r) = 0 \tag{7.44}$$

with operators

$$A_r = \frac{1}{r}\frac{d}{dr}, \quad B_r = \frac{d}{dr}r.$$

These operators satisfy the commutation relation

$$[A_r, B_r] = 2A_r,$$

and this implies

$$A_r\left(A_r B_r + k^2\right) = A_r\left(B_r A_r + 2A_r + k^2\right) = \left(A_r B_r + 2A_r + k^2\right) A_r,$$

$$A_r^\ell\left(A_r B_r + k^2\right) = \left(A_r B_r + 2\ell A_r + k^2\right) A_r^\ell. \tag{7.45}$$

This yields

$$f_{k,\ell}(r) \propto \frac{1}{r}\frac{d}{dr}f_{k,\ell-1}(r) \propto \left(\frac{1}{r}\frac{d}{dr}\right)^{\ell} f_{k,0}(r) = \sqrt{\frac{2}{\pi}}\left(\frac{1}{r}\frac{d}{dr}\right)^{\ell}\frac{\sin(kr)}{kr}, \tag{7.46}$$

$$\psi_{k,\ell}(r) \propto \sqrt{\frac{2}{\pi}}r^{\ell}\left(\frac{1}{r}\frac{d}{dr}\right)^{\ell}\frac{\sin(kr)}{kr} = (-)^{\ell}k^{\ell}\sqrt{\frac{2}{\pi}}j_{\ell}(kr). \tag{7.47}$$

Here we used the definition of the spherical Bessel functions

$$j_{\ell}(x) = (-x)^{\ell}\left(\frac{1}{x}\frac{d}{dx}\right)^{\ell}\frac{\sin x}{x} = \sqrt{\frac{\pi}{2x}}J_{\ell+\frac{1}{2}}(x).$$

The asymptotic expansion of $j_{\ell}(x)$ is

$$j_{\ell}(x)\Big|_{x\gg 1} \approx \frac{1}{x}\sin\left(x - \frac{\pi\ell}{2}\right).$$

Therefore the properly normalized radial eigenfunctions are

$$\psi_{k,\ell}(r) = \sqrt{\frac{2}{\pi}}\mathrm{i}^{\ell}j_{\ell}(kr) = \sqrt{\frac{2}{\pi}}\left(\frac{r}{\mathrm{i}k}\right)^{\ell}\left(\frac{1}{r}\frac{d}{dr}\right)^{\ell}\frac{\sin(kr)}{kr}, \tag{7.48}$$

and the free spherical waves with sharp angular momenta $\boldsymbol{M}^2$, M_z are

$$\langle \boldsymbol{r}|k,\ell,m\rangle = \sqrt{\frac{2}{\pi}}\mathrm{i}^{\ell}j_{\ell}(kr)Y_{\ell,m}(\vartheta,\varphi) = \frac{\mathrm{i}^{\ell}}{\sqrt{kr}}J_{\ell+\frac{1}{2}}(kr)Y_{\ell,m}(\vartheta,\varphi). \tag{7.49}$$

Our conventions for the phase and the normalization of the radial wave function are motivated by the expansion of plane waves in terms of spherical harmonics. If we define

$$\langle \boldsymbol{k}|k',\ell,m\rangle = \frac{1}{kk'}\delta(k-k')Y_{\ell,m}(\hat{\boldsymbol{k}}),$$

we automatically get the expansion of plane waves in terms of spherical harmonics,

$$\begin{aligned}\langle \boldsymbol{r}|\boldsymbol{k}\rangle &= \frac{1}{\sqrt{2\pi}^3}\exp(\mathrm{i}\boldsymbol{k}\cdot\boldsymbol{r}) = \sqrt{\frac{2}{\pi}}\sum_{\ell=0}^{\infty}\sum_{m=-\ell}^{\ell}\mathrm{i}^{\ell}j_{\ell}(kr)Y_{\ell,m}(\hat{\boldsymbol{r}})Y^{+}_{\ell,m}(\hat{\boldsymbol{k}}) \\ &= \frac{1}{\sqrt{kr}}\sum_{\ell=0}^{\infty}\sum_{m=-\ell}^{\ell}\mathrm{i}^{\ell}J_{\ell+\frac{1}{2}}(kr)Y_{\ell,m}(\hat{\boldsymbol{r}})Y^{+}_{\ell,m}(\hat{\boldsymbol{k}}).\end{aligned} \tag{7.50}$$

This expansion is also particularly useful for $\exp(\mathrm{i}kz)$. We have $P_{\ell}^{m}(1) = \delta_{m,0}$ and therefore $Y_{\ell,m}(\boldsymbol{e}_z) = Y_{\ell,m}(\vartheta = 0) = \sqrt{(2\ell+1)/4\pi}$. This yields

$$\exp(\mathrm{i}kz) = \sum_{\ell=0}^{\infty}(2\ell+1)\mathrm{i}^{\ell}j_{\ell}(kr)P_{\ell}(\cos\vartheta). \tag{7.51}$$

The radial wave functions (7.48) of the free spherical waves (7.49) satisfy completeness relations on the half-line

$$\int_0^\infty dr\,r^2\,\psi_{k,\ell}(r)\psi_{k',\ell}(r) = \frac{1}{k^2}\delta(k-k'),$$

$$\int_0^\infty dk\,k^2\,\psi_{k,\ell}(r)\psi_{k,\ell}(r') = \frac{1}{r^2}\delta(r-r'). \qquad (7.52)$$

If our discussion above does not refer to motion of a single particle with mass μ, but to relative motion of two non-interacting particles at locations

$$\boldsymbol{x}_1 = \boldsymbol{R} + \frac{m_2}{m_1+m_2}\boldsymbol{r}, \quad \boldsymbol{x}_2 = \boldsymbol{R} - \frac{m_1}{m_1+m_2}\boldsymbol{r}$$

we can write a full two-particle wave function with sharp angular momentum quantum numbers for the relative motion as

$$\langle \boldsymbol{R},\boldsymbol{r}|\boldsymbol{K},k,\ell,m\rangle = \frac{\mathrm{i}^\ell}{2\pi^2}\exp(\mathrm{i}\boldsymbol{K}\cdot\boldsymbol{R})j_\ell(kr)Y_{\ell,m}(\hat{\boldsymbol{r}}),$$

or we could also require sharp angular momentum quantum numbers L, M for the center or mass motion[5],

$$\langle \boldsymbol{R},\boldsymbol{r}|K,L,M,k,\ell,m\rangle = \frac{2}{\pi}\mathrm{i}^{L+\ell}j_L(KR)j_\ell(kr)Y_{L,M}(\hat{\boldsymbol{R}})Y_{\ell,m}(\hat{\boldsymbol{r}}).$$

7.8 Bound energy eigenstates of the hydrogen atom

The solution for the hydrogen atom was reported by Schrödinger in 1926 in the same paper where he introduced the time-independent Schrödinger equation[6].

We recall that separation of the wave function in equation (7.12)

$$\psi(\boldsymbol{r}) = \psi(r)Y_{\ell,m}(\vartheta,\varphi) \qquad (7.53)$$

and use of $\boldsymbol{M}^2|\ell,m\rangle = \hbar^2\ell(\ell+1)|\ell,m\rangle$ yields the radial Schrödinger equation

$$-\frac{\hbar^2}{2\mu}\frac{1}{r}\frac{d^2}{dr^2}r\psi(r) + \left(\frac{\hbar^2\ell(\ell+1)}{2\mu r^2} - \frac{e^2}{4\pi\epsilon_0 r}\right)\psi(r) = E\psi(r), \qquad (7.54)$$

[5]...or we could use total angular momentum, i.e. quantum numbers $K, k, j \in \{|L-\ell|,\ldots,L+\ell\}, m_j = M+m, L, \ell$.

[6]E. Schrödinger, Annalen Phys. 384, 361 (1926).

where the attractive Coulomb potential between charges e and $-e$ has been inserted. This yields asymptotic equations for small r,

$$-r^2\frac{d^2}{dr^2}r\psi(r) + \ell(\ell+1)r\psi(r) = 0, \tag{7.55}$$

and for large r,

$$-\frac{d^2}{dr^2}r\psi(r) = \frac{2\mu E}{\hbar^2}r\psi(r). \tag{7.56}$$

The Euler type differential equation (7.55) has basic solutions $r\psi(r) = Ar^{\ell+1} + Br^{-\ell}$, but with $\ell \geq 0$ only the first solution $r\psi(r) \propto r^{\ell+1}$ will yield a finite probability density $|\psi(\boldsymbol{r})|^2$ near the origin.

The normalizable solution of (7.56) for $E < 0$ is

$$r\psi(r) \propto \exp\!\left(-\sqrt{-2\mu E}r/\hbar\right). \tag{7.57}$$

We combine the asymptotic solutions with a polynomial $w(r) = \sum_{\nu\geq 0} c_\nu r^\nu$,

$$r\psi(r) = r^{\ell+1}w(r)\exp(-\kappa r), \quad \kappa = \sqrt{-2\mu E}r/\hbar.$$

Substitution in (7.54) yields the condition

$$r\frac{d^2}{dr^2}w(r) + 2(\ell+1-\kappa r)\frac{d}{dr}w(r) + \left(\frac{\mu e^2}{2\pi\epsilon_0\hbar^2} - 2\kappa(\ell+1)\right)w(r) = 0,$$

which in turn yields a recursion relation for the coefficients in the polynomial $w(r)$,

$$c_{\nu+1} = c_\nu\frac{2\kappa(\nu+\ell+1) - \frac{\mu e^2}{2\pi\epsilon_0\hbar^2}}{(\nu+1)(\nu+2\ell+2)}. \tag{7.58}$$

Normalizability of the solution requires termination of the polynomial $w(r)$ with a maximal power $N \equiv \max(\nu) \geq 0$ of r, i.e. $c_{N+1} = 0$ and therefore

$$\kappa \equiv \frac{\sqrt{-2\mu E}}{\hbar} = \frac{\mu e^2}{4\pi\epsilon_0\hbar^2(N+\ell+1)}. \tag{7.59}$$

This implies energy quantization for the bound states in the form

$$E_n = -\frac{\mu e^4}{32\pi^2\epsilon_0^2\hbar^2}\frac{1}{n^2} = -\frac{\alpha^2}{2}\mu c^2\frac{1}{n^2} \tag{7.60}$$

with the principal quantum number $n \equiv N + \ell + 1$. Note that $N \geq 0$ implies the relation $n \geq \ell + 1$ between the principal and the magnetic quantum number.

We used the definition

$$\alpha = \frac{e^2}{4\pi\epsilon_0\hbar c} = 7.29735\ldots \times 10^{-3} = \frac{1}{137.036\ldots}. \tag{7.61}$$

of Sommerfeld's *fine structure constant* in (7.60).

We will also use equation (7.59) in the form $\kappa = (na)^{-1}$ with the Bohr radius

$$a \equiv \frac{4\pi\epsilon_0\hbar^2}{\mu e^2} = \frac{\hbar}{\alpha\mu c}. \tag{7.62}$$

The recursion relation is then

$$c_{\nu+1} = c_\nu \frac{2}{na}\frac{\nu+\ell+1-n}{(\nu+1)(\nu+2\ell+2)}, \quad 0 \leq \nu \leq N \equiv n-\ell-1. \tag{7.63}$$

This defines all coefficients c_ν in $w(r)$ in terms of the coefficient c_0, which finally must be determined from normalization. The factor $2/na$ in the recursion relation will generate a power $(2/na)^\nu$ in c_ν, such that $w(r)$ will be a polynomial in $2r/na$. The factor $(\nu+1)^{-1}$ will generate a factor $1/\nu!$ in c_ν, and the factor $(\nu+\alpha)/(\nu+\beta)$ with $\alpha = \ell+1-n$, $\beta = 2\ell+2$ will finally yield a polynomial of the form

$$w(r) = c_0\left[1 + \frac{\alpha}{\beta}\frac{2r}{na} + \frac{1}{2!}\frac{\alpha(\alpha+1)}{\beta(\beta+1)}\left(\frac{2r}{na}\right)^2 \right.$$
$$\left. + \frac{1}{3!}\frac{\alpha(\alpha+1)(\alpha+2)}{\beta(\beta+1)(\beta+2)}\left(\frac{2r}{na}\right)^3 + \ldots\right] = c_0 \times {}_1F_1(\alpha;\beta;2r/na).$$

As indicated in this equation, the series for $c_0 = 1$ defines the confluent hypergeometric function ${}_1F_1(\alpha;\beta;x) \equiv M(\alpha;\beta;x)$ (also known as Kummer's function [1]). For $-\alpha \in \mathbb{N}_0$ and $\beta \in \mathbb{N}$ this function can also be expressed as an associated Laguerre polynomial. The normalized radial wave functions can then be written as

$$\psi_{n,\ell}(r) = \frac{2}{n^2}\sqrt{\frac{(n+\ell)!}{(n-\ell-1)!a^3}}\frac{{}_1F_1(-n+\ell+1;2\ell+2;2r/na)}{(2\ell+1)!}$$
$$\times\left(\frac{2r}{na}\right)^\ell \exp\left(-\frac{r}{na}\right)$$
$$= \frac{2}{n^2}\sqrt{\frac{(n-\ell-1)!}{(n+\ell)!a^3}}\left(\frac{2r}{na}\right)^\ell L^{2\ell+1}_{n-\ell-1}\left(\frac{2r}{na}\right)\exp\left(-\frac{r}{na}\right). \tag{7.64}$$

Substitution of the explicit series representation for $w(r)$ shows that the radial wave functions are products of a polynomial in $2r/na$ of order $n-1$ with $n-\ell$ terms, multiplied with the exponential function $\exp(-r/na)$,

$$\psi_{n,\ell}(r) = \frac{2}{n^2}(-)^\ell\sqrt{\frac{(n+\ell)!(n-\ell-1)!}{a^3}}\exp\left(-\frac{r}{na}\right)$$
$$\times\sum_{k=\ell}^{n-1}\frac{(-2r/na)^k}{(k-\ell)!(n-k-1)!(k+\ell+1)!}. \tag{7.65}$$

The representation (7.64) in terms of the associated Laguerre polynomials differs from older textbook representations by a factor $(n+\ell)!$ due to the modern definition of the normalization of associated Laguerre polynomials,

$$L_n^m(x) = \frac{(-)^m}{(n+m)!}\frac{d^m}{dx^m}\left(\exp(x)\frac{d^{n+m}}{dx^{n+m}}\left[x^{n+m}\exp(-x)\right]\right)$$
$$= \frac{(m+n)!}{n!\cdot m!}{}_1F_1(-n;m+1;x),$$

which is also used in symbolic calculation programs. The normalization follows from

$$\int_0^\infty dx\,\exp(-x)x^{m+1}[L_n^m(x)]^2 = (2n+m+1)\frac{(n+m)!}{n!}, \tag{7.66}$$

but their standard orthogonality relation is

$$\int_0^\infty dx\,\exp(-x)x^m L_n^m(x)L_{n'}^m(x) = \frac{(n+m)!}{n!}\delta_{n,n'}. \tag{7.67}$$

Since they appear as eigenstates of the hydrogen Hamiltonian, the normalized bound radial wave functions must satisfy the orthogonality relation

$$\int_0^\infty dr\,r^2\,\psi_{n,\ell}(r)\psi_{n',\ell}(r) = \delta_{n,n'}. \tag{7.68}$$

This implies that the associated Laguerre polynomials must also satisfy a peculiar additional orthogonality relation which generalizes (7.66),

$$\int_0^\infty dx\,\exp\!\left(-\frac{(n+n'+m+1)x}{(2n+m+1)(2n'+m+1)}\right)x^{m+1}L_n^m\!\left(\frac{x}{2n+m+1}\right)$$
$$\times L_{n'}^m\!\left(\frac{x}{2n'+m+1}\right) = (2n+m+1)^{m+3}\frac{(n+m)!}{n!}\delta_{n,n'}. \tag{7.69}$$

Squares $\psi_{n,\ell}^2(r)$ of the radial wave functions are plotted for low lying values of n and ℓ in Figures 7.1–7.6.

For the meaning of the radial wave function, recall that the full three-dimensional wave function is

$$\psi_{n,\ell,m}(\boldsymbol{r}) = \psi_{n,\ell}(r)Y_{\ell,m}(\vartheta,\varphi).$$

This implies that $\psi_{n,\ell}^2(r)$ is a radial profile of the probability density $|\psi_{n,\ell,m}(\boldsymbol{r})|^2$ to find the particle (or rather the quasiparticle which describes relative motion in the hydrogen atom) in the location $\boldsymbol{r}$, but note that in each particular direction (ϑ,φ) the

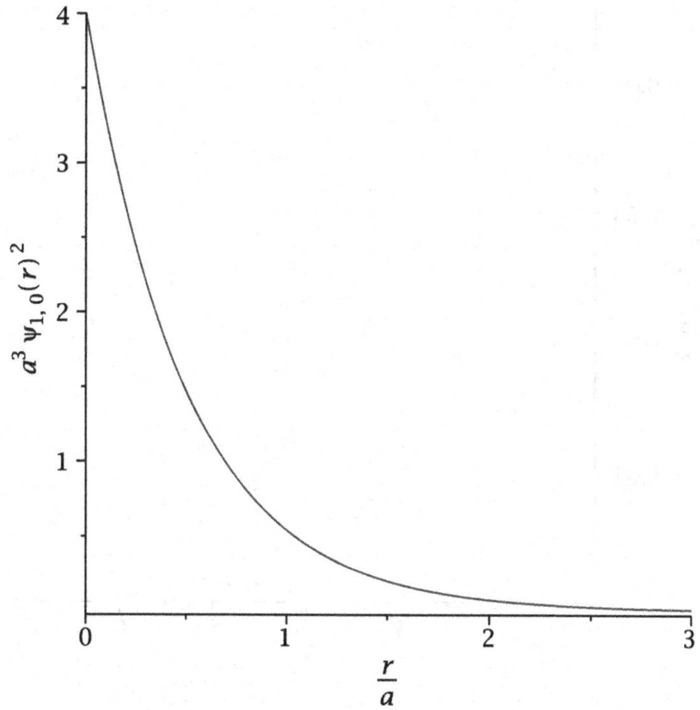

Fig. 7.1 The function $a^3\psi_{1,0}^2(r)$

radial profile is scaled by the factor $Y_{\ell,m}^2(\vartheta,\varphi)$ to give the actual radial profile of the probability density in that direction. Furthermore, note that the probability density for finding the electron-proton pair with separation between r and $r+dr$ is

$$\int_0^\pi d\vartheta \int_0^{2\pi} d\varphi\, r^2 \sin\vartheta\; |\psi_{n,\ell,m}(\boldsymbol{r})|^2 = r^2\psi_{n,\ell}^2(r).$$

The function $\psi_{n,\ell}^2(r)$ is proportional to the radial probability density in fixed directions, while $r^2\psi_{n,\ell}^2(r)$ samples the full spherical shell between r and $r+dr$ in all directions, and therefore the latter probability density is scaled by the geometric size factor r^2 for thin spherical shells.

Nowadays radial expectation values

$$\langle r^h\rangle_{n,\ell} = \int_0^\infty dr\, r^{h+2}\psi_{n,\ell}^2(r)$$

are readily calculated with symbolic computation programs. One finds in particular

$$\langle r\rangle_{n,\ell} = \frac{3n^2-\ell(\ell+1)}{2}a, \quad \langle r^2\rangle_{n,\ell} = \frac{n^2}{2}[5n^2+1-3\ell(\ell+1)]a^2.$$

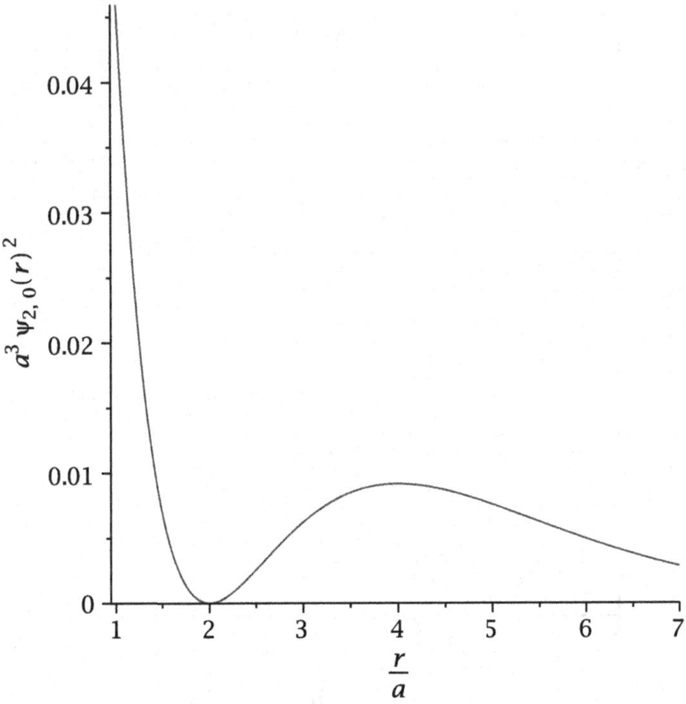

Fig. 7.2 The function $a^3\psi^2_{2,0}(r)$ for $r > a$

The resulting uncertainty in distance between the proton and the electron

$$(\Delta r)_{n,\ell} = \langle r^2\rangle_{n,\ell} - \langle r\rangle^2_{n,\ell} = \frac{a}{2}\sqrt{n^2(n^2+2) - \ell^2(\ell+1)^2}$$

is relatively large for most states in the sense that $(\Delta r/\langle r\rangle)_{n,\ell}$ is not small, except for large n states with large angular momentum. For example, we have $(\Delta r/\langle r\rangle)_{n,0} = \sqrt{1 + (2/n^2)/3} > 1/3$ but $(\Delta r/\langle r\rangle)_{n,n-1} = 1/\sqrt{2n+1}$. However, even for large n and ℓ, the particle could still have magnetic quantum number $m = 0$, whence its probability density would be uniformly spread over directions (ϑ, φ). This means that a hydrogen atom with sharp energy generically cannot be considered as consisting of a well localized electron near a well localized proton. This is just another illustration of the fact that simple particle pictures make no sense at the quantum level.

We also note from (7.64) or (7.65) that the bound eigenstates $\psi_{n,\ell,m}(\boldsymbol{r}) = \psi_{n,\ell}(r)Y_{\ell,m}(\vartheta, \varphi)$ have a typical linear scale

$$na = n\frac{4\pi\epsilon_0\hbar^2}{\mu Ze^2} \propto n\frac{\mu^{-1}}{Ze^2}. \tag{7.70}$$

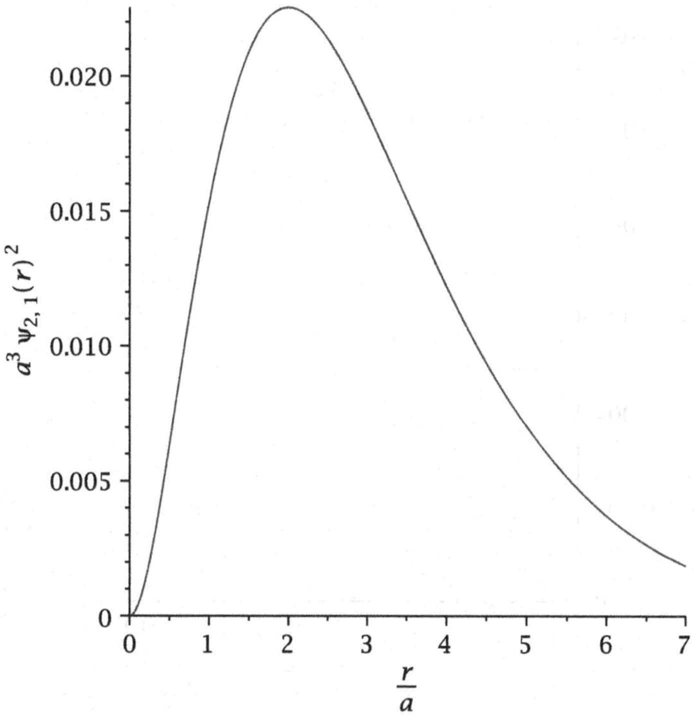

Fig. 7.3 The function $a^3\psi_{2,1}^2(r)$

Here we have generalized the definition of the Bohr radius a to the case of an electron in the field of a nucleus of charge Ze. Equation (7.70) is another example of the competition between the kinetic term $p^2/2\mu$ driving wave packets apart, and an attractive potential, here $V(r) = -Ze^2/4\pi\epsilon_0 r$, trying to collapse the wave function into a point. Metaphorically speaking, pressure from kinetic terms stabilizes the wave function. For given ratio of force constant Ze^2 and kinetic parameter μ^{-1} the attractive potential cannot compress the wave packet to sizes smaller than a, and therefore there is no way for the system to release any more energy. Superficially, there seems to exist a classical analog to the quantum mechanical competition between kinetic energy and attractive potentials in the Schrödinger equation. In classical mechanics, competition between centrifugal terms and attractive potentials can yield stable bound systems. However, the classical analogy is incomplete in a crucial point. The centrifugal term for $\ell \neq 0$ is also there in equation (7.54) exactly as in the classical Coulomb or Kepler problems. However, what stabilizes the wave function against core collapse in the crucial lowest energy case with $\ell = 0$ is the radial kinetic term, whereas in the classical case bound Coulomb or Kepler systems with vanishing angular momentum always collapse. To understand the quantum mechanical stabilization of atoms against collapse a little better, let us repeat equation (7.54) for $\ell = 0$ and nuclear charge Ze, and for low values of r,

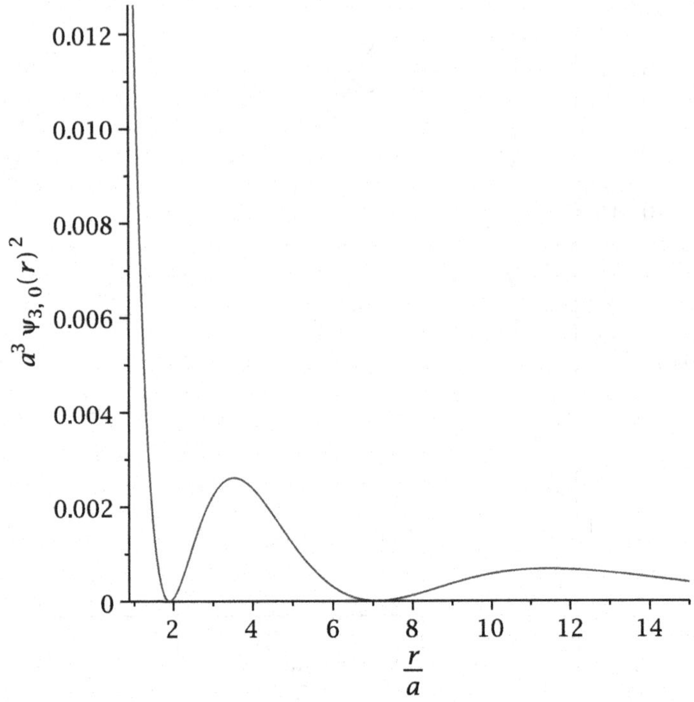

Fig. 7.4 The function $a^3\psi^2_{3,0}(r)$ for $r > a$

where we can assume $\psi(r) \neq 0$:

$$\frac{\hbar^2}{2\mu}\frac{1}{\psi(r)}\frac{d^2}{dr^2}r\psi(r) = -Er - \frac{Ze^2}{4\pi\epsilon_0}. \tag{7.71}$$

The radial probability amplitude $r\psi(r)$ must satisfy $\psi^{-1}(r)d^2(r\psi(r))/dr^2 < 0$ near the origin, to bend the function around to eventually yield $\lim_{r\to\infty} r\psi(r) = 0$, which is necessary for normalizability of $r^2\psi^2(r)$ on the half-axis $r > 0$. But near $r = 0$, the only term that bends the wave function in the right direction for normalizability is essentially the ratio Ze^2/μ^{-1},

$$\frac{1}{\psi(r)}\frac{d^2}{dr^2}r\psi(r) \simeq -\frac{Ze^2\mu}{2\pi\epsilon_0\hbar^2}.$$

If we want to concentrate more and more of the wave function near the origin $r \simeq 0$, we have to bend it around already very close to $r = 0$ to reach small values $ar^2\psi^2(r) \ll 1$ very early. But the only parameter that bends the wave function near the origin $r \simeq 0$ is the ratio between attractive force constant and kinetic parameter, Ze^2/μ^{-1}. This limits the minimal spatial extension of the wave function

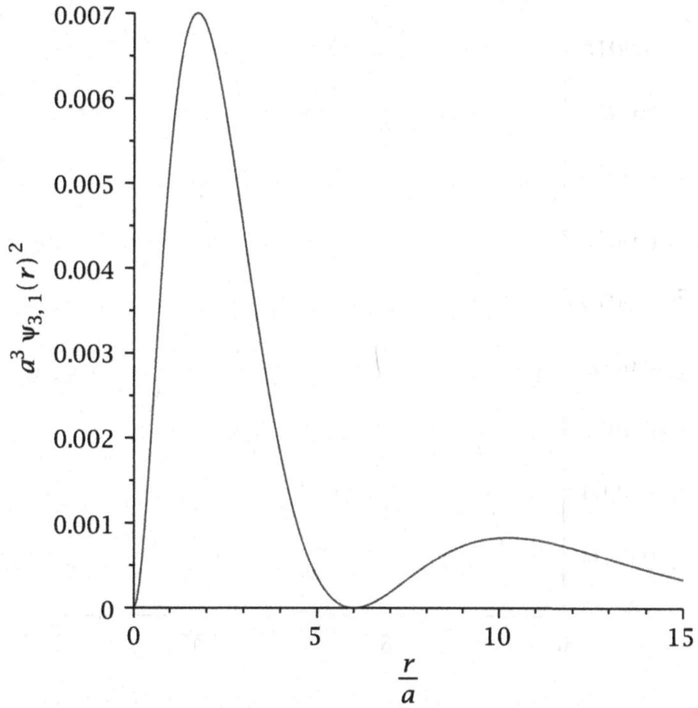

Fig. 7.5 The function $a^3\psi_{3,1}^2(r)$

and therefore prevents the classically inevitable core collapse in the bound Coulomb system with vanishing angular momentum $\boldsymbol{M} = \boldsymbol{0}$. In a nutshell, there is only so much squeezing of the wave function that Ze^2/μ^{-1} can do. See also Problem 7.15 for squeezing or stretching of a hydrogen atom near its ground state.

The radial probability amplitude $r\psi_{1,0}(r)$ for the ground state is plotted in Figure 7.7.

7.9 Spherical Coulomb waves

Now we assume $E > 0$. Recall that the asymptotic solutions for $r\psi(r)$ for $r \to 0$ were of the form $Ar^{\ell+1} + Br^{-\ell}$. Let us initially focus on the solutions which remain regular in the origin.

The symptotic behavior for large r *seems* to correspond to outgoing and incoming radial waves

$$r\psi_{\pm}(r) \to A_{\pm}\exp(\pm ikr), \quad k = \sqrt{2\mu E}/\hbar.$$

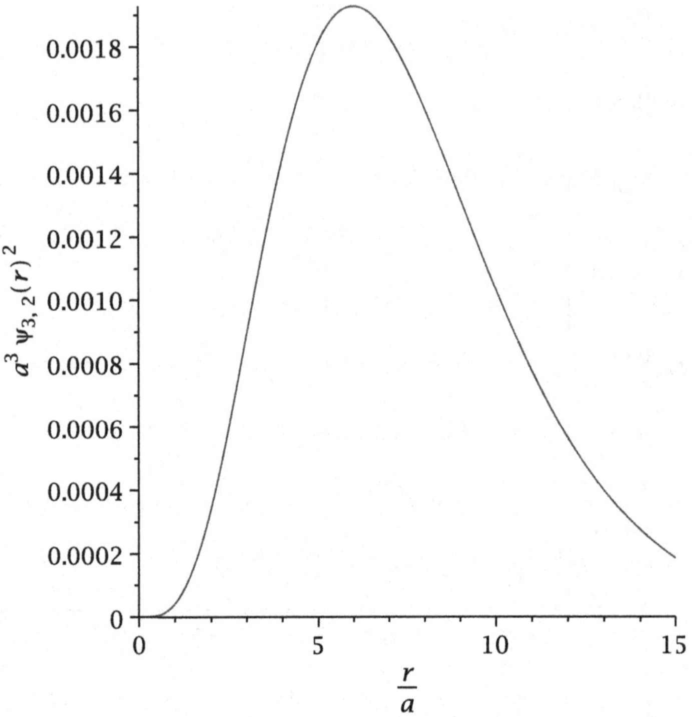

Fig. 7.6 The function $a^3\psi_{3,2}^2(r)$

Therefore we try an *ansatz*

$$r\psi_\pm(r) = w_\pm(r)r^{\ell+1}\exp(\pm ikr), \quad w_\pm(r) = \sum_{\nu\geq 0} c_{\pm,\nu}r^\nu.$$

Instead of the recursion relation (7.58) we now find

$$c_{\pm,\nu+1} = -c_{\pm,\nu}\frac{2}{\nu+1}\,\frac{\frac{1}{a}\pm ik(\nu+\ell+1)}{\nu+2\ell+2} \tag{7.72}$$

and therefore

$$\begin{aligned} w_\pm(r) &\propto 1 \mp \frac{\ell+1\mp\frac{\mathrm{i}}{ka}}{2\ell+2}2\mathrm{i}kr + \frac{\left(\ell+1\mp\frac{\mathrm{i}}{ka}\right)\left(\ell+2\mp\frac{\mathrm{i}}{ka}\right)}{(2\ell+2)(2\ell+3)}\frac{(2\mathrm{i}kr)^2}{2!} \\ &\quad\mp \frac{\left(\ell+1\mp\frac{\mathrm{i}}{ka}\right)\left(\ell+2\mp\frac{\mathrm{i}}{ka}\right)\left(\ell+3\mp\frac{\mathrm{i}}{ka}\right)}{(2\ell+2)(2\ell+3)(2\ell+4)}\frac{(2\mathrm{i}kr)^3}{3!} + \ldots \\ &= {}_1F_1(\ell+1\mp(\mathrm{i}/ka); 2\ell+2; \mp 2\mathrm{i}kr). \end{aligned}$$

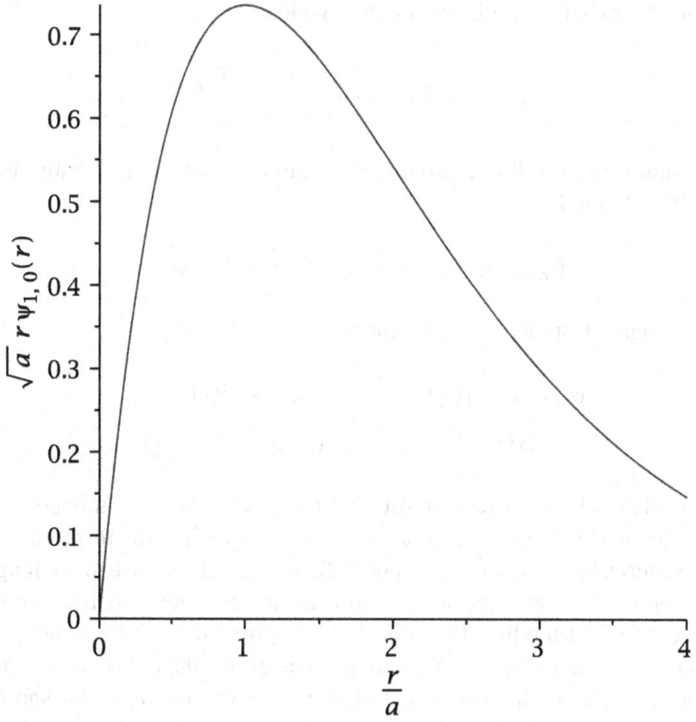

Fig. 7.7 The function $\sqrt{a}r\psi_{1,0}(r)$

However, the confluent hypergeometric function satisfies Kummer's identity

$$ {}_1F_1(\alpha;\beta;z) = \exp(z)\,{}_1F_1(\beta-\alpha;\beta;-z), $$

and this implies in particular

$$ \begin{aligned} &\exp(\mathrm{i}kr)\,{}_1F_1(\ell+1-(\mathrm{i}/ka);2\ell+2;-2\mathrm{i}kr) \\ &\quad= \exp(-\mathrm{i}kr)\,{}_1F_1(\ell+1+(\mathrm{i}/ka);2\ell+2;2\mathrm{i}kr), \end{aligned} $$

i.e. there is only one regular solution for given quantum numbers (k,ℓ), and it corresponds neither to an outgoing nor to an incoming spherical wave, but is apparently rather a superposition of incoming and outgoing waves. For applications in scattering theory one often also has to look at solutions which are irregular in the origin,

$$ r\psi(r) = r^{-\ell}v_\pm(r)\exp(\pm\mathrm{i}kr), \quad v_\pm(r) = \sum_{\nu\geq 0} d_{\pm,\nu}r^\nu. $$

In this case, the radial Schrödinger equation yields

$$d_{\pm,\nu+1} = \mp d_{\pm,\nu} \frac{2ik}{\nu+1} \frac{\nu - \ell \mp \frac{i}{ka}}{\nu - 2\ell}. \tag{7.73}$$

This recursion relation follows also directly from (7.72) with the substitution $\ell \to -\ell - 1$. The solution is

$$v_\pm(r) \propto {}_1F_1(-\ell \mp (i/ka); -2\ell; \mp 2ikr),$$

and we have again from Kummer's identity

$$\begin{aligned}&\exp(ikr)\,{}_1F_1(-\ell - (i/ka); -2\ell; -2ikr)\\ &= \exp(-ikr)\,{}_1F_1(-\ell + (i/ka); -2\ell; 2ikr),\end{aligned}$$

i.e. there is also only one irregular solution for given quantum numbers (k, ℓ), as expected. Regular and irregular solutions can be combined to form outgoing or incoming spherical waves, see e.g. [1] or [28]. This is relevant when the long range Coulomb potential is combined with a short range scattering potential, because the short range part will modify the short distance properties of the states and both the regular and irregular spherical Coulomb waves are then needed to model the asymptotic behavior of incoming and scattered waves far from the short range scattering potential. This is relevant for scattering in nuclear physics, when short range scattering is effected by nuclear forces.

However, for us the regular solutions are more interesting because together with the bound states $\psi_{n,\ell,m}(\boldsymbol{r}) = \psi_{n,\ell}(r)Y_{\ell,m}(\vartheta,\varphi)$, the regular unbound states $\psi_{k,\ell,m}(\boldsymbol{r}) = \psi_{k,\ell}(r)Y_{\ell,m}(\vartheta,\varphi)$ form a complete set in Hilbert space.

We use the normalization[7]

$$\begin{aligned}\psi_{k,\ell}(r) = \sqrt{\frac{2}{\pi}} \exp\!\left(\frac{\pi}{2ak}\right) \frac{\left|\Gamma\!\left(\ell+1+\frac{i}{ka}\right)\right|}{(2\ell+1)!}(2kr)^\ell \exp(-ikr)\\ \times {}_1F_1(\ell+1+(i/ka); 2\ell+2; 2ikr).\end{aligned} \tag{7.74}$$

With this normalization the regular spherical Coulomb waves become the free waves with sharp angular momentum (7.48) in the limit of vanishing Coulomb potential $e^2 \to 0 \;\Rightarrow\; a \to \infty$, or if the energy $E = \hbar^2k^2/2\mu$ of the spherical Coulomb waves is much larger than the binding energy $E_B = -E_1 = \hbar^2/2\mu a^2$ of hydrogen, $E \gg E_B \;\Rightarrow\; ka \gg 1$.

Apart from the normalization, the spherical Coulomb waves $\psi_{k,\ell}(r)$ become the radial bound state wave functions $\psi_{n,\ell}(r)$ through the substitution $ik \to (na)^{-1}$. This is expected since this substitution takes the positive energy Schrödinger equation into the negative energy Schrödinger equation.

[7] W. Gordon, Annalen Phys. 394, 1031 (1929); M. Stobbe, Annalen Phys. 399, 661 (1930), see also [3]. Gordon and Stobbe normalized in the k scale, i.e. to $\delta(k-k')$ instead of $\delta(k-k')/k^2$.

The regular spherical Coulomb waves satisfy the orthogonality properties (cf. (7.68))

$$\int_0^\infty dr\, r^2\, \psi_{k,\ell}(r)\psi_{k',\ell}(r) = \frac{1}{k^2}\delta(k-k'), \quad \int_0^\infty dr\, r^2\, \psi_{k,\ell}(r)\psi_{n,\ell}(r) = 0,$$

and together with the radial bound state wave functions they satisfy the completeness relation[8]

$$\sum_{n=\ell+1}^{\infty} \psi_{n,\ell}(r)\psi_{n,\ell}(r') + \int_0^\infty dk\, k^2 \psi_{k,\ell}(r)\psi_{k,\ell}(r') = \frac{1}{r^2}\delta(r-r').$$

Together with the completeness relation (7.39) for the spherical harmonics, this implies completeness of the regular hydrogen states,

$$\sum_{\ell=0}^{\infty}\sum_{m=-\ell}^{\ell}\left(\sum_{n=\ell+1}^{\infty} \psi_{n,\ell,m}(\boldsymbol{r})\psi^{+}_{n,\ell,m}(\boldsymbol{r}') + \int_0^\infty dk\, k^2 \psi_{k,\ell,m}(\boldsymbol{r})\psi^{+}_{k,\ell,m}(\boldsymbol{r}')\right)$$
$$= \delta(\boldsymbol{r}-\boldsymbol{r}'). \tag{7.75}$$

For calculations of transitions between free and bound states, e.g. for electron-proton recombination cross sections, one needs free eigenstates which are not radially symmetric but approximate plane waves at large separations. To construct such a state from the spherical Coulomb waves, we can use that equation (7.51) tells us the decomposition of the plane wave $\exp(ikz)$ in terms of the free states of sharp angular momentum (7.48),

$$\exp(ikz) = \sqrt{\frac{\pi}{2}}\sum_{\ell=0}^{\infty}(2\ell+1)P_\ell(\cos\vartheta)\psi^{(e^2=0)}_{k,\ell}(r).$$

However, one needs to adjust the phase factors in the sum if one wants to get an asymptotic superposition of plane waves and outgoing spherical waves. The superposition of spherical Coulomb waves[9] (7.74),

$$\langle \boldsymbol{r}|\boldsymbol{k}\rangle_{MG} = \sqrt{\frac{\pi}{2}}\sum_{\ell=0}^{\infty}(2\ell+1)P_\ell(\cos\vartheta)\frac{\Gamma\!\left(\ell+1+\frac{\mathrm{i}}{ka}\right)}{\left|\Gamma\!\left(\ell+1+\frac{\mathrm{i}}{ka}\right)\right|}\psi_{k,\ell}(r), \tag{7.76}$$

will correspond to a free energy eigenstate of hydrogen with energy $\hbar^2k^2/2\mu$ which up to logarithmic corrections approximates a superposition of a plane wave $\exp(ikz)$ with outgoing radial waves.

[8]See e.g. N. Mukunda, Amer. J. Phys. 46, 910 (1978).

[9]N.F. Mott, Proc. Roy. Soc. London A 118, 542 (1928); W. Gordon, Z. Phys. 48, 180 (1928).

7.10 Problems

7.1. Show that the transformations (7.3, 7.4) imply

$$\boldsymbol{K}\cdot\boldsymbol{R}+\boldsymbol{k}\cdot\boldsymbol{r}=\boldsymbol{k}_1\cdot\boldsymbol{x}_1+\boldsymbol{k}_2\cdot\boldsymbol{x}_2. \tag{7.77}$$

What is then the proper boundary condition for $\lim_{r\to\infty}\psi(\boldsymbol{r})$ for an unbound 2-particle state of the form (7.7) if we assume that the two particles have asymptotic momenta $\hbar\boldsymbol{k}_1$ and $\hbar\boldsymbol{k}_2$ for large separation?

7.2. How large is the minimal value of the product $\Delta M_x \Delta M_y$ of uncertainties of angular momentum components in a state $|\ell, m_\ell\rangle$?

7.3. Why does equation (7.45) imply that there is no other choice but (7.46) for the regular solution $f_{k,\ell}(r)$ of (7.44)?

7.4. A simple spherical model for a color center or a quantum dot consists of an electron confined to a sphere of radius R. Inside the sphere the electron can move freely because the potential energy vanishes there, $V(r) = 0$ for $r < R$. The wave function in the sphere for given angular momentum quantum numbers will therefore have the form $\psi(\boldsymbol{r}) \propto j_\ell(kr)Y_{\ell,m}(\vartheta,\varphi)$.

Which energy quantization conditions will we get from the condition that the wave function vanishes for $r \geq R$? How large is the radius R if the electron absorbs photons of energy 2.3 eV?

Zeros $x_{n,\ell}$ of spherical Bessel functions, $j_\ell(x_{n,\ell}) = 0$, $n = 1, 2, \ldots$ can be found e.g. in Chapter 10 of [1].

Which relation between R and lattice constant d follows from Mollwo's relation $\nu d^2 = 5.02 \times 10^{-5}\,\mathrm{m^2\,Hz}$?

In hindsight, color centers could be considered as the first realization of atomic scale quantum dots.

7.5. Show that the radial density profile $\psi^2_{n,\ell}(r)$ for bound states of hydrogen has maxima at the extrema of the radial wave function $\psi_{n,\ell}(r)$.

Remark. $\psi_{n,\ell}(r)$ and $d\psi_{n,\ell}(r)/dr$ have no common zeros, because this would contradict the radial Schrödinger equation.

7.6. Calculate the radius $r_{max}^{(n,n-1)}$ where the radial wave function $\psi_{n,n-1}(r)$ has a maximum. Compare your result to $\langle r\rangle_{n,n-1} \pm (\Delta r)_{n,n-1}$.

7.7. For $n \geq 2$ calculate the radius $r_{max}^{(n,n-2)}$ where the radial wave function $\psi_{n,n-2}(r)$ has a maximum.

7.8. As a rule of thumb, quantum systems tend to approach classical behavior for large quantum numbers. We have seen that for large quantum number n the radial wave function $\psi_{n,n-1}(r)$ is localized in a spherical shell $\langle r\rangle_{n,n-1} \pm (\Delta r)_{n,n-1}$ which is "thin" in the sense of $(\Delta r/\langle r\rangle)_{n,n-1} = 1/\sqrt{2n+1} \to 0$.

For sharp energy $E_{n,\ell,m}$, could we ever hope to find an approximately localized electron in a hydrogen atom?

7.9. Calculate the radial expectation values $\langle r^{-1}\rangle_{n,\ell}$, $\langle r^{-2}\rangle_{n,\ell}$ and $\langle r^{-3}\rangle_{n,\ell}$. These expectation values are relevant for interaction energies within the atom.

7.10. We have seen how the expectation value $\langle r\rangle$ for the separation between the electron and the proton depends on the quantum numbers n and ℓ. How large are the corresponding expectation values for the distances of the two particles from the center of mass of the hydrogen atom?

7.11. An electric field along the z axis shifts the potential energy of a hydrogen atom by a perturbation $\Delta V \propto z$. For which combinations of quantum numbers are the matrix elements $\langle n_1', n_2', m'|z|n_1, n_2, m\rangle$ different from zero?

7.12. Calculate the probability density to find the momentum $\boldsymbol{p}$ in the relative motion in a hydrogen atom in its ground state.

7.13. Calculate the probability densities to find the momentum $\boldsymbol{p}$ in the relative motion in a hydrogen atom for the $n = 2$ states $|2, \ell, m_\ell\rangle$.

7.14. We cannot construct energy eigenstates of the hydrogen atom which separate in the coordinates $\boldsymbol{x}_e$ and $\boldsymbol{x}_p$ of the electron and the proton. If we want to have a representation which factorizes in electron and proton wave functions, the best we can do is to expand the energy eigenstates $\Psi_{\boldsymbol{K},n,\ell,m}(\boldsymbol{R},\boldsymbol{r})$ in terms of complete sets of functions $f_e(\boldsymbol{x}_e)g_p(\boldsymbol{x}_p)$ which arise from complete sets of functions $f(\boldsymbol{x})$, $g(\boldsymbol{x})$ for single particle states. Expand the ground state of a hydrogen atom with center of mass momentum $\hbar\boldsymbol{K}$,

$$\Psi_{\boldsymbol{K},1,0,0}(\boldsymbol{R},\boldsymbol{r}) = \frac{1}{\sqrt{2a}^3\pi^2}\exp\!\left(\mathrm{i}\boldsymbol{K}\cdot\boldsymbol{R} - \frac{r}{a}\right)$$

in terms of the complete basis of factorized plane electron and proton waves,

$$\langle \boldsymbol{x}_e, \boldsymbol{x}_p|\boldsymbol{k}_e, \boldsymbol{k}_p\rangle = \frac{1}{(2\pi)^3}\exp\!\left(\mathrm{i}\boldsymbol{k}_e\cdot\boldsymbol{x}_e + \mathrm{i}\boldsymbol{k}_p\cdot\boldsymbol{x}_p\right).$$

Solution. Fourier transformation of

$$\langle \boldsymbol{x}_e, \boldsymbol{x}_p|1,0,0;\boldsymbol{K}\rangle = \frac{1}{\sqrt{2a}^3\pi^2}\exp\!\left(\mathrm{i}\boldsymbol{K}\cdot\frac{m_e\boldsymbol{x}_e + m_p\boldsymbol{x}_p}{m_e+m_p} - \frac{|\boldsymbol{x}_e - \boldsymbol{x}_p|}{a}\right)$$

yields

$$\langle \boldsymbol{k}_e, \boldsymbol{k}_p|1,0,0;\boldsymbol{K}\rangle = \frac{\sqrt{2a}^3}{\pi}\delta(\boldsymbol{K} - \boldsymbol{k}_e - \boldsymbol{k}_p)$$
$$\times\left[1 + \frac{a^2}{(m_e+m_p)^2}\left|m_p\boldsymbol{k}_e - m_e\boldsymbol{k}_p\right|^2\right]^{-2}, \qquad (7.78)$$

i.e. the decomposition of the 1s hydrogen state with center of mass momentum $\hbar\boldsymbol{K}$ in terms of electron and proton plane wave states is

$$|1,0,0;\boldsymbol{K}\rangle = \frac{\sqrt{2a}^3}{\pi}\int d^3k_e\,|\boldsymbol{k}_e\rangle \otimes |\boldsymbol{K}-\boldsymbol{k}_e\rangle \times \left[1 + a^2\left|\boldsymbol{k}_e - \frac{m_e}{m_e+m_p}\boldsymbol{K}\right|^2\right]^{-2}. \tag{7.79}$$

Hint for the Fourier transformation: It is advantageous to use center of mass and relative coordinates for the calculation of the Fourier integrals, $d^3\boldsymbol{x}_e \wedge d^3\boldsymbol{x}_p = d^3\boldsymbol{r} \wedge d^3\boldsymbol{R}$.

7.15. Suppose we force a hydrogen atom into a 1s type state

$$\psi(\boldsymbol{r},t) = \frac{1}{\sqrt{\pi b^3}}\exp(-r/b)\exp(-\mathrm{i}E_1 t/\hbar), \tag{7.80}$$

where

$$E_1 = -\frac{\mu e^4}{32\pi^2\epsilon_0^2\hbar^2} = -\frac{e^2}{8\pi\epsilon_0 a}$$

is the ground state energy of the hydrogen atom, but the length parameter b is *not* the Bohr radius a.

7.15a. How do the expectation values for kinetic, potential and total energy in the state (7.80) compare to the corresponding values in the ground state of the unperturbed hydrogen atom?

7.15b. How do we have to change the potential energy of the system to force the hydrogen atom into the state (7.80)? Show that the change in potential energy can be written as

$$\Delta V = \frac{\hbar^2}{\mu}\left(\frac{1}{a}-\frac{1}{b}\right)\left(\frac{1}{r}-\frac{1}{2a}-\frac{1}{2b}\right).$$

7.16. Solve the differential equation (6.6) for the harmonic oscillator not by the operator method, but by the same methods which we have used to solve the radial equation (7.54) for the hydrogen atom.

7.17. The proposal of Bohmian mechanics[10] asserts that quantum mechanics with the Born probability interpretation should be replaced by a pilot wave theory. The wave function would still satisfy the Schrödinger equation. However, instead of

[10]D. Bohm, Phys. Rev. 85, 166 & 180 (1952).

serving as a probability amplitude for the outcome of single measurements, the wave function provides a pilot wave for particles in the sense that an N particle wave function determines the velocity field for the particles through the equation

$$\frac{d\boldsymbol{x}_I(t)}{dt} = \frac{\hbar}{2im_I\,|\psi(\boldsymbol{x}_1(t),\ldots\boldsymbol{x}_N(t);t)|^2} \times\left(\psi^+(\boldsymbol{x}_1(t),\ldots\boldsymbol{x}_N(t);t)\stackrel{\leftrightarrow}{\nabla}_I\psi(\boldsymbol{x}_1(t),\ldots\boldsymbol{x}_N(t);t)\right), \quad (7.81)$$

where $\psi^+\stackrel{\leftrightarrow}{\nabla}\psi \equiv \psi^+\nabla\psi - (\nabla\psi^+)\psi$.

It has been claimed that this leads to predicitions which are indistinguishable from quantum mechanics, at least as long as we are only concerned with motion of non-relativistic particles.

We consider a hydrogen atom with center of mass velocity $\boldsymbol{V} = \hbar\boldsymbol{K}/(m_e + m_p)$. Which velocities would equation (7.81) predict for the velocities of the proton and the electron in the ground state of the atom? How would the proton and the electron then be arranged in the ground state of a hydrogen atom?

Solution. The ground state wave function in terms of electron and proton coordinates is

$$\langle\boldsymbol{x}_e,\boldsymbol{x}_p|1,0,0;\boldsymbol{K}(t)\rangle = \exp\!\left(\mathrm{i}\boldsymbol{K}\cdot\frac{m_e\boldsymbol{x}_e + m_p\boldsymbol{x}_p}{m_e+m_p}\right)\exp\!\left(-\frac{|\boldsymbol{x}_e-\boldsymbol{x}_p|}{a}\right) \times\frac{1}{\pi^2\sqrt{2a}^3}\exp\!\left(-\frac{\mathrm{i}\hbar}{2(m_e+m_p)}\boldsymbol{K}^2t - \frac{\mathrm{i}}{\hbar}E_1t\right).$$

Equation (7.81) then yields

$$\boldsymbol{v}_e = \boldsymbol{v}_p = \boldsymbol{V}.$$

This result agrees with the corresponding expectation values for particle velocities in quantum mechanics. However, here we assume that both the electron and the proton have well defined (although not individually observable) trajectories, and their velocities are sharply defined. Therefore the electron and the proton would both move with the constant center of mass velocity $\boldsymbol{V}$ along straight lines. Motion with a fixed distance between the two particles seems hardly compatible with their electromagnetic attraction, but Bohmian mechanics explains this in terms of an additional quantum potential generated by the wave function,

$$V_\psi(\boldsymbol{x}) = -\frac{\hbar^2}{2m|\psi(\boldsymbol{x})|}\Delta|\psi(\boldsymbol{x})|,$$

i.e. the wave function would also induce an additional force field in Bohmian mechanics.

However, motion with fixed separation between the electron and the proton should imply observation of an electric dipole moment for individual hydrogen atoms, contrary to ordinary quantum mechanics.

On the other hand, motion of the electron and the proton right on top of each other is an appealing classical picture, but is incompatible with the positive nucleus plus negative electron hull structure of atoms that follows e.g. from the van der Waals equation of state for gases (which gives atomic radii between 1 and 2 Å) and Rutherford scattering (which tells us two things: only the positive charge is concentrated in the nucleus, and the nucleus has only a radius of a few femtometers).

To avoid this negative verdict, we might argue that we should rather consider a cold gas of Bohmian hydrogen atoms to understand the implications of the Bohmian interpretation for the ground state wave function.

In a cold gas of Bohmian hydrogen atoms the static distance between the electron and the proton would be distributed according to $|\langle \boldsymbol{x}_e - \boldsymbol{x}_p|1,0,0\rangle|^2$. There would be many hydrogen atoms with the electron sitting right on top of the proton, but there would also be a lot of hydrogen atoms with a large separation and a corresponding static electric dipole moment $\boldsymbol{d} = e(\boldsymbol{x}_p - \boldsymbol{x}_e)$. Standard quantum mechanics in Born's interpretation does not predict an electric dipole moment in any of the hydrogen atoms, because an electron would only appear to have a particular location if we specifically perform a measurement asking for the location. However, in Bohmian mechanics, the electron and proton would exist as particles at all times with fixed relative location, and therefore there should be an average dipole moment per atom in the ground state with magnitude $\langle|\boldsymbol{d}|\rangle = e\langle r\rangle = 3ea/2$.

These dipole moments might be randomly distributed and therefore we might not observe a macroscopic dipole moment. However, we could align these dipole moments with a weak static external electric field. The field strength would be much weaker than the internal field strength in hydrogen, to ensure that the ground state wave function is not perturbed. In addition to any induced electric dipole moment in the Bohmian hydrogen atoms (which would also exist in the same way for the standard quantum mechanical hydrogen atoms) there would be a macroscopic dipole moment from orientation polarization. This would be a real difference from the standard quantum mechanical cold hydrogen gas. Therefore I disagree with claims that Bohmian mechanics is just a different ontological interpretation of non-relativistic quantum mechanics. Trying to make pilot wave theories work is certainly tempting, but I cannot consider Bohmian mechanics as a serious competitor to standard quantum mechanics with the Born interpretation of quantum states.

Chapter 8
Spin and Addition of Angular Momentum Type Operators

We have seen in Section 7.4 that representations of the angular momentum Lie algebra (7.21) are labelled by a quantum number ℓ which can take half-integer or integer values. However, we have also seen in Section 7.5 that ℓ is limited to integer values when the operators $\boldsymbol{M}$ actually refer to angular momentum, because the wave functions[1] $\langle \boldsymbol{x}|n, \ell, m_\ell\rangle$ or $\langle \boldsymbol{x}|k, \ell, m_\ell\rangle$ for angular momentum eigenstates must be single valued. It was therefore very surprising when Stern, Gerlach, Goudsmit, Uhlenbeck and Pauli in the 1920s discovered that half-integer values of ℓ are also realized in nature, although in that case ℓ cannot be related to an angular momentum any more. Half-integer values of ℓ arise in nature because leptons and quarks carry a representation of the "covering group" SU(2) of the proper rotation group SO(3), where SU(2) stands for the group which can be represented by special unitary 2×2 matrices[2]. The designation "special" refers to the fact that the matrices are also required to have determinant 1. The generators of the groups SU(2) and SO(3) satisfy the same Lie algebra (7.21), but for every rotation matrix $\underline{R}(\varphi) = \underline{R}(\varphi + 2\pi\hat{\varphi})$ there are two unitary 2×2 matrices $\underline{U}(\varphi) = -\underline{U}(\varphi + 2\pi\hat{\varphi})$. In that sense SU(2) provides a double cover of SO(3).

We will use the notations $\boldsymbol{l}$ and $\boldsymbol{M}$ for angular momenta, and $\boldsymbol{s}$ or $\boldsymbol{S}$ for spins.

---

[1] We denote the magnetic quantum number with m_ℓ in this chapter because m will denote the mass of a particle.

[2] Ultimately, all particles carry representations of the covering group SL(2,ℂ) of the group SO(1,3) of proper orthochronous Lorentz transformations, see Appendices B and H.

R. Dick, *Advanced Quantum Mechanics*, Graduate Texts in Physics,
DOI 10.1007/978-3-319-25675-7_8

8.1 Spin and magnetic dipole interactions

A particle of charge q and mass m which moves with angular momentum $\boldsymbol{l}$ through a constant magnetic field $\boldsymbol{B}$ has its energy levels shifted through a Zeeman term in the Hamiltonian,

$$H_Z = -\frac{q}{2m}\boldsymbol{l}\cdot\boldsymbol{B}. \tag{8.1}$$

We will explore the origin of this term in Chapter 15, see Problem 15.2, but for now we can think of it as a magnetic dipole term with a dipole moment

$$\boldsymbol{\mu}_l = \frac{q}{2m}\boldsymbol{l}.$$

The relation between $\boldsymbol{\mu}_l$ and $\boldsymbol{l}$ can be motivated from electrodynamics, but is actually a consequence of the coupling to magnetic vector potentials in the Schrödinger equation.

The quantization $\langle \ell, m_\ell | l_z | \ell, m_\ell \rangle = \hbar m_\ell$ for angular momentum components in a fixed direction yields a Zeeman shift

$$\Delta E = -\frac{q\hbar}{2m}Bm_\ell, \quad -\ell \leq m_\ell \leq \ell,$$

of the energy levels of a charged particle in a magnetic field. For orbital momentum the resulting number $2\ell + 1$ of energy levels is odd. However, the observation of motion of Ag atoms through an inhomogeneous field by Stern and Gerlach in 1921 revealed a split of energy levels of these atoms into two levels in a magnetic field. This complies with a split into $2s + 1$ levels only if the angular momentum like quantum number s is $1/2$. This additional angular momentum type quantum number is denoted as spin. Spin behaves in many respects similar to angular momentum, but it cannot be an orbital angular momentum because that would exclude half-integer values for s. Another major difference to angular momentum concerns the fact that the spectroscopically observed splitting of energy levels due to spin complies with a magnetic dipole type interaction only if the corresponding Zeeman type term is increased by a factor g_s,

$$H = -\boldsymbol{\mu}_s\cdot\boldsymbol{B}, \quad \boldsymbol{\mu}_s = g_s\frac{q}{2m}\boldsymbol{s}.$$

This "anomalous g factor" is in very good approximation $g_s \simeq 2$. The relation between $\boldsymbol{\mu}_s$ and $\boldsymbol{s}$ is a consequence of relativistic quantum mechanics and will be explained in Section 21.5.

The important observation for now is that there exist operators which satisfy the angular momentum Lie algebra (7.21),

$$[S_i, S_j] = i\hbar\epsilon_{ijk}S_k,$$

and therefore have representations of the form (7.26–7.31),

$$S_z|s,m_s\rangle = \hbar m_s|s,m_s\rangle, \tag{8.2}$$

$$S_\pm|s,m_s\rangle = \hbar\sqrt{s(s+1)-m_s(m_s\pm 1)}|s,m_s\pm 1\rangle, \tag{8.3}$$

$$\mathbf{S}^2|s,m_s\rangle = \hbar^2 s(s+1)|s,m_s\rangle.$$

However, these operators are not related to orbital angular momentum and therefore can have half-integer values of the quantum number s in their representations.

Our previous calculations of matrix representations of the rotation group in Section 7.4 imply that spin is related to transformation properties of particle wave functions under rotations. However, before we can elaborate on this, we have to take a closer look at the representations with $s = 1/2$.

In the following mapping between matrices we use an index mapping for the magnetic quantum numbers $m_s = \pm 1/2$ to indices

$$a(m_s) = (3/2) - m_s, \tag{8.4}$$

i.e. $m_s = 1/2 \to a(m_s) = 1$, $m_s = -1/2 \to a(m_s) = 2$.

Substitution of $s = 1/2$ in equations (8.2, 8.3) yields

$$\langle 1/2, m_s'|S_3|1/2, m_s\rangle = \hbar m_s \delta_{m_s',m_s} = \frac{\hbar}{2}(\sigma_3)_{a(m_s'),a(m_s)}, \tag{8.5}$$

$$\langle 1/2, m_s'|S_1|1/2, m_s\rangle = \frac{\hbar}{2}\left(\delta_{m_s',m_s+1} + \delta_{m_s',m_s-1}\right) = \frac{\hbar}{2}(\sigma_1)_{a(m_s'),a(m_s)}, \tag{8.6}$$

and

$$\langle 1/2, m_s'|S_2|1/2, m_s\rangle = \frac{\hbar}{2\mathrm{i}}\left(\delta_{m_s',m_s+1} - \delta_{m_s',m_s-1}\right) = \frac{\hbar}{2}(\sigma_2)_{a(m_s'),a(m_s)}, \tag{8.7}$$

with the Pauli matrices

$$\underline{\sigma}_1 = \begin{pmatrix} 0 & 1 \\ 1 & 0 \end{pmatrix}, \quad \underline{\sigma}_2 = \begin{pmatrix} 0 & -\mathrm{i} \\ \mathrm{i} & 0 \end{pmatrix}, \quad \underline{\sigma}_3 = \begin{pmatrix} 1 & 0 \\ 0 & -1 \end{pmatrix}. \tag{8.8}$$

The Pauli matrices provide a basis for hermitian traceless 2×2 matrices and satisfy the relation

$$\underline{\sigma}_i \cdot \underline{\sigma}_j = \delta_{ij}\underline{1} + \mathrm{i}\epsilon_{ijk}\underline{\sigma}_k. \tag{8.9}$$

The index mapping $m_s \to a(m_s)$ is employed in the notation of spin states as $|1/2, m_s\rangle \to |a(m_s)\rangle$ such that a general $s = 1/2$ state is

$$|\psi\rangle = \sum_{m_s=1/2}^{-1/2} |1/2, m_s\rangle\langle 1/2, m_s|\psi\rangle = \sum_{a=1}^{2} |a\rangle\langle a|\psi\rangle \tag{8.10}$$

Knowledge of a spin 1/2 state $|\psi\rangle$ is equivalent to the knowledge of its two components $\langle 1/2, 1/2|\psi\rangle \equiv \langle 1|\psi\rangle \equiv \psi_1$, $\langle 1/2, -1/2|\psi\rangle \equiv \langle 2|\psi\rangle \equiv \psi_2$. In column notation this corresponds to the *2-spinor* ψ

$$\psi = \begin{pmatrix} \psi_1 \\ \psi_2 \end{pmatrix}, \tag{8.11}$$

such that application of a spin operator S_i

$$\langle 1/2, m_s|\psi'\rangle = \langle 1/2, m_s|S_i|\psi\rangle = \sum_{m_s'=1/2}^{-1/2} \langle 1/2, m_s|S_i|1/2, m_s'\rangle\langle 1/2, m_s'|\psi\rangle$$

corresponds to the matrix multiplication

$$\psi' = \frac{\hbar}{2}\underline{\sigma}_i \cdot \psi. \tag{8.12}$$

For example, a general electron state $|\psi\rangle$ corresponds to a superposition of spin orientations $\pm 1/2$ and a superposition of $\boldsymbol{x}$ eigenstates,

$$|\psi\rangle = \int d^3\boldsymbol{x} \sum_{m_s=1/2}^{-1/2} |\boldsymbol{x}; m_s\rangle\langle \boldsymbol{x}; m_s|\psi\rangle \equiv \int d^3\boldsymbol{x} \sum_{m_s=1/2}^{-1/2} |\boldsymbol{x}; m_s\rangle \psi_{a(m_s)}(\boldsymbol{x}),$$

and is given in 2-spinor notation (listing all common index conventions) as

$$\psi(\boldsymbol{x}) = \begin{pmatrix} \psi_1(\boldsymbol{x}) \\ \psi_2(\boldsymbol{x}) \end{pmatrix} \equiv \begin{pmatrix} \psi_{1/2}(\boldsymbol{x}) \\ \psi_{-1/2}(\boldsymbol{x}) \end{pmatrix} \equiv \begin{pmatrix} \psi_+(\boldsymbol{x}) \\ \psi_-(\boldsymbol{x}) \end{pmatrix} \equiv \begin{pmatrix} \psi_\uparrow(\boldsymbol{x}) \\ \psi_\downarrow(\boldsymbol{x}) \end{pmatrix}. \tag{8.13}$$

The normalization is

$$\int d^3\boldsymbol{x} \left(|\psi_1(\boldsymbol{x})|^2 + |\psi_2(\boldsymbol{x})|^2\right) = 1.$$

The probability densities for finding the electron with spin up or down in the location $\boldsymbol{x}$ are $|\psi_1(\boldsymbol{x})|^2$ and $|\psi_2(\boldsymbol{x})|^2$, respectively, while the probability density to find the electron in the location $\boldsymbol{x}$ in any spin orientation is $|\psi_1(\boldsymbol{x})|^2 + |\psi_2(\boldsymbol{x})|^2$. Note that these three probability densities can have maxima in three different locations, which reminds us how questionable the concept of a particle is in quantum mechanics.

8.2 Transformation of scalar, spinor, and vector wave functions under rotations

The commutation relations between angular momentum $\boldsymbol{M} = \mathbf{x} \times \mathbf{p}$ and $\mathbf{x}$,

$$[M_i, \mathrm{x}_j] = i\hbar\epsilon_{ijk}\mathrm{x}_k$$

imply with the rotation generators $(L_i)_{jk} = \epsilon_{ijk}$ and the rotation matrices from Section 7.4

$$\exp\left(\frac{\mathrm{i}}{\hbar}\boldsymbol{\varphi}\cdot\boldsymbol{M}\right)\mathbf{x}\exp\left(-\frac{\mathrm{i}}{\hbar}\boldsymbol{\varphi}\cdot\boldsymbol{M}\right) = \exp(-\boldsymbol{\varphi}\cdot\boldsymbol{L})\cdot\mathbf{x} = \underline{R}(-\boldsymbol{\varphi})\cdot\mathbf{x},$$

and therefore

$$\langle\boldsymbol{x}|\exp\left(\frac{\mathrm{i}}{\hbar}\boldsymbol{\varphi}\cdot\boldsymbol{M}\right) = \langle\underline{R}(-\boldsymbol{\varphi})\cdot\boldsymbol{x}|.$$

Rotation of a state

$$|\psi(t)\rangle \to |\psi'(t)\rangle = \exp\left(\frac{\mathrm{i}}{\hbar}\boldsymbol{\varphi}\cdot\boldsymbol{M}\right)|\psi(t)\rangle$$

therefore implies for the rotated wave function

$$\langle\boldsymbol{x}'|\psi'(t)\rangle = \langle\underline{R}(\boldsymbol{\varphi})\cdot\boldsymbol{x}|\psi'(t)\rangle = \langle\boldsymbol{x}|\psi(t)\rangle, \tag{8.14}$$

where

$$\boldsymbol{x}' = \underline{R}(\boldsymbol{\varphi})\cdot\boldsymbol{x}$$

is the rotated coordinate vector.

A transformation behavior like (8.14) tells us that the transformed wave function at the transformed set of coordinates is the same as the original wave function at the original set of coordinates. Such a transformation behavior is denoted as a *scalar* transformation law, and the corresponding wave functions are scalar functions.

On the other hand, spinor wave functions have two components which denote probability amplitudes for spin orientation along a given spatial axis, conventionally chosen as the z axis. The z' axis of the rotated frame will generically have a direction which is different from the z axis, and the probability amplitudes for spin along the z' direction will be different from the probability amplitudes along the z direction.

The rotated 2-spinor state

$$|\psi(t)\rangle \to |\psi'(t)\rangle = \exp\left(\frac{\mathrm{i}}{\hbar}\boldsymbol{\varphi}\cdot(\boldsymbol{M}+\boldsymbol{S})\right)|\psi(t)\rangle \tag{8.15}$$

has components

$$\begin{aligned}\langle\boldsymbol{x}',a|\psi'(t)\rangle \equiv \psi'_a(\boldsymbol{x}',t) &= \langle\underline{R}(\boldsymbol{\varphi})\cdot\boldsymbol{x},a|\exp\left(\frac{\mathrm{i}}{\hbar}\boldsymbol{\varphi}\cdot(\boldsymbol{M}+\boldsymbol{S})\right)|\psi(t)\rangle\\ &= \langle\boldsymbol{x},a|\exp\left(\frac{\mathrm{i}}{\hbar}\boldsymbol{\varphi}\cdot\boldsymbol{S}\right)|\psi(t)\rangle = \left[\exp\left(\frac{\mathrm{i}}{2}\boldsymbol{\varphi}\cdot\underline{\boldsymbol{\sigma}}\right)\right]_{ab}\langle\boldsymbol{x},b|\psi(t)\rangle\end{aligned}$$

or in terms of the column 2-spinor (8.13),

$$\psi'(\mathbf{x}', t) = \exp\left(\frac{\mathrm{i}}{2}\boldsymbol{\varphi} \cdot \underline{\boldsymbol{\sigma}}\right) \cdot \psi(\mathbf{x}, t). \tag{8.16}$$

For comparison, we also give the result if we use the representation (8.2, 8.3) with $s = 1$ for the spin operators $\mathbf{S}$ on wave functions. In that case the matrix correspondence

$$\langle s = 1/2, m_s' | \mathbf{S} | s = 1/2, m_s \rangle = \hbar \boldsymbol{\sigma}_{a(m_s'),a(m_s)}/2$$

is replaced in a first step by

$$\langle s = 1, m_s' | \mathbf{S} | s = 1, m_s \rangle = \hbar \boldsymbol{\Sigma}_{j(m_s'),j(m_s)}$$

with $j(m_s) = 2 - m_s$,

$$\underline{\Sigma}_1 = \frac{1}{\sqrt{2}} \begin{pmatrix} 0 & 1 & 0 \\ 1 & 0 & 1 \\ 0 & 1 & 0 \end{pmatrix}, \quad \underline{\Sigma}_2 = \frac{\mathrm{i}}{\sqrt{2}} \begin{pmatrix} 0 & -1 & 0 \\ 1 & 0 & -1 \\ 0 & 1 & 0 \end{pmatrix},$$

$$\underline{\Sigma}_3 = \begin{pmatrix} 1 & 0 & 0 \\ 0 & 0 & 0 \\ 0 & 0 & -1 \end{pmatrix}. \tag{8.17}$$

However, this is still not the standard matrix representation for spin $s = 1$. The connection with the conventional representation (7.18) of vector rotation operators is achieved through the similarity transformation

$$\underline{L} = \frac{\mathrm{i}}{\hbar} \underline{M} = \mathrm{i} \underline{A} \cdot \underline{\boldsymbol{\Sigma}} \cdot \underline{A}^{-1} \tag{8.18}$$

with the unitary matrix

$$\underline{A} = \frac{1}{\sqrt{2}} \begin{pmatrix} -1 & 0 & 1 \\ -\mathrm{i} & 0 & -\mathrm{i} \\ 0 & \sqrt{2} & 0 \end{pmatrix}, \quad \underline{A}^{-1} = \frac{1}{\sqrt{2}} \begin{pmatrix} -1 & \mathrm{i} & 0 \\ 0 & 0 & \sqrt{2} \\ 1 & \mathrm{i} & 0 \end{pmatrix}.$$

The transformation law for *vector wave functions* $\langle \mathbf{x}, i | A(t) \rangle \equiv A_i(\mathbf{x}, t)$ under rotations is then given in terms of the same rotation matrices $\underline{R}(\boldsymbol{\varphi}) = \exp(\boldsymbol{\varphi} \cdot \underline{\mathbf{L}})$ which effect rotations of the vector $\mathbf{x}$,

$$\mathbf{x}' = \exp(\boldsymbol{\varphi} \cdot \underline{\mathbf{L}}) \cdot \mathbf{x}, \quad \mathbf{A}'(\mathbf{x}', t) = \exp(\boldsymbol{\varphi} \cdot \underline{\mathbf{L}}) \cdot \mathbf{A}(\mathbf{x}, t). \tag{8.19}$$

We will see in Chapter 18 that photons are described by vector wave functions.

8.3 Addition of angular momentum like quantities

In classical mechanics, angular momentum is an additive vector quantity which is conserved in rotationally symmetric systems. Furthermore, the transformation equation (8.15) for spinor states involved addition of two different operators which both satisfy the angular momentum Lie algebra (7.21). However, before immersing ourselves into the technicalities of how angular momentum type operators are combined in quantum mechanics, it is worthwhile to point out that interactions in atoms and materials provide another direct physical motivation for addition of angular momentum like quantities.

We have seen in Section 7.1 that relative motion of two interacting particles with an interaction potential $V(\boldsymbol{x}_1 - \boldsymbol{x}_2)$ can be described in terms of effective single particle motion of a (quasi)particle with location $\boldsymbol{r}(t) = \boldsymbol{x}_1(t) - \boldsymbol{x}_2(t)$, mass $m = m_1 m_2/(m_1+m_2)$, momentum $\boldsymbol{p} = (m_2\boldsymbol{p}_1 - m_1\boldsymbol{p}_2)/(m_1+m_2)$ and angular momentum $\boldsymbol{l} = \boldsymbol{r}\times\boldsymbol{p}$.

Furthermore, if $m_2 \gg m_1$, but the charge q_2 is not much larger than q_1 and the spin $|\boldsymbol{s}_2|$ is not much larger than $|\boldsymbol{s}_1|$, then we can assign a charge[3] $q = q_1$ and a spin $\boldsymbol{s} = \boldsymbol{s}_1$ to the quasiparticle with mass $m \simeq m_1$.

A particle of charge $-e$ and mass m with angular momentum operators $\boldsymbol{l}$ and spin $\boldsymbol{s}$ experiences a contribution to its energy levels from an interaction term

$$H_{l\cdot s} = \frac{\mu_0 e^2}{8\pi m^2 r^3}\boldsymbol{l}\cdot\boldsymbol{s} \tag{8.20}$$

in its Hamiltonian, if it is moving in the electric field $\boldsymbol{E} = \hat{\boldsymbol{r}}e/(4\pi\epsilon_0 r^2)$ of a much heavier particle of charge e. One can think of $H_{l\cdot s}$ as a magnetic dipole-dipole interaction $(\mu_0/4\pi r^3)\boldsymbol{\mu}_l\cdot\boldsymbol{\mu}_s$, but finally it arises as a consequence of a relativistic generalization of the Schrödinger equation. We will see this in Chapter 21, in particular equation (21.117). However, for the moment we simply accept the existence of terms like (8.20) as an experimental fact. These terms contribute to the fine structure of spectral lines. The term (8.20) is known as a *spin-orbit coupling* term or *ls* coupling term, and applies in this particular form to the energy levels of the quasiparticle which describes relative motion in a two-particle system. However, if there are many charged particles like in a many-electron atom, then there will also be interaction terms between angular momenta and spins of different particles in the system, i.e. we will have terms of the form

$$H_{j_1\cdot j_2} = f(r_{12})\boldsymbol{j}_1\cdot\boldsymbol{j}_2, \tag{8.21}$$

where $\boldsymbol{j}_i$ are angular momentum like operators. We will superficially denote all these operators (including spin) simply as angular momentum operators in the following.

[3]We will return to the question of assignment of charge and spin to the quasiparticle for relative motion in Section 18.4.

Diagonalization of Hamiltonians like (8.20) or (8.21) requires us to combine two operators to a new operator according to $\boldsymbol{j} = \boldsymbol{l} + \boldsymbol{s}$ or $\boldsymbol{j} = \boldsymbol{j}_1 + \boldsymbol{j}_1$, respectively. From the perspective of spectroscopy, terms like (8.20) or (8.21) are the very reason why we have to know how to combine two angular momentum type operators in quantum mechanics. Diagonalization of (8.20) and (8.21) is important for understanding the spectra of atoms and molecules, and spin-orbit coupling also affects energy levels in materials. Furthermore, Hamiltonians of the form $-2J\boldsymbol{s}_1 \cdot \boldsymbol{s}_2$ provide an effective description of interactions in magnetic materials, see Section 17.7, and they are important for spin entanglement and spintronics. The advantage of introducing the combined angular momentum operator $\boldsymbol{j} = \boldsymbol{l} + \boldsymbol{s}$ is that it also satisfies angular momentum commutation rules (7.21) $[j_a, j_b] = \mathrm{i}\hbar\epsilon_{abc}j_c$ and therefore should have eigenstates $|j, m_j\rangle$,

$$\boldsymbol{j}^2|j, m_j\rangle = \hbar^2 j(j+1)|j, m_j\rangle, \quad j_z|j, m_j\rangle = \hbar m_j|j, m_j\rangle. \tag{8.22}$$

However, $\boldsymbol{j}$ commutes with $\boldsymbol{l}^2$ and $\boldsymbol{s}^2$, $[j_a, \boldsymbol{l}^2] = [j_a, \boldsymbol{s}^2] = 0$, and therefore we can try to construct the states in (8.22) such that they also satisfy the properties

$$\boldsymbol{l}^2|j, m_j, \ell, s\rangle = \hbar^2\ell(\ell+1)|j, m_j, \ell, s\rangle,$$

$$\boldsymbol{s}^2|j, m_j, \ell, s\rangle = \hbar^2 s(s+1)|j, m_j, \ell, s\rangle.$$

The advantage of these states is that they are eigenstates of the coupling operator (8.20),

$$\begin{aligned}\boldsymbol{l}\cdot\boldsymbol{s}|j, m_j, \ell, s\rangle &= \frac{\boldsymbol{j}^2 - \boldsymbol{l}^2 - \boldsymbol{s}^2}{2}|j, m_j, \ell, s\rangle \\ &= \hbar^2\frac{j(j+1) - \ell(\ell+1) - s(s+1)}{2}|j, m_j, \ell, s\rangle,\end{aligned} \tag{8.23}$$

and therefore the energy shifts from spin-orbit coupling in these states are

$$\Delta E = \frac{\mu_0 e^2\hbar^2}{16\pi m^2}\langle r^{-3}\rangle\left[j(j+1) - \ell(\ell+1) - s(s+1)\right]. \tag{8.24}$$

The states that we know for the operators $\boldsymbol{l}$ and $\boldsymbol{s}$ are the eigenstates $|\ell, m_\ell\rangle$ for $\boldsymbol{l}^2$ and l_z, and $|s, m_s\rangle$ for $\boldsymbol{s}^2$ and s_z, respectively. We can combine these states into states

$$|\ell, m_\ell\rangle \otimes |s, m_s\rangle \equiv |\ell, m_\ell; s, m_s\rangle \tag{8.25}$$

which will be denoted as a tensor product basis of angular momentum states. The understanding in the tensor product notation is that $\boldsymbol{l}$ only acts on the first factor and $\boldsymbol{s}$ only on the second factor. Strictly speaking the combined angular momentum operator should be written as

$$\boldsymbol{j} = \boldsymbol{l} \otimes 1 + 1 \otimes \boldsymbol{s},$$

which automatically ensures the correct rule

$$\boldsymbol{j}(|\ell, m_\ell\rangle \otimes |s, m_s\rangle) = \boldsymbol{l}|\ell, m_\ell\rangle \otimes |s, m_s\rangle + |\ell, m_\ell\rangle \otimes \boldsymbol{s}|s, m_s\rangle,$$

but we will continue with the standard physics notation $\boldsymbol{j} = \boldsymbol{l} + \boldsymbol{s}$.

The main problem for combination of angular momenta is how to construct the eigenstates $|j, m_j, \ell, s\rangle$ for total angular momentum from the tensor products (8.25) of eigenstates of the initial angular momenta,

$$|j, m_j, \ell, s\rangle = \sum_{m_\ell, m_s} |\ell, m_\ell; s, m_s\rangle\langle\ell, m_\ell; s, m_s|j, m_j, \ell, s\rangle. \tag{8.26}$$

We will denote the states $|j, m_j, \ell, s\rangle$ as the combined angular momentum states.

There is no summation over indices $\ell' \neq \ell$ or $s' \neq s$ on the right hand side because all states involved are eigenstates of $\boldsymbol{l}^2$ and $\boldsymbol{s}^2$ with the same eigenvalues $\hbar^2\ell(\ell+1)$ or $\hbar^2 s(s+1)$, respectively.

The components $\langle\ell, m_\ell; s, m_s|j, m_j, \ell, s\rangle$ of the transformation matrix from the initial angular momenta states to the combined angular momentum states are known as *Clebsch-Gordan coefficients* or *vector addition coefficients*. The notation $\langle\ell, m_\ell; s, m_s|j, m_j, \ell, s\rangle$ is logically satisfactory by explicitly showing that the Clebsch-Gordan coefficients can also be thought of as the representation of the combined angular momentum states $|j, m_j, \ell, s\rangle$ in the basis of tensor product states $|\ell, m_\ell; s, m_s\rangle$. However, the notation is also redundant in terms of the quantum numbers ℓ and s, and a little clumsy. It is therefore convenient to abbreviate the notation by setting

$$\langle\ell, m_\ell; s, m_s|j, m_j, \ell, s\rangle \equiv \langle\ell, m_\ell; s, m_s|j, m_j\rangle.$$

The new angular momentum eigenstates must also be normalizable and orthogonal for different eigenvalues, i.e. the transformation matrix must be unitary,

$$\sum_{m_\ell, m_s} \langle j, m_j|\ell, m_\ell; s, m_s\rangle\langle\ell, m_\ell; s, m_s|j', m_j'\rangle = \delta_{j,j'}\delta_{m_j,m_j'}, \tag{8.27}$$

$$\sum_{j, m_j} \langle\ell, m_\ell; s, m_s|j, m_j\rangle\langle j, m_j|\ell, m_\ell'; s, m_s'\rangle = \delta_{m_\ell,m_\ell'}\delta_{m_s,m_s'}. \tag{8.28}$$

The hermiticity properties

$$j_z = (l_z + s_z)^+, \quad j_\pm = (l_\mp + s_\mp)^+$$

imply with the definition (4.31) of adjoint operators the relations

$$m_j\langle\ell, m_\ell; s, m_s|j, m_j\rangle = (m_\ell + m_s)\,\langle\ell, m_\ell; s, m_s|j, m_j\rangle \tag{8.29}$$

and

$$\begin{aligned}
&\sqrt{j(j+1)-m_j(m_j\pm 1)}\langle\ell,m_\ell;s,m_s|j,m_j\pm 1\rangle\\
&=\sqrt{\ell(\ell+1)-m_\ell(m_\ell\mp 1)}\langle\ell,m_\ell\mp 1;s,m_s|j,m_j\rangle\\
&\quad+\sqrt{s(s+1)-m_s(m_s\mp 1)}\langle\ell,m_\ell;s,m_s\mp 1|j,m_j\rangle.
\end{aligned}\tag{8.30}$$

Equation (8.29) yields

$$\langle\ell,m_\ell;s,m_s|j,m_j\rangle=\delta_{m_\ell+m_s,m_j}\langle\ell,m_\ell;s,m_s|j,m_\ell+m_s\rangle.$$

The highest occurring value of m_j which is also the highest occurring value for j is therefore $\ell+s$, and there is only one such state. This determines the state $|\ell+s,\ell+s,\ell,s\rangle$ up to a phase factor to

$$|\ell+s,\ell+s,\ell,s\rangle=|\ell,\ell;s,s\rangle,\tag{8.31}$$

i.e. we choose the phase factor as

$$\langle\ell,\ell;s,s|\ell+s,\ell+s\rangle=1.$$

Repeated application of $j_-=l_-+s_-$ on the state (8.31) then yields all the remaining states of the form $|\ell+s,m_j,\ell,s\rangle$ or equivalently the remaining Clebsch-Gordan coefficients of the form $\langle\ell,m_\ell;s,m_s|\ell+s,m_j=m_\ell+m_s\rangle$ with $-\ell-s\leq m_j<\ell+s$. For example, the next two lower states with $j=\ell+s$ are given by

$$\begin{aligned}
j_-|\ell+s,\ell+s,\ell,s\rangle&=\sqrt{2(\ell+s)}|\ell+s,\ell+s-1,\ell,s\rangle\\
&=\sqrt{2\ell}|\ell,\ell-1;s,s\rangle+\sqrt{2s}|\ell,\ell;s,s-1\rangle
\end{aligned}$$

and

$$\begin{aligned}
j_-^2|\ell+s,\ell+s,\ell,s\rangle&=2\sqrt{\ell+s}\sqrt{2(\ell+s)-1}|\ell+s,\ell+s-2,\ell,s\rangle\\
&=2\sqrt{\ell(2\ell-1)}|\ell,\ell-2;s,s\rangle+4\sqrt{\ell s}|\ell,\ell-1;s,s-1\rangle\\
&\quad+2\sqrt{s(2s-1)}|\ell,\ell;s,s-2\rangle.
\end{aligned}\tag{8.32}$$

However, we have two states in the $|\ell,m_\ell;s,m_s\rangle$ basis with total magnetic quantum number $\ell+s-1$, but so far discovered only one state in the $|j,m_j,\ell,s\rangle$ basis with this magnetic quantum number. We can therefore construct a second state with $m_j=\ell+s-1$, which is orthogonal to the state $|\ell+s,\ell+s-1,\ell,s\rangle$,

$$|\ell+s-1,\ell+s-1,\ell,s\rangle=\sqrt{\frac{s}{\ell+s}}|\ell,\ell-1;s,s\rangle-\sqrt{\frac{\ell}{\ell+s}}|\ell,\ell;s,s-1\rangle.\tag{8.33}$$

Application of $\vec{j}^2$ would show that this state has $j = \ell + s - 1$, which was already anticipated in the notation. Repeated application of the lowering operator j_- on this state would then yield all remaining states of the form $|\ell + s - 1, m_j, \ell, s\rangle$ with $1 - \ell - s \le m_j < \ell + s - 1$, e.g.

$$\sqrt{\ell+s-1}|\ell+s-1,\ell+s-2,\ell,s\rangle = \sqrt{s\frac{2\ell-1}{\ell+s}}|\ell,\ell-2;s,s\rangle$$
$$-\sqrt{\ell\frac{2s-1}{\ell+s}}|\ell,\ell;s,s-2\rangle + \frac{s-\ell}{\sqrt{\ell+s}}|\ell,\ell-1;s-1,s\rangle. \quad (8.34)$$

We have three states with $m_j = \ell + s - 2$ in the direct product basis, *viz.* $|\ell, \ell - 2; s, s\rangle$, $|\ell, \ell; s, s - 2\rangle$ and $|\ell, \ell - 1; s - 1, s\rangle$, but so far we have only constructed two states in the combined angular momentum basis with $m_j = \ell + s - 2$, *viz.* $|\ell + s, \ell + s - 2, \ell, s\rangle$ and $|\ell + s - 1, \ell + s - 2, \ell, s\rangle$. We can therefore construct a third state in the combined angular momentum basis which is orthogonal to the other two states,

$$\begin{aligned}|\ell+s-2,\ell+s-2,\ell,s\rangle &\propto |\ell,\ell-1;s-1,s\rangle\\ &-|\ell+s,\ell+s-2,\ell,s\rangle\langle\ell+s,\ell+s-2,\ell,s|\ell,\ell-1;s-1,s\rangle\\ &-|\ell+s-1,\ell+s-2,\ell,s\rangle\langle\ell+s-1,\ell+s-2,\ell,s|\ell,\ell-1;s-1,s\rangle.\end{aligned}$$

Substitution of the states and Clebsch-Gordan coefficients from (8.32) and (8.34) and normalization yields

$$|\ell+s-2,\ell+s-2,\ell,s\rangle = \sqrt{\frac{(2\ell-1)(2s-1)}{(2\ell+2s-1)(\ell+s-1)}}|\ell,\ell-1;s-1,s\rangle$$
$$+\frac{\sqrt{\ell(2\ell-1)}|\ell,\ell;s,s-2\rangle - \sqrt{s(2s-1)}|\ell,\ell-2;s,s\rangle}{\sqrt{(2\ell+2s-1)(\ell+s-1)}}. \quad (8.35)$$

Application of j_- then yields the remaining states of the form $|\ell + s - 2, m_j, \ell, s\rangle$. This process of repeated applications of j_- and forming new states with lower j through orthogonalization to the higher j states terminates when j reaches a minimal value $j = |\ell - s|$, when all $(2\ell + 1)(2s + 1)$ states $|\ell, m_\ell; s, m_s\rangle$ have been converted into the same number of states of the form $|j, m_j, \ell, s\rangle$. In particular, we observe that there are $2 \times \min(\ell, s) + 1$ allowed values for j,

$$j \in \{|\ell - s|, |\ell - s| + 1, \ldots, \ell + s - 1, \ell + s\}. \quad (8.36)$$

The procedure to reduce the state space in terms of total angular momentum eigenstates $|j, m_j, \ell, s\rangle$ through repeated applications of j_- and orthogonalizations is lengthy when the number of states $(2\ell + 1)(2s + 1)$ is large, and the reader will certainly appreciate that Wigner [42] and Racah[4] have derived expressions for general Clebsch-Gordan coefficients. Racah derived in particular the following expression (see also [9, 34])

$$\langle \ell, m_\ell; s, m_s | j, m_j \rangle = \delta_{m_\ell + m_s, m_j}$$
$$\times \sum_{\nu=\nu_1}^{\nu_2} (-)^\nu \left(\frac{\sqrt{(2j+1)\cdot(\ell+s-j)!\cdot(j+\ell-s)!\cdot(j+s-\ell)!}}{\sqrt{(j+\ell+s+1)!}\cdot\nu!\cdot(\ell-m_\ell-\nu)!\cdot(s+m_s-\nu)!}\right.$$
$$\left.\times \frac{\sqrt{(\ell+m_\ell)!\cdot(\ell-m_\ell)!\cdot(s+m_s)!\cdot(s-m_s)!\cdot(j+m_j)!\cdot(j-m_j)!}}{(j-s+m_\ell+\nu)!\cdot(j-\ell-m_s+\nu)!\cdot(\ell+s-j-\nu)!}\right). \tag{8.37}$$

The boundaries of the summation are determined by the requirements

$$\max[0, s-m_\ell-j, \ell+m_s-j] \leq \nu \leq \min[\ell+s-j, \ell-m_\ell, s+m_s].$$

Even if we decide to follow the standard convention of using real Clebsch-Gordan coefficients, there are still sign ambiguities for every particular value of j in $|\ell-s| \leq j \leq \ell+s$. This arises from the ambiguity of constructing the next orthogonal state when going from completed sets of states $|j', m_{j'}, \ell, s\rangle$, $j < j' \leq \ell+s$ to the next lower level j, because a sign ambiguity arises in the construction of the next orthogonal state $|j, j, \ell, s\rangle$. For example, Racah's formula (8.37) would give us the state $|\ell+s-2, \ell+s-2, \ell, s\rangle$ constructed before in equation (8.35), but with an overall minus sign.

Tables of Clebsch-Gordan coefficients had been compiled in the olden days, but nowadays these coefficients are implemented in commercial mathematical software programs for numerical and symbolic calculation, and there are also free online applets for the calculation of Clebsch-Gordan coefficients.

8.4 Problems

8.1. Calculate the spinor rotation matrix

$$\underline{U}(\varphi) = \exp\left(\frac{\mathrm{i}}{2}\varphi\cdot\underline{\sigma}\right).$$

Hint: Use the expansion of the exponential function and consider odd and even powers of the exponent separately.

[4]G. Racah, Phys. Rev. 62, 438 (1942).

Verify the property

$$\underline{U}(\boldsymbol{\varphi}) = -\underline{U}(\boldsymbol{\varphi} + 2\pi\hat{\boldsymbol{\varphi}}).$$

8.2. We perform a rotation of the reference frame by an angle φ around the x-axis. How does this change the coordinates of the vector $\boldsymbol{x}$? Suppose we have a spinor which has only a spin up component in the old reference frame. How large are the spin up and spin down components of the spinor with respect to the rotated z axis?

8.3. The Cartesian coordinates $\{x, y, z\}$ transform under rotations according to

$$\boldsymbol{x} \to \boldsymbol{x}' = \exp\left(\boldsymbol{\varphi} \cdot \underline{\boldsymbol{L}}\right) \cdot \boldsymbol{x}.$$

Construct coordinates $\{X, Y, Z\}$ which transform with the matrices (8.17) under rotations,

$$\boldsymbol{X} \to \boldsymbol{X}' = \exp\left(\mathrm{i}\boldsymbol{\varphi} \cdot \underline{\boldsymbol{\Sigma}}\right) \cdot \boldsymbol{X}.$$

8.4. Construct the matrices $\langle s, m_s'|\boldsymbol{S}|s, m_s\rangle = \hbar\boldsymbol{\Sigma}_{j(m_s'),j(m_s)}$ for $s = 3/2$. Choose the index mapping $m_s \to j(m_s)$ such that

$$\underline{\Sigma}_3 = \frac{1}{2}\begin{pmatrix} 3 & 0 & 0 & 0 \\ 0 & 1 & 0 & 0 \\ 0 & 0 & -1 & 0 \\ 0 & 0 & 0 & -3 \end{pmatrix}.$$

Suppose we have an excited Lithium atom in a spin $s = 3/2$ state, which is described by the 4-component wave function $\Psi_j(\boldsymbol{x}_1, \boldsymbol{x}_2, \boldsymbol{x}_3)$, $1 \leq j \leq 4$. How does this wave function transform under a rotation around the x axis by an angle $\varphi = \pi/2$?

8.5. Construct all the states $|j, m_j, \ell = 1, s = 1/2\rangle$ as linear combinations of the tensor product states $|\ell = 1, m_\ell; s = 1/2, m_s\rangle$, using either the recursive construction from the state $|j = 3/2, m_j = 3/2, \ell = 1, s = 1/2\rangle = |\ell = 1, m_\ell = 1; s = 1/2, m_s = 1/2\rangle$ or Racah's formula (8.37). Compare with the results from a symbolic computation program or an online applet for the calculation of Clebsch-Gordan coefficients.

Chapter 9
Stationary Perturbations in Quantum Mechanics

We denote a quantum system with a time-independent Hamiltonian H_0 as *solvable* (or sometimes also as *exactly solvable*) if we can calculate the energy eigenvalues and eigenstates of H_0 analytically. The harmonic oscillator and the hydrogen atom provide two examples of solvable quantum systems. Exactly solvable systems provide very useful models for quantum behavior in physical systems. The harmonic oscillator describes systems near a stable equilibrium, while the Hamiltonian with a Coulomb potential is an important model system for atomic physics and for every quantum system which is dominated by Coulomb interactions. However, in many cases the Schrödinger equation will not be solvable, and we have to go beyond solvable model systems to calculate quantitative properties. In these cases we have to resort to the calculation of approximate solutions. The methods developed in the present chapter are applicable to perturbations of discrete energy levels by time-independent perturbations V of the Hamiltonian, $H_0 \to H = H_0 + V$.

9.1 Time-independent perturbation theory without degeneracies

We consider a perturbation of a solvable time-independent Hamiltonian H_0 by a time-independent term V, and for bookkeeping purposes we extract a coupling constant λ from the perturbation,

$$H = H_0 + V \to H = H_0 + \lambda V.$$

After the relevant expressions for shifts of states and energy levels have been calculated to the desired order in λ, we usually subsume λ again in V, such that e.g. $\lambda\langle\phi^{(0)}|V|\psi^{(0)}\rangle \to \langle\phi^{(0)}|V|\psi^{(0)}\rangle$.

R. Dick, *Advanced Quantum Mechanics*, Graduate Texts in Physics,
DOI 10.1007/978-3-319-25675-7_9

We know the unperturbed energy levels and eigenstates of the solvable Hamiltonian H_0,

$$H_0|\psi_j^{(0)}\rangle = E_j^{(0)}|\psi_j^{(0)}\rangle.$$

In the present section we assume that the energy levels $E_j^{(0)}$ are not degenerate, and we want to calculate in particular approximations for the energy level E_i which arises from the unperturbed energy level $E_i^{(0)}$ due to the presence of the perturbation V. We will see below that consistency of the formalism requires that the differences $|E_i^{(0)} - E_j^{(0)}|$ for $j \neq i$ must have a positive minimal value, i.e. the unperturbed energy level $E_i^{(0)}$ for which we want to calculate corrections has to be discrete[1].

Orthogonality of eigenstates for different energy eigenvalues implies

$$\langle\psi_i^{(0)}|\psi_j^{(0)}\rangle = \delta_{ij}.$$

In the most common form of time-independent perturbation theory we try to find an approximate solution to the equation

$$H|\psi_i\rangle = E_i|\psi_i\rangle$$

in terms of power series expansions in the coupling constant λ,

$$|\psi_i\rangle = \sum_{n\geq 0}\lambda^n|\psi_i^{(n)}\rangle,\ \ \langle\psi_i^{(0)}|\psi_i^{(n\geq 1)}\rangle = 0,\ \ E_i = \sum_{n\geq 0}\lambda^n E_i^{(n)}. \tag{9.1}$$

Depending on the properties of V, these series may converge for small values of $|\lambda|$, or they may only hold as asymptotic expansions for $|\lambda| \to 0$. The book by Kato [21] provides results and resources on convergence and applicability properties of the perturbation series. Here we will focus on the commonly used first and second order expressions for wave functions and energy levels.

We can require

$$\langle\psi_i^{(0)}|\psi_i^{(n)}\rangle = \delta_{n,0} \tag{9.2}$$

because the recursion equation (9.3) below, which is derived without the assumption (9.2), does not determine these particular coefficients. One way to understand this is to observe that we can decompose $|\psi_i^{(n\geq 1)}\rangle$ into terms parallel and orthogonal to $|\psi_i^{(0)}\rangle$,

$$|\psi_i^{(n\geq 1)}\rangle = |\psi_i^{(0)}\rangle\langle\psi_i^{(0)}|\psi_i^{(n\geq 1)}\rangle + |\psi_i^{(n\geq 1)}\rangle - |\psi_i^{(0)}\rangle\langle\psi_i^{(0)}|\psi_i^{(n\geq 1)}\rangle.$$

[1] This condition is not affected by a possible degeneracy of $E_i^{(0)}$, as will be shown in Section 9.2.

Inclusion of the parallel part $|\psi_i^{(0)}\rangle\langle\psi_i^{(0)}|\psi_i^{(n\geq 1)}\rangle$ in the zeroth order term, followed by a rescaling by

$$\left(1+\langle\psi_i^{(0)}|\psi_i^{(n\geq 1)}\rangle\right)^{-1} = 1-\langle\psi_i^{(0)}|\psi_i^{(n\geq 1)}\rangle + \mathcal{O}(\lambda^{2n})$$

to restore a coefficient 1 in the zeroth order term, affects only terms of order λ^{n+1} or higher in the perturbation series. This implies that if we have solved the Schrödinger equation to order λ^{n-1} with the constraint

$$\langle\psi_i^{(0)}|\psi_i^{(m)}\rangle = \delta_{m,0}, \quad 0\leq m\leq n-1,$$

then ensuring that constraint also to order λ^n preserves the constraint for the lower order terms. Therefore we can fulfill the constraint (9.2) to any desired order in which we wish to calculate the perturbation series.

Substitution of the perturbative expansions into the Schrödinger equation $H|\psi_i\rangle = E_i|\psi_i\rangle$ yields

$$\begin{aligned}\sum_{n\geq 0}\lambda^n H_0|\psi_i^{(n)}\rangle + \sum_{n\geq 0}\lambda^{n+1}V|\psi_i^{(n)}\rangle &= \sum_{m,n\geq 0}\lambda^{m+n}E_i^{(m)}|\psi_i^{(n)}\rangle \\ &= \sum_{n\geq 0}\sum_{m=0}^{n}\lambda^n E_i^{(m)}|\psi_i^{(n-m)}\rangle.\end{aligned}$$

This equation is automatically fulfilled at zeroth order. Isolation of terms of order λ^{n+1} for $n\geq 0$ yields

$$H_0|\psi_i^{(n+1)}\rangle + V|\psi_i^{(n)}\rangle = \sum_{m=0}^{n+1}E_i^{(m)}|\psi_i^{(n-m+1)}\rangle,$$

and projection of this equation onto $|\psi_j^{(0)}\rangle$ yields

$$\begin{aligned}E_j^{(0)}\langle\psi_j^{(0)}|\psi_i^{(n+1)}\rangle + \langle\psi_j^{(0)}|V|\psi_i^{(n)}\rangle = &\sum_{m=0}^{n}E_i^{(m)}\langle\psi_j^{(0)}|\psi_i^{(n-m+1)}\rangle \\ &+E_i^{(n+1)}\delta_{ij}.\end{aligned} \tag{9.3}$$

We can first calculate the first order corrections for energy levels and wave functions from this equation, and then solve it recursively to any desired order.

First order corrections to the energy levels and eigenstates

The first order corrections are found from equation (9.3) for $n=0$. Substitution of $j=i$ implies for the first order shifts of the energy levels the result

$$E_i^{(1)} = \langle\psi_i^{(0)}|V|\psi_i^{(0)}\rangle, \tag{9.4}$$

and $j \neq i$ yields with $E_i^{(0)} \neq E_j^{(0)}$ the first order shifts of the energy eigenstates

$$\langle\psi_j^{(0)}|\psi_i^{(1)}\rangle = \frac{\langle\psi_j^{(0)}|V|\psi_i^{(0)}\rangle}{E_i^{(0)} - E_j^{(0)}}. \tag{9.5}$$

Recursive solution of equation (9.3) for n ≥ 1

We first observe that $j = i$ in equation (9.3) implies with the condition (9.2)

$$E_i^{(n+1)} = \langle\psi_i^{(0)}|V|\psi_i^{(n)}\rangle - \sum_{m=1}^{n} E_i^{(m)}\langle\psi_i^{(0)}|\psi_i^{(n-m+1)}\rangle = \langle\psi_i^{(0)}|V|\psi_i^{(n)}\rangle, \tag{9.6}$$

and $i \neq j$ yields

$$\left(E_i^{(0)} - E_j^{(0)}\right)\langle\psi_j^{(0)}|\psi_i^{(n+1)}\rangle = \langle\psi_j^{(0)}|V|\psi_i^{(n)}\rangle - \sum_{m=1}^{n} E_i^{(m)}\langle\psi_j^{(0)}|\psi_i^{(n-m+1)}\rangle. \tag{9.7}$$

The right hand side of both equations depends only on lower order shifts of energy levels and eigenstates. Therefore these equations can be used for the recursive solution of equation (9.3) to arbitrary order.

Second order corrections to the energy levels and eigenstates

Substitution of $n = 1$ into equation (9.6) yields with (9.5) and

$$\sum\!\!\!\!\!\!\int_k |\psi_k^{(0)}\rangle\langle\psi_k^{(0)}| = 1$$

the second order shift

$$E_i^{(2)} = \sum\!\!\!\!\!\!\int_{k\neq i} \frac{\langle\psi_i^{(0)}|V|\psi_k^{(0)}\rangle\langle\psi_k^{(0)}|V|\psi_i^{(0)}\rangle}{E_i^{(0)} - E_k^{(0)}} = \sum\!\!\!\!\!\!\int_{k\neq i} \frac{|\langle\psi_i^{(0)}|V|\psi_k^{(0)}\rangle|^2}{E_i^{(0)} - E_k^{(0)}}. \tag{9.8}$$

States in the continuous part of the spectrum of H_0 will also contribute to the shifts in energy levels and eigenstates. It is only required that the energy level $E_i^{(0)}$, for which we want to calculate the corrections, is discrete and does not overlap with any continuous energy levels.

Note that equation (9.8) implies that the second order correction to the ground state energy is always negative.

For the eigenstates, equation (9.7) yields with the first order results (9.4, 9.5) the equation (recall $i \neq j$ in (9.7))

$$\langle\psi_j^{(0)}|\psi_i^{(2)}\rangle = \sum_{k\neq i}\frac{\langle\psi_j^{(0)}|V|\psi_k^{(0)}\rangle\langle\psi_k^{(0)}|V|\psi_i^{(0)}\rangle}{(E_i^{(0)}-E_j^{(0)})(E_i^{(0)}-E_k^{(0)})} - \frac{\langle\psi_i^{(0)}|V|\psi_i^{(0)}\rangle\langle\psi_j^{(0)}|V|\psi_i^{(0)}\rangle}{(E_i^{(0)}-E_j^{(0)})^2}. \tag{9.9}$$

Now we can explain why it is important that our original unperturbed energy level $E_i^{(0)}$ is discrete. To ensure that the n-th order corrections to the energy levels and eigenstates in equations (9.1) are really of order λ^n (or smaller than all previous terms), the matrix elements $|\langle\psi_j^{(0)}|V|\psi_k^{(0)}\rangle|$ of the perturbation operator should be at most of the same order of magnitude as the energy differences $|E_i^{(0)} - E_j^{(0)}|$ between the unperturbed level $E_i^{(0)}$ and the other unperturbed energy levels in the system. This implies in particular that the minimal absolute energy difference between $E_i^{(0)}$ and the other unperturbed energy levels must not vanish, i.e. $E_i^{(0)}$ must be a discrete energy level.

Equations (9.4) and (9.8) (and their counterparts (9.16) and (9.24) in degenerate perturbation theory below) used to be the most frequently employed equations of time-independent perturbation theory, because historically many experiments were concerned with spectroscopic determinations of energy levels. However, measurements e.g. of local electron densities or observations of wave functions (e.g. in scanning tunneling microscopes or through X-ray scattering using synchrotrons) are very common nowadays, and therefore the corrections to the states are also directly relevant for the interpretation of experimental data.

Summary of non-degenerate perturbation theory in second order

If we include λ with V, the states and energy levels in second order are

$$|\psi_i\rangle = |\psi_i^{(0)}\rangle + |\psi_i^{(1)}\rangle + |\psi_i^{(2)}\rangle = |\psi_i^{(0)}\rangle + \sum_{j\neq i}|\psi_j^{(0)}\rangle\frac{\langle\psi_j^{(0)}|V|\psi_i^{(0)}\rangle}{E_i^{(0)}-E_j^{(0)}} + \sum_{j,k\neq i}|\psi_j^{(0)}\rangle\frac{\langle\psi_j^{(0)}|V|\psi_k^{(0)}\rangle\langle\psi_k^{(0)}|V|\psi_i^{(0)}\rangle}{(E_i^{(0)}-E_j^{(0)})(E_i^{(0)}-E_k^{(0)})} - \sum_{j\neq i}|\psi_j^{(0)}\rangle\frac{\langle\psi_j^{(0)}|V|\psi_i^{(0)}\rangle\langle\psi_i^{(0)}|V|\psi_i^{(0)}\rangle}{(E_i^{(0)}-E_j^{(0)})^2} \tag{9.10}$$

and

$$E_i = E_i^{(0)} + \langle\psi_i^{(0)}|V|\psi_i^{(0)}\rangle + \sum_{j\neq i} \frac{|\langle\psi_i^{(0)}|V|\psi_j^{(0)}\rangle|^2}{E_i^{(0)} - E_j^{(0)}}. \tag{9.11}$$

The second order states $|\psi_i\rangle$ are not normalized any more,

$$\langle\psi_i|\psi_j\rangle = \delta_{ij} + \mathcal{O}(\lambda^2)\delta_{ij}.$$

Normalization is preserved in first order due to

$$\langle\psi_i^{(0)}|\psi_j^{(1)}\rangle + \langle\psi_i^{(1)}|\psi_j^{(0)}\rangle = 0,$$

but in second order we have

$$\langle\psi_i^{(0)}|\psi_j^{(2)}\rangle + \langle\psi_i^{(1)}|\psi_j^{(1)}\rangle + \langle\psi_i^{(2)}|\psi_j^{(0)}\rangle = \sum_{k\neq i} \frac{|\langle\psi_i^{(0)}|V|\psi_k^{(0)}\rangle|^2}{\left(E_i^{(0)} - E_k^{(0)}\right)^2}\delta_{ij}.$$

However, we can add to the leading term $|\psi_i^{(0)}\rangle$ in $|\psi_i\rangle$ a term of the form $|\psi_i^{(0)}\rangle\mathcal{O}(\lambda^2)$ and still preserve the master equation (9.3) to second order. We can therefore rescale (9.10) by a factor $[1 + \mathcal{O}(\lambda^2)]^{-1/2}$ to a normalized second order state

$$\begin{aligned}
|\psi_i\rangle = {} & |\psi_i^{(0)}\rangle - \frac{1}{2}|\psi_i^{(0)}\rangle \sum_{j\neq i} \frac{|\langle\psi_i^{(0)}|V|\psi_j^{(0)}\rangle|^2}{\left(E_i^{(0)} - E_j^{(0)}\right)^2} + \sum_{j\neq i} |\psi_j^{(0)}\rangle \frac{\langle\psi_j^{(0)}|V|\psi_i^{(0)}\rangle}{E_i^{(0)} - E_j^{(0)}} \\
& + \sum_{j,k\neq i} |\psi_j^{(0)}\rangle \frac{\langle\psi_j^{(0)}|V|\psi_k^{(0)}\rangle\langle\psi_k^{(0)}|V|\psi_i^{(0)}\rangle}{\left(E_i^{(0)} - E_j^{(0)}\right)\left(E_i^{(0)} - E_k^{(0)}\right)} \\
& - \sum_{j\neq i} |\psi_j^{(0)}\rangle \frac{\langle\psi_j^{(0)}|V|\psi_i^{(0)}\rangle\langle\psi_i^{(0)}|V|\psi_i^{(0)}\rangle}{\left(E_i^{(0)} - E_j^{(0)}\right)^2}.
\end{aligned} \tag{9.12}$$

Now the second order shift is not orthogonal to $|\psi_i^{(0)}\rangle$ any more, but we still have a solution of equation (9.3) to second order.

9.2 Time-independent perturbation theory with degenerate energy levels

Now we admit degeneracy of energy levels of our unperturbed Hamiltonian H_0. Time-independent perturbation theory in the previous section repeatedly involved division by energy differences $[E_i^{(0)} - E_j^{(0)}]_{i\neq j}$. This will not be possible any more for pairs of degenerate energy levels, and we have to carefully reconsider each step in the previous derivation if degeneracies are involved.

The full Hamiltonian and the 0-th order results are now

$$H = H_0 + \lambda V,\ H_0|\psi_{j\alpha}^{(0)}\rangle = E_j^{(0)}|\psi_{j\alpha}^{(0)}\rangle,$$

where Greek indices denote sets of degeneracy indices. For example, if H_0 would correspond to a hydrogen atom, the quantum number j would correspond to the principal quantum number n of a bound state or the wave number k of a spherical Coulomb wave, and the degeneracy index α would correspond to the set of angular momentum quantum number, magnetic quantum number, and spin projection, $\alpha = \{\ell, m_\ell, m_s\}$. For the same reasons as in equation (9.9), the energy level for which we wish to calculate an approximation must be discrete, i.e. the techniques developed in this chapter can be used to study perturbations of the bound states of hydrogen atoms, but not perturbations of Coulomb waves.

We denote the degeneracy subspace to the energy level $E_j^{(0)}$ as $\mathcal{E}_j$ and the projector on $\mathcal{E}_j$ is

$$\mathcal{P}_j^{(0)} = \sum_\alpha |\psi_{j\alpha}^{(0)}\rangle\langle\psi_{j\alpha}^{(0)}|.$$

As in the previous section, we wish to calculate approximations for the energy level $E_{i\alpha}$ and corresponding eigenstates $|\psi_{i\alpha}\rangle$, $H|\psi_{i\alpha}\rangle = E_{i\alpha}|\psi_{i\alpha}\rangle$, which arise from the energy level $E_i^{(0)}$ and the eigenstates $|\psi_{i\alpha}^{(0)}\rangle$ due to the perturbation V. The energy level $E_i^{(0)}$ may split into several energy levels $E_{i\alpha}$ because the perturbation might lift the degeneracy of $E_i^{(0)}$. We will actually assume that the perturbation V lifts the degeneracy of the energy level $E_i^{(0)}$ already at first order, $E_{i\alpha}^{(1)} \neq E_{i\beta}^{(1)}$ if $\alpha \neq \beta$.

The Rayleigh-Ritz-Schrödinger *ansatz* is

$$|\psi_{i\alpha}\rangle = \sum_{n\geq 0} \lambda^n |\psi_{i\alpha}^{(n)}\rangle,\ \langle\psi_{i\alpha}^{(0)}|\psi_{i\alpha}^{(n\geq 1)}\rangle = 0,\ E_{i\alpha} = \sum_{n\geq 0} \lambda^n E_{i\alpha}^{(n)}. \tag{9.13}$$

Substitution into the full time-independent Schrödinger equation yields

$$\begin{aligned}\sum_{n\geq 0} \lambda^n H_0|\psi_{i\alpha}^{(n)}\rangle + \sum_{n\geq 0} \lambda^{n+1} V|\psi_{i\alpha}^{(n)}\rangle &= \sum_{m,n\geq 0} \lambda^{m+n} E_{i\alpha}^{(m)}|\psi_{i\alpha}^{(n)}\rangle \\ &= \sum_{n\geq 0}\sum_{m=0}^{n} \lambda^n E_{i\alpha}^{(m)}|\psi_{i\alpha}^{(n-m)}\rangle.\end{aligned}$$

This is yields in $(n + 1)$-st order for $n \geq 0$

$$H_0|\psi_{i\alpha}^{(n+1)}\rangle + V|\psi_{i\alpha}^{(n)}\rangle = \sum_{m=0}^{n+1} E_{i\alpha}^{(m)}|\psi_{i\alpha}^{(n-m+1)}\rangle. \tag{9.14}$$

We determine the corrections $|\psi_{i\alpha}^{(n\geq 1)}\rangle$ to the wave functions through their projections $\langle\psi_{j\beta}^{(0)}|\psi_{i\alpha}^{(n\geq 1)}\rangle$ onto the basis of unperturbed states. Projection of equation (9.14) yields

$$E_j^{(0)}\langle\psi_{j\beta}^{(0)}|\psi_{i\alpha}^{(n+1)}\rangle + \langle\psi_{j\beta}^{(0)}|V|\psi_{i\alpha}^{(n)}\rangle = \sum_{m=0}^{n} E_{i\alpha}^{(m)}\langle\psi_{j\beta}^{(0)}|\psi_{i\alpha}^{(n-m+1)}\rangle + E_{i\alpha}^{(n+1)}\delta_{ij}\delta_{\alpha\beta}. \tag{9.15}$$

First order corrections to the energy levels

The first order equations ($n = 0$ in equation (9.15)) yield for $j = i$ and $\beta = \alpha$ the equation

$$E_{i\alpha}^{(1)} = \langle\psi_{i\alpha}^{(0)}|V|\psi_{i\alpha}^{(0)}\rangle, \tag{9.16}$$

while $j = i$, $\alpha \neq \beta$ imposes a consistency condition on the choice of basis of unperturbed states,

$$\langle\psi_{i\beta}^{(0)}|V|\psi_{i\alpha}^{(0)}\rangle\Big|_{\beta\neq\alpha} = 0, \tag{9.17}$$

This condition means that we have to diagonalize V first *within each degeneracy subspace* $\mathcal{E}_i$ in the sense

$$V|\psi_{i\alpha}^{(0)}\rangle = E_{i\alpha}^{(1)}|\psi_{i\alpha}^{(0)}\rangle + \sum\!\!\!\!\!\!\int_{j\neq i}\sum_{\beta}|\psi_{j\beta}^{(0)}\rangle\langle\psi_{j\beta}^{(0)}|V|\psi_{i\alpha}^{(0)}\rangle, \tag{9.18}$$

before we can use the perturbation *ansatz* (9.13), and according to (9.16) the first order energy corrections $E_{i\alpha}^{(1)}$ are the corresponding eigenvalues in the i-th degeneracy subspace. If the first order energy corrections $E_{i\alpha}^{(1)}$ are all we care about, this means that we can calculate them from the eigenvalue conditions

$$\det\Big[\langle\psi_{i\beta}^{(0)}|V|\psi_{i\alpha}^{(0)}\rangle - E_{i\alpha}^{(1)}\delta_{\alpha\beta}\Big] = 0, \tag{9.19}$$

using any initial choice of unperturbed orthogonal energy eigenstates. But that would achieve only a very limited objective.

As also indicated in equation (9.18), diagonalization within the subspaces means only diagonalization of the operators $\mathcal{P}_i^{(0)}V\mathcal{P}_i^{(0)}$, which does *not* amount to total diagonalization of V,

$$\sum_i \mathcal{P}_i^{(0)}V\mathcal{P}_i^{(0)} \neq V = \sum_{i,j}\mathcal{P}_i^{(0)}V\mathcal{P}_j^{(0)}.$$

We still will have non-vanishing transition matrix elements $\langle\psi_{j\beta}^{(0)}|V|\psi_{i\alpha}^{(0)}\rangle \neq 0$ between different degeneracy subspaces $i \neq j$.

First order corrections to the energy eigenstates

Setting $i \neq j$ in equation (9.15) yields a part of the first order corrections to the wave functions,

$$\langle \psi_{j\beta}^{(0)} | \psi_{i\alpha}^{(1)} \rangle = \frac{\langle \psi_{j\beta}^{(0)} | V | \psi_{i\alpha}^{(0)} \rangle}{E_i^{(0)} - E_j^{(0)}}. \tag{9.20}$$

However, this yields only the projections $\langle \psi_{j\beta}^{(0)} | \psi_{i\alpha}^{(1)} \rangle$ of the first order corrections $|\psi_{i\alpha}^{(1)}\rangle$ onto the unperturbed states for $j \neq i$. We need to use $j = i$ in the second order equations to calculate the missing terms $\langle \psi_{i\beta}^{(0)} | \psi_{i\alpha}^{(1)} \rangle$, $(\beta \neq \alpha)$, for the first order corrections.

Equation (9.15) yields for $n = 1, j = i$ and $\beta \neq \alpha$ the equation

$$\langle \psi_{i\beta}^{(0)} | V | \psi_{i\alpha}^{(1)} \rangle \Big|_{\beta \neq \alpha} = E_{i\alpha}^{(1)} \, \langle \psi_{i\beta}^{(0)} | \psi_{i\alpha}^{(1)} \rangle \Big|_{\beta \neq \alpha}$$

and after substitution of equations (9.16, 9.17, 9.20)

$$\begin{aligned}\left(E_{i\alpha}^{(1)} - E_{i\beta}^{(1)}\right) \langle \psi_{i\beta}^{(0)} | \psi_{i\alpha}^{(1)} \rangle \Big|_{\beta \neq \alpha} &= \sum\!\!\!\!\!\!\!\int_{j \neq i} \sum_{\gamma} \langle \psi_{i\beta}^{(0)} | V | \psi_{j\gamma}^{(0)} \rangle \langle \psi_{j\gamma}^{(0)} | \psi_{i\alpha}^{(1)} \rangle \\ &= \sum\!\!\!\!\!\!\!\int_{j \neq i} \sum_{\gamma} \frac{\langle \psi_{i\beta}^{(0)} | V | \psi_{j\gamma}^{(0)} \rangle \langle \psi_{j\gamma}^{(0)} | V | \psi_{i\alpha}^{(0)} \rangle}{E_i^{(0)} - E_j^{(0)}},\end{aligned}$$

i.e. we find the missing pieces of the first order corrections to the states

$$\begin{aligned}\langle \psi_{i\beta}^{(0)} | \psi_{i\alpha}^{(1)} \rangle \Big|_{\beta \neq \alpha} &= \frac{1}{\langle \psi_{i\alpha}^{(0)} | V | \psi_{i\alpha}^{(0)} \rangle - \langle \psi_{i\beta}^{(0)} | V | \psi_{i\beta}^{(0)} \rangle} \\ &\quad \times \sum\!\!\!\!\!\!\!\int_{j \neq i} \sum_{\gamma} \frac{\langle \psi_{i\beta}^{(0)} | V | \psi_{j\gamma}^{(0)} \rangle \langle \psi_{j\gamma}^{(0)} | V | \psi_{i\alpha}^{(0)} \rangle}{E_i^{(0)} - E_j^{(0)}}\end{aligned} \tag{9.21}$$

if V has removed the degeneracy between $|\psi_{i\alpha}\rangle$ and $|\psi_{i\beta}\rangle$ in first order, $E_{i\alpha}^{(1)} \neq E_{i\beta}^{(1)}$.

Recursive solution of equation (9.15) for $n \geq 1$

We first rewrite equation (9.15) by inserting

$$1 = \sum\!\!\!\!\!\!\!\int_{k,\gamma} |\psi_{k\gamma}^{(0)}\rangle \langle \psi_{k\gamma}^{(0)}|$$

in the matrix element of V, and using equations (9.16, 9.17):

$$E_j^{(0)}\langle\psi_{j\beta}^{(0)}|\psi_{i\alpha}^{(n+1)}\rangle + E_{j\beta}^{(1)}\langle\psi_{j\beta}^{(0)}|\psi_{i\alpha}^{(n)}\rangle + \sum\!\!\!\!\!\!\!\int_{k\neq j}\sum_{\gamma}\langle\psi_{j\beta}^{(0)}|V|\psi_{k\gamma}^{(0)}\rangle\langle\psi_{k\gamma}^{(0)}|\psi_{i\alpha}^{(n)}\rangle$$

$$= E_i^{(0)}\langle\psi_{j\beta}^{(0)}|\psi_{i\alpha}^{(n+1)}\rangle + E_{i\alpha}^{(1)}\langle\psi_{j\beta}^{(0)}|\psi_{i\alpha}^{(n)}\rangle + \Theta(n\geq 2)\sum_{m=2}^{n}E_{i\alpha}^{(m)}\langle\psi_{j\beta}^{(0)}|\psi_{i\alpha}^{(n-m+1)}\rangle$$

$$+E_{i\alpha}^{(n+1)}\delta_{ij}\delta_{\alpha\beta}. \tag{9.22}$$

Substitution of $j = i$ and $\beta = \alpha$ yields

$$E_{i\alpha}^{(n+1)} = \sum\!\!\!\!\!\!\!\int_{k\neq i}\sum_{\gamma}\langle\psi_{i\alpha}^{(0)}|V|\psi_{k\gamma}^{(0)}\rangle\langle\psi_{k\gamma}^{(0)}|\psi_{i\alpha}^{(n)}\rangle, \tag{9.23}$$

where equations (9.13, 9.17) have been used. The second order correction is in particular with equation (9.20):

$$E_{i\alpha}^{(2)} = \sum\!\!\!\!\!\!\!\int_{j\neq i}\sum_{\beta}\frac{|\langle\psi_{j\beta}^{(0)}|V|\psi_{i\alpha}^{(0)}\rangle|^2}{E_i^{(0)}-E_j^{(0)}}. \tag{9.24}$$

We find again that the second order correction to the ground state energy is always negative.

For the higher order shifts of the states we find for $j \neq i$ in equation (9.22)

$$\left(E_i^{(0)}-E_j^{(0)}\right)\langle\psi_{j\beta}^{(0)}|\psi_{i\alpha}^{(n+1)}\rangle = \langle\psi_{j\beta}^{(0)}|V|\psi_{i\alpha}^{(n)}\rangle - \sum_{m=1}^{n}E_{i\alpha}^{(m)}\langle\psi_{j\beta}^{(0)}|\psi_{i\alpha}^{(n-m+1)}\rangle$$

$$= \langle\psi_{j\beta}^{(0)}|V|\psi_{i\alpha}^{(n)}\rangle - \langle\psi_{i\alpha}^{(0)}|V|\psi_{i\alpha}^{(0)}\rangle\langle\psi_{j\beta}^{(0)}|\psi_{i\alpha}^{(n)}\rangle$$

$$-\Theta(n\geq 2)\sum_{m=1}^{n-1}\sum\!\!\!\!\!\!\!\int_{k\neq i}\sum_{\gamma}\langle\psi_{i\alpha}^{(0)}|V|\psi_{k\gamma}^{(0)}\rangle\langle\psi_{k\gamma}^{(0)}|\psi_{i\alpha}^{(m)}\rangle\langle\psi_{j\beta}^{(0)}|\psi_{i\alpha}^{(n-m)}\rangle, \tag{9.25}$$

which gives us the contributions $\langle\psi_{j\beta}^{(0)}|\psi_{i\alpha}^{(n+1)}\rangle\Big|_{j\neq i}$ to the $(n+1)$-st order wave function corrections.

Substitution of $i = j$, $\alpha \neq \beta$ yields finally

$$\left(E_{i\alpha}^{(1)}-E_{i\beta}^{(1)}\right)\langle\psi_{i\beta}^{(0)}|\psi_{i\alpha}^{(n\geq 1)}\rangle\Big|_{\beta\neq\alpha} = \sum\!\!\!\!\!\!\!\int_{k\neq i}\sum_{\gamma}\langle\psi_{i\beta}^{(0)}|V|\psi_{k\gamma}^{(0)}\rangle\langle\psi_{k\gamma}^{(0)}|\psi_{i\alpha}^{(n)}\rangle$$

$$-\Theta(n\geq 2)\sum_{m=2}^{n}E_{i\alpha}^{(m)}\langle\psi_{i\beta}^{(0)}|\psi_{i\alpha}^{(n-m+1)}\rangle = \sum\!\!\!\!\!\!\!\int_{k\neq i}\sum_{\gamma}\langle\psi_{i\beta}^{(0)}|V|\psi_{k\gamma}^{(0)}\rangle\langle\psi_{k\gamma}^{(0)}|\psi_{i\alpha}^{(n)}\rangle$$

$$-\Theta(n\geq 2)\sum_{m=1}^{n-1}\sum\!\!\!\!\!\!\!\int_{k\neq i}\sum_{\gamma}\langle\psi_{i\alpha}^{(0)}|V|\psi_{k\gamma}^{(0)}\rangle\langle\psi_{k\gamma}^{(0)}|\psi_{i\alpha}^{(m)}\rangle\langle\psi_{i\beta}^{(0)}|\psi_{i\alpha}^{(n-m)}\rangle. \tag{9.26}$$

This gives us the missing pieces $\langle\psi_{i\beta}^{(0)}|\psi_{i\alpha}^{(n)}\rangle\Big|_{\beta\neq\alpha}$ of the n-th order wave function correction for $E_{i\alpha}^{(1)} \neq E_{i\beta}^{(1)}$.

Summary of first order shifts of the level $E_i^{(0)}$ if the perturbation lifts the degeneracy of the level

We must diagonalize the perturbation operator V within the degeneracy subspace $\mathcal{E}_i$ in the sense of (9.18), i.e. we must choose the unperturbed eigenstates $|\psi_{i\alpha}^{(0)}\rangle$ such that the equation

$$\langle\psi_{i\alpha}^{(0)}|V|\psi_{i\beta}^{(0)}\rangle = E_{i\alpha}^{(1)}\delta_{\alpha\beta} \tag{9.27}$$

also holds for $\alpha \neq \beta$.

The projections of the first order shifts of the energy eigenstates onto states in other degeneracy sectors are

$$\langle\psi_{j\beta}^{(0)}|\psi_{i\alpha}^{(1)}\rangle\Big|_{j\neq i} = \frac{\langle\psi_{j\beta}^{(0)}|V|\psi_{i\alpha}^{(0)}\rangle}{E_i^{(0)} - E_j^{(0)}}, \tag{9.28}$$

and the projections within the degeneracy sector are

$$\langle\psi_{i\beta}^{(0)}|\psi_{i\alpha}^{(1)}\rangle\Big|_{\beta\neq\alpha} = \frac{1}{\langle\psi_{i\alpha}^{(0)}|V|\psi_{i\alpha}^{(0)}\rangle - \langle\psi_{i\beta}^{(0)}|V|\psi_{i\beta}^{(0)}\rangle}$$
$$\times \sum_{j\neq i}\sum_{\gamma} \frac{\langle\psi_{i\beta}^{(0)}|V|\psi_{j\gamma}^{(0)}\rangle\langle\psi_{j\gamma}^{(0)}|V|\psi_{i\alpha}^{(0)}\rangle}{E_i^{(0)} - E_j^{(0)}}. \tag{9.29}$$

This requires that the first order shifts have completely removed the degeneracies in the i-th energy level, $E_{i\beta}^{(1)} \neq E_{i\alpha}^{(1)}$ for $\beta \neq \alpha$.

9.3 Problems

9.1. A one-dimensional harmonic oscillator is perturbed by a term

$$V = \lambda[(a^{+})^2 + a^2]^2.$$

Calculate the first and second order corrections to the ground state energy and wave function.

9.2. An atom on a surface is prevented from moving along the surface through a two-dimensional potential

$$V(x,y)=\frac{1}{2}m\omega^2(x^2+y^2)+Ax^4+By^4,\quad A\geq 0,\quad B\geq 0.$$

Find an approximation H_0 for the Hamiltonian of the atom where you can write down exact energy levels and eigenstates for the atom.

Use the remaining terms in $H-H_0$ to calculate first order corrections to the energy levels and eigenstates of the atom.

9.3. Which results do you get for the perturbed system from 9.2 in second order perturbation theory?

9.4. Suppose that the perturbation V has removed all degeneracies in all energy levels of an unperturbed system. Show that all the first order states $|\psi_{i\alpha}^{(0)}\rangle+|\psi_{i\alpha}^{(1)}\rangle$ are orthonormal in first order.

9.5. A hydrogen atom is perturbed by a static electric field $\boldsymbol{E}=\mathcal{E}\boldsymbol{e}_z$ in z direction. This field induces an extra potential

$$V=-e\Phi=e\mathcal{E}z \tag{9.30}$$

in the Hamiltonian for relative motion.

9.5a. Calculate the shift of the ground state energy up to second order in $\mathcal{E}$.

9.5b. Calculate the shift of the ground state wave function up to second order in $\mathcal{E}$.

9.5c. Which constraints on $\mathcal{E}$ do you find from the requirement of applicability of perturbation theory?

9.6. Calculate the first order shifts of the $n=2$ level of hydrogen under the perturbation (9.30).

9.7. A two-level system has two energy eigenstates $|E_\pm\rangle$ with energies

$$H_0|E_\pm\rangle=\left(E_0\pm\frac{\Delta E}{2}\right)|E_\pm\rangle,\quad \Delta E\neq 0.$$

We can use 2-spinor notation such that a general state in the two-level system is

$$|\psi\rangle=\sum_\pm|E_\pm\rangle\langle E_\pm|\psi\rangle\to\psi=\begin{pmatrix}\psi_1\\ \psi_2\end{pmatrix},\quad \psi_1=\langle E_+|\psi\rangle,\quad \psi_2=\langle E_-|\psi\rangle.$$

The Hamiltonian in 2-spinor notation is

$$H_0=E_0\underline{1}+\frac{\Delta E}{2}\underline{\sigma}_3.$$

We now perturb the Hamiltonian $H_0 \to H = H_0 + V$ through a term

$$V = \frac{V_1}{2}\underline{\sigma}_1 + \frac{V_2}{2}\underline{\sigma}_2.$$

9.7a. Calculate the first order corrections to the energy levels and eigenstates due to the perturbation V.

9.7b. Calculate the second order corrections to the energy levels and eigenstates due to the perturbation V.

9.7c. The Hamiltonian H is a hermitian 2×2 matrix which can be diagonalized exactly.

Calculate the exact energy levels and eigenstates of H. Compare with the perturbative results from 9.7a and 9.7b.

9.8. Which consistency conditions in the degeneracy subspace $\mathcal{E}_i$ would you find if the perturbation V does *not* remove the degeneracy in that subspace in first order?

Solution. The derivation of (9.21) shows that if we still have $E^{(1)}_{i\alpha} = E^{(1)}_{i\beta}$ for all degeneracy indices in $\mathcal{E}_i$, consistency of the second order equation requires that not just the operator $V^{(1)}_i = \mathcal{P}^{(0)}_i V \mathcal{P}^{(0)}_i$ is diagonal, but also that the operator

$$V^{(2)}_i = \mathcal{P}^{(0)}_i V \frac{1 - \mathcal{P}^{(0)}_i}{E^{(0)}_i - H_0} V \mathcal{P}^{(0)}_i$$

is diagonal. However, consistency of the simultaneous diagonalization of $V^{(1)}_i$ and $V^{(2)}_i$ then also implies the condition

$$\left[V^{(1)}_i, V^{(2)}_i\right] = \mathcal{P}^{(0)}_i V \left(\mathcal{P}^{(0)}_i V \frac{1 - \mathcal{P}^{(0)}_i}{E^{(0)}_i - H_0} - \frac{1 - \mathcal{P}^{(0)}_i}{E^{(0)}_i - H_0} V \mathcal{P}^{(0)}_i\right) V \mathcal{P}^{(0)}_i = 0.$$

If V preserves the degeneracy in $\mathcal{E}_i$, but these consistency conditions cannot be met, then $H = H_0 + V$ apparently does not have a complete set of eigenstates which scale analytically under scaling $V \to \lambda V$ of the perturbation.

Chapter 10
Quantum Aspects of Materials I

Quantum mechanics is indispensable for the understanding of materials. In return, solid state physics provides beautiful illustrations for the impact of quantum dynamics on allowed energy levels in a system, for wave-particle duality, and for applications of perturbation theory.

In the present chapter we will focus on Bloch's theorem, the duality between Bloch and Wannier states, the emergence of energy bands in crystals, and the emergence of effective mass in *kp* perturbation theory. We will do this for one-dimensional lattices, since this captures the essential ideas. Students who would like to follow up on our introductory exposition and understand the profound impact of quantum mechanics on every physical property of materials at a deeper level should consult the monographs of Callaway [5], Ibach and Lüth [17], Kittel [22] or Madelung [25], or any of the other excellent texts on condensed matter physics - and they should include courses on condensed matter physics in their curriculum!

10.1 Bloch's theorem

Electrons in solid materials provide a particularly beautiful realization of wave-particle duality. Bloch's theorem covers the wave aspects of this duality. From a practical perspective, Bloch's theorem implies that we can discuss electrons in terms of states which sample the whole lattice of ion cores in a solid material. This has important implications for the energy levels of electrons in materials, and therefore for all physical properties of materials.

It is useful to recall the theory of discrete Fourier transforms as a preparation for the proof of Bloch's theorem. We write the discrete Fourier expansion for functions $f(x)$ with periodicity a as

$$f(x) = \sum_{n=-\infty}^{\infty} f_n \exp\!\left(2\pi \mathrm{i}\frac{nx}{a}\right). \tag{10.1}$$

R. Dick, *Advanced Quantum Mechanics*, Graduate Texts in Physics,
DOI 10.1007/978-3-319-25675-7_10

The orthogonality relation

$$\frac{1}{a}\int_0^a dx\,\exp\!\left(2\pi\mathrm{i}\frac{mx}{a}\right)\exp\!\left(-2\pi\mathrm{i}\frac{nx}{a}\right)=\delta_{mn} \tag{10.2}$$

yields the inversion

$$f_n=\frac{1}{a}\int_0^a dx f(x)\exp\!\left(-2\pi\mathrm{i}\frac{nx}{a}\right),$$

and substituting this back into equation (10.1) yields a representation of the δ-function in a finite interval of length a,

$$\frac{1}{a}\sum_{n=-\infty}^{\infty}\exp\!\left(2\pi\mathrm{i}n\frac{x-x'}{a}\right)=\delta(x-x'), \tag{10.3}$$

or equivalently

$$\sum_{n=-\infty}^{\infty}\exp(\mathrm{i}n\xi)=2\pi\delta(\xi). \tag{10.4}$$

Equation (10.3) is the completeness relation for the Fourier monomials on an interval of length a.

The Hamiltonian for electrons in a lattice with periodicity a is

$$H=\frac{\mathrm{p}^2}{2m}+V(\mathrm{x}),$$

where the potential operator has the periodicity of the lattice,

$$V(\mathrm{x})=V(\mathrm{x}+a)=\exp\!\left(\frac{\mathrm{i}}{\hbar}a\mathrm{p}\right)V(\mathrm{x})\exp\!\left(-\frac{\mathrm{i}}{\hbar}a\mathrm{p}\right).$$

This implies

$$\exp\!\left(\frac{\mathrm{i}}{\hbar}a\mathrm{p}\right)H=H\exp\!\left(\frac{\mathrm{i}}{\hbar}a\mathrm{p}\right),$$

and therefore eigenspace of H with eigenvalue E_n can be decomposed into eigenspaces of the lattice translation operator

$$T(a)=\exp\!\left(\frac{\mathrm{i}}{\hbar}a\mathrm{p}\right). \tag{10.5}$$

The eigenvalues of this unitary operator must be pure phase factors[1],

$$\exp\left(\frac{\mathrm{i}}{\hbar}a\mathrm{p}\right)|E_n,k\rangle = \exp(\mathrm{i}ka)\,|E_n,k\rangle. \tag{10.6}$$

Let us repeat this result in the x-representation:

$$\begin{aligned}\langle x|\exp\left(\frac{\mathrm{i}}{\hbar}a\mathrm{p}\right)|E_n,k\rangle &= \exp\left(a\frac{d}{dx}\right)\langle x|E_n,k\rangle = \langle x+a|E_n,k\rangle \\ &= \exp(\mathrm{i}ka)\,\langle x|E_n,k\rangle. \end{aligned} \tag{10.7}$$

This means that the energy eigenstate $\langle x|E_n,k\rangle \equiv \psi_n(k,x)$ has exactly the same periodicity properties under lattice translations as the plane wave $\langle x|k\rangle = \exp(\mathrm{i}kx)/\sqrt{2\pi}$. The ratio $\psi_n(k,x)/\langle x|k\rangle$ must therefore be a periodic function! This is Bloch's theorem in solid state physics[2]:

Energy eigenstates in a periodic lattice can always be written as the product of a periodic function $u_n(k,x+a) = u_n(k,x)$ with a plane wave,

$$\psi_n(k,x) = \sqrt{\frac{a}{2\pi}}\exp(\mathrm{i}kx)u_n(k,x). \tag{10.8}$$

The quasiperiodicity parameter k (multiplied with $\hbar$) has momentum-like properties, but it is not the momentum $\langle E_n,k|\mathrm{p}|E_n,k\rangle$ in the state $|E_n,k\rangle$. Therefore it is often denoted as a quasimomentum or a pseudomomentum.

Periodicity of the modulation factor $u_n(k,x)$ implies the expansions

$$u_n(k,x) = \sum_{\ell\in\mathbb{Z}} u_{n;\ell}(k)\exp\left(2\pi\mathrm{i}\frac{\ell x}{a}\right),$$

$$u_{n;\ell}(k) = \frac{1}{a}\int_0^a dx\, u_n(k,x)\exp\left(-2\pi\mathrm{i}\frac{\ell x}{a}\right).$$

We denote the eigenfunctions $\psi_n(k,x) \equiv \langle x|E_n,k\rangle$ of the lattice Hamiltonian as *Bloch functions* or equivalently as the x representation of the *Bloch states* $|E_n,k\rangle$. The corresponding periodic functions $u_n(k,x) \equiv \langle x|u_n(k)\rangle$ will be denoted as *Bloch factors*. Equation (10.8) for the Bloch state $|E_n,k\rangle$ reads in basis free notation

$$|E_n,k\rangle = \sqrt{\frac{a}{2\pi}}\exp(\mathrm{i}k\mathrm{x})|u_n(k)\rangle. \tag{10.9}$$

---

[1]This is a consequence of Schur's Lemma in group theory: Abelian symmetry groups have one-dimensional irreducible representations.

[2]F. Bloch, Z. Phys. 52, 555 (1929). As a mathematical theorem in the theory of differential equations it is known as Floquet's theorem due to G. Floquet, Ann. sci. de l'É.N.S., 2e série, 12, 47 (1883).

For arbitrary $\ell \in \mathbb{Z}$ the eigenvalues of the lattice translations satisfy

$$\exp(\mathrm{i}ka) = \exp\left[\mathrm{i}\left(k + \frac{2\pi\ell}{a}\right)a\right],$$

and therefore the quasimomentum k can be restricted to the region

$$-\frac{\pi}{a} < k \leq \frac{\pi}{a}, \qquad (10.10)$$

which is denoted as the first Brillouin zone of the (rescaled) dual lattice.

The quasiperiodicity property will impact the possible eigenstates of the lattice Hamiltonian H. The parameter k will therefore also impact the corresponding eigenvalues E_n, i.e. the eigenvalues will be functions $E_n(k)$ of the pseudomomentum. The index n enumerates different energy levels for each value of k in the first Brillouin zone. The functions $E_n(k)$ are known as energy bands, and we will see below that there are enumerably many energy bands in a lattice.

Orthogonality of the periodic Bloch factors

The orthogonality relation for the energy eigenstates $\psi_n(k,x) = \langle x|E_n,k\rangle$ implies an orthogonality property for the periodic Bloch factors $u_n(k,x) = \langle x|u_n(k)\rangle$. We have

$$\begin{aligned}\delta_{mn}\delta(k-k') &= \langle E_m,k'|E_n,k\rangle \\ &= \frac{a}{2\pi}\int_{-\infty}^{\infty} dx\,\exp\left[\mathrm{i}(k-k')x\right]u_m^+(k',x)u_n(k,x) \\ &= \frac{a}{2\pi}\sum_{\ell\in\mathbb{Z}}\exp\left[\mathrm{i}(k-k')\ell a\right] \\ &\quad\times\int_0^a dx\,\exp\left[\mathrm{i}(k-k')x\right]u_m^+(k',x)u_n(k,x).\end{aligned} \qquad (10.11)$$

Equation (10.4) implies

$$\sum_{\ell\in\mathbb{Z}}\exp\left[\mathrm{i}(k-k')\ell a\right] = \frac{2\pi}{a}\delta(k-k'), \qquad (10.12)$$

and substitution of this into equation (10.11) yields the orthogonality relations for the periodic Bloch factors,

$$\int_0^a dx\,u_m^+(k,x)u_n(k,x) = a\sum_{\ell}u_{m;\ell}^+(k)u_{n;\ell}(k) = \delta_{mn}. \qquad (10.13)$$

Note that this in turn also implies a normalization of the Bloch functions in the lattice cell,

$$\int_0^a dx\, |\psi_n(k,x)|^2 = \frac{a}{2\pi}.$$

The plane wave normalization in (10.11) implies length dimension 0 for the Bloch functions and length dimension -1/2 for the Bloch factors.

We remark that the completeness of the energy eigenstates yields

$$\begin{aligned}\delta(x-x') &= \sum_n \int_{-\pi/a}^{\pi/a} dk\, \langle x|E_n,k\rangle\langle E_n,k|x'\rangle \\ &= \frac{a}{2\pi}\sum_n \int_{-\pi/a}^{\pi/a} dk\, \exp\big[ik(x-x')\big]\, u_n^+(k,x')u_n(k,x), \qquad (10.14)\end{aligned}$$

but we cannot read off a separate relation for the Bloch factors from this.

10.2 Wannier states

The Bloch functions $\psi_n(k,x)$ are plane waves with a periodic modulation factor $u_n(k,x) = u_n(k,x+a)$, and therefore extend over the full lattice in x-space. However, due to

$$\exp(\mathrm{i}\,[k + (2\pi\ell/a)]\,a) = \exp(\mathrm{i}ka)$$

the quasiperiodicity parameter k was restricted to the first Brillouin zone

$$-\frac{\pi}{a} < k \leq \frac{\pi}{a}, \qquad (10.15)$$

i.e. as a function of k, $\psi_n(k,x)$ is only defined in the finite interval (10.15) (or equivalently has periodicity under shifts of k by multiples of $2\pi/a$). This implies the expansions[3]

$$\psi_n(k,x) = \sqrt{\frac{a}{2\pi}}\exp(\mathrm{i}kx)u_n(k,x) = \sqrt{\frac{a}{2\pi}}\sum_{\nu\in\mathbb{Z}} w_{n,\nu}(x)\exp(\mathrm{i}\nu ka), \qquad (10.16)$$

$$u_n(k,x) = \sum_{\nu\in\mathbb{Z}} w_{n,\nu}(x)\exp[-\mathrm{i}k(x-\nu a)]. \qquad (10.17)$$

[3] This is exactly as in (10.1), only with periodicity $2\pi/a$.

The functions $w_{n,\nu}(x) = \langle x|w_{n,\nu}\rangle$ are apparently Fourier transforms of the Bloch functions $\psi_n(k,x)$ with respect to the quasiperiodicity parameter k. These functions are known as *Wannier functions*[4].

The inversion of the expansion is

$$w_{n,\nu}(x) = \sqrt{\frac{a}{2\pi}}\int_{-\pi/a}^{\pi/a} dk\, \psi_n(k,x)\exp(-\mathrm{i}\nu ka)$$
$$= \frac{a}{2\pi}\int_{-\pi/a}^{\pi/a} dk\, u_n(k,x)\exp[\mathrm{i}k(x-\nu a)]. \tag{10.18}$$

The corresponding states

$$|w_{n,\nu}\rangle = \sqrt{\frac{a}{2\pi}}\int_{-\pi/a}^{\pi/a} dk\, \exp(-\mathrm{i}\nu ka)|E_n,k\rangle \tag{10.19}$$

are *Wannier states.*

The periodicity $u_n(k,x-\mu a) = u_n(k,x)$ of the Bloch functions also yields

$$w_{n,\nu}(x-\mu a) = w_{n,\nu+\mu}(x). \tag{10.20}$$

This implies in particular a localization property of Wannier functions,

$$w_{n,\nu}(x) = w_{n,0}(x-\nu a). \tag{10.21}$$

Determining all the functions $w_{n,\nu}(x)$ in one cell of the direct lattice is equivalent to finding the function $w_{n,0}(x)$ over the full lattice. Furthermore, $w_{n,\nu}(x)$ depends only on $x - \nu a$, i.e. it is attached to a lattice cell[5].

Wannier states satisfy completeness relations as a consequence of the completeness relations of the Bloch states. The relations are

$$\int_{-\infty}^{\infty} dx\, w^+_{m,\mu}(x)w_{n,\nu}(x) = \frac{a}{2\pi}\int_{-\pi/a}^{\pi/a} dk'\int_{-\pi/a}^{\pi/a} dk\int_{-\infty}^{\infty} dx\, \psi^+_m(k',x)\psi_n(k,x)$$
$$\times\exp[\mathrm{i}(\mu k' - \nu k)a]$$
$$= \delta_{mn}\frac{a}{2\pi}\int_{-\pi/a}^{\pi/a} dk\, \exp[\mathrm{i}(\mu-\nu)ka] = \delta_{mn}\delta_{\mu\nu},$$

[4]G.H. Wannier, Phys. Rev. 52, 191 (1937).

[5]It is tempting to conclude that the Wannier functions $w_{n,\nu}(x)$ should be centered around the lattice site $x = \nu a$, but this is not what generically happens. The Wannier function $w_{n,\nu}(x)$ is usually large in a unit cell containing the lattice site $x = \nu a$, but localization around the lattice site requires inclusion of extra phase factors $\exp[\mathrm{i}\varphi(n,k)]$ in the Bloch functions, see W. Kohn, Phys. Rev. 115, 809 (1959) and F.B. Pedersen, G.T. Einevoll, P.C. Hemmer, Phys. Rev. B 44, 5470 (1991).

and

$$\sum_{n,\nu} w_{n,\nu}(x)w^{+}_{n,\nu}(x') = \int_{-\pi/a}^{\pi/a} dk \int_{-\pi/a}^{\pi/a} dk' \frac{a}{2\pi} \sum_{\nu} \exp[-\mathrm{i}\nu(k-k')a]$$
$$\times \sum_{n} \psi_n(k,x)\psi_n^{+}(k',x')$$
$$= \int_{-\pi/a}^{\pi/a} dk \int_{-\pi/a}^{\pi/a} dk'\, \delta(k-k') \sum_{n} \psi_n(k,x)\psi_n^{+}(k',x')$$
$$= \int_{-\pi/a}^{\pi/a} dk \sum_{n} \psi_n(k,x)\psi_n^{+}(k,x') = \delta(x-x'). \qquad (10.22)$$

The periodicity of the Bloch functions in the dual lattice

$$\psi_n(k,x) = \psi_n\!\left(k + \frac{2\pi}{a}, x\right)$$

implies for the Bloch factors the quasiperiodicity

$$u_n\!\left(k + \frac{2\pi}{a}, x\right) = \exp\!\left(-2\pi\mathrm{i}\frac{x}{a}\right) u_n(k,x), \quad u_{n,\ell}\!\left(k + \frac{2\pi}{a}\right) = u_{n,\ell+1}(k),$$

and in particular

$$u_{n,\ell}(k) = u_{n,0}\!\left(k + \frac{2\pi\ell}{a}\right). \qquad (10.23)$$

This property of the Fourier coefficients of the Bloch factors in the dual lattice is dual to the property (10.21) of the Wannier functions in the direct lattice. Knowing all the Fourier coefficients $u_{n,\ell}(k)$ of the Bloch factors in a Brillouin zone is equivalent to knowing the Fourier coefficients $u_{n,0}(k)$ throughout the dual lattice. We can think of the functions $u_{n,-\ell}(k)$ as dual Wannier functions in k space. Indeed, these functions are related through Fourier transforms,

$$w_{n,0}(x) = \frac{a}{2\pi} \int_{-\infty}^{\infty} dk\, u_{n,0}(k) \exp(\mathrm{i}kx),$$
$$u_{n,0}(k) = \frac{1}{a} \int_{-\infty}^{\infty} dx\, w_{n,0}(x) \exp(-\mathrm{i}kx), \qquad (10.24)$$

see Problem 10.3. Knowing any particular Wannier function $w_{n,\nu}(x)$ in the whole lattice, or any particular dual Wannier function $u_{n,\ell}(k)$ everywhere in the dual lattice completely determines the Wannier and Bloch functions, and the Wannier and Bloch states for given band index n.

We can summarize the periodicity properties of the Bloch functions and the Bloch factors in the assertions that the Bloch function

$$\begin{aligned}\psi_n(k,x) &= \sqrt{\frac{a}{2\pi}}\exp(\mathrm{i}kx)u_n(k,x)\\ &= \sqrt{\frac{a}{2\pi}}\sum_{\ell\in\mathbb{Z}} u_{n,0}\left(k+\frac{2\pi\ell}{a}\right)\exp\left[\mathrm{i}\left(k+\frac{2\pi\ell}{a}\right)x\right]\\ &= \sqrt{\frac{a}{2\pi}}\sum_{\nu\in\mathbb{Z}} w_{n,0}(x+\nu a)\exp(-\mathrm{i}\nu ka)\end{aligned} \tag{10.25}$$

is periodic in dual space and quasiperiodic in direct space, whereas the Bloch factor

$$\begin{aligned}u_n(k,x) &= \sum_{\ell\in\mathbb{Z}} u_{n,0}\left(k+\frac{2\pi\ell}{a}\right)\exp\left(2\pi\mathrm{i}\frac{\ell x}{a}\right)\\ &= \sum_{\nu\in\mathbb{Z}} w_{n,0}(x+\nu a)\exp[-\mathrm{i}k(x+\nu a)]\end{aligned} \tag{10.26}$$

is quasiperiodic in dual space and periodic in direct space.

10.3 Time-dependent Wannier states

The usual stationary Wannier states (10.18) do not satisfy the time-independent Schrödinger equation in the crystal because they are linear combinations of stationary solutions for different eigenvalues $E_n(k)$. However, the solutions

$$\psi_n(k,x,t) = \psi_n(k,x)\exp(-\mathrm{i}E_n(k)t/\hbar)$$

of the time-dependent Schrödinger equation satisfy the same periodicity properties in the dual lattice as $\psi_n(k,x)$ because the energy bands $E_n(k)$ are periodic in the dual lattice. Therefore we can write down expansions

$$\psi_n(k,x,t) = \sqrt{\frac{a}{2\pi}}\sum_{\nu\in\mathbb{Z}} w_{n,\nu}(x,t)\exp(\mathrm{i}\nu ka), \tag{10.27}$$

$$\begin{aligned}w_{n,\nu}(x,t) &= \sqrt{\frac{a}{2\pi}}\int_{-\pi/a}^{\pi/a} dk\,\psi_n(k,x,t)\exp(-\mathrm{i}\nu ka)\\ &= \frac{a}{2\pi}\int_{-\pi/a}^{\pi/a} dk\,u_n(k,x)\exp[\mathrm{i}k(x-\nu a)]\exp\left(-\frac{\mathrm{i}}{\hbar}E_n(k)t\right).\end{aligned} \tag{10.28}$$

The time-dependent Wannier states $w_{n,\nu}(x,t) = \langle x|w_{n,\nu}(t)\rangle$ also satisfy completeness relations,

$$\int_{-\infty}^{\infty} dx\, w^{+}_{m,\mu}(x,t) w_{n,\nu}(x,t) = \delta_{mn}\delta_{\mu\nu} \tag{10.29}$$

and

$$\sum_{n,\nu} w_{n,\nu}(x,t) w^{+}_{n,\nu}(x',t) = \delta(x-x'), \tag{10.30}$$

and the periodicity $u_n(k, x - \mu a) = u_n(k,x)$ of the Bloch functions also implies localization of the time-dependent states,

$$w_{n,\nu}(x,t) = w_{n,0}(x - \nu a, t). \tag{10.31}$$

These states are therefore still associated with individual lattice sites, but contrary to the states (10.19), the states (10.28) are solutions of a Schrödinger equation in the lattice,

$$\mathrm{i}\hbar \frac{d}{dt}|w_{n,\nu}(t)\rangle = \left(\frac{\mathrm{p}^2}{2m} + V(\mathrm{x})\right)|w_{n,\nu}(t)\rangle.$$

We can think of this as a manifestation of wave-particle duality for electrons in a crystal. We can describe electrons as waves penetrating the whole crystal, or as particles associated with particular lattice sites.

10.4 The Kronig-Penney model

The Kronig-Penney model[6] discusses motion of non-relativistic particles in a periodic piecewise constant potential. It provides a beautiful explanation for the emergence of energy bands in materials by demonstrating that only certain energy ranges in a periodic potential can yield electron states which comply with Bloch's theorem. We discuss the simplified version where the periodic potential is a series of δ-peaks at distance a,

$$V(x) = V_0 b \sum_{\nu\in\mathbb{Z}} \delta(x - \nu a). \tag{10.32}$$

[6] R. de L. Kronig, W.G. Penney, Proc. Roy. Soc. London A 130, 449 (1931).

V_0 is a constant energy, while $b > 0$ is a constant length to make the equation dimensionally correct. In a model with finite width, V_0 would be the height of a barrier and b the width.

Since we have vanishing potential for $-a < x < 0$, the energy eigenstates

$$\langle x|E,k\rangle = \sqrt{\frac{a}{2\pi}}\exp(ikx)u(k,x)$$

in this region must satisfy

$$-\frac{\hbar^2}{2m}\frac{d^2}{dx^2}\langle x|E,k\rangle = E\langle x|E,k\rangle, \quad -a < x < 0,$$

i.e. they must be combinations of plane waves $\exp(\pm\mathrm{i}Kx)$ with wave vector $K = \sqrt{2mE}/\hbar$ for $E > 0$, or combinations of real exponentials $\exp(\pm Kx)$ with $K = \sqrt{-2mE}/\hbar$ for $E < 0$. Solution with positive energy exist both for $V_0 > 0$ and $V_0 < 0$, but negative energy solutions exist only for $V_0 < 0$.

We discuss the positive energy solutions first. Once we know the energy eigenstates in one interval of length a, we know them everywhere, because we know from Bloch's theorem that whenever we proceed by a length a the wave function only changes by a factor $\exp(ika)$. For the intervals $-a < x < 0$ and $0 < x < a$ this implies in particular (with $E > 0$)

$$\langle x|E,k\rangle = \begin{cases} A\exp[\mathrm{i}Kx] + B\exp[-\mathrm{i}Kx], & -a \le x \le 0, \\ A\exp[\mathrm{i}(Kx - Ka + ka)] + B\exp[\mathrm{i}(Ka - Kx + ka)], \\ \quad 0 \le x \le a, \end{cases}$$

and for $(\nu - 1)a < x < \nu a$:

$$\langle x|E,k\rangle = A\exp[\mathrm{i}(Kx - \nu Ka + \nu ka)] + B\exp[\mathrm{i}(\nu Ka - Kx + \nu ka)].$$

The junction conditions following from the full Schrödinger equation

$$\frac{d^2}{dx^2}\langle x|E,k\rangle + \frac{2mE}{\hbar^2}\langle x|E,k\rangle = \frac{2m}{\hbar^2}V_0 b\sum_{\nu\in\mathbb{Z}}\delta(x-\nu a)\langle x|E,k\rangle$$

read

$$\lim_{\epsilon\to 0}\left(\langle \nu a+\epsilon|E,k\rangle - \langle \nu a-\epsilon|E,k\rangle\right) = 0,$$

$$\lim_{\epsilon\to +0}\left(\frac{d}{dx}\langle x|E,k\rangle\Big|_{x=\nu a+\epsilon} - \frac{d}{dx}\langle x|E,k\rangle\Big|_{x=\nu a-\epsilon}\right) = 2\frac{u}{a}\langle \nu a|E,k\rangle,$$

where the new constant $u = mV_0ab/\hbar^2$ was introduced for convenience. The resulting junction conditions are identical at all lattice points $x = \nu a$,

$$A\exp[\mathrm{i}(k-K)a] + B\exp[\mathrm{i}(k+K)a] = A + B, \tag{10.33}$$

$$\mathrm{i}KA\Big(\exp[\mathrm{i}(k-K)a] - 1\Big) - \mathrm{i}KB\Big(\exp[\mathrm{i}(k+K)a] - 1\Big) = 2\frac{u}{a}(A+B). \tag{10.34}$$

The requirement for existence of a non-trivial solution of these equations yields the condition

$$\begin{vmatrix} \exp[\mathrm{i}(k-K)a]-1 & \exp[\mathrm{i}(k+K)a]-1 \\ \mathrm{i}K(\exp[\mathrm{i}(k-K)a]-1)-(2u/a) & -\mathrm{i}K(\exp[\mathrm{i}(k+K)a]-1)-(2u/a) \end{vmatrix}$$
$$=0,$$

i.e.

$$\cos(Ka)+\frac{u}{Ka}\sin(Ka)=\cos(ka), \tag{10.35}$$

which in turn implies a condition for the allowed energy values $E=\hbar^2K^2/2m$,

$$\left|\cos(Ka)+\frac{u}{Ka}\sin(Ka)\right|\leq 1. \tag{10.36}$$

The limit $E\to 0+$ is allowed if and only if $|1+u|\leq 1$, or equivalently if and only if $-2\leq u\leq 0$.

The function on the left hand side of equation (10.35) is plotted for $u=5$ in Figure 10.1 and for negative values of u in Figures 10.6 and 10.8 below. For $0>u>-2$ the lowest energy band has both positive and negative energy values.

Negative energy solutions $E<0$ might exist for $V_0<0$. The Schrödinger equation for $-a<x<0$ and the Bloch theorem imply

$$\langle x|E,k\rangle=\begin{cases} A\exp[Kx]+B\exp[-Kx], & -a\leq x\leq 0, \\ A\exp[Kx-Ka+\mathrm{i}ka]+B\exp[Ka-Kx+\mathrm{i}ka], & \\ \quad 0\leq x\leq a, & \end{cases}$$

with $K=\sqrt{-2mE}/\hbar$.

The matching conditions at $x=0$ (and for any $x=\nu a$) are

$$A\exp[(\mathrm{i}k-K)a]+B\exp[(\mathrm{i}k+K)a]=A+B,$$

$$KA(\exp[(\mathrm{i}k-K)a]-1)-KB(\exp[(\mathrm{i}k+K)a]-1)=-\frac{2m}{\hbar^2}|V_0|b(A+B),$$

and the condition for existence of non-trivial solutions is with $u=mV_0ab/\hbar^2<0$,

$$\begin{vmatrix} \exp[(\mathrm{i}k-K)a]-1 & \exp[(\mathrm{i}k+K)a]-1 \\ K(\exp[(\mathrm{i}k-K)a]-1)-(2u/a) & -K(\exp[(\mathrm{i}k+K)a]-1)-(2u/a) \end{vmatrix}$$
$$=0.$$

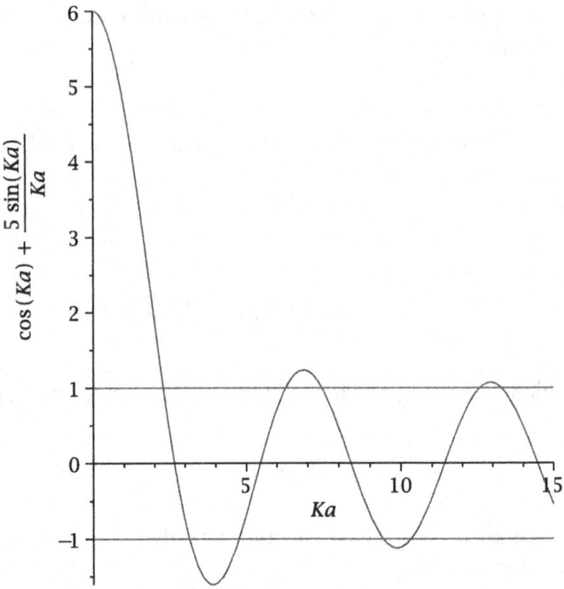

Fig. 10.1 The function $f(Ka) = \cos(Ka) + (u/Ka)\sin(Ka)$ for $u = 5$

This yields

$$\cosh(Ka) + \frac{u}{Ka}\sinh(Ka) = \cos(ka) \tag{10.37}$$

The limit $E \to 0-$ exists if and only if $-2 \leq u \leq 0$, i.e. in the same range for V_0 which was found for $E \to 0+$.

For $V_0 < 0$ one always finds one negative energy band (besides the positive energy bands), see e.g. Figures 10.5 and 10.7. This negative energy band goes from a minimum at $k = 0$ to positive maxima at $k = \pm\pi/a$ if $-2 \leq u \leq 0$, by joining to a positive energy branch at some intermediate values $\pm k_0$. The intermediate value k_0 with $E(k_0) = 0$ satisfies

$$\cos(k_0 a) = 1 + u.$$

For $u < -2$ the lowest band is entirely in the negative energy range, but still with the minimum at $k = 0$ and the maxima at $k = \pm\pi/a$.

It is useful to plot the functions $f(Ka) = \cos(Ka) + (u/Ka)\sin(Ka)$, and for negative u also $g(Ka) = \cosh(Ka) + (u/Ka)\sinh(Ka)$ to analyze the implications of the conditions (10.35) and (10.37). We will do this for $u = 5$.

Increasing Ka in Figure 10.1 corresponds to increasing energy $E = \hbar^2K^2/2m$. Due to the condition (10.35) there are no allowed energies for $0 \leq Ka < K_1a \simeq 2.284$ where $f(Ka) > 1$. At K_1a we have the lowest allowed energy value $\hbar^2K_1^2/2m$ with corresponding pseudomomentum $ka = 0$. Between $K_1a \leq Ka \leq K_2a = \pi$

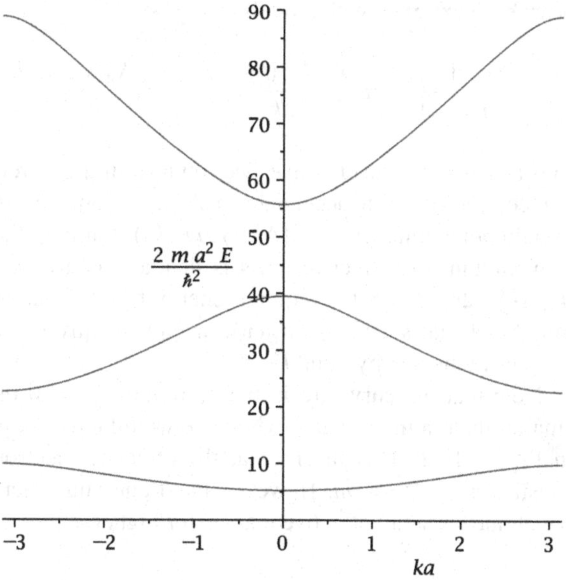

Fig. 10.2 The three lowest energy bands $E_0(k)$, $E_1(k)$ and $E_2(k)$ for $u = 5$

the energy increases to $\hbar^2\pi^2/2ma^2$ and the values of $|ka|$ increase to the boundaries $|ka| = \pi$ of the Brillouin zone. The width of this lowest energy band is

$$W = \frac{\hbar^2}{2m}\left(K_2^2 - K_1^2\right) \simeq \frac{\hbar^2}{2ma^2}\left(\pi^2 - 2.284^2\right).$$

For $\pi < Ka < K_3a \simeq 4.761$ there are again no allowed energy values, i.e. there is an energy gap of width

$$\Delta E_g = \frac{\hbar^2}{2m}\left(K_3^2 - K_2^2\right) \simeq \frac{\hbar^2}{2ma^2}\left(4.761^2 - \pi^2\right)$$

between the lowest energy band and the next energy band. Between $K_3a \leq Ka \leq K_4a = 2\pi$ the energy increases from $\hbar^2K_3^2/2m$ to $\hbar^2K_4^2/2m$, while $|ka|$ decreases from $|ka| = \pi$ to $ka = 0$. This behavior occurs over and over again, with decreasing energy gaps ΔE_g between adjacent bands. The three lowest energy bands $E_n(k)$ for $u = 5$ are plotted in Figure 10.2.

The energy bands have extrema in the center and at the boundaries of the Brillouin zone,

$$\left.\frac{dE_n(k)}{dk}\right|_{k=0,\pm\pi/a} = \left.\frac{\hbar^2K_n(k)}{m}\frac{dK_n(k)}{dk}\right|_{k=0,\pm\pi/a} = 0.$$

Therefore we can use parabolic approximations

$$E_n(k) \simeq E_n(k_0) \pm \frac{\hbar^2}{2m_n(k_0)}(k - k_0)^2 \tag{10.38}$$

near those extrema, with effective masses

$$\frac{1}{m_n(k_0)} = \pm\frac{1}{\hbar^2}\left.\frac{dE_n(k)}{dk}\right|_{k=k_0} = \pm\frac{K_n(k_0)}{m}\left.\frac{d^2K_n(k)}{dk^2}\right|_{k=k_0}. \tag{10.39}$$

This is denoted as an effective electron mass if the extremum is a minimum, because the required energy to accelerate an electron from the band minimum $\hbar k_0$ to a nearby pseudomomentum $\hbar k$ is $\hbar^2(\Delta k)^2/2m_n(k_0)$. On the other hand, if the extremum is a maximum, the effective mass is denoted as a hole mass, because in that case $\hbar^2(\Delta k)^2/2m_n(k_0)$ is the required energy to move an electron from a nearby point k to k_0 (if that state was vacant), or equivalently move the vacant state (or hole) from k_0 to the nearby point k.

Note that the curvature in the minimum $k_0 = 0$ of the lowest energy band is smaller than in the vacuum, where we would have the parabola $2mEa^2/\hbar^2 = (ka)^2$ in Figure 10.2. This means that the effective electron mass in the lowest band satisfies $m_{n=0}(0) > m$. However, band curvature increases for the higher bands, which means small effective masses for higher n.

10.5 *kp* perturbation theory and effective mass

The combination of the Bloch theorem with second order perturbation theory provides another beautiful introduction to the concept of effective electron or hole mass in materials. This is a little more technical, but also more general in the sense that it does not rely on a particular potential model.

The starting point for *kp* perturbation theory is an effective Schrödinger equation for the Bloch factors $u_n(k,x) = \langle x|u_n(k)\rangle$. The identity

$$\mathrm{p}^2|E_n,k\rangle = \sqrt{\frac{a}{2\pi}}\exp(ik\mathrm{x})(\mathrm{p}+\hbar k)^2|u_n(k)\rangle$$

implies the following effective Schrödinger equation for the Bloch factors:

$$\left(\frac{\mathrm{p}^2}{2m} + \frac{\hbar}{m}k\mathrm{p} + V(\mathrm{x})\right)|u_n(k)\rangle = \left(E_n(k) - \frac{\hbar^2k^2}{2m}\right)|u_n(k)\rangle. \tag{10.40}$$

Now suppose that we know the Bloch factors and energy levels at a point k_0 in the Brillouin zone, and we take these solutions of

$$\left(\frac{\mathrm{p}^2}{2m} + \frac{\hbar}{m}k_0\mathrm{p} + V(\mathrm{x})\right)|E_n(k_0)\rangle = \left(E_n(k_0) - \frac{\hbar^2k_0^2}{2m}\right)|E_n(k_0)\rangle \tag{10.41}$$

as 0-th order approximation to the perturbative solution of equation (10.40),

$$\left(\frac{\mathrm{p}^2}{2m}+\frac{\hbar}{m}k_0\mathrm{p}+V(\mathrm{x})+\frac{\hbar}{m}(k-k_0)\mathrm{p}\right)|u_n(k)\rangle=\left(E_n(k)-\frac{\hbar^2k^2}{2m}\right)|u_n(k)\rangle,$$

i.e. the perturbatively sought states and eigenvalues are $|u_n(k)\rangle$ and $E_n(k)-(\hbar^2k^2/2m)$, and the perturbation operator is $\hbar(k-k_0)\mathrm{p}/m$. The energy levels in second order perturbation theory are therefore

$$\begin{aligned}E_n(k)&=E_n(k_0)+\frac{\hbar^2}{2m}(k^2-k_0^2)+\frac{\hbar}{m}(k-k_0)\langle E_n(k_0)|\mathrm{p}|E_n(k_0)\rangle\\&\quad+\frac{\hbar^2}{m^2}(k-k_0)^2\sum_{m\neq n}\frac{|\langle E_m(k_0)|\mathrm{p}|E_n(k_0)\rangle|^2}{E_n(k_0)-E_m(k_0)},\end{aligned}$$

and the effective mass near an extremum k_0 in the n-th band is then in second order perturbation theory

$$\frac{1}{m_n(k_0)}=\frac{1}{\hbar^2}\frac{d^2}{dk^2}E_n(k)\bigg|_{k=k_0}=\frac{1}{m}+\frac{2}{m^2}\sum_{m\neq n}\frac{|\langle E_m(k_0)|\mathrm{p}|E_n(k_0)\rangle|^2}{E_n(k_0)-E_m(k_0)}.$$

If there appear degeneracies between different bands at $k=k_0$, we should split the band indices $n\to i,\alpha$, and we have to apply the result (9.24) to find

$$\begin{aligned}E_{i,\alpha}(k)&=E_i(k_0)+\frac{\hbar^2}{2m}(k^2-k_0^2)+\frac{\hbar}{m}(k-k_0)\langle u_{i,\alpha}(k_0)|\mathrm{p}|u_{i,\alpha}(k_0)\rangle\\&\quad+\frac{\hbar^2}{m^2}(k-k_0)^2\sum_{j\neq i}\sum_{\beta}\frac{|\langle u_{j,\beta}(k_0)|\mathrm{p}|u_{i,\alpha}(k_0)\rangle|^2}{E_i(k_0)-E_j(k_0)},\end{aligned}$$

$$\frac{1}{m_{i,\alpha}(k_0)}=\frac{1}{\hbar^2}\frac{d^2}{dk^2}E_{i,\alpha}(k)\bigg|_{k=k_0}=\frac{1}{m}+\frac{2}{m^2}\sum_{j\neq i}\sum_{\beta}\frac{|\langle u_{j,\beta}(k_0)|\mathrm{p}|u_{i,\alpha}(k_0)\rangle|^2}{E_i(k_0)-E_j(k_0)}.$$

These results indicate that the effective mass in the lowest energy band is always larger than the free electron mass m, in agreement with our observation from the Kronig-Penney model.

10.6 Problems

10.1. Show that the lattice translation operator (10.5) acts on the Wannier states

$$|w_{n,\nu}(t)\rangle=\sqrt{\frac{a}{2\pi}}\int_{-\pi/a}^{\pi/a}dk\,|E_n,k\rangle\exp(-\mathrm{i}\nu ka)\exp[-\mathrm{i}E_n(k)t/\hbar]$$

according to

$$T(a)|w_{n,\nu}(t)\rangle = |w_{n,\nu-1}(t)\rangle.$$

10.2. Show that the momentum per lattice site in the Bloch function $\psi_n(k,x)$,

$$\langle p\rangle_{n,k} = \frac{2\pi}{a}\int_{\nu a}^{(\nu+1)a} dx\,\psi_n^+(k,x)\frac{\hbar}{\mathrm{i}}\frac{\partial}{\partial x}\psi_n(k,x).$$

is independent of ν and is given in terms of a weighted sum over equivalent sites in the dual lattice,

$$\langle p\rangle_{n,k} = a\sum_\ell \hbar\left(k+\frac{2\pi}{a}\ell\right)\left|u_{n,0}\left(k+\frac{2\pi\ell}{a}\right)\right|^2.$$

10.3. Show that the Wannier functions $w_{n,\nu}(x) = w_{n,0}(x-\nu a)$ in the lattice and the dual Wannier functions $u_{n,-\ell}(k) = u_{n,0}(k-(2\pi/a)\ell)$ in the dual lattice are related according to

$$w_{n,\nu}(x) = \frac{a}{2\pi}\int_{-\pi/a}^{\pi/a} dk\sum_\ell u_{n,-\ell}(k)\exp\left[\mathrm{i}\left(k-\frac{2\pi}{a}\ell\right)(x-\nu a)\right],$$

$$u_{n,-\ell}(k) = \frac{1}{a}\int_0^a dx\sum_\nu w_{n,\nu}(x)\exp\left[-\mathrm{i}\left(k-\frac{2\pi}{a}\ell\right)(x-\nu a)\right].$$

Furthermore, show that these relations are equivalent to (10.24).

10.4. Show that we can write the (non-normalized) positive energy Bloch functions for the Kronig-Penney model in Section 10.4 in the form

$$\begin{aligned}\psi_K(k,x) = \langle x|E,k\rangle &= \sqrt{\frac{a}{2\pi}}\sum_{q\in\mathbb{Z}}\Theta(x+a-qa)\Theta(qa-x)\exp[\mathrm{i}(q-1)ka]\\ &\times\Big(\exp[\mathrm{i}K(x-qa)]-\exp[\mathrm{i}K(x-qa)]\exp[\mathrm{i}(k+K)a]\\ &-\exp[-\mathrm{i}K(x-qa)]+\exp[-\mathrm{i}K(x-qa)]\exp[\mathrm{i}(k-K)a]\Big).\end{aligned} \tag{10.42}$$

We use the label $K\equiv K(n,k)=\sqrt{2mE_n(k)}/\hbar$ instead of the energy band index n. Up to an undetermined phase factor $\exp[\mathrm{i}\varphi_K(k)]$, the omitted normalization factor N_K is given by

$$\begin{aligned}N_K^{-2} &= 4a[1-\cos(Ka)\cos(ka)]+\frac{4}{K}\sin(Ka)[\cos(ka)-\cos(Ka)]\\ &= 4a\sin(Ka)\left[\sin(Ka)\left(1+\frac{u}{K^2a^2}\right)-\frac{u}{Ka}\cos(Ka)\right].\end{aligned} \tag{10.43}$$

Determine the Bloch factor $u_K(k,x)$ and the dual Wannier states $u_{K,\ell}(k)$ for the Bloch function (10.42).

10.5. We consider the Kronig-Penney model in the limit of vanishing potential $u = 0$.

Show that the solutions for the energy levels for k in the first Brillouin zone are given by

$$\frac{\sqrt{2mE_n(k)}}{\hbar} \equiv K_n(k) = |k| + \frac{2\pi}{a}n, \quad n = 0, 1, \ldots$$

Using (10.42) with the normalization factor included, construct the Bloch functions $\psi_n(k,x)$ and Bloch factors $u_n(k,x)$ for k in the first Brillouin zone.

Show that the Bloch factors in the whole dual lattice are given by

$$u_n(k,x) = \frac{1}{\mathrm{i}\sqrt{a}} \sum_{\ell\in\mathbb{Z}} \left[\Theta\!\left(k - \frac{2\pi}{a}\ell\right) \Theta\!\left((2\ell+1)\frac{\pi}{a} - k\right) \exp\!\left(2\pi\mathrm{i}\frac{n-\ell}{a}x\right) + \Theta\!\left(k - (2\ell-1)\frac{\pi}{a}\right) \Theta\!\left(\frac{2\pi}{a}\ell - k\right) \exp\!\left(-2\pi\mathrm{i}\frac{n+\ell}{a}x\right)\right].$$

Show that Wannier functions are given by

$$w_{n,0}(x) = \frac{\sqrt{a}}{\mathrm{i}\pi} \frac{\sin[(2n+1)\pi x/a] - \sin[2n\pi x/a]}{x}.$$

Examples are shown in Figures 10.3 and 10.4.

10.6. We consider the Kronig-Penney model in the limit of vanishing potential $u = 0$. Show that the time-dependent Bloch functions in the whole dual lattice are given by

$$\begin{aligned}\psi_n(k,x,t) = \frac{\exp(\mathrm{i}kx)}{\mathrm{i}\sqrt{2\pi}} \sum_{\ell\in\mathbb{Z}} \Bigg[&\Theta\!\left(k - \frac{2\pi}{a}\ell\right) \Theta\!\left((2\ell+1)\frac{\pi}{a} - k\right) \\ &\times \exp\!\left(2\pi\mathrm{i}\frac{n-\ell}{a}x\right) \exp\!\left[-\frac{\mathrm{i}\hbar t}{2m}\left(k + 2\pi\frac{n-\ell}{a}\right)^2\right] \\ &+ \Theta\!\left(k - (2\ell-1)\frac{\pi}{a}\right) \Theta\!\left(\frac{2\pi}{a}\ell - k\right) \exp\!\left(-2\pi\mathrm{i}\frac{n+\ell}{a}x\right) \\ &\times \exp\!\left[-\frac{\mathrm{i}\hbar t}{2m}\left(k - 2\pi\frac{n+\ell}{a}\right)^2\right]\Bigg].\end{aligned}$$

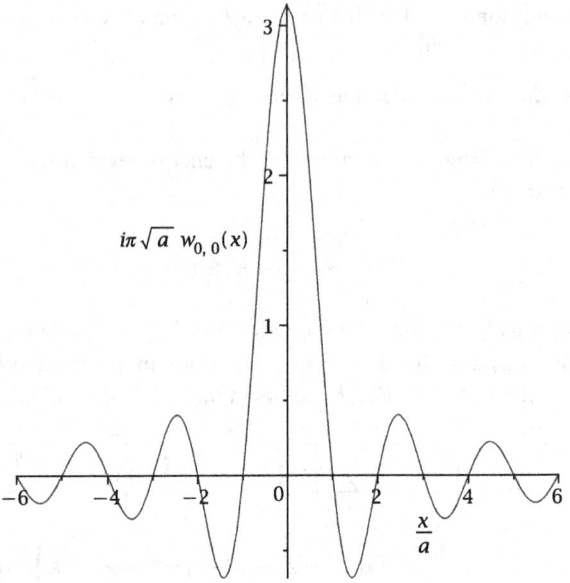

Fig. 10.3 The function $w_{0,0}(x)$ for $u = 0$

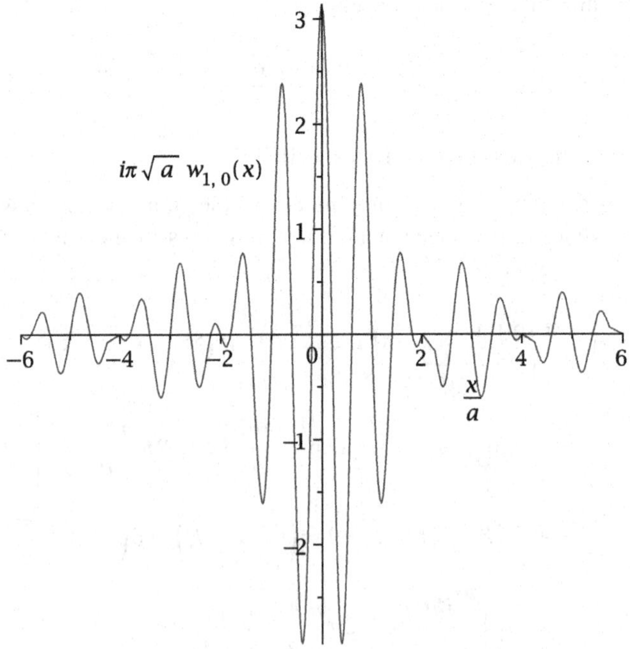

Fig. 10.4 The function $w_{1,0}(x)$ for $u = 0$

Show that the time-dependent Wannier functions (10.28) are given by

$$
\begin{aligned}
w_{n,0}(x,t) = & -\sqrt{\frac{\mathrm{i}ma}{8\pi\hbar t}}\exp\left(\mathrm{i}\frac{mx^2}{2\hbar t}\right)\left[\mathrm{erf}\left(-\sqrt{\mathrm{i}\frac{mx^2}{2\hbar t}}-2n\pi\sqrt{\frac{\mathrm{i}\hbar t}{2ma^2}}\right)\right. \\
& -\mathrm{erf}\left(-\sqrt{\mathrm{i}\frac{mx^2}{2\hbar t}}-(2n+1)\pi\sqrt{\mathrm{i}\frac{\hbar t}{2ma^2}}\right) \\
& +\mathrm{erf}\left(-\sqrt{\mathrm{i}\frac{mx^2}{2\hbar t}}+(2n+1)\pi\sqrt{\frac{\mathrm{i}\hbar t}{2ma^2}}\right) \\
& \left.-\,\mathrm{erf}\left(-\sqrt{\mathrm{i}\frac{mx^2}{2\hbar t}}+2n\pi\sqrt{\frac{\mathrm{i}\hbar t}{2ma^2}}\right)\right],
\end{aligned}
$$

where $\sqrt{\mathrm{i}} = (1+\mathrm{i})/\sqrt{2}$.

10.7. Figures 10.5 and 10.6 illustrate the conditions (10.37) and (10.35) for existence of negative or positive energies for $u = -3$.

Note that for the negative energies increasing K corresponds to decreasing E. Therefore the energy minimum in the negative energy band arises from $g(Ka) = 1$, $ka = 0$, and the maximum in the negative energy band arises from $g(Ka) = -1$, $ka = \pm\pi$ in Figure 10.5. Analyze the band structure in this model similar to the analysis of Figure 10.1. Contrary to the case of positive u, there are also negative energy values possible for $u = -3$. How many negative energy bands are there for $u = -3$?

10.8. Figures 10.7 and 10.8 illustrate the conditions (10.37) and (10.35) for existence of negative or positive energies for $u = -1.5$.

Analyze the band structure in this model similar to the analysis of Problem 10.7. Contrary to the case $u < -2$, we reach the value $E = 0$ for $Ka = 0$ in Figure 10.7 for $0 < |ka| < \pi$. At this point we go into the positive energies corresponding to the values $0 \leq Ka \lesssim 1.689$ in Figure 10.8, i.e. the lowest energy band contains both negative and positive energies in this case. For which value of ka is $E = 0$?

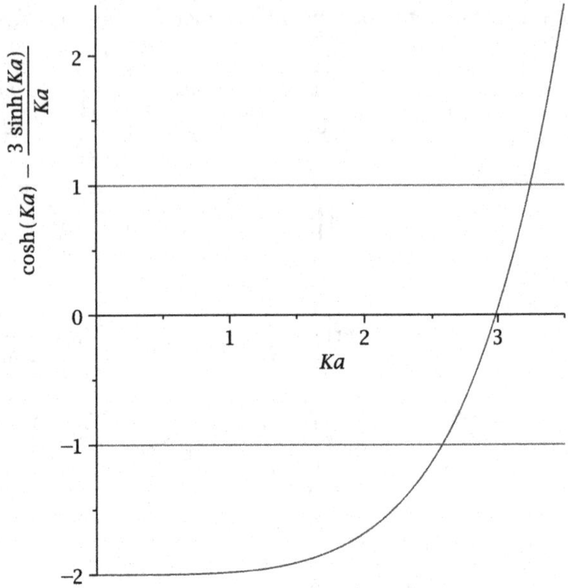

Fig. 10.5 The function $g(Ka) = \cosh(Ka) - (3/Ka)\sinh(Ka)$. Only values of Ka with $|g(Ka)| \leq 1$ correspond to allowed energy values $E = -\hbar^2 K^2/2m$ in the Kronig-Penney model with $u = -3$

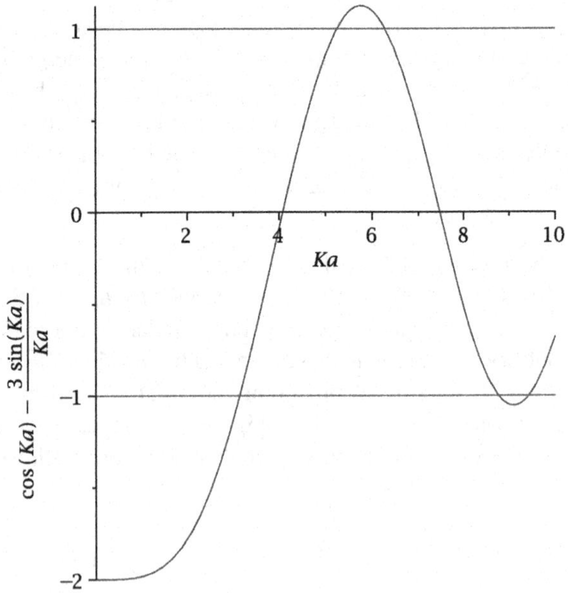

Fig. 10.6 The function $f(Ka) = \cos(Ka) - (3/Ka)\sin(Ka)$. Only values of Ka with $|f(Ka)| \leq 1$ correspond to allowed energy values $E = \hbar^2 K^2/2m$ in the Kronig-Penney model with $u = -3$

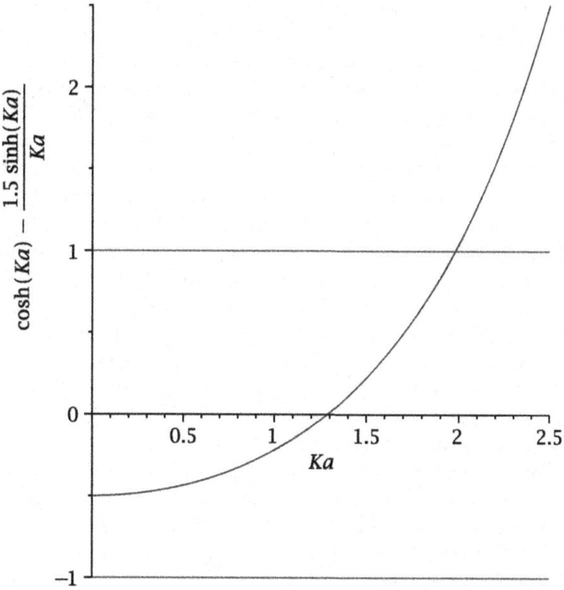

Fig. 10.7 The function $g(Ka) = \cosh(Ka) - (1.5/Ka)\sinh(Ka)$. Only values of Ka with $|g(Ka)| \le 1$ correspond to allowed energy values $E = -\hbar^2 K^2/2m$ in the Kronig-Penney model with $u = -1.5$

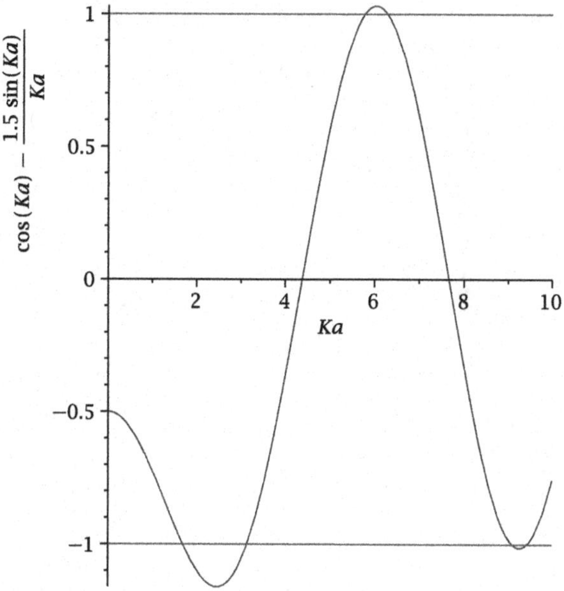

Fig. 10.8 The function $f(Ka) = \cos(Ka) - (1.5/Ka)\sin(Ka)$. Only values of Ka with $|f(Ka)| \le 1$ correspond to allowed energy values $E = \hbar^2 K^2/2m$ in the Kronig-Penney model with $u = -1.5$

Chapter 11
Scattering Off Potentials

Most two-particle interaction potentials $V(\boldsymbol{x}_1 - \boldsymbol{x}_2)$ assume a finite value V_∞ if $|\boldsymbol{x}_1 - \boldsymbol{x}_2| \to \infty$. If the relative motion of the two-particle system has an energy $E > V_\infty$ the particles can have arbitrary large distance. In particular, we can imagine a situation where the two particles approach each other from an initially large separation and after reaching some minimal distance move away from each other. The force between the two particles will influence the trajectories of the two particles, and this influence will be strongest when the particles are close together. The deflection of particle trajectories due to interaction forces is denoted as scattering. This is denoted as potential scattering if the interaction forces between the particles can be expressed through a potential. We have seen in Section 7.1 that the motion of two particles with an interaction potential of the form $V(\boldsymbol{r}) = V(\boldsymbol{x}_1 - \boldsymbol{x}_2)$ can be separated into center of mass motion and relative motion, $\Psi(\boldsymbol{x}_1, \boldsymbol{x}_2) = \psi(\boldsymbol{r}) \exp(\mathrm{i}\boldsymbol{K} \cdot \boldsymbol{R})/\sqrt{2\pi}^3$, where the factor $\psi(\boldsymbol{r})$ for relative motion of the two particles satisfies

$$E\psi(\boldsymbol{r}) = -\frac{\hbar^2}{2m}\Delta\psi(\boldsymbol{r}) + V(\boldsymbol{r})\psi(\boldsymbol{r}). \tag{11.1}$$

E is the contribution from relative motion to the total energy (7.9) of the two-particle system, and I wrote m for the reduced mass of the two-particle system.

Equation (11.1) with $E > V_\infty$ does not only describe two-particle scattering, but also scattering of a particle of mass m off a potential with fixed center $r = 0$, e.g. because the source of the potential is fixed by forces which do not affect the scattered particle.

Scattering is an important technique for the determination of properties of a physical system. Within the framework of potential scattering, observations of deflections of particle trajectories in a potential can be used to determine the strength and functional dependence $V(\boldsymbol{r})$ of a scattering potential.

R. Dick, *Advanced Quantum Mechanics*, Graduate Texts in Physics,
DOI 10.1007/978-3-319-25675-7_11

Suppose that we wish to determine a scattering potential $V(\boldsymbol{r})$ through scattering of non-relativistic particles of momentum $\hbar\boldsymbol{k}$ off the potential. The deflected particles will have momenta $\boldsymbol{k}' \neq \boldsymbol{k}$, and one observable that we should certainly be able to measure is the number $dn(\Omega)/dt$ of particles per time which are deflected into a small solid angle $d\Omega = d\vartheta d\varphi \sin\vartheta$ in the direction $\hat{\boldsymbol{x}} = (\sin\vartheta\cos\varphi, \sin\vartheta\sin\varphi, \cos\vartheta)$. According to the definition of particle current densities j, this number will be given by

$$\frac{dn(\Omega)}{dt} = \lim_{r\to\infty} j_{out}(\boldsymbol{k}')r^2 d\Omega,$$

where $j_{out}(\boldsymbol{k}')$ is the number of deflected particles per area and per time which are moving in the direction $\hat{\boldsymbol{k}}' = \hat{\boldsymbol{x}}$ with momentum $\hbar\boldsymbol{k}'$. We are taking the limit $r \to \infty$ because we are interested in measuring $dn(\Omega)/dt$ far from the scattering center (or for large separation of particles in particle-particle scattering), to make sure that the scattering potential does not deflect the scattered particles any further.

The number of particles $dn(\Omega)/dt$ which are scattered into the direction $\hat{\boldsymbol{k}}'$ is of course proportional to the number $j_{in}(\boldsymbol{k})$ of particles per area and per time which are incident on the scattering center, and it is also proportional to the width $d\Omega$ of the solid angle over which we sum the scattered particles. Therefore we expect that the observable which may really tell us something about the scattering potential is gotten by dividing out the trivial dependence on j_{in} and $d\Omega$, i.e. we define

$$\frac{d\sigma}{d\Omega} = \frac{1}{j_{in}(\boldsymbol{k})}\frac{dn(\Omega)}{d\Omega dt} = \lim_{r\to\infty} r^2\frac{j_{out}(\boldsymbol{k}')}{j_{in}(\boldsymbol{k})}. \tag{11.2}$$

The quantity $d\sigma(\Omega) = (dn(\Omega)/dt)/j_{in}$ has the dimension of an area and is therefore known as a *differential scattering cross section*. Differential scattering cross sections are the observables or primary interest in potential scattering.

If we integrate over all possible scattering directions, we get the *scattering cross section*

$$\sigma = \int d\sigma = \frac{1}{j_{in}(\boldsymbol{k})}\int d\Omega\frac{dn(\Omega)}{dt} = \frac{1}{j_{in}(\boldsymbol{k})}\frac{dn}{dt},$$

i.e. the scattering cross section is the total number of scattered particles per time, dn/dt, divided by the current density of incident particles.

For an explanation of the name *cross section* we also remark that the calculation of scattering of particles off a hard sphere of radius R in *classical mechanics* yields a scattering cross section $\sigma = \pi R^2$ which equals the cross section of the sphere. We will see below in Section 11.3 that quantum mechanics actually yields a larger scattering cross section of a sphere, e.g. $\sigma = 4\pi R^2$ for scattering of very low energetic particles. Scattering of low energy particles off a sphere could be considered as a most basic illustration of measuring properties of a scattering center. Measuring the number dn/dt of low energy particles per time which are scattered off the sphere and dividing by the incident particle current density provides a measurement of the radius $R = \sqrt{\sigma/4\pi} = \sqrt{(dn/dt)/4\pi j_{in}}$ of the scattering center.

The particle current density of particles described by a wave function $\psi(\boldsymbol{x}, t)$ will be proportional to the corresponding probability current density

$$j = \frac{\hbar}{2im}\left(\psi^+ \cdot \nabla\psi - \nabla\psi^+ \cdot \psi\right),$$

and therefore we can use probability current densities in the calculation of the ratio in (11.2).

11.1 The free energy-dependent Green's function

Many applications of quantum mechanics require the calculation of the inverse (or resolvent) $\mathcal{G}(E)$ of the operator $E - H_0 = E - (\mathbf{p}^2/2m)$,

$$(E - H_0)\mathcal{G}(E) = 1. \tag{11.3}$$

E.g. if we consider a time-independent potential V, the time-independent Schrödinger equation

$$(E - H_0)|\psi(E)\rangle = V|\psi(E)\rangle \tag{11.4}$$

is satisfied if the energy eigenstate $|\psi(E)\rangle$ satisfies a *Lippmann-Schwinger equation*[1]

$$|\psi(E)\rangle = |\psi_0(E)\rangle + \mathcal{G}(E)V|\psi(E)\rangle, \tag{11.5}$$

where $|\psi_0(E)\rangle$ satisfies the condition $(E - H_0)|\psi_0(E)\rangle = 0$. Iteration of (11.5) for $|\psi_0(E)\rangle \neq 0$ then yields the perturbation series

$$|\psi(E)\rangle = \sum_{n=0}^{\infty} [\mathcal{G}(E)V]^n \, |\psi_0(E)\rangle. \tag{11.6}$$

Equations (11.5, 11.6) with $|\psi_0(E)\rangle \neq 0$ are also sometimes written as

$$|\psi(E)\rangle = \frac{1}{1 - \mathcal{G}(E)V}|\psi_0(E)\rangle. \tag{11.7}$$

[1]B.A. Lippmann, J. Schwinger, Phys. Rev. 79, 469 (1950).

Contrary to equation (11.5), the equations (11.6, 11.7) assume that E is an eigenvalue of *both* H *and* H_0. A necessary condition for equations (11.6, 11.7) is therefore $E \geq 0$ because we use $H_0 = \mathbf{p}^2/2m$. We also assume that the series on the right hand side of (11.6) converges in a suitable sense (e.g. such that we can integrate it with square integrable functions).

Equation (11.5) is more general than equations (11.6, 11.7) because if E is not an eigenvalue of H_0 we have $|\psi_0(E)\rangle = 0$ and equation (11.5) simply states that the solutions of (11.4) are eigenstates of $\mathcal{G}(E)V$ with eigenvalue 1, whereas equations (11.6, 11.7) become singular if $\mathcal{G}(E)V$ has eigenvalue 1.

For potential scattering theory it is customary to rescale $\mathcal{G}(E)$ by a factor $-\hbar^2/2m$, such that the zero energy Green's function $G(0)$ is the inverse of the negative Laplace operator. The equation

$$\mathcal{G}(E) = -\frac{2m}{\hbar^2}G(E) = \frac{1}{E - H_0 + \mathrm{i}\epsilon}, \quad \epsilon \to +0, \tag{11.8}$$

is then in $\boldsymbol{k}$-representation

$$\langle \boldsymbol{k}|G(E)|\boldsymbol{k}'\rangle = \frac{\delta(\boldsymbol{k} - \boldsymbol{k}')}{\boldsymbol{k}^2 - (2mE/\hbar^2) - \mathrm{i}\epsilon} = G(E, \boldsymbol{k})\delta(\boldsymbol{k} - \boldsymbol{k}'). \tag{11.9}$$

The condition (11.3) for the energy-dependent Green's function in $\boldsymbol{x}$-representation

$$\begin{aligned}\langle \boldsymbol{x}|G(E)|\boldsymbol{x}'\rangle &= \int d^3\boldsymbol{k} \int d^3\boldsymbol{k}'\, \langle \boldsymbol{x}|\boldsymbol{k}\rangle\langle \boldsymbol{k}|G(E)|\boldsymbol{k}'\rangle\langle \boldsymbol{k}'\boldsymbol{x}'\rangle \\ &= \frac{1}{(2\pi)^3}\int d^3\boldsymbol{k}\, \exp[\mathrm{i}\boldsymbol{k}\cdot(\boldsymbol{x} - \boldsymbol{x}')]G(E, \boldsymbol{k}) \\ &= G(E, \boldsymbol{x} - \boldsymbol{x}')\end{aligned} \tag{11.10}$$

is

$$\Delta G(E, \boldsymbol{x} - \boldsymbol{x}') + \frac{2m}{\hbar^2}EG(E, \boldsymbol{x} - \boldsymbol{x}') = -\delta(\boldsymbol{x} - \boldsymbol{x}'). \tag{11.11}$$

The shift $\mathrm{i}\epsilon \to +\mathrm{i}0$ in equation (11.8) defines the *retarded Green's function* for the Schrödinger equation. The reason for this terminology is that the corresponding Green's function in the time domain (cf. (5.11)),

$$\mathcal{G}(t) = \frac{1}{2\pi\hbar}\int_{-\infty}^{\infty} dE\, \mathcal{G}(E)\exp\!\left(-\frac{\mathrm{i}}{\hbar}Et\right) = \frac{\Theta(t)}{\mathrm{i}\hbar}\exp\!\left(-\frac{\mathrm{i}}{\hbar}H_0 t\right), \tag{11.12}$$

satisfies the conditions

$$\mathrm{i}\hbar\frac{\partial \mathcal{G}(t)}{\partial t} - H_0\mathcal{G}(t) = \delta(t), \quad \mathcal{G}(t)\Big|_{t<0} = 0.$$

This implies that $\mathcal{G}$ propagates time-dependent perturbations

$$\mathrm{i}\hbar\frac{\partial}{\partial t}|\psi(t)\rangle - H_0|\psi(t)\rangle = V(t)|\psi(t)\rangle$$

forward in time,

$$|\psi(t)\rangle = |\psi_0(t)\rangle + \int_{-\infty}^{\infty} dt'\, \mathcal{G}(t-t')V(t')|\psi(t')\rangle$$
$$= |\psi_0(t)\rangle - \frac{\mathrm{i}}{\hbar}\int_{-\infty}^{t} dt'\, \exp\!\left(-\frac{\mathrm{i}}{\hbar}H_0(t-t')\right) V(t')|\psi(t')\rangle. \quad (11.13)$$

We will revisit time-dependent perturbations in Chapter 13 and focus on scattering due to time-independent perturbations (11.4–11.6) for now.

We first calculate the Green's function $G(E, \boldsymbol{x})$ for $E > 0$. The equations (11.9) and (11.10) yield

$$G(E,\boldsymbol{x}) = \frac{1}{(2\pi)^3}\int d^3\boldsymbol{k}\,\frac{\exp(\mathrm{i}\boldsymbol{k}\cdot\boldsymbol{x})}{k^2-(2mE/\hbar^2)-\mathrm{i}\epsilon}$$
$$= \frac{1}{(2\pi)^2}\int_0^{\infty}dk\int_{-1}^{1}d\xi\,k^2\frac{\exp(\mathrm{i}kr\xi)}{k^2-(2mE/\hbar^2)-\mathrm{i}\epsilon}$$
$$= \frac{1}{(2\pi)^2\mathrm{i}r}\int_0^{\infty}dk\,k\frac{\exp(\mathrm{i}kr)-\exp(-\mathrm{i}kr)}{k^2-(2mE/\hbar^2)-\mathrm{i}\epsilon}$$
$$= \frac{1}{(2\pi)^2\mathrm{i}r}\int_{-\infty}^{\infty}dk\,k\frac{\exp(\mathrm{i}kr)}{k^2-(2mE/\hbar^2)-\mathrm{i}\epsilon}. \quad (11.14)$$

Due to $r > 0$ we can add a semi-circle with radius $|k| \to \infty$ in the upper half of the complex k plane to the integration contour, see Figure 11.1. This additional

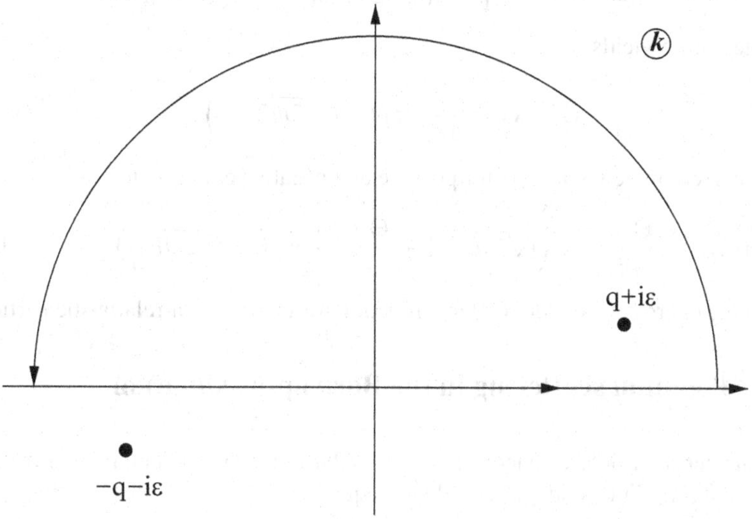

Fig. 11.1 Location of the poles and integration contour in the complex k plane. Here $\mathrm{q} \equiv \sqrt{2mE}/\hbar$

path segment $\cap$ will not change the integral in (11.14) because we have with $k = |k|\exp(\mathrm{i}\phi)$

$$\lim_{|k|\to\infty}\int_\cap dk\,k\frac{\exp(\mathrm{i}kr)}{k^2-(2mE/\hbar^2)-\mathrm{i}\epsilon}$$
$$=\lim_{|k|\to\infty}\int_0^\pi d\phi\,\frac{\mathrm{i}|k|^2\exp(2\mathrm{i}\phi)\exp[\mathrm{i}|k|r\cos(\phi)]}{|k|^2\exp(2\mathrm{i}\phi)-(2mE/\hbar^2)}\exp[-|k|r\sin(\phi)]=0.$$

However, adding the semi-circle to the integration contour allows us to use the residue theorem. Decomposing the denominator into its simple poles

$$\frac{1}{k^2-(2mE/\hbar^2)-\mathrm{i}\epsilon}=\frac{1}{[k-(\sqrt{2mE}/\hbar)-\mathrm{i}\epsilon][k+(\sqrt{2mE}/\hbar)+\mathrm{i}\epsilon]}$$

then yields

$$G(E,\boldsymbol{x}) = \frac{1}{2\pi r}\frac{k\exp(\mathrm{i}kr)}{k+(\sqrt{2mE}/\hbar)}\bigg|_{k=\sqrt{2mE}/\hbar}$$
$$= \frac{1}{4\pi r}\exp\!\left(\mathrm{i}\sqrt{2mE}r/\hbar\right), \tag{11.15}$$

i.e. the retardation requirement $\mathcal{G}(t) \propto \Theta(t)$ yields only outgoing spherical waves for positive energy.

If we perform the same calculation for $E < 0$ we find a denominator

$$\frac{1}{k^2-(2mE/\hbar^2)}=\frac{1}{[k-\mathrm{i}(\sqrt{-2mE}/\hbar)][k+\mathrm{i}(\sqrt{-2mE}/\hbar)]},$$

and integration yields

$$G(E,\boldsymbol{x}) = \frac{1}{4\pi r}\exp\!\left(-\sqrt{-2mE}r/\hbar\right). \tag{11.16}$$

We can combine the results for positive and negative energy into

$$G(E,\boldsymbol{x}) = \frac{\Theta(E)}{4\pi r}\exp\!\left(\mathrm{i}\sqrt{2mE}\frac{r}{\hbar}\right)+\frac{\Theta(-E)}{4\pi r}\exp\!\left(-\sqrt{-2mE}\frac{r}{\hbar}\right). \tag{11.17}$$

This is the energy-dependent Green's function for the free non-relativistic particle.

11.2 Potential scattering in the Born approximation

We consider a particle of energy $E = \hbar^2k^2/2m$ in a static potential $V(\boldsymbol{x})$ of finite range. The time-independent Schrödinger equation

$$(\Delta + k^2)\psi(\boldsymbol{x}) = \frac{2m}{\hbar^2}V(\boldsymbol{x})\psi(\boldsymbol{x}) \tag{11.18}$$

can be converted into an integral equation using the Green's function (11.15),

$$\psi(\boldsymbol{x}) = \frac{\exp(\mathrm{i}\boldsymbol{k}\cdot\boldsymbol{x})}{(2\pi)^{3/2}} - \frac{m}{2\pi\hbar^2}\int d^3x'\,\frac{\exp(\mathrm{i}k|\boldsymbol{x}-\boldsymbol{x}'|)}{|\boldsymbol{x}-\boldsymbol{x}'|}V(\boldsymbol{x}')\psi(\boldsymbol{x}'). \tag{11.19}$$

Note that the Lippmann-Schwinger equation (11.5) is the representation free operator version of this equation for static potential and plane waves as unperturbed states.

First order iteration of (11.19) and neglect of the irrelevant normalization factor $(2\pi)^{-3/2}$ yields

$$\psi(\boldsymbol{x}) \approx \exp(\mathrm{i}\boldsymbol{k}\cdot\boldsymbol{x}) - \frac{m}{2\pi\hbar^2}\int d^3x'\,\frac{V(\boldsymbol{x}')}{|\boldsymbol{x}-\boldsymbol{x}'|}\exp(\mathrm{i}k|\boldsymbol{x}-\boldsymbol{x}'| + \mathrm{i}\boldsymbol{k}\cdot\boldsymbol{x}'), \tag{11.20}$$

The overall normalization is irrelevant, because finally we are only interested in the ratio of the different parts of the wave function.

For $r \gg r'$ we have

$$|\boldsymbol{x}-\boldsymbol{x}'| \approx \sqrt{r^2 - 2rr'\cos\theta} \approx r - r'\cos\theta = r - \frac{1}{r}\boldsymbol{x}\cdot\boldsymbol{x}' = r - \hat{\boldsymbol{x}}\cdot\boldsymbol{x}'.$$

We need the expansion to this order in the exponent of the Green's function in equation (11.20). However, for the denominator the expansion

$$\frac{1}{|\boldsymbol{x}-\boldsymbol{x}'|} \approx \frac{1}{r}$$

will suffice, because the subleading term $(r'/r^2)\cos\theta$ will not contribute to the differential scattering cross section (11.2) due to the limit $\lim_{r\to\infty} r^2 j_{out}$.

Substitution of the approximations yields the *Born approximation*

$$\begin{aligned}\psi(\boldsymbol{x}) &= \exp(\mathrm{i}\boldsymbol{k}\cdot\boldsymbol{x}) - \frac{m}{2\pi\hbar^2}\frac{1}{r}\exp(\mathrm{i}kr)\int d^3x'\,\exp[\mathrm{i}(\boldsymbol{k}-k\hat{\boldsymbol{x}})\cdot\boldsymbol{x}']V(\boldsymbol{x}')\\ &= \exp(\mathrm{i}\boldsymbol{k}\cdot\boldsymbol{x}) + f(k\hat{\boldsymbol{x}}-\boldsymbol{k})\frac{1}{r}\exp(\mathrm{i}kr) = \psi^{(\mathrm{in})}(\boldsymbol{x}) + \psi^{(\mathrm{out})}(\boldsymbol{x}),\end{aligned} \tag{11.21}$$

with the *scattering amplitude*

$$f(\Delta\boldsymbol{k}) = -\frac{m}{2\pi\hbar^2}\int d^3x\,\exp(-\mathrm{i}\Delta\boldsymbol{k}\cdot\boldsymbol{x})V(\boldsymbol{x}). \tag{11.22}$$

Note that $\hbar\Delta\boldsymbol{k} \equiv \hbar(k\hat{\boldsymbol{x}}-\boldsymbol{k}) = \hbar\boldsymbol{k}' - \hbar\boldsymbol{k}$ is the momentum transfer imparted on the particle which is detected in the direction $\hat{\boldsymbol{x}}$, i.e. the scattering amplitude is up to a factor the Fourier transformed scattering potential evaluated at the momentum transfer. For later reference, we note that $f(\Delta\boldsymbol{k})$ can also be written as a transition matrix element of the operator $V(\mathbf{x})$,

$$f(\Delta\boldsymbol{k}) = -(2\pi)^2\frac{m}{\hbar^2}\langle\boldsymbol{k}'|V|\boldsymbol{k}\rangle. \tag{11.23}$$

Like in the prototype one-dimensional scattering event described in Section 3.1, the monochromatic asymptotic wave function (11.21) describes both the incident and the scattered particles simultaneously, for the same reasons as in Section 3.1. The current density j_{out} of scattered particles is therefore calculated from the outgoing spherical wave component $\psi^{(out)}(\boldsymbol{x})$ in the wave function (11.21), and we only need the leading term for $r \to \infty$,

$$\boldsymbol{j}_{out} = \frac{\hbar}{2im}\left(\psi^{(out)+}\nabla\psi^{(out)} - \nabla\psi^{(out)+}\cdot\psi^{(out)}\right)\Big|_{\text{leading term for } r\to\infty}$$
$$= \frac{\hbar k}{m}\frac{\hat{\boldsymbol{x}}}{r^2}|f(k\hat{\boldsymbol{x}} - \boldsymbol{k})|^2. \tag{11.24}$$

The incoming current density j_{in} is calculated from the incoming plane wave component $\psi^{(in)}(\boldsymbol{x})$,

$$j_{in} = \frac{\hbar k}{m}. \tag{11.25}$$

Both j_{in} and j_{out} come in units of cm/s instead of the expected cm^{-2}/s for particle or probability current densities. The reason for this is the use of plane or spherical wave states in $\boldsymbol{k}$ space which are dimensionless in $\boldsymbol{x}$ representation, see Section 5.3. Therefore the current densities (11.24) and (11.25) are actually current densities per $\boldsymbol{k}$ space volume[2]. The normalization to $\boldsymbol{k}$ space volume cancels in the ratio j_{out}/j_{in}, and substitution of equations (11.24, 11.25) into (11.2) yields

$$\frac{d\sigma_k}{d\Omega} = |f(k\hat{\boldsymbol{x}} - \boldsymbol{k})|^2. \tag{11.27}$$

The scattering amplitude (11.22) for a spherically symmetric potential is

$$f(\Delta k) \equiv f_k(\theta) = -\frac{2m}{\hbar^2\Delta k}\int_0^\infty dr\, r\sin(\Delta kr)V(r), \tag{11.28}$$

where $\Delta k = 2k\sin(\theta/2)$ and θ is the scattering angle, see Figure 11.2.

In agreement with the observation that the energy-dependent wave function describes both the incoming and the scattered particles, we have split the wave function $\psi(\boldsymbol{r})$ into the components $\psi^{(in)}(\boldsymbol{r})$ and $\psi^{(out)}(\boldsymbol{r})$, and then calculated separate current densities $\boldsymbol{j}_{in}$ and $\boldsymbol{j}_{out}$ from both contributions rather than calculate a total current density $\boldsymbol{j}$ for $\psi(\boldsymbol{r})$. On the other hand, probability conservation implies for stationary states $\nabla\cdot\boldsymbol{j} = 0$, but only for the full current density including the

[2] We could more appropriately write $d\boldsymbol{j}(\boldsymbol{k})/d^3\boldsymbol{k} = \hbar\boldsymbol{k}/m$, and the current density from a volume $\tilde{V}$ in $\boldsymbol{k}$ space is

$$\boldsymbol{j}(\tilde{V}) = \int_{\tilde{V}} d^3\boldsymbol{k}\,\frac{\hbar\boldsymbol{k}}{m}. \tag{11.26}$$

Note that this has the correct units cm^{-2}/s for a current density.

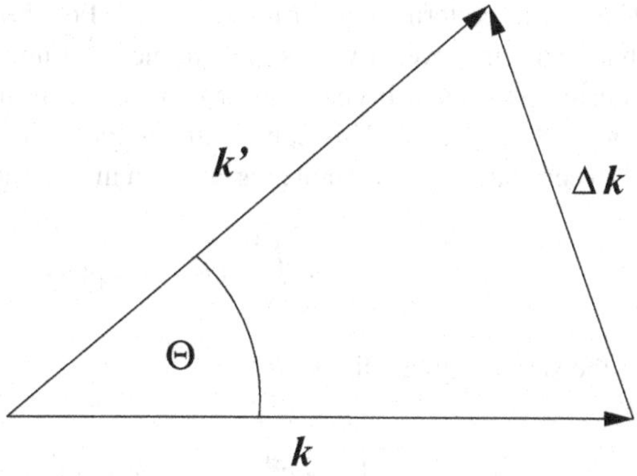

Fig. 11.2 Energy conservation implies $k \equiv |\boldsymbol{k}| = |\boldsymbol{k}'|$. Δk is therefore related to k and the scattering angle θ according to $\Delta k = 2k\sin(\theta/2)$

interference terms $\boldsymbol{j} - \boldsymbol{j}_{in} - \boldsymbol{j}_{out}$ between the incoming and scattered parts of the wave function. Therefore the interference terms will describe reduction of the current of incoming particles due to scattering.

When we discuss this effect on the basis of the wave function (11.21), we have to keep in mind that this is only a large distance approximation which was justified by the observation that we are interested in the large distance limit of $r^2 j_{out}$. Furthermore, the wave function (11.21) will only yield components j_r and j_θ in spherical coordinates, and only j_r will be relevant for the detailed balance between incoming and scattered particles. The radial current density from (11.21) in the large distance approximation, i.e. for $kr \gg 1$ and under neglect of terms which drop off faster than r^{-2}, is

$$\frac{m}{\hbar} j_r = \Im\left(\psi^+ \frac{\partial}{\partial r}\psi\right) \simeq k\cos\theta + k\frac{|f_k(\theta)|^2}{r^2} + \frac{k}{r}$$
$$\times \Re\big(f_k(\theta)\exp[\mathrm{i}kr(1-\cos\theta)] + f_k^+(\theta)\cos\theta\exp[-\mathrm{i}kr(1-\cos\theta)]\big).$$

Conservation of particles requires $\int d\Omega\, r^2 j_r = 0$. The first term $k\cos\theta$ yields null after integration over the sphere. The remaining terms yield with $u = 1 - \cos\theta$, $F_k(u) = f_k(\theta)$,

$$k\sigma_k + 2\pi kr\int_0^2 du\, \Re\big[F_k(u)\exp(\mathrm{i}kru) + F_k^+(u)(1-u)\exp(-\mathrm{i}kru)\big] = 0. \qquad (11.29)$$

Here $\sigma_k \equiv \int d\Omega\, d\sigma_k/d\Omega$ is the total scattering cross section.

Two-fold integration by parts yields

$$kr\int_0^2 du\, F_k(u)\exp(\mathrm{i}kru) = \mathrm{i}F_k(0) - \mathrm{i}F_k(2)\exp(2\mathrm{i}kr) - \frac{1}{kr}F_k'(0)$$
$$+ \frac{1}{kr}F_k'(2)\exp(2\mathrm{i}kr) - \frac{1}{kr}\int_0^2 du\, F_k''(u)\exp(\mathrm{i}kru).$$

The last three terms vanish for $kr \to \infty$. For the term $F_k(2)\exp(2ikr)$ we observe that averaging over a very small momentum uncertainty $\Delta k/k = \pi/kr \ll 1$ also yields a null result because it corresponds to an integration in k space over a range $-\pi/r \leq k \leq \pi/r$. This can be understood physically as destructive interference between states with a minute variation in momentum. Therefore we find for $kr \to \infty$

$$kr \int_0^2 du\, F_k(u) \exp(ikru) \to \mathrm{i}f_k(0).$$

In the same way one finds

$$kr \int_0^2 du\, F_k^+(u)(1-u) \exp(-ikru) \to -\mathrm{i}f_k^+(0),$$

and equation (11.29) yields in the large kr limit the *optical theorem*

$$\sigma_k = \frac{4\pi}{k}\Im f_k(0) \tag{11.30}$$

between the total scattering cross section and the imaginary part of the scattering amplitude in forward direction.

11.3 Scattering off a hard sphere

The hard sphere of radius R corresponds to the $V_0 \to \infty$ limit of a potential $V(r) = V_0\Theta(R-r)$. This reduces to the solution of the free Schrödinger equation for $r > R$ and a boundary condition on the surface of the sphere,

$$\psi(\boldsymbol{r})\Big|_{r=R} = 0. \tag{11.31}$$

We recall that the radial Schrödinger equation for a free particle with fixed angular momentum M_z, $\boldsymbol{M}^2$ and energy $E = \hbar^2k^2/2m$ yields the radial equation (7.43),

$$\left(\frac{d^2}{dr^2} - \frac{\ell(\ell+1)}{r^2} + k^2\right) r\psi(r) = 0. \tag{11.32}$$

We have seen in Section 7.7 that the regular solutions for arbitrary ℓ can be gotten through repeated application of $r^{-1}d/dr$ on the regular solution for $\ell = 0$,

$$\psi^{(\mathrm{in})}_{\ell,k}(r) \propto j_\ell(kr), \quad j_\ell(x) = (-x)^\ell \left(\frac{1}{x}\frac{d}{dx}\right)^\ell j_0(x), \quad j_0(x) = \frac{\sin x}{x}.$$

We denote the regular solutions $j_\ell(kr)$ as $\psi^{(\mathrm{in})}_{\ell,k}(r)$, because we can superimpose those functions according to equation (7.51) to form an incoming plane wave.

However, equation (11.32) also has an outgoing radial wave as a solution for $\ell = 0$,

$$\psi^{(\mathrm{out})}_{0,k}(r) \propto \frac{\exp(\mathrm{i}kr)}{kr} = -\,\mathrm{i}h^{(1)}_0(kr).$$

The reasoning leading to equations (7.46, 7.47) also implies that repeated application of $r^{-1}d/dr$ on the outgoing radial wave solution leads to solutions for higher ℓ,

$$\psi^{(\mathrm{out})}_{\ell,k}(r) \propto -\,\mathrm{i}h^{(1)}_\ell(kr), \quad h^{(1)}_\ell(x) = (-x)^\ell \left(\frac{1}{x}\frac{d}{dx}\right)^\ell h^{(1)}_0(x).$$

In leading order in $1/r$, these are again outgoing radial waves,

$$h^{(1)}_\ell(kr) \simeq (-)^\ell \frac{\exp(\mathrm{i}kr)}{\mathrm{i}kr}. \tag{11.33}$$

The functions $h^{(1)}_\ell(x)$ are known as spherical Hankel functions of the first kind.

We can use the spherical Bessel and Hankel functions to form solutions of equation (11.32) which satisfy the condition (11.31) and contain an outgoing spherical wave in the asymptotic limit,

$$\psi_{\ell,k}(r) = \psi^{(\mathrm{in})}_{\ell,k}(r) + \psi^{(\mathrm{out})}_{\ell,k}(r) \propto j_\ell(kr) - h^{(1)}_\ell(kr)\frac{j_\ell(kR)}{h^{(1)}_\ell(kR)}.$$

However, equation (7.51) then tells us how to write down a solution to the free Schrödinger equation for energy $E = \hbar^2k^2/2m$ outside of the hard sphere, which satisfies the boundary condition (11.31) and contains both a plane wave and an outgoing spherical wave,

$$\begin{aligned}\psi_k(\boldsymbol{r}) &= \sum_{\ell=0}^{\infty}(2\ell+1)\mathrm{i}^\ell\left(j_\ell(kr) - h^{(1)}_\ell(kr)\frac{j_\ell(kR)}{h^{(1)}_\ell(kR)}\right)P_\ell(\cos\theta)\\ &= \exp(\mathrm{i}kz) - \sum_{\ell=0}^{\infty}(2\ell+1)\mathrm{i}^\ell h^{(1)}_\ell(kr)\frac{j_\ell(kR)}{h^{(1)}_\ell(kR)}P_\ell(\cos\theta)\\ &= \psi^{(\mathrm{in})}(\boldsymbol{r}) + \psi^{(\mathrm{out})}(\boldsymbol{r}).\end{aligned} \tag{11.34}$$

From our previous experience in Sections 3.1 and 11.2 we already anticipated that the monochromatic wave function will describe both incoming and scattered particles. According to equation (11.33), the asymptotic expansion of the wave function for large r is

$$\psi_k(\boldsymbol{r}) \simeq \exp(\mathrm{i}kz) + f_k(\theta)\frac{\exp(\mathrm{i}kr)}{r},$$

$$f_k(\theta) = -\frac{1}{k}\sum_{\ell=0}^{\infty}(2\ell+1)(-\mathrm{i})^{\ell+1}P_\ell(\cos\theta)\frac{j_\ell(kR)}{h^{(1)}_\ell(kR)}.$$

The resulting expression for the differential scattering cross section of the hard sphere is a little unwieldy,

$$\frac{d\sigma_k}{d\Omega} = |f_k(\theta)|^2 = \frac{1}{k^2}\sum_{\ell,\ell'=0}^{\infty}(2\ell+1)(2\ell'+1)\mathrm{i}^{\ell-\ell'}P_\ell(\cos\theta)P_{\ell'}(\cos\theta)$$
$$\times\frac{j_\ell^+(kR)j_{\ell'}(kR)}{h_\ell^{(1)+}(kR)h_{\ell'}^{(1)}(kR)}. \tag{11.35}$$

However, for the scattering cross section the orthogonality property of Legendre polynomials

$$\int_0^\pi d\theta\,\sin\theta\,P_\ell(\cos\theta)P_{\ell'}(\cos\theta) = \frac{2}{2\ell+1}\delta_{\ell,\ell'}$$

yields a much simpler result,

$$\sigma_k = \frac{4\pi}{k^2}\sum_{\ell=0}^{\infty}(2\ell+1)\left|\frac{j_\ell(kR)}{h_\ell^{(1)}(kR)}\right|^2. \tag{11.36}$$

This is shown in Figure 11.3. The asymptotic behavior for small arguments kR, $j_\ell(kR) \simeq (kR)^\ell/(2\ell+1)!!$ and $h_\ell^{(1)}(kR) \simeq -\mathrm{i}(2\ell-1)!!(kR)^{-\ell-1}$, imply for the low energy or long wavelength limit $\lambda \gg R$ that only the $\ell = 0$ contribution survives with

$$\lim_{kR\to 0}\sigma_k = 4\pi R^2.$$

The quantum mechanical scattering cross section drops continuously from $\lim_{kR\to 0}\sigma_k = 4\pi R^2$ to $\lim_{kR\to\infty}\sigma_k = 2\pi R^2$, i.e. it always exceeds the classical value $\sigma_{cl} = \pi R^2$ by more than a factor of 2. In terms of the variables k and R, σ_k seems to be independent of $\hbar$ and one might naively expect that this is the reason for absence of a classical limit for scattering off a hard sphere, but this is wrong for two reasons. If one compares with classical results one should use the same variables as in classical mechanics, and if in terms of the classical variables the quantum mechanical result is independent of $\hbar$, we rather expect to find the same result as in classical mechanics. Furthermore, when one calculates classical scattering cross sections, one uses the momentum $p = \hbar k$ for the incident particles as a variable besides the radius R of the sphere, i.e. in terms of classical variables the cross section σ_k does depend on $\hbar$, and the classical limit should correspond to $pR \gg \hbar$, $kR \gg 1$. However, the classical limit fails because there is an important difference between the classical calculation and quantum mechanical scattering. The classical calculation requires particles to hit the hard sphere with an impact parameter $b = |\boldsymbol{M}|/|\boldsymbol{p}|$ which is limited by the requirement that all scattered particles must actually hit the sphere, $b \leq R$. This corresponds to a classical angular momentum cutoff $|\boldsymbol{M}| \leq pR$. However, quantum mechanically, particles with arbitrary high angular momentum still feel the presence of the hard sphere

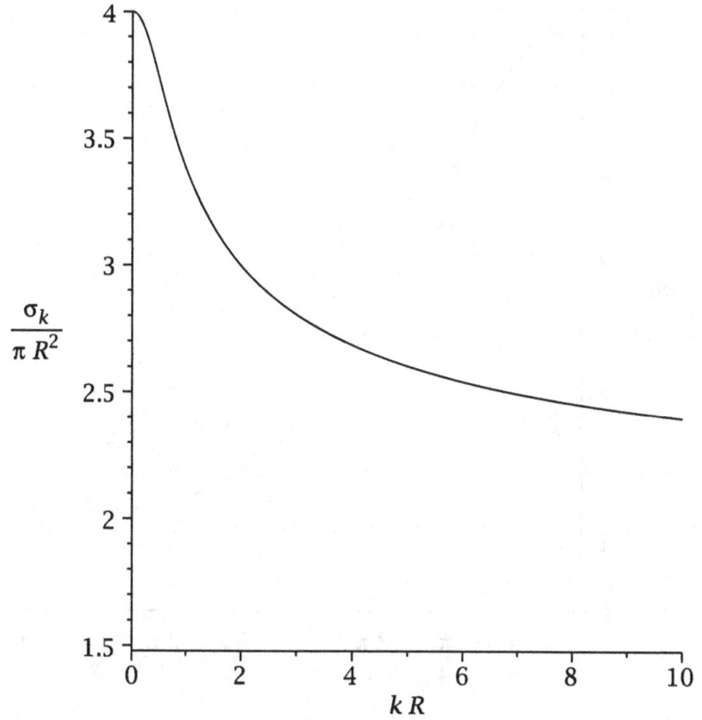

Fig. 11.3 The scattering cross section of a hard sphere normalized to the classical scattering cross section $\sigma_{classical} = \pi R^2$

and can be scattered similar to classical wave diffraction. This regime of deviation between the classical and quantum picture concerns large angular momenta and small deflection angles, i.e. the forward scattering region. In the classical picture the forward scattered particles are considered as missing the sphere and therefore ignored in the classical scattering cross section. Therefore the classical cross section is always smaller than the quantum mechanical cross section, even in the classical limit $kR \gg 1$. The increasing concentration of scattering in forward direction with increasing kR is demonstrated in Figures 11.4 and 11.5.

An approximate evaluation of the θ dependence of the extra "non-classical" part of the differential cross section for $kR \gg 1$ in terms of shadow forming waves is given in [29]. However, the Figures 11.4 and 11.5 use the exact result (11.35). Either way, the ultimate reason for the discrepancy between the quantum result and the classical result in the classical limit $kR \gg 1$ is different accounting of scattered versus unscattered particles in the forward scattering region.

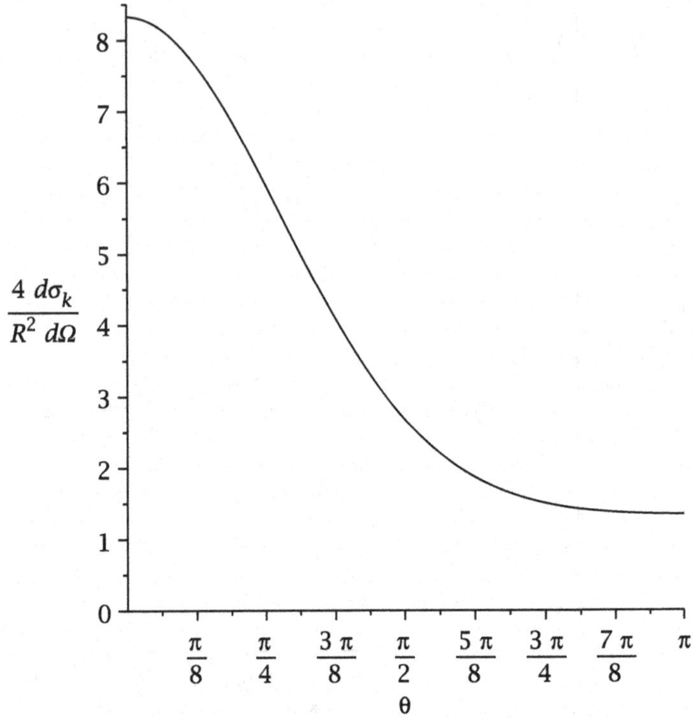

Fig. 11.4 The differential scattering cross section $d\sigma_k/d\Omega$ of a hard sphere of radius R normalized to the classical differential scattering cross section $(d\sigma_k/d\Omega)_{classical} = R^2/4$ for $kR = 1$

11.4 Rutherford scattering

Axial symmetry often plays a role in atoms which interact with their surroundings. External fields will often have axial symmetry, and this motivated Schrödinger to solve the hydrogen problem in parabolic coordinates for his perturbative analysis of the Stark effect[3]. Furthermore, if a hydrogen atom is formed through electron-proton recombination, the initial plane wave state describing the mutual approach of the electron and the proton will also break the rotational symmetry to axial symmetry and the calculation of recombination cross sections can be performed in terms of parabolic coordinates [3]. Maybe the best known application of parabolic coordinates concerns the calculation of Rutherford scattering. The incident plane wave $\psi^{(\mathrm{in})} \sim \exp(ikz)$ breaks the rotational symmetry of the problem down to an axial symmetry, but we still expect an outgoing spherical wave $\psi^{(\mathrm{out})} \sim \exp(ikr)/r$. The reconciliation of axial symmetry with the use of r makes parabolic coordinates more useful than cylinder coordinates for the study of scattering in rotationally

[3]E. Schrödinger, Annalen Phys. 385, 437 (1926).

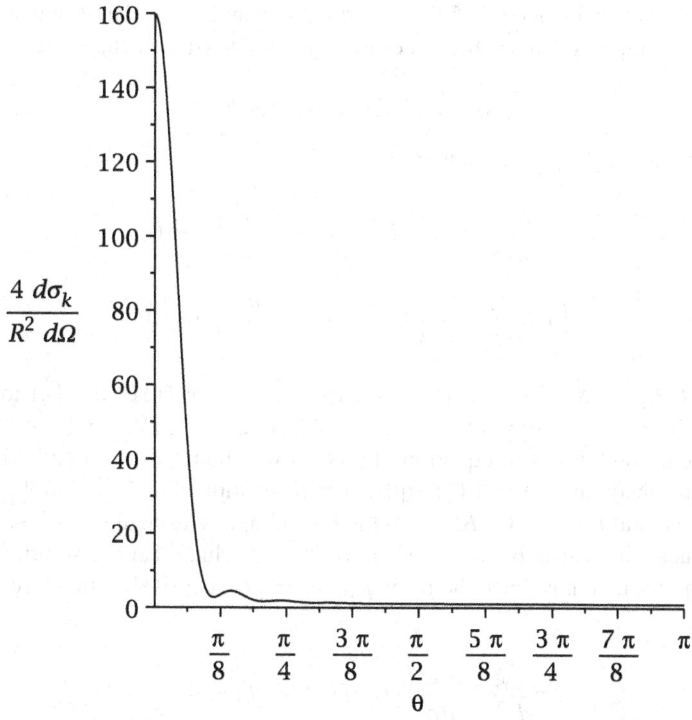

Fig. 11.5 The differential scattering cross section $d\sigma_k/d\Omega$ of a hard sphere of radius R normalized to the classical differential scattering cross section $(d\sigma_k/d\Omega)_{classical} = R^2/4$ for $kR = 10$

symmetric potentials beyond the Born approximation, and can occasionally render them also more useful than spherical coordinates. The separability of the Coulomb problem in parabolic coordinates makes them particularly useful for the study of Rutherford scattering.

We define parabolic coordinates through the following relations[4],

$$x = 2\sqrt{\xi\eta}\cos\varphi, \quad y = 2\sqrt{\xi\eta}\sin\varphi, \quad z = \xi - \eta,$$
$$2\xi = r + z, \quad 2\eta = r - z, \quad \varphi = \arctan\frac{y}{x}. \tag{11.37}$$

Using the methods developed in Section 5.4, one finds the Schrödinger equation for motion of a particle of energy $E = \hbar^2k^2/2\mu$ in the Coulomb potential $V = q_1q_2/4\pi\epsilon_0 r = \hbar^2K/2\mu(\xi + \eta)$ in parabolic coordinates,s

$$\frac{1}{\xi+\eta}\left[\frac{\partial}{\partial\xi}\left(\xi\frac{\partial}{\partial\xi}\right) + \frac{\partial}{\partial\eta}\left(\eta\frac{\partial}{\partial\eta}\right)\right]\psi + \frac{1}{4\xi\eta}\frac{\partial^2\psi}{\partial^2\varphi} + k^2\psi - \frac{K}{\xi+\eta}\psi = 0. \tag{11.38}$$

[4] Please note that the definition used here differs by factors of 2 from the definition used by Schrödinger, $\lambda_1 \equiv \xi_S = 2\xi$, $\lambda_2 \equiv \eta_S = 2\eta$.

The arguments in Section 5.5, in particular concerning Hamiltonians of the form (5.36), imply that the solutions of this equation must have the form

$$\psi(\xi,\eta,\varphi) = f(\xi)g(\eta)\exp(im\varphi),$$

with the remaining separated equations

$$\left(\xi\frac{d}{d\xi}\right)^2 f + \left(k^2\xi^2 - K_1\xi - \frac{m^2}{4}\right)f = 0, \tag{11.39}$$

$$\left(\eta\frac{d}{d\eta}\right)^2 g + \left(k^2\eta^2 - K_2\eta - \frac{m^2}{4}\right)g = 0, \tag{11.40}$$

with $K_1 + K_2 = K$. We want $\psi^{(\mathrm{in})} \sim \exp(ikz) = \exp(ik\xi)\exp(-ik\eta)$ to be the dominant term in the solution near the half-axis $z < 0$, i.e. for $\xi \to 0$. This requirement complies with equation (11.39) if we choose $m = 0$ and $K_1 = ik$. If we then substitute $f(\xi) = F(\xi)\exp(ik\xi)$ into equation (11.39) to find the second solution, we find $F(\xi) = A + BE_i(-2ik\xi)$, which implies a singularity of the second solution near the half-axis $z < 0$. Therefore we conclude that the solution to our scattering problem must have the form $\psi(\xi,\eta) = g(\eta)\exp(ik\xi)$ with the remaining condition

$$\eta\frac{d^2g}{d\eta^2} + \frac{dg}{d\eta} + \left(k^2\eta - K + \mathrm{i}k\right)g = 0. \tag{11.41}$$

Comparison with equation (11.39) for $m = 0$ and $K_1 = ik$ tells us that $g(\eta) = \exp(-ik\eta)$ is the regular solution of equation (11.41) if $K = 0$. This is also clear from the physical point of view. If there is no scattering potential, the plane wave $\exp(ikz) = \exp(ik\xi)\exp(-ik\eta)$ that we imposed near the half-axis $z < 0$ must persist everywhere. This motivates a substitution $g(\eta) = h(\eta)\exp(-ik\eta)$ in (11.41),

$$\eta\frac{d^2h}{d\eta^2} + (1 - 2\mathrm{i}k\eta)\frac{dh}{d\eta} - Kh = 0. \tag{11.42}$$

Substitution of $h(\eta) = \sum_{n\geq 0} c_n\eta^n$ yields

$$c_{n+1} = \frac{K + 2\mathrm{i}kn}{(n+1)^2}c_n$$

and therefore

$$\begin{aligned} h(\eta) &= c_0\sum_{n=0}^{\infty}\frac{K(K+2\mathrm{i}k)\ldots(K+2\mathrm{i}k(n-1))}{n!}\frac{\eta^n}{n!} \\ &= c_0\,{}_1F_1(-\mathrm{i}K/2k;1;2\mathrm{i}k\eta) = c_0\exp(2\mathrm{i}k\eta)\,{}_1F_1(1+\mathrm{i}K/2k;1;2\mathrm{i}k\eta). \end{aligned}$$

The wave function for our scattering problem is therefore up to normalization[5]

$$\begin{aligned}\psi(\boldsymbol{r}) &= \exp[ik(\xi-\eta)]\,{}_1F_1(-\mathrm{i}K/2k;1;2ik\eta)\\ &= \exp(ikz)\,{}_1F_1(-\mathrm{i}K/2k;1;ik(r-z))\\ &= \exp(ikr)\,{}_1F_1(1+\mathrm{i}K/2k;1;ik(r-z)).\end{aligned} \tag{11.43}$$

The normalization factor c_0 is irrelevant because it cancels in the calculation of the cross section.

Identification of the incoming and scattered components in the wave function requires asymptotic expansion for large values of the argument $2k\eta = k(r-z)$. The asymptotic expansion of confluent hypergeometric functions ${}_1F_1(a;b;\zeta)$ for large $|\zeta|$ [1] yields the leading order terms

$$\begin{aligned}{}_1F_1(-\mathrm{i}K/2k;1;ik(r-z)) &\simeq \exp\!\left(\frac{\pi K}{4k}\right)\left[\frac{2k}{\mathrm{i}K}\Gamma^{-1}\!\left(\frac{\mathrm{i}K}{2k}\right)\right.\\ &\times\exp\!\left(\frac{\mathrm{i}K}{2k}\ln[k(r-z)]\right)\left(1+\frac{K^2}{4ik^3(r-z)}\right)+\frac{\exp[ik(r-z)]}{ik(r-z)}\Gamma^{-1}\!\left(\frac{K}{2ik}\right)\\ &\left.\times\exp\!\left(-\frac{\mathrm{i}K}{2k}\ln[k(r-z)]\right)\right].\end{aligned}$$

After neglecting another irrelevant overall factor we find the asymptotic form

$$\begin{aligned}\psi(\boldsymbol{r}) &\simeq \exp\!\left(ikz+\frac{\mathrm{i}K}{2k}\ln[k(r-z)]\right)\left(1+\frac{K^2}{4ik^3r(1-\cos\theta)}\right)\\ &+\frac{\Gamma(\mathrm{i}K/2k)}{\Gamma(-\mathrm{i}K/2k)}\frac{K}{2k^2r(1-\cos\theta)}\exp\!\left(ikr-\frac{\mathrm{i}K}{2k}\ln[k(r-z)]\right)\\ &= \psi^{(\mathrm{in})}(\boldsymbol{r})+\psi^{(\mathrm{out})}(\boldsymbol{r}),\end{aligned} \tag{11.44}$$

where $\theta = \arccos(z/r)$ is the scattering angle.

This yields a differential scattering cross section

$$\frac{d\sigma}{d\Omega} = \lim_{r\to\infty}\frac{r^2 j_{out}}{j_{in}} = \left(\frac{K}{4k^2\sin^2(\theta/2)}\right)^2 = \left(\frac{q_1q_2}{16\pi\epsilon_0 E}\right)^2\frac{1}{\sin^4(\theta/2)}, \tag{11.45}$$

which equals exactly the corresponding cross section calculated in classical mechanics and used by Rutherford in 1911 to infer the existence of a tiny positively charged nucleus in atoms. The cross section (11.45) is an example of a quantum mechanical result which is independent of $\hbar$ when expressed in terms of classical variables, and therefore it must agree with the classical result.

---

[5] W. Gordon, Z. Phys. 48, 180 (1928).

Use of the asymptotic expansion of the hypergeometric function ${}_1F_1(a;b;\zeta)$ for large $|\zeta|$ in the present case implies the requirement $kr(1-\cos\theta) = 2kr\sin^2(\theta/2) \gg 1$, i.e. the Rutherford formula is only applicable for scattering angles $\theta \gg \sqrt{2/kr} = \sqrt{\lambda/\pi r}$. This limitation is usually irrelevant, because the values of λ e.g. in the experiments of Geiger and Marsden were only a few femtometers.

11.5 Problems

11.1. Show that the Lippmann-Schwinger equations (11.6, 11.7) can also be written in the form

$$|\psi(E)\rangle = (E - H_0 + \mathrm{i}\epsilon)\frac{1}{E - H + \mathrm{i}\epsilon}|\psi_0(E)\rangle.$$

11.2. The free time-dependent retarded Green's function in $\boldsymbol{x}$ representation has to satisfy the conditions

$$\left(\mathrm{i}\hbar\frac{\partial}{\partial t} + \frac{\hbar^2}{2m}\Delta\right)\mathcal{G}(\boldsymbol{x},t) = \delta(\boldsymbol{x})\delta(t), \quad \mathcal{G}(\boldsymbol{x},t)\Big|_{t<0} = 0.$$

Show that this function satisfies the following equations,

$$\begin{aligned}\mathcal{G}(\boldsymbol{x},t) &= \frac{1}{(2\pi)^4\hbar}\int d^3\boldsymbol{k}\int_{-\infty}^{\infty} d\omega\, \frac{\exp[\mathrm{i}(\boldsymbol{k}\cdot\boldsymbol{x} - \omega t)]}{\omega - (\hbar\boldsymbol{k}^2/2m) + \mathrm{i}\epsilon} \\ &= -\frac{m}{\pi\hbar^3}\int_{-\infty}^{\infty} dE\, G(\boldsymbol{x},E)\exp(-\mathrm{i}Et/\hbar) \\ &= \frac{\Theta(t)}{(2\pi)^3\mathrm{i}\hbar}\int d^3\boldsymbol{k}\,\exp\left[\mathrm{i}\left(\boldsymbol{k}\cdot\boldsymbol{x} - \frac{\hbar t}{2m}\boldsymbol{k}^2\right)\right] \\ &= \frac{\Theta(t)}{\mathrm{i}\hbar}\sqrt{\frac{m}{2\pi\mathrm{i}\hbar t}}^{\,3}\exp\left(\mathrm{i}\frac{m\boldsymbol{x}^2}{2\hbar t}\right). \end{aligned} \tag{11.46}$$

This also corresponds to the relation

$$\mathcal{G}(\boldsymbol{x},t) = \frac{\Theta(t)}{\mathrm{i}\hbar}U(\boldsymbol{x},t) = \frac{\Theta(t)}{\mathrm{i}\hbar}\langle\boldsymbol{x}|\exp\left(-\frac{\mathrm{i}t}{2m}\mathbf{p}^2\right)|\mathbf{0}\rangle$$

between the retarded Green's function and the propagator for the free Schrödinger equation.

11.3. Show that transformation of equation (11.13) into the frequency domain (5.12, 5.13) yields

$$|\psi(\omega)\rangle = |\psi_0(\omega)\rangle + \frac{1}{\sqrt{2\pi}}\int_{-\infty}^{\infty} d\omega'\,\mathcal{G}(\hbar\omega)V(\omega-\omega')|\psi(\omega')\rangle. \tag{11.47}$$

Show also that this reduces to the Lippmann-Schwinger equation (11.5) if the perturbation V is time-independent.

11.4. Calculate the differential scattering cross sections for the following potentials in Born approximation.

11.4a. $V(r) = V_0\Theta(R - r)$,

11.4b. $V(r) = V_0 \exp(-r/R)$,

11.4c. $V(r) = V_0 \exp(-r^2/R^2)$.

11.5. Calculate the total cross sections for the potentials from Problem 11.4 in Born approximation.

11.6. Calculate the differential cross section for Rutherford scattering in Born approximation. Compare with the exact result.

11.7. Calculate the differential and total scattering cross sections for Rutherford scattering with screened electromagnetic interactions in Born approximation. Use the following models for the screened interactions,

11.7a. $V(r) = (qQ/4\pi\epsilon_0 r)\exp(-r/R)$,

11.7b. $V(r) = (qQ/4\pi\epsilon_0 r)\exp(-r^2/R^2)$.

11.8. Show that the scattering amplitude in Born approximation (11.22) satisfies $\Im f_k(0) = 0$. Why does this not contradict the optical theorem (11.30)?

Hint: Split the potential into parity even and odd parts, $V(\boldsymbol{x}) = V_+(\boldsymbol{x}) + V_-(\boldsymbol{x})$, $V_\pm(\boldsymbol{x}) = [V(\boldsymbol{x}) \pm V(-\boldsymbol{x})]/2$. Consider powers of V (or equivalently of coupling constants) in (11.30).

Chapter 12
The Density of States

Many applications of quantum mechanics require the concept of density of states. The notion of density of states is not entirely unique. Depending on the context and the requirements of the problem at hand, it most often refers to the number of quantum states per volume and per unit of energy, or to the number of states in a volume unit $d^3\boldsymbol{k}$ in $\boldsymbol{k}$ space, and for both notions there are several variants of the density of states. Therefore the purpose of this chapter is not only to introduce the concept of density of states, but also to enumerate all the different definitions which are commonly used in physics.

Various forms of the density of states appear in numerous places in physics, e.g. in thermodynamics and optics we need the density of photon states in the derivation of Planck's law, in solid state physics the density of electron states appears in the integral of energy dependent functions over the Brillouin zone, in statistical physics we need it to calculate energy densities in physical systems, and in quantum mechanics we need it to calculate transition probabilities involving states in an energy continuum, e.g. to calculate electron emission probabilities for ionization or for the photoelectric effect, or to calculate scattering cross sections. Transition probabilities involving quantum states in an energy continuum (e.g. unbound states or states in an energy band in a solid) involve the density of states per particle as the number of states dn per unit of volume in $\boldsymbol{k}$ space,

$$dn = d^3\boldsymbol{k}. \tag{12.1}$$

More precisely, this is a density of states per spin or polarization or helicity states of a particle. Otherwise it would have to be multiplied by the number g of spin or helicity states.

The densities of electron states, photon states, and all kinds of quasiparticle states in materials are also very important quantities in materials science. These densities determine the momentum and energy distributions of (quasi)particles in materials, and the number of available states e.g. for charge or momentum transport,

R. Dick, *Advanced Quantum Mechanics*, Graduate Texts in Physics,
DOI 10.1007/978-3-319-25675-7_12

or for excitation of electrons or phonons. Densities of states therefore have profound impacts on electric and thermal conductivity and on optical properties of materials. We will see that there exist several ways to justify equation (12.1), and we will also explore the many different, but related definitions of the density of states.

12.1 Counting of oscillation modes

The basic notion of density of states concerns the $\boldsymbol{k}$ space density of linearly independent oscillation modes in a homogeneous volume. This is a very basic quantity in physics from which more advanced notions like local densities of states can be inferred. There are two basic ways to derive the $\boldsymbol{k}$ space density of states in a finite volume V. One of the derivations is more intuitive and the other one is slightly more formal, but the density of states is such an important concept that it is worthwhile to discuss both derivations.

The reasoning with periodic boundary conditions in a finite volume

The simplest derivation of (12.1) counts the number of independent oscillation modes in a rectangular cavity with periodic boundary conditions. A general wave vector can always be written in the form

$$\boldsymbol{k} = \frac{2\pi}{\lambda}\hat{\boldsymbol{k}} = \frac{2\pi}{\lambda}\sum_i \cos\theta_i\,\boldsymbol{e}_i,$$

where $\cos\theta_i$, $\sum_i \cos\theta_i^2 = 1$, are the directional cosines of the vector.

Suppose that the wave has to be periodic with periodicity L_i in direction $\boldsymbol{e}_i$. In that case the length L_i must be an integer multiple of the projection $\lambda_i = \lambda/\cos\theta_i$ of the wavelength onto the direction $\boldsymbol{e}_i$:

$$L_i = n_i\lambda_i = n_i\frac{\lambda}{\cos\theta_i}, \quad n_i \geq 0, \tag{12.2}$$

see Figure 12.1.

Equation (12.2) can be written in terms of the components of the wave vector $\boldsymbol{k}$,

$$k_i = \frac{2\pi}{\lambda}\cos\theta_i = \frac{2\pi}{\lambda_i} = 2\pi\frac{n_i}{L_i}.$$

The volume of a single state in $\boldsymbol{k}$-space is therefore (with g spin or helicity states per wave)

$$\left[\Delta^3\boldsymbol{k}\right]_{\text{single state}} = \frac{(2\pi)^3}{gV},$$

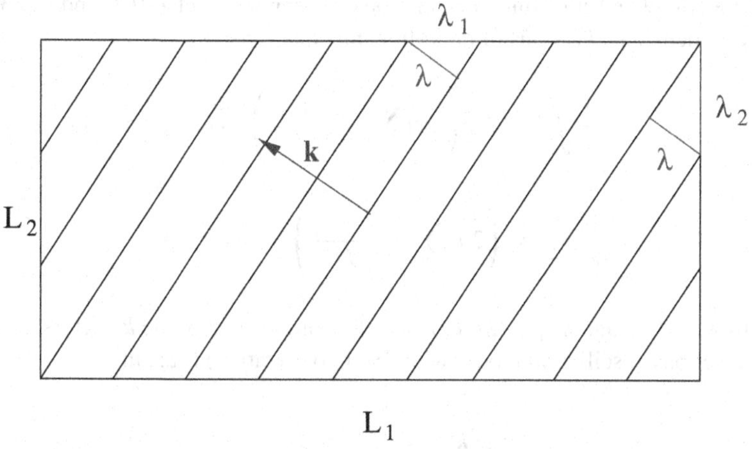

Fig. 12.1 A standing wave in a cavity with periodic boundary conditions

since g helicity states reside in a cell of volume $(2\pi)^3/V$ in $\boldsymbol{k}$-space. This yields the proportionality factor between the measure for the number of states dn and the volume measure in $\boldsymbol{k}$-space: The number dn of states in a volume $d^3\boldsymbol{k}$ in $\boldsymbol{k}$ space is

$$dn = d^3\boldsymbol{k} \Big/ \left[\Delta^3\boldsymbol{k}\right]_{\text{single state}} = \frac{gV}{8\pi^3} d^3\boldsymbol{k}. \tag{12.3}$$

Inclusion of the factor g corresponds to a summation over all possible polarizations or helicities. This version of the density of states is often employed in thermodynamics and statistical physics. In quantum mechanics and scattering theory we often need the density of states with a given polarization or helicity,

$$dn = \frac{V}{8\pi^3} d^3\boldsymbol{k}. \tag{12.4}$$

The quantity $dn/d^3\boldsymbol{k} = V/(2\pi)^3$ is the density of polarized states (if the system can have polarization) in $\boldsymbol{k}$ space, and $dn/(Vd^3\boldsymbol{k}) = 1/(2\pi)^3$ is the corresponding density of states in phase space. We will mostly use the density (12.4) or its continuum limit, i.e. we will usually count states with a given polarization or helicity, e.g. states of spin up electrons, states of photons of given polarization, etc.

The reasoning based on the completeness of plane wave states

A slightly more formal reasoning is based on the completeness of Fourier monomials. The Fourier monomials

$$\langle \boldsymbol{x}|\boldsymbol{n}\rangle = \frac{1}{\sqrt{V}} \exp(\mathrm{i}\boldsymbol{k}\cdot\boldsymbol{x}) = \frac{1}{\sqrt{V}} \exp\left(2\pi\mathrm{i}\sum_i \frac{n_i x_i}{L_i}\right) \tag{12.5}$$

provide a complete set of functions in a box of lengths L_i, cf. (10.2) and (10.3) for the one-dimensional versions of the following equations:

$$\frac{1}{V}\int_V d^3x\,\exp\!\left(2\pi\mathrm{i}\sum_i \frac{n_i - n_i'}{L_i}x_i\right) = \delta_{n,n'},$$

$$\frac{1}{V}\sum_n \exp\!\left(2\pi\mathrm{i}\sum_i n_i\frac{x_i - x_i'}{L_i}\right) = \delta(x - x').$$

Therefore we find again $k_i = 2\pi n_i/L_i$ for the components of the k vectors, and the volume per base oscillation mode (with fixed polarization) is again

$$\Delta^3 k\Big|_{\text{single mode}} = \frac{8\pi^3}{V}.$$

This yields again the equation (12.4),

$$dn = \frac{V}{8\pi^3}d^3k.$$

We remark that the measure dn for the number of states in k-space (12.4) can also be written in terms of the wave numbers $\tilde{\nu}_i \equiv 1/\lambda_i = k_i/2\pi$,

$$dn = \frac{V}{8\pi^3}d^3k = Vd^3\tilde{\nu} = Vd^3\frac{1}{\lambda}.$$

If we also replace the volume V in position space with the volume measure d^3x, we find a particularly intuitive and suggestive form for the corresponding measure of states in phase space,

$$dn = d^3x\,d^3\tilde{\nu} = d^3x\,d^3\frac{1}{\lambda} = \frac{d^3x\,d^3p}{h^3}.$$

Here $\lambda = \lambda\hat{k}$, and we define the "inverse" vector as $1/\lambda \equiv \lambda/\lambda^2$.

12.2 The continuum limit

In the limit $V \to \infty$, the discrete enumerable set of normalized plane waves in a cubic volume V, $\exp(\mathrm{i}k\cdot x)/\sqrt{V}$, $k = 2\pi n/L = 2\pi n/V^{1/3}$, is replaced by the continuous non-enumerable set $\exp(\mathrm{i}k\cdot x)/\sqrt{2\pi}^3$. We have derived the completeness relation for the continuous Fourier monomials in Section 2.1, see equations (2.2) and (2.9). However, for the discussion of the continuum limit of (12.4) it is useful to revisit the completeness relation of the continuous Fourier monomials as a continuum limit of the completeness relation of Fourier monomials in finite volume.

We are using $\Delta^3\boldsymbol{n} = 1$ for the volume of a triplet of integers in $\mathbb{Z}^3$. The completeness relation for the Fourier monomials in a cubic box can then be written as

$$\begin{aligned}
\delta(\boldsymbol{x}-\boldsymbol{x}') &= \frac{1}{V}\sum_{\boldsymbol{n}}\exp\!\left(\frac{2\pi\mathrm{i}}{V^{1/3}}\boldsymbol{n}\cdot(\boldsymbol{x}-\boldsymbol{x}')\right) \\
&= \frac{1}{V}\sum_{\boldsymbol{n}}\Delta^3\boldsymbol{n}\,\exp\!\left(\frac{2\pi\mathrm{i}}{V^{1/3}}\boldsymbol{n}\cdot(\boldsymbol{x}-\boldsymbol{x}')\right) \\
&= \frac{1}{V}\sum_{\boldsymbol{k}}\frac{V}{(2\pi)^3}\Delta^3\boldsymbol{k}\,\exp\!\left[\mathrm{i}\boldsymbol{k}\cdot(\boldsymbol{x}-\boldsymbol{x}')\right] \\
&\to \frac{1}{(2\pi)^3}\int d^3\boldsymbol{k}\,\exp\!\left[\mathrm{i}\boldsymbol{k}\cdot(\boldsymbol{x}-\boldsymbol{x}')\right],
\end{aligned} \tag{12.6}$$

where we have used the fact that the volume cancels to take the continuum limit in the final step.

This corresponds to the substitution $V \Rightarrow (2\pi)^3$ in the plane waves, and the corresponding substitution for the measure for the number of states is indeed

$$dn = \frac{V}{(2\pi)^3}d^3\boldsymbol{k} \;\Rightarrow\; d^3\boldsymbol{k}. \tag{12.7}$$

Note that either way, the density of states per volume V of a particle with fixed helicity or spin is

$$\frac{dn}{V} = \frac{d^3\boldsymbol{k}}{(2\pi)^3}, \tag{12.8}$$

irrespective of whether we have taken the continuum limit or not. However, please also note that in the continuum limit both the differential number of states $dn = d^3\boldsymbol{k}$ and the number of states per volume $dn/V = d^3\boldsymbol{k}/(2\pi)^3$ come with dimensions length^{-3}.

The "dimensionally wrong" density of states $dn/d^3\boldsymbol{k} = 1$ in $\boldsymbol{k}$ space is a consequence of the cancellation of V in (12.6) and the ensuing interpretation of $\exp(\mathrm{i}\boldsymbol{k}\cdot\boldsymbol{x})/\sqrt{2\pi}^3$ as the properly normalized plane wave states in the continuum limit. This has shifted the length dimension from the states

$$\dim\!\left[\exp(\mathrm{i}\boldsymbol{k}\cdot\boldsymbol{x})/\sqrt{V}\right] = \ell^{-3/2} \to \dim\!\left[\exp(\mathrm{i}\boldsymbol{k}\cdot\boldsymbol{x})/\sqrt{2\pi}^3\right] = 1$$

into the $\boldsymbol{k}$ space measure of states,

$$\dim\left[dn = d^3\boldsymbol{k}V/(2\pi)^3\right] = 1 \to \dim\left[dn = d^3\boldsymbol{k}\right] = \ell^{-3}.$$

We have to keep this in mind when we are using dimensional analysis of quantum mechanical transition amplitudes in time-dependent perturbation theory in Chapters 13 and 18.

Another reasoning for the continuum limit

We consider the matrix element of the time evolution operator $U_D(t,t')$ between a bound hydrogen state $|n,\ell,m_\ell\rangle$ and a plane wave $|\boldsymbol{k}\rangle$. The definition of U_D and the motivation for considering its matrix elements will be given in the following chapter. For now we only need to know that it is a unitary operator which describes e.g. how bound hydrogen eigenstates are scattered into other states under a time-dependent perturbation of the hydrogen atom.

Unitarity of $U_D(t,t')$ and the completeness relation for plane waves imply

$$\int d^3\boldsymbol{k}\left|\langle\boldsymbol{k}|U_D(t,t')|n,\ell,m_\ell\rangle\right|^2 = 1.$$

This tells us that we can interprete $|\langle\boldsymbol{k}|U_D(t,t')|n,\ell,m_\ell\rangle|^2$ as a probability density for the system to end up in a plane wave state $|\boldsymbol{k}\rangle$, and $d^3\boldsymbol{k}$ as a measure for the number of states, such that the probability for the system to end up in a region $\mathcal{K}$ in $\boldsymbol{k}$ space is

$$P_{n,\ell,m_\ell\to\mathcal{K}}(t,t') = \int_{\mathcal{K}} d^3\boldsymbol{k}\left|\langle\boldsymbol{k}|U_D(t,t')|n,\ell,m_\ell\rangle\right|^2.$$

This confirms yet again that $dn = d^3\boldsymbol{k}$ is the correct density of states in $\boldsymbol{k}$ space in the continuum limit.

Different forms of the density of states in a homogeneous medium

We may or may not include the number g of helicity or spin states in the density of states, we can normalize to finite volume V or take the continuum limit $V\to\infty$, and we may also use the number of states per $\boldsymbol{k}$ space volume and per direct volume V (i.e. calculate the density of states in phase space). All these simple alternatives amount to eight basic options for the density of states in $\boldsymbol{k}$ space,

$$dn = [g]\left[\frac{[V]}{8\pi^3}\right]d^3\boldsymbol{k}.$$

The first term in square brackets is included if we sum over all possible polarizations of the particle, and the fraction $V/(8\pi^3)$ is included if we use box normalization. The volume factor is not included if the density of states is also counted per volume in position space, dn/V. The fraction $V/(8\pi^3)$ in dn disappears in the continuum limit.

12.3 The density of states in the energy scale

In solid state physics (and in variants of time-dependent perturbation theory and scattering theory) one is often interested in transforming $d^3\boldsymbol{k}$ to variables $d^2\boldsymbol{k}_\parallel$ parallel to surfaces of constant energy $E(\boldsymbol{k})$ in $\boldsymbol{k}$ space and the energy E, which increases orthogonal to the surfaces of constant energy. The normalized unit vector in the direction of increasing E is

$$\hat{\boldsymbol{k}}_\perp = \frac{\partial E(\boldsymbol{k})/\partial \boldsymbol{k}}{|\partial E(\boldsymbol{k})/\partial \boldsymbol{k}|} = \frac{\boldsymbol{v}(\boldsymbol{k})}{|\boldsymbol{v}(\boldsymbol{k})|}$$

(recall that $\boldsymbol{v}(\boldsymbol{k}) = \hbar^{-1}\partial E(\boldsymbol{k})/\partial \boldsymbol{k}$ is the group velocity). Therefore we have

$$dk_\perp = d\boldsymbol{k} \cdot \hat{\boldsymbol{k}}_\perp = \frac{d\boldsymbol{k} \cdot \partial E(\boldsymbol{k})/\partial \boldsymbol{k}}{|\partial E(\boldsymbol{k})/\partial \boldsymbol{k}|} = \frac{dE}{|\partial E(\boldsymbol{k})/\partial \boldsymbol{k}|} = \frac{dE}{\hbar|\boldsymbol{v}(\boldsymbol{k})|}$$

and

$$d^3\boldsymbol{k} = d^2\boldsymbol{k}_\parallel \frac{dE}{|\partial E(\boldsymbol{k})/\partial \boldsymbol{k}|}.$$

Here $d^2\boldsymbol{k}_\parallel$ is some appropriate measure for coordinates along the constant energy surfaces.

An isotropic dispersion relation, $E(\boldsymbol{k}) = E(k)$, yields

$$d^3\boldsymbol{k} = d^2\Omega_k k^2 \frac{dE}{dE/dk}.$$

The corresponding number of states is then

$$dn = [g]\left[\frac{[V]}{8\pi^3}\right] d^2\Omega_k k^2 \frac{dE}{|dE/dk|} = \varrho(E) dE d^2\Omega_k, \tag{12.9}$$

with a density of states *per energy* or density of states *in the energy scale*

$$\varrho(E) = [g]\left[\frac{[V]}{8\pi^3}\right] \frac{k^2}{|dE/dk|}. \tag{12.10}$$

Here the absolute value $|dE/dk|$ is taken in the denominator, because in cases where $dE/dk < 0$, the convention is to substitute an integral in positive dk direction with an integral in positive dE direction in the summation over states,

$$dk = \frac{dE}{dE/dk} \to \frac{dE}{|dE/dk|}.$$

In isotropic problems the angular variables are often integrated over, and one uses the convention

$$dn \to \varrho(E)dE$$

with the factor 4π included in ϱ. Altogether this leaves us with the following sixteen possibilities for the density of states in the energy scale,

$$\varrho(E) = [g]\left[\frac{[V]}{8\pi^3}\right][4\pi]\frac{k^2}{|dE/dk|}. \tag{12.11}$$

We remark that generalization of the previous arguments to d spatial dimensions yields the following results for the density of states,

$$dn = [g]\left[\frac{[V]}{(2\pi)^d}\right]d^d\boldsymbol{k}, \quad \varrho_d(E) = [g]\left[\frac{[V]}{(2\pi)^d}\right]\left[\frac{2\sqrt{\pi}^d}{\Gamma(d/2)}\right]\frac{k^{d-1}}{|dE/dk|}, \tag{12.12}$$

where $S_{d-1} = 2\sqrt{\pi}^d/\Gamma(d/2)$ is the $(d-1)$-dimensional hyper-area of a unit sphere in d dimensions.

12.4 Density of states for free non-relativistic particles and for radiation

The free non-relativistic particle satisfies $E = \hbar^2k^2/2m$, and equation (12.12) yields the following forms of the density of states in the energy scale in d dimensions,

$$\varrho_d(E) = \Theta(E)\,[g]\left[\frac{[V]}{(2\pi)^d}\right]\left[\frac{2\sqrt{\pi}^d}{\Gamma(d/2)}\right]\left(\frac{\sqrt{m}}{\hbar}\right)^d\sqrt{2E}^{d-2}.$$

The most commonly used version gives the density of free non-relativistic states per volume and per energy in d dimensions as

$$\varrho_d(E) = g\Theta(E)\sqrt{\frac{m}{2\pi}}^d\frac{\sqrt{E}^{d-2}}{\Gamma(d/2)\hbar^d}. \tag{12.13}$$

For $d = 3$ this yields the density of states in a free electron model for metals. For other materials this equation can be used to calculate the density of electron states near the minimum of an energy band or the density of hole states near the maximum of an energy band if we replace E with the difference to the local minimum or maximum in energy: $E \to E - E_{min}$ or $E \to E_{max} - E$. In these cases m is the effective electron or hole mass near E_{min} or E_{max}, respectively. Equation (12.13) is also often employed for $d = 1$ and $d = 2$ to estimate the density of states in quantum wires or quantum wells.

The energy of a photon of frequency f is $E = hf = \hbar ck$ and we have $g = 2$ independent polarization states. Equations (12.3) or (12.11) therefore yield the following expressions for the density of photon states per volume and per unit of energy,

$$\varrho(E) = \frac{dn}{VdE} = \frac{2}{8\pi^3}4\pi k^2\frac{dk}{dE} = \frac{E^2}{\pi^2(\hbar c)^3},$$

or in d dimensions (with $g = d - 1$ polarizations),

$$\varrho_d(E) = \frac{(d-1)E^{d-1}}{2^{d-1}\pi^{d/2}\Gamma(d/2)(\hbar c)^d}.$$

The density of photon states (per volume V) in the frequency scale follows as

$$\varrho_d(f) = \frac{dn}{Vdf} = \frac{2(d-1)\pi^{d/2}}{\Gamma(d/2)}\frac{f^{d-1}}{c^d}.$$

For $d = 3$ this is equation (1.6) which we have used in the derivation of Planck's law.

12.5 The density of states for other quantum systems

It is also useful to note that we can express the density of states per volume in plane waves trivially through the corresponding wave functions,

$$\frac{dn}{V} = \frac{d^3\boldsymbol{k}}{(2\pi)^3} = d^3\boldsymbol{k}\,|\langle\boldsymbol{x}|\boldsymbol{k}\rangle|^2.$$

This suggests the following identification of the number of states per volume in terms of quantum states which are labeled through a set of quantum numbers α,

$$\frac{dn}{V}(\boldsymbol{x}) = d\alpha\,|\langle\boldsymbol{x}|\alpha\rangle|^2. \tag{12.14}$$

From this observation we can infer a generalization of the density of states per volume and per unit of energy which also holds for discrete spectra. Suppose the Hamiltonian H has a discrete spectrum E_n and continuous spectra in ranges $E_{b1} \le E \le E_{b2}$. We use $\alpha = (E, \nu)$ for the set of quantum numbers, where ν is a set of degeneracy indices. Then the previous identification yields the density of states per volume and per energy as

$$\varrho(E,\boldsymbol{x}) = \sum_n \delta(E - E_n) \sum\!\!\!\!\!\!\int d\nu_n \, |\langle \boldsymbol{x}|E_n, \nu_n\rangle|^2$$

$$+ \sum_b \Theta(E - E_{b1})\Theta(E_{b2} - E) \sum\!\!\!\!\!\!\int d\nu(E) \, |\langle \boldsymbol{x}|E, \nu(E)\rangle|^2 . \quad (12.15)$$

E.g. the density of states per volume and per energy for the hydrogen atom would be (with factors of 2 from summation over spins)

$$\varrho(E,\boldsymbol{x}) = 2\sum_{n=1}^{\infty} \delta(E - E_n) \sum_{\ell=0}^{n-1} \sum_{m_\ell=-\ell}^{\ell} |\langle \boldsymbol{x}|n, \ell, m_\ell\rangle|^2$$

$$+ \Theta(E) \sum_{\ell=0}^{\infty} \sum_{m_\ell=-\ell}^{\ell} \frac{1}{\hbar^3} \sqrt{(2m)^3 E} \, |\langle \boldsymbol{x}|k, \ell, m_\ell\rangle|^2 , \quad (12.16)$$

where $\langle \boldsymbol{x}|k, \ell, m_\ell\rangle$ are the Coulomb waves $\psi_{k,\ell,m_\ell}(\boldsymbol{x})$ from Section 7.9 and $\sqrt{(2m)^3 E}/\hbar^3 = 2k^2 dk/dE$.

A short hand version of equation (12.15) is

$$\varrho(E,\boldsymbol{x}) = \sum\!\!\!\!\!\!\int dE' \, d\nu(E') \, \delta(E - E') \left|\langle \boldsymbol{x}|E', \nu(E')\rangle\right|^2 . \quad (12.17)$$

Note that for each quantum system, the total number of single-particle states per volume diverges in a very specific way,

$$\int \frac{dn}{V} = g\langle \boldsymbol{x}|\boldsymbol{x}\rangle = g\delta(\boldsymbol{0}).$$

12.6 Problems

12.1. We consider a free gas of spin 1/2 fermions in a finite volume $V = L^3$ with periodic boundary conditions. This implies the constraints

$$k_i = \frac{2\pi}{L} n_i, \quad n_i \in \mathbb{Z}$$

on the components of the wave vector.

Our fermion gas contains $N \gg 1$ particles, and we assume it to be in the state of minimal energy. How large is the maximal momentum $p_F = \hbar k_F$ (the *Fermi momentum*) in the fermi gas?

You have to take into account that only two fermions can have the same momentum.

Solution. With $N \gg 1$ we have $\lambda_F \ll L$ or $k_F \gg \frac{2\pi}{L}$. The number of states with momenta $p \leq p_F$ is then

$$2\frac{4\pi}{3}k_F^3\frac{L^3}{8\pi^3} = \frac{1}{3\pi^2}k_F^3L^3 = N,$$

and therefore

$$k_F = \frac{1}{L}\left(3\pi^2 N\right)^{\frac{1}{3}} = \left(3\pi^2 n\right)^{\frac{1}{3}}, \quad p_F = \hbar k_F, \tag{12.18}$$

where $n = N/V$ is the particle density.

12.2. The equation (12.12) for the density of states in d dimensions holds for isotropic dispersion relations $E = E(|\boldsymbol{k}|)$. We used $k \equiv |\boldsymbol{k}|$ in (12.12).

For one-dimensional models that equation yields the density of states in the energy scale per volume $V \equiv a$ (a lattice constant) and per helicity state as

$$\varrho_1(E) = \frac{1}{ag}\frac{dn}{dE} = \frac{1}{2\pi} \times 2 \times \left|\frac{d|k|}{dE}\right|. \tag{12.19}$$

The factor of 2 comes from the "volume" $S_0 = [2\sqrt{\pi}^d/\Gamma(d/2)]_{d=1}$ of the zero-dimensional unit sphere. This sphere consists of the two points 1 and -1. Is equation (12.19) correct? Or should we abandon the factor of 2?

12.3. Equation (12.14) for the local density of states yields for one-dimensional lattices with volume $V = a$ and Bloch states (10.8) the local density of states in the k scale as

$$\varrho(k,x) = \frac{dn}{agdk}(x) = \sum_n |\psi_n(k,x)|^2.$$

Note that we also divided out the number g of spin or helicity states, which is included as a discrete parameter in the set of quantum numbers α in (12.14).

Show that transformation to the energy scale and spatial averaging reproduces the isotropic result (12.19),

$$\varrho(E) = \frac{1}{a}\int_0^a dx\, \varrho(E,x) = \frac{1}{\pi}\left|\frac{d|k|}{dE}\right|.$$

12.4. We consider the Kronig-Penney model from Section 10.4.

Show for $E > 0$ that the spatially averaged one-dimensional density of states in the energy scale,

$$\varrho(E) = \frac{dn}{agdE} = \frac{1}{\pi}\left|\frac{d|k|}{dE}\right|,$$

is given by

$$\varrho(E) = \frac{m}{\pi\hbar^2 K}\left|\left(1+\frac{u}{(Ka)^2}\right)\sin(Ka) - \frac{u}{Ka}\cos(Ka)\right| \times\left[\left(1-\frac{u^2}{(Ka)^2}\right)\sin^2(Ka) - \frac{u}{Ka}\sin(2Ka)\right]^{-1/2}, \quad (12.20)$$

with $K = \sqrt{2mE}/\hbar$. This equation only applies where states exist, i.e. where the condition (10.35) is met.

The resulting density of states for $u = 5$ in the region of the first two energy bands is plotted in Figure 12.2 for a lattice constant $a = 3.5$ Å.

12.5. Calculate the density of states in the energy scale for free unpolarized electrons in three dimensions if you *cannot* assume that the kinetic energy of the electrons is much smaller than their rest energy.

Which result do you get in the non-relativistic limit?

Derive the corresponding results also in d spatial dimensions.

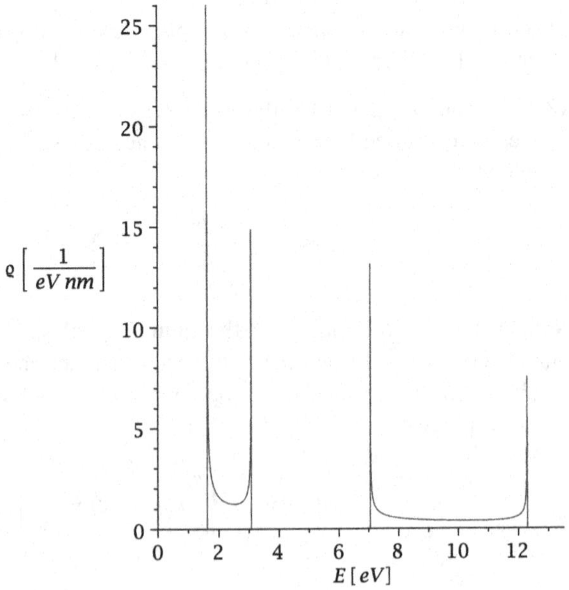

Fig. 12.2 The one-dimensional density of states in the energy scale (12.20) for $u = 5$ and a lattice constant $a = 3.5$ Å. The energy scale covers the first two energy bands $E_0(k)$ and $E_1(k)$, cf. Figure 10.2

Result. The relativistic dispersion relation $E = \hbar c\sqrt{k^2 + (mc/\hbar)^2}$ yields for particles with g helicity (or spin or polarization) states the density of states per volume and in the energy scale

$$\begin{aligned}\varrho(E) &= \frac{dn}{dVdE} = \frac{g}{(2\pi)^d}\frac{2\sqrt{\pi}^d}{\Gamma(d/2)}k^{d-1}\frac{dk}{dE} \\ &= \frac{2g\Theta(E-mc^2)}{(2\sqrt{\pi}\hbar c)^d\Gamma(d/2)}E\sqrt{E^2-m^2c^4}^{\,d-2}. \end{aligned} \quad (12.21)$$

For the comparison with the non-relativistic limit we should write this in terms of the kinetic energy $K = E - mc^2$, because K is usually denoted as the energy of the particle in the non-relativistic limit,

$$\varrho(K) = \frac{2g\Theta(K)}{(2\sqrt{\pi}\hbar c)^d\Gamma(d/2)}(K+mc^2)\sqrt{K(K+2mc^2)}^{\,d-2}. \quad (12.22)$$

This yields the non-relativistic result (12.13) (with the substitution $E_{non-rel.} = K$) in the limit $K \ll mc^2$.

Note that use of $E = \pm\hbar c\sqrt{k^2 + (mc/\hbar)^2}$ yields a symmetric density of states which includes the anti-particles as negative energy states,

$$\hat{\varrho}(E) = \varrho(E) + \overline{\varrho}(\overline{E}) = \frac{2g\Theta(E^2-m^2c^4)}{(2\sqrt{\pi}\hbar c)^d\Gamma(d/2)}|E|\sqrt{E^2-m^2c^4}^{\,d-2}. \quad (12.23)$$

Here $\overline{E} = -E > mc^2$ corresponds to the energy of the anti-particles. As for the factor g, we just remark for completeness that a massive vector field in $d+1$ space-time dimensions has $g_d = d$ possible polarizations. Furthermore, a spinor in $d+1$ space-time dimensions has $2^{\lfloor (d+1)/2\rfloor}$ components which describe both particles and anti-particles, and therefore an electron in $d+1$ space-time dimensions has $g_d = 2^{\lfloor (d-1)/2\rfloor}$ spin components. The floor function used here yields $\lfloor n/2\rfloor = n/2$ if n is even, $\lfloor n/2\rfloor = (n-1)/2$ if n is odd. Please see Chapter 21 and Appendix G (note that d denotes the number of space-time dimensions in Appendix G).

Chapter 13
Time-dependent Perturbations in Quantum Mechanics

The development of time-dependent perturbation theory was initiated by Paul Dirac's early work on the semi-classical description of atoms interacting with electromagnetic fields[1]. Dirac, Wheeler, Heisenberg, Feynman and Dyson developed it into a powerful set of techniques for studying interactions and time evolution in quantum mechanical systems which cannot be solved exactly. It is used for the quantitative description of phenomena as diverse as proton-proton scattering, photo-ionization of materials, scattering of electrons off lattice defects in a conductor, scattering of neutrons off nuclei, electric susceptibilities of materials, neutron absorption cross sections in a nuclear reactor etc. The list is infinitely long. Time-dependent perturbation theory is an extremely important tool for calculating properties of any physical system.

So far all the Hamiltonians which we had studied were time-independent. This property was particularly important for the time-energy Fourier transformation from the time-dependent Schrödinger equation to a time-independent Schrödinger equation. Time-independence of H also ensures conservation of energy, as will be discussed in detail in Chapter 16. Time-dependent perturbation theory, on the other hand, is naturally also concerned with time-dependent Hamiltonians $H(t)$ (although it provides very useful results also for time-independent Hamiltonians, and we will see later that most of its applications in quantum field theory concern systems with time-independent Hamiltonians). We will therefore formulate all results in this chapter for time-dependent Hamiltonians, and only specify to time-independent cases where it is particularly useful for applications.

[1] P.A.M. Dirac, Proc. Roy. Soc. London A 112, 661 (1926).

R. Dick, *Advanced Quantum Mechanics*, Graduate Texts in Physics,
DOI 10.1007/978-3-319-25675-7_13

13.1 Pictures of quantum dynamics

As a preparation for the discussion of time-dependent perturbation theory (and of field quantization later on), we now enter the discussion of different *pictures* of quantum dynamics.

The picture which we have used so far is the *Schrödinger picture* of quantum dynamics: The time evolution of a system is encoded in its states $|\psi(t)\rangle$ which have to satisfy a Schrödinger equation $i\hbar d|\psi_S(t)\rangle/dt = H(t)|\psi_S(t)\rangle$. However, every transformation on states and operators $|\psi\rangle \to U|\psi\rangle, A \to U \cdot A \cdot U^+$ with a unitary operator U leaves the matrix elements $\langle\phi|A|\psi\rangle$ and therefore the observables of a system invariant.

If U is in particular a time-dependent unitary operator, then this changes the time-evolution of the states and operators without changing the time-evolution of the observables. Application of a time-dependent $U(t)$ corresponds to a change of the picture of quantum dynamics, and two important cases besides the Schrödinger picture are the *Heisenberg picture* and the *interaction (or Dirac) picture*. In the Heisenberg picture all time dependence is cast from the states onto the operators, whereas in the Dirac picture the operators follow a "free" (or better: exactly solvable) time evolution, while the interaction (non-solvable) part of the Hamiltonian determines the time evolution of the states.

There are essentially two reasons for introducing the Heisenberg picture. The less important of these reasons is that the Hamilton-Poisson formulation of the classical limit of quantum systems is related to the Heisenberg picture. The really important reason is that quantum field theory in Chapter 17 appears first in the Heisenberg picture.

The rationale for introducing the Dirac picture is that time-dependent perturbation theory automatically leads to the calculation of matrix elements of the time evolution operator in the Dirac picture. As soon as we want to calculate transition probabilities in a quantum system under the influence of time-dependent perturbations, we automatically encounter the Dirac picture.

Before immersing ourselves into the discussion of the Heisenberg and Dirac pictures, we have to take a closer look at time evolution in the Schrödinger picture.

Time evolution in the Schrödinger picture

In the Schrödinger picture the basic operators Φ_S (like $\mathbf{x}$ or $\mathbf{p}$) are time-independent, $d\Phi_S/dt = 0$, and all the time evolution from the dynamics is carried by the states. The differential equation

$$i\hbar\frac{d}{dt}|\psi_S(t)\rangle = H(t)|\psi_S(t)\rangle$$

yields an equivalent integral equation

$$|\psi_S(t)\rangle = |\psi_S(t_0)\rangle - \frac{\mathrm{i}}{\hbar}\int_{t_0}^{t} d\tau\, H(\tau)|\psi_S(\tau)\rangle,$$

and iteration of this equation yields

$$|\psi_S(t)\rangle = U(t,t_0)|\psi_S(t_0)\rangle$$

with the *time evolution operator*[2]

$$\begin{aligned} U(t,t_0) &= \sum_n \frac{1}{(\mathrm{i}\hbar)^n}\int_{t_0}^{t} d\tau_1 \int_{t_0}^{\tau_1} d\tau_2 \ldots \int_{t_0}^{\tau_{n-1}} d\tau_n\, H(\tau_1)H(\tau_2)\ldots H(\tau_n) \\ &= \sum_n \frac{1}{(\mathrm{i}\hbar)^n}\int_{t_0}^{t} d\tau_n \int_{\tau_n}^{t} d\tau_{n-1} \ldots \int_{\tau_2}^{t} d\tau_1\, H(\tau_1)H(\tau_2)\ldots H(\tau_n) \\ &= \mathrm{T}\exp\!\left(-\frac{\mathrm{i}}{\hbar}\int_{t_0}^{t} d\tau\, H(\tau)\right). \end{aligned} \tag{13.1}$$

Taking the adjoint switches t with t_0 in the argument of the time evolution operator,

$$\begin{aligned} U^+(t,t_0) &= \sum_n \left(\frac{\mathrm{i}}{\hbar}\right)^n \int_{t_0}^{t} d\tau_1 \int_{t_0}^{\tau_1} d\tau_2 \ldots \int_{t_0}^{\tau_{n-1}} d\tau_n\, H(\tau_n)H(\tau_{n-1})\ldots H(\tau_1) \\ &= \sum_n \frac{1}{(\mathrm{i}\hbar)^n}\int_{t}^{t_0} d\tau_n \int_{t}^{\tau_n} d\tau_{n-1} \ldots \int_{t}^{\tau_2} d\tau_1\, H(\tau_n)H(\tau_{n-1})\ldots H(\tau_1) \\ &= \mathrm{T}\exp\!\left(-\frac{\mathrm{i}}{\hbar}\int_{t}^{t_0} d\tau\, H(\tau)\right) = U(t_0,t). \end{aligned} \tag{13.2}$$

This and the composition law (13.7) below imply unitarity of the time evolution operator.

Please note that the *time ordering operator* T in equations (13.1) and (13.2) always ensures that the Hamiltonians are ordered from right to left such that their time arguments go from closer to the *lower integration boundary* (t_0 in equation (13.1), t in equation (13.2)) to the *upper integration boundary* (t in equation (13.1), t_0 in equation (13.2)), *irrespective of whether the upper integration boundary is larger or smaller than the lower integration boundary*, e.g. if $t > t_0$ in equation (13.1) then of course $t_0 < t$ in equation (13.2). Apparently, the identification of "lower" and "upper" integration boundary in the previous statement implies the convention that the integrand in the exponent is $-\mathrm{i}H(t)/\hbar$. Otherwise the statement would be ambiguous.

The re-ordering of integrations in the second lines of equations (13.1, 13.2) is trivial for the 0th and 1st order terms. For the higher order terms e.g. in equation (13.1) we can recursively use for any consecutive pair of integrations

[2]F.J. Dyson, Phys. Rev. 75, 1736 (1949). Equation (13.1) gives three different representations of the time evolution operator. Equivalence of these representations is demonstrated in equations (13.3, 13.4) and in Problem 13.1.

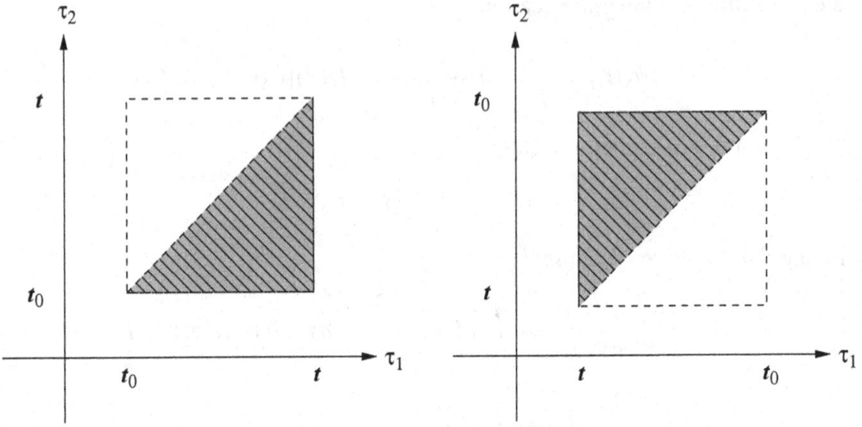

Fig. 13.1 The integration domain in equation (13.3) is shown in green. The left panel is for $t > t_0$ (forward evolution by $U(t, t_0)$), the right panel is for $t < t_0$ (backward evolution by $U(t, t_0)$). In either case re-arranging of the order of integration over the same domain yields equation (13.3)

$$\int_{t_0}^{t} d\tau_1 \int_{t_0}^{\tau_1} d\tau_2\, A(\tau_1, \tau_2) = \int_{t_0}^{t} d\tau_2 \int_{\tau_2}^{t} d\tau_1\, A(\tau_1, \tau_2), \tag{13.3}$$

which proves the re-ordering for $n = 2$, see also Figure 13.1. For higher n we can perform an induction step,

$$\begin{aligned}
&\int_{t_0}^{t} d\tau_1 \int_{t_0}^{\tau_1} d\tau_2 \ldots \int_{t_0}^{\tau_{n-1}} d\tau_n \int_{t_0}^{\tau_n} d\tau_{n+1}\, H(\tau_1)H(\tau_2)\ldots H(\tau_n)H(\tau_{n+1})\\
&= \int_{t_0}^{t} d\tau_1 \int_{t_0}^{\tau_1} d\tau_{n+1} \int_{\tau_{n+1}}^{\tau_1} d\tau_n \ldots \int_{\tau_3}^{\tau_1} d\tau_2\, H(\tau_1)H(\tau_2)\ldots H(\tau_n)H(\tau_{n+1})\\
&= \int_{t_0}^{t} d\tau_{n+1} \int_{\tau_{n+1}}^{t} d\tau_1 \int_{\tau_{n+1}}^{\tau_1} d\tau_n \ldots \int_{\tau_3}^{\tau_1} d\tau_2\, H(\tau_1)H(\tau_2)\ldots H(\tau_n)H(\tau_{n+1})\\
&= \int_{t_0}^{t} d\tau_{n+1} \int_{\tau_{n+1}}^{t} d\tau_n \int_{\tau_n}^{t} d\tau_1 \int_{\tau_n}^{\tau_1} d\tau_{n-1} \ldots \int_{\tau_3}^{\tau_1} d\tau_2\, H(\tau_1)H(\tau_2)\ldots H(\tau_{n+1})\\
&= \int_{t_0}^{t} d\tau_{n+1} \int_{\tau_{n+1}}^{t} d\tau_n \ldots \int_{\tau_3}^{t} d\tau_1 \int_{\tau_3}^{\tau_1} d\tau_2\, H(\tau_1)H(\tau_2)\ldots H(\tau_n)H(\tau_{n+1})\\
&= \int_{t_0}^{t} d\tau_{n+1} \int_{\tau_{n+1}}^{t} d\tau_n \ldots \int_{\tau_3}^{t} d\tau_2 \int_{\tau_2}^{t} d\tau_1\, H(\tau_1)H(\tau_2)\ldots H(\tau_n)H(\tau_{n+1}),
\end{aligned} \tag{13.4}$$

which concludes the proof.

Time evolution operators satisfy several important properties which include Schrödinger type operator equations, unitarity, and a simple composition property. We begin with the Schrödinger type differential equations satisfied by $U(t,t')$.

The derivative with respect to the first time argument of the time evolution operator is most easily calculated using the representation in the first line of (13.1),

$$\mathrm{i}\hbar\frac{\partial}{\partial t}U(t,t_0) = H(t)U(t,t_0), \tag{13.5}$$

while the derivative with respect to the second argument follows most easily from the second line in (13.1),

$$\mathrm{i}\hbar\frac{\partial}{\partial t}U(t',t) = -\,U(t',t)H(t). \tag{13.6}$$

Taking the adjoint of (13.5) or using (13.2) yields

$$\mathrm{i}\hbar\frac{\partial}{\partial t}U^{+}(t,t_0) = -\,U^{+}(t,t_0)H(t).$$

The time evolution operator is the unique solution of these differential equations with initial condition $U(t_0,t_0) = 1$. The differential equations together with the initial condition also imply the integral equations

$$U(t,t') = 1 - \frac{\mathrm{i}}{\hbar}\int_{t'}^{t} d\tau\, H(\tau)U(\tau,t') = 1 - \frac{\mathrm{i}}{\hbar}\int_{t'}^{t} d\tau\, U(t,\tau)H(\tau).$$

Another important property of the time evolution operator is the composition law

$$U(t',t)U(t,t_0) = U(t',t_0). \tag{13.7}$$

Proving this through multiplication of the left hand side and sorting out the n-th order term is clumsy, due to the need to prove that the sum over $n+1$ n-fold integrals on the left hand side really produces the n-th order term on the right hand side. However, we can find a much more elegant proof by observing that $U(t',t)U(t,t_0)$ is actually independent of t due to equations (13.5, 13.6),

$$\frac{\partial}{\partial t}U(t',t)U(t,t_0) = 0,$$

and therefore

$$U(t',t)U(t,t_0) = U(t',t')U(t',t_0) = U(t',t_0).$$

The composition law yields in particular

$$U(t_0,t)U(t,t_0) = U(t_0,t_0) = 1, \quad U(t_0,t) = U^{-1}(t,t_0),$$

and combined with (13.2) this implies unitarity of the time evolution operator,

$$U^{+}(t,t_0) = U(t_0,t) = U^{-1}(t,t_0), \tag{13.8}$$

i.e. time evolution preserves the norm of states.

The time evolution operator for the harmonic oscillator

The time evolution operator for time-independent Hamiltonians H is invariant under time translations,

$$U(t-t_0) = \exp\!\left(-\frac{\mathrm{i}}{\hbar}H(t-t_0)\right).$$

The matrix elements in x space can then be written in terms of the wave functions of energy eigenstates $H|E,\nu\rangle = E|E,\nu\rangle$, where ν is a set of degeneracy indices. There are no degeneracy indices in one dimension and the expansion takes the form

$$\langle x|U(t)|x'\rangle = \sum\!\!\!\!\!\!\int dE\,\exp\!\left(-\frac{\mathrm{i}}{\hbar}Et\right)\langle x|E\rangle\langle E|x'\rangle.$$

E.g. the time evolution operator of the harmonic oscillator

$$U(t) = \exp(-\mathrm{i}\omega a^+ a t)\exp(-\mathrm{i}\omega t/2)$$

has matrix elements

$$\begin{aligned}\langle x|\exp(-\mathrm{i}\omega a^+ a t)\,|x'\rangle &= \sum_{n=0}^{\infty}\langle x|n\rangle\langle n|x'\rangle\exp(-\mathrm{i}n\omega t)\\ &= \sqrt{\frac{m\omega}{\pi\hbar}}\sum_{n=0}^{\infty}\frac{\exp(-\mathrm{i}n\omega t)}{2^n n!}H_n\!\left(\sqrt{\frac{m\omega}{\hbar}}x\right)\\ &\quad\times H_n\!\left(\sqrt{\frac{m\omega}{\hbar}}x'\right)\exp\!\left(-\frac{m\omega}{2\hbar}\left(x^2+x'^2\right)\right).\end{aligned}$$

Use of the Mehler formula (D.8) yields

$$\begin{aligned}\langle x|U(t)|x'\rangle &= \langle x|\exp(-\mathrm{i}\omega a^+ a t)\,|x'\rangle\exp(-\mathrm{i}\omega t/2)\\ &= \sqrt{\frac{m\omega}{2\pi\mathrm{i}\hbar\sin(\omega t)}}\exp\!\left(\mathrm{i}\frac{m\omega}{2\hbar}\frac{\left(x^2+x'^2\right)\cos(\omega t)-2xx'}{\sin(\omega t)}\right).\end{aligned} \tag{13.9}$$

To use the Mehler formula we should take $\omega \to \omega - \mathrm{i}\epsilon$ for $t > 0$. This complies with the shifts $E' \to E' - \mathrm{i}\epsilon$,

$$\mathcal{G}(E) = \frac{1}{E-H+\mathrm{i}\epsilon} = \sum\!\!\!\!\!\!\int dE'\,\frac{|E'\rangle\langle E'|}{E-E'+\mathrm{i}\epsilon},$$

which define retarded Green's functions in the energy representation, see e.g. (11.8, 20.14). The time-dependent retarded Green's function for the oscillator is related to the propagator (13.9) in the standard way

$$\langle x|\mathcal{G}(t)|x'\rangle = \frac{\Theta(t)}{i\hbar}\langle x|U(t)|x'\rangle.$$

The Heisenberg picture

In the Heisenberg picture we use the unitary time evolution operator $U(t, t_0)$ to cast the time dependence from the states onto the operators,

$$|\psi_H\rangle = |\psi_S(t_0)\rangle = U^+(t, t_0)|\psi_S(t)\rangle,$$

$$\Phi_H(t) = U^+(t, t_0)\Phi_S U(t, t_0).$$

For the time evolution of the operators in the Heisenberg picture we observe that

$$\begin{aligned}\Phi_H(t) &= U^+(t, t_0)\Phi_S U(t, t_0) = U^+(t, t_0)\Phi_S U^+(t_0, t)\\ &= U^+(t, t')U^+(t', t_0)\Phi_S U^+(t_0, t')U^+(t', t) = U^+(t, t')\Phi_H(t')U(t, t'),\end{aligned}$$

and the Heisenberg evolution equation

$$\begin{aligned}i\hbar\frac{d}{dt}\Phi_H(t) &= -U^+(t, t_0)[H(t)\Phi_S - \Phi_S H(t)]U(t, t_0)\\ &= -U^+(t, t_0)H(t)U(t, t_0)U^+(t, t_0)\Phi_S U(t, t_0)\\ &\quad + U^+(t, t_0)\Phi_S U(t, t_0)U^+(t, t_0)H(t)U(t, t_0)\\ &= -[H_H(t), \Phi_H(t)].\end{aligned} \tag{13.10}$$

In the last equation, $H_H(t)$ is the Hamiltonian written in terms of operators $\Phi_H(t)$ in the Heisenberg picture.

For time-dependent $\Phi_S(t)$ we have

$$\frac{d}{dt}\Phi_H(t) = U^+(t, t_0)\left(\frac{i}{\hbar}[H(t), \Phi_S(t)] + \frac{d}{dt}\Phi_S(t)\right)U(t, t_0).$$

13.2 The Dirac picture

For the Dirac or interaction picture we split the Schrödinger picture Hamiltonian $H(t)$ into a "free" (or rather: solvable) part $H_0(t)$ and an "interaction" (or rather: perturbation) part $V(t)$,

$$H(t) = H_0(t) + V(t), \tag{13.11}$$

and define the "free" time evolution operator

$$U_0(t,t_0) = \mathrm{T}\exp\!\left(-\frac{\mathrm{i}}{\hbar}\int_{t_0}^{t} d\tau\, H_0(\tau)\right).$$

The common terminology of denoting $H_0(t)$ and $U_0(t, t_0)$ as "free" Hamiltonian and time evolution operator while $V(t)$ is the "interaction" part is motivated from scattering theory, which is one of the most common applications of time-dependent perturbation theory. However, we should always keep in mind that $H_0(t)$ does not really need to be a free particle Hamiltonian. E.g. for a hydrogen atom under the influence of an external electromagnetic field with wavelength $\lambda \gg a_0$, the "free" part H_0 would actually be the hydrogen Hamiltonian including the Coulomb interaction between the proton and the electron, while $V(t)$ would describe the effective coupling of the electromagnetic field to the quasiparticle which describes relative motion in the hydrogen atom. We will discuss this case in detail in Chapter 18, and in particular in Section 18.4.

The interaction picture splits off the solvable part of the time evolution from the states,

$$\begin{aligned} |\psi_D(t)\rangle &= U_0^+(t,t_0)|\psi_S(t)\rangle = U_0^+(t,t_0)U(t,t')|\psi_S(t')\rangle \\ &= U_0^+(t,t_0)U(t,t')U_0(t',t_0)|\psi_D(t')\rangle = U_D(t,t')|\psi_D(t')\rangle, \end{aligned} \quad (13.12)$$

where the last line identifies the time evolution operator $U_D(t, t')$ acting on the states in the interaction picture.

The solvable part of the time evolution is cast onto the operators

$$\Phi_D(t) = U_0^+(t,t_0)\Phi_S U_0(t,t_0) \quad (13.13)$$

to preserve the time evolution of matrix elements and expectation values in the interaction picture.

Substituting the composition law for time evolution operators confirms that $\Phi_D(t)$ evolves freely between different times,

$$\begin{aligned} \Phi_D(t) &= U_0^+(t,t_0)\Phi_S U_0^+(t_0,t) = U_0^+(t,t')U_0^+(t',t_0)\Phi_S U_0^+(t_0,t')U_0^+(t',t) \\ &= U_0^+(t,t')\Phi_D(t')U_0(t,t'), \end{aligned} \quad (13.14)$$

and substituting $\Phi_H(t)$ for Φ_S shows that $\Phi_D(t)$ is related to the operator in the Heisenberg picture through the particular variant $U_D(t, t_0)$ of the interaction picture evolution operator $U_D(t, t')$ (13.12),

$$\begin{aligned} \Phi_D(t) &= U_0^+(t,t_0)U(t,t_0)\Phi_H(t)U^+(t,t_0)U_0(t,t_0) \\ &= U_D(t,t_0)\Phi_H(t)U_D^+(t,t_0). \end{aligned} \quad (13.15)$$

The differential equation for time evolution of the operators is

$$\begin{aligned}
i\hbar\frac{d}{dt}\Phi_D(t) &= -\,U_0^+(t,t_0)[H_0(t)\Phi_S - \Phi_S H_0(t)]U_0(t,t_0)\\
&= -\,U_0^+(t,t_0)H_0(t)U_0(t,t_0)U_0^+(t,t_0)\Phi_S U_0(t,t_0)\\
&\quad + U_0^+(t,t_0)\Phi_S U_0(t,t_0)U_0^+(t,t_0)H_0(t)U_0(t,t_0)\\
&= -\,[H_{0,D}(t),\Phi_D(t)],
\end{aligned}$$

where in the last equation (similar to the previous remark for the Heisenberg picture) $H_{0,D}(t)$ is written in terms of operators $\Phi_D(t)$ in the Dirac picture.

The interactions are encoded in the time evolution of the states,

$$\begin{aligned}
i\hbar\frac{d}{dt}|\psi_D(t)\rangle &= U_0^+(t,t_0)[H(t) - H_0(t)]|\psi_S(t)\rangle\\
&= U_0^+(t,t_0)V(t)U_0(t,t_0)|\psi_D(t)\rangle = H_D(t)|\psi_D(t)\rangle,
\end{aligned}\tag{13.16}$$

where again $U_0^+(t,t_0)V(t)U_0(t,t_0) = V_D(t) \equiv H_D(t)$ due to the operator transition $\Phi_S \to \Phi_D$ in the Hamiltonians.

Conversion of equation (13.16) into the equivalent integral equation gives us another equation for the time evolution operator $U_D(t,t')$ for the states in the Dirac picture,

$$U_D(t,t') = U_0^+(t,t_0)U(t,t')U_0(t',t_0) = \mathrm{T}\exp\!\left(-\frac{\mathrm{i}}{\hbar}\int_{t'}^{t} d\tau\, H_D(\tau)\right). \tag{13.17}$$

This evolution operator apparently satisfies

$$i\hbar\frac{\partial}{\partial t}U_D(t,t') = H_D(t)U_D(t,t'),\quad i\hbar\frac{\partial}{\partial t'}U_D(t,t') = -\,U_D(t,t')H_D(t').$$

We have split the time evolution asymmetrically between states and operators, and therefore there are two Hamiltonians and related time evolution operators in the interaction picture: the "free" Hamiltonian $H_0(t)$ for the evolution of the operators and the interaction Hamiltonian $H_D(t)$ for the evolution of the states (and then there is the third Hamiltonian $H(t)$ and its time evolution operator appearing in the derivation of the interaction picture).

If we substitute[3]

$$H_D(t) = U_0^+(t,t_0)V(t)U_0(t,t_0) = U_0(t_0,t)V(t)U_0(t,t_0)$$

into equation (13.17) and use the composition property for time evolution operators

$$U_0(\tau,t_0)U_0(t_0,\tau') = U_0(\tau,\tau'),$$

[3]The transformation law for operators from the Schrödinger picture into the interaction picture implies $H_D(t) \equiv V_D(t)$. The notation $V_D(t)$ is therefore also often used for $H_D(t)$.

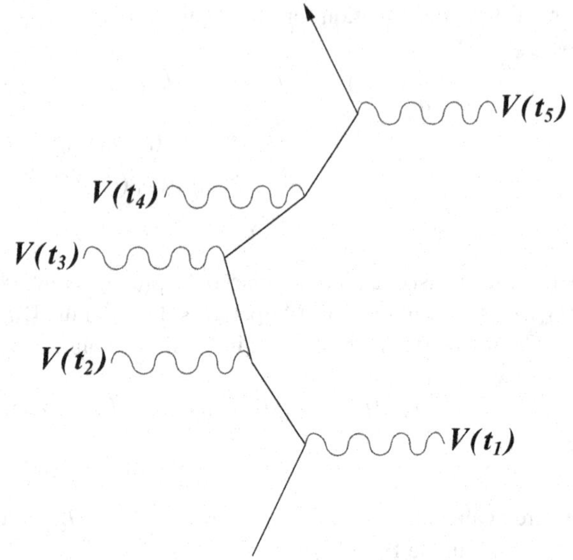

Fig. 13.2 Scattering off a time-dependent perturbation

we find

$$\begin{aligned} U_D(t,t') &= \mathrm{T}\exp\!\left(-\frac{\mathrm{i}}{\hbar}\int_{t'}^{t} d\tau\, H_D(\tau)\right) \\ &= \sum_{n=0}^{\infty}\frac{1}{(\mathrm{i}\hbar)^n}\int_{t'}^{t} d\tau_1\int_{t'}^{\tau_1} d\tau_2\ldots\int_{t'}^{\tau_{n-1}} d\tau_n\, H_D(\tau_1)H_D(\tau_2)\ldots H_D(\tau_n) \\ &= \sum_{n=0}^{\infty}\frac{1}{(\mathrm{i}\hbar)^n}\int_{t'}^{t} d\tau_1\int_{t'}^{\tau_1} d\tau_2\ldots\int_{t'}^{\tau_{n-1}} d\tau_n\, U_0(t_0,\tau_1)V(\tau_1)U_0(\tau_1,\tau_2) \\ &\quad\times V(\tau_2)U_0(\tau_2,\tau_3)\ldots U_0(\tau_{n-1},\tau_n)V(\tau_n)U_0(\tau_n,t_0). \end{aligned} \tag{13.18}$$

The n-th term in the sum can be interpreted as n scatterings at the perturbation $V(t)$, with "free" time evolution under the Hamiltonian $H_0(t)$ between any two scattering events, see Figure 13.2. In the end everything is evolved again to the fiducial time t_0. Equation (13.21) below will show that this as a consequence of the fact that we will express transition probability amplitudes in terms of states at some fixed time t_0.

Dirac picture for constant H_0

We have $H_0 = H_{0,D}$ if H_0 is a time-independent operator in the Schrödinger picture, because H_0 and $U_0(t,t_0) = \exp[-\mathrm{i}H_0(t-t_0)/\hbar]$ commute.

The Hamiltonian $H_D(t)$ acting on the states in the interaction picture is related to the Hamiltonian with the ordinary operators $\mathbf{p}$, $\mathbf{x}$,... of the Schrödinger picture via

$$H_D(t) = \exp\!\left(\frac{\mathrm{i}}{\hbar}H_0(t-t_0)\right) V(t) \exp\!\left(-\frac{\mathrm{i}}{\hbar}H_0(t-t_0)\right).$$

The time evolution operator for the states in the interaction picture is then

$$\begin{aligned} U_D(t,t') &= \mathrm{T}\exp\!\left(-\frac{\mathrm{i}}{\hbar}\int_{t'}^{t} d\tau\, H_D(\tau)\right) \\ &= \sum_{n=0}^{\infty} \frac{1}{(\mathrm{i}\hbar)^n}\int_{t'}^{t} d\tau_1 \int_{t'}^{\tau_1} d\tau_2 \ldots \int_{t'}^{\tau_{n-1}} d\tau_n\, \exp\!\left(-\frac{\mathrm{i}}{\hbar}H_0(t_0-\tau_1)\right) \\ &\quad \times V(\tau_1)\exp\!\left(-\frac{\mathrm{i}}{\hbar}H_0(\tau_1-\tau_2)\right) V(\tau_2)\exp\!\left(-\frac{\mathrm{i}}{\hbar}H_0(\tau_2-\tau_3)\right)\ldots \\ &\quad \times \exp\!\left(-\frac{\mathrm{i}}{\hbar}H_0(\tau_{n-1}-\tau_n)\right) V(\tau_n)\exp\!\left(-\frac{\mathrm{i}}{\hbar}H_0(\tau_n-t_0)\right). \end{aligned} \quad (13.19)$$

The case of time-independent unperturbed operators H_0 is the most common case in applications of time-dependent perturbation theory. Equation (13.19) therefore shows the most commonly employed form of $U_D(t,t')$ for the evaluation of the transition amplitudes or scattering matrix elements which will be introduced in Section 13.3.

13.3 Transitions between discrete states

We are now in a position to discuss transitions in a quantum system under the influence of time-dependent perturbations. We are still operating in the framework of "ordinary" quantum mechanics ("first quantized theory"), and at this stage time-dependent perturbations of a quantum system arise from time dependence of the parameters in the Schrödinger equation.

We will denote states as *discrete states* if they can be characterized by a set of discrete quantum numbers, e.g. the bound energy eigenstates $|n,\ell,m_\ell,m_s\rangle$ of hydrogen or the states $|n_1,n_2,n_3\rangle$ of a three-dimensional harmonic oscillator are discrete. States which require at least one continuous quantum number for their labeling are denoted as *continuous states*. Momentum eigenstates $|\boldsymbol{k}\rangle$ are examples of continuous states. Quantum mechanical transitions involving continuous states require special considerations. Therefore we will first discuss transitions between discrete states, e.g. transitions between atomic or molecular bound states.

We consider a system with an unperturbed Hamiltonian H_0 under the influence of a perturbation $V(t)$:

$$H(t) = H_0 + V(t).$$

The perturbation operator will in general be a function of the operators $\mathbf{p}$ and $\mathbf{x}$, $V(t) \equiv V(\mathbf{p}, \mathbf{x}, t)$. We will see later that in many applications $V(t)$ has the form

$$V(\mathbf{p}, \mathbf{x}, t) = V_1(\mathbf{x}, t) + \mathbf{p} \cdot V_2(\mathbf{x}, t). \tag{13.20}$$

In this section we assume that all states under consideration can be normalized to 1.

For the calculation of transition probabilities in the system, recall that the expansion of a general state $|\phi\rangle$ in terms of an orthonormal complete set of states $|\psi_n\rangle$ is

$$|\phi\rangle = \sum_n |\psi_n\rangle\langle\psi_n|\phi\rangle,$$

and therefore the probability $P_n(\phi)$ of finding the state $|\psi_n\rangle$ in a measurement performed on the state $|\phi\rangle$ is $P_n(\phi) = |\langle\psi_n|\phi\rangle|^2$. We can also understand this as the expectation value of the projection operator $|\psi_n\rangle\langle\psi_n|$ in the state $|\phi\rangle$.

Now assume that the state $|\phi\rangle$ is a state $|\psi_{in}(t)\rangle$, where the state at an earlier time $t' < t$ was an unperturbed state $|\psi_{in}^{(0)}(t')\rangle$, typically an eigenstate of H_0. Then we know that the state at time t is

$$|\psi_{in}(t)\rangle = U(t, t')|\psi_{in}^{(0)}(t')\rangle,$$

and since the state now evolved with the full Hamiltonian including the perturbation $V(t)$, it will not be an unperturbed state any more, but a superposition of unperturbed states. If at time t a measurement is performed on the state $|\psi_{in}(t)\rangle$, the probability to measure a certain unperturbed state $|\psi_{out}^{(0)}(t)\rangle$ will be $|\langle\psi_{out}^{(0)}(t)|\psi_{in}(t)\rangle|^2$.

Therefore the probability amplitude for transition from an unperturbed state $|\psi_{in}^{(0)}(t')\rangle$ to an unperturbed state $|\psi_{out}^{(0)}(t)\rangle$ between times t' and t is

$$\begin{aligned}
\langle\psi_{out}^{(0)}(t)|\psi_{in}(t)\rangle &= \langle\psi_{out}^{(0)}(t)|U(t, t')|\psi_{in}^{(0)}(t')\rangle \\
&= \langle\psi_{out}^{(0)}(t_0)|U_0^+(t, t_0)U(t, t')U_0(t', t_0)|\psi_{in}^{(0)}(t_0)\rangle \\
&= \langle\psi_{out}^{(0)}(t_0)|U_D(t, t')|\psi_{in}^{(0)}(t_0)\rangle.
\end{aligned} \tag{13.21}$$

The transition probability amplitudes between unperturbed states are matrix elements of the time evolution operator in the interaction picture, where the unperturbed states are taken at some arbitrary fixed time.

The Schrödinger equations for the unperturbed states $|\psi^{(0)}(t_0)\rangle$ and the free evolution operators $U_0(t',t_0)$ and $U_0^+(t,t_0)$ imply

$$\frac{\partial}{\partial t_0}\langle\psi_{out}^{(0)}(t_0)|U_0^+(t,t_0)U(t,t')U_0(t',t_0)|\psi_{in}^{(0)}(t_0)\rangle = 0, \tag{13.22}$$

i.e. the choice of the parameter t_0 is (of course) irrelevant for the transition matrix element. We set $t_0 = 0$ in the following.

If we substitute the expansion (13.18) for the time evolution operator in the interaction picture we get a series

$$\begin{aligned}
\langle\psi_{out}^{(0)}(0)|U_D(t,t')|\psi_{in}^{(0)}(0)\rangle &= \langle\psi_{out}^{(0)}(0)|\mathrm{T}\exp\!\left(-\frac{\mathrm{i}}{\hbar}\int_{t'}^{t}d\tau H_D(\tau)\right)|\psi_{in}^{(0)}(0)\rangle \\
&= \sum_{n=0}^{\infty}\frac{1}{(\mathrm{i}\hbar)^n}\int_{t'}^{t}d\tau_1\int_{t'}^{\tau_1}d\tau_2\ldots\int_{t'}^{\tau_{n-1}}d\tau_n\,\langle\psi_{out}^{(0)}(0)|\exp\!\left(\frac{\mathrm{i}}{\hbar}H_0\tau_1\right) \\
&\times V(\tau_1)\exp\!\left(-\frac{\mathrm{i}}{\hbar}H_0(\tau_1-\tau_2)\right)V(\tau_2)\exp\!\left(-\frac{\mathrm{i}}{\hbar}H_0(\tau_2-\tau_3)\right) \\
&\times\ldots\exp\!\left(-\frac{\mathrm{i}}{\hbar}H_0(\tau_{n-1}-\tau_n)\right)V(\tau_n)\exp\!\left(-\frac{\mathrm{i}}{\hbar}H_0\tau_n\right)|\psi_{in}^{(0)}(0)\rangle.
\end{aligned} \tag{13.23}$$

Now we assume that our unperturbed states are energy eigenstates

$$\begin{aligned}
|\psi_{out}^{(0)}(0)\rangle &= |\psi_n(0)\rangle = |n\rangle,\ H_0|\psi_n(0)\rangle = E_n|\psi_n(0)\rangle, \\
|\psi_{in}^{(0)}(0)\rangle &= |\psi_m(0)\rangle = |m\rangle,\ H_0|\psi_m(0)\rangle = E_m|\psi_m(0)\rangle
\end{aligned}$$

of the unperturbed Hamiltonian. Equation (13.23) then yields for the transition probability amplitude between eigenstates of H_0 (see also equation (13.19)),

$$\begin{aligned}
\langle n|U_D(t,t')|m\rangle &= \delta_{n,m} - \frac{\mathrm{i}}{\hbar}\int_{t'}^{t}d\tau\ \exp(\mathrm{i}\omega_{nm}\tau)\langle n|V(\tau)|m\rangle \\
&\quad -\frac{1}{\hbar^2}\sum_l\int_{t'}^{t}d\tau_1\int_{t'}^{\tau_1}d\tau_2\ \exp(\mathrm{i}\omega_{nl}\tau_1)\langle n|V(\tau_1)|l\rangle \\
&\quad \times\exp(\mathrm{i}\omega_{lm}\tau_2)\langle l|V(\tau_2)|m\rangle + \ldots,
\end{aligned} \tag{13.24}$$

with the transition frequencies $\omega_{nm} = (E_n - E_m)/\hbar$.

The transition probability from a discrete state $|m\rangle$ to a discrete state $|n\rangle$ is then

$$P_{m\to n}(t,t') = \left|\langle n|U_D(t,t')|m\rangle\right|^2. \tag{13.25}$$

Equation (13.24) assumes that we use eigenstates of H_0 for the initial and final states, but equation (13.25) holds for arbitrary discrete initial and final states, and we even do not have to require the same basis for the decomposition of the initial and the final state, i.e. equation (13.25) also holds if m and n are discrete quantum numbers referring to different bases of states.

$P_{m\to n}(t,t')$ is a dimensionless positive number if both the initial and final states are discrete states, i.e. dimensionless states (see the discussion of dimensions of states in Section 5.3), and due to the unitarity of $U_D(t,t')$ it is also properly normalized as a probability,

$$\sum_n P_{m\to n}(t,t') = \sum_n \langle m|U_D^+(t,t')|n\rangle\langle n|U_D(t,t')|m\rangle = \langle m|m\rangle = 1.$$

As a corollary, this observation also implies $0 \le |\langle n|U_D(t,t')|m\rangle|^2 \le 1$, as required for a probability.

We will denote the *transition probability amplitude* $\langle n|U_D(t,t')|m\rangle$ also as a *scattering matrix element* or *S matrix element*,

$$\begin{aligned} S_{nm}(t,t') &= \langle n|U_D(t,t')|m\rangle = \langle n|\mathrm{T}\exp\!\left(-\frac{\mathrm{i}}{\hbar}\int_{t'}^{t} d\tau\, H_D(\tau)\right)|m\rangle \\ &= \langle m|U_D^+(t,t')|n\rangle^* = \langle m|U_D(t',t)|n\rangle^* = \left(S_{mn}^{-1}(t,t')\right)^* \\ &= \left(S^{-1+}(t,t')\right)_{nm}. \end{aligned} \tag{13.26}$$

In the literature this definition is more commonly employed with default values $t\to\infty$, $t'\to-\infty$ for the initial and final times, $S_{nm} \equiv S_{nm}(\infty,-\infty)$. It is also usually reserved for transitions with two particles in the initial state (to be discussed in Chapter 17 and following chapters), but here we are still dealing with a single particle perturbed by a potential $V(t)$, or an effective single particle description of relative motion of two particles. The connection with many particle scattering theory later on is easier if we introduce the scattering matrix already for single particle problems, and it is also useful to have this notion available for arbitrary initial and final times.

Møller operators

At this point it is also interesting to note a factorized representation of the time evolution operator in the interaction picture, which is applicable if both H and H_0 do not depend on time. In this case we have with $t_0 = 0$,

$$U_D(t,t') = \exp\!\left(\frac{\mathrm{i}}{\hbar}H_0 t\right)\exp\!\left(-\frac{\mathrm{i}}{\hbar}H(t-t')\right)\exp\!\left(-\frac{\mathrm{i}}{\hbar}H_0 t'\right) = \Omega^+(t)\Omega(t')$$

with the Møller operator

$$\Omega(t) = \exp\!\left(\frac{\mathrm{i}}{\hbar}Ht\right)\exp\!\left(-\frac{\mathrm{i}}{\hbar}H_0 t\right).$$

Let us repeat the basic equation (13.21) and substitute this definition,

$$\begin{aligned}\langle\psi_{out}^{(0)}(t)|\psi_{in}(t)\rangle &= \langle\psi_{out}^{(0)}(t)|U(t,t')|\psi_{in}^{(0)}(t')\rangle \\ &= \langle\psi_{out}^{(0)}|U_0^+(t)U(t,t')U_0(t')|\psi_{in}^{(0)}\rangle = \langle\psi_{out}^{(0)}|U_D(t,t')|\psi_{in}^{(0)}\rangle \\ &= \langle\psi_{out}^{(0)}|\Omega^+(t)\Omega(t')|\psi_{in}^{(0)}\rangle = \langle\Psi_{out}\{t\}|\Psi_{in}\{t'\}\rangle.\end{aligned}$$

Here we have introduced states

$$\begin{aligned}|\Psi\{t\}\rangle = \Omega(t)|\psi^{(0)}\rangle &= \exp\!\left(\frac{\mathrm{i}}{\hbar}Ht\right)\exp\!\left(-\frac{\mathrm{i}}{\hbar}H_0t\right)|\psi^{(0)}\rangle \\ &= \exp\!\left(\frac{\mathrm{i}}{\hbar}Ht\right)|\psi^{(0)}(t)\rangle. \qquad (13.27)\end{aligned}$$

For the interpretation of these states we notice

$$\exp\!\left(-\frac{\mathrm{i}}{\hbar}Ht\right)|\Psi\{t\}\rangle = |\psi^{(0)}(t)\rangle,$$

i.e. $|\Psi\{t\}\rangle$ is the fictitious interacting state at time $t_0 = 0$ which yields the unperturbed state $|\psi^{(0)}(t)\rangle$ at time t under *full* time evolution from $t_0 = 0$ to t.

In the framework or quantum mechanics, the case that both H and H_0 are time-independent would often be dealt with in the framework of time-independent perturbation theory or potential scattering theory. However, we will see later that in the framework of quantum field theory, time-independent H and H_0 is very common in applications of time-dependent perturbation theory.

First order transition probability between discrete energy eigenstates

For $n \neq m$, the first order result for S_{nm} is the matrix element of the Fourier component $V(\omega_{nm})$,

$$\begin{aligned}S_{nm} &= -\frac{\mathrm{i}}{\hbar}\int_{-\infty}^{\infty}dt\,\exp(\mathrm{i}\omega_{nm}t)\langle n|V(t)|m\rangle \\ &= -\frac{\mathrm{i}}{\hbar}\sqrt{2\pi}\,\langle n|V(\omega_{nm})|m\rangle. \qquad (13.28)\end{aligned}$$

If the time dependence of the perturbation $V(t)$ is such that the Fourier transform $V(\omega)$ exists in the sense of standard Fourier theory (i.e. if $V(\omega)$ is a sufficiently well behaved function, which is the case e.g. if $V(t)$ is absolutely integrable or square integrable with respect to t), then the first order scattering matrix (13.28) provides us with finite first order approximations for transition probabilities

$$P_{m\to n} = |S_{nm}|^2 = \frac{2\pi}{\hbar^2}\,|\langle n|V(\omega_{nm})|m\rangle|^2. \tag{13.29}$$

Note that the Fourier transform

$$V(\omega) = \frac{1}{\sqrt{2\pi}}\int dt\,\exp(\mathrm{i}\omega t)V(t)$$

of a potential $V(t)$ has the dimension energy×time. Therefore $P_{m\to n}$ is a dimensionless number, as it should be. Furthermore, the probability interpretation and the use of first order perturbation theory entail that we should have $|\langle n|V(\omega_{nm})|m\rangle| < \hbar/\sqrt{2\pi}$. Otherwise first order perturbation theory is not applicable and higher order terms must be included to estimate transition probabilities.

The first order transition probability between discrete states requires existence of a regular Fourier transform $V(\omega)$ of the perturbation $V(t)$. This condition is not satisfied in the important case of monochromatic perturbations like $V(t) = W\exp(-\mathrm{i}\omega t)$, which have a δ function as Fourier transform,

$$V(t) = W\exp(-\mathrm{i}\omega t), \quad V(\omega_{nm}) = \sqrt{2\pi}\,W\delta(\omega_{nm}-\omega).$$

Consistent treatment of this case requires that at least one of the states involved is part of a continuum of states, as discussed in Sections 13.4 and 13.5. If both the initial and final atomic or molecular state are discrete, then the perturbation $V(t) = W\exp(-\mathrm{i}\omega t)$ must be treated as arising from a quantized field which comes with its own continuum of states. Monochromatic perturbations $V(t) = W\exp(\pm\mathrm{i}\omega t)$ typically arise from photon absorption or emission, and the previous statement simply means that the consistent treatment of transitions between bound states due to monochromatic perturbations requires the full quantum theory of the photon, see Section 18.6. See also Problem 13.6 for an explanation why the Golden Rule of first order perturbation theory, which is discussed in the next section, cannot be used for transitions between discrete states.

13.4 Transitions from discrete states into continuous states: Ionization or decay rates

Ionization of atoms or molecules, transitions from discrete donor states into conduction bands in n-doped semiconductors, or disintegration of nuclei are processes where particles make a transition from discrete states into states in a continuum.

We assume that the unperturbed Hamiltonian H_0 contains an attractive radially symmetric potential which generates bound states $|n,\ell,m\rangle$, where ℓ and m are the usual angular momentum quantum numbers for the bound states and the quantum number n labels the energy levels. The free states for H_0 are usually given in terms of hypergeometric functions, e.g. the Coulomb waves $|k,\ell,m\rangle$ from Section 7.9.

Here we initially use plane wave states instead and ask for the probability for the system to go from a bound state $|n, \ell, m\rangle$ into a plane wave state $|\boldsymbol{k}\rangle$ under the influence of a perturbation $V(t)$. This is a simplification, but the prize that we pay is that the transition matrix elements from a bound state into plane waves do not necessarily tell us something about ionization or decay of a bound system, because those transition matrix elements will also not vanish for perturbations which primarily generate another bound state since the bound states can also be written as superpositions of plane waves, see e.g. Problem 13.7. Therefore the transition matrix elements into plane wave states generically correspond to a mixture of transitions into bound states and free states. However, the focus in this preliminary discussion is not the calculation of actual ionization or decay rates, but to explain how continuous final states affect the interpretation of transition matrix elements.

For continuous final states like $|\boldsymbol{k}\rangle$, the appropriate projection of $U_D(t,t')|\psi_{in}^{(0)}\rangle$ is onto the dimensionless combination $\sqrt{d^3k}\langle\boldsymbol{k}|$ (recall from $\langle\boldsymbol{k}|\boldsymbol{k}'\rangle = \delta(\boldsymbol{k}-\boldsymbol{k}')$ that the plane wave states $|\boldsymbol{k}\rangle$ in three dimensions have length dimension length$^{3/2}$, see Section 5.3). This means that in a transition from a discrete state $|n, \ell, m\rangle$ into a momentum eigenstate $\boldsymbol{k}$, the dimensionless quantity

$$\sqrt{d^3k}\, S_{\boldsymbol{k};n,\ell,m}(t,t') = \sqrt{d^3k}\langle\boldsymbol{k}|U_D(t,t')|n,\ell,m\rangle$$

is a *differential transition probability amplitude*, in the sense that

$$dP_{n,\ell,m\to\boldsymbol{k}}(t,t') = d^3k\, \left|\langle\boldsymbol{k}|U_D(t,t')|n,\ell,m\rangle\right|^2$$

is a *differential transition probability* for the transition from the discrete state into a volume element d^3k around the vector $\boldsymbol{k}$ in momentum space. The meaning of this statement is that

$$P_{n,\ell,m\to\mathcal{K}}(t,t') = \int_{\mathcal{K}} d^3k\, \left|\langle\boldsymbol{k}|U_D(t,t')|n,\ell,m\rangle\right|^2 \tag{13.30}$$

is the *transition probability* from the discrete state $|n, \ell, m\rangle$ into a volume $\mathcal{K}$ in $\boldsymbol{k}$-space. Another way to say this is to denote the quantity with the dimension length^3

$$\mathcal{P}_{n,\ell,m\to\boldsymbol{k}}(t,t') = \frac{dP_{n,\ell,m\to\boldsymbol{k}}(t,t')}{d^3k} = \left|\langle\boldsymbol{k}|U_D(t,t')|n,\ell,m\rangle\right|^2$$

as the *transition probability density* per $\boldsymbol{k}$-space volume. The S matrix element

$$S_{\boldsymbol{k};n,\ell,m}(t,t') = \langle\boldsymbol{k}|U_D(t,t')|n,\ell,m\rangle$$

is then a *transition probability density amplitude* (just like a wave function $\langle\boldsymbol{x}|\psi(t)\rangle$ is a *probability density amplitude* rather than a probability amplitude, but for obvious reasons neither of these designations are ever used).

With this interpretation, the transition amplitudes into continuous states yield correctly normalized probabilities, e.g. for plane waves,

$$\begin{aligned}\int d^3\boldsymbol{k}\,\mathcal{P}_{n,\ell,m\to\boldsymbol{k}}(t,t') &= \int d^3\boldsymbol{k}\;\left|\langle\boldsymbol{k}|U_D(t,t')|n,\ell,m\rangle\right|^2\\ &= \int d^3\boldsymbol{k}\,\langle n,\ell,m|U_D^+(t,t')|\boldsymbol{k}\rangle\langle\boldsymbol{k}|U_D(t,t')|n,\ell,m\rangle\\ &= \langle n,\ell,m|U_D^+(t,t')U_D(t,t')|n,\ell,m\rangle\\ &= \langle n,\ell,m|n,\ell,m\rangle = 1.\end{aligned}$$

The important conclusion from this is that transition matrix elements of $U_D(t,t')$ from discrete states into continuous final states yield transition probability *densities*, which have to be integrated to yield transition probabilities. We will also rediscover this in the framework of the spherical Coulomb waves in the following subsection.

Ionization probabilities for hydrogen

Now that we have clarified the meaning of transition amplitudes from discrete states into continuous states with the familiar basis of plane wave states, let us come back to the ionization or decay problems, i.e. transitions from the discrete bound spectrum of an unperturbed Hamiltonian H_0 into the continuum of unbound states. We will use hydrogen states as an example, but the derivations go through in the same way for any Hamiltonian H_0 with discrete and continuous states.

The unperturbed Hamiltonian for hydrogen is

$$H_0 = \frac{\mathbf{p}^2}{2\mu} - \frac{e^2}{4\pi\epsilon_0|\mathbf{r}|}, \tag{13.31}$$

and the ionization problem concerns transitions from bound states $|n,\ell,m\rangle$ into Coulomb waves $|k,\ell,m\rangle$ under the influence of a time-dependent perturbation[4] $V(t)$. The contribution from Coulomb waves to the decomposition of unity in terms of hydrogen states came with a measure k^2dk (7.75),

[4]If the perturbation $V(t)$ contains directional information (e.g. polarization of an incoming photon or the direction of an electric field), then we might also like to calculate probabilities for the direction of dissociation of the hydrogen atom. This direction would be given by the $\boldsymbol{k}$ vector of relative motion between the electron and the proton after separation. For the calculation of directional information we would have to combine the spherical Coulomb waves $|k,\ell,m\rangle$ into states which approximate plane wave states $|\boldsymbol{k}\rangle$ at infinity, similar to the construction of incoming approximate plane wave states in Section 13.5, see also the discussion of the photoeffect in [3].

$$1 = \sum_{\ell=0}^{\infty}\sum_{m=-\ell}^{\ell}\left(\sum_{n=\ell+1}^{\infty}|n,\ell,m\rangle\langle n,\ell,m| + \int_0^\infty dk\, k^2|k,\ell,m\rangle\langle k,\ell,m|\right)$$
$$= \sum_{\ell=0}^{\infty}\sum_{m=-\ell}^{\ell}\left(\sum_{n=\ell+1}^{\infty}|n,\ell,m\rangle\langle n,\ell,m| + \int_0^\infty dE\,|E,\ell,m\rangle\varrho(E)\langle E,\ell,m|\right),$$

where we also introduced an energy representation for the spherical Coulomb waves, $|E,\ell,m\rangle = |k,\ell,m\rangle$,

$$E = \frac{\hbar^2k^2}{2\mu} = \frac{\hbar^2k^2}{2}\left(\frac{1}{m_e} + \frac{1}{m_p}\right),$$

and the corresponding density of spherical Coulomb waves in the energy scale,

$$\varrho(E) = \Theta(E)k^2\frac{dk}{dE} = \frac{\Theta(E)}{\hbar^3}\sqrt{2\mu^3E}. \tag{13.32}$$

This differs from (12.13) for $d = 3$ by missing a factor $g/2\pi^2 = g4\pi/8\pi^3$. The spin factor is $g = 1$, because spin flips can usually be neglected in ionization transitions. Inclusion of spin quantum numbers m_s and m_s' for the initial and final states would therefore result in a factor $\delta_{m_s,m_s'}$. There is no factor 4π because the angular directions in $\boldsymbol{k}$ space have been discretized in terms of angular momentum quantum numbers (ℓ, m), and there is no factor $(2\pi)^{-3}$ because the density $\varrho(E)$ in equation (13.32) is a number of states per unit of energy, but it is *not* a number of states per energy and *volume* (remember $V \to (2\pi)^3$ in the continuum limit). It comes in units $\mathrm{cm}^{-3}\mathrm{eV}^{-1}$ because the projector $|k,\ell,m\rangle\langle k,\ell,m|$ for spherical Coulomb waves has dimension length^3, and therefore scattering matrix elements $|\langle E,\ell',m'|U_D(\infty,-\infty)|n,\ell,m\rangle|^2$ from bound states into ionized states come in units of cm^3. Please also recall the remark after equation (12.8).

Suppose we start with an unperturbed bound state $|n,\ell,m\rangle$. We can calculate two kinds of scattering matrix elements, *viz.* for transitions into bound states,

$$S_{n',\ell',m';n,\ell,m} = \langle n',\ell',m'|U_D(\infty,-\infty)|n,\ell,m\rangle,$$

and into ionized states

$$S_{E,\ell',m';n,\ell,m} = \langle E,\ell',m'|U_D(\infty,-\infty)|n,\ell,m\rangle, \quad E > 0.$$

For the sums of the absolute squares of these scattering matrix elements, we observe from the completeness relation for hydrogen states, the unitarity of $U_D(\infty,-\infty)$, and $\langle n,\ell,m|n,\ell,m\rangle = 1$ that

$$\sum_{\ell'=0}^{\infty}\sum_{m'=-\ell'}^{\ell'}\left(\sum_{n'=\ell'+1}^{\infty}|\langle n',\ell',m'|U_D(\infty,-\infty)|n,\ell,m\rangle|^2\right.$$
$$\left.+\int_0^\infty dE\,\varrho(E)\,|\langle E,\ell',m'|U_D(\infty,-\infty)|n,\ell,m\rangle|^2\right)=1.$$

This confirms $0\leq|\langle n',\ell',m'|U_D(\infty,-\infty)|n,\ell,m\rangle|^2\leq 1$, as is required for transition probabilities between bound states, but it also tells us that

$$P_{n,\ell,m\to E>0}=1-P_{n,\ell,m\to E<0}$$
$$=\sum_{\ell'=0}^{\infty}\sum_{m'=-\ell'}^{\ell'}\int_0^\infty dE\,\varrho(E)\,|\langle E,\ell',m'|U_D(\infty,-\infty)|n,\ell,m\rangle|^2 \quad (13.33)$$

must be the ionization probability due to the perturbation $V(t)$, since the sum over all transition probabilities into bound states is

$$P_{n,\ell,m\to E<0}=\sum_{\ell'=0}^{\infty}\sum_{m'=-\ell'}^{\ell'}\sum_{n'=\ell'+1}^{\infty}|\langle n',\ell',m'|U_D(\infty,-\infty)|n,\ell,m\rangle|^2.$$

This confirms again that absolute squares of scattering matrix elements into continuous final states must be integrated against final state densities to yield transition probabilities, where the appropriate density of final states follows from the completeness relation of the unperturbed system.

If we want to know the probability for the hydrogen atom to ionize into a state with energy $0<E_1\leq E\leq E_2$ for the relative motion between proton and electron, we have to calculate

$$P_{n,\ell,m\to[E_1,E_2]}=\sum_{\ell'=0}^{\infty}\sum_{m'=-\ell'}^{\ell'}\int_{E_1}^{E_2} dE\,\varrho(E)\,|\langle E,\ell',m'|U_D(\infty,-\infty)|n,\ell,m\rangle|^2.$$

On the other hand, if we only know the energy level E_n of the initial bound state, we would calculate the ionization probability of the atom as a weighted average

$$P_{E_n\to E>0}=\frac{1}{n^2}\sum_{\ell=0}^{n-1}\sum_{m=-\ell}^{\ell}P_{n,\ell,m\to E>0}. \quad (13.34)$$

The first order results for the ionization probabilities follow from the first order scattering matrix elements

$$
\begin{aligned}
S^{(1)}_{k,\ell',m';n,\ell,m} &= \langle k,\ell',m'|U_D(\infty,-\infty)|n,\ell,m\rangle^{(1)} \\
&= -\frac{i}{\hbar}\int_{-\infty}^{\infty} dt\, \exp\!\left(\frac{i}{\hbar}(E_k-E_n)t\right)\langle k,\ell',m'|V(t)|n,\ell,m\rangle \\
&= -i\frac{\sqrt{2\pi}}{\hbar}\langle k,\ell',m'|V(\omega_{kn})|n,\ell,m\rangle, \qquad (13.35)
\end{aligned}
$$

with the transition frequency $\omega_{kn} = (E_k - E_n)/\hbar$. This assumes that the Fourier transformed operator $V(\omega_{kn})$ exists in the sense of standard Fourier theory. The case of a monochromatic perturbation, for which the Fourier transform is a δ function in frequency space, requires special treatment and is discussed in the following subsection.

Even for well behaved Fourier transform $V(\omega_{kn})$, use of the first order result (13.35) to estimate the ionization probability,

$$
P^{(1)}_{n,\ell,m\to E>0} = \frac{2\pi}{\hbar^2}\sum_{\ell'=0}^{\infty}\sum_{m'=-\ell'}^{\ell'}\int_0^{\infty} dk\, k^2 \left|\langle k,\ell',m'|V(\omega_{kn})|n,\ell,m\rangle\right|^2,
$$

can only make sense for $P^{(1)}_{n,\ell,m\to E>0} \leq 1$.

The Golden Rule for transitions from discrete states into a continuum of states

Now assume that we perturb a hydrogen atom in the initial bound state $|n,\ell,m\rangle$ with a monochromatic perturbation[5]

$$
V(t) = W\exp(-i\omega t) + W^{+}\exp(i\omega t), \qquad (13.36)
$$

$$
V(\omega') = \sqrt{2\pi}W\delta(\omega'-\omega) + \sqrt{2\pi}W^{+}\delta(\omega'+\omega). \qquad (13.37)
$$

The corresponding scattering matrix element for transition into the ionized state $|E,\ell',m'\rangle$ is in first order

$$
\begin{aligned}
S_{E,\ell',m';n,\ell,m} = &-\frac{i}{\hbar}2\pi\langle E,\ell',m'|W|n,\ell,m\rangle\delta\!\left(\frac{E-E_n}{\hbar}-\omega\right) \\
&-\frac{i}{\hbar}2\pi\langle E,\ell',m'|W^{+}|n,\ell,m\rangle\delta\!\left(\frac{E-E_n}{\hbar}+\omega\right). \qquad (13.38)
\end{aligned}
$$

[5]Recall that the notation tacitly implies dependence of the operators V and W on $\mathbf{x}$ and $\mathbf{p}$ (just like we usually write H instead of $H(\mathbf{x},\mathbf{p})$ for a Hamilton operator).

The cross multiplication terms in $|S_{E,\ell',m';n,\ell,m}|^2$ cancel for $\omega \neq 0$ due to the incompatibility of the δ functions, and therefore we can focus in the following discussion only on the first terms in equations (13.36–13.38), i.e. we continue with

$$S_{E,\ell',m';n,\ell,m} = -\frac{\mathrm{i}}{\hbar}2\pi\langle E,\ell',m'|W|n,\ell,m\rangle\delta\left(\frac{E-E_n}{\hbar}-\omega\right).$$

The square of this S matrix element yields a factor

$$\delta(0) = \lim_{\omega\to 0}\delta(\omega) = \lim_{\omega\to 0}\lim_{T\to\infty}\frac{1}{2\pi}\int_{-T/2}^{T/2}dt\,\exp(\mathrm{i}\omega t) = \lim_{T\to\infty}\frac{T}{2\pi}$$

in $dP_{n,\ell,m\to E,\ell',m'}/dE = \varrho(E)|S_{E,\ell',m';n,\ell,m}|^2$. Dividing by the factor T provides us with a differential transition rate into a final state energy interval $[E, E+dE]$,

$$\begin{aligned}
dw_{n,\ell,m\to E,\ell',m'} &= \frac{1}{T}dP_{n,\ell,m\to E,\ell',m'} = dE\,\varrho(E)\frac{1}{T}\left|S_{E,\ell',m';n,\ell,m}\right|^2 \\
&= dE\,\varrho(E)\frac{2\pi}{\hbar^2}\left|\langle E,\ell',m'|W|n,\ell,m\rangle\right|^2\delta\left(\frac{E-E_n}{\hbar}-\omega\right) \\
&= dE\,\varrho(E)\frac{2\pi}{\hbar}\left|\langle E,\ell',m'|W|n,\ell,m\rangle\right|^2\delta(E-E_n-\hbar\omega).
\end{aligned}$$

Integration over the final state energy E then yields an expression for the transition rate,

$$w_{n,\ell,m\to E,\ell',m'} = \frac{2\pi}{\hbar}\left|\langle E,\ell',m'|W|n,\ell,m\rangle\right|^2\varrho(E)\Big|_{E=E_n+\hbar\omega}, \qquad (13.39)$$

which is commonly referred to as the *Golden Rule*.

The total first order ionization rate of the state $|n,\ell,m\rangle$ under the perturbation (13.36) is then

$$w_{n,\ell,m} = \frac{2\pi}{\hbar}\sum_{\ell'=0}^{\infty}\sum_{m'=-\ell'}^{\ell'}\left|\langle E,\ell',m'|W|n,\ell,m\rangle\right|^2\varrho(E)\Bigg|_{E=E_n+\hbar\omega}. \qquad (13.40)$$

The standard expression for the Golden Rule for the transition rate from a discrete state $|m\rangle$ into a continuous state $|n\rangle$ due to the perturbation $V = W\exp(-\mathrm{i}\omega t)$ is

$$w_{m\to n} = \int_{-\infty}^{\infty}dE\,\frac{dw_{m\to n}}{dE} = \frac{2\pi}{\hbar}\varrho(E_n)|\langle n|W|m\rangle|^2\Big|_{E_n=E_m+\hbar\omega}. \qquad (13.41)$$

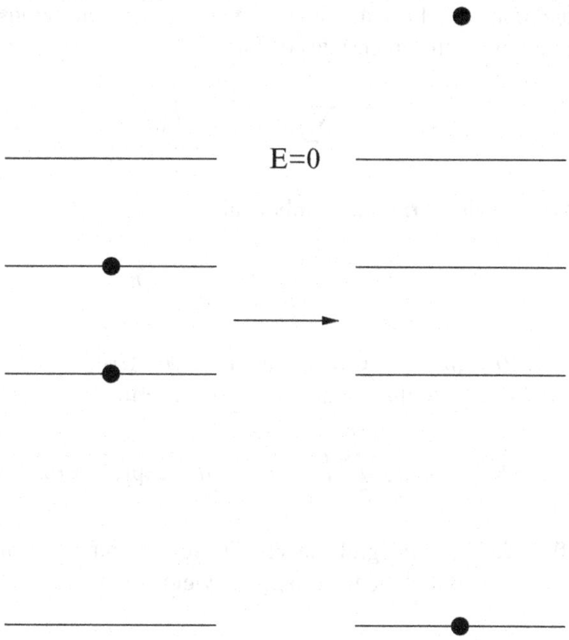

Fig. 13.3 Energy schematics for an Auger process. The initial bound state of the two electrons has the same energy as the final continuous state of an ion and a free electron

This is also particularly popular for time-independent V,

$$w_{m\to n} = \frac{2\pi}{\hbar}\varrho(E_n)|\langle n|V|m\rangle|^2\Big|_{E_n=E_m}. \tag{13.42}$$

Quantum systems can have degeneracy between states $|m\rangle$ which are labelled by discrete quantum numbers and states $|n\rangle$ with continuous quantum numbers. Metastable states, or excited bound states in many-electron atoms provide examples for this, and equation (13.42) would be the first order expression for the decay rate of these states. An example for this is the Auger effect, which is electron emission from atoms due to Coulomb repulsion. The perturbation operator[6] $V = e^2/4\pi|\mathbf{x}_1 - \mathbf{x}_2|$ is time-independent, and energy conservation is fulfilled because the discrete bound state of two electrons in an excited atom can exceed the sum of ground state energy and ionization energy, see Figure 13.3.

Time-dependent perturbation theory in second order and the Golden Rule #1

We will discuss a time-independent perturbation V,

$$H = H_0 + V,$$

[6]G. Wentzel, Z. Phys. 43, 524 (1927).

and transition from a discrete state $|m\rangle$ into a continuous state $|n\rangle$. The completeness relation for the eigenstates of H_0 is

$$\sum_m |m\rangle\langle m| + \int dE_n \varrho(E_n)|n\rangle\langle n| = 1.$$

We will also write this symbolically as

$$\sum\!\!\!\!\!\!\int |l\rangle\langle l| = 1.$$

If $\langle n|V|m\rangle = 0$, the leading order term for the scattering matrix element $\langle n|U_D(\infty, -\infty)|m\rangle$ is the second order term

$$S_{nm}^{(2)} = -\frac{1}{\hbar^2}\sum\!\!\!\!\!\!\int \int_{-\infty}^{\infty} d\tau \int_{-\infty}^{\tau} d\tau'\, \exp(\mathrm{i}\omega_{nl}\tau)\langle n|V|l\rangle \exp(\mathrm{i}\omega_{lm}\tau')\langle l|V|m\rangle.$$

To make the τ' integral convergent, we add a small negative imaginary part to $\omega_{lm} \to \omega_{lm} - \mathrm{i}\epsilon$, so that the time integrals yield

$$\begin{aligned}
\int_{-\infty}^{\infty} d\tau \int_{-\infty}^{\tau} d\tau'\, \exp(\mathrm{i}\omega_{nl}\tau)\exp\!\big(\mathrm{i}\omega_{lm}\tau' + \epsilon\tau'\big) &= \frac{1}{\mathrm{i}\omega_{lm} + \epsilon}\int_{-\infty}^{\infty} d\tau\, \exp(\mathrm{i}\omega_{nm}\tau) \\
&= \frac{2\pi}{\mathrm{i}\omega_{lm} + \epsilon}\delta(\omega_{nm}).
\end{aligned} \tag{13.43}$$

This yields the second order scattering matrix element

$$\begin{aligned}
S_{nm}^{(2)} &= \frac{2\pi\mathrm{i}}{\hbar^2}\delta(\omega_{nm})\sum\!\!\!\!\!\!\int \frac{\langle n|V|l\rangle\langle l|V|m\rangle}{\omega_{lm} - \mathrm{i}\epsilon} \\
&= 2\pi\mathrm{i}\delta(E_n - E_m)\sum\!\!\!\!\!\!\int \frac{\langle n|V|l\rangle\langle l|V|m\rangle}{E_l - E_m - \mathrm{i}\epsilon},
\end{aligned}$$

and the differential transition rate

$$\begin{aligned}
dw_{m\to n} &= dE_n\varrho(E_n)\frac{1}{T}|S_{nm}|^2 \\
&= dE_n\varrho(E_n)\frac{2\pi}{\hbar}\delta(E_n - E_m)\left|\sum\!\!\!\!\!\!\int \frac{\langle n|V|l\rangle\langle l|V|m\rangle}{E_l - E_m - \mathrm{i}\epsilon}\right|^2.
\end{aligned} \tag{13.44}$$

Integration yields the second order expression for the transition rate,

$$w_{m\to n} = \frac{2\pi}{\hbar}\varrho(E_n)\left|\sum\!\!\!\!\!\!\int \frac{\langle n|V|l\rangle\langle l|V|m\rangle}{E_l - E_m - \mathrm{i}\epsilon}\right|^2\Bigg|_{E_n=E_m}. \tag{13.45}$$

Equation (13.45) tells us how transitions through virtual intermediate states can generate the transition from $|m\rangle$ to $|n\rangle$ even if the direct transition is forbidden due to a selection rule $\langle n|V|m\rangle = 0$.

In his famous lectures on nuclear physics at the University of Chicago in 1949, Fermi coined the phrase "Golden Rule #2" for the first order transition rate (13.41, 13.42). He denoted the corresponding second order expression for transition rates as "Golden Rule #1", because it is important for nuclear reactions through intermediate compound nuclei [30].

13.5 Transitions from continuous states into discrete states: Capture cross sections

Transitions from continuous to discrete states arise e.g. in the capture of electrons by ions, in the absorption of an electron from a valence band into an acceptor state in a p-doped semiconductor, in neutron capture by nuclei etc. Consider e.g. the process $|k,\ell,m\rangle \to |n,\ell',m'\rangle$ of absorption of an electron by an H^+ ion, where we still assume that the hydrogen Hamiltonian (13.31) for relative motion is perturbed by addition of an operator $V(t)$. From our previous experience, we expect that the transition matrix element

$$S_{n,\ell',m';k,\ell,m} = \langle n,\ell',m'|U_D(\infty,-\infty)|k,\ell,m\rangle$$

yields a measure of probability for the absorption in the form of a transition probability density

$$\mathcal{P}_{k,\ell,m\to n,\ell',m'} = \left|S_{n,\ell',m';k,\ell,m}\right|^2. \tag{13.46}$$

Indeed, the dimensionless number

$$\begin{aligned} P_{n,\ell',m'} &= \sum_{\ell=0}^{\infty}\sum_{m=-\ell}^{\ell}\int_0^{\infty} dk\,k^2 \left|\langle n,\ell',m'|U_D(\infty,-\infty)|k,\ell,m\rangle\right|^2 \\ &= 1 - \sum_{\ell=0}^{\infty}\sum_{m=-\ell}^{\ell}\sum_{n'=\ell+1}^{\infty} \left|\langle n,\ell',m'|U_D(\infty,-\infty)|n',\ell,m\rangle\right|^2 \end{aligned}$$

is the probability that the state $|n,\ell',m'\rangle$ emerged from some capture ($p^+ + e^- \to H$) event rather than from an internal transition in the hydrogen atom. This assumes again that the perturbation $V(t)$ has a well behaved Fourier transform $V(\omega)$ such that the time integrals in the perturbation series can be defined as classical functions. However, a more common use of transition matrix elements from continuous initial states is the calculation of cross sections due to monochromatic perturbations. One possibility to calculate capture or absorption due to a Coulomb potential is to use parabolic coordinates because the incoming asymptotic plane wave can be described in parabolic coordinates, just like in Rutherford scattering [3]. However, radial coordinates are just as convenient for this problem.

Calculation of the capture cross section

We will outline how to calculate the first order cross section for the reaction $p^+ + e^- \to H$ due to a monochromatic perturbation[7] $V(t) = W\exp(i\omega t)$. For a judicious choice of the operator $W \equiv W(\mathbf{p}, \mathbf{x})$ this describes electron-proton recombination due to emission of a photon with energy $\hbar\omega$. We will discuss these operators in Chapter 18, but here we do not specify the operator W further. Our present focus is rather to develop the formalism for calculating the capture or recombination cross section for a general perturbation $W(\mathbf{p}, \mathbf{x})\exp(i\omega t)$. We should also mention that perturbations $V(t) = V(\mathbf{x}, t)$ due to interactions with additional nearby electrons or ions are much more efficient and therefore more important for electron capture than direct radiative recombination due to photon emission.

The wave function for the approach between a free electron and a proton in the effective single particle description for relative motion is given by the wave function $\langle \boldsymbol{x}|k\rangle_{MG}$ which was constructed by Mott and Gordon in 1928 (7.76). The normalization factor is irrelevant because it cancels in the cross section. For convenience, it was chosen in equation (7.76) such that the asymptotic incoming current density is

$$j_{in} = \frac{\hbar k}{\mu}, \tag{13.47}$$

where μ is the reduced mass of the two-particle system. This current density has units of $\mathrm{cm/s} = \mathrm{cm}^{-2}\mathrm{s}^{-1}/\mathrm{cm}^{-3}$ because it is actually a current density $dj_{in}/d^3\boldsymbol{k}$ per unit of volume in $\boldsymbol{k}$ space, which is a consequence of the use of an asymptotic plane wave state in its calculation. A current density per $\boldsymbol{k}$ space volume is the correct notion for the calculation of the electron-proton recombination cross section, because the S matrix element

$$S_{n,\ell,m;k} = \frac{2\pi}{i\hbar}\langle n,\ell,m|W|k\rangle_{MG}\delta(\omega_{nk} + \omega)$$

yields a transition probability density per $\boldsymbol{k}$ space volume

$$\mathcal{P}_{k\to n,\ell,m} = |S_{n,\ell,m;k}|^2$$

[7] See the discussion after equation (13.38) for an explanation why we can deal with monochromatic perturbations as abridged non-hermitian operators.

which comes in units of cm^3, again due to the use of an asymptotic plane wave state as incoming state.

$\mathcal{P}_{k\to n,\ell,m}$ contains the factor

$$\delta(0)=\lim_{\omega\to 0}\delta(\omega)=\lim_{\omega\to 0}\lim_{T\to\infty}\frac{1}{2\pi}\int_{-T/2}^{T/2}dt\,\exp(i\omega t)=\lim_{T\to\infty}\frac{T}{2\pi}.$$

We can use this to calculate a transition rate density per $\boldsymbol{k}$ space volume

$$\mathcal{W}_{k\to n,\ell,m}=\frac{1}{T}\left|S_{n,\ell,m;k}\right|^2=\frac{2\pi}{\hbar^2}\left|\langle n,\ell,m|W|k\rangle_{MG}\right|^2\delta(\omega_{nk}+\omega). \tag{13.48}$$

The transition rate is certainly proportional to the asymptotic current density j_{in}, and therefore we divide the transition rate density by this current density to get a measure for the probability of the absorption process $|k\rangle_{MG}\to|n,\ell,m\rangle$. This yields the absorption cross section

$$\sigma_{k\to n,\ell,m}=\frac{\mathcal{W}_{k\to n,\ell,m}}{j_{in}}=\frac{2\pi\mu}{\hbar^3 k}\left|\langle n,\ell,m|W|k\rangle_{MG}\right|^2\delta(\omega_{nk}+\omega) \tag{13.49}$$

with units of cm^2. The total absorption cross section due to the perturbation operator $W(\mathbf{p},\mathbf{x})\exp(i\omega t)$ is then

$$\sigma_k=\sum_{\ell=0}^{\infty}\sum_{m=-\ell}^{\ell}\sum_{n=\ell+1}^{\infty}\sigma_{k\to n,\ell,m}.$$

The capture cross section enters into the calculations of rate coefficients $(\sigma v)_{av}$, where the notation indicates averaging over the distribution of relative particle velocities in a plasma of ions and electrons. The rate coefficients go into the balance equations for electron and ion densities,

$$\frac{d\rho_e}{dt}=\frac{d\rho_p}{dt}=-(\sigma v)_{av}\rho_p\rho_e,$$

where in general additional terms due to collisional relaxation and ionization have to be included. Due to (13.47) the rate coefficients are directly related to the transition rates per $\boldsymbol{k}$ space volume calculated in the state (7.76),

$$v\sigma_k=j_{in}\sigma_k=\sum_{\ell=0}^{\infty}\sum_{m=-\ell}^{\ell}\sum_{n=\ell+1}^{\infty}\mathcal{W}_{k\to n,\ell,m}.$$

Calculations of radiative capture cross sections for electron-proton recombination into arbitrary hydrogen shells were performed in parabolic coordinates by

Oppenheimer[8] and by Bethe and Salpeter [3]. Calculations in polar coordinates had been performed by Wessel, Stückelberg and Morse, and Stobbe[9]. All these authors had noticed that the electron capture cross sections for ions from radiative recombination were much too small to explain the experimental values, and it was eventually recognized that collisional relaxation due to interactions with spectator particles dominated the observed recombination rates. Therefore modern calculations of electron-ion recombination rates focus on collisional relaxation, which means that the relevant perturbation operators V are not determined by photon emission but by Coulomb interactions in a plasma, and the spectator particles also have to be taken into account in the initial and final states. Electron-ion recombination rates are particularly important for plasma physics and astrophysics.

13.6 Transitions between continuous states: Scattering

For transitions between continuous states, e.g. $|\boldsymbol{k}\rangle \to |\boldsymbol{k}'\rangle$, the S matrix element

$$S_{\boldsymbol{k}',\boldsymbol{k}} = \langle \boldsymbol{k}'|U_D(\infty,-\infty)|\boldsymbol{k}\rangle$$

is a quantity with the dimension length^3, because both external states have dimension length$^{3/2}$. We know from Section 13.4 how to make sense of transition matrix elements with continuous final states, *viz.* as transition probability densities $d^3\boldsymbol{k}'\,|S_{\boldsymbol{k}',\boldsymbol{k}}|^2$ in the final state space. We also know from the discussion in Section 13.5 that a continuous initial state in the scattering matrix will yield a transition probability density in the space of initial states if $V(\omega)$ is a classical function. In that case

$$P_{\mathcal{K}\to\mathcal{K}'} = \int_{\mathcal{K}'} d^3\boldsymbol{k}' \int_{\mathcal{K}} d^3\boldsymbol{k}\,|S_{\boldsymbol{k}',\boldsymbol{k}}|^2$$

will tell us the probability for transitions between states in $\boldsymbol{k}$ space volumes $\mathcal{K}$ and $\mathcal{K}'$ due to the perturbation $V(t)$.

However, just like in Section 13.5, the most important applications of scattering matrix elements with continuous initial states concern the calculation of cross sections due to monochromatic perturbations. We know from Sections 13.4 and 13.5 that monochromatic perturbations call for normalization of $|S_{\boldsymbol{k}',\boldsymbol{k}}|^2$ by the reaction time T, and we have learned in Section 13.5 that continuous initial states under the influence of a monochromatic perturbation require normalization of the transition rate with the current density j_{in} of incident particles to calculate a cross section

[8] J.R. Oppenheimer, Z. Phys. 55, 725 (1929).

[9] W. Wessel, Annalen Phys. 397, 611 (1930); E.C.G. Stückelberg, P.M. Morse, Phys. Rev. 36, 16 (1930); M. Stobbe, Annalen Phys. 399, 661 (1930).

for the quantum mechanical reaction described by the S matrix element, see equation (13.49). Our previous experience with initial or final continuous states therefore motivates the definition of the *differential scattering cross section*

$$d\sigma_{k\to k'} = d^3k' \frac{|S_{k',k}|^2}{Tj_{in}}. \tag{13.50}$$

This has again the dimension length^2, because the incident current density j_{in} for plane waves has units of cm/s, see equation (13.47) and the following discussion.

The notion of a differential scattering cross section is sufficiently important to warrant rederivation of equation (13.50) in simple steps in the next paragraph.

Cross section for scattering off a periodic perturbation

We apply the transition probability between continuous states to calculate the scattering cross section for a monochromatic perturbation

$$V(t) = W\exp(-\mathrm{i}\omega t).$$

Our Hamiltonian is

$$H = \frac{\mathbf{p}^2}{2m} + W(\mathbf{x})\exp(-\mathrm{i}\omega t),$$

and our unperturbed states are plane waves $|\boldsymbol{k}\rangle$.

The first order result for the scattering matrix $S_{k',k} = \langle \boldsymbol{k}'|U_D(\infty, -\infty)|\boldsymbol{k}\rangle$ is

$$\begin{aligned} S_{k',k} &= -\frac{\mathrm{i}}{\hbar}\int_{-\infty}^{\infty} dt\, \exp\!\left[\mathrm{i}\left(\hbar\frac{k'^2-k^2}{2m} - \omega\right)t\right] \langle \boldsymbol{k}'|W|\boldsymbol{k}\rangle \\ &= -\frac{2\pi\mathrm{i}}{\hbar}\langle \boldsymbol{k}'|W|\boldsymbol{k}\rangle \delta\!\left(\hbar\frac{k'^2-k^2}{2m} - \omega\right) \\ &= -\mathrm{i}\mathcal{M}_{k',k}\delta\!\left(\hbar\frac{k'^2-k^2}{2m} - \omega\right). \end{aligned} \tag{13.51}$$

The factor $\mathcal{M}_{k',k}$ in the scattering matrix element is also denoted as a *scattering amplitude*.

The transition probability density

$$\mathcal{P}_{k\to k'} = \left|S_{k',k}\right|^2 = \left|\mathcal{M}_{k',k}\right|^2 \delta(0)\delta\!\left(\hbar\frac{k'^2-k^2}{2m} - \omega\right)$$

contains the factor

$$\delta(0) = \lim_{\omega\to 0}\delta(\omega) = \lim_{\omega\to 0}\lim_{T\to\infty}\frac{1}{2\pi}\int_{-T/2}^{T/2} dt\, \exp(\mathrm{i}\omega t) = \lim_{T\to\infty}\frac{T}{2\pi},$$

and we can calculate a transition rate density

$$\mathcal{W}_{k\to k'} = \frac{1}{T}\left|S_{k',k}\right|^2 = \frac{\left|\mathcal{M}_{k',k}\right|^2}{2\pi}\delta\left(\hbar\frac{k'^2-k^2}{2m}-\omega\right)$$
$$= \frac{2\pi}{\hbar^2}\left|\langle k'|W|k\rangle\right|^2\delta\left(\hbar\frac{k'^2-k^2}{2m}-\omega\right).$$

The corresponding differential transition rate into the final state volume d^3k' is

$$dw_{k\to k'} = d^3k'\,\mathcal{W}_{k\to k'} = d^3k'\,\frac{1}{T}\left|S_{k',k}\right|^2.$$

However, this still comes in units of cm^3/s instead of s^{-1}, due to the initial plane wave state. For initial continuous states, we do not apply a volume measure (here d^3k) in the space of initial states, but normalize by the current density of the incident particles. This yields a differential cross section for scattering of momentum eigenstates,

$$d\sigma_{k\to k'} = \frac{dw_{k\to k'}}{j_{in}} = d^3k'\,\frac{2\pi}{\hbar^2 j_{in}}\left|\langle k'|W|k\rangle\right|^2\delta\left(\hbar\frac{k'^2-k^2}{2m}-\omega\right). \tag{13.52}$$

The motivation for dividing out the current density j_{in} of incoming particles from the scattering rate is the trivial dependence of the scattering rate on this parameter: if we double the number of incoming particles per second or per cm^2, we will trivially double the number of scattering events per second. Therefore all the interesting physics is in the proportionality factor $d\sigma$ between j_{in} and dw. This proportionality factor has the dimension of an area, and in classical mechanics, integration of $d\sigma$ over d^3p' for scattering of classical particles off a hard sphere of radius r yields the cross section area of the sphere $\sigma = \int d\sigma = \pi r^2$. Therefore the name *differential scattering cross section* for $d\sigma$.

The current density $\boldsymbol{j} = (\hbar/2\mathrm{i}m)(\psi^+\nabla\psi - \nabla\psi^+\cdot\psi)$ for a plane wave, $\boldsymbol{j} = \hbar\boldsymbol{k}/(2\pi)^3 m$, is actually a current density per unit of volume in $\boldsymbol{k}$-space. This is the correct current density to be used in (13.52), because $dw_{k\to k'}$ is a transition rate per unit of volume in $\boldsymbol{k}$-space, and the ratio yields a *bona fide* differential cross section[10]. Expressed in terms of continuum plane wave matrix elements, the differential scattering cross section is

[10] Alternatively, we could have used box normalization for the incoming plane waves, $\langle \boldsymbol{x}|\boldsymbol{k}\rangle = \exp(\mathrm{i}\boldsymbol{k}\cdot\boldsymbol{x})/\sqrt{V}$ both in $dw_{k\to k'}$ and in $\boldsymbol{j}$ ($\Rightarrow \boldsymbol{j} = \hbar\boldsymbol{k}/(mV) = \boldsymbol{v}/V$), or we could have rescaled both $dw_{k\to k'}$ and $\boldsymbol{j}$ with the conversion factor $8\pi^3/V$ to make both quantities separately dimensionally correct, $[dw_{k\to k'}] = \mathrm{s}^{-1}$, $[j] = \mathrm{cm}^{-2}\mathrm{s}^{-1}$. All three methods yield the same result for the scattering cross section, of course.

$$d\sigma_{\boldsymbol{k}\to\boldsymbol{k}'} = d^3\boldsymbol{k}'\,\frac{(2\pi)^4 m}{\hbar^3 k}\left|\langle\boldsymbol{k}'|W|\boldsymbol{k}\rangle\right|^2\delta\!\left(\hbar\frac{k'^2-k^2}{2m}-\omega\right)$$
$$= d^3\boldsymbol{k}'\,(2\pi)^4\frac{2m^2}{\hbar^4 k}\left|\langle\boldsymbol{k}'|W|\boldsymbol{k}\rangle\right|^2\delta\!\left(k'^2-k^2-\frac{2m}{\hbar}\omega\right). \tag{13.53}$$

We can use the δ-function in (13.53) to integrate over k'. This leaves us with a differential cross section per unit of solid angle,

$$\frac{d\sigma}{d\Omega} = (2\pi)^4\frac{m^2}{\hbar^4}\sqrt{1+\frac{2m\omega}{\hbar k^2}}\left|\langle\boldsymbol{k}'|W|\boldsymbol{k}\rangle\right|^2\bigg|_{k'=\sqrt{k^2+(2m\omega/\hbar)}}. \tag{13.54}$$

The corresponding result for $\omega = 0$ (scattering off a static potential) can also be derived within the framework of the time-independent Schrödinger equation, see Chapter 11. For the comparison note that we can write the differential scattering cross section (13.54) as

$$\frac{d\sigma}{d\Omega} = \sqrt{1+\frac{2m\omega}{\hbar k^2}}\,|f(\Delta\boldsymbol{k})|^2\bigg|_{k'=\sqrt{k^2+(2m\omega/\hbar)}}, \tag{13.55}$$

with the scattering amplitude

$$f(\Delta\boldsymbol{k}) = -(2\pi)^2\frac{m}{\hbar^2}\langle\boldsymbol{k}'|W|\boldsymbol{k}\rangle = -2\pi\frac{m}{\hbar}\mathcal{M}_{\boldsymbol{k}',\boldsymbol{k}}, \tag{13.56}$$

cf. (11.23), i.e. equation (13.55) reduces to (11.27) for scattering off a static potential if $\omega = 0$. The potential scattering formalism could be extended to time-dependent perturbations by using the asymptotic expansion of the time-dependent retarded Green's function (11.46). However, the equivalent scattering matrix formalism is more convenient.

Scattering theory in second order

We will discuss scattering off the time-independent potential V in second order. The Hamiltonian is

$$H = \frac{\mathbf{p}^2}{2m} + V.$$

If $\boldsymbol{k}' \neq \boldsymbol{k}$ and $\langle \boldsymbol{k}'|V|\boldsymbol{k}\rangle = 0$, the leading order term for the S-matrix is the second order term

$$S_{k',k} = -\frac{1}{\hbar^2}\int d^3\boldsymbol{q}\int_{-\infty}^{\infty}d\tau\int_{-\infty}^{\tau}d\tau'\,\exp\!\left(\frac{\mathrm{i}\hbar}{2m}\left(k'^2-q^2\right)\tau\right)$$
$$\times\langle \boldsymbol{k}'|V|\boldsymbol{q}\rangle\exp\!\left(\frac{\mathrm{i}\hbar}{2m}\left(q^2-k^2\right)\tau'\right)\langle \boldsymbol{q}|V|\boldsymbol{k}\rangle.$$

To make the τ' integral convergent, we add a term $\epsilon\tau'$ in the exponent, so that the time integrals yield

$$\int_{-\infty}^{\infty}d\tau\int_{-\infty}^{\tau}d\tau'\,\exp\!\left(\frac{\mathrm{i}\hbar}{2m}\left(k'^2-q^2\right)\tau\right)\exp\!\left(\frac{\mathrm{i}\hbar}{2m}\left(q^2-k^2\right)\tau'+\epsilon\tau'\right)$$
$$=\frac{1}{\frac{\mathrm{i}\hbar}{2m}\left(q^2-k^2\right)+\epsilon}\int_{-\infty}^{\infty}d\tau\,\exp\!\left(\frac{\mathrm{i}\hbar}{2m}\left(k'^2-k^2\right)\tau\right)$$
$$=\frac{2\pi}{\frac{\mathrm{i}\hbar}{2m}\left(q^2-k^2\right)+\epsilon}\delta\!\left(\frac{\hbar}{2m}\left(k'^2-k^2\right)\right)$$

and

$$S_{k',k} = \frac{2\pi\mathrm{i}}{\hbar^2}\delta\big(\omega(k')-\omega(k)\big)\int d^3\boldsymbol{q}\,\frac{\langle \boldsymbol{k}'|V|\boldsymbol{q}\rangle\langle \boldsymbol{q}|V|\boldsymbol{k}\rangle}{\omega(q)-\omega(k)-\mathrm{i}\epsilon}$$
$$=2\pi\mathrm{i}\delta\big(E(k')-E(k)\big)\int d^3\boldsymbol{q}\,\frac{\langle \boldsymbol{k}'|V|\boldsymbol{q}\rangle\langle \boldsymbol{q}|V|\boldsymbol{k}\rangle}{E(q)-E(k)-\mathrm{i}\epsilon}.$$

The corresponding differential transition rate is

$$dw_{\boldsymbol{k}\to\boldsymbol{k}'} = d^3\boldsymbol{k}'\frac{1}{T}\left|S_{k',k}\right|^2.$$
$$=d^3\boldsymbol{k}'\frac{2\pi}{\hbar}\delta\big(E(k')-E(k)\big)\left|\int d^3\boldsymbol{q}\,\frac{\langle \boldsymbol{k}'|V|\boldsymbol{q}\rangle\langle \boldsymbol{q}|V|\boldsymbol{k}\rangle}{E(q)-E(k)-\mathrm{i}\epsilon}\right|^2,$$

and the differential cross section for scattering of momentum eigenstates in second order is

$$d\sigma_{\boldsymbol{k}\to\boldsymbol{k}'} = \frac{dw_{\boldsymbol{k}\to\boldsymbol{k}'}}{j_{in}}$$
$$=d^3\boldsymbol{k}'\frac{(2\pi)^4m}{\hbar^2k}\delta\big(E(k')-E(k)\big)\left|\int d^3\boldsymbol{q}\,\frac{\langle \boldsymbol{k}'|V|\boldsymbol{q}\rangle\langle \boldsymbol{q}|V|\boldsymbol{k}\rangle}{E(q)-E(k)-\mathrm{i}\epsilon}\right|^2$$
$$=d^3\boldsymbol{k}'\frac{(2\pi)^4m^2}{\hbar^4kk'}\delta\big(k'-k\big)\left|\int d^3\boldsymbol{q}\,\frac{\langle \boldsymbol{k}'|V|\boldsymbol{q}\rangle\langle \boldsymbol{q}|V|\boldsymbol{k}\rangle}{E(q)-E(k)-\mathrm{i}\epsilon}\right|^2. \quad (13.57)$$

Integration over k' yields the differential scattering cross section per unit of solid angle in second order,

$$\frac{d\sigma}{d\Omega} = (2\pi)^4 \frac{m^2}{\hbar^4} \left| \int d^3\boldsymbol{q}\, \frac{\langle \boldsymbol{k}'|V|\boldsymbol{q}\rangle\langle \boldsymbol{q}|V|\boldsymbol{k}\rangle}{E(q) - E(k) - \mathrm{i}\epsilon} \right|^2 \Bigg|_{k'=k}. \tag{13.58}$$

Equations (13.54) and (13.58) could be denoted as Fermi's Golden Rules #2 and #1 for scattering theory.

13.7 Expansion of the scattering matrix to higher orders

For time-independent perturbation V we can write the expansion of the scattering matrix in the form

$$\begin{aligned}
S_{fi}(t,t') &= \langle f|U_D(t,t')|i\rangle = \langle f|\mathrm{T}\exp\!\left(-\frac{\mathrm{i}}{\hbar}\int_{t'}^{t} d\tau\, H_D(\tau)\right)|i\rangle \\
&= \sum_{n=0}^{\infty} \frac{1}{(\mathrm{i}\hbar)^n} \int_{t'}^{t} d\tau_1 \int_{t'}^{\tau_1} d\tau_2 \ldots \int_{t'}^{\tau_{n-1}} d\tau_n\, \langle f|U_0(t_0,\tau_1)VU_0(\tau_1,\tau_2) \\
&\quad \times VU_0(\tau_2,\tau_3)\ldots U_0(\tau_{n-1},\tau_n)VU_0(\tau_n,t_0)|i\rangle \\
&= \sum_{n=0}^{\infty} \sum\!\!\!\!\!\!\int_{j_1,\ldots j_{n-1}} \frac{1}{(\mathrm{i}\hbar)^n} \int_{t'}^{t} d\tau_1 \int_{t'}^{\tau_1} d\tau_2 \ldots \int_{t'}^{\tau_{n-1}} d\tau_n\, \exp\!\left(-\frac{\mathrm{i}}{\hbar}E_f t_0\right) \\
&\quad \times \exp\!\left(\frac{\mathrm{i}}{\hbar}E_f\tau_1\right) V_{fj_1} \exp\!\left(-\frac{\mathrm{i}}{\hbar}E_{j_1}\tau_1\right) \exp\!\left(\frac{\mathrm{i}}{\hbar}E_{j_1}\tau_2\right) V_{j_1j_2} \\
&\quad \times \exp\!\left(-\frac{\mathrm{i}}{\hbar}E_{j_2}\tau_2\right) \exp\!\left(\frac{\mathrm{i}}{\hbar}E_{j_2}\tau_3\right) \ldots \exp\!\left(\frac{\mathrm{i}}{\hbar}E_{j_{m-1}}\tau_m\right) V_{j_{m-1}j_m} \\
&\quad \times \exp\!\left(-\frac{\mathrm{i}}{\hbar}E_{j_m}\tau_m\right) \ldots \exp\!\left(\frac{\mathrm{i}}{\hbar}E_{j_{n-2}}\tau_{n-1}\right) V_{j_{n-2}j_{n-1}} \\
&\quad \times \exp\!\left(-\frac{\mathrm{i}}{\hbar}E_{j_{n-1}}\tau_{n-1}\right) \exp\!\left(\frac{\mathrm{i}}{\hbar}E_{j_{n-1}}\tau_n\right) V_{j_{n-1}i} \exp\!\left(-\frac{\mathrm{i}}{\hbar}E_i\tau_n\right) \\
&\quad \times \exp\!\left(\frac{\mathrm{i}}{\hbar}E_i t_0\right).
\end{aligned}$$

Taking the limits $t' \to -\infty$ and $t \to \infty$, we find the equation

$$\begin{aligned}
S_{fi} &= \delta_{fi} - 2\pi\mathrm{i}\delta(E_f - E_i)V_{fi} - 2\pi\mathrm{i}\delta(E_f - E_i) \\
&\quad \times \sum_{n=2}^{\infty} \sum\!\!\!\!\!\!\int_{j_1,\ldots j_{n-1}} V_{fj_1}V_{j_1j_2}\ldots V_{j_{n-2}j_{n-1}}V_{j_{n-1}i}\big[(E_i - E_{j_1} + \mathrm{i}\epsilon) \\
&\quad \times (E_i - E_{j_2} + \mathrm{i}\epsilon)\ldots(E_i - E_{j_{n-2}} + \mathrm{i}\epsilon)(E_i - E_{j_{n-1}} + \mathrm{i}\epsilon)\big]^{-1}.
\end{aligned} \tag{13.59}$$

However, we can also use

$$-2\pi i\delta(E_f - E_i) = -\frac{i}{\hbar}\lim_{\epsilon\to 0, t\to\infty, \epsilon t\to 0}\int_{-\infty}^{t} d\tau\, \exp\!\left(\frac{i(E_f - E_i) + \epsilon}{\hbar}\tau\right)$$
$$= \lim_{\epsilon\to 0, t\to\infty, \epsilon t\to 0}\frac{1}{E_i - E_f + i\epsilon}\exp\!\left(\frac{i(E_f - E_i)t + \epsilon t}{\hbar}\right).$$

This yields a form which resembles expressions for the shifts of wave functions in time-independent perturbation theory,

$$S_{fi} = \lim_{\epsilon\to 0, t\to\infty, \epsilon t\to 0}\exp\!\left(\frac{i(E_f - E_i)t + \epsilon t}{\hbar}\right)\left(\delta_{fi} + \frac{V_{fi}}{E_i - E_f + i\epsilon}\right.$$
$$+ \sum_{n=2}^{\infty}\sum\!\!\!\!\!\!\int_{j_1,\dots j_{n-1}} V_{fj_1}V_{j_1j_2}\dots V_{j_{n-2}j_{n-1}}V_{j_{n-1}i}\big[(E_i - E_f + i\epsilon)$$
$$\left.\times (E_i - E_{j_1} + i\epsilon)\dots(E_i - E_{j_{n-2}} + i\epsilon)(E_i - E_{j_{n-1}} + i\epsilon)\big]^{-1}\right). \quad (13.60)$$

If the initial state is continuous, $|S_{fi}|^2$ will enter into the calculation of cross sections. If only the final state is continuous, $|S_{fi}|^2$ will enter into the calculation of decay rates. If both external states are discrete, the perturbation V should be treated as arising from a quantum field, see the remarks at the end of Section 13.3.

We can write the result (13.60) also as

$$S_{fi} = \lim_{\epsilon\to 0, t\to\infty, \epsilon t\to 0}\langle\psi_f|\exp\!\left(-i\frac{E_i - H_0 + i\epsilon}{\hbar}t\right)|\Psi_i^{(\epsilon)}\rangle \quad (13.61)$$

with the state

$$|\Psi_i^{(\epsilon)}\rangle = \sum_{n=0}^{\infty}\left(\frac{1}{E_i - H_0 + i\epsilon}V\right)^n|\psi_i\rangle. \quad (13.62)$$

This state satisfies the Lippmann-Schwinger equation (11.5)

$$|\Psi_i^{(\epsilon)}\rangle = |\psi_i\rangle + \frac{1}{E_i - H_0 + i\epsilon}V|\Psi_i^{(\epsilon)}\rangle \quad (13.63)$$

and therefore also

$$\lim_{\epsilon\to 0}(E_i - H + i\epsilon)|\Psi_i^{(\epsilon)}\rangle = 0.$$

Indeed, one of the objectives of the original work of Lippmann and Schwinger was to relate states of the form (13.62) to the scattering matrix, and it was thought that they relate to the Møller states (13.27). However, we now see that they instead appear as stated in equation (13.61).

We can also write the result (13.61) in even neater form

$$S_{fi} = \lim_{\epsilon\to 0, t\to\infty, \epsilon t\to 0} \langle \psi_f | \Psi_i^{(\epsilon)}(t)\rangle \tag{13.64}$$

with the state

$$|\Psi_i^{(\epsilon)}(t)\rangle = \sum_{n=0}^{\infty} \left(\frac{1}{E_i - H_0 + \mathrm{i}\epsilon} H_D(t) \right)^n |\psi_i\rangle. \tag{13.65}$$

This state satisfies the equation

$$|\Psi_i^{(\epsilon)}(t)\rangle = |\psi_i\rangle + \frac{1}{E_i - H_0 + \mathrm{i}\epsilon} H_D(t) |\Psi_i^{(\epsilon)}(t)\rangle \tag{13.66}$$

and therefore also

$$\lim_{\epsilon\to 0} (E_i - H_0 + \mathrm{i}\epsilon)|\Psi_i^{(\epsilon)}(t)\rangle = \lim_{\epsilon\to 0} H_D(t)|\Psi_i^{(\epsilon)}(t)\rangle. \tag{13.67}$$

13.8 Energy-time uncertainty

We are now finally in a position to address the origin of energy-time uncertainty in a more formal way. Energy conservation in each term of the scattering matrix (13.59) came from the final time integral over τ_1, which in symmetric form for the initial and final time limits can be written as

$$\begin{aligned} \lim_{\Delta t\to\infty} \frac{1}{\hbar} \int_{-\Delta t/2}^{\Delta t/2} dt\, \exp(\mathrm{i}\Delta E_{fi} t/\hbar) &= 2 \lim_{\Delta t\to\infty} \frac{\sin(\Delta E_{fi}\Delta t/2\hbar)}{\Delta E_{fi}} \\ &= 2\pi\delta(\Delta E_{fi}). \end{aligned} \tag{13.68}$$

However, this tells us that if we allocate only a finite time window Δt to observe the evolution of the system, or if the system is forced to make the transition within a time window Δt, then we will observe violations of energy conservation of order

$$|\Delta E_{fi}| \simeq \frac{4\hbar}{\Delta t}. \tag{13.69}$$

Here we used that the sinc function $\sin(x)/x$ is rather broad with half maximum near $x = \pm 2$.

How can that be? The theorem of energy conservation for time-independent Hamiltonian $H = H_0 + V$ in a static spacetime holds in quantum mechanics just as in any other physical field theory. We will see this in Section 16.2. However, by allocating a finite time window Δt for our measurement device to observe the system, or by constraining the system to make the transition within the fixed finite time window, we apparently introduce a time-dependent perturbation into the system that results in an energy uncertainty in excess of $\hbar/\Delta t$ in the final state.

13.9 Problems

13.1. The last lines in equations (13.1, 13.2) require the property

$$\mathrm{T}\frac{1}{n!}\left(\int_{t_0}^{t} d\tau\, H(\tau)\right)^n \equiv \frac{1}{n!}\mathrm{T}\int_{t_0}^{t} d\tau_1 \int_{t_0}^{t} d\tau_2 \ldots \int_{t_0}^{t} d\tau_n\, H(\tau_1)H(\tau_2)\ldots H(\tau_n)$$
$$= \int_{t_0}^{t} d\tau_1 \int_{t_0}^{\tau_1} d\tau_2 \ldots \int_{t_0}^{\tau_{n-1}} d\tau_n\, H(\tau_1)H(\tau_2)\ldots H(\tau_n). \tag{13.70}$$

Prove this property.

Hints: The equation for $n = 0$ and $n = 1$ is trivial, and for $n = 2$ it can easily be demonstrated from equation (13.3). This motivates a proof by induction with respect to n, which can easily be accomplished using the property

$$\mathrm{T}H(\tau_1)H(\tau_2)\ldots H(\tau_n)H(\tau_{n+1}) = \mathrm{T}[\mathrm{T}H(\tau_1)H(\tau_2)\ldots H(\tau_n)]H(\tau_{n+1}).$$

A more direct way to prove (13.70) is to express the ordering of operators through appropriate Θ functions under the assumption $t > t_0$ (forward evolution) or $t < t_0$ (backward evolution).

13.2. Use Fourier transformation to calculate the matrix elements $\langle x|U(t)|x'\rangle$ for the free time evolution operator in one dimension. Compare with the result (13.9) for the harmonic oscillator.

13.3. Calculate the annihilation and creation operators $a(t)$ and $a^+(t)$ of the harmonic oscillator in the Heisenberg picture.

Use the previous results to calculate the operators $\mathrm{x}(t)$ and $\mathrm{p}(t)$ for the harmonic oscillator in the Heisenberg picture.

13.4. Start from the definition

$$U_D(t,t') = U_0^+(t,t_0)U(t,t')U_0(t',t_0)$$

of the time evolution operator of states in the interaction picture to prove that

$$i\hbar\frac{\partial}{\partial t}U_D(t,t') = H_D(t)U_D(t,t'), \quad i\hbar\frac{\partial}{\partial t'}U_D(t,t') = -U_D(t,t')H_D(t').$$

13.5. Calculate the first order transition probability for the transition 1s→2p for a hydrogen atom which is perturbed by a potential

$$V(t) = P\frac{z}{2\tau}\exp(-|t|/\tau).$$

P and τ are constants. What is the meaning of P in the limit $\tau \to 0$?

13.6. The Golden Rule #2 for the first order transition rate (13.41) is often abused for the discussion of transitions between discrete states. In this problem you will be asked to figure out where the derivation of the Golden Rule #2 for transitions between discrete states breaks down.

13.6a. Calculate the first order transition probability for transitions between discrete energy eigenstates $|m\rangle \to |n\rangle$ under the influence of a monochromatic perturbation $V(t) = W\exp(-\mathrm{i}\omega t)$ which only acts between times t' and t. Which consistency requirements do you find from the condition that the first order result describes a transition probability $P_{m\to n}(t,t')$? Calculate also the transition rate $w_{m\to n}(t,t') = dP_{m\to n}(t,t')/dt$.

13.6b. Try to take the limit $t - t' \to \infty$ to derive the Golden Rule #2. Does this comply with the consistency requirements from 13.6a?

13.6c. Why do the inconsistencies of 13.6b not appear if the final state $|n\rangle$ is a continuous state?

Solution for Problem 13.6. For a periodic perturbation $V(t) = W\exp(-\mathrm{i}\omega t)$ the first order transition amplitude between times t' and t, and between different eigenstates of H_0 becomes

$$\begin{aligned}\langle n|U_D(t,t')|m\rangle &\approx -\frac{\mathrm{i}}{\hbar}\int_{t'}^{t} d\tau\, \exp[\mathrm{i}(\omega_{nm}-\omega)\tau]\langle n|W|m\rangle \\ &= \frac{\exp[\mathrm{i}(\omega_{nm}-\omega)t'] - \exp[\mathrm{i}(\omega_{nm}-\omega)t]}{\hbar(\omega_{nm}-\omega)}\langle n|W|m\rangle.\end{aligned}$$

The resulting transition probability is

$$\begin{aligned}P_{m\to n}(t-t') &= 2\frac{1-\cos[(\omega_{nm}-\omega)(t-t')]}{\hbar^2(\omega_{nm}-\omega)^2}|\langle n|W|m\rangle|^2 \\ &= \left(\frac{2\sin[(\omega_{nm}-\omega)(t-t')/2]}{\hbar(\omega_{nm}-\omega)}|\langle n|W|m\rangle|\right)^2,\end{aligned} \tag{13.71}$$

and the rate of change of the transition probability follows as

$$w_{m\to n}(\Delta t) = \frac{d}{dt}P_{m\to n}(\Delta t) = \frac{2}{\hbar^2}\frac{\sin[(\omega_{nm}-\omega)\Delta t]}{\omega_{nm}-\omega}|\langle n|W|m\rangle|^2. \tag{13.72}$$

Equations (13.71) and (13.72) yield perfectly well behaved, dimensionally correct expressions for the first order transition probability and transition rate between discrete states. Consistency with the probability interpretation for the extreme case $\omega_{nm} - \omega = 0$ requires

$$\Delta t = t - t' \leq \frac{\hbar}{|\langle n|W|m\rangle|}, \tag{13.73}$$

or alternatively, consistency of (13.71) with the probability interpretation for arbitrary Δt requires

$$|\omega_{nm} - \omega| \geq \frac{2}{\hbar}|\langle n|W|m\rangle|. \tag{13.74}$$

The problem arises with the limit $\Delta t \to \infty$, which would transform the transition rate from an ordinary function of frequencies into a δ function,

$$\begin{aligned} w_{m\to n} &= \lim_{\Delta t\to\infty} w_{m\to n}(\Delta t) = \frac{2\pi}{\hbar^2}|\langle n|W|m\rangle|^2\delta(\omega_{nm} - \omega) \\ &= \frac{2\pi}{\hbar}|\langle n|W|m\rangle|^2\delta(E_n - E_m - \hbar\omega). \end{aligned} \tag{13.75}$$

Here we used

$$\begin{aligned} \lim_{\Delta t\to\infty} \frac{\sin[(\omega_{nm} - \omega)\Delta t]}{\omega_{nm} - \omega} &= \lim_{\Delta t\to\infty} \frac{1}{2}\int_{-\Delta t}^{\Delta t} d\tau\, \exp[\mathrm{i}(\omega_{nm} - \omega)\tau] \\ &= \pi\delta(\omega_{nm} - \omega) = \pi\hbar\delta(E_n - E_m - \hbar\omega). \end{aligned}$$

Taking the limit $\Delta t \to \infty$ violates either the condition (13.73), or the condition (13.74) through its result $\omega_{nm} - \omega \to 0$ for a transition. From this point of view (and ignoring the fact that we should have at least one continuous external state when properly taking into account photons, see Section 18.6 and Problem 18.11), the resolution of the paradox of emergence of a δ function between discrete states in the limit $\Delta t \to \infty$ is that in the region of frequencies (13.74) where the first order result might be applicable, the first order result becomes subdominant for large Δt and (at the very least) higher order terms would have to be included to get estimates of transition probabilities and transition rates, or perturbation theory is just not suitable any more to get reliable estimates for those parameters.

These problems do not arise for continuous final states, because in these cases $P_{m\to n}(t, t') \to dP_{m\to n}(t, t') = dE_n\varrho(E_n)|S_{n,m}(t, t')|^2$ are not transition probabilities any more (which would be bounded by 1), but only transition probability densities for which only the integral over the energy scale with measure factor $\varrho(E_n)$ is bounded.

13.7. Calculate the representation of the ground state of hydrogen as a superposition of plane waves.

Solution. The $\boldsymbol{x}$ representation of the ground state of hydrogen is

$$\langle \boldsymbol{x}|1,0,0\rangle = \frac{1}{\sqrt{\pi a^3}}\exp(-r/a),$$

where a is the Bohr radius (7.62).

Fourier transformation to $\boldsymbol{k}$ space yields

$$\begin{aligned}\langle \boldsymbol{k}|1,0,0\rangle &= \frac{1}{\pi\sqrt{2a^3}}\int_0^\infty dr \int_{-1}^{1} d\xi\, r^2 \exp(-ikr\xi)\exp(-r/a)\\ &= \frac{1}{ik\pi\sqrt{2a^3}}\int_0^\infty dr\, r\exp(-r/a)[\exp(ikr)-\exp(-ikr)]\\ &= \frac{1}{ik\pi\sqrt{2a^3}}\left[\left.\frac{1}{\kappa^2}\right|_{\kappa=(1/a)-ik} - \left.\frac{1}{\kappa^2}\right|_{\kappa=(1/a)+ik}\right]\\ &= \frac{\sqrt{2a}^3}{\pi}\frac{1}{[1+(ka)^2]^2}.\end{aligned}$$

The representation in terms of plane waves is therefore

$$\langle \boldsymbol{x}|1,0,0\rangle = \sqrt{\frac{a}{\pi}}^3\int d^3\boldsymbol{k}\,\frac{\exp(i\boldsymbol{k}\cdot\boldsymbol{x})}{\pi[1+(ka)^2]^2},$$

i.e. the ground state is an isotropic superposition of plane waves which is dominated by small wave numbers $k \lesssim 1/a$ or large wavelengths $\lambda \gtrsim 2\pi a$. This problem was included in this chapter to drive home the point that calculation of transition rates into plane wave states does not necessarily tell us something about scattering or ionization in a system with bound states, unless the energy of the final plane wave state is large compared to the binding energies of the bound states.

13.8. Calculate the first order ionization rate for particles which are trapped in a one-dimensional δ-function potential (Section 3.3), if the particles are perturbed by a potential $V(t) = F_0 \mathrm{x}\exp(-i\omega t)$. What is the meaning of the constant F_0?

13.9. Calculate the first order capture cross section for free particles with wave function (3.18) which can become trapped in a one-dimensional δ-function potential, if the particles are perturbed by a potential $V(t) = F_0\mathrm{x}\exp(i\omega t)$. Recall that the normalization of initial states does not matter in the calculation of cross sections since it cancels in the ratio of capture rate to current density.

13.10. Calculate the cross section for recombination of an electron and a proton with energy $\hbar^2k^2/2\mu$ (in their relative motion) into the ground state of hydrogen. Perform the calculation both in parabolic and in polar coordinates.

13.11. Calculate the differential and total scattering cross sections for particles with initial momentum $\hbar\boldsymbol{k}$ which are scattered off the time-dependent potential

$$V(t) = \frac{A}{r}\Theta(R - r)\exp(-\mathrm{i}\omega t).$$

13.12. In 1984 Michael Berry published a paper studying (among other things) the following interesting question: Suppose $H(t)$ is a time-dependent Hamiltonian with the property that for each value of t there is a discrete spectrum $\mathcal{E}_n(t)$ such that

$$H(t)|\mathcal{E}_n(t)\rangle = \mathcal{E}_n(t)|\mathcal{E}_n(t)\rangle, \tag{13.76}$$

$$\langle\mathcal{E}_m(t)|\mathcal{E}_n(t)\rangle = \delta_{mn}, \quad \sum_n |\mathcal{E}_n(t)\rangle\langle\mathcal{E}_n(t)| = \underline{1}. \tag{13.77}$$

We can relate the eigenstates $|\mathcal{E}_n(t)\rangle$ of $H(t)$ to the states $|\psi_n(t)\rangle$ of the physical system described by $H(t)$ simply through the completeness relation (13.77),

$$|\psi_n(t)\rangle = \sum_m |\mathcal{E}_m(t)\rangle\langle\mathcal{E}_m(t)|\psi_n(t)\rangle. \tag{13.78}$$

However, assume that we start with an eigenstate $|\mathcal{E}_n(0)\rangle$ at time $t = 0$, i.e. we are seeking a solution of the initial value problem

$$\mathrm{i}\hbar\frac{d}{dt}|\psi_n(t)\rangle = H(t)|\psi_n(t)\rangle, \quad |\psi_n(0)\rangle = |\mathcal{E}_n(0)\rangle. \tag{13.79}$$

Can we directly relate $|\psi_n(t)\rangle$ to $|\mathcal{E}_n(t)\rangle$ without invoking a superposition (13.78) of all the eigenstates $|\mathcal{E}_m(t)\rangle$?

13.12a. Can you give an example of a time-dependent Hamiltonian satisfying the requirements (13.76, 13.77)?

13.12b. Since the states $|\psi_n(t)\rangle$ and $|\mathcal{E}_n(t)\rangle$ are both normalized they could only differ by a time-dependent phase if they are directly related,

$$|\psi_n(t)\rangle = \exp[-\mathrm{i}\Omega_n(t)]|\mathcal{E}_n(t)\rangle. \tag{13.80}$$

Which conditions would the phase $\Omega_n(t)$ have to fulfill for $|\psi_n(t)\rangle$ to satisfy the initial value problem (13.79)?

13.12c. What would be the solution for $\Omega_n(t)$ if a solution exists?
Which condition does $|\mathcal{E}_n(t)\rangle$ have to satisfy for existence of $\Omega_n(t)$?

Solution for 13.12b and 13.12c. Substitution of (13.80) into the time-dependent Schrödinger equation (13.79) and taking into account (13.76) yields

$$\mathrm{i}\hbar\frac{d}{dt}|\psi_n(t)\rangle - H(t)|\psi_n(t)\rangle = \exp[-\mathrm{i}\Omega(t)]$$
$$\times\left(\hbar\frac{d\Omega_n(t)}{dt}|\mathcal{E}_n(t)\rangle + \mathrm{i}\hbar\frac{d}{dt}|\mathcal{E}_n(t)\rangle - \mathcal{E}_n(t)|\mathcal{E}_n(t)\rangle\right),$$

i.e. we would need to satisfy the conditions

$$\hbar\frac{d\Omega_n(t)}{dt}|\mathcal{E}_n(t)\rangle + \mathrm{i}\hbar\frac{d}{dt}|\mathcal{E}_n(t)\rangle - \mathcal{E}_n(t)|\mathcal{E}_n(t)\rangle = 0 \tag{13.81}$$

and $\Omega(0) = 0$ to ensure that the *Ansatz* (13.80) satisfies the initial value problem (13.79). *If* the condition (13.81) is consistent, it is equivalent to

$$\hbar\frac{d\Omega_n(t)}{dt} = \mathcal{E}_n(t) - \mathrm{i}\hbar\langle\mathcal{E}_n(t)|\frac{d}{dt}|\mathcal{E}_n(t)\rangle \tag{13.82}$$

with solution

$$\Omega_n(t) = \int_0^t d\tau\left(\frac{\mathcal{E}_n(\tau)}{\hbar} - \mathrm{i}\langle\mathcal{E}_n(\tau)|\frac{d}{d\tau}|\mathcal{E}_n(\tau)\rangle\right). \tag{13.83}$$

However, the condition (13.81) will usually *not* be consistent and $\Omega_n(t)$ will not exist in many cases. The problem is that condition (13.81) requires that $d|\mathcal{E}_n(t)\rangle/dt \propto |\mathcal{E}_n(t)\rangle$,

$$\frac{d}{dt}|\mathcal{E}_n(t)\rangle = \mathrm{i}\left(\frac{d\Omega_n(t)}{dt} - \frac{\mathcal{E}_n(t)}{\hbar}\right)|\mathcal{E}_n(t)\rangle, \tag{13.84}$$

which would yield with (13.83)

$$\begin{aligned}|\mathcal{E}_n(t)\rangle &= \exp\left(\int_0^t d\tau\,\langle\mathcal{E}_n(\tau)|\frac{d}{d\tau}|\mathcal{E}_n(\tau)\rangle\right)|\mathcal{E}_n(0)\rangle\\ &= \exp[-\mathrm{i}\beta_n(t)]|\mathcal{E}_n(0)\rangle,\end{aligned} \tag{13.85}$$

with the *Berry phase*[11]

$$\beta_n(t) = \mathrm{i}\int_0^t d\tau\,\langle\mathcal{E}_n(\tau)|\frac{d}{d\tau}|\mathcal{E}_n(\tau)\rangle. \tag{13.86}$$

The constraints on the existence of physical states of the form (13.80) and on the usefulness of the Berry phase can most easily be seen from the fact that (13.85) is equivalent to

$$\frac{d}{dt}|\mathcal{E}_n(t)\rangle = |\mathcal{E}_n(t)\rangle\langle\mathcal{E}_n(t)|\frac{d}{dt}|\mathcal{E}_n(t)\rangle, \tag{13.87}$$

[11]M.V. Berry, Proc. Roy. Soc. London A 392, 45 (1984).

whereas generically time-dependence of the Hamiltonian mixes its eigenstates under time-evolution,

$$\frac{d}{dt}|\mathcal{E}_n(t)\rangle = \sum_m |\mathcal{E}_m(t)\rangle\langle\mathcal{E}_m(t)|\frac{d}{dt}|\mathcal{E}_n(t)\rangle. \tag{13.88}$$

Stated differently, the condition for existence of $\Omega_n(t)$ is

$$\langle\mathcal{E}_m(t)|\frac{d}{dt}|\mathcal{E}_n(t)\rangle = \delta_{m,n}\langle\mathcal{E}_n(t)|\frac{d}{dt}|\mathcal{E}_n(t)\rangle, \tag{13.89}$$

or the condition for (13.80, 13.83) as an *approximate* solution of the time-dependent Schrödinger equation is that for every $m \neq n$

$$\left|\langle\mathcal{E}_m(t)|\frac{d}{dt}|\mathcal{E}_n(t)\rangle\right| \ll \left|\langle\mathcal{E}_n(t)|\frac{d}{dt}|\mathcal{E}_n(t)\rangle\right|. \tag{13.90}$$

Note that if $\Omega_n(t)$ exists, the solution (13.80, 13.83) can also be written as

$$|\psi_n(t)\rangle = \exp\!\left(-\frac{\mathrm{i}}{\hbar}\int_0^t d\tau\,\mathcal{E}_n(\tau)\right)|\mathcal{E}_n(0)\rangle. \tag{13.91}$$

Of course, the Berry phase always exists in the sense of the definition (13.86), and in the same manner one might simply adopt (13.83) as a definition of $\Omega_n(t)$. The problem is whether or not they are related to the evolution of the states $|\psi_n(t)\rangle$ of the system described by the Hamiltonian $H(t)$. We have found (13.90) as a condition for the usefulness of the Berry phase. Comparison of (13.91) with the exact evolution formula

$$|\psi_n(t)\rangle = \mathrm{T}\exp\!\left(-\frac{\mathrm{i}}{\hbar}\int_0^t d\tau\,H(\tau)\right)|\mathcal{E}_n(0)\rangle \tag{13.92}$$

shows that the condition for usefulness of the Berry phase for the approximate description of the evolution of the system between times 0 and t can also be expressed in the form

$$H(\tau)|\mathcal{E}_n(0)\rangle \simeq \mathcal{E}_n(\tau)|\mathcal{E}_n(0)\rangle \tag{13.93}$$

for $0 \leq \tau \leq t$ or $t \leq \tau \leq 0$.

Chapter 14
Path Integrals in Quantum Mechanics

Path integrals provide in many instances an elegant complementary description of quantum mechanics and also for the quantization of fields, which we will study from a canonical point of view in Chapter 17 and following chapters. Path integrals are particularly popular in scattering theory, because the techniques of path integration were originally developed in the study of time evolution operators. Other areas where path integrals are used include statistical physics and the description of dissipative systems.

Path integration is based on a beautiful intuitive description of the quantum mechanical time evolution of particles or wave functions from initial to final states. The prize for the intuitive elegance in the description of time evolution is that the description of bound systems and the identification of the corresponding states is often cumbersome with path integral methods. On the other hand, path integration and canonical quantization complement each other particularly well in relativistic scattering theory, where canonical methods are needed for unitarity of the scattering matrix, for the normalization of the scattering states, and also for the correct choice of propagators in perturbation theory, while the path integral formulation provides an elegant tool for the development of rules for covariant perturbation theory.

Path integrals had been developed by Richard Feynman as a tool for understanding the role of the classical action in quantum mechanics, and had then evolved into a basis for covariant perturbation theory in relativistic field theories[1]. Our introductory exposition will focus on the use of path integrals in scattering theory. The first authoritative textbook on path integrals was co-authored by Feynman himself [10]. Extensive discussions and many applications of path integrals can be found in [13] and [23]. The use of path integrals in perturbative relativistic quantum field theory from a particle physics perspective is discussed e.g. in [18, 31, 41].

[1]R.P. Feynman, Ph.D. thesis, Princeton University 1942; Rev. Mod. Phys. 20, 367 (1948).

R. Dick, *Advanced Quantum Mechanics*, Graduate Texts in Physics,
DOI 10.1007/978-3-319-25675-7_14

14.1 Correlation and Green's functions for free particles

Before we enter the discussion of free particle motion and potential scattering in terms of path integrals, it is useful to discuss Green's functions for the Newton equation and canonical correlation functions for free particles.

The equation of motion of a classical non-relativistic particle under the influence of a force $\boldsymbol{F}(t)$ is directly integrable,

$$\boldsymbol{x}(t) = \boldsymbol{x}_i + \boldsymbol{v}_i(t - t_i) + \frac{1}{m}\int_{t_i}^{t} dt' \int_{t_i}^{t'} dt''\, \boldsymbol{F}(t''). \tag{14.1}$$

Partial integration of the acceleration term yields a Green's function representation

$$\begin{aligned}\boldsymbol{x}(t) &= \boldsymbol{x}_i + \boldsymbol{v}_i(t - t_i) + \frac{1}{m}\int_{t_i}^{t} dt'\,(t - t')\boldsymbol{F}(t') \\ &= \boldsymbol{x}_i + \boldsymbol{v}_i(t - t_i) + \frac{1}{m}\int_{-\infty}^{\infty} dt'\, G_i(t,t')\boldsymbol{F}(t'),\end{aligned} \tag{14.2}$$

with a Green's function which satisfies homogeneous initial conditions,

$$\begin{aligned}G_i(t,t') &= (t - t')\left[\Theta(t - t') - \Theta(t_i - t')\right] \\ &= (t - t')\left[\Theta(t' - t_i) - \Theta(t' - t)\right],\end{aligned} \tag{14.3}$$

$$\frac{\partial^2}{\partial t^2}G_i(t,t') = \delta(t - t'), \quad \frac{\partial}{\partial t}G_i(t,t')\Big|_{t=t_i} = 0, \quad G_i(t_i,t') = 0.$$

If we determine the velocity $\boldsymbol{v}_i$ such that $\boldsymbol{x}(t_f) = \boldsymbol{x}_f$, we find another Green's function representation

$$\begin{aligned}\boldsymbol{x}(t) &= \boldsymbol{x}_i\frac{t - t_f}{t_i - t_f} + \boldsymbol{x}_f\frac{t - t_i}{t_f - t_i} + \int_{t_f}^{t} dt'\,\frac{tt_f + t't_i}{t_f - t_i}\frac{\boldsymbol{F}(t')}{m} \\ &\quad + \int_{t_i}^{t} dt'\,\frac{tt_i + t't_f}{t_i - t_f}\frac{\boldsymbol{F}(t')}{m} + \int_{t_i}^{t_f} dt'\,\frac{tt' + t_it_f}{t_f - t_i}\frac{\boldsymbol{F}(t')}{m} \\ &= \boldsymbol{x}_i\frac{t - t_f}{t_i - t_f} + \boldsymbol{x}_f\frac{t - t_i}{t_f - t_i} + \frac{1}{m}\int_{-\infty}^{\infty} dt'\, G_{fi}(t,t')\boldsymbol{F}(t'),\end{aligned} \tag{14.4}$$

with a Green's function which satisfies homogeneous boundary conditions,

$$\begin{aligned}G_{fi}(t,t') &= \frac{tt_f + t't_i}{t_f - t_i}\left[\Theta(t - t') - \Theta(t_f - t')\right] \\ &\quad + \frac{tt_i + t't_f}{t_i - t_f}\left[\Theta(t - t') - \Theta(t_i - t')\right]\end{aligned}$$

$$
\begin{aligned}
&+\frac{tt'+t_it_f}{t_f-t_i}\left[\Theta(t_f-t')-\Theta(t_i-t')\right]\\
&=\Theta(t_f-t')\frac{(t-t_i)(t'-t_f)}{t_f-t_i}+\Theta(t_i-t')\frac{(t-t_f)(t_i-t')}{t_f-t_i}\\
&+(t-t')\Theta(t-t'),
\end{aligned}
\tag{14.5}
$$

$$
\frac{\partial^2}{\partial t^2}G_{fi}(t,t')=\delta(t-t'),\quad G_{fi}(t_f,t')=0,\quad G_{fi}(t_i,t')=0.
$$

The most general form of the Green's function for the Newton equation is

$$
G(t,t')=\frac{|t-t'|}{2}+\alpha(t')t+\beta(t'). \tag{14.6}
$$

A particular of these Green's functions appears also in canonical quantum mechanics in the time ordered two-point correlation function of the Heisenberg position operator

$$
\mathbf{x}(t)=\exp\!\left(\frac{it}{2m\hbar}\mathbf{p}^2\right)\mathbf{x}\exp\!\left(-\frac{it}{2m\hbar}\mathbf{p}^2\right)=\mathbf{x}+\frac{t}{m}\mathbf{p}.
$$

In general we can define N-point correlation functions without or with time-ordering,

$$
\boldsymbol{g}_{fi}^{(N)}(t_N,t_{N-1},\ldots,t_1)=\langle\boldsymbol{x}_f,t_f|\mathbf{x}(t_N)\otimes\mathbf{x}(t_{N-1})\otimes\ldots\otimes\mathbf{x}(t_1)|\boldsymbol{x}_i,t_i\rangle, \tag{14.7}
$$

$$
\boldsymbol{G}_{fi}^{(N)}(t_N,t_{N-1},\ldots,t_1)=\langle\boldsymbol{x}_f,t_f|\mathrm{T}_+\mathbf{x}(t_1)\otimes\mathbf{x}(t_2)\otimes\ldots\otimes\mathbf{x}(t_N)|\boldsymbol{x}_i,t_i\rangle. \tag{14.8}
$$

Here the time ordering operator T_+ arranges the Heisenberg operators from right to left by increasing time, but does not affect the times t_i and t_f, e.g.

$$
\boldsymbol{G}_{fi}^{(2)}(t,t')=\Theta(t-t')\boldsymbol{g}_{fi}^{(2)}(t,t')+\Theta(t'-t)\boldsymbol{g}_{fi}^{(2)}(t',t).
$$

The states $|\boldsymbol{x},t\rangle$ are the position eigenkets in the Heisenberg picture,

$$
|\boldsymbol{x},t\rangle=\exp\!\left(\frac{it}{2m\hbar}\mathbf{p}^2\right)|\boldsymbol{x}\rangle,\quad \mathbf{x}(t)|\boldsymbol{x},t\rangle=\boldsymbol{x}|\boldsymbol{x},t\rangle,
$$

and their products coincide with the position representation of the free non-relativistic particle propagator which we have encountered on several occasions before, see e.g. (3.33) and (4.45),

$$g_{fi}^{(0)} = \langle \boldsymbol{x}_f, t_f | \boldsymbol{x}_i, t_i \rangle = \langle \boldsymbol{x}_f | \exp\!\Big(-\mathrm{i}\frac{t_f - t_i}{2m\hbar}\mathbf{p}^2\Big) |\boldsymbol{x}_i\rangle = \langle \boldsymbol{x}_f | U_0(t_f, t_i | \boldsymbol{x}_i \rangle$$
$$= \sqrt{\frac{m}{2\pi\mathrm{i}\hbar(t_f - t_i)}}^{\,-3} \exp\left[-\frac{m\,(\boldsymbol{x}_f - \boldsymbol{x}_i)^2}{2\mathrm{i}\hbar(t_f - t_i)}\right]. \tag{14.9}$$

A small imaginary shift $t_f - t_i \to t_f - t_i - \mathrm{i}\epsilon$ is implied for convergence properties of Gaussian integrals which appear in the evaluation of $\langle \boldsymbol{x}_f, t_f | \boldsymbol{x}_i, t_i \rangle$.

Note that $\ddot{\mathbf{x}}(t) = -[H_0, [H_0, \mathbf{x}(t)]]/\hbar^2 = 0$ and therefore the second order time derivatives of $g_{fi}^{(N)}(t_N, t_{N-1}, \ldots, t_1)$ with respect to the time arguments t_I of the Heisenberg operators vanish. However, this implies that the time ordered two-point function contains a Green's function of the Newton equation on the diagonal (no summation over the index pair aa),

$$\frac{\partial^2}{\partial t^2} G_{fi,aa}^{(2)}(t, t') = \delta(t - t')\frac{1}{m}\langle \boldsymbol{x}_f, t_f | [p_a, x_a(t')] | \boldsymbol{x}_i, t_i \rangle$$
$$= \delta(t - t')\frac{\hbar}{\mathrm{i}m}\langle \boldsymbol{x}_f, t_f | \boldsymbol{x}_i, t_i \rangle. \tag{14.10}$$

The functions $g_{fi}^{(N)}(t_N, t_{N-1}, \ldots, t_1)$ will generically not be symmetric in their time arguments. They are easily evaluated by observing that the relation

$$\exp\!\Big(-\frac{\mathrm{i}t_f}{2m\hbar}\mathbf{p}^2\Big)\,\mathbf{x}(t)\exp\!\Big(\frac{\mathrm{i}t_f}{2m\hbar}\mathbf{p}^2\Big) = \mathbf{x} + \frac{t - t_f}{m}\mathbf{p} = \mathbf{x}(t - t_f)$$

implies the recursion relation

$$g_{fi}^{(N)}(t_N, t_{N-1}, \ldots, t_1) = \Big(\boldsymbol{x}_f - \frac{\mathrm{i}\hbar}{m}(t_N - t_f)\frac{\partial}{\partial \boldsymbol{x}_f}\Big)\, g_{fi}^{(N-1)}(t_{N-1}, \ldots, t_1).$$

The one-point function is in particular

$$g_{fi}^{(1)}(t) \equiv \boldsymbol{G}_{fi}^{(1)}(t) = \langle \boldsymbol{x}_f, t_f | \mathbf{x}(t) | \boldsymbol{x}_i, t_i \rangle$$
$$= \langle \boldsymbol{x}_f | \Big(\mathbf{x} + \frac{t - t_f}{m}\mathbf{p}\Big) \exp\!\Big(-\mathrm{i}\frac{t_f - t_i}{2m\hbar}\mathbf{p}^2\Big) |\boldsymbol{x}_i\rangle$$
$$= \Big(\boldsymbol{x}_f - \frac{\mathrm{i}\hbar}{m}(t - t_f)\frac{\partial}{\partial \boldsymbol{x}_f}\Big) \langle \boldsymbol{x}_f | U_0(t_f, t_i) | \boldsymbol{x}_i \rangle$$
$$= \Big(\boldsymbol{x}_f + (\boldsymbol{x}_f - \boldsymbol{x}_i)\,\frac{t - t_f}{t_f - t_i}\Big) \langle \boldsymbol{x}_f | U_0(t_f, t_i) | \boldsymbol{x}_i \rangle, \tag{14.11}$$

i.e. the ratio between the one-point function and the zero-point function is the free classical path which passes through $\boldsymbol{x}_i$ and $\boldsymbol{x}_f$,

$$\frac{g_{fi}^{(1)}(t)}{g_{fi}^{(0)}} = \boldsymbol{x}_f \frac{t - t_i}{t_f - t_i} + \boldsymbol{x}_i \frac{t_f - t}{t_f - t_i}.$$

The canonical two-point function is

$$\begin{aligned} g^{(2)}_{fi}(t,t') &= \langle x_f, t_f|\mathbf{x}(t)\otimes\mathbf{x}(t')|x_i, t_i\rangle \\ &= \left(x_f - \frac{i\hbar}{m}(t-t_f)\frac{\partial}{\partial x_f}\right)\otimes\left(x_f - \frac{i\hbar}{m}(t'-t_f)\frac{\partial}{\partial x_f}\right)\langle x_f|U_0(t_f,t_i)|x_i\rangle \\ &= \left(x_f\frac{t-t_i}{t_f-t_i} + x_i\frac{t_f-t}{t_f-t_i}\right)\otimes\left(x_f\frac{t'-t_i}{t_f-t_i} + x_i\frac{t_f-t'}{t_f-t_i}\right)\langle x_f|U_0(t_f,t_i)|x_i\rangle \\ &+ \frac{i\hbar}{m}\frac{(t_f-t)(t'-t_i)}{t_f-t_i}\underline{1}\langle x_f|U_0(t_f,t_i)|x_i\rangle. \end{aligned} \tag{14.12}$$

The relation (14.10) for the diagonal entries of the time-ordered two-point function is easily confirmed.

The primary use of N-point functions in quantum mechanics concerns the perturbative evaluation of scattering amplitudes in analytic scattering potentials. We will see this in Section 14.3.

14.2 Time evolution in the path integral formulation

The standard formulation of path integrals derives from the time evolution of states in the x representation,

$$\langle x|\psi(t)\rangle = \langle x|U(t,t_0)|\psi(t_0)\rangle,$$

We can also write this as

$$\langle x|\psi(t)\rangle = \langle x,t,t_0|\psi(t_0)\rangle$$

if we define the time-dependent states

$$|x,t,t_0\rangle = U^+(t,t_0)|x\rangle = \mathrm{T}\exp\!\left(-\frac{i}{\hbar}\int_t^{t_0}d\tau\, H(\tau)\right)|x\rangle. \tag{14.13}$$

Recall the definition of the time ordering operator T which was given following equation (13.2).

The parameter t_0 is usually suppressed in the notation of states, $|x,t,t_0\rangle \equiv |x,t\rangle$, $|\psi(t_0)\rangle \equiv |\psi\rangle$. The time-dependent basis states (14.13) are just the eigenstates of the Heisenberg picture operator

$$\mathbf{x}(t) = U^+(t,t_0)\mathbf{x}U(t,t_0), \quad \mathbf{x}(t)|x,t\rangle = x|x,t\rangle, \tag{14.14}$$

and the time parameter t_0 is the time parameter where the Schrödinger picture and the Heisenberg picture coincide.

The Heisenberg picture eigenstates satisfy the completeness relation

$$\int d^3x\,|x,t\rangle\langle x,t| = 1 \tag{14.15}$$

as a consequence of the completeness relation of the **x** eigenstates $|x\rangle$ and the unitarity of the time evolution operators. Furthermore, the composition property (13.7) of time evolution operators implies that the products of the Heisenberg picture eigenstates yield the x representation of the time evolution operator,

$$\langle x,t|x',t'\rangle = \langle x|U(t,t')|x'\rangle. \tag{14.16}$$

The properties (14.15) and (14.16) imply the following representation of the time evolution of a state,

$$\begin{aligned}\langle x,t|\psi\rangle &= \langle x,t|\left(\prod_{n=1}^{N}\int d^3x_n\,|x_n,t_n\rangle\langle x_n,t_n|\right)|\psi\rangle\\ &= \int d^3x_N\ldots\int d^3x_1\,\langle x|U(t,t_N)|x_N\rangle\langle x_N|U(t_N,t_{N-1})|x_{N-1}\rangle\ldots\\ &\quad\times\langle x_2|U(t_2,t_1)|x_1\rangle\langle x_1|U(t_1,t_0)|\psi\rangle.\end{aligned} \tag{14.17}$$

Equivalently, we could also have arrived at this equation directly from the composition property (13.7) of the time evolution operator and the completeness of the Schrödinger picture eigenkets $|x\rangle$.

Equation (14.17) implies in particular for the initial state $|\psi(t_0)\rangle = |x_0\rangle \equiv |x_0,t_0\rangle$ the evolution equation

$$\begin{aligned}\langle x,t|x_0\rangle &= \langle x|U(t,t_0)|x_0\rangle\\ &= \int d^3x_N\ldots\int d^3x_1\,\langle x|U(t,t_N)|x_N\rangle\langle x_N|U(t_N,t_{N-1})|x_{N-1}\rangle\ldots\\ &\quad\times\langle x_2|U(t_2,t_1)|x_1\rangle\langle x_1|U(t_1,t_0)|x_0\rangle.\end{aligned} \tag{14.18}$$

Intuitively the formula (14.18) can be considered as an integration over the set of all paths that a particle can take from an initial location x_0 at time t_0 to the location x at time t. In particular, if we use

$$\begin{aligned}\langle x|U(t,t_0)|x_0\rangle &= \langle x|\exp\!\left[-\mathrm{i}\frac{t-t_0}{\hbar}\left(\frac{\mathbf{p}^2}{2m}+V(\mathbf{x})\right)\right]|x_0\rangle\\ &= \langle x|\lim_{N\to\infty}\left[1-\mathrm{i}\frac{t-t_0}{N\hbar}\left(\frac{\mathbf{p}^2}{2m}+V(\mathbf{x})\right)\right]^N|x_0\rangle\end{aligned} \tag{14.19}$$

and substitute the following peculiar decomposition of unity,

$$\begin{aligned} 1 &= \int d^3x \int d^3p\, |x\rangle\langle x|p\rangle\langle p| \\ &= \int d^3x \int \frac{d^3p}{\sqrt{2\pi\hbar}^3}\, |x\rangle \exp\!\left(\frac{\mathrm{i}}{\hbar} p\cdot x\right) \langle p| \end{aligned} \tag{14.20}$$

between any two factors in the product (14.19), we find

$$\begin{aligned} \langle x, t|x_0\rangle &= \langle x|U(t,t_0)|x_0\rangle \\ &= \lim_{N\to\infty} \left(\prod_{I=1}^{N} \int \frac{d^3x_I d^3p_I}{\sqrt{2\pi\hbar}^3}\right) \exp\!\left(\frac{\mathrm{i}}{\hbar}\sum_{J=1}^{N} p_J\cdot x_J\right) \langle x|x_N\rangle\langle p_N| \\ &\quad\times \left[1 - \mathrm{i}\frac{t-t_0}{N\hbar}\left(\frac{\mathbf{p}^2}{2m} + V(\mathbf{x})\right)\right] |x_{N-1}\rangle\langle p_{N-1}| \ldots \\ &\quad\times |x_2\rangle\langle p_2| \left[1 - \mathrm{i}\frac{t-t_0}{N\hbar}\left(\frac{\mathbf{p}^2}{2m} + V(\mathbf{x})\right)\right] |x_1\rangle\langle p_1| \\ &\quad\times \left[1 - \mathrm{i}\frac{t-t_0}{N\hbar}\left(\frac{\mathbf{p}^2}{2m} + V(\mathbf{x})\right)\right] |x_0\rangle. \end{aligned} \tag{14.21}$$

The momentum integrals are

$$\begin{aligned} &\int \frac{d^3p_I}{\sqrt{2\pi\hbar}^3} \exp\!\left(\frac{\mathrm{i}}{\hbar} p_I\cdot x_I\right) \langle p_N| \left[1 - \mathrm{i}\frac{t-t_0}{N\hbar}\left(\frac{\mathbf{p}^2}{2m} + V(\mathbf{x})\right)\right] |x_{I-1}\rangle \\ &= \int \frac{d^3p_I}{(2\pi\hbar)^3} \left[1 - \mathrm{i}\frac{t-t_0}{N\hbar}\left(V(x_{I-1}) - \frac{\hbar^2}{2m}\frac{\partial^2}{\partial x_{I-1}^2}\right)\right] \\ &\quad\times \exp\!\left(\frac{\mathrm{i}}{\hbar} p_I\cdot (x_I - x_{I-1})\right) \\ &= \left[1 - \mathrm{i}\frac{t-t_0}{N\hbar}\left(V(x_{I-1}) - \frac{\hbar^2}{2m}\frac{\partial^2}{\partial x_{I-1}^2}\right)\right] \delta\left(x_I - x_{I-1}\right), \end{aligned} \tag{14.22}$$

and this exactly returns equation (14.19) if we would have substituted N copies of

$$1 = \int d^3x\, |x\rangle\langle x|$$

instead of (14.20). This is exactly as it should be. However, if we substitute instead

$$\left[1 - \mathrm{i}\frac{t-t_0}{N\hbar}\left(\frac{p_I^2}{2m} + V(x_{I-1})\right)\right] \simeq \exp\!\left[-\mathrm{i}\frac{t-t_0}{N\hbar}\left(\frac{p_I^2}{2m} + V(x_{I-1})\right)\right]$$

in (14.22), we find that the momentum integrals are

$$\int\frac{d^3p_I}{(2\pi\hbar)^3}\exp\!\left[-\mathrm{i}\frac{t-t_0-\mathrm{i}\epsilon}{2m\hbar N}\left(p_I-mN\frac{x_I-x_{I-1}}{t-t_0-\mathrm{i}\epsilon}\right)^2+\mathrm{i}N\frac{m}{2\hbar}\frac{(x_I-x_{I-1})^2}{t-t_0}\right]$$
$$=\sqrt{\frac{mN}{2\pi\mathrm{i}\hbar(t-t_0-\mathrm{i}\epsilon)}}^{\,-3}\exp\!\left[\frac{\mathrm{i}}{\hbar}\frac{m}{2}\left(N\frac{x_I-x_{I-1}}{t-t_0}\right)^2\frac{t-t_0}{N}\right].$$

This motivates the following formula for the matrix elements of the time evolution operator,

$$\begin{aligned}\langle x,t|x_0\rangle &= \langle x|U(t,t_0)|x_0\rangle\\ &\simeq \lim_{N\to\infty}\exp\!\left(\frac{\mathrm{i}}{\hbar}\sum_{J=1}^{N}\left[\frac{m}{2}\left(N\frac{x_J-x_{J-1}}{t-t_0}\right)^2-V(x_{J-1})\right]\frac{t-t_0}{N}\right)\\ &\quad\times\sqrt{\frac{mN}{2\pi\mathrm{i}\hbar(t-t_0)}}^{\,-3N}\left(\prod_{I=1}^{N}\int d^3x_I\right)\delta(x-x_N).\end{aligned}\tag{14.23}$$

The exponent is a discretized version of the action integral of a non-relativistic particle, and this motivates the further short hand notation

$$\begin{aligned}\langle x|U(t,t_0)|x_0\rangle &= \int_{x(t_0)=x_0}^{x(t)=x}D^3x(t')\,\exp\!\left[\frac{\mathrm{i}}{\hbar}\int_{t_0}^{t}dt'\,\left(\frac{m}{2}\dot{x}^2(t')-V(x(t'))\right)\right]\\ &= \int_{x(t_0)=x_0}^{x(t)=x}D^3x(t')\,\exp\!\left(\frac{\mathrm{i}}{\hbar}S[x(t')]\right),\end{aligned}\tag{14.24}$$

where $S[x(t')]$ is the action functional of the particle (see Appendix A). Please note that this standard notation for path integrals is misleading with regard to the length dimension or units of the path integral. The x matrix elements of the time evolution operator have dimension length^{-3} in agreement with the dimension length$^{-3/2}$ of $\mathbf{x}$ eigenstates in three dimensions, see Section 5.3. This of course agrees with the discretized version on the right hand side of equation (14.23). The three-dimensional path integral therefore has dimension length^{-3}, but the notation $\int D^3x\exp(\mathrm{i}S[x]/\hbar)$ suggests length dimension length^3. A dimensionally correct, but also more awkward notation would be

$$\begin{aligned}\langle x|U(t,t_0)|x_0\rangle &= \delta\big(x(t_0)-x_0\big)\delta\big(x(t)-x\big)\\ &\quad\times\int D^3x(t')\,\exp\!\left(\frac{\mathrm{i}}{\hbar}S[x(t')]\right),\end{aligned}\tag{14.25}$$

where the end point integration of the path would implement the boundary point constraints. We will continue to use the standard notation (14.24), but keep the fact in mind that this notation is not dimensionally correct.

Equation (14.24) defines the path integral representation of the propagator in configuration ($\boldsymbol{x}$) space. Note that nothing in the derivation required forward evolution $t > t_0$ in time. Of course, the same results apply for backward evolution. However, the discretization into time steps $(t - t_0)/N$ imply that consecutive steps are either always later or always earlier depending on $t > t_0$ or $t < t_0$, respectively. Therefore path integrals with factors like $\boldsymbol{x}(t_1)\boldsymbol{x}(t_2)$ in the integrand correspond to time ordered matrix elements in canonical quantization, but whether time ordering refers to later times or earlier times depends on whether we are studying forward or backward evolution in time. Usually we are interested in forward evolution, i.e. we assume $t > t_0$ in the following.

A virtue of the path integral is that it explains the principle of stationary action of classical paths as a consequence of dominant contributions from those trajectories where small fluctuations of the path do not yield cancellation of the integral from phase fluctuations.

As a relatively simple exercise, let us see how this reproduces the $\boldsymbol{x}$ representation (4.45) of the free propagator.

The integrations in (14.23) for $V(\boldsymbol{x}) = 0$ include a set of $N-1$ Gaussian integrals. The first integral over $d^3\boldsymbol{x}_1$ yields

$$\sqrt{\frac{mN}{2\pi i\hbar(t-t_0-i\epsilon)}}^{\,3}\exp\!\left[-\frac{mN}{2i\hbar(t-t_0-i\epsilon)}\frac{1}{2}(\boldsymbol{x}_2-\boldsymbol{x}_0)^2\right]$$
$$\times\int d^3\boldsymbol{x}_1\,\exp\!\left[-\frac{mN}{2i\hbar(t-t_0-i\epsilon)}2\left(\boldsymbol{x}_1-\frac{\boldsymbol{x}_2+\boldsymbol{x}_0}{2}\right)^2\right]$$
$$=\frac{1}{\sqrt{2}^{\,3}}\exp\!\left[-\frac{mN}{2i\hbar(t-t_0-i\epsilon)}\frac{1}{2}(\boldsymbol{x}_2-\boldsymbol{x}_0)^2\right].$$

Next we evaluate the $\boldsymbol{x}_2$ integral and then work consecutively through all the integrals. This reproduces always a similar result with minor variations. One can show by induction with respect to I that the $\boldsymbol{x}_I$ integral yields

$$\sqrt{\frac{mN}{2\pi i\hbar(t-t_0-i\epsilon)I}}^{\,3}\exp\!\left[-\frac{mN}{2i\hbar(t-t_0-i\epsilon)}\frac{1}{I+1}(\boldsymbol{x}_{I+1}-\boldsymbol{x}_0)^2\right]$$
$$\times\int d^3\boldsymbol{x}_I\,\exp\!\left[-\frac{mN}{2i\hbar(t-t_0-i\epsilon)}\frac{I+1}{I}\left(\boldsymbol{x}_I-\frac{I}{I+1}\left(\boldsymbol{x}_{I+1}+\frac{\boldsymbol{x}_0}{I}\right)\right)^2\right]$$
$$=\frac{1}{\sqrt{I+1}^{\,3}}\exp\!\left[-\frac{mN}{2i\hbar(t-t_0-i\epsilon)}\frac{1}{I+1}(\boldsymbol{x}_{I+1}-\boldsymbol{x}_0)^2\right].$$

After the final integrations over $\boldsymbol{x}_{N-1}$ and $\boldsymbol{x}_N$ (which is trivial due to the δ function in (14.23)), we are left with

$$\begin{aligned}\langle \boldsymbol{x}, t|\boldsymbol{x}_0\rangle &= \langle \boldsymbol{x}|U(t,t_0)|\boldsymbol{x}_0\rangle \\ &= \sqrt{\frac{m}{2\pi \mathrm{i}\hbar(t-t_0-\mathrm{i}\epsilon)}}^{\,3} \exp\left[-\frac{m\,(\boldsymbol{x}-\boldsymbol{x}_0)^2}{2\mathrm{i}\hbar(t-t_0-\mathrm{i}\epsilon)}\right],\end{aligned}$$

which is indeed the $\boldsymbol{x}$ representation (4.45) of the free propagator.

Note that the classical trajectory of the particle from the location $\boldsymbol{x}_0$ at time t_0 to the location $\boldsymbol{x}$ at time t is given by

$$\boldsymbol{x}_{cl}(t') = \boldsymbol{x}_0 + \frac{\boldsymbol{x}-\boldsymbol{x}_0}{t-t_0}(t'-t_0) = \boldsymbol{x}\frac{t'-t_0}{t-t_0} + \boldsymbol{x}_0\frac{t'-t}{t_0-t},$$

and therefore the factor in the exponent of the free propagator is just the action functional evaluated on the classical trajectory,

$$\frac{m}{2}\frac{(\boldsymbol{x}-\boldsymbol{x}_0)^2}{t-t_0} = S[\boldsymbol{x}_{cl}(t')].$$

This holds in general for propagators where the Lagrange function contains at most second order terms in particle velocities and locations, and the path integral formulation is particularly well suited to prove this. If the Lagrange function contains at most second order terms in $\dot{\boldsymbol{x}}$ and $\boldsymbol{x}$, then due to fixed initial and final points $\boldsymbol{x}(t_0) \equiv \boldsymbol{x}_0$ and $\boldsymbol{x}(t) \equiv \boldsymbol{x}$, the action functional for all admissible paths $\boldsymbol{x}(t')$ is exactly

$$\begin{aligned}S[\boldsymbol{x}(t')] = S[\boldsymbol{x}_{cl}(t')] &+ \frac{1}{2}\int_{t_0}^{t} dt'' \int_{t_0}^{t} dt' \left(\boldsymbol{x}(t'') - \boldsymbol{x}_{cl}(t'')\right) \\ &\times \frac{\delta^2 S}{\delta \boldsymbol{x}(t'')\delta \boldsymbol{x}(t')} \cdot \left(\boldsymbol{x}(t') - \boldsymbol{x}_{cl}(t')\right),\end{aligned} \tag{14.26}$$

see Problem 14.1. Functional integration over $\exp(\mathrm{i}S[\boldsymbol{x}(t')]/\hbar)$ then yields a constant from the Gaussian integral over the fluctuations $\boldsymbol{x}(t') - \boldsymbol{x}_{cl}(t')$, and a remnant exponential factor,

$$\langle \boldsymbol{x}|U(t,t_0)|\boldsymbol{x}_0\rangle \sim \exp\left(\frac{\mathrm{i}}{\hbar}S[\boldsymbol{x}_{cl}(t')]\right).$$

However, note that this requires vanishing fluctuations at the boundaries, $\boldsymbol{x}(t_{(0)}) - \boldsymbol{x}_{cl}(t_{(0)}) = \boldsymbol{0}$. Otherwise boundary terms involving $\boldsymbol{x}(t_{(0)}) - \boldsymbol{x}_{cl}(t_{(0)})$ will appear in the exponent. This is important e.g. in scattering theory in the following section, when we are really concerned with fixed initial and final momenta rather than locations.

14.3 Path integrals in scattering theory

We have seen in Chapter 13 that the calculation of transition probabilities or scattering cross section from an initial state $|\psi_i(t')\rangle$ to a final state $|\psi_f(t)\rangle$ requires the calculation of the scattering matrix element

$$S_{fi}(t,t') = \langle\psi_f(t)|U(t,t')|\psi_i(t')\rangle = \langle\psi_f|U_D(t,t')|\psi_i\rangle,$$

where

$$U_D(t,t') = \exp\!\left(\frac{\mathrm{i}}{\hbar}H_0 t\right)\mathrm{T}\exp\!\left(-\frac{\mathrm{i}}{\hbar}\int_{t'}^{t}d\tau\,H(\tau)\right)\exp\!\left(-\frac{\mathrm{i}}{\hbar}H_0 t'\right)$$

is the time evolution operator on the states in the interaction picture. We also recall that the usual default definition of the scattering matrix involves $t\to\infty$, $t'\to-\infty$, $S_{fi}\equiv S_{fi}(\infty,-\infty)$. For the following discussion it is convenient to relabel initial and final times as $t'\to t_i$, $t\to t_f$. Equation (14.24) then implies a connection between scattering matrix elements and path integrals,

$$S_{fi} = \lim_{t_i\to-\infty,t_f\to\infty}\int d^3\boldsymbol{x}_f\int d^3\boldsymbol{x}_i\,\langle\psi_f|\exp\!\left(\frac{\mathrm{i}}{\hbar}H_0 t_f\right)|\boldsymbol{x}_f\rangle$$
$$\times\int_{\boldsymbol{x}(t_i)=\boldsymbol{x}_i}^{\boldsymbol{x}(t_f)=\boldsymbol{x}_f} D^3\boldsymbol{x}(t)\,\exp\!\left(\frac{\mathrm{i}}{\hbar}S[\boldsymbol{x}(t)]\right)\langle\boldsymbol{x}_i|\exp\!\left(-\frac{\mathrm{i}}{\hbar}H_0 t_i\right)|\psi_i\rangle. \quad (14.27)$$

This is still a mixed formula involving both canonical operators and a path integral. We now assume that our initial and final states are momentum eigenstates $|\psi_i\rangle = |\boldsymbol{p}_i\rangle$ and $|\psi_f\rangle = |\boldsymbol{p}_f\rangle$, and we also assume that the scattering potential $V(\boldsymbol{x},t)$ is analytic with finite range. The free Hamiltonian for the free-free scattering problem is $H_0 = \mathbf{p}^2/2m$. The resulting scattering matrix element is then

$$S_{fi} = \lim_{t_i\to-\infty,t_f\to\infty}\int d^3\boldsymbol{x}_f\int d^3\boldsymbol{x}_i\int_{\boldsymbol{x}(t_i)=\boldsymbol{x}_i}^{\boldsymbol{x}(t_f)=\boldsymbol{x}_f} D^3\boldsymbol{x}(t)\,\exp\!\left(\frac{\mathrm{i}}{\hbar}S[\boldsymbol{x}(t)]\right)$$
$$\times\frac{1}{(2\pi\hbar)^3}\exp\!\left[\frac{\mathrm{i}}{\hbar}\left(\frac{\boldsymbol{p}_f^2 t_f-\boldsymbol{p}_i^2 t_i}{2m}+\boldsymbol{p}_i\cdot\boldsymbol{x}_i-\boldsymbol{p}_f\cdot\boldsymbol{x}_f\right)\right]. \quad (14.28)$$

For the perturbative evaluation of (14.28) we introduce an auxiliary external force $\boldsymbol{F}(t)$, such that the Lagrange function including the scattering potential $V(\boldsymbol{x},t)$ takes the form

$$L = \frac{m}{2}\dot{\boldsymbol{x}}^2(t) - V(\boldsymbol{x}(t),t) + \boldsymbol{F}(t)\cdot\boldsymbol{x}(t).$$

The path integral in (14.28) then takes the form

$$\int D^3x(t)\,\exp\!\left(\frac{\mathrm{i}}{\hbar}S[x(t)]\right)=\int D^3x(t)\sum_{n=0}^{\infty}\frac{1}{(\mathrm{i}\hbar)^n n!}\int_{t_i}^{t_f}dt_1\ldots\int_{t_i}^{t_f}dt_n$$
$$\times V(\boldsymbol{x}(t_1),t_1)\ldots V(\boldsymbol{x}(t_n),t_n)\exp\!\left[\frac{\mathrm{i}}{\hbar}\int_{t_i}^{t_f}dt\left(\frac{m}{2}\dot{\boldsymbol{x}}^2(t)+\boldsymbol{F}(t)\cdot\boldsymbol{x}(t)\right)\right]$$
$$=\int D^3x(t)\sum_{n=0}^{\infty}\frac{1}{(\mathrm{i}\hbar)^n n!}\int_{t_i}^{t_f}dt_1\ldots\int_{t_i}^{t_f}dt_n\,V\!\left(\frac{\hbar}{\mathrm{i}}\frac{\delta}{\delta\boldsymbol{F}(t_1)},t_1\right)\ldots$$
$$\times V\!\left(\frac{\hbar}{\mathrm{i}}\frac{\delta}{\delta\boldsymbol{F}(t_n)},t_n\right)\exp\!\left[\frac{\mathrm{i}}{\hbar}\int_{t_i}^{t_f}dt\left(\frac{m}{2}\dot{\boldsymbol{x}}^2(t)+\boldsymbol{F}(t)\cdot\boldsymbol{x}(t)\right)\right]$$
$$=\int D^3x(t)\,\exp\!\left[-\frac{\mathrm{i}}{\hbar}\int_{t_i}^{t_f}dt'\,V\!\left(\frac{\hbar}{\mathrm{i}}\frac{\delta}{\delta\boldsymbol{F}(t')},t'\right)\right]$$
$$\times\exp\!\left[\frac{\mathrm{i}}{\hbar}\int_{t_i}^{t_f}dt\left(\frac{m}{2}\dot{\boldsymbol{x}}^2(t)+\boldsymbol{F}(t)\cdot\boldsymbol{x}(t)\right)\right].\tag{14.29}$$

Evaluation of the Gaussian integrals as in equation (14.23) for $V(\boldsymbol{x})=0$ reproduces the canonical perturbation series (13.18). However, a different representation is gotten if we pull the variational derivative operators $V(-\mathrm{i}\hbar\delta/\delta\boldsymbol{F}(t),t)$ out of the path integral,

$$\int D^3x(t)\,\exp\!\left(\frac{\mathrm{i}}{\hbar}S[x(t)]\right)=\exp\!\left[-\frac{\mathrm{i}}{\hbar}\int_{t_i}^{t_f}dt'\,V\!\left(\frac{\hbar}{\mathrm{i}}\frac{\delta}{\delta\boldsymbol{F}(t')},t'\right)\right]Z[\boldsymbol{F}],$$
$$Z[\boldsymbol{F}]=\int D^3x(t)\,\exp\!\left[\frac{\mathrm{i}}{\hbar}\int_{t_i}^{t_f}dt\left(\frac{m}{2}\dot{\boldsymbol{x}}^2(t)+\boldsymbol{F}(t)\cdot\boldsymbol{x}(t)\right)\right].\tag{14.30}$$

It is useful to have a convolution notation for the following calculations. We define

$$(G\circ\boldsymbol{F})(t)\equiv\int_{-\infty}^{\infty}dt'\,G(t,t')\boldsymbol{F}(t')$$

and

$$(\dot{G}\circ\boldsymbol{F})(t)\equiv\int_{-\infty}^{\infty}dt'\,\frac{\partial}{\partial t}G(t,t')\boldsymbol{F}(t').$$

Partial integration yields the following representation of the action of a particle under the influence of a force $\boldsymbol{F}(t)$ for every Green's function (14.6),

$$
\begin{aligned}
S[\boldsymbol{x},\boldsymbol{F}] &= \int_{t_i}^{t_f} dt\left(\frac{m}{2}\dot{\boldsymbol{x}}^2(t)+\boldsymbol{F}(t)\cdot\boldsymbol{x}(t)\right)\\
&= \frac{m}{2}\int_{t_i}^{t_f} dt\left(\dot{\boldsymbol{x}}(t)-\frac{(\dot{G}\circ\boldsymbol{F})(t)}{m}\right)^2+\frac{1}{2m}\int_{t_i}^{t_f} dt\,\boldsymbol{F}(t)\cdot(G\circ\boldsymbol{F})(t)\\
&\quad+\left(\boldsymbol{x}(t_f)-\frac{(G\circ\boldsymbol{F})(t_f)}{2m}\right)\cdot(\dot{G}\circ\boldsymbol{F})(t_f)\\
&\quad-\left(\boldsymbol{x}(t_i)-\frac{(G\circ\boldsymbol{F})(t_i)}{2m}\right)\cdot(\dot{G}\circ\boldsymbol{F})(t_i).
\end{aligned}
\tag{14.31}
$$

The trajectory $\boldsymbol{x}(t)$ between $\boldsymbol{x}_i$ and $\boldsymbol{x}_f$ appears only in the free particle action for the trajectory

$$
\boldsymbol{X}(t)=\boldsymbol{x}(t)-\frac{1}{m}(G\circ\boldsymbol{F})(t),
\tag{14.32}
$$

which classically satisfies $\ddot{\boldsymbol{X}}(t)=\boldsymbol{0}$. Therefore the path integral (14.30) can be evaluated in terms of the result for the free particle,

$$
\begin{aligned}
Z[\boldsymbol{F}] &= \sqrt{\frac{m}{2\pi\mathrm{i}\hbar(t_f-t_i)}}^{\,3}\exp\!\left[\frac{\mathrm{i}}{\hbar}\left(\boldsymbol{X}_f\cdot(\dot{G}\circ\boldsymbol{F})(t_f)-\boldsymbol{X}_i\cdot(\dot{G}\circ\boldsymbol{F})(t_i)\right)\right]\\
&\quad\times\exp\!\left(\frac{\mathrm{i}}{2m\hbar}\left[(G\circ\boldsymbol{F})(t_f)\cdot(\dot{G}\circ\boldsymbol{F})(t_f)-(G\circ\boldsymbol{F})(t_i)\cdot(\dot{G}\circ\boldsymbol{F})(t_i)\right]\right)\\
&\quad\times\exp\!\left(\mathrm{i}m\frac{(\boldsymbol{X}_f-\boldsymbol{X}_i)^2}{2\hbar(t_f-t_i)}+\frac{\mathrm{i}}{2m\hbar}\int_{t_i}^{t_f} dt\,\boldsymbol{F}(t)\cdot(G\circ\boldsymbol{F})(t)\right)\\
&= \langle\boldsymbol{X}_f|U_0(t_f,t_i)|\boldsymbol{X}_i\rangle\exp\!\left(\frac{\mathrm{i}}{2m\hbar}\int_{t_i}^{t_f} dt\,\boldsymbol{F}(t)\cdot(G\circ\boldsymbol{F})(t)\right)\\
&\quad\times\exp\!\left(\frac{\mathrm{i}}{2m\hbar}\left[(G\circ\boldsymbol{F})(t_f)\cdot(\dot{G}\circ\boldsymbol{F})(t_f)-(G\circ\boldsymbol{F})(t_i)\cdot(\dot{G}\circ\boldsymbol{F})(t_i)\right]\right)\\
&\quad\times\exp\!\left[\frac{\mathrm{i}}{\hbar}\left(\boldsymbol{X}_f\cdot(\dot{G}\circ\boldsymbol{F})(t_f)-\boldsymbol{X}_i\cdot(\dot{G}\circ\boldsymbol{F})(t_i)\right)\right].
\end{aligned}
\tag{14.33}
$$

We can summarize our results in the equations

$$
S_{fi}=\lim_{t_i\to-\infty,t_f\to\infty}\exp\!\left[-\frac{\mathrm{i}}{\hbar}\int_{t_i}^{t_f} dt\,V\!\left(\frac{\hbar}{\mathrm{i}}\frac{\delta}{\delta\boldsymbol{F}(t)},t\right)\right]S_{fi}[\boldsymbol{F}]\Big|_{\boldsymbol{F}=\boldsymbol{0}},
\tag{14.34}
$$

$$S_{fi}[\boldsymbol{F}] = \frac{1}{(2\pi\hbar)^3}\int d^3\boldsymbol{X}_f \int d^3\boldsymbol{X}_i\, Z[\boldsymbol{F}](\boldsymbol{X}_f, t_f; \boldsymbol{X}_i, t_i)$$
$$\times \exp\left(\mathrm{i}\frac{\boldsymbol{p}_f^2 t_f - \boldsymbol{p}_i^2 t_i}{2m\hbar}\right)\exp\left[\frac{\mathrm{i}}{\hbar}\boldsymbol{p}_i\cdot\left(\boldsymbol{X}_i + \frac{1}{m}(G\circ\boldsymbol{F})(t_i)\right)\right] \quad (14.35)$$
$$\times \exp\left[-\frac{\mathrm{i}}{\hbar}\boldsymbol{p}_f\cdot\left(\boldsymbol{X}_f + \frac{1}{m}(G\circ\boldsymbol{F})(t_f)\right)\right].$$

The integrals over $\boldsymbol{X}_f$ and $\boldsymbol{X}_i$ amount to a Gaussian integral involving $\boldsymbol{X}_f - \boldsymbol{X}_i$ and an integral over a Fourier monomial involving $\boldsymbol{X}_i$. Evaluation of the integrals yields

$$S_{fi}[\boldsymbol{F}] = \exp\left(\frac{\mathrm{i}}{2m\hbar}\int_{t_i}^{t_f} dt\,\boldsymbol{F}(t)\cdot(G\circ\boldsymbol{F})(t)\right)$$
$$\times \exp\left(\frac{\mathrm{i}}{2m\hbar}\left[2\boldsymbol{p}_f - (\dot{G}\circ\boldsymbol{F})(t_f)\right]\cdot\left[t_f(\dot{G}\circ\boldsymbol{F})(t_f) - (G\circ\boldsymbol{F})(t_f)\right]\right)$$
$$\times \exp\left(-\frac{\mathrm{i}}{2m\hbar}\left[2\boldsymbol{p}_i - (\dot{G}\circ\boldsymbol{F})(t_i)\right]\cdot\left[t_i(\dot{G}\circ\boldsymbol{F})(t_i) - (G\circ\boldsymbol{F})(t_i)\right]\right)$$
$$\times\delta\big(\boldsymbol{p}_f - (\dot{G}\circ\boldsymbol{F})(t_f) - \boldsymbol{p}_i + (\dot{G}\circ\boldsymbol{F})(t_i)\big). \quad (14.36)$$

For consistency we note that this reproduces the correct result $S_{fi} = \delta\big(\boldsymbol{p}_f - \boldsymbol{p}_i\big)$ for the free particle. The δ function implies conservation of the free momentum $\boldsymbol{P} = \boldsymbol{p}(t) - (\dot{G}\circ\boldsymbol{F})(t)$, or equivalently matching of the external momenta under evolution with the force $\boldsymbol{F}(t)$,

$$\boldsymbol{p}_f = \boldsymbol{p}_i + \int_{t_i}^{t_f} dt\,\boldsymbol{F}(t). \quad (14.37)$$

Please note that it is not possible to impose simultaneous boundary conditions

$$t_f\,\frac{\partial}{\partial t}G(t,t')\bigg|_{t=t_f} = G(t_f, t')$$

and

$$t_i\,\frac{\partial}{\partial t}G(t,t')\bigg|_{t=t_i} = G(t_i, t'),$$

because such a Green's function does not exist. As a consequence it is not possible to eliminate the initial and final state dependent exponentials in the scattering matrix through a clever choice of the Green's function. This is of course as it should be, because the scattering amplitude $\mathcal{M}_{fi} = \mathrm{i}(S_{fi} - \delta_{fi})/\delta(\boldsymbol{P}_f - \boldsymbol{P}_i)$ generically must depend on the initial and final states.

The functionals $S[\boldsymbol{x},\boldsymbol{F}]$ (14.31), $Z[\boldsymbol{F}]$ (14.33) and $S_{fi}[\boldsymbol{F}]$ (14.36) are all independent of the boundary functions $\alpha(t')$ and $\beta(t')$ in the general Green's function (14.6). The easiest way to show this is by observing that the functionals are invariant under shifts

$$(G\circ\boldsymbol{F})(t)\to(G\circ\boldsymbol{F})(t)+\boldsymbol{A}t+\boldsymbol{B}$$

with constant vectors $\boldsymbol{A}$ and $\boldsymbol{B}$. For $Z[\boldsymbol{F}]$ the demonstration has to take into account that $\boldsymbol{X}_f$ and $\boldsymbol{X}_i$ contain $(G\circ\boldsymbol{F})(t_f)$ or $(G\circ\boldsymbol{F})(t_i)$ according to (14.32).

We are therefore free to use e.g. the Green's functions $G_i(t,t')$ (14.3) or $G_{fi}(t,t')$ (14.5), or the retarded Green's function $G_{ret}(t,t')=(t-t')\Theta(t-t')$ or a Stückelberg-Feynman type Green's function with equal contributions from retarded and advanced components, $G_{SF}(t,t')=|t-t'|/2$, or any other Green's function of the form (14.6).

The limit $t_i\to-\infty$, $t_f\to\infty$ in equation (14.36) yields the following representation of the S-matrix element for scattering due to the external force $\boldsymbol{F}(t)$,

$$\begin{aligned}S_{fi}[\boldsymbol{F}]=&\,\delta\left(\boldsymbol{p}_f-\boldsymbol{p}_i-\int_{-\infty}^{\infty}dt\,\boldsymbol{F}(t)\right)\exp\left(\mathrm{i}\frac{\boldsymbol{p}_f+\boldsymbol{p}_i}{2m\hbar}\cdot\int_{-\infty}^{\infty}dt\,t\boldsymbol{F}(t)\right)\\&\times\exp\left(\frac{\mathrm{i}}{4m\hbar}\int_{-\infty}^{\infty}dt\int_{-\infty}^{\infty}dt'\,|t-t'|\boldsymbol{F}(t)\cdot\boldsymbol{F}(t')\right).\end{aligned}\tag{14.38}$$

In the next steps we will compare the correlation functions between the canonical and the path integral formalism.

The calculation of the one-point function from the path integral (14.33) has to take into account that the generic Green's function (14.6) shifts $\boldsymbol{x}_{f/i}$ to $\boldsymbol{X}_{f/i}$ according to equation (14.32). This implies for the one-point function in the path integral formalism

$$\begin{aligned}\langle\boldsymbol{x}_f,t_f|\boldsymbol{x}(t)|\boldsymbol{x}_i,t_i\rangle=&-\mathrm{i}\hbar\,\frac{\delta}{\delta\boldsymbol{F}(t)}Z[\boldsymbol{F}]\bigg|_{F=0}=\langle\boldsymbol{x}_f|U_0(t_f,t_i)|\boldsymbol{x}_i\rangle\\&\times\left(\frac{\boldsymbol{x}_f+\boldsymbol{x}_i}{2}+\alpha(t)\left(\boldsymbol{x}_f-\boldsymbol{x}_i\right)+m\frac{\delta}{\delta\boldsymbol{F}(t)}\left(\boldsymbol{X}_f-\boldsymbol{X}_i\right)\cdot\frac{\boldsymbol{x}_f-\boldsymbol{x}_i}{t_f-t_i}\right).\end{aligned}$$

However, we have

$$m\frac{\delta}{\delta\boldsymbol{F}(t)}\left(\boldsymbol{X}_f-\boldsymbol{X}_i\right)=\alpha(t)(t_i-t_f)+t-\frac{t_f+t_i}{2},$$

and therefore the path integral result for the one-point function is indeed independent on the gauge functions $\alpha(t')$ and $\beta(t')$, as was already clear from the cancellation of those terms in $Z[\boldsymbol{F}]$,

$$\langle \boldsymbol{x}_f, t_f|\boldsymbol{x}(t)|\boldsymbol{x}_i, t_i\rangle = -\mathrm{i}\hbar \left.\frac{\delta}{\delta \boldsymbol{F}(t)} Z[\boldsymbol{F}]\right|_{\boldsymbol{F}=\boldsymbol{0}} = \langle \boldsymbol{x}_f|U_0(t_f, t_i)|\boldsymbol{x}_i\rangle \times \left(\boldsymbol{x}_f \frac{t-t_i}{t_f-t_i} + \boldsymbol{x}_i \frac{t_f-t}{t_f-t_i}\right), \tag{14.39}$$

i.e. we do find the same result (14.11) as in the canonical formalism.

For the calculation of the two-point functions in the functional formalism

$$\langle \boldsymbol{x}_f, t_f|\boldsymbol{x}(t_2)\otimes\boldsymbol{x}(t_1)|\boldsymbol{x}_i, t_i\rangle = -\hbar^2 \left.\frac{\delta^2 Z[\boldsymbol{F}]}{\delta\boldsymbol{F}(t_2)\otimes\delta\boldsymbol{F}(t_1)}\right|_{\boldsymbol{F}=\boldsymbol{0}}$$

it is useful to observe that

$$\frac{1}{Z[\boldsymbol{F}]}\frac{\delta^2 Z[\boldsymbol{F}]}{\delta\boldsymbol{F}(t_2)\otimes\delta\boldsymbol{F}(t_1)} = \frac{\delta^2 \ln Z[\boldsymbol{F}]}{\delta\boldsymbol{F}(t_2)\otimes\delta\boldsymbol{F}(t_1)} + \frac{\delta \ln Z[\boldsymbol{F}]}{\delta\boldsymbol{F}(t_2)} \otimes \frac{\delta \ln Z[\boldsymbol{F}]}{\delta\boldsymbol{F}(t_1)}.$$

The factors in the last term were evaluated at $\boldsymbol{F} = \boldsymbol{0}$ in (14.39) and reproduce the tensor product of one-point functions in (14.12). The second order variational derivative of $\ln Z[\boldsymbol{F}]$ yields for $t_i \le t_1 \le t_2 \le t_f$ (but *only* in that case)

$$-\hbar^2 \frac{\delta^2 \ln Z[\boldsymbol{F}]}{\delta\boldsymbol{F}(t_2)\otimes\delta\boldsymbol{F}(t_1)} = \frac{\mathrm{i}\hbar}{m}\frac{(t_f-t_2)(t_1-t_i)}{t_f-t_i}\underline{1}.$$

The general result for $t_i < t_f$ is

$$-\hbar^2 \frac{\delta^2 \ln Z[\boldsymbol{F}]}{\delta\boldsymbol{F}(t_2)\otimes\delta\boldsymbol{F}(t_1)} = \frac{\mathrm{i}\hbar}{m(t_f-t_i)}\Big[\Theta(t_f-t_2)\Theta(t_2-t_1)\Theta(t_1-t_i) \times(t_f-t_2)(t_1-t_i) + \Theta(t_f-t_1)\Theta(t_1-t_2)\Theta(t_2-t_i)(t_f-t_1)(t_2-t_i)\Big]\underline{1}.$$

Therefore we cannot in general simply identify $-\hbar^2\delta^2 Z[\boldsymbol{F}]/(\delta\boldsymbol{F}(t_2)\otimes\delta\boldsymbol{F}(t_1))$ at $\boldsymbol{F} = \boldsymbol{0}$ with either $\boldsymbol{g}^{(2)}_{fi}(t_2, t_1)$ or $\boldsymbol{G}^{(2)}_{fi}(t_2, t_1)$, but we have

$$-\hbar^2 \left.\frac{\delta^2 Z[\boldsymbol{F}]}{\delta\boldsymbol{F}(t_2)\otimes\delta\boldsymbol{F}(t_1)}\right|_{\boldsymbol{F}=\boldsymbol{0}} = \boldsymbol{g}^{(2)}_{fi}(t_2, t_1) = \boldsymbol{G}^{(2)}_{fi}(t_2, t_1)$$

if $t_i \le t_1 \le t_2 \le t_f$.

It seems surprising that substitution of (14.38) into equation (14.34) and setting $\boldsymbol{F} = \boldsymbol{0}$ after evaluation of the functional derivatives yields scattering from the potential V. However, equations (14.34, 14.38) compare to the practically useful relation (13.18) (or the equivalent relation (14.29)) for the scattering matrix elements like the representation

$$\langle \boldsymbol{x}|U_0(t-t')|\boldsymbol{x}'\rangle = \exp\left(\mathrm{i}\hbar\frac{t-t'-\mathrm{i}\epsilon}{2m}\frac{\partial^2}{\partial \boldsymbol{x}^2}\right)\delta(\boldsymbol{x}-\boldsymbol{x}')$$

for the $\boldsymbol{x}$ matrix elements of the free time evolution operator compares to the practically more useful representation (4.45).

Recasting the perturbation series in terms of the operator $V(-\mathrm{i}\hbar\delta/\delta\boldsymbol{F}(t),t)$ instead of $V(\boldsymbol{x},t)$ does not yield a more efficient or practical representation for potential scattering theory. However, recasting interactions in terms of functional derivatives is useful when interactions are expressed in terms of higher order products of wave functions instead of potentials. Therefore we used the transcription of potential scattering theory in terms of functional derivatives with respect to auxiliary forces as an illustration for functional methods in perturbation theory.

14.4 Problems

14.1. Verify equation (14.26) for the general second order particle action

$$S[\boldsymbol{x}(t')] = \int_{t_0}^{t} dt'\left(\frac{1}{2}\dot{\boldsymbol{x}}(t')\cdot\underline{M}\cdot\dot{\boldsymbol{x}}(t') + \frac{1}{2}\boldsymbol{x}(t')\cdot\underline{F}\cdot\dot{\boldsymbol{x}}(t')\right.$$
$$\left. - \frac{1}{2}\boldsymbol{x}(t')\cdot\underline{\Omega}^2\cdot\boldsymbol{x}(t') + \boldsymbol{F}\cdot\boldsymbol{x}(t')\right),$$
$$\underline{F}^T = -\underline{F},\quad \boldsymbol{x}(t_0) = \boldsymbol{x}_{cl}(t_0) = \boldsymbol{x}_0,\quad \boldsymbol{x}(t) = \boldsymbol{x}_{cl}(t) = \boldsymbol{x}.$$

14.2. Which exponential factor in the propagator $\langle \boldsymbol{x}|U(t,t_0)|\boldsymbol{x}_0\rangle$ do you find for a harmonic oscillator?

14.3. Derive the particular Green's functions (14.3) and (14.5) from the general form (14.6).

14.4. Show that the terms

$$\mathcal{A}[\boldsymbol{F}] = \frac{1}{2m}\left[(G\circ\boldsymbol{F})(t_f)\cdot(\dot{G}\circ\boldsymbol{F})(t_f) - (G\circ\boldsymbol{F})(t_i)\cdot(\dot{G}\circ\boldsymbol{F})(t_i)\right]$$
$$+\frac{1}{2m}\int_{t_i}^{t_f} dt\,\boldsymbol{F}(t)\cdot(G\circ\boldsymbol{F})(t)$$

in the exponent in (14.33) are actually the action $S[\boldsymbol{x},\boldsymbol{F}]$ for the classical trajectory $\boldsymbol{x}(t) = (G\circ\boldsymbol{F})(t)/m$.

Why is $Z[\boldsymbol{F}]$ not just given by $\langle\boldsymbol{0}|U_0(t_f,t_i)|\boldsymbol{0}\rangle\exp(\mathrm{i}\mathcal{A}[\boldsymbol{F}]/\hbar)$, in spite of what you might have expected from (14.26)?

14.5. Calculate the time ordered three-point function

$$\langle \boldsymbol{x}_f, t_f|\mathbf{x}(t_3)\otimes\mathbf{x}(t_2)\otimes\mathbf{x}(t_1)|\boldsymbol{x}_i, t_i\rangle, \quad t_i < t_1 < t_2 < t_3 < t_f,$$

both in the canonical formalism and in the path integral formalism.

14.6. The functional $S_{fi}[\boldsymbol{F}] \equiv S(\boldsymbol{p}_f, \boldsymbol{p}_i)[\boldsymbol{F}]$ (14.38) is a scattering matrix element between momentum eigenstates. The functional $Z[\boldsymbol{F}]$ (14.33) on the other hand is *not* a scattering matrix element in position space because it does *not* correspond to a position space matrix element of an *interaction picture time evolution operator*. Instead it corresponds to the path integral result for the position matrix element of the *full time evolution operator* of a free particle under the influence of a spatially homogeneous force $\boldsymbol{F}(t)$. However, we can derive a position space scattering matrix element through Fourier transformation of $S(\boldsymbol{p}_f, \boldsymbol{p}_i)[\boldsymbol{F}]$. Show that

$$\begin{aligned}S(\boldsymbol{x}_f, \boldsymbol{x}_i)[\boldsymbol{F}] &= \int d^3p_f \int d^3p_i\, \exp\!\left(\frac{\mathrm{i}}{\hbar}\left(\boldsymbol{p}_f\cdot\boldsymbol{x}_f - \boldsymbol{p}_i\cdot\boldsymbol{x}_i\right)\right)\frac{S(\boldsymbol{p}_f, \boldsymbol{p}_i)[\boldsymbol{F}]}{(2\pi\hbar)^3}\\ &= \delta\!\left(\boldsymbol{x}_f - \boldsymbol{x}_i + \frac{1}{m}\int_{-\infty}^{\infty} dt\, t\boldsymbol{F}(t)\right)\exp\!\left(\mathrm{i}\frac{\boldsymbol{x}_f + \boldsymbol{x}_i}{2\hbar}\cdot\int_{-\infty}^{\infty} dt\, \boldsymbol{F}(t)\right)\\ &\quad\times\exp\!\left(\frac{\mathrm{i}}{4m\hbar}\int_{-\infty}^{\infty} dt \int_{-\infty}^{\infty} dt'\, |t-t'|\boldsymbol{F}(t)\cdot\boldsymbol{F}(t')\right).\end{aligned} \tag{14.40}$$

Show also that with the conditions

$$\lim_{t\to\pm\infty} t\int_{-\infty}^{t} dt'\, \boldsymbol{F}(t') = \boldsymbol{0} \tag{14.41}$$

the δ function implies

$$\boldsymbol{x}_f = \boldsymbol{x}_i + \frac{1}{m}\int_{-\infty}^{\infty} dt \int_{-\infty}^{t} dt'\, \boldsymbol{F}(t'), \tag{14.42}$$

while the conditions

$$\lim_{t\to\pm\infty} t\int_{t}^{\infty} dt'\, \boldsymbol{F}(t') = \boldsymbol{0} \tag{14.43}$$

yield with the δ function the relation

$$\boldsymbol{x}_i = \boldsymbol{x}_f + \frac{1}{m}\int_{\infty}^{-\infty} dt \int_{\infty}^{t} dt'\, \boldsymbol{F}(t'). \tag{14.44}$$

Equation (14.42) describes the asymptotic solution of a classical trajectory of a non-relativistic particle which started out at rest in $\boldsymbol{x}_i$ at $t \to -\infty$, while (14.44) reconstructs the initial location for $t \to -\infty$ of a particle which comes to rest in $\boldsymbol{x}_f$ in the limit $t \to \infty$. The conditions (14.41) or (14.43) do not generate extra restrictions on the physical motion of the particle, but are mathematical conditions for convergence of the time integrals in (14.42) or (14.44), respectively, i.e. they are necessary for existence of solutions of the Newton equation for $t \to \pm\infty$.

Chapter 15
Coupling to Electromagnetic Fields

Electromagnetism is the most important interaction for the study of atoms, molecules and materials. It determines most of the potentials or perturbation operators V which are studied in practical applications of quantum mechanics, and it also serves as a basic example for the implementation of other, more complicated interactions in quantum mechanics. Therefore the primary objective of the current chapter is to understand how electromagnetic fields are introduced in the Schrödinger equation.

15.1 Electromagnetic couplings

The introduction of electromagnetic fields into the Schrödinger equation for a particle of mass m and electric charge q can be inferred from the description of the particle in classical Lagrangian mechanics.

The Lagrange function for the particle in electromagnetic fields

$$\boldsymbol{E}(\boldsymbol{x},t) = -\nabla\Phi(\boldsymbol{x},t) - \frac{\partial \boldsymbol{A}(\boldsymbol{x},t)}{\partial t}, \quad \boldsymbol{B}(\boldsymbol{x},t) = \nabla \times \boldsymbol{A}(\boldsymbol{x},t)$$

is

$$L = \frac{m}{2}\dot{\boldsymbol{x}}(t)^2 + q\dot{\boldsymbol{x}}(t)\cdot\boldsymbol{A}(\boldsymbol{x}(t),t) - q\Phi(\boldsymbol{x}(t),t). \tag{15.1}$$

Let us check (or review) that equation (15.1) is indeed the correct Lagrange function for the particle. The electromagnetic potentials in the Lagrange function depend on the time t both explicitly and implicitly through the time dependence $\boldsymbol{x}(t)$ of the trajectory of the particle. The time derivative of the conjugate momentum

$$\boldsymbol{p} = \frac{\partial L}{\partial \dot{\boldsymbol{x}}} = m\dot{\boldsymbol{x}} + q\boldsymbol{A} \tag{15.2}$$

R. Dick, *Advanced Quantum Mechanics*, Graduate Texts in Physics,
DOI 10.1007/978-3-319-25675-7_15

is therefore

$$\frac{d\boldsymbol{p}}{dt} = m\ddot{\boldsymbol{x}} + q\dot{x}_i\frac{\partial \boldsymbol{A}}{\partial x_i} + q\frac{\partial \boldsymbol{A}}{\partial t}.$$

According to the Euler-Lagrange equations (cf. Appendix A), this must equal

$$\frac{\partial L}{\partial \boldsymbol{x}} = q\dot{x}_i\nabla A_i - q\nabla\Phi.$$

The property (7.14) of the ϵ tensor implies

$$\boldsymbol{e}_i\left(\dot{x}_j\partial_i A_j - \dot{x}_j\partial_j A_i\right) = \boldsymbol{e}_i\epsilon_{ijk}\epsilon_{klm}\dot{x}_j\partial_l A_m = \dot{\boldsymbol{x}} \times \boldsymbol{B},$$

and therefore the Euler-Lagrange equation yields the Lorentz force law

$$m\ddot{\boldsymbol{x}} = q(\boldsymbol{E} + \boldsymbol{v} \times \boldsymbol{B})\,, \tag{15.3}$$

as required.

The classical Hamiltonian for the particle follows as

$$H = \boldsymbol{p}\cdot\dot{\boldsymbol{x}} - L = \frac{1}{2m}(\boldsymbol{p} - q\boldsymbol{A})^2 + q\Phi = \frac{m}{2}\dot{\boldsymbol{x}}^2 + q\Phi. \tag{15.4}$$

The Hamilton operator of the charged particle therefore becomes

$$H = \frac{1}{2m}[\mathbf{p} - q\boldsymbol{A}(\mathbf{x},t)]^2 + q\Phi(\mathbf{x},t), \tag{15.5}$$

and the Schrödinger equation in $\boldsymbol{x}$ representation is

$$\mathrm{i}\hbar\frac{\partial}{\partial t}\Psi(\boldsymbol{x},t) = -\frac{1}{2m}[\hbar\nabla - \mathrm{i}q\boldsymbol{A}(\boldsymbol{x},t)]^2\Psi(\boldsymbol{x},t) + q\Phi(\boldsymbol{x},t)\Psi(\boldsymbol{x},t). \tag{15.6}$$

This is the Schrödinger equation for a charged particle in electromagnetic fields. If we write this in the form

$$\mathrm{i}\hbar\frac{\partial}{\partial t}\Psi - q\Phi\Psi = \frac{1}{2m}(\mathrm{i}\hbar\nabla + q\boldsymbol{A})^2\Psi$$

we also recognize that this arises from the free Schrödinger equation through the substitutions

$$\mathrm{i}\hbar\nabla \to \mathrm{i}\hbar\nabla + q\boldsymbol{A}, \quad \mathrm{i}\hbar\frac{\partial}{\partial t} = \mathrm{i}\hbar c\partial_0 \to \mathrm{i}\hbar c\partial_0 - q\Phi. \tag{15.7}$$

These equations can be combined in 4-vector notation with $p_0 = -E/c$, $A_0 = -\Phi/c$,

$$p_\mu = -\mathrm{i}\hbar\partial_\mu \to p_\mu - qA_\mu = -\mathrm{i}\hbar\partial_\mu - qA_\mu.$$

This observation is useful for recognizing a peculiar symmetry property of equation (15.6). Classical electromagnetism is invariant under gauge transformations of the electromagnetic potentials (here we use $f(x) \equiv f(\boldsymbol{x}, t)$),

$$\Phi(x) \to \Phi'(x) = \Phi(x) - c\partial_0\varphi(x), \quad \boldsymbol{A}(x) \to \boldsymbol{A}'(x) = \boldsymbol{A}(x) + \nabla\varphi(x), \tag{15.8}$$

where the arbitrary function $\varphi(x)$ has the dimension of a magnetic flux, i.e. it comes in units of Vs. The Schrödinger equation (15.6) should respect this invariance of classical electromagnetism to comply with classical limits, and indeed it does. If we also transform the wave function according to

$$\Psi(x) \to \Psi'(x) = \exp\!\left(\mathrm{i}\frac{q}{\hbar}\varphi(x)\right)\Psi(x), \tag{15.9}$$

then the Schrödinger equation in the transformed fields and wave functions has exactly the same form as the Schrödinger equation in the original fields, because the linear transformation property

$$\begin{aligned}
&\mathrm{i}\hbar\frac{\partial}{\partial t}\Psi' - q\Phi'\Psi' + \frac{1}{2m}(\hbar\nabla - \mathrm{i}q\boldsymbol{A}')^2\Psi' \\
&= \exp\!\left(\mathrm{i}\frac{q}{\hbar}\varphi(x)\right)\left[\mathrm{i}\hbar\frac{\partial}{\partial t}\Psi - q\Phi\Psi + \frac{1}{2m}(\hbar\nabla - \mathrm{i}q\boldsymbol{A})^2\Psi\right]
\end{aligned}$$

implies that

$$\mathrm{i}\hbar\frac{\partial}{\partial t}\Psi' - q\Phi'\Psi' + \frac{1}{2m}(\hbar\nabla - \mathrm{i}q\boldsymbol{A}')^2\Psi' = 0$$

holds in the transformed fields if and only if the Schrödinger equation also holds in the original fields,

$$\mathrm{i}\hbar\frac{\partial}{\partial t}\Psi - q\Phi\Psi + \frac{1}{2m}(\hbar\nabla - \mathrm{i}q\boldsymbol{A})^2\Psi = 0.$$

The reason for the linear transformation law is

$$\begin{aligned}
\partial_\mu - \mathrm{i}\frac{q}{\hbar}A'_\mu &= \partial_\mu - \mathrm{i}\frac{q}{\hbar}A_\mu - \mathrm{i}\frac{q}{\hbar}(\partial_\mu\varphi) \\
&= \exp\!\left(\mathrm{i}\frac{q}{\hbar}\varphi\right)\left(\partial_\mu - \mathrm{i}\frac{q}{\hbar}A_\mu\right)\exp\!\left(-\mathrm{i}\frac{q}{\hbar}\varphi\right),
\end{aligned}$$

which implies that the *covariant derivatives*

$$D_\mu\Psi = \left(\partial_\mu - \mathrm{i}\frac{q}{\hbar}A_\mu\right)\Psi$$

transform exactly like the fields,

$$\Psi(x) \to \Psi'(x) = \exp\!\left(\mathrm{i}\frac{q}{\hbar}\varphi(x)\right)\Psi(x),$$

$$D_\mu\Psi(x) \to D'_\mu\Psi'(x) = \exp\!\left(\mathrm{i}\frac{q}{\hbar}\varphi(x)\right)D_\mu\Psi(x),$$

$$D_\mu D_\nu \ldots D_\rho\Psi(x) \to D'_\mu D'_\nu \ldots D'_\rho\Psi'(x) = \exp\!\left(\mathrm{i}\frac{q}{\hbar}\varphi(x)\right)D_\mu D_\nu \ldots D_\rho\Psi(x).$$

This implies preservation of every partial differential equation which like the Schrödinger equation uses only covariant derivatives,

$$\mathrm{i}\hbar c D_0\Psi(x) = -\frac{\hbar^2}{2m}\boldsymbol{D}^2\Psi(x) \quad \Leftrightarrow \quad \mathrm{i}\hbar c D'_0\Psi'(x) = -\frac{\hbar^2}{2m}\boldsymbol{D}'^2\Psi'(x).$$

Coupling of matter wave functions to electromagnetic potentials through covariant derivatives is known as *minimal coupling*.

Observables are gauge invariant, too. For example, the mechanical momentum of the charged particle in electromagnetic fields is

$$\begin{aligned} m\frac{d}{dt}\langle\boldsymbol{x}\rangle(t) &= \int d^3\boldsymbol{x}\,\Psi^+(\boldsymbol{x},t)\left[-\mathrm{i}\hbar\nabla - q\boldsymbol{A}(\boldsymbol{x},t)\right]\Psi(\boldsymbol{x},t) \\ &= \int d^3\boldsymbol{x}\,\Psi'^+(\boldsymbol{x},t)\left[-\mathrm{i}\hbar\nabla - q\boldsymbol{A}'(\boldsymbol{x},t)\right]\Psi'(\boldsymbol{x},t). \end{aligned} \quad (15.10)$$

Electromagnetic interactions ensure local phase invariance of nature. We can rotate the wave function with an arbitrary local phase factor without changing the dynamics or observables of a physical system, due to the presence of the electromagnetic potentials. In hindsight, we should consider this as the reason for the peculiar coupling of the electromagnetic potentials in the Schrödinger equation (15.6).

The presence of the electromagnetic potentials in the observables will of course affect conservation laws. E.g. the mechanical momentum (15.10) of the particle will generically not be conserved, because it can exchange momentum with the electromagnetic field which carries momentum $\boldsymbol{p}_{em}(t) = \epsilon_0\int d^3\boldsymbol{x}\,\boldsymbol{E}(\boldsymbol{x},t)\times\boldsymbol{B}(\boldsymbol{x},t)$. The conserved momentum[1] of the coupled system of non-relativistic charged

[1]The canonical momentum $\langle\boldsymbol{p}\rangle(t) = -\mathrm{i}\hbar\int d^3\boldsymbol{x}\,\Psi^+(\boldsymbol{x},t)\boldsymbol{\nabla}\Psi(\boldsymbol{x},t) = m(d\langle\boldsymbol{x}\rangle(t)/dt) + q\langle\boldsymbol{A}(\boldsymbol{x},t)\rangle$ is also generically not conserved, except if the particle moves in a spatially homogeneous electric field $\boldsymbol{E}(t) = -d\boldsymbol{A}(t)/dt$, e.g. in a plate capacitor. However, note that this is an artifact of the gauge $\Phi = 0$.

particle and electromagnetic fields is

$$\boldsymbol{P} = m\frac{d}{dt}\langle\boldsymbol{x}\rangle(t) + \boldsymbol{p}_{em}(t) = \int d^3x\,\mathcal{P}(\boldsymbol{x},t),$$

where the momentum density with symmetrized action of the derivatives on the wave functions is given by

$$\begin{aligned}\mathcal{P}(\boldsymbol{x},t) &= \frac{\hbar}{2\mathrm{i}}\left[\Psi^+(\boldsymbol{x},t)\cdot\nabla\Psi(\boldsymbol{x},t) - \nabla\Psi^+(\boldsymbol{x},t)\cdot\Psi(\boldsymbol{x},t)\right] \\ &\quad - q\Psi^+(\boldsymbol{x},t)\boldsymbol{A}(\boldsymbol{x},t)\Psi(\boldsymbol{x},t) + \epsilon_0\boldsymbol{E}(\boldsymbol{x},t)\times\boldsymbol{B}(\boldsymbol{x},t).\end{aligned} \tag{15.11}$$

The derivation of momentum conservation for the classical particle-field system in the full relativistic setting can be found in Appendix B, see in particular equation (B.29). Systematic derivations of momentum densities in the coupled system of charged particles and electromagnetic fields in the framework of relativistic spinor quantum electrodynamics (QED) can be found in Sections 21.4 and 21.5, see in particular equations (21.90) and (21.106). Problem 21.5 and equation (21.133) provide the corresponding results in relativistic scalar QED.

Both the non-relativistic limits for bosons and fermions lead to (15.11) for the conserved momentum density in non-relativistic QED, which is also known as *quantum electronics*.

Multipole moments

In many applications of quantum mechanics, simplifications of the electromagnetic coupling terms in equation (15.6) can be employed if the electromagnetic fields have large wavelengths compared to the wave functions in the Schrödinger equation. The leading order and most common approximation is related to the electric dipole moment of charge distributions, and therefore we will briefly discuss the origin of multipole moments in electromagnetism.

Suppose that we probe the electromagnetic potential of a charge q which is located at $\boldsymbol{x}$. We are interested in the potential at location $\boldsymbol{r}$, where $|\boldsymbol{r}| \gg |\boldsymbol{x}|$. Second order Taylor expansion of the Coulomb term in the variables $\boldsymbol{x}$ yields

$$\frac{q}{|\boldsymbol{r}-\boldsymbol{x}|} \approx \frac{q}{r} + q\frac{\boldsymbol{r}\cdot\boldsymbol{x}}{r^3} + q\frac{3(\boldsymbol{r}\cdot\boldsymbol{x})^2 - r^2\boldsymbol{x}^2}{2r^5} = \frac{q}{r} + \frac{\boldsymbol{r}\cdot\boldsymbol{d}}{r^3} + \frac{1}{2r^5}\boldsymbol{r}\cdot\underline{\mathrm{Q}}\cdot\boldsymbol{r}$$

with the dipole and quadrupole terms

$$\boldsymbol{d} = q\boldsymbol{x}, \quad \underline{\mathrm{Q}} = q\left(3\boldsymbol{x}\otimes\boldsymbol{x} - \boldsymbol{x}^2\underline{1}\right).$$

For an extended charge distribution $\varrho(\boldsymbol{x})$ this implies at large distance a representation of the potential

$$4\pi\epsilon_0\Phi(\boldsymbol{r}) = \int d^3x\,\frac{\varrho(\boldsymbol{x})}{|\boldsymbol{r}-\boldsymbol{x}|} \approx \frac{q}{r} + \frac{\boldsymbol{r}\cdot\boldsymbol{d}}{r^3} + \frac{1}{2r^5}\boldsymbol{r}\cdot\underline{\mathrm{Q}}\cdot\boldsymbol{r}$$

in terms of the monopole, dipole, and quadrupole moments

$$q = \int d^3x\,\varrho(\boldsymbol{x}),\quad \boldsymbol{d} = \int d^3x\,\varrho(\boldsymbol{x})\boldsymbol{x},\quad \underline{\mathrm{Q}} = \int d^3x\,\varrho(\boldsymbol{x})[3\boldsymbol{x}\otimes\boldsymbol{x} - \boldsymbol{x}^2\underline{1}].$$

We will find that the leading order coupling of long wavelength electromagnetic fields to charges appears through electric dipole moments of the charges.

Semiclassical treatment of the matter-radiation system in the dipole approximation

In the semiclassical treatment the electromagnetic fields are considered as external classical fields with which the quantum mechanical matter (atom, nucleus, molecule, solid) interacts.

If we consider e.g. an atom with an internal (average or effective) potential $V_{int}(\boldsymbol{x})$ experienced by the electrons, the Schrödinger equation for these electrons in the external electromagnetic fields is

$$\mathrm{i}\hbar\frac{\partial}{\partial t}\Psi = -\frac{1}{2m}(\hbar\nabla - \mathrm{i}q\boldsymbol{A})^2\Psi + (q\Phi + V_{int})\Psi. \tag{15.12}$$

If the electromagnetic fields vary weakly over the extension a of the wave functions (corresponding to approximately homogeneous field over the extension of the atom or molecule under consideration), we can effectively assume a spatially homogeneous field $\boldsymbol{E} = \boldsymbol{E}(t)$ corresponding to a potential $\Phi(\boldsymbol{x},t) = -\boldsymbol{E}(t)\cdot\boldsymbol{x}$. If we assume that our material probes range over length scales from 1 Å (corresponding to the size of atoms) to several Å (corresponding to molecules containing e.g. several Benzene rings), electromagnetic fields with wavelengths larger than 100 nm or photon energies smaller than 12 eV can be considered as approximately spatially homogeneous over the size of the probe. Furthermore the magnetic field in the electromagnetic wave satisfies

$$\boldsymbol{B}(t) = \frac{1}{2}\nabla\times(\boldsymbol{B}(t)\times\boldsymbol{x}) \tag{15.13}$$

and

$$|\boldsymbol{B}| = \frac{1}{c}|\boldsymbol{E}|,\quad |\dot{\boldsymbol{B}}| = \frac{\omega}{c}|\boldsymbol{E}| = \frac{2\pi}{\lambda}|\boldsymbol{E}|.$$

We have

$$\left|\frac{\partial \boldsymbol{A}}{\partial t}\right| = \frac{1}{2}\left|\dot{\boldsymbol{B}} \times \boldsymbol{x}\right| \simeq \frac{\pi a}{\lambda}\,|\boldsymbol{E}| \ll |\boldsymbol{E}|\,,$$

and therefore the description of $\boldsymbol{E}$ only through the electric potential,

$$\boldsymbol{E}(t) = -\nabla\Phi(\boldsymbol{x},t) = \nabla(\boldsymbol{E}(t)\cdot\boldsymbol{x}),$$

is justified for $\lambda \gg a$. Furthermore, the magnitudes of magnetic contributions to the Schrödinger equation are of order

$$\frac{q\hbar}{m}\,|\boldsymbol{A}\cdot\nabla\Psi| \simeq \frac{q\hbar}{2mc}\,|\boldsymbol{E}|\,|\boldsymbol{x}|\,|\nabla\Psi| \simeq \frac{q\hbar}{2mc}\,|\boldsymbol{E}|\,|\Psi|\,, \tag{15.14}$$

$$\frac{q^2}{2m}\boldsymbol{A}^2\,|\Psi| \simeq \frac{q^2}{8mc^2}\boldsymbol{E}^2\boldsymbol{x}^2\,|\Psi| \simeq \frac{q^2a^2}{8mc^2}\boldsymbol{E}^2\,|\Psi|\,. \tag{15.15}$$

For comparison, the electric contribution has a magnitude of order

$$q\,|\boldsymbol{E}|\,|\boldsymbol{x}|\,|\Psi| \simeq qa\,|\boldsymbol{E}|\,|\Psi|\,.$$

The ratio of the linear magnetic term (15.14) to the electric term is $\hbar/(2mca)$. If we use the electron mass for m, we find

$$\frac{\hbar}{2mca} \leq 2\times 10^{-3} \times \frac{1\,\text{Å}}{a},$$

i.e. the linear magnetic term is often negligible compared to the electric term.

The ratio of the second magnetic term (15.15) to the electric term is approximately $qa\,|\boldsymbol{E}|\,/8mc^2$. Validity of the non-relativistic approximation requires that the electrostatic energy $qa\,|\boldsymbol{E}|$ due to the electric field should be small compared to mc^2. Therefore we also find that the second magnetic term should be negligible compared to the electric term. Quantitatively, if we assume $mc^2 = 511\,\text{keV}$, we have $ea\,|\boldsymbol{E}|\,/8mc^2 \ll 1$ for

$$|\boldsymbol{E}| \ll \frac{8mc^2}{ea} = 4\times 10^{16}\,\frac{\text{V}}{\text{m}} \times \frac{1\,\text{Å}}{a}.$$

For comparison, the internal field strength in hydrogen is of order $e/(4\pi\epsilon_0 a_0^2) \simeq 5\times 10^{11}$ V/m.

We conclude that for $\lambda \gg a$ the effect of external electromagnetic fields can be approximated by the addition of a term

$$\Delta V(\mathbf{x},t) = q\Phi(\mathbf{x},t) = -q\boldsymbol{E}(t)\cdot\mathbf{x} = -\boldsymbol{d}\cdot\boldsymbol{E}(t) \tag{15.16}$$

in the Schrödinger equation. The approximation of spatially homogeneous external field yields a perturbation proportional to the dipole operator and is therefore denoted as *dipole approximation*.

Two cautionary remarks are in order at this point. The term *dipole approximation* is nowadays more widely used for the long wavelength approximation $\exp(\mathrm{i}\boldsymbol{k}\cdot\boldsymbol{x}) \simeq 1$ in matrix elements irrespective of whether the perturbation operator has the dipole form (15.16) or is given in terms of the coupling to the vector potential $\boldsymbol{A}(\mathbf{x}, t)$ in (15.12).

Furthermore, if we describe electromagnetic interactions at the level of photon-matter interactions, the dipole approximation (15.16) is generically limited to first order perturbation theory, and holds in second order perturbation theory only if additional conditions are met, see Section 18.9 and Problem 18.10.

Dipole selection rules

The first order scattering matrix elements in dipole approximation are given by $S_{fi} = \sqrt{2\pi}\mathrm{i}q\boldsymbol{E}(\omega_{fi})\cdot\langle f|\mathbf{x}|i\rangle/\hbar$, i.e. only transitions $|i\rangle \to |f\rangle$ with non-vanishing dipole matrix elements $q\langle f|\mathbf{x}|i\rangle$ are allowed in this approximation. This yields straightforward selection rules for states which are eigenstates of $\boldsymbol{M}^2$ and M_z. The commutator relation $[M_z, \mathrm{z}] = 0$ implies

$$\langle n', \ell', m'|[M_z, \mathrm{z}]|n, \ell, m\rangle = \hbar\langle n', \ell', m'|\mathrm{z}|n, \ell, m\rangle(m' - m) = 0, \tag{15.17}$$

and therefore an electric field component in z direction can only induce transitions between states with the same magnetic quantum number.

In the same way, the commutators $[M_z, \mathrm{x} \pm \mathrm{iy}] = \pm\hbar(\mathrm{x} \pm \mathrm{iy})$ imply

$$\langle n', \ell', m'|\mathrm{x} \pm \mathrm{iy}|n, \ell, m\rangle(m' - m \mp 1) = 0, \tag{15.18}$$

such that electric field components in the (x, y) plane can only induce transitions which increase or decrease the magnetic quantum number by one unit.

Finally, the fairly complicated relation $[\boldsymbol{M}^2, [\boldsymbol{M}^2, \mathbf{x}]] = 2\hbar^2\{\boldsymbol{M}^2, \mathbf{x}\}$ yields

$$\langle n', \ell', m'|\mathbf{x}|n, \ell, m\rangle(\ell + \ell')(\ell + \ell' + 2)[(\ell - \ell')^2 - 1] = 0. \tag{15.19}$$

This implies that the matrix element can be non-vanishing only if $\ell' = \ell \pm 1$. $\ell' = \ell = 0$ is not a solution, because in this case the wave functions depend only on r and the angular integrations then show that the matrix element vanishes.

Equations (15.17–15.19) imply the dipole selection rules $\Delta\ell = \pm 1$ and $\Delta m = 0, \pm 1$.

15.2 Stark effect and static polarizability tensors

Polarizability tensors characterize the response of a quantum system to an external electric field $\boldsymbol{E}$. The calculation of polarizability tensors is another example of applications of second order perturbation theory in materials science. It also illustrates the role of perturbation theory in derivations of quantum mechanical expressions for measurable physical quantities, which were first introduced in classical electrodynamics and were initially approximated by means of simple mechanical models.

The calculation of polarizabilities generically involves many particles and related dipole operators $V(t) = -\sum_{i=1}^{N} q_i\boldsymbol{E}(t)\cdot\mathbf{x}_i$, where it is assumed that all particles are confined to a region which is still small compared to the wavelength of the electric field. We will develop the theory in a single-particle approximation in the sense that we only use the single charged (quasi)particle operator $V(t) = -q\boldsymbol{E}(t)\cdot\mathbf{x}$. In the present section we will do this for time-independent external field, where we can use the techniques of time-independent perturbation theory. The case of dynamical polarizability for time-dependent external fields will be discussed in Section 15.3.

Linear Stark effect

Before we jump into the second order calculation of the response to an electric field, we consider the implications of first order perturbation theory for the dipole approximation.

An external static electric field shifts the Hamilton operator according to

$$H_0 \to H = H_0 + V = H_0 - q\boldsymbol{E}\cdot\mathbf{x}.$$

Time-independent perturbation theory tells us that the first order shifts of atomic or molecular energy levels due to the external field have to be determined as the eigenvalues of the matrix

$$\langle\Psi_{n,\alpha}^{(0)}|V(\mathbf{x})|\Psi_{n,\beta}^{(0)}\rangle = -q\langle\Psi_{n,\alpha}^{(0)}|\mathbf{x}|\Psi_{n,\beta}^{(0)}\rangle\cdot\boldsymbol{E},$$

and when the n-th degeneracy subspace has been internally diagonalized with respect to $V(\mathbf{x})$, the first order shifts are

$$E_{n,\alpha}^{(1)} = -q\langle\Psi_{n,\alpha}^{(0)}|\mathbf{x}|\Psi_{n,\alpha}^{(0)}\rangle\cdot\boldsymbol{E} = -\boldsymbol{d}_{n,\alpha}\cdot\boldsymbol{E},$$

with the *intrinsic* dipole moment in the state $|\Psi_{n,\alpha}^{(0)}\rangle$

$$\boldsymbol{d}_{n,\alpha} = q\langle\Psi_{n,\alpha}^{(0)}|\mathbf{x}|\Psi_{n,\alpha}^{(0)}\rangle. \tag{15.20}$$

The perturbation V has odd parity under $\boldsymbol{x} \to -\boldsymbol{x}$, while atomic states of opposite parity are usually not degenerate. Therefore in systems which are symmetric under the parity transformation $\boldsymbol{x} \to -\boldsymbol{x}$, the states in the n-th energy level usually satisfy

$$\langle \Psi^{(0)}_{n,\alpha}|\mathbf{x}|\Psi^{(0)}_{n,\beta}\rangle = 0$$

because the integrand is odd under the parity transformation. Usually this implies absence of a linear Stark effect in atoms, and the same remark applies to molecules with parity symmetry. An important exception is the hydrogen atom due to ℓ-degeneracy of its energy levels (if the matrix elements of V are larger than the fine structure of the hydrogen levels). States with angular momentum quantum number ℓ have parity $(-1)^\ell$, so that the n-th hydrogen level with $n > 1$ contains degenerate states of opposite parity. Diagonalization of V in that degeneracy subspace then yields states $|\Psi^{(0)}_{n,\alpha}\rangle$ with $\langle \Psi^{(0)}_{n,\alpha}|\mathbf{x}|\Psi^{(0)}_{n,\alpha}\rangle \neq 0$.

Quadratic Stark effect and the static polarizability tensor

Second order perturbation theory yields the following corrections to discrete atomic or molecular energy levels,

$$E^{(2)}_{n\alpha} = \sum\!\!\!\!\!\!\int_{m\neq n} \sum_\beta \frac{|\langle \Psi^{(0)}_{m\beta}|V|\Psi^{(0)}_{n\alpha}\rangle|^2}{E^{(0)}_n - E^{(0)}_m}$$

$$= q^2\boldsymbol{E}\cdot \sum\!\!\!\!\!\!\int_{m\neq n} \sum_\beta \frac{\langle \Psi^{(0)}_{n\alpha}|\mathbf{x}|\Psi^{(0)}_{m\beta}\rangle\langle \Psi^{(0)}_{m\beta}|\mathbf{x}|\Psi^{(0)}_{n\alpha}\rangle}{E^{(0)}_n - E^{(0)}_m}\cdot\boldsymbol{E}.$$

The notation takes into account that the intermediate levels can be continuous, but degeneracy indices are always discrete.

We can write the second order shifts in the form

$$E^{(2)}_{n\alpha} = -\frac{1}{2}\boldsymbol{d}_{(n\alpha)}\cdot\boldsymbol{E} = -\frac{1}{2}\boldsymbol{E}\cdot\underline{\alpha}_{(n\alpha)}\cdot\boldsymbol{E},$$

where

$$\boldsymbol{d}_{(n\alpha)} = \underline{\alpha}_{(n\alpha)}\cdot\boldsymbol{E}$$

is the *induced* dipole moment and $\underline{\alpha}_{(n\alpha)}$ is the static electronic polarizability tensor in the state $|\Psi^{(0)}_{n\alpha}\rangle$,

$$\underline{\alpha}_{(n\alpha)} = -q^2 \sum\!\!\!\!\!\!\int_{m\neq n} \sum_\beta \frac{1}{E^{(0)}_n - E^{(0)}_m}\Big(\langle \Psi^{(0)}_{n\alpha}|\mathbf{x}|\Psi^{(0)}_{m\beta}\rangle\otimes\langle \Psi^{(0)}_{m\beta}|\mathbf{x}|\Psi^{(0)}_{n\alpha}\rangle$$
$$+ \langle \Psi^{(0)}_{m\beta}|\mathbf{x}|\Psi^{(0)}_{n\alpha}\rangle\otimes\langle \Psi^{(0)}_{n\alpha}|\mathbf{x}|\Psi^{(0)}_{m\beta}\rangle\Big). \qquad (15.21)$$

Note that in the ground state $\alpha_{ii} > 0$ (no summation convention), i.e. in second order perturbation theory, which usually should capture all linear contributions from a weak external electric field to the induced dipole moment, there is no electronic dia-electricity for the ground state.

15.3 Dynamical polarizability tensors

We cannot use time-independent perturbation theory if the perturbation operator $V(t) = -q\mathbf{x}\cdot\boldsymbol{E}(t)$ varies with time. Application of our results from Chapter 13 for time-dependent perturbations implies that the first order transition probability from a state $|m\rangle$ into a state $|n\rangle$ under the action of the electric field $\boldsymbol{E}(t)$ between times t' and t is proportional to[2]

$$P^{(1)}_{m\to n}(t,t') = \left|\frac{q}{\hbar}\int_{t'}^{t} d\tau\, \exp(\mathrm{i}\omega_{nm}\tau)\boldsymbol{E}(\tau)\cdot\langle n|\mathbf{x}|m\rangle\right|^2,$$

where

$$\omega_{nm} = \frac{1}{\hbar}(E_n - E_m).$$

For $t' \to -\infty$, $t \to \infty$, this becomes in particular (see our previous results (13.28, 13.29))

$$P^{(1)}_{m\to n} = 2\pi\left|\frac{q}{\hbar}\boldsymbol{E}(\omega_{nm})\cdot\langle n|\mathbf{x}|m\rangle\right|^2,$$

i.e. long term action of an external electric field can induce a transition in first order between energy levels E_m and E_n only if the field contains a Fourier component of the corresponding frequency ω_{nm}.

However, at this time we are interested in the problem how equation (15.21) can be generalized to a dynamical polarizability in the presence of a time-dependent external field $\boldsymbol{E}(t)$.

Suppose the system was in the state $|\Psi^{(0)}_{n,\alpha}(0)\rangle \equiv |\Psi^{(0)}_{n,\alpha}\rangle$ at $t = 0$, when it begins to experience the effect of the electric field. The shift of the wave function $|\Psi^{(0)}_{n,\alpha}(t)\rangle$ under the influence of the external field is

$$\begin{aligned}|\Psi_{n,\alpha}(t)\rangle - |\Psi^{(0)}_{n,\alpha}(t)\rangle &= \Theta(t)\left[U(t) - U_0(t)\right]|\Psi^{(0)}_{n,\alpha}\rangle\\ &= \Theta(t)U_0(t)\left[U_0^+(t)U(t)U_0(0) - 1\right]|\Psi^{(0)}_{n,\alpha}\rangle\\ &= \Theta(t)U_0(t)\left[U_D(t) - 1\right]|\Psi^{(0)}_{n,\alpha}\rangle,\end{aligned}$$

[2] Recall that $|S_{nm}|^2$ is a true transition probability only if the initial and final state are discrete, while otherwise it enters into decay rates or cross sections.

and the first order shift is therefore

$$
\begin{aligned}
|\Psi^{(1)}_{n,\alpha}(t)\rangle &= -\frac{\mathrm{i}}{\hbar}\Theta(t)U_0(t)\int_0^t d\tau\, H_D(\tau)|\Psi^{(0)}_{n,\alpha}\rangle \\
&= -\frac{\mathrm{i}}{\hbar}\Theta(t)U_0(t)\int_0^t d\tau\, U_0^+(\tau)V(\tau)U_0(\tau)|\Psi^{(0)}_{n,\alpha}\rangle \\
&= -\frac{\mathrm{i}}{\hbar}\Theta(t)\int_0^t d\tau\, U_0(t-\tau)V(\tau)U_0(\tau)|\Psi^{(0)}_{n,\alpha}\rangle.
\end{aligned}
$$

The *induced* dipole moment in the state $|\Psi^{(0)}_{n,\alpha}\rangle$ is then given in leading order by the first order terms (recall that the 0th order term corresponds to the *intrinsic* dipole moment (15.20))

$$
\begin{aligned}
\boldsymbol{d}_{(n\alpha)}(t) &= \langle\Psi^{(0)}_{n,\alpha}(t)|q\mathbf{x}|\Psi^{(1)}_{n,\alpha}(t)\rangle + \langle\Psi^{(1)}_{n,\alpha}(t)|q\mathbf{x}|\Psi^{(0)}_{n,\alpha}(t)\rangle \\
&= q^2\frac{\mathrm{i}}{\hbar}\Theta(t)\int_0^t d\tau\,\langle\Psi^{(0)}_{n,\alpha}|U_0^+(t)\mathbf{x}U_0(t-\tau)\mathbf{x}\cdot\boldsymbol{E}(\tau)U_0(\tau)|\Psi^{(0)}_{n,\alpha}\rangle \\
&\quad - q^2\frac{\mathrm{i}}{\hbar}\Theta(t)\int_0^t d\tau\,\langle\Psi^{(0)}_{n,\alpha}|U_0^+(\tau)\mathbf{x}\cdot\boldsymbol{E}(\tau)U_0^+(t-\tau)\mathbf{x}U_0(t)|\Psi^{(0)}_{n,\alpha}\rangle.
\end{aligned}
$$

This becomes after insertion of complete sets of unperturbed states in $U_0(t-\tau)=U_0(t)U_0^+(\tau)$ and $U_0^+(t-\tau)=U_0(\tau)U_0^+(t)$

$$
\begin{aligned}
\boldsymbol{d}_{(n\alpha)}(t) &= q^2\frac{\mathrm{i}}{\hbar}\Theta(t)\sum\!\!\!\!\!\!\int_{m,\beta}\int_0^t d\tau\,\exp[\mathrm{i}\omega_{nm}(t-\tau)]\langle\Psi^{(0)}_{n,\alpha}|\mathbf{x}|\Psi^{(0)}_{m,\beta}\rangle \\
&\quad\times\langle\Psi^{(0)}_{m,\beta}|\mathbf{x}\cdot\boldsymbol{E}(\tau)|\Psi^{(0)}_{n,\alpha}\rangle - q^2\frac{\mathrm{i}}{\hbar}\Theta(t)\sum\!\!\!\!\!\!\int_{m,\beta}\int_0^t d\tau\,\exp[-\mathrm{i}\omega_{nm}(t-\tau)] \\
&\quad\times\langle\Psi^{(0)}_{n,\alpha}|\mathbf{x}\cdot\boldsymbol{E}(\tau)|\Psi^{(0)}_{m,\beta}\rangle\langle\Psi^{(0)}_{m,\beta}|\mathbf{x}|\Psi^{(0)}_{n,\alpha}\rangle \\
&= \int_0^\infty d\tau\,\underline{\alpha}_{(n\alpha)}(t-\tau)\cdot\boldsymbol{E}(\tau),
\end{aligned}
\tag{15.22}
$$

with a dynamical polarizability tensor

$$
\begin{aligned}
\underline{\alpha}_{(n\alpha)}(t) &= q^2\frac{\mathrm{i}}{\hbar}\Theta(t)\left(\sum\!\!\!\!\!\!\int_{m,\beta}\exp(\mathrm{i}\omega_{nm}t)\langle\Psi^{(0)}_{n,\alpha}|\mathbf{x}|\Psi^{(0)}_{m,\beta}\rangle\otimes\langle\Psi^{(0)}_{m,\beta}|\mathbf{x}|\Psi^{(0)}_{n,\alpha}\rangle\right. \\
&\quad\left.-\sum\!\!\!\!\!\!\int_{m,\beta}\exp(-\mathrm{i}\omega_{nm}t)\langle\Psi^{(0)}_{m,\beta}|\mathbf{x}|\Psi^{(0)}_{n,\alpha}\rangle\otimes\langle\Psi^{(0)}_{n,\alpha}|\mathbf{x}|\Psi^{(0)}_{m,\beta}\rangle\right).
\end{aligned}
\tag{15.23}
$$

Now we assume harmonic time dependence of an electric field which is switched on at $t=0$,

$$
\boldsymbol{E}(\tau)\equiv\boldsymbol{E}_\omega(\tau)=\boldsymbol{E}\Theta(t)\sin(\omega\tau)=\boldsymbol{E}\Theta(t)\,\frac{\exp(\mathrm{i}\omega\tau)-\exp(-\mathrm{i}\omega\tau)}{2\mathrm{i}}.
$$

The time integrals in the two terms for $\boldsymbol{d}_{(n\alpha)}(t)$ then yield

$$\begin{aligned}
&\pm\frac{q^2}{2\hbar}\int_0^t d\tau\Big(\exp[\pm\mathrm{i}\omega_{nm}(t-\tau)+\mathrm{i}\omega\tau]-\exp[\pm\mathrm{i}\omega_{nm}(t-\tau)-\mathrm{i}\omega\tau]\Big)\\
&=\pm\frac{q^2}{2\mathrm{i}\hbar}\left(\frac{\exp(\mathrm{i}\omega t)-\exp(\pm\mathrm{i}\omega_{nm}t)}{\omega\mp\omega_{nm}}+\frac{\exp(-\mathrm{i}\omega t)-\exp(\pm\mathrm{i}\omega_{nm}t)}{\omega\pm\omega_{nm}}\right)\\
&=\pm\frac{q^2}{\mathrm{i}\hbar}\frac{\omega\cos(\omega t)\pm\mathrm{i}\omega_{nm}\sin(\omega t)-\omega\exp(\pm\mathrm{i}\omega_{nm}t)}{\omega^2-\omega_{nm}^2}.
\end{aligned}\tag{15.24}$$

We also assume slowly oscillating field in the sense $\omega \ll |\omega_{nm}|$ for all quantum numbers m which correspond to large matrix elements $|\langle\Psi^{(0)}_{m,\beta}|\mathbf{x}|\Psi^{(0)}_{n,\alpha}\rangle|$. This means that the external field is not likely to induce direct transitions between different energy levels. Under these conditions, the contribution from the integrals in equation (15.24) to $\boldsymbol{d}_{(n\alpha)}(t)$ will be dominated by the term which is in phase with the external field,

$$\begin{aligned}
&\pm\frac{q^2}{2\hbar}\int_0^t d\tau\Big(\exp[\pm\mathrm{i}\omega_{nm}(t-\tau)+\mathrm{i}\omega\tau]-\exp[\pm\mathrm{i}\omega_{nm}(t-\tau)-\mathrm{i}\omega\tau]\Big)\\
&\to\frac{q^2}{\hbar}\frac{\omega_{nm}\sin(\omega t)}{\omega^2-\omega_{nm}^2},
\end{aligned}$$

and the induced dipole moment in this approximation is

$$\begin{aligned}
\boldsymbol{d}_{(n\alpha)\omega}(t)=\frac{q^2}{\hbar}\sum\!\!\!\!\!\!\!\int_{m,\beta}\frac{\omega_{mn}}{\omega_{mn}^2-\omega^2}&\Big(\langle\Psi^{(0)}_{n,\alpha}|\mathbf{x}|\Psi^{(0)}_{m,\beta}\rangle\langle\Psi^{(0)}_{m,\beta}|\mathbf{x}\cdot\boldsymbol{E}_\omega(t)|\Psi^{(0)}_{n,\alpha}\rangle\\
&+\langle\Psi^{(0)}_{m,\beta}|\mathbf{x}|\Psi^{(0)}_{n,\alpha}\rangle\langle\Psi^{(0)}_{n,\alpha}|\mathbf{x}\cdot\boldsymbol{E}_\omega(t)|\Psi^{(0)}_{m,\beta}\rangle\Big).
\end{aligned}$$

This can also be written as

$$\boldsymbol{d}_{(n\alpha)\omega}(t)=\underline{\alpha}_{(n\alpha)}(\omega)\cdot\boldsymbol{E}_\omega(t)$$

with the frequency dependent polarizability tensor for the state $|\Psi^{(0)}_{n,\alpha}\rangle$ (usually the ground state)

$$\begin{aligned}
\underline{\alpha}_{(n\alpha)}(\omega)=\frac{q^2}{\hbar}\sum\!\!\!\!\!\!\!\int_{m,\beta}\frac{\omega_{mn}}{\omega_{mn}^2-\omega^2}&\Big(\langle\Psi^{(0)}_{n,\alpha}|\mathbf{x}|\Psi^{(0)}_{m,\beta}\rangle\otimes\langle\Psi^{(0)}_{m,\beta}|\mathbf{x}|\Psi^{(0)}_{n,\alpha}\rangle\\
&+\langle\Psi^{(0)}_{m,\beta}|\mathbf{x}|\Psi^{(0)}_{n,\alpha}\rangle\otimes\langle\Psi^{(0)}_{n,\alpha}|\mathbf{x}|\Psi^{(0)}_{m,\beta}\rangle\Big).
\end{aligned}\tag{15.25}$$

The zero frequency polarizability tensor $\underline{\alpha}_{(n\alpha)}(0)$ is the static tensor (15.21), as expected.

The frequency dependent polarizability tensor is not only relevant for slowly oscillating fields, but appears implicitly already in the equations (15.22, 15.23), which do not include a restriction to slowly oscillating external field. If we agree to shift the denominator in (15.25) by small imaginary numbers according to

$$\underline{\alpha}_{(n\alpha)}(\omega) = \frac{q^2}{\hbar}\left(\sum\!\!\!\!\!\!\int_{m,\beta} \frac{\omega_{nm}}{\omega^2 - \omega_{nm}^2 - \mathrm{i}\epsilon}\langle\Psi^{(0)}_{n,\alpha}|\mathbf{x}|\Psi^{(0)}_{m,\beta}\rangle \otimes \langle\Psi^{(0)}_{m,\beta}|\mathbf{x}|\Psi^{(0)}_{n,\alpha}\rangle\right.$$
$$\left. + \sum\!\!\!\!\!\!\int_{m,\beta} \frac{\omega_{nm}}{\omega^2 - \omega_{nm}^2 + \mathrm{i}\epsilon}\langle\Psi^{(0)}_{m,\beta}|\mathbf{x}|\Psi^{(0)}_{n,\alpha}\rangle \otimes \langle\Psi^{(0)}_{n,\alpha}|\mathbf{x}|\Psi^{(0)}_{m,\beta}\rangle\right), \quad (15.26)$$

we find that the dynamical polarizability tensors in equations (15.23) and (15.26) are related via

$$\underline{\alpha}_{(n\alpha)}(t) = \frac{\Theta(t)}{\pi}\int_{-\infty}^{\infty} d\omega\, \underline{\alpha}_{(n\alpha)}(\omega)\exp(-\mathrm{i}\omega t). \quad (15.27)$$

Oscillator strength

Equation (15.26) yields an averaged polarizability

$$\alpha_{(n\alpha)}(\omega) = \frac{1}{3}\mathrm{tr}\,\underline{\alpha}_{(n\alpha)}(\omega) = \frac{2q^2}{3\hbar}\sum\!\!\!\!\!\!\int_{m,\beta} \frac{\omega_{mn}}{\omega_{mn}^2 - \omega^2}|\langle\Psi^{(0)}_{m,\beta}|\mathbf{x}|\Psi^{(0)}_{n,\alpha}\rangle|^2$$
$$= \frac{q^2}{m}\sum\!\!\!\!\!\!\int_{m,\beta} \frac{f_{m,\beta;n,\alpha}}{\omega_{mn}^2 - \omega^2}$$

with the *oscillator strength* for the transition $|\Psi^{(0)}_{n,\alpha}\rangle \to |\Psi^{(0)}_{m,\beta}\rangle$:

$$f_{m,\beta;n,\alpha} = \frac{2m}{3\hbar}\omega_{mn}|\langle\Psi^{(0)}_{m,\beta}|\mathbf{x}|\Psi^{(0)}_{n,\alpha}\rangle|^2 = -f_{n,\alpha;m,\beta}. \quad (15.28)$$

We use m both for the mass of the charged (quasi)particle which has its wave functions shifted due to the external field, and as a label for the intermediate states. Since mass never appears as an index in equation (15.28) or the following equations, this should not cause confusion.

The polarizability is also often averaged over degenerate initial states. If the degeneracy of the n-th energy level is g_n, then

$$\alpha_n(\omega) = \frac{1}{g_n}\sum_\alpha \alpha_{(n\alpha)}(\omega) = \frac{q^2}{m}\sum\!\!\!\!\!\!\int_m \frac{f_{m|n}}{\omega_{mn}^2 - \omega^2}$$

with an effective oscillator strength which is averaged over degenerate initial states and summed over degenerate final states,

$$f_{m|n} = \frac{1}{g_n}\sum_{\alpha,\beta} f_{m,\beta;n,\alpha} = -\frac{g_m}{g_n}f_{n|m}. \quad (15.29)$$

With these conventions, positive oscillator strength corresponds to absorption and negative oscillator strength corresponds to emission. Oscillator strengths are sometimes also defined through absolute values, but for the f-sum rules below it plays a role that emission transitions contribute with negative sign.

For an explanation of the name *oscillator strength* for $f_{m,\beta;n,\alpha}$, we observe that a classical isotropic harmonic oscillator model for polarizability

$$m\ddot{\boldsymbol{x}}(t) + m\omega_0^2\boldsymbol{x}(t) = q\boldsymbol{E}\sin(\omega t)$$

yields an induced dipole moment

$$\boldsymbol{d}_\omega(t) = q\boldsymbol{x}(t) = \frac{q^2}{m}\frac{1}{\omega_0^2 - \omega^2}\boldsymbol{E}\sin(\omega t) = \alpha(\omega)\boldsymbol{E}_\omega(t)$$

with the polarizability

$$\alpha(\omega) = \frac{q^2}{m}\frac{1}{\omega_0^2 - \omega^2},$$

i.e. every virtual transition $|\Psi^{(0)}_{n,\alpha}\rangle \to |\Psi^{(0)}_{m,\beta}\rangle$ contributes effectively like an oscillator of frequency $|\omega_{mn}| = \left|E^{(0)}_m - E^{(0)}_n\right|/\hbar$ to the polarizability $\alpha_{(n\alpha)}(\omega)$ of the state $|\Psi^{(0)}_{n,\alpha}\rangle$, but the contribution of that transition is weighted with the oscillator strength (15.28).

Thomas-Reiche-Kuhn sum rule (f-sum rule) for the oscillator strength

Kuhn, Reiche and Thomas found a sum rule for the oscillator strength already in the framework of old quantum theory[3]. The quantum mechanical proof is based on the fact that the Hamiltonian operator $H = (\mathbf{p}^2/2m) + V(\mathbf{x})$ yields a commutator

$$[H, \mathbf{x}] = \frac{\hbar\mathbf{p}}{im}. \tag{15.30}$$

This implies for a discrete normalized state $|\Psi^{(0)}_{n,\alpha}\rangle$

$$\begin{aligned}\sum\!\!\!\!\!\!\int_{m,\beta} f_{m,\beta;n,\alpha} &= \frac{2m}{3\hbar}\sum\!\!\!\!\!\!\int_{m,\beta} \omega_{mn}\langle\Psi^{(0)}_{n,\alpha}|\mathbf{x}|\Psi^{(0)}_{m,\beta}\rangle\cdot\langle\Psi^{(0)}_{m,\beta}|\mathbf{x}|\Psi^{(0)}_{n,\alpha}\rangle \\ &= \frac{2m}{3\hbar^2}\sum\!\!\!\!\!\!\int_{m,\beta}\left(E^{(0)}_m - E^{(0)}_n\right)\langle\Psi^{(0)}_{n,\alpha}|\mathbf{x}|\Psi^{(0)}_{m,\beta}\rangle\cdot\langle\Psi^{(0)}_{m,\beta}|\mathbf{x}|\Psi^{(0)}_{n,\alpha}\rangle\end{aligned}$$

[3] W. Kuhn, Z. Phys. 33, 408 (1925); F. Reiche, W. Thomas, Z. Phys. 34, 510 (1925).

$$= \frac{m}{3\hbar^2} \sum\!\!\!\!\!\!\int_{m,\beta} \Big(\langle \Psi^{(0)}_{n,\alpha} | \mathbf{x} | \Psi^{(0)}_{m,\beta} \rangle \cdot \langle \Psi^{(0)}_{m,\beta} | [H_0, \mathbf{x}] | \Psi^{(0)}_{n,\alpha} \rangle$$

$$- \langle \Psi^{(0)}_{n,\alpha} | [H_0, \mathbf{x}] | \Psi^{(0)}_{m,\beta} \rangle \cdot \langle \Psi^{(0)}_{m,\beta} | \mathbf{x} | \Psi^{(0)}_{n,\alpha} \rangle \Big)$$

$$= \frac{1}{3i\hbar} \sum\!\!\!\!\!\!\int_{m,\beta} \Big(\langle \Psi^{(0)}_{n,\alpha} | \mathbf{x} | \Psi^{(0)}_{m,\beta} \rangle \cdot \langle \Psi^{(0)}_{m,\beta} | \mathbf{p} | \Psi^{(0)}_{n,\alpha} \rangle$$

$$- \Big\langle \Psi^{(0)}_{n,\alpha} | \mathbf{p} | \Psi^{(0)}_{m,\beta} \rangle \cdot \langle \Psi^{(0)}_{m,\beta} | \mathbf{x} | \Psi^{(0)}_{n,\alpha} \rangle \Big)$$

$$= \frac{1}{3i\hbar} \langle \Psi^{(0)}_{n,\alpha} | (\mathbf{x} \cdot \mathbf{p} - \mathbf{p} \cdot \mathbf{x}) | \Psi^{(0)}_{n,\alpha} \rangle = 1.$$

This is the[4] *Thomas-Reiche-Kuhn sum rule*,

$$\sum\!\!\!\!\!\!\int_{m,\beta} f_{m,\beta;n,\alpha} = 1. \tag{15.31}$$

Averaging over initial degeneracy indices (15.29) then also yields

$$\sum\!\!\!\!\!\!\int_{m} f_{m|n} = 1. \tag{15.32}$$

Equation (15.30) implies a further relation which connects matrix elements of $\mathbf{x}$ and $\mathbf{p}$,

$$\omega_{mn} \langle \Psi^{(0)}_{m,\beta} | \mathbf{x} | \Psi^{(0)}_{n,\alpha} \rangle = \frac{1}{im} \langle \Psi^{(0)}_{m,\beta} | \mathbf{p} | \Psi^{(0)}_{n,\alpha} \rangle.$$

This yields an alternative representation of the oscillator strength

$$f_{m,\beta;n,\alpha} = \frac{2}{3m\hbar\omega_{mn}} |\langle \Psi^{(0)}_{m,\beta} | \mathbf{p} | \Psi^{(0)}_{n,\alpha} \rangle|^2, \tag{15.33}$$

which is known as the *velocity form* of the oscillator strength, while equation (15.28) is denoted as the *length form* of the oscillator strength. Yet another common definition in atomic, molecular and optical physics is

$$f_{m,\beta;n,\alpha} = \frac{2m\omega_{mn}}{3\hbar q^2} S_{m,\beta;n,\alpha}, \quad f_{m|n} = \frac{2m\omega_{mn}}{3\hbar q^2} S_{m,n},$$

[4] If the wave functions are N-particle wave functions and the potential V is the corresponding sum of dipole operators, the number on the right hand side of the sum rules becomes N.

with the *electric dipole line strength* of the transition $|\Psi_{n,\alpha}^{(0)}\rangle \to |\Psi_{m,\beta}^{(0)}\rangle$

$$\mathcal{S}_{m,\beta;n,\alpha} = |\langle\Psi_{m,\beta}^{(0)}|q\mathbf{x}|\Psi_{n,\alpha}^{(0)}\rangle|^2 = \left|\frac{q}{m\omega_{mn}}\langle\Psi_{m,\beta}^{(0)}|\mathbf{p}|\Psi_{n,\alpha}^{(0)}\rangle\right|^2,$$

$$\mathcal{S}_{m,n} = \frac{1}{g_n}\sum_{\alpha,\beta}\mathcal{S}_{m,\beta;n,\alpha} = \frac{g_m}{g_n}\mathcal{S}_{n,m}.$$

Tensorial oscillator strengths and sum rules

We can define oscillator strength tensors through the relations

$$\underline{\alpha}_{(n\alpha)}(\omega) = \frac{q^2}{m}\sum\!\!\!\!\!\!\int_{m,\beta}\frac{\underline{f}_{m,\beta;n,\alpha}}{\omega_{mn}^2-\omega^2},$$

$$\underline{\alpha}_n(\omega) = \frac{1}{g_n}\sum_\alpha\underline{\alpha}_{(n\alpha)}(\omega) = \frac{q^2}{m}\sum_m\frac{\underline{f}_{m,n}}{\omega_{mn}^2-\omega^2},$$

i.e. we have representations for oscillator strength tensors

$$\begin{aligned}
\underline{f}_{m,\beta;n,\alpha} &= \frac{m}{\hbar}\omega_{mn}\Big(\langle\Psi_{n,\alpha}^{(0)}|\mathbf{x}|\Psi_{m,\beta}^{(0)}\rangle\otimes\langle\Psi_{m,\beta}^{(0)}|\mathbf{x}|\Psi_{n,\alpha}^{(0)}\rangle \\
&\quad+\langle\Psi_{m,\beta}^{(0)}|\mathbf{x}|\Psi_{n,\alpha}^{(0)}\rangle\otimes\langle\Psi_{n,\alpha}^{(0)}|\mathbf{x}|\Psi_{m,\beta}^{(0)}\rangle\Big) \\
&= \frac{m}{2\hbar^2}\Big(\langle\Psi_{n,\alpha}^{(0)}|\mathbf{x}|\Psi_{m,\beta}^{(0)}\rangle\otimes\langle\Psi_{m,\beta}^{(0)}|[H_0,\mathbf{x}]|\Psi_{n,\alpha}^{(0)}\rangle \\
&\quad-\langle\Psi_{n,\alpha}^{(0)}|[H_0,\mathbf{x}]|\Psi_{m,\beta}^{(0)}\rangle\otimes\langle\Psi_{m,\beta}^{(0)}|\mathbf{x}|\Psi_{n,\alpha}^{(0)}\rangle \\
&\quad+\langle\Psi_{m,\beta}^{(0)}|[H_0,\mathbf{x}]|\Psi_{n,\alpha}^{(0)}\rangle\otimes\langle\Psi_{n,\alpha}^{(0)}|\mathbf{x}|\Psi_{m,\beta}^{(0)}\rangle \\
&\quad-\langle\Psi_{m,\beta}^{(0)}|\mathbf{x}|\Psi_{n,\alpha}^{(0)}\rangle\otimes\langle\Psi_{n,\alpha}^{(0)}|[H_0,\mathbf{x}]|\Psi_{m,\beta}^{(0)}\rangle\Big) \\
&= \frac{1}{2\mathrm{i}\hbar}\Big(\langle\Psi_{n,\alpha}^{(0)}|\mathbf{x}|\Psi_{m,\beta}^{(0)}\rangle\otimes\langle\Psi_{m,\beta}^{(0)}|\mathbf{p}|\Psi_{n,\alpha}^{(0)}\rangle-\langle\Psi_{n,\alpha}^{(0)}|\mathbf{p}|\Psi_{m,\beta}^{(0)}\rangle\otimes\langle\Psi_{m,\beta}^{(0)}|\mathbf{x}|\Psi_{n,\alpha}^{(0)}\rangle \\
&\quad+\langle\Psi_{m,\beta}^{(0)}|\mathbf{p}|\Psi_{n,\alpha}^{(0)}\rangle\otimes\langle\Psi_{n,\alpha}^{(0)}|\mathbf{x}|\Psi_{m,\beta}^{(0)}\rangle-\langle\Psi_{m,\beta}^{(0)}|\mathbf{x}|\Psi_{n,\alpha}^{(0)}\rangle\otimes\langle\Psi_{n,\alpha}^{(0)}|\mathbf{p}|\Psi_{m,\beta}^{(0)}\rangle\Big) \\
&= -\underline{f}_{n,\alpha;m,\beta},
\end{aligned} \tag{15.34}$$

and reduced oscillator strength tensors

$$\underline{f}_{m|n} = \frac{1}{g_n}\sum_{\alpha,\beta}\underline{f}_{m,\beta;n,\alpha} = -\frac{g_m}{g_n}\underline{f}_{n|m}. \tag{15.35}$$

This yields tensorial f-sum rules,

$$\sum_{m,\beta} \underline{f}_{m,\beta;n,\alpha} = \frac{1}{2i\hbar}\langle\Psi^{(0)}_{n,\alpha}|\left([\mathrm{x}_i, \mathrm{p}_j] - [\mathrm{p}_i, \mathrm{x}_j]\right)|\Psi^{(0)}_{n,\alpha}\rangle \boldsymbol{e}_i \otimes \boldsymbol{e}_j = \underline{1} = \sum_m \underline{f}_{m|n}.$$

For comparison, we note that the polarization tensor of an isotropic classical oscillator is easily shown to be

$$\underline{\alpha}(\omega) = \frac{q^2}{m}\frac{1}{\omega_0^2 - \omega^2}\underline{1}.$$

The standard oscillator strength is related to the oscillator strength tensor via

$$f_{m,\beta;n,\alpha} = \frac{1}{3}\mathrm{tr}\underline{f}_{m,\beta;n,\alpha}.$$

15.4 Problems

15.1. Show that the probability current density in the presence of electromagnetic potentials is given by

$$\boldsymbol{j} = \frac{\hbar}{2im}\left(\Psi^+\nabla\Psi - \nabla\Psi^+\cdot\Psi - 2i\frac{q}{\hbar}\Psi^+\boldsymbol{A}\Psi\right).$$

Is this expression gauge invariant?

15.2. Suppose that a particle is moving in a spatially homogeneous magnetic field $\boldsymbol{B}(t)$. Show that in order $|q\boldsymbol{B}|$ this yields the Zeeman term

$$\langle\boldsymbol{x}|H_Z(t)|\Psi(t)\rangle = -\langle\boldsymbol{x}|\frac{q}{2m}\boldsymbol{B}(t)\cdot\boldsymbol{l}|\Psi(t)\rangle = i\frac{q\hbar}{2m}\boldsymbol{B}(t)\cdot(\boldsymbol{x}\times\nabla)\Psi(\boldsymbol{x},t)$$

in the Schrödinger equation (15.6).

Hint: You can use equation (15.13).

15.3. A hydrogen atom is initially in its ground state when it is excited by an external electric field $\boldsymbol{E}(t)$.

15.3a. Show through direct evaluation of the matrix elements that the dipole term $V(t) = -q\mathbf{x}\cdot\boldsymbol{E}(t)$ in first order only excites higher level p states.

15.3b. The external field is

$$\boldsymbol{E}(t) = \boldsymbol{e}_z\mathcal{E}\exp(-t^2/\tau^2). \tag{15.36}$$

How large are the first order transition probabilities $P_{1\to n}$ into excited bound energy levels?

15.4. How large is the ionization probability for a hydrogen atom in the electric field (15.36) in leading order perturbation theory?

15.5. Calculate the ionization probability (13.40) for a hydrogen atom in its ground state which is perturbed by an oscillating electric field in z direction,

$$V(t) = ez\mathcal{E}\cos(\omega t).$$

15.6. Calculate the linear Stark effect for the first excited level of hydrogen due to a homogeneous static electric field $\boldsymbol{E}$.

15.7. Calculate the static polarizability tensor in the ground state of hydrogen.

15.8. Calculate the oscillator strengths $f_{n';n} = 2m\omega_{n'n}|\langle n'|x|n\rangle|^2/\hbar$ for a one-dimensional oscillator. Why does the equation for the one-dimensional oscillator strength differ by a factor 3 from the three-dimensional oscillator strength (15.28)?

15.9. Calculate the oscillator strengths $f_{n,\ell,m;1,0,0}$ for the hydrogen atom. How large is the sum

$$\sum_{\ell=0}^{\infty}\sum_{m=-\ell}^{\ell}\int_0^\infty dk\, k^2 f_{k,\ell,m;1,0,0}$$

of the oscillator strengths into Coulomb waves?

15.10. We consider the transition $|m\rangle \to |n\rangle$ due to an external electric field $\boldsymbol{E}(t)$. Show that the square of the corresponding first order scattering matrix element is related to the oscillator strength tensor of the transition through

$$|S_{nm}|^2 = \frac{\pi q^2}{m\hbar\omega_{nm}}\boldsymbol{E}(\omega_{nm})\cdot \underline{f}_{n;m}\cdot \boldsymbol{E}(\omega_{nm}).$$

15.11. Show that normalizable energy eigenstates, $\langle n|n\rangle = 1$, have vanishing momentum expectation values,

$$\langle n|\mathbf{p}|n\rangle = 0.$$

Why does this equation not hold for plane wave states?

15.12. Prove the *Bethe sum rule*[5],

$$\frac{2m}{\hbar}\sum\!\!\!\!\!\!\!\int_{m,\beta}\omega_{mn}\left|\langle\Psi^{(0)}_{m,\beta}|\exp(\mathrm{i}\boldsymbol{k}\cdot\mathbf{x})|\Psi^{(0)}_{n,\alpha}\rangle\right|^2 = \boldsymbol{k}^2.$$

[5]H. Bethe, Annalen Phys. 397, 325 (1930).

Chapter 16
Principles of Lagrangian Field Theory

The replacement of Newton's equation by quantum mechanical wave equations in the 1920s implied that by that time all known fundamental degrees of freedom in physics were described by fields like $\boldsymbol{A}(\boldsymbol{x},t)$ or $\Psi(\boldsymbol{x},t)$, and their dynamics was encoded in wave equations. However, all the known fundamental wave equations can be derived from a field theory version of Hamilton's principle[1], i.e. the concept of the Lagrange function $L(q(t),\dot{q}(t))$ and the related action $S=\int dt\,L$ generalizes to a Lagrange density $\mathcal{L}(\phi(\boldsymbol{x},t),\dot{\phi}(\boldsymbol{x},t),\nabla\phi(\boldsymbol{x},t))$ with related action $S=\int dt\int d^3\boldsymbol{x}\,\mathcal{L}$, such that all fundamental wave equations can be derived from the variation of an action,

$$\frac{\partial\mathcal{L}}{\partial\phi}-\partial_\mu\frac{\partial\mathcal{L}}{\partial(\partial_\mu\phi)}=0.$$

This formulation of dynamics is particularly useful for exploring the connection between symmetries and conservation laws of physical systems, and it also allows for a systematic approach to the quantization of fields, which allows us to describe creation and annihilation of particles.

16.1 Lagrangian field theory

Irrespective of whether we work with relativistic or non-relativistic field theories, it is convenient to use four-dimensional notation for coordinates and partial derivatives,

$$x^\mu=\{x^0,\boldsymbol{x}\}\equiv\{ct,\boldsymbol{x}\},\quad \partial_\mu=\frac{\partial}{\partial x^\mu}=\{\partial_0,\nabla\}.$$

[1]Please review Appendix A if you are not familiar with Lagrangian mechanics, or if you need a reminder.

R. Dick, *Advanced Quantum Mechanics*, Graduate Texts in Physics,
DOI 10.1007/978-3-319-25675-7_16

We proceed by first deriving the general field equations following from a Lagrangian $\mathcal{L}(\partial\phi_I, \phi_I)$ which depends on a set of fields $\phi_I(x) \equiv \phi_I(\boldsymbol{x}, t)$ and their first order derivatives $\partial_\mu\phi_I(x)$. These fields will be the Schrödinger field $\Psi(\boldsymbol{x}, t)$ and its complex conjugate field $\Psi^+(\boldsymbol{x}, t)$ in Chapter 17, but in Chapter 18 we will also deal with the wave function $\boldsymbol{A}(x)$ of the photon.

We know that the equations of motion for the variables $\boldsymbol{x}(t)$ of classical mechanics follow from action principles $\delta S = \delta \int dt L(\dot{\boldsymbol{x}}, \boldsymbol{x}) = 0$ in the form of the Euler-Lagrange equations

$$\frac{\partial L}{\partial x_i} - \frac{d}{dt}\frac{\partial L}{\partial \dot{x}_i} = 0.$$

The variation of a field dependent action functional

$$S[\phi] = \frac{1}{c}\int_{\mathcal{V}} d^4x\, \mathcal{L}(\partial\phi_I, \phi_I)$$

for fields $\phi_I(x)$ proceeds in the same way as in classical mechanics, the only difference being that we apply the Gauss theorem for the partial integrations.

To elucidate this, we require that arbitrary first order variation

$$\phi_I(x) \to \phi_I(x) + \delta\phi_I(x)$$

with fixed fields at initial and final times t_0 and t_1,

$$\delta\phi_I(\boldsymbol{x}, t_0) = 0, \quad \delta\phi_I(\boldsymbol{x}, t_1) = 0,$$

leaves the action $S[\phi]$ in first order invariant. We also assume that the fields and their variations vanish at spatial infinity.

The first order variation of the action between the times t_0 and t_1 is

$$\begin{aligned}
\delta S[\phi] &= S[\phi + \delta\phi] - S[\phi] \\
&= \int d^3\boldsymbol{x} \int_{t_0}^{t_1} dt\, [\mathcal{L}(\partial\phi_I + \partial\delta\phi_I, \phi_I + \delta\phi_I) - \mathcal{L}(\partial\phi_I, \phi_I)] \\
&= \int d^3\boldsymbol{x} \int_{t_0}^{t_1} dt \left(\delta\phi_I \frac{\partial\mathcal{L}}{\partial\phi_I} + \frac{\partial\mathcal{L}}{\partial(\partial_\mu\phi_I)}\partial_\mu\delta\phi_I\right).
\end{aligned}$$

Partial integration in the last term yields

$$\delta S[\phi] = \int d^3\boldsymbol{x} \int_{t_0}^{t_1} dt\, \delta\phi_I \left(\frac{\partial\mathcal{L}}{\partial\phi_I} - \partial_\mu \frac{\partial\mathcal{L}}{\partial(\partial_\mu\phi_I)}\right), \tag{16.1}$$

where the boundary terms vanish because of the vanishing variations at spatial infinity and at t_0 and t_1.

Equation (16.1) implies that we can have $\delta S[\phi] = 0$ for arbitrary variations $\delta\phi_I(x)$ between t_0 and t_1 if and only if the equations

$$\frac{\partial\mathcal{L}}{\partial\phi_I} - \partial_\mu\frac{\partial\mathcal{L}}{\partial(\partial_\mu\phi_I)} = 0 \tag{16.2}$$

hold for all the fields $\phi_I(x)$. These are the Euler-Lagrange equations for Lagrangian field theory.

The derivation of equation (16.2) does not depend on the number of four spacetime dimensions, $\mu \in \{0, 1, 2, 3\}$. It would just as well go through in any number d of dimensions, where d could be a number of spatial dimensions if we study equilibrium or static phenomena in field theory, or d can be $d - 1$ spatial and one time dimension. Relevant cases for observations include $d = 1$ (mechanics or equilibrium in one-dimensional systems), $d = 2$ (equilibrium phenomena on interfaces or surfaces, time-dependent phenomena in one-dimensional systems), $d = 3$ (equilibrium phenomena in three dimensions, time-dependent phenomena on interfaces or surfaces), and $d = 4$ (time-dependent phenomena in observable spacetime). In particular, classical particle mechanics can be considered as a field theory in one spacetime dimension.

The Lagrange density for the Schrödinger field

An example is provided by the Lagrange density for the Schrödinger field,

$$\mathcal{L} = \frac{i\hbar}{2}\left(\Psi^+\cdot\frac{\partial\Psi}{\partial t} - \frac{\partial\Psi^+}{\partial t}\cdot\Psi\right) - \frac{\hbar^2}{2m}\nabla\Psi^+\cdot\nabla\Psi - \Psi^+\cdot V\cdot\Psi. \tag{16.3}$$

In the notation of the previous paragraph, this corresponds to fields $\phi_1(x) = \Psi^+(x)$ and $\phi_2(x) = \Psi(x)$, or we could also denote the real and imaginary parts of Ψ as the two fields.

We have the following partial derivatives of the Lagrange density,

$$\frac{\partial\mathcal{L}}{\partial\Psi^+} = \frac{i\hbar}{2}\frac{\partial\Psi}{\partial t} - V\Psi, \quad \frac{\partial\mathcal{L}}{\partial(\partial_t\Psi^+)} = -\frac{i\hbar}{2}\Psi, \quad \frac{\partial\mathcal{L}}{\partial(\partial_i\Psi^+)} = -\frac{\hbar^2}{2m}\partial_i\Psi,$$

and the corresponding adjoint equations. The Euler-Lagrange equation from variation of the action with respect to Ψ^+,

$$\frac{\partial\mathcal{L}}{\partial\Psi^+} - \partial_t\frac{\partial\mathcal{L}}{\partial(\partial_t\Psi^+)} - \partial_i\frac{\partial\mathcal{L}}{\partial(\partial_i\Psi^+)} = 0,$$

is the Schrödinger equation

$$i\hbar\frac{\partial}{\partial t}\Psi + \frac{\hbar^2}{2m}\Delta\Psi - V\Psi = 0.$$

The Euler-Lagrange equation from variation with respect to Ψ in turn yields the complex conjugate Schrödinger equation for Ψ^+. This is of course required for consistency, and is a consequence of $\mathcal{L} = \mathcal{L}^+$.

The Schrödinger field is slightly unusual in that variation of the action with respect to $\phi_1(x) = \Psi^+(x)$ yields the equation for $\phi_2(x) = \Psi(x)$ and vice versa. Generically, variation of the action with respect to a field $\phi_I(x)$ yields the equation of motion for that field[2]. However, the important conclusion from this section is that Schrödinger's quantum mechanics is a Lagrangian field theory with a Lagrange density (16.3).

16.2 Symmetries and conservation laws

We consider an action with fields ϕ (ϕ_I, $1 \leq I \leq N$) in a d-dimensional space or spacetime:

$$S = \frac{1}{c}\int d^d x\, \mathcal{L}(\phi, \partial\phi). \tag{16.4}$$

To reveal the connection between symmetries and conservation laws, we calculate the first order change of the action S (16.4) if we perform transformations of the coordinates,

$$x'(x) = x - \epsilon(x). \tag{16.5}$$

This transforms the integration measure in the action as

$$d^d x' = d^d x\left(1 - \partial_\mu \epsilon^\mu\right),$$

and partial derivatives transform according to

$$\partial'_\mu = \partial_\mu + \left(\partial_\mu \epsilon^\nu\right)\partial_\nu. \tag{16.6}$$

We also include transformations of the fields,

$$\phi'(x') = \phi(x) + \delta\phi(x). \tag{16.7}$$

[2]The unconventional behavior for the Schrödinger field can be traced back to how it arises from the Klein-Gordon or Dirac fields in the non-relativistic limit, see Chapter 21.

Coordinate transformations often also imply transformations of the fields, e.g. if ϕ is a tensor field of n-th order with components $\phi_{\alpha\ldots\nu}(x)$, the transformation induced by the coordinate transformation $x \to x'(x) = x - \epsilon(x)$ is

$$\phi'_{\alpha'\ldots\nu'}(x') = \partial_{\alpha'}x^{\alpha} \cdot \partial_{\beta'}x^{\beta} \ldots \partial_{\nu'}x^{\nu} \cdot \phi_{\alpha\beta\ldots\nu}(x).$$

This yields is in first order

$$\begin{aligned}\delta\phi_{\alpha\beta\ldots\nu}(x) &= \phi'_{\alpha\ldots\nu}(x') - \phi_{\alpha\ldots\nu}(x)\\ &= \partial_{\alpha}\epsilon^{\sigma} \cdot \phi_{\sigma\beta\ldots\nu}(x) + \partial_{\beta}\epsilon^{\sigma} \cdot \phi_{\alpha\sigma\ldots\nu}(x) + \ldots + \partial_{\nu}\epsilon^{\sigma} \cdot \phi_{\alpha\beta\ldots\sigma}(x).\end{aligned}$$

Fields can also transform without a coordinate transformation, e.g. through a phase transformation.

We denote the transformations (16.5, 16.7) as a *symmetry* of the Lagrangian field theory (16.4) if they leave the volume form $d^d x\, \mathcal{L}$ invariant,

$$d^d x'\, \mathcal{L}(\phi', \partial'\phi'; x') = d^d x\, \mathcal{L}(\phi, \partial\phi; x). \tag{16.8}$$

Here we also allow for an explicit dependence of the Lagrange density on the coordinates x besides the implicit coordinate dependence through the dependence on the fields $\phi(x)$. If we define a transformed Lagrange density from the requirement of invariance of the action S under the transformations (16.5, 16.7),

$$\mathcal{L}'(\phi', \partial'\phi'; x') = \det(\partial' x)\mathcal{L}(\phi, \partial\phi; x), \tag{16.9}$$

the symmetry condition (16.8) amounts to form invariance of the Lagrange density.

The equations (16.6) and (16.7) imply the following first order change of partial derivative terms:

$$\delta\left(\partial_{\mu}\phi\right) = \partial_{\mu}\delta\phi + \left(\partial_{\mu}\epsilon^{\nu}\right)\partial_{\nu}\phi. \tag{16.10}$$

The resulting first order change of the volume form is (with the understanding that we sum over all fields in all multiplicative terms where the field ϕ appears twice):

$$\begin{aligned}\delta(d^d x\, \mathcal{L}) &= d^d x\left[\left(1 - \partial_{\mu}\epsilon^{\mu}\right)\left(\mathcal{L} + \delta\phi\frac{\partial\mathcal{L}}{\partial\phi} + \delta\left(\partial_{\rho}\phi\right)\frac{\partial\mathcal{L}}{\partial(\partial_{\rho}\phi)} - \epsilon^{\sigma}\delta_{\sigma}\mathcal{L}\right) - \mathcal{L}\right]\\ &= d^d x\left[(\partial_{\mu}\epsilon^{\nu})\left(\partial_{\nu}\phi \cdot \frac{\partial\mathcal{L}}{\partial(\partial_{\mu}\phi)} - \eta_{\nu}{}^{\mu}\mathcal{L}\right) + \partial_{\mu}\left(\delta\phi\frac{\partial\mathcal{L}}{\partial(\partial_{\mu}\phi)}\right)\right.\\ &\quad + \delta\phi\left(\frac{\partial\mathcal{L}}{\partial\phi} - \partial_{\mu}\frac{\partial\mathcal{L}}{\partial(\partial_{\mu}\phi)}\right)\\ &\quad \left. - \epsilon^{\mu}\left(\partial_{\mu}\mathcal{L} - \partial_{\mu}\phi \cdot \frac{\partial\mathcal{L}}{\partial\phi} - \partial_{\mu}\partial_{\nu}\phi \cdot \frac{\partial\mathcal{L}}{\partial(\partial_{\nu}\phi)}\right)\right]\end{aligned}$$

$$= d^d x \left\{ \partial_\mu \left[\epsilon^\nu \left(\partial_\nu \phi \cdot \frac{\partial \mathcal{L}}{\partial(\partial_\mu \phi)} - \eta_\nu{}^\mu \mathcal{L} \right) + \delta\phi \frac{\partial \mathcal{L}}{\partial(\partial_\mu \phi)} \right] \right.$$
$$\left. + (\delta\phi + \epsilon^\nu \partial_\nu \phi) \left(\frac{\partial \mathcal{L}}{\partial \phi} - \partial_\mu \frac{\partial \mathcal{L}}{\partial(\partial_\mu \phi)} \right) \right\}. \tag{16.11}$$

Here

$$\delta_\mu \mathcal{L} = \partial_\mu \mathcal{L} - \partial_\mu \phi \cdot \frac{\partial \mathcal{L}}{\partial \phi} - \partial_\mu \partial_\nu \phi \cdot \frac{\partial \mathcal{L}}{\partial(\partial_\nu \phi)}$$

is the partial derivative of $\mathcal{L}$ with respect to any *explicit* coordinate dependence.

If we have off-shell $\delta(d^d x\, \mathcal{L}) = 0$ for the proposed transformations ϵ, $\delta\phi$, we find a local on-shell conservation law

$$\partial_\mu j^\mu = 0 \tag{16.12}$$

with the current density

$$j^\mu = \epsilon^\nu \left(\eta_\nu{}^\mu \mathcal{L} - \partial_\nu \phi \cdot \frac{\partial \mathcal{L}}{\partial(\partial_\mu \phi)} \right) - \delta\phi \frac{\partial \mathcal{L}}{\partial(\partial_\mu \phi)}. \tag{16.13}$$

The corresponding charge in a d-dimensional spacetime

$$Q = \frac{1}{c} \int d^{d-1}\boldsymbol{x}\, j^0(\boldsymbol{x}, t) = \int d^{d-1}\boldsymbol{x}\, \varrho(\boldsymbol{x}, t) \tag{16.14}$$

is conserved if no charges are escaping or entering at $|\boldsymbol{x}| \to \infty$:

$$\lim_{|\boldsymbol{x}| \to \infty} \int d^{d-2}\Omega\, |\boldsymbol{x}|^{d-3} \boldsymbol{x} \cdot \boldsymbol{j}(\boldsymbol{x}, t) = 0.$$

Here $d^{d-2}\Omega = d\theta_1 \ldots d\theta_{d-2} \sin^{d-3}\theta_1 \ldots \sin\theta_{d-3}$ is the measure on the $(d-2)$-dimensional sphere in the $d-1$ spatial dimensions, see also (J.22) (note that in (J.22) the number of spatial dimensions is denoted as d).

If the off-shell variation of $d^d x \mathcal{L}$ satisfies $\delta(d^d x\, \mathcal{L}) \equiv d^d x\, \partial_\mu K^\mu$, the on-shell conserved current is $J^\mu = j^\mu + K^\mu$ and the charge is the spatial integral over J^0/c.

Symmetry transformations which only transform the fields, but leave the coordinates invariant ($\delta\phi \neq 0$, $\epsilon = 0$), are denoted as *internal symmetries*. Symmetry transformations involving coordinate transformations are denoted as *external symmetries*.

The connection between symmetries and conservation laws was developed by Emmy Noether[3] and is known as *Noether's theorem*.

[3] E. Noether, Nachr. König. Ges. Wiss. Göttingen, Math.-phys. Klasse, 235 (1918), see also arXiv:physics/0503066.

Energy-momentum tensors

We now specialize to inertial (i.e. pseudo-Cartesian) coordinates in Minkowski spacetime. If the coordinate shift in (16.5) is a constant translation, $\partial_\mu\epsilon^\nu = 0$, all fields transform like scalars, $\delta\phi = 0$, and the conserved current becomes

$$j^\mu = \epsilon^\nu \left(\eta_\nu{}^\mu \mathcal{L} - \partial_\nu\phi \cdot \frac{\partial\mathcal{L}}{\partial(\partial_\mu\phi)}\right) = \epsilon^\nu \Theta_\nu{}^\mu.$$

Omitting the d irrelevant constants ϵ^ν leaves us with d conserved currents ($0 \leq \nu \leq d-1$)

$$\partial_\mu \Theta_\nu{}^\mu = 0, \tag{16.15}$$

with components

$$\Theta_\nu{}^\mu = \eta_\nu{}^\mu \mathcal{L} - \partial_\nu\phi \cdot \frac{\partial\mathcal{L}}{\partial(\partial_\mu\phi)}. \tag{16.16}$$

The corresponding conserved charges

$$p_\nu = \frac{1}{c}\int d^{d-1}\boldsymbol{x}\, \Theta_\nu{}^0 \tag{16.17}$$

are the components of the four-dimensional energy-momentum vector of the physical system described by the Lagrange density $\mathcal{L}$, and the tensor with components $\Theta_\nu{}^\mu$ is therefore denoted as an *energy-momentum tensor*.

The spatial components Θ^{ij} of the energy-momentum tensor have dual interpretations in terms of momentum current densities and forces. To explain the meaning of Θ^{ij}, we pick an arbitrary (but stationary) spatial volume V. Since we are talking about fields, part of the fields will reside in V. From equation (16.17), the fields in V will carry a part of the total momentum $\boldsymbol{p}$ which is

$$\boldsymbol{p}_V = \boldsymbol{e}_i \frac{1}{c}\int_V d^{d-1}\boldsymbol{x}\, \Theta^{i0}.$$

The equations (16.15) and (16.17) imply that the change of $\boldsymbol{p}_V$ is given by

$$\frac{d}{dt}\boldsymbol{p}_V = \boldsymbol{e}_i \int_V d^{d-1}\boldsymbol{x}\, \partial_0\Theta^{i0} = -\boldsymbol{e}_i \oint_{\partial V} d^{d-2}S_j\, \Theta^{ij}, \tag{16.18}$$

where the Gauss theorem in $d-1$ spatial dimensions was employed and $d^{d-2}S_j$ is the outward bound surface element on the boundary ∂V of the volume.

This equation tells us that the component Θ^{ij} describes the flow of the momentum component p^i through the plane with normal vector $\boldsymbol{e}_j$, i.e. Θ^{ij} is the flow of momentum p^i in the direction $\boldsymbol{e}_j$ and $\boldsymbol{j}^i = \Theta^{ij}\boldsymbol{e}_j$ is the corresponding current density. In the dual interpretation, we read equation (16.18) with the relation $\boldsymbol{F}_V = d\boldsymbol{p}_V/dt$ between force and momentum change in mind. In this interpretation, $\boldsymbol{F}_V$ *is the force exerted on the fields in the fixed volume* V, because it describes the rate of change of momentum of the fields in V. $-\boldsymbol{F}_V$ *is the force exerted by the fields in the fixed volume* V. The component Θ^{ij} is then the force in direction $\boldsymbol{e}_i$ per area with normal vector $\boldsymbol{e}_j$. This represents strain or pressure for $i = j$ and stress for $i \neq j$. The energy-momentum tensor is therefore also known as *stress-energy tensor*.

There is another equation for the energy-momentum tensor in general relativity, which agrees with equation (16.16) for scalar fields, but not for vector or relativistic spinor fields. Both definitions yield the same conserved energy and momentum of a system, but improvement terms have to be added to the tensor from equation (16.16) in relativistic field theories to get the correct expressions for local densities for energy and momentum. We will discuss the necessary modifications of $\Theta_\nu{}^\mu$ for the Maxwell field (photons) in Section 18.1 and for relativistic fermions in Section 21.4.

16.3 Applications to Schrödinger field theory

The energy-momentum tensor for the Schrödinger field is found by substituting (16.3) into equation (16.16). The corresponding energy density is usually written as a Hamiltonian density $\mathcal{H}$,

$$\mathcal{H} = c\mathcal{P}^0 = -\Theta_0{}^0 = \frac{\hbar^2}{2m}\nabla\Psi^+\cdot\nabla\Psi + \Psi^+\cdot V\cdot\Psi, \tag{16.19}$$

and the momentum density is

$$\boldsymbol{\mathcal{P}} = \frac{1}{c}\boldsymbol{e}_i\Theta^{i0} = \frac{\hbar}{2\mathrm{i}}\left(\Psi^+\cdot\nabla\Psi - \nabla\Psi^+\cdot\Psi\right). \tag{16.20}$$

The energy current density for the Schrödinger field follows as

$$\boldsymbol{j}_{\mathcal{H}} = -c\Theta_0{}^i\boldsymbol{e}_i = -\frac{\hbar^2}{2m}\left(\nabla\Psi^+\cdot\frac{\partial\Psi}{\partial t} + \frac{\partial\Psi^+}{\partial t}\cdot\nabla\Psi\right). \tag{16.21}$$

The energy $E = \int d^3\boldsymbol{x}\,\mathcal{H}$ and momentum $\boldsymbol{p} = \int d^3\boldsymbol{x}\,\boldsymbol{\mathcal{P}}$ agree with the corresponding expectation values of the Schrödinger wave function in quantum mechanics. The results of the previous section, or direct application of the Schrödinger equation, tell us that E is conserved if the potential is time-independent, $V = V(\boldsymbol{x})$, and the momentum component e.g. in x-direction is conserved if the momentum does not depend on x, $V = V(y, z)$.

Probability and charge conservation from invariance under phase rotations

The Lagrange density (16.3) is invariant under phase rotations of the Schrödinger field,

$$\delta\Psi(\boldsymbol{x},t) = \mathrm{i}\frac{q}{\hbar}\varphi\Psi(\boldsymbol{x},t), \quad \delta\Psi^+(\boldsymbol{x},t) = -\mathrm{i}\frac{q}{\hbar}\varphi\Psi^+(\boldsymbol{x},t).$$

We wrote the constant phase in the peculiar form $q\varphi/\hbar$ in anticipation of the connection to local gauge transformations (15.8, 15.9), which will play a recurring role later on. However, for now we note that substitution of the phase transformations into the equation (16.13) yields after division by the irrelevant constant $q\varphi$ the density

$$\varrho = \frac{j^0}{c} = -\frac{1}{q\varphi}\left(\delta\Psi\frac{\partial\mathcal{L}}{\partial(\partial_t\Psi)} + \delta\Psi^+\frac{\partial\mathcal{L}}{\partial(\partial_t\Psi^+)}\right) = \Psi^+\Psi = \frac{1}{q}\varrho_q \tag{16.22}$$

and the related current density

$$\begin{aligned}\boldsymbol{j} &= -\frac{1}{q\varphi}\left(\delta\Psi\frac{\partial\mathcal{L}}{\partial(\nabla\Psi)} + \delta\Psi^+\frac{\partial\mathcal{L}}{\partial(\nabla\Psi^+)}\right) = \frac{\hbar}{2\mathrm{i}m}\left(\Psi^+\cdot\nabla\Psi - \nabla\Psi^+\cdot\Psi\right)\\ &= \frac{1}{q}\boldsymbol{j}_q.\end{aligned} \tag{16.23}$$

Comparison with equations (1.17) and (1.18) shows that probability conservation in Schrödinger theory can be considered as a consequence of invariance under global phase rotations.

Had we not divided out the charge q, we would have drawn the same conclusion for conservation of electric charge with $\varrho_q = q\Psi^+\Psi$ as the charge density and $\boldsymbol{j}_q = q\boldsymbol{j}$ as the electric current density. The coincidence of the conservation laws for probability and electric charge in Schrödinger theory arises because it is a theory for non-relativistic particles. Only charge conservation will survive in the relativistic limit, but probability conservation for particles will not hold any more, because $\varrho_q(\boldsymbol{x},t)/q$ will not be positive definite any more and therefore will not yield a quantity that could be considered as a probability density to find a particle in the location $\boldsymbol{x}$ at time t.

Comparison with equation (16.20) tells us that $\boldsymbol{j}$ is also proportional to the momentum density,

$$\boldsymbol{j}(\boldsymbol{x},t) = \frac{1}{m}\mathcal{P}(\boldsymbol{x},t), \tag{16.24}$$

which tells us that the probability current density of the Schrödinger field is also a velocity density.

16.4 Problems

16.1. Show that addition of any derivative term $\partial_\mu \mathcal{F}(\phi_I)$ to the Lagrange density $\mathcal{L}(\phi_I, \partial\phi_I)$ does not change the Euler-Lagrange equations.

16.2. We consider classical particle mechanics with a Lagrangian $L(q_I, \dot{q}_I)$.

16.2a. Suppose the action is invariant under constant shifts δq_J of the coordinate $q_J(t)$. Which conserved quantity do you find from equation (16.13)? Which condition must L fulfill to ensure that the action is not affected by the constant shift δq_J?

16.2b. Now we assume that the action is invariant under constant shifts $\delta t = -\epsilon$ of the internal coordinate t. Which conserved quantity do you find from equation (16.13) in this case?

16.3. Use the Schrödinger equation to confirm that the energy density (16.19) and the energy current density (16.21) indeed satisfy the local conservation law

$$\frac{\partial}{\partial t}\mathcal{H} = -\nabla \cdot \boldsymbol{j}_{\mathcal{H}}$$

if the potential is time-independent, $V = V(\boldsymbol{x})$.

How does E change if $V = V(\boldsymbol{x}, t)$ is time-dependent?

16.4. We have only evaluated the components $\Theta_0{}^0$, $\Theta_i{}^0$ and $\Theta_0{}^i$ of the energy-momentum tensor of the Schrödinger field in equations (16.19)–(16.21). Which momentum current densities $\boldsymbol{j}^i_{\mathcal{P}}$ do you find from the energy-momentum tensor of the Schrödinger field?

16.5. Schrödinger fields can have different transformation properties under coordinate rotations $\delta\boldsymbol{x} = -\boldsymbol{\varphi} \times \boldsymbol{x}$, see Section 8.2. In this problem we analyze a Schrödinger field which transforms like a scalar under rotations,

$$\delta\Psi(\boldsymbol{x}, t) = \Psi'(\boldsymbol{x}', t) - \Psi(\boldsymbol{x}, t) = 0.$$

The Lagrange density (16.3) is invariant under rotations if $V = V(r, t)$. Which conserved quantity do you find from this observation?

Solution. Equation (16.13) yields with $\boldsymbol{\epsilon} = \boldsymbol{\varphi} \times \boldsymbol{x}$ a conserved charge density

$$\varrho = \frac{j^0}{c} = -(\boldsymbol{\varphi} \times \boldsymbol{x}) \cdot \left(\nabla\Psi \frac{\partial\mathcal{L}}{\partial(\partial_t\Psi)} + \nabla\Psi^+ \frac{\partial\mathcal{L}}{\partial(\partial_t\Psi^+)}\right)$$
$$= -\frac{\mathrm{i}\hbar}{2}\boldsymbol{\varphi} \cdot \left[\boldsymbol{x} \times \left(\Psi^+ \cdot \nabla\Psi - \nabla\Psi^+ \cdot \Psi\right)\right] = \boldsymbol{\varphi} \cdot \mathcal{M},$$

with an angular momentum density

$$\mathcal{M} = \frac{\hbar}{2\mathrm{i}}\boldsymbol{x} \times \left(\Psi^+ \cdot \nabla\Psi - \nabla\Psi^+ \cdot \Psi\right) = \boldsymbol{x} \times \mathcal{P}. \tag{16.25}$$

Since the constant parameters φ are arbitrary, we find three linearly independent conserved quantities, *viz.* the angular momentum

$$M = \int d^3x\, \mathcal{M} = \langle \mathbf{x} \times \mathbf{p} \rangle$$

of the scalar Schrödinger field.

16.6. Now we assume that our Schrödinger field is a 2-spinor with the transformation property

$$\delta\Psi = \frac{\mathrm{i}}{2}(\varphi \cdot \underline{\sigma}) \cdot \Psi, \quad \delta\Psi^+ = -\frac{\mathrm{i}}{2}\Psi^+ \cdot (\varphi \cdot \underline{\sigma}).$$

Show that the corresponding density of "total angular momentum" of the Schrödinger field in this case consists of an orbital and a spin part,

$$\begin{aligned} \mathcal{J} &= \frac{\hbar}{2\mathrm{i}} \boldsymbol{x} \times \left(\Psi^+ \cdot \nabla\Psi - \nabla\Psi^+ \cdot \Psi\right) + \frac{\hbar}{2}\Psi^+ \cdot \underline{\sigma} \cdot \Psi \\ &= \boldsymbol{x} \times \mathcal{P} + \Psi^+ \cdot \underline{S} \cdot \Psi = \mathcal{M} + \mathcal{S}. \end{aligned} \tag{16.26}$$

Rotational invariance implies only conservation of the total angular momentum $\boldsymbol{J} = \int d^3x\, \mathcal{J}$. However, on the level of the Lagrange density (16.3), which does not contain spin-orbit interaction terms (8.20), the orbital and spin parts are preserved separately. We will see in Section 21.5 that spin-orbit coupling is a consequence of relativity.

16.7. Suppose the Hamiltonian has the spin-orbit coupling form $H = \alpha \boldsymbol{M} \cdot \boldsymbol{S}$, where M_i and S_i are angular momentum and spin operators. How do these operators evolve in the Heisenberg picture?

16.7a. Show that the Heisenberg evolution equations for the operators yield

$$\dot{\boldsymbol{M}} = \alpha \boldsymbol{S} \times \boldsymbol{M}, \quad \dot{\boldsymbol{S}} = \alpha \boldsymbol{M} \times \boldsymbol{S}. \tag{16.27}$$

16.7b. Show that $\boldsymbol{J} \equiv \boldsymbol{M} + \boldsymbol{S}$, $\boldsymbol{M}^2$, $\boldsymbol{S}^2$ and $\boldsymbol{M} \cdot \boldsymbol{S}$ are all constant.

16.7c. Show that the evolution equations (16.27) are solved by

$$\boldsymbol{M}(t) = \exp(-\alpha \boldsymbol{J} \cdot \underline{L} t - \mathrm{i}\hbar\alpha t) \cdot \boldsymbol{M} \tag{16.28}$$

and

$$\boldsymbol{S}(t) = \exp(-\alpha \boldsymbol{J} \cdot \underline{L} t - \mathrm{i}\hbar\alpha t) \cdot \boldsymbol{S}, \tag{16.29}$$

where $\boldsymbol{M} \equiv \boldsymbol{M}(0)$, $\boldsymbol{S} \equiv \boldsymbol{S}(0)$, and $\underline{\boldsymbol{L}} = (\underline{L}_1, \underline{L}_2, \underline{L}_3)$ is the vector of matrices with components $(\underline{L}_i)_{jk} = \epsilon_{ijk}$, see equation (7.18).

16.7d. Except for the phase rotations, the equations (16.28, 16.29) seem to suggest that $\boldsymbol{M}(t)$ and $\boldsymbol{S}(t)$ are rotating around the direction of the vector $\boldsymbol{J}$ with angular velocity $\boldsymbol{\omega} = \alpha\boldsymbol{J}$. This suggestive picture of coupled angular momentum type operators rotating around the total angular momentum vector is often denoted as the *vector model* of spin-orbit type couplings. However, note that the total angular momentum vector is acting on tensor products of eigenstates and in fully explicit notation has the form

$$\boldsymbol{J} = \boldsymbol{M} \otimes \underline{1} + \underline{1} \otimes \boldsymbol{S}.$$

That does not mean that the results (16.27–16.29) or the conservation laws expressed in 16.7b are incorrect, but we must beware of simple interpretations in terms of vectors living within one and the same vector space.

Repeat the previous problems 16.7a–c in terms of the explicit tensor product notation using the Hamiltonian

$$H = \alpha \boldsymbol{M} \overset{\cdot}{\otimes} \boldsymbol{S} = \alpha M_i \otimes S_i.$$

16.8. Show that the Lagrange density

$$\mathcal{L} = \frac{\mathrm{i}\hbar}{2}\left(\Psi^+ \cdot \frac{\partial \Psi}{\partial t} - \frac{\partial \Psi^+}{\partial t} \cdot \Psi\right) - q\Psi^+ \cdot \Phi \cdot \Psi$$
$$- \frac{\hbar^2}{2m}\left(\nabla\Psi^+ + \mathrm{i}\frac{q}{\hbar}\Psi^+\boldsymbol{A}\right) \cdot \left(\nabla\Psi - \mathrm{i}\frac{q}{\hbar}\boldsymbol{A}\Psi\right). \quad (16.30)$$

yields the equations of motion for the Schrödinger field in external electromagnetic fields

$$\boldsymbol{E}(\boldsymbol{x},t) = -\nabla\Phi(\boldsymbol{x},t) - \frac{\partial}{\partial t}\boldsymbol{A}(\boldsymbol{x},t), \quad \boldsymbol{B}(\boldsymbol{x},t) = \nabla \times \boldsymbol{A}(\boldsymbol{x},t).$$

16.9. Derive the electric charge and current densities for the Schrödinger field in electromagnetic fields from the phase invariance of (16.30).

Answers. The charge density is

$$\varrho_q = q\Psi^+\Psi. \quad (16.31)$$

The current density is

$$\boldsymbol{j}_q = \frac{q\hbar}{2\mathrm{i}m}\left(\Psi^+ \cdot \nabla\Psi - \nabla\Psi^+ \cdot \Psi\right) - \frac{q^2}{m}\Psi^+\boldsymbol{A}\Psi. \quad (16.32)$$

Are the charge and current densities gauge invariant?

Chapter 17
Non-relativistic Quantum Field Theory

Quantum mechanics, as we know it so far, deals with invariant particle numbers,

$$\frac{d}{dt}\langle\Psi(t)|\Psi(t)\rangle = 0.$$

However, at least one of the early indications of wave-particle duality implies disappearance of a particle, *viz.* absorption of a photon in the photoelectric effect. This reminds us of two deficiencies of Schrödinger's wave mechanics: it cannot deal with absorption or emission of particles, and it cannot deal with relativistic particles.

In the following sections we will deal with the problem of absorption and emission of particles in the non-relativistic setting, i.e. for slow electrons, protons, neutrons, or nuclei, or quasiparticles in condensed matter physics. The strategy will be to follow a quantization procedure that works for the promotion of classical mechanics to quantum mechanics, but this time for Schrödinger theory. The correspondences are summarized in Table 17.1.

The key ingredient is promotion of the "classical" variables $\boldsymbol{x}$ or $\Psi(\boldsymbol{x}, t)$ to operators through "canonical (anti-)commutation relations", as outlined in the last two lines of Table 17.1. This procedure of promoting classical variables to operators by imposing canonical commutation or anti-commutation relations is called *canonical quantization*. Canonical quantization of fields is denoted as *field quantization*. Since the fields are often wave functions (like the Schrödinger wave function) which arose from the quantization of $\boldsymbol{x}$ and $\boldsymbol{p}$, field quantization is sometimes also called *second quantization*. A quantum theory that involves quantized fields is denoted as a quantum field theory.

Indeed, quantum field theory is essentially as old as Schrödinger's wave mechanics, because it was clear right after the inception of quantum mechanics that the formalism was not yet capable of the description of quantum effects for photons. This led to the rapid invention of field quantization in several steps between

R. Dick, *Advanced Quantum Mechanics*, Graduate Texts in Physics,
DOI 10.1007/978-3-319-25675-7_17

Table 17.1 Correspondence between first and second quantization

Classical mechanics	Schrödinger's wave mechanics
Independent variable t	Independent variables $\boldsymbol{x}, t$
Dependent variables $\boldsymbol{x}(t)$	Dependent variables $\Psi(\boldsymbol{x}, t)$, $\Psi^+(\boldsymbol{x}, t)$
Newton's equation $m\ddot{\boldsymbol{x}} = -\boldsymbol{\nabla} V(\boldsymbol{x})$	Schrödinger's equation $\mathrm{i}\hbar\frac{\partial}{\partial t}\Psi = -\frac{\hbar^2}{2m}\Delta\Psi + V\Psi$
Lagrangian $L = \frac{m}{2}\dot{\boldsymbol{x}}^2 - V(\boldsymbol{x})$	Lagrangian $\mathcal{L} = \frac{\mathrm{i}\hbar}{2}\left(\Psi^+ \cdot \frac{\partial}{\partial t}\Psi - \frac{\partial}{\partial t}\Psi^+ \cdot \Psi\right)$ $-\frac{\hbar^2}{2m}\boldsymbol{\nabla}\Psi^+ \cdot \boldsymbol{\nabla}\Psi - \Psi^+ \cdot V \cdot \Psi$
Conjugate momenta $p_i(t) = \partial L/\partial\dot{x}_i(t) = m\dot{x}_i(t)$	Conjugate momenta $\Pi_\Psi(\boldsymbol{x}, t) = \partial\mathcal{L}/\partial\dot{\Psi}(\boldsymbol{x}, t) = \frac{\mathrm{i}\hbar}{2}\Psi^+(\boldsymbol{x}, t)$, $\Pi_\Psi^+(\boldsymbol{x}, t) = -\frac{\mathrm{i}\hbar}{2}\Psi(\boldsymbol{x}, t)$
Canonical commutators $[x_i(t), p_j(t)] = \mathrm{i}\hbar\delta_{ij}$, $[x_i(t), x_j(t)] = 0$, $[p_i(t), p_j(t)] = 0$	Canonical (anti-)commutators $[\Psi(\boldsymbol{x}, t), \Psi^+(\boldsymbol{x}', t)]_\mp = \delta(\boldsymbol{x} - \boldsymbol{x}')$, $[\Psi(\boldsymbol{x}, t), \Psi(\boldsymbol{x}', t)]_\mp = 0$

1925 and 1928. Key advancements[1] were the formulation of a quantum field as a superposition of infinitely many oscillation operators by Born, Heisenberg and Jordan in 1926, the application of infinitely many oscillation operators by Dirac in 1927 for photon emission and absorption, and the introduction of anti-commutation relations for fermionic field operators by Jordan and Wigner in 1928. Path integration over fields was introduced by Feynman in the 1940s.

17.1 Quantization of the Schrödinger field

We will now start to perform the program of canonical quantization of Schrödinger's wave mechanics. First steps will involve the promotion of wave functions like $\Psi(\boldsymbol{x}, t)$ and $\Psi^+(\boldsymbol{x}, t)$ to *field operators* or *quantum fields* through the proposition of canonical commutation or anti-commutation relations, and the identification of related composite field operators like the Hamiltonian, momentum and charge operators. The composite operators will then help us to reveal the physical meaning of the Schrödinger quantum fields $\Psi(\boldsymbol{x}, t)$ and $\Psi^+(\boldsymbol{x}, t)$ as annihilation and creation operators for particles.

The Lagrange density (16.3) yields the canonically conjugate momenta

$$\Pi_\Psi = \frac{\partial\mathcal{L}}{\partial\dot{\Psi}} = \frac{\mathrm{i}\hbar}{2}\Psi^+, \quad \Pi_{\Psi^+} = \frac{\partial\mathcal{L}}{\partial\dot{\Psi}^+} = -\frac{\mathrm{i}\hbar}{2}\Psi,$$

[1]M. Born, W. Heisenberg, P. Jordan, Z. Phys. 35, 557 (1926); P.A.M. Dirac, Proc. Roy. Soc. London A 114, 243 (1927); P. Jordan, E. Wigner, Z. Phys. 47, 631 (1928).

and the canonical commutation relations[2] translate for fermions (with the upper signs corresponding to anti-commutators) and bosons (with the lower signs corresponding to commutators) into

$$\begin{aligned}[\Psi(\boldsymbol{x},t),\Psi^+(\boldsymbol{x}',t)]_\pm &\equiv \Psi(\boldsymbol{x},t)\Psi^+(\boldsymbol{x}',t) \pm \Psi^+(\boldsymbol{x}',t)\Psi(\boldsymbol{x},t)\\ &= \delta(\boldsymbol{x}-\boldsymbol{x}'),\end{aligned} \tag{17.1}$$

$$[\Psi(\boldsymbol{x},t),\Psi(\boldsymbol{x}',t)]_\pm = 0, \quad [\Psi^+(\boldsymbol{x},t),\Psi^+(\boldsymbol{x}',t)]_\pm = 0.$$

Whether the quantum field for a particle should be quantized using commutation or anti-commutation relations depends on the spin of the particle, i.e. on the transformation properties of the field under rotations, see Chapter 8. Bosons have integer spin and are quantized through commutation relations while fermions have half-integer spin and are quantized through anti-commutation relations. Therefore we should include spin labels (which were denoted as m_s or a in Chapter 8) with the quantum fields, e.g. $\Psi_{m_s}(\boldsymbol{x},t)$, $m_s \in \{-s, -s+1, \ldots, s\}$, for a field describing particles of spin s and spin projection m_s. We will explicitly include spin labels in Section 17.5, but for now we will not clutter the equations any more than necessary, since spin labels can usually be ignored as long as dipole approximation $\lambda \gg a_0$ applies. Here a_0 is the Bohr radius and λ is the wavelength of photons which might interact with the Schrödinger field. Spin-flipping transitions are suppressed roughly by a factor a_0^2/λ^2 relative to spin-preserving transitions in dipole approximation. See the remarks after equation (18.106).

The commutation relations (17.1) in the bosonic case are like the commutation relations $[a_i, a_j^+] = \delta_{ij}$ etc. for oscillator operators. Therefore we can think of the field operators $\Psi(\boldsymbol{x},t)$ and $\Psi^+(\boldsymbol{x}',t)$ as annihilation and creation operators for each point in spacetime. We will explicitly confirm this interpretation below by showing that the corresponding Fourier transformed operators $a(\boldsymbol{k})$ and $a^+(\boldsymbol{k})$ (in the Schrödinger picture) annihilate or create particles of momentum $\hbar\boldsymbol{k}$, respectively. We will also see how linear superpositions of the operators $\psi^+(\boldsymbol{x}) = \Psi^+(\boldsymbol{x},0)$ act on the vacuum to generate e.g. states $|n,\ell,m_\ell\rangle$ which correspond to hydrogen eigenstates.

Note that $\Psi(\boldsymbol{x},t)$ and $\Psi^+(\boldsymbol{x},t)$ are now *time-dependent operators* and their time evolution is determined by the full dynamics of the system. Therefore they are operators in the *Heisenberg picture* of the second quantized theory, i.e. what had been representations of states in the Schrödinger picture of the first quantized theory has become field operators in the Heisenberg picture of the second quantized theory.

The elevation of wave functions to operators implies that functions or functionals of the wave functions that we had encountered in quantum mechanics now also

[2] Recall the canonical commutation relations $[x_i(t), p_j(t)] = i\hbar\delta_{ij}$, $[x_i(t), x_j(t)] = 0$, $[p_i(t), p_j(t)] = 0$ in the Heisenberg picture of quantum mechanics. It is customary to dismiss a factor of 2 in the (anti-)commutation relations (17.1), which otherwise would simply reappear in different places of the quantized Schrödinger theory.

become operators. Particularly important cases of functionals of wave functions include expectation values for observables like energy, momentum, and charge, and these will all become operators in the second quantized theory. E.g. the Hamiltonian density is related to the Lagrange density through a Legendre transformation (cf. $H = \sum_i p_i \dot{q}_i - L$ in mechanics), $\mathcal{H} = \Pi_\Psi \dot{\Psi} + \dot{\Psi}^+ \Pi_{\Psi^+} - \mathcal{L}$. This yields the Hamiltonian $H = \int d^3x\, \mathcal{H}$ in the form

$$H = \int d^3x \left(\frac{\hbar^2}{2m} \nabla \Psi^+(x,t) \cdot \nabla \Psi(x,t) + \Psi^+(x,t) V(x) \Psi(x,t) \right). \tag{17.2}$$

We have also found the Hamiltonian density in equation (16.19) from the energy-momentum tensor of the Schrödinger field, which in addition gave us the momentum

$$\begin{aligned} \boldsymbol{P}(t) &= \int d^3x\, \mathcal{P}(x,t) \\ &= \int d^3x\, \frac{\hbar}{2\mathrm{i}} \left(\Psi^+(x,t) \cdot \nabla \Psi(x,t) - \nabla \Psi^+(x,t) \cdot \Psi(x,t) \right). \end{aligned} \tag{17.3}$$

We can just as well use the equivalent expressions

$$H = \int d^3x \left(-\frac{\hbar^2}{2m} \Psi^+(x,t) \Delta \Psi(x,t) + \Psi^+(x,t) \cdot V(x) \cdot \Psi(x,t) \right)$$

and $\boldsymbol{P}(t) = -\mathrm{i}\hbar \int d^3x\, \Psi^+(x,t) \nabla \Psi(x,t)$, which can be motivated from the corresponding equations for the energy and momentum expectation values in the first quantized Schrödinger theory.

Other frequently used composite operators[3] include the number and charge operators N and Q, cf. (16.23),

$$N = \int d^3x\, \varrho(x,t) = \int d^3x\, \Psi^+(x,t) \Psi(x,t) = \frac{1}{q} Q. \tag{17.4}$$

Before we continue with the demonstration that $\Psi(x,t)$ and $\Psi^+(x',t)$ are annihilation and creation operators, we should confirm our suspicion that they are indeed operators in the Heisenberg picture of quantum field theory. We will do this next.

[3] For another composite operator we can also define an integrated current density through $\boldsymbol{I}_q(t) = \int d^3x\, \boldsymbol{j}_q(x,t) = q\boldsymbol{P}(t)/m$, where the last equation follows from (16.24). However, recall that $\boldsymbol{j}_q(x,t)$ is a current density, but it is *not* a current per volume, and therefore $\boldsymbol{I}_q(t)$ is not an electric current but comes in units of e.g. Ampère meter. It is related to charge transport like momentum $\boldsymbol{P}(t)$ is related to mass transport.

Time evolution of the field operators

Very useful identities for commutators involving products of operators are

$$\begin{aligned}[AB, C] &= ABC - CAB = ABC + ACB - ACB - CAB \\ &= A[B, C]_{\pm} - [C, A]_{\pm}B, \\ [A, BC] &= ABC - BCA = ABC + BAC - BAC - BCA \\ &= [A, B]_{\pm}C - B[C, A]_{\pm}. \end{aligned} \tag{17.5}$$

These relations and the canonical (anti-)commutation relations between the field operators imply that both bosonic and fermionic field operators $\Psi(\boldsymbol{x}, t)$ satisfy the Heisenberg evolution equations,

$$\frac{\partial}{\partial t}\Psi(\boldsymbol{x}, t) = \mathrm{i}\frac{\hbar}{2m}\Delta\Psi(\boldsymbol{x}, t) - \frac{\mathrm{i}}{\hbar}V(\boldsymbol{x})\Psi(\boldsymbol{x}, t) = \frac{\mathrm{i}}{\hbar}[H, \Psi(\boldsymbol{x}, t)], \tag{17.6}$$

$$\frac{\partial}{\partial t}\Psi^+(\boldsymbol{x}, t) = -\mathrm{i}\frac{\hbar}{2m}\Delta\Psi^+(\boldsymbol{x}, t) + \frac{\mathrm{i}}{\hbar}V(\boldsymbol{x})\Psi^+(\boldsymbol{x}, t) = \frac{\mathrm{i}}{\hbar}[H, \Psi^+(\boldsymbol{x}, t)]. \tag{17.7}$$

However, then we also get (note that here the time-independence of $V(\boldsymbol{x})$ is important)

$$\frac{d}{dt}H = \frac{\mathrm{i}}{\hbar}[H, H] = 0,$$

which was already anticipated in the notation by writing H rather than $H(t)$. The relations (17.6, 17.7) confirm the Heisenberg picture interpretation of the Schrödinger field operators $\Psi(\boldsymbol{x}, t)$ and $\Psi^+(\boldsymbol{x}, t)$.

k-space representation of quantized Schrödinger theory

In quantum mechanics, we used wave functions in $\boldsymbol{k}$-space both for scattering theory and for the calculation of the time evolution of free wave packets. The $\boldsymbol{k}$-space representation becomes even more important in quantum field theory because ensembles of particles have additive quantum numbers like total momentum and total kinetic energy which depend on the wave vector $\boldsymbol{k}$ of a particle, and this will help us to reveal the meaning of the Schrödinger field operators.

The mode expansion in the Heisenberg picture

$$\Psi(\boldsymbol{x}, t) = \frac{1}{\sqrt{2\pi}^3}\int d^3\boldsymbol{k}\, a(\boldsymbol{k}, t)\exp(\mathrm{i}\boldsymbol{k}\cdot\boldsymbol{x}), \tag{17.8}$$

$$a(\boldsymbol{k}, t) = \frac{1}{\sqrt{2\pi}^3}\int d^3\boldsymbol{x}\, \Psi(\boldsymbol{x}, t)\exp(-\mathrm{i}\boldsymbol{k}\cdot\boldsymbol{x}) \tag{17.9}$$

implies with (17.1) the (anti-)commutation relations for the field operators in $\boldsymbol{k}$-space,

$$[a(\boldsymbol{k},t),a^{+}(\boldsymbol{k}',t)]_{\pm}=\delta(\boldsymbol{k}-\boldsymbol{k}'),$$

$$[a(\boldsymbol{k},t),a(\boldsymbol{k}',t)]_{\pm}=0,\quad [a^{+}(\boldsymbol{k},t),a^{+}(\boldsymbol{k}',t)]_{\pm}=0.$$

Furthermore, substitution of equation (17.8) into the charge, momentum and energy operators yields

$$Q=qN=q\int d^{3}\boldsymbol{k}\,a^{+}(\boldsymbol{k},t)a(\boldsymbol{k},t), \tag{17.10}$$

$$\boldsymbol{P}(t)=\int d^{3}\boldsymbol{k}\,\hbar\boldsymbol{k}\,a^{+}(\boldsymbol{k},t)a(\boldsymbol{k},t) \tag{17.11}$$

and

$$H=H_{0}(t)+V(t), \tag{17.12}$$

with the kinetic and potential operators

$$H_{0}(t)=\int d^{3}\boldsymbol{k}\,\frac{\hbar^{2}\boldsymbol{k}^{2}}{2m}a^{+}(\boldsymbol{k},t)a(\boldsymbol{k},t) \tag{17.13}$$

and

$$V(t)=\int d^{3}\boldsymbol{k}\int d^{3}\boldsymbol{q}\,a^{+}(\boldsymbol{k}+\boldsymbol{q},t)V(\boldsymbol{q})a(\boldsymbol{k},t). \tag{17.14}$$

Here we used the following normalization for the Fourier transform of single particle potentials,

$$V(\boldsymbol{x})=\int d^{3}\boldsymbol{q}\,V(\boldsymbol{q})\exp(\mathrm{i}\boldsymbol{q}\cdot\boldsymbol{x}),$$

$$V(\boldsymbol{q})=V^{+}(-\boldsymbol{q})=\frac{1}{(2\pi)^{3}}\int d^{3}\boldsymbol{x}\,V(\boldsymbol{x})\exp(-\mathrm{i}\boldsymbol{q}\cdot\boldsymbol{x}).$$

Field operators in the Schrödinger picture and the Fock space for the Schrödinger field

The relations in the Heisenberg picture

$$\frac{\partial}{\partial t}\Psi(\boldsymbol{x},t)=\frac{\mathrm{i}}{\hbar}[H,\Psi(\boldsymbol{x},t)],\quad \frac{\partial}{\partial t}a(\boldsymbol{k},t)=\frac{\mathrm{i}}{\hbar}[H,a(\boldsymbol{k},t)],\quad \frac{d}{dt}H=0$$

imply

$$\Psi(\boldsymbol{x},t)=\exp\!\left(\frac{\mathrm{i}}{\hbar}Ht\right)\psi(\boldsymbol{x})\exp\!\left(-\frac{\mathrm{i}}{\hbar}Ht\right),$$

$$a(\boldsymbol{k},t)=\exp\!\left(\frac{\mathrm{i}}{\hbar}Ht\right)a(\boldsymbol{k})\exp\!\left(-\frac{\mathrm{i}}{\hbar}Ht\right).$$

The time-independent operators $\psi(\boldsymbol{x}) = \Psi(\boldsymbol{x},0)$, $a(\boldsymbol{k}) = a(\boldsymbol{k},0)$ are the corresponding operators in the *Schrödinger picture* of the quantum field theory[4]. Having time-independent operators in the Schrödinger picture comes at the expense of time-dependent states

$$|\boldsymbol{\Phi}(t)\rangle = \exp\!\left(-\frac{\mathrm{i}}{\hbar}Ht\right)|\boldsymbol{\Phi}(0)\rangle,$$

to preserve the time dependence of matrix elements and observables. Here we use a boldface bra-ket notation $\langle\boldsymbol{\Phi}|$ and $|\boldsymbol{\Phi}\rangle$ for states in the second quantized theory to distinguish them from the states $\langle\Phi|$ and $|\Phi\rangle$ in the first quantized theory.

The canonical (anti-)commutation relations for the Heisenberg picture operators imply canonical (anti-)commutation relations for the Schrödinger picture operators,

$$[\psi(\boldsymbol{x}),\psi^+(\boldsymbol{x}')]_\pm = \delta(\boldsymbol{x}-\boldsymbol{x}'), \tag{17.15}$$

$$[\psi(\boldsymbol{x}),\psi(\boldsymbol{x}')]_\pm = 0, \quad [\psi^+(\boldsymbol{x}),\psi^+(\boldsymbol{x}')]_\pm = 0,$$

$$[a(\boldsymbol{k}),a^+(\boldsymbol{k}')]_\pm = \delta(\boldsymbol{k}-\boldsymbol{k}'), \tag{17.16}$$

$$[a(\boldsymbol{k}),a(\boldsymbol{k}')]_\pm = 0, \quad [a^+(\boldsymbol{k}),a^+(\boldsymbol{k}')]_\pm = 0.$$

These are oscillator like commutation or anti-commutation relations, and to figure out what they mean we will look at all the composite operators of the Schrödinger field that we had constructed before.

Time-independence of the full Hamiltonian implies that we can express H in terms of the field operators $\Psi(\boldsymbol{x},t)$ in the Heisenberg picture or the field operators $\psi(\boldsymbol{x})$ in the Schrödinger picture,

$$\begin{aligned} H &= \int d^3\boldsymbol{x}\left(\frac{\hbar^2}{2m}\nabla\Psi^+(\boldsymbol{x},t)\cdot\nabla\Psi(\boldsymbol{x},t) + \Psi^+(\boldsymbol{x},t)\cdot V(\boldsymbol{x})\cdot\Psi(\boldsymbol{x},t)\right) \\ &= \int d^3\boldsymbol{x}\left(\frac{\hbar^2}{2m}\nabla\psi^+(\boldsymbol{x})\cdot\nabla\psi(\boldsymbol{x}) + \psi^+(\boldsymbol{x})\cdot V(\boldsymbol{x})\cdot\psi(\boldsymbol{x})\right) \\ &= \int d^3\boldsymbol{k}\,\frac{\hbar^2\boldsymbol{k}^2}{2m}a^+(\boldsymbol{k})a(\boldsymbol{k}) + \int d^3\boldsymbol{k}\int d^3\boldsymbol{q}\,a^+(\boldsymbol{k}+\boldsymbol{q})V(\boldsymbol{q})a(\boldsymbol{k}). \end{aligned} \tag{17.17}$$

However, the free Hamiltonians in the Heisenberg picture and in the Schrödinger picture depend in the same way on the respective field operators, but they are different operators if $V \neq 0$,

$$\begin{aligned} H_0 &= \exp\!\left(-\frac{\mathrm{i}}{\hbar}Ht\right)H_0(t)\exp\!\left(\frac{\mathrm{i}}{\hbar}Ht\right) \\ &= \exp\!\left(-\frac{\mathrm{i}}{\hbar}Ht\right)\int d^3\boldsymbol{x}\,\frac{\hbar^2}{2m}\nabla\Psi^+(\boldsymbol{x},t)\cdot\nabla\Psi(\boldsymbol{x},t)\exp\!\left(\frac{\mathrm{i}}{\hbar}Ht\right) \\ &= \int d^3\boldsymbol{x}\,\frac{\hbar^2}{2m}\nabla\psi^+(\boldsymbol{x})\cdot\nabla\psi(\boldsymbol{x}) = \int d^3\boldsymbol{k}\,\frac{\hbar^2\boldsymbol{k}^2}{2m}a^+(\boldsymbol{k})a(\boldsymbol{k}). \end{aligned} \tag{17.18}$$

[4]For convenience, we have chosen the time when both pictures coincide as $t_0 = 0$.

The number and charge operators in the Schrödinger picture are

$$N = \int d^3x\, \varrho(x) = \int d^3x\, \psi^+(x)\psi(x) = \int d^3k\, a^+(k)a(k) = \frac{1}{q}Q,$$

and the momentum operator is

$$\begin{aligned} P &= \int d^3x\, \frac{\hbar}{2\mathrm{i}}\left(\psi^+(x)\cdot\nabla\psi(x) - \nabla\psi^+(x)\cdot\psi(x)\right) \\ &= \int d^3k\, \hbar k\, a^+(k)a(k). \end{aligned} \tag{17.19}$$

The momentum operator $P(t)$ in the Heisenberg picture (17.11) is related to the momentum operator P in the Schrödinger picture through the standard transformation between Schrödinger picture and Heisenberg picture,

$$P(t) = \exp\left(\frac{\mathrm{i}}{\hbar}Ht\right)P\exp\left(-\frac{\mathrm{i}}{\hbar}Ht\right),$$

and the same similarity transformation applies to all the other operators. However, we did not write $N(t)$ or $Q(t)$ in equations (17.4, 17.10), because $[H, N] = 0$ for the single particle Hamiltonian (17.17).

We are now fully prepared to identify the meaning of the operators $a(k)$ and $a^+(k)$. The commutation relations

$$[H_0, a(k)] = -\frac{\hbar^2 k^2}{2m}a(k), \quad [H_0, a^+(k)] = \frac{\hbar^2 k^2}{2m}a^+(k), \tag{17.20}$$

$$[P, a(k)] = -\hbar k a(k), \quad [P, a^+(k)] = \hbar k a^+(k), \tag{17.21}$$

$$[Q, a(k)] = -q a(k), \quad [Q, a^+(k)] = q a^+(k), \tag{17.22}$$

$$[N, a(k)] = -a(k), \quad [N, a^+(k)] = a^+(k), \tag{17.23}$$

imply that $a(k)$ annihilates a particle with energy $\hbar^2 k^2/2m$, momentum $\hbar k$, mass m and charge q, while $a^+(k)$ generates such a particle. This follows exactly in the same way as the corresponding proof for energy annihilation and creation for the harmonic oscillator (6.11–6.13). Suppose e.g. that $|K\rangle$ is an eigenstate of the momentum operator,

$$P|K\rangle = \hbar K|K\rangle.$$

The commutation relation (17.21) then implies

$$Pa^+(k)|K\rangle = a^+(k)\left(P + \hbar k\right)|K\rangle = \hbar\left(K + k\right)a^+(k)|K\rangle,$$

i.e.

$$a^+(k)|K\rangle \propto |K + k\rangle,$$

while (17.20) implies

$$a^+(\boldsymbol{k})|E\rangle \propto |E + (\hbar^2\boldsymbol{k}^2/2m)\rangle.$$

The Hamilton operator (17.18) therefore corresponds to an infinite number of harmonic oscillators with frequencies $\omega(\boldsymbol{k}) = \hbar\boldsymbol{k}^2/2m$, and there must exist a lowest energy state $|0\rangle$ which must be annihilated by the lowering operators,

$$a(\boldsymbol{k})|0\rangle = 0.$$

The general state then corresponds to linear superpositions of states of the form

$$|\{n_{\boldsymbol{k}}\}\rangle = \prod_{\boldsymbol{k}} \frac{a^+(\boldsymbol{k})^{n_{\boldsymbol{k}}}}{\sqrt{n_{\boldsymbol{k}}!}}|0\rangle.$$

This vector space of states is denoted as a *Fock space*.

The particle annihilation and creation interpretation of $a(\boldsymbol{k})$ and $a^+(\boldsymbol{k})$ then also implies that the Fourier component $V(\boldsymbol{q})$ in the potential term of the full Hamiltonian (17.17) shifts the momentum of a particle by $\Delta\boldsymbol{p} = \hbar\boldsymbol{q}$ by replacing a particle with momentum $\hbar\boldsymbol{k}$ with a particle of momentum $\hbar\boldsymbol{k} + \hbar\boldsymbol{q}$.

Time-dependence of H_0

The free Hamiltonian H_0 (17.18) is time-independent in the Schrödinger picture (and also in the Dirac picture introduced below), but not in the Heisenberg picture if $[H_0, H] \neq 0$. The transformation from the Schrödinger picture into the Heisenberg picture,

$$H_0(t) = \int d^3\boldsymbol{x}\, \frac{\hbar^2}{2m}\nabla\Psi^+(\boldsymbol{x}, t)\cdot\nabla\Psi(\boldsymbol{x}, t) = \exp\!\left(\frac{\mathrm{i}}{\hbar}Ht\right) H_0 \exp\!\left(-\frac{\mathrm{i}}{\hbar}Ht\right),$$

implies the evolution equation

$$\begin{aligned}\frac{dH_0(t)}{dt} &= \frac{\mathrm{i}}{\hbar}[H, H_0(t)] = \frac{\mathrm{i}}{\hbar}[V(t), H_0(t)]\\ &= \frac{\mathrm{i}}{\hbar}\exp\!\left(\frac{\mathrm{i}}{\hbar}Ht\right)[V, H_0]\exp\!\left(-\frac{\mathrm{i}}{\hbar}Ht\right),\end{aligned} \tag{17.24}$$

The operator

$$V(t) = \int d^3\boldsymbol{x}\, \Psi^+(\boldsymbol{x}, t)V(\boldsymbol{x})\Psi(\boldsymbol{x}, t) = \exp\!\left(\frac{\mathrm{i}}{\hbar}Ht\right) V \exp\!\left(-\frac{\mathrm{i}}{\hbar}Ht\right)$$

is the potential operator in the Heisenberg picture, while the potential operator in the Schrödinger picture is

$$V = \int d^3\boldsymbol{x}\, \psi^+(\boldsymbol{x})V(\boldsymbol{x})\psi(\boldsymbol{x}) = \int d^3\boldsymbol{k} \int d^3\boldsymbol{q}\, a^+(\boldsymbol{k}+\boldsymbol{q})V(\boldsymbol{q})a(\boldsymbol{k}).$$

The commutator in the Schrödinger picture follows from the canonical commutators or anti-commutators of the field operators as

$$[V, H_0] = \int d^3\boldsymbol{x}\, \frac{\hbar^2}{2m} \left(\psi^+(\boldsymbol{x})\cdot\nabla\psi(\boldsymbol{x}) - \nabla\psi^+(\boldsymbol{x})\cdot\psi(\boldsymbol{x})\right)\cdot\nabla V(\boldsymbol{x}) \quad (17.25)$$

$$= -\int d^3\boldsymbol{k} \int d^3\boldsymbol{q}\, \frac{\hbar^2}{2m} \left(\boldsymbol{q}^2 + 2\boldsymbol{k}\cdot\boldsymbol{q}\right) a^+(\boldsymbol{k}+\boldsymbol{q})V(\boldsymbol{q})a(\boldsymbol{k}). \quad (17.26)$$

The integral in equation (17.25) contains the current density (1.18, 16.23) of the Schrödinger field. The commutator can therefore be written as

$$[V, H_0] = \mathrm{i}\hbar \int d^3\boldsymbol{x}\boldsymbol{j}(\boldsymbol{x})\cdot\nabla V(\boldsymbol{x}),$$

and substitution into the Heisenberg picture evolution equations for $H_0(t)$ (17.24) yields

$$\frac{d}{dt}H_0(t) = -\int d^3\boldsymbol{x}\boldsymbol{j}(\boldsymbol{x},t)\cdot\nabla V(\boldsymbol{x}). \quad (17.27)$$

However, we have also identified $\boldsymbol{j}(\boldsymbol{x},t)$ as a velocity density operator for the Schrödinger field, cf. (16.24). The classical analog of equation (17.27) is therefore the equation for the change of the kinetic energy of a classical non-relativistic particle moving under the influence of the force $\boldsymbol{F}(\boldsymbol{x}) = -\nabla V(\boldsymbol{x})$,

$$\frac{d}{dt}K(t) = -\boldsymbol{v}(t)\cdot\nabla V(\boldsymbol{x}).$$

17.2 Time evolution for time-dependent Hamiltonians

The generic case in quantum field theory are time-independent Hamilton operators in the Heisenberg and Schrödinger pictures. We will see the reason for this below, after discussing the general case of a Heisenberg picture Hamiltonian $H(t) \equiv H_H(t)$ which could depend on time.

Integration of equation (17.6) yields in the general case of time-dependent $H(t)$

$$\Psi(t) = \Psi(t_0) + \frac{\mathrm{i}}{\hbar}\int_{t_0}^{t} d\tau\, [H(\tau), \Psi(\tau)] = \tilde{U}(t,t_0)\Psi(t_0)\tilde{U}^+(t,t_0),$$

with the unitary operator

$$\tilde{U}(t,t_0) = \tilde{\mathrm{T}}\exp\!\left(\frac{\mathrm{i}}{\hbar}\int_{t_0}^{t} d\tau\, H(\tau)\right).$$

Here $\tilde{\mathrm{T}}$ locates the Hamiltonians near the upper time integration boundary leftmost, but for the factor $+\mathrm{i}$ in front of the integral.

Recall that in the Heisenberg picture, we have all time dependence in the operators, but time-independent states. To convert to the Schrödinger picture, we remove the time dependence from the operators and cast it onto the states such that matrix elements remain the same, $\langle\Phi(t_0)|\Psi(t)|\Phi(t_0)\rangle = \langle\Phi(t)|\Psi(t_0)|\Phi(t)\rangle$. The time evolution of the states in the Schrödinger picture is therefore given by

$$|\Phi(t)\rangle = \tilde{U}(t_0,t)|\Phi(t_0)\rangle. \tag{17.28}$$

This implies a Schrödinger equation

$$\begin{aligned}\mathrm{i}\hbar\frac{d}{dt}|\Phi(t)\rangle &= \tilde{U}(t_0,t)H_H(t)|\Phi(t_0)\rangle = \tilde{U}(t_0,t)H_H(t)\tilde{U}(t,t_0)|\Phi(t)\rangle\\ &= H_S(t)|\Phi(t)\rangle.\end{aligned}$$

Therefore we also have

$$|\Phi(t)\rangle = U(t,t_0)|\Phi(t_0)\rangle = \mathrm{T}\exp\!\left(-\frac{\mathrm{i}}{\hbar}\int_{t_0}^{t} d\tau\, H_S(\tau)\right)|\Phi(t_0)\rangle,$$

i.e.

$$\begin{aligned}\tilde{U}(t_0,t) &= \tilde{\mathrm{T}}\exp\!\left(\frac{\mathrm{i}}{\hbar}\int_{t}^{t_0} d\tau\, H_H(\tau)\right) = U(t,t_0)\\ &= \mathrm{T}\exp\!\left(-\frac{\mathrm{i}}{\hbar}\int_{t_0}^{t} d\tau\, H_S(\tau)\right),\end{aligned} \tag{17.29}$$

where

$$H_S(t) = \tilde{U}(t_0,t)H_H(t)\tilde{U}(t,t_0), \quad H_H(t) = U(t_0,t)H_S(t)U(t,t_0).$$

The Hamiltonian in the Schrödinger picture depends only on the t-independent field operators $\Psi(t_0)$, i.e. any time dependence of H_S can only result from an explicit time dependence of any parameter, e.g. if a coupling constant or mass would somehow depend on time. If such a time dependence through a parameter is not there, then $U(t,t_0) = \exp[-\mathrm{i}H_S(t-t_0)/\hbar]$ and $H_H(t) = H_S$, i.e. H_S is time-independent if and only if H_H is time-independent, and then $H_S = H_H$.

This explains why time-independent Hamiltonians $H_S = H_H$ are the generic case in quantum field theory. Usually, if we would discover any kind of time dependence in any parameter $\lambda = \lambda(t)$ in H_S, we would suspect that there must be a dynamical explanation in terms of a corresponding field, i.e. we would promote $\lambda(t)$ to a full dynamical field operator besides all the other field operators in H_S, including a kinetic term for $\lambda(t)$, and then the new Hamiltonian would again be time-independent.

Occasionally, we might prefer to treat a dynamical field as a given time-dependent parameter, e.g. include electric fields in a semi-classical approximation instead of dealing with the quantized photon operators. This is standard practice in the "first quantized" theory, and therefore time dependence of the Schrödinger and Heisenberg Hamiltonians plays a prominent role there. However, once we go through the hassle of field quantization, we may just as well do the same for all the fields in the theory, including electromagnetic fields, and therefore semi-classical approximations and ensuing time dependence through parameters is not as important in the second quantized theory.

17.3 The connection between first and second quantized theory

For a single particle first and second quantized theory should yield the same expectation values, i.e. matrix elements in the 1-particle sector should agree:

$$\langle\Phi|\Psi\rangle = \langle\Phi|\Psi\rangle. \tag{17.30}$$

For the states

$$|\boldsymbol{x}\rangle = \psi^+(\boldsymbol{x})|0\rangle, \quad |\boldsymbol{k}\rangle = a^+(\boldsymbol{k})|0\rangle,$$

equation (17.30) is fulfilled due to the standard Fourier transformation relation between the operators in $\boldsymbol{x}$-space and $\boldsymbol{k}$-space. The relations

$$\psi^+(\boldsymbol{x}) = \int d^3\boldsymbol{k}\, a^+(\boldsymbol{k})\langle\boldsymbol{k}|\boldsymbol{x}\rangle, \quad \psi(\boldsymbol{x}) = \int d^3\boldsymbol{k}\,\langle\boldsymbol{x}|\boldsymbol{k}\rangle a(\boldsymbol{k}),$$

$$a^+(\boldsymbol{k}) = \int d^3\boldsymbol{x}\,\psi^+(\boldsymbol{x})\langle\boldsymbol{x}|\boldsymbol{k}\rangle, \quad a(\boldsymbol{k}) = \int d^3\boldsymbol{x}\,\langle\boldsymbol{k}|\boldsymbol{x}\rangle\psi(\boldsymbol{x}),$$

yield

$$\langle\boldsymbol{x}|\boldsymbol{k}\rangle = \langle 0|\psi(\boldsymbol{x})a^+(\boldsymbol{k})|0\rangle = \int d^3\boldsymbol{k}'\,\langle\boldsymbol{x}|\boldsymbol{k}'\rangle\langle 0|a(\boldsymbol{k}')a^+(\boldsymbol{k})|0\rangle$$

$$= \int d^3\boldsymbol{k}'\,\langle\boldsymbol{x}|\boldsymbol{k}'\rangle\langle 0|[a(\boldsymbol{k}'), a^+(\boldsymbol{k})]_\pm|0\rangle = \langle\boldsymbol{x}|\boldsymbol{k}\rangle = \frac{1}{\sqrt{2\pi}^3}\exp(\mathrm{i}\boldsymbol{k}\cdot\boldsymbol{x}).$$

To explore this connection further, we will use superscripts (1) and (2) to designate operators in first and second quantized theory. E.g. the 1-particle Hamiltonians in first and second quantized theory can be written as

$$H^{(1)} = \int d^3\boldsymbol{x}\, |\boldsymbol{x}\rangle \left(-\frac{\hbar^2}{2m}\Delta + V(\boldsymbol{x})\right) \langle \boldsymbol{x}|, \tag{17.31}$$

$$H^{(2)} = \int d^3\boldsymbol{x}\, \psi^+(\boldsymbol{x})\left(-\frac{\hbar^2}{2m}\Delta + V(\boldsymbol{x})\right) \psi(\boldsymbol{x}). \tag{17.32}$$

We can rewrite $H^{(2)}$ as

$$H^{(2)} = \int d^3\boldsymbol{x}' \int d^3\boldsymbol{x}'' \int d^3\boldsymbol{x}\, \psi^+(\boldsymbol{x}')\delta(\boldsymbol{x}'-\boldsymbol{x}'')\left(-\frac{\hbar^2}{2m}\Delta'' + V(\boldsymbol{x}'')\right)$$
$$\times\delta(\boldsymbol{x}''-\boldsymbol{x})\psi(\boldsymbol{x}) = \int d^3\boldsymbol{x}' \int d^3\boldsymbol{x}\, \psi^+(\boldsymbol{x}')\langle \boldsymbol{x}'|H^{(1)}|\boldsymbol{x}\rangle\psi(\boldsymbol{x}),$$

and again we have exact correspondence between 1-particle matrix elements in the first and second quantized theory,

$$\langle \boldsymbol{x}'|H^{(1)}|\boldsymbol{x}\rangle = \langle \boldsymbol{x}'|H^{(2)}|\boldsymbol{x}\rangle. \tag{17.33}$$

This works in general. For an operator $K^{(1)}$ from first quantized theory, the requirement of equality of 1-particle matrix elements

$$\langle \boldsymbol{k}'|K^{(2)}|\boldsymbol{k}\rangle = \langle \boldsymbol{k}'|K^{(1)}|\boldsymbol{k}\rangle, \quad \langle \boldsymbol{x}'|K^{(2)}|\boldsymbol{x}\rangle = \langle \boldsymbol{x}'|K^{(1)}|\boldsymbol{x}\rangle \tag{17.34}$$

can be solved by

$$K^{(2)} = \int d^3\boldsymbol{k}' \int d^3\boldsymbol{k}\, a^+(\boldsymbol{k}')\langle \boldsymbol{k}'|K^{(1)}|\boldsymbol{k}\rangle a(\boldsymbol{k})$$
$$= \int d^3\boldsymbol{x}' \int d^3\boldsymbol{x}\, \psi^+(\boldsymbol{x}')\langle \boldsymbol{x}'|K^{(1)}|\boldsymbol{x}\rangle\psi(\boldsymbol{x}).$$

General 1-particle states and corresponding annihilation and creation operators in second quantized theory

The equivalence of first and second quantized theory in the single-particle sector also allows us to derive the equations for 1-particle states and corresponding annihilation and creation operators in second quantization. Suppose $|m\rangle$ and $|n\rangle$ are two states of the first quantized theory. The corresponding matrix element of the Hamiltonian in the first quantized theory is

$$\begin{aligned}
\langle m|H^{(1)}|n\rangle &= \int d^3x\,\langle m|x\rangle\left(-\frac{\hbar^2}{2m}\Delta + V(x)\right)\langle x|n\rangle \\
&= \int d^3x\int d^3x'\int d^3x''\,\langle m|x''\rangle\delta(x''-x)\left(-\frac{\hbar^2}{2m}\Delta + V(x)\right) \\
&\quad\times\delta(x-x')\langle x'|n\rangle \\
&= \int d^3x\int d^3x'\int d^3x''\,\langle m|x''\rangle\langle 0|\psi(x'')\psi^+(x)\left(-\frac{\hbar^2}{2m}\Delta + V(x)\right) \\
&\quad\times\psi(x)\psi^+(x')|0\rangle\langle x'|n\rangle \\
&= \int d^3x'\int d^3x''\,\langle m|x''\rangle\langle 0|\psi(x'')H^{(2)}\psi^+(x')|0\rangle\langle x'|n\rangle,
\end{aligned} \tag{17.35}$$

where we used the identity

$$\delta(x''-x)\delta(x-x') = \langle 0|\psi(x'')\psi^+(x)\psi(x)\psi^+(x')|0\rangle$$

to write the matrix element of the 1st quantized theory as a matrix element of the 2nd quantized theory.

We can interprete the result (17.35) as equality of single particle matrix elements,

$$\langle m|H^{(1)}|n\rangle = \langle m|H^{(2)}|n\rangle$$

if we define the 1-particle states

$$|n\rangle = \int d^3x\,\psi^+(x)|0\rangle\langle x|n\rangle = \int d^3x\,|x\rangle\langle x|n\rangle. \tag{17.36}$$

This also motivates the definition of corresponding creation and annihilation operators

$$a_n^+ \equiv \psi_n^+ = \int d^3x\,\psi^+(x)\langle x|n\rangle = \int d^3k\,a^+(k)\langle k|n\rangle, \tag{17.37}$$

$$a_n \equiv \psi_n = \int d^3x\,\psi(x)\langle n|x\rangle = \int d^3k\,a(k)\langle n|k\rangle. \tag{17.38}$$

E.g. the operator

$$a_{n,\ell,m}^+ \equiv \psi_{n,\ell,m}^+ = \int d^3x\,\psi^+(x)\langle x|n,\ell,m\rangle = \int d^3k\,a^+(k)\langle k|n,\ell,m\rangle$$

will create an electron (or more precisely, the corresponding quasiparticle for relative motion of the electron and the proton) in the $|n,\ell,m\rangle$ state of hydrogen.

The canonical relations for the field operators $\psi(\boldsymbol{x})$ or $a(\boldsymbol{k})$ imply that the operators for multiplets of quantum numbers n also satisfy (anti-)commutation relations

$$[\psi_m, \psi_n^+]_\pm = \delta_{m,n}, \quad [\psi_m, \psi_n]_\pm = 0, \quad [\psi_m^+, \psi_n^+]_\pm = 0. \tag{17.39}$$

Substituting $\langle \boldsymbol{x}|n\rangle = \langle \boldsymbol{x}|n\rangle$ in equation (17.36) also shows the completeness relation in the single particle sector of the Fock space,

$$\int d^3\boldsymbol{x}\, |\boldsymbol{x}\rangle\langle\boldsymbol{x}| = 1. \tag{17.40}$$

Time evolution of 1-particle states in second quantized theory

According to our previous observations, a state in the Schrödinger picture evolves according to

$$|\Phi(t)\rangle = \exp\!\left(-\frac{\mathrm{i}}{\hbar}H^{(2)}t\right)|\Phi(0)\rangle. \tag{17.41}$$

On the other hand, according to equation (17.36), a single particle state at time $t = 0$ should be given in terms of the corresponding first quantized state $|\Phi(0)\rangle$,

$$|\Phi(0)\rangle = \int d^3\boldsymbol{x}\, \psi^+(\boldsymbol{x})|0\rangle\langle\boldsymbol{x}|\Phi(0)\rangle.$$

Here we wish to show that this relation is preserved under time evolution.

We find from equations (17.41), (17.40) and (17.33)

$$\begin{aligned}|\Phi(t)\rangle &= \int d^3\boldsymbol{x}\, |\boldsymbol{x}\rangle\langle\boldsymbol{x}|\exp\!\left(-\frac{\mathrm{i}}{\hbar}H^{(2)}t\right)|\Phi(0)\rangle \\ &= \int d^3\boldsymbol{x}\, \psi^+(\boldsymbol{x})|0\rangle\langle\boldsymbol{x}|\exp\!\left(-\frac{\mathrm{i}}{\hbar}H^{(1)}t\right)|\Phi(0)\rangle \\ &= \int d^3\boldsymbol{x}\, \psi^+(\boldsymbol{x})|0\rangle\langle\boldsymbol{x}|\Phi(t)\rangle,\end{aligned} \tag{17.42}$$

i.e. the equation (17.36) is indeed preserved under time evolution of the states.

We can write the Schrödinger state $|\Phi(t)\rangle$ also in the form

$$|\Phi(t)\rangle = \Phi^+(t)|0\rangle$$

with the creation operator of the particle in the first quantized state $|\Phi(t)\rangle$,

$$\Phi^+(t) = \int d^3\boldsymbol{x}\, \psi^+(\boldsymbol{x})\langle\boldsymbol{x}|\Phi(t)\rangle = \int d^3\boldsymbol{k}\, a^+(\boldsymbol{k})\langle\boldsymbol{k}|\Phi(t)\rangle. \tag{17.43}$$

Note that $\Phi^+(t)$ is an operator in the Schrödinger picture of the theory. The time-dependence arises only because it is a superposition of Schrödinger picture operators with time-dependent amplitudes. The corresponding Heisenberg picture operator is given in equation (17.47) below.

Other equivalent forms of the representation of states in the Schrödinger picture involve linear combinations of Heisenberg picture field operators, e.g.

$$\begin{aligned} |\Phi(t)\rangle &= \exp\!\left(-\frac{\mathrm{i}}{\hbar}H^{(2)}t\right)|\Phi(0)\rangle \\ &= \int d^3\boldsymbol{x}\,\exp\!\left(-\frac{\mathrm{i}}{\hbar}H^{(2)}t\right)\psi^+(\boldsymbol{x})|0\rangle\langle\boldsymbol{x}|\Phi(0)\rangle \\ &= \int d^3\boldsymbol{x}\,\Psi^+(\boldsymbol{x},-t)|0\rangle\langle\boldsymbol{x}|\Phi(0)\rangle \end{aligned} \tag{17.44}$$

and

$$\langle\boldsymbol{x}|\Phi(t)\rangle = \langle\boldsymbol{x}|\exp\!\left(-\frac{\mathrm{i}}{\hbar}H^{(2)}t\right)|\Phi(0)\rangle = \langle\boldsymbol{x},t|\Phi(0)\rangle,$$

with moving base kets

$$|\boldsymbol{x},t\rangle = \exp\!\left(\frac{\mathrm{i}}{\hbar}H^{(2)}t\right)|\boldsymbol{x}\rangle = \exp\!\left(\frac{\mathrm{i}}{\hbar}H^{(2)}t\right)\psi^+(\boldsymbol{x})|0\rangle = \Psi^+(\boldsymbol{x},t)|0\rangle,$$

$$|\boldsymbol{k},t\rangle = a^+(\boldsymbol{k},t)|0\rangle.$$

At first sight, the time-dependence of the creation operator in (17.44) may not be what one naively might have expected, but as we have seen it is implied by the correspondence of single particle matrix elements between the second and first quantized theory. In a slightly different way, the correctness of the time-dependence in (17.44) can also be confirmed by verifying that it is exactly the time-dependence which ensures that the Heisenberg evolution equation (17.7) is equivalent to the Schrödinger equation on the single particle wave function, see Problem 17.10.

The Heisenberg picture state corresponding to $|\Phi(t)\rangle$ is

$$\begin{aligned} |\Phi_H\rangle &= \exp\!\left(\frac{\mathrm{i}}{\hbar}H^{(2)}t\right)|\Phi(t)\rangle = |\Phi(0)\rangle \\ &= \int d^3\boldsymbol{x}\,\psi^+(\boldsymbol{x})|0\rangle\langle\boldsymbol{x}|\Phi(0)\rangle. \end{aligned} \tag{17.45}$$

Note that substitution of (17.42) into the first equation in (17.45) implies that we can write this state also in the form

$$|\Phi_H\rangle = \int d^3\boldsymbol{x}\,\Psi^+(\boldsymbol{x},t)|0\rangle\langle\boldsymbol{x}|\Phi(t)\rangle = \Phi_H^+(t)|0\rangle \tag{17.46}$$

with the Heisenberg picture operator

$$\Phi_H^+(t) = \exp\left(\frac{\mathrm{i}}{\hbar}H^{(2)}t\right)\Phi^+(t)\exp\left(-\frac{\mathrm{i}}{\hbar}H^{(2)}t\right)$$
$$= \int d^3x\,\Psi^+(x,t)\langle x|\Phi(t)\rangle. \quad (17.47)$$

The time-independence of the Heisenberg picture state $|\Phi_H\rangle$ is manifest in (17.45) but appears rather suspicious in (17.46). However, the representation (17.46) directly leads back to (17.45) if we use the correspondence of single particle matrix elements,

$$\langle x|\Phi(t)\rangle = \langle x|\Phi(t)\rangle = \langle x|\exp\left(-\frac{\mathrm{i}}{\hbar}H^{(2)}t\right)|\Phi(0)\rangle,$$

and the completeness relation in the single particle sector,

$$\exp\left(\frac{\mathrm{i}}{\hbar}H^{(2)}t\right)\int d^3x\,|x\rangle\langle x|\exp\left(-\frac{\mathrm{i}}{\hbar}H^{(2)}t\right) = 1.$$

There is a subtle point underlying the discussion in this section that students who go through their first iteration of learning quantum field theory would not notice, because we have not yet discussed interacting quantum field theories. However, I should point out that the equivalence of first and second quantization in the single particle sector holds if the single particle states cannot spontaneously absorb another particle or decay into two or more particles. This property also holds in interacting quantum field theories like quantum electronics or quantum electrodynamics, because conservation laws prevent e.g. single charged particles from spontaneously absorbing or radiating photons. These theories require at least two particles in both the initial and final states (or semi-classical inclusion of a second particle in the form of an external potential) for particle number changing processes. Quantum field theory can also describe inherently unstable particles which decay into two or more particles. This could be mapped back to a corresponding first quantized theory in terms of coupled many particle wave equations for N-particle wave functions. However, that would yield an unwieldy and inefficient formalism. Quantum fields are much more convenient than wave functions when it comes to the description of particle number changing processes.

17.4 The Dirac picture in quantum field theory

Although our Hamiltonians in the Heisenberg and Schrödinger pictures are usually time-independent in quantum field theory, time-dependent perturbation theory is still used for the calculation of transition rates even with time-independent perturbations V. This will lead again to the calculation of scattering matrix elements $S_{fi} = \langle f|U_D(\infty, -\infty)|i\rangle$ of the time-evolution operator in the interaction picture. Therefore we will automatically encounter field operators in the Dirac picture, which are gotten from the time-independent field operators of the Schrödinger picture through application of an unperturbed Hamiltonian $H_0 = H - V$. In many cases this will be the free Schrödinger picture Hamiltonian

$$H_0 = \int d^3x \, \frac{\hbar^2}{2m} \nabla\psi^+(x) \cdot \nabla\psi(x) = \int d^3k \, \frac{\hbar^2 k^2}{2m} a^+(k)a(k).$$

Please note that the free Hamilton operator in the Heisenberg picture (we set again $t_0 = 0$ for the time when the two pictures coincide)

$$\begin{aligned} H_{0,H}(t) &= \exp\left(\frac{\mathrm{i}}{\hbar}Ht\right) H_0 \exp\left(-\frac{\mathrm{i}}{\hbar}Ht\right) \\ &= \int d^3x \, \frac{\hbar^2}{2m} \nabla\Psi^+(x,t) \cdot \nabla\Psi(x,t) = \int d^3k \, \frac{\hbar^2 k^2}{2m} a^+(k,t)a(k,t) \end{aligned}$$

usually differs from H_0, because generically

$$[H, H_0] = [V, H_0] \neq 0.$$

Transformation of the basic field operators from the Schrödinger picture into the Dirac picture yields

$$\begin{aligned} a_D(k,t) &= \exp\left(\frac{\mathrm{i}}{\hbar}H_0 t\right) a(k) \exp\left(-\frac{\mathrm{i}}{\hbar}H_0 t\right) = a(k) \exp\left(-\frac{\mathrm{i}\hbar}{2m}k^2 t\right) \\ &= \frac{1}{\sqrt{2\pi}^3} \int d^3x \, \psi(x,t) \exp(-\mathrm{i}k \cdot x), \end{aligned} \tag{17.48}$$

$$\begin{aligned} \psi(x,t) &= \exp\left(\frac{\mathrm{i}}{\hbar}H_0 t\right) \psi(x) \exp\left(-\frac{\mathrm{i}}{\hbar}H_0 t\right) \\ &= \frac{1}{\sqrt{2\pi}^3} \int d^3k \, a_D(k,t) \exp(\mathrm{i}k \cdot x) \\ &= \frac{1}{\sqrt{2\pi}^3} \int d^3k \, a(k) \exp\left(\mathrm{i}k \cdot x - \frac{\mathrm{i}\hbar}{2m}k^2 t\right). \end{aligned} \tag{17.49}$$

Due to the simple relation (17.48) $a_D(k,t)$ is always substituted with $a(k)$ in applications of the Dirac picture.

We summarize the conventions for the notation for basic field operators in Schrödinger field theory in Table 17.2.

The Hamiltonian and the corresponding time evolution operator on the states, as well as the transition amplitudes are derived in exactly the same way as in the first quantized theory. However, these topics are important enough to warrant

Table 17.2 Conventions for basic field operators in different pictures of Schrödinger field theory

Heisenberg picture	Schrödinger picture	Dirac picture
$\Psi(\boldsymbol{x}, t)$	$\psi(\boldsymbol{x})$	$\psi(\boldsymbol{x}, t)$
$\Psi^{+}(\boldsymbol{x}, t)$	$\psi^{+}(\boldsymbol{x})$	$\psi^{+}(\boldsymbol{x}, t)$
$a(\boldsymbol{k}, t)$	$a(\boldsymbol{k})$	$a_D(\boldsymbol{k}, t)$
$a^{+}(\boldsymbol{k}, t)$	$a^{+}(\boldsymbol{k})$	$a_D^{+}(\boldsymbol{k}, t)$

repetition in the framework of the second quantized theory. This time we can limit the discussion to the simpler case of time-independent Hamiltonians H and H_0 in the Schrödinger picture.

The states in the Schrödinger picture of quantum field theory satisfy the Schrödinger equation

$$\mathrm{i}\hbar\frac{d}{dt}|\Phi(t)\rangle = H|\Phi(t)\rangle,$$

which implies

$$|\Phi(t)\rangle = \exp\!\left(-\frac{\mathrm{i}}{\hbar}H(t-t')\right)|\Phi(t')\rangle.$$

The transformation (17.49) $\psi(\boldsymbol{x}) \to \psi(\boldsymbol{x}, t)$ into the Dirac picture implies for the states the transformation

$$|\Phi(t)\rangle \to |\Phi_D(t)\rangle = \exp\!\left(\frac{\mathrm{i}}{\hbar}H_0 t\right)|\Phi(t)\rangle.$$

The time evolution of the states in the Schrödinger picture then determines the time evolution of the states in the Dirac picture

$$\begin{aligned}|\Phi_D(t)\rangle &= \exp\!\left(\frac{\mathrm{i}}{\hbar}H_0 t\right)\exp\!\left(-\frac{\mathrm{i}}{\hbar}H(t-t')\right)\exp\!\left(-\frac{\mathrm{i}}{\hbar}H_0 t'\right)|\Phi_D(t')\rangle\\ &= U_D(t,t')|\Phi_D(t')\rangle\end{aligned}$$

with the time evolution operator on the states[5]

$$U_D(t,t') = \exp\!\left(\frac{\mathrm{i}}{\hbar}H_0 t\right)\exp\!\left(-\frac{\mathrm{i}}{\hbar}H(t-t')\right)\exp\!\left(-\frac{\mathrm{i}}{\hbar}H_0 t'\right).$$

[5]Recall that there are two time evolution operators in the Dirac picture. The free time evolution operator $U_0(t-t')$ evolves the operators $\psi(\boldsymbol{x}, t) = U_0^{+}(t-t')\psi(\boldsymbol{x}, t')U_0(t-t')$, while $U_D(t,t')$ evolves the states.

This operator satisfies the initial condition $U_D(t',t') = 1$ and the differential equations

$$\begin{aligned}
\mathrm{i}\hbar\frac{\partial}{\partial t}U_D(t,t') &= \exp\left(\frac{\mathrm{i}}{\hbar}H_0 t\right)(H-H_0)\exp\left(-\frac{\mathrm{i}}{\hbar}H(t-t')\right)\exp\left(-\frac{\mathrm{i}}{\hbar}H_0 t'\right)\\
&= \exp\left(\frac{\mathrm{i}}{\hbar}H_0 t\right)V\exp\left(-\frac{\mathrm{i}}{\hbar}H_0 t\right)U_D(t,t') = H_D(t)U_D(t,t'),
\end{aligned}$$

$$\mathrm{i}\hbar\frac{\partial}{\partial t'}U_D(t,t') = -U_D(t,t')H_D(t'),$$

and can therefore also be written as

$$U_D(t,t') = \mathrm{T}\exp\left(-\frac{\mathrm{i}}{\hbar}\int_{t'}^{t} d\tau\, H_D(\tau)\right).$$

The states in the Dirac picture therefore satisfy the Schrödinger equation

$$\mathrm{i}\hbar\frac{d}{dt}|\Phi_D(t)\rangle = H_D(t)|\Phi_D(t)\rangle$$

with the Hamiltonian

$$H_D(t) \equiv V_D(t) = \exp\left(\frac{\mathrm{i}}{\hbar}H_0 t\right)V\exp\left(-\frac{\mathrm{i}}{\hbar}H_0 t\right).$$

The transition amplitude from an initial unperturbed state $|\Phi_i(t')\rangle$ at time t' to a final state $|\Phi_f(t)\rangle$ at time t is

$$\begin{aligned}
S_{fi}(t,t') &= \langle\Phi_f(t)|\Phi_i(t)\rangle = \langle\Phi_f(t)|\exp\left(-\frac{\mathrm{i}}{\hbar}H(t-t')\right)|\Phi_i(t')\rangle\\
&= \langle\Phi_f(0)|\exp\left(\frac{\mathrm{i}}{\hbar}H_0 t\right)\exp\left(-\frac{\mathrm{i}}{\hbar}H(t-t')\right)\exp\left(-\frac{\mathrm{i}}{\hbar}H_0 t'\right)|\Phi_i(0)\rangle,
\end{aligned}$$

or with $|f\rangle \equiv |\Phi_f(0)\rangle$

$$S_{fi}(t,t') = \langle f|U_D(t,t')|i\rangle = \langle f|\mathrm{T}\exp\left(-\frac{\mathrm{i}}{\hbar}\int_{t'}^{t} d\tau\, H_D(\tau)\right)|i\rangle. \tag{17.50}$$

The scattering matrix $S_{fi} = \langle f|U_D(\infty,-\infty)|i\rangle$ contains information about all processes which take a physical system e.g. from an initial state $|i\rangle$ with n_i particles to a final state $|f\rangle$ with n_f. This includes in particular also processes where the interactions in $H_D(t)$ generate virtual intermediate particles which do not couple to any of the external particles. These vacuum processes need to be subtracted from the scattering matrix in each order of perturbation theory, which amounts to simply neglecting them in the evaluation of the scattering matrix. The vacuum processes also appear in the vacuum to vacuum amplitude, and the subtraction in each order of perturbation theory can also be understood as dividing the vacuum to vacuum amplitude out of the scattering matrix,

$$S_{fi} = \frac{\langle f|U_D(\infty, -\infty)|i\rangle}{\langle 0|U_D(\infty, -\infty)|0\rangle}. \tag{17.51}$$

However, unitarity of the time evolution operator $U_D(\infty, -\infty)$ implies unitarity of the scattering matrix $S_{fi} = \langle f|U_D(\infty, -\infty)|i\rangle$ as defined earlier,

$$\begin{aligned}
\sum_i S_{fi}S^+_{if'} &= \sum_i S_{fi}S^*_{f'i} = \sum_i \langle f|U_D(\infty, -\infty)|i\rangle\langle f'|U_D(\infty, -\infty)|i\rangle^* \\
&= \sum_i \langle f|U_D(\infty, -\infty)|i\rangle\langle i|U^+_D(\infty, -\infty)|f'\rangle \\
&= \langle f|U_D(\infty, -\infty)U^+_D(\infty, -\infty)|f'\rangle = \delta_{ff'}.
\end{aligned}$$

Therefore division by the vacuum to vacuum matrix element $\langle 0|U_D(\infty, -\infty)|0\rangle$ in the alternative definition (17.51) can only yield a unitary scattering matrix if the amplitude $\langle 0|U_D(\infty, -\infty)|0\rangle$ is a phase factor. We can understand this in the following way. Conservation laws prevent spontaneous decay of the vacuum into any excited states $|N\rangle$,

$$\langle N \neq 0|U_D(\infty, -\infty)|0\rangle = 0.$$

The completeness relation

$$|0\rangle\langle 0| + \sum_{N\neq 0} |N\rangle\langle N| = 1$$

and unitarity of the time evolution operator then implies

$$\begin{aligned}
|\langle 0|U_D(\infty, -\infty)|0\rangle|^2 &= |\langle 0|U_D(\infty, -\infty)|0\rangle|^2 + \sum_{N\neq 0} |\langle N|U_D(\infty, -\infty)|0\rangle|^2 \\
&= \langle 0|U^+_D(\infty, -\infty)U_D(\infty, -\infty)|0\rangle = 1,
\end{aligned} \tag{17.52}$$

thus confirming that the vacuum to vacuum amplitude is a phase factor. We will continue to use the simpler notation $S_{fi} = \langle f|U_D(\infty, -\infty)|i\rangle$ for the scattering matrix with the understanding that we can neglect vacuum processes.

17.5 Inclusion of spin

The wave functions of particles with spin s have $2s + 1$ components $\Psi_\sigma(\boldsymbol{x}, t) \equiv \langle \boldsymbol{x}, \sigma|\Psi(t)\rangle$, $\sigma \equiv m_s \in \{-s, -s+1, \ldots, s\}$, and the normalization condition is

$$\sum_{\sigma=-s}^{s} \int d^3x\, |\Psi_\sigma(\boldsymbol{x}, t)|^2 = 1,$$

see e.g. Section 8.1. The Schrödinger equation with spin-dependent local interaction potentials between the different components,

$$i\hbar\frac{\partial}{\partial t}\Psi_\sigma(\boldsymbol{x},t) = -\frac{\hbar^2}{2m}\Delta\Psi_\sigma(\boldsymbol{x},t) + \sum_{\sigma'} V_{\sigma\sigma'}(\boldsymbol{x})\Psi_{\sigma'}(\boldsymbol{x},t)$$

follows from a Lagrange density

$$\mathcal{L} = \frac{i\hbar}{2}\sum_\sigma\left(\Psi_\sigma^+\cdot\frac{\partial\Psi_\sigma}{\partial t} - \frac{\partial\Psi_\sigma^+}{\partial t}\cdot\Psi_\sigma\right) - \frac{\hbar^2}{2m}\sum_\sigma \nabla\Psi_\sigma^+\cdot\nabla\Psi_\sigma$$
$$-\sum_{\sigma,\sigma'}\Psi_\sigma^+\cdot V_{\sigma\sigma'}\cdot\Psi_{\sigma'}. \tag{17.53}$$

Canonical quantization then yields the (anti-)commutation relations for the Heisenberg picture field operators,

$$[\Psi_\sigma(\boldsymbol{x},t),\Psi_{\sigma'}^+(\boldsymbol{x}',t)]_\pm = \delta_{\sigma\sigma'}\delta(\boldsymbol{x}-\boldsymbol{x}'),$$
$$[\Psi_\sigma(\boldsymbol{x},t),\Psi_{\sigma'}(\boldsymbol{x}',t)]_\pm = 0, \quad [\Psi_\sigma^+(\boldsymbol{x},t),\Psi_{\sigma'}^+(\boldsymbol{x}',t)]_\pm = 0,$$

with commutators for bosons (integer spin) and anti-commutators for fermions (half-integer spin).

The charge, Hamiltonian and momentum operators for particles with spin follow from (17.53) using the methods of Sections (16.2,16.3),

$$Q = qN = q\int d^3\boldsymbol{x}\sum_\sigma \Psi_\sigma^+(\boldsymbol{x},t)\Psi_\sigma(\boldsymbol{x},t), \tag{17.54}$$

$$H = H_0(t) + V(t),$$

$$H_0(t) = \int d^3\boldsymbol{x}\sum_\sigma \frac{\hbar^2}{2m}\nabla\Psi_\sigma^+(\boldsymbol{x},t)\cdot\nabla\Psi_\sigma(\boldsymbol{x},t), \tag{17.55}$$

$$V(t) = \int d^3\boldsymbol{x}\sum_{\sigma,\sigma'}\Psi_\sigma^+(\boldsymbol{x},t)V_{\sigma\sigma'}(\boldsymbol{x})\Psi_{\sigma'}(\boldsymbol{x},t), \tag{17.56}$$

and

$$\boldsymbol{P}(t) = \int d^3\boldsymbol{x}\sum_\sigma \frac{\hbar}{2i}\left(\Psi_\sigma^+(\boldsymbol{x},t)\cdot\nabla\Psi_\sigma(\boldsymbol{x},t) - \nabla\Psi_\sigma^+(\boldsymbol{x},t)\cdot\Psi_\sigma(\boldsymbol{x},t)\right).$$

The transition to the Schrödinger picture field operators then proceeds in the standard way,

$$\psi_\sigma(\boldsymbol{x}) = \exp\left(-\frac{\mathrm{i}}{\hbar}Ht\right)\Psi_\sigma(\boldsymbol{x},t)\exp\left(\frac{\mathrm{i}}{\hbar}Ht\right),$$

$$[\psi_\sigma(\boldsymbol{x}),\psi^+_{\sigma'}(\boldsymbol{x}')]_\pm = \delta_{\sigma\sigma'}\delta(\boldsymbol{x}-\boldsymbol{x}'),$$

$$[\psi_\sigma(\boldsymbol{x}),\psi_{\sigma'}(\boldsymbol{x}')]_\pm = 0, \quad [\psi^+_\sigma(\boldsymbol{x}),\psi^+_{\sigma'}(\boldsymbol{x}')]_\pm = 0.$$

The most common case of non-vanishing spin in non-relativistic quantum mechanics is $s = 1/2$, and then the common conventions for assigning values for the spin label σ are $1/2, +, \uparrow$ for $s_z = \hbar/2$, and $-1/2, -, \downarrow$ for $s_z = -\hbar/2$. Higher spin values can arise within non-relativistic quantum mechanics in nuclei, atoms, and molecules.

Since we now use $\Psi_\sigma(\boldsymbol{x},t)$ to denote quantum fields in the Heisenberg picture, we will denote the time-dependent states in the Schrödinger picture and the corresponding wave function components by $\Phi_\sigma(\boldsymbol{x},t)$. Within the framework of the "first quantized theory", a single particle state for a particle with spin s is then given by

$$|\Phi(t)\rangle = \sum_{\sigma=-s}^{s}\int d^3\boldsymbol{x}\,|\boldsymbol{x},\sigma\rangle\langle\boldsymbol{x},\sigma|\Phi(t)\rangle, \tag{17.57}$$

such that $|\langle\boldsymbol{x},\sigma|\Phi(t)\rangle|^2 = |\Phi_\sigma(\boldsymbol{x},t)|^2$ is the probability density to find the particle with spin projection $\hbar\sigma$ in the location $\boldsymbol{x}$ at time t, and the normalization condition is

$$\sum_{\sigma=-s}^{s}\int d^3\boldsymbol{x}\,|\langle\boldsymbol{x},\sigma|\Phi(t)\rangle|^2 = 1.$$

The Fock space creation and annihilation operators for particles in first quantized particle states $|\Phi(t)\rangle$ are then in direct generalization of (17.43)

$$\Phi^+(t) = \sum_\sigma\int d^3\boldsymbol{x}\,\psi^+_\sigma(\boldsymbol{x})\langle\boldsymbol{x},\sigma|\Phi(t)\rangle = \sum_\sigma\int d^3\boldsymbol{k}\,a^+_\sigma(\boldsymbol{k})\langle\boldsymbol{k},\sigma|\Phi(t)\rangle \tag{17.58}$$

and

$$\Phi(t) = \sum_\sigma\int d^3\boldsymbol{x}\,\psi_\sigma(\boldsymbol{x})\langle\Phi(t)|\boldsymbol{x},\sigma\rangle = \sum_\sigma\int d^3\boldsymbol{k}\,a_\sigma(\boldsymbol{k})\langle\Phi(t)|\boldsymbol{k},\sigma\rangle. \tag{17.59}$$

A single particle wave function with a set n of orbital quantum numbers and definite spin projection σ is e.g. $\langle\boldsymbol{x},\sigma'|\Phi_{n,\sigma}(t)\rangle = \langle\boldsymbol{x}|\Phi_n(t)\rangle\delta_{\sigma\sigma'}$, and the corresponding single particle state in the quantized field theory is

$$|\Phi_{n,\sigma}(t)\rangle = \Phi^+_{n,\sigma}(t)|0\rangle = \int d^3\boldsymbol{x}\,\psi^+_\sigma(\boldsymbol{x})|0\rangle\langle\boldsymbol{x}|\Phi_n(t)\rangle. \tag{17.60}$$

It is appropriate to add a remark on notation for the general single particle field operators. Following the conventions of equation (17.37), we like to write the creation operator for a particle with a set n of orbital quantum numbers and spin projection $\hbar\sigma$ as $\psi^+_{n,\sigma}(t)$ or $a^+_{n,\sigma}(t)$, but if no quantum numbers are specified and we just talk about an abstract single particle state $|\Phi(t)\rangle$, the notation $\Phi^+(t)$ from equation (17.58) is more suitable. The three notations for given sets of orbital quantum numbers and spin projection are therefore

$$\Phi^+_{n,\sigma}(t) = \psi^+_{n,\sigma}(t) = a^+_{n,\sigma}(t) = \int d^3\boldsymbol{x}\,\psi^+_\sigma(\boldsymbol{x})\langle\boldsymbol{x}|\Phi_n(t)\rangle$$
$$= \int d^3\boldsymbol{k}\,a^+_\sigma(\boldsymbol{k})\langle\boldsymbol{k}|\Phi_n(t)\rangle.$$

Note that these time-dependent operators are operators in the *Schrödinger picture* of the theory. Their time-dependence arises only because they are time-dependent superpositions of the Schrödinger picture operators $\psi^+_\sigma(\boldsymbol{x})$ or $a^+_\sigma(\boldsymbol{k})$. We can easily verify that the state (17.60) is a Schrödinger picture state by using the correspondence of single particle matrix elements between first and second quantized theory,

$$\begin{aligned}\langle\boldsymbol{x}|\Phi_n(t)\rangle\delta_{\sigma\sigma'} &= \langle\boldsymbol{x},\sigma'|\Phi_{n,\sigma}(t)\rangle = \langle\boldsymbol{x},\sigma'|\exp(-\mathrm{i}H^{(1)}t/\hbar)|\Phi_{n,\sigma}(0)\rangle\\ &= \langle\boldsymbol{x},\sigma'|\exp(-\mathrm{i}Ht/\hbar)|\Phi_{n,\sigma}(0)\rangle,\end{aligned} \tag{17.61}$$

$$\begin{aligned}|\Phi_{n,\sigma}(t)\rangle &= \sum_{\sigma'}\int d^3\boldsymbol{x}\,|\boldsymbol{x},\sigma'\rangle\langle\boldsymbol{x},\sigma'|\exp(-\mathrm{i}Ht/\hbar)\,|\Phi_{n,\sigma}(0)\rangle\\ &= \exp(-\mathrm{i}Ht/\hbar)\,|\Phi_{n,\sigma}(0)\rangle.\end{aligned} \tag{17.62}$$

The operators $\Phi_{n,\sigma}(t)$ and $\Phi^+_{n,\sigma}(t)$ satisfy canonical commutation or anti-commutation relations as a consequence of the corresponding relations for $\psi_\sigma(\boldsymbol{x})$ and $\psi^+_\sigma(\boldsymbol{x})$, see Problem 17.4.

The Heisenberg picture operators for a state with quantum numbers n and σ are

$$\Phi^+_{H,n,\sigma}(t) = \exp\!\left(\frac{\mathrm{i}}{\hbar}Ht\right)\Phi^+_{n,\sigma}(t)\exp\!\left(-\frac{\mathrm{i}}{\hbar}Ht\right) = \int d^3\boldsymbol{x}\,\Psi^+_\sigma(\boldsymbol{x},t)\langle\boldsymbol{x}|\Phi_n(t)\rangle.$$

The Schrödinger picture state (17.60) yields the correct expectation value for the kinetic energy of the particle if evaluated with the time-independent kinetic Hamiltonian H_0 in the Schrödinger picture, but the Heisenberg picture state

$$\begin{aligned}|\Phi_{n,\sigma}\rangle &= \exp\!\left(\frac{\mathrm{i}}{\hbar}Ht\right)|\Phi_{n,\sigma}(t)\rangle = \Phi^+_{H,n,\sigma}(t)|0\rangle\\ &= \int d^3\boldsymbol{x}\,\Psi^+_\sigma(\boldsymbol{x},t)|0\rangle\langle\boldsymbol{x}|\Phi_n(t)\rangle\\ &= |\Phi_{n,\sigma}(0)\rangle = \int d^3\boldsymbol{x}\,\psi^+_\sigma(\boldsymbol{x})|0\rangle\langle\boldsymbol{x}|\Phi_n(0)\rangle.\end{aligned} \tag{17.63}$$

has to be evaluated with the generically time-dependent kinetic Hamiltonian $H_0(t)$ in the Heisenberg picture to yield the kinetic energy of the particle.

The actual time-independence of $\int d^3x\,\Psi_\sigma^+(x,t)|0\rangle\langle x|\Phi_n(t)\rangle$ follows already from the first line of equation (17.63), but we can also verify it again from the correspondence (17.61) of single particle matrix elements,

$$\int d^3x\,\Psi_\sigma^+(x,t)|0\rangle\langle x|\Phi_n(t)\rangle = \exp\!\left(\frac{\mathrm{i}}{\hbar}Ht\right)\sum_{\sigma'}\int d^3x\,|x,\sigma'\rangle\langle x,\sigma'|$$

$$\times\exp\!\left(-\frac{\mathrm{i}}{\hbar}Ht\right)|\Phi_{n,\sigma}(0)\rangle = |\Phi_{n,\sigma}(0)\rangle = \int d^3x\,\psi_\sigma^+(x)|0\rangle\langle x|\Phi_n(0)\rangle.$$

We will mostly use Schrödinger picture states and operators in the remainder of this chapter.

A general two-particle state with particle species a and a' (e.g. an electron and a proton or two electrons) will have the form

$$|\Phi_{a,a'}(t)\rangle = \frac{1}{\sqrt{1+\delta_{a,a'}}}\sum_{\sigma,\sigma'}\int d^3x\int d^3x'\,\psi_{a,\sigma}^+(x)\psi_{a',\sigma'}^+(x')|0\rangle$$

$$\times\langle x,\sigma;x',\sigma'|\Phi_{a,a'}(t)\rangle. \tag{17.64}$$

For identical particles it makes sense to require the symmetry property

$$\langle x,\sigma;x',\sigma'|\Phi_{a,a}(t)\rangle = \mp\,\langle x',\sigma';x,\sigma|\Phi_{a,a}(t)\rangle, \tag{17.65}$$

with the upper sign applying to fermions and the lower sign for bosons.

In the ideal case of a completely normalizable system (e.g. two particles trapped in an oscillator potential or a box), the quantity $|\langle x,\sigma;x',\sigma'|\Phi_{a,a'}(t)\rangle|^2$ is a probability density for finding one particle at x with spin projection σ and the second particle at x' with spin projection σ', and we should have

$$\sum_{\sigma,\sigma'}\int d^3x\int d^3x'\,|\langle x,\sigma;x',\sigma'|\Phi_{a,a'}(t)\rangle|^2 = 1$$

if we know that there is exactly one particle of kind a and one particle of kind a' in the system. It then follows with (17.65) that the state (17.64) is also normalized,

$$\langle\Phi_{a,a'}(t)|\Phi_{a,a'}(t)\rangle = 1.$$

For an example we consider a state where a particle of type a has orbital quantum numbers n and spin projection σ, and a particle of type a' has orbital quantum numbers n' and spin projection σ'. The two-particle amplitude

$$\langle x,\rho;x',\rho'|\Phi_{a,n,\sigma;a',n',\sigma'}(t)\rangle = \langle x,\rho;x',\rho'|\Phi_{a,n,\sigma}(t),\Phi_{a',n',\sigma'}(t)\rangle$$

$$\equiv \frac{\delta_{\rho\sigma}\delta_{\rho'\sigma'}\langle \boldsymbol{x}|\Phi_{a,n}(t)\rangle\langle \boldsymbol{x}'|\Phi_{a',n'}(t)\rangle}{\sqrt{1+\delta_{aa'}}\sqrt{1+\delta_{aa'}\delta_{nn'}\delta_{\sigma\sigma'}}}$$
$$\mp\delta_{aa'}\frac{\delta_{\rho'\sigma}\delta_{\rho\sigma'}\langle \boldsymbol{x}'|\Phi_{a,n}(t)\rangle\langle \boldsymbol{x}|\Phi_{a',n'}(t)\rangle}{\sqrt{2(1+\delta_{nn'}\delta_{\sigma\sigma'})}} \tag{17.66}$$

yields the tensor product state

$$|\Phi_{a,n,\sigma}(t),\Phi_{a',n',\sigma'}(t)\rangle = \int d^3\boldsymbol{x}\int d^3\boldsymbol{x}'\,\psi^+_{a,\sigma}(\boldsymbol{x})\psi^+_{a',\sigma'}(\boldsymbol{x}')|0\rangle$$
$$\times\frac{\langle \boldsymbol{x}|\Phi_{a,n}(t)\rangle\langle \boldsymbol{x}'|\Phi_{a',n'}(t)\rangle}{\sqrt{1+\delta_{aa'}\delta_{\sigma\sigma'}\delta_{nn'}}}. \tag{17.67}$$

A two-particle state will generically *not* have the factorized form (17.67) because this form is incompatible with interactions between the two particles. We can explain this with the case of a system of two different particles that we have solved in Chapter 7. A two-particle state of a proton with quantum numbers N and an electron with quantum numbers n and definite spin projections of the two particles could be written in the form

$$|\phi_{n,\sigma}(t),\Phi_{N,\Sigma}(t)\rangle = \int d^3\boldsymbol{x}\int d^3\boldsymbol{x}'\,\psi^+_{e,\sigma}(\boldsymbol{x})\psi^+_{p,\Sigma}(\boldsymbol{x}')|0\rangle\langle \boldsymbol{x}|\phi_n(t)\rangle\langle \boldsymbol{x}'|\Phi_N(t)\rangle,$$

but we had seen in Chapter 7 that no such electron-proton state is compatible with the Coulomb interaction of the two particles. There is no solution of the Schrödinger equation for the two particles which factorizes into the product of an electron wave function with a proton wave function. We did find factorized solutions of the form

$$\langle \boldsymbol{x},\boldsymbol{x}'|\Phi_{\boldsymbol{K},n,\ell,m}(t)\rangle = \langle \boldsymbol{R}|\Phi_{\boldsymbol{K}}(t)\rangle\langle \boldsymbol{r}|\Phi_{n,\ell,m}(t)\rangle,$$

where the first factor

$$\langle \boldsymbol{R}|\Phi_{\boldsymbol{K}}(t)\rangle = \frac{1}{\sqrt{2\pi}^3}\exp\!\left(\mathrm{i}\boldsymbol{K}\cdot\boldsymbol{R}-\mathrm{i}\frac{\hbar}{2m}\boldsymbol{K}^2t\right)$$

describes center of mass motion, and the second factor

$$\langle \boldsymbol{r}|\Phi_{n,\ell,m}(t)\rangle = \langle \boldsymbol{r}|n,\ell,m\rangle\exp(-\mathrm{i}E_nt/\hbar)$$

describes relative motion. Therefore we can write down a two-particle state for the electron-proton system in the form

$$|\Phi_{\boldsymbol{K},n,\ell,m;\sigma,\Sigma}(t)\rangle = \int d^3\boldsymbol{x}\int d^3\boldsymbol{x}'\,\psi^+_{e,\sigma}(\boldsymbol{x})\psi^+_{p,\Sigma}(\boldsymbol{x}')|0\rangle\langle \boldsymbol{x}-\boldsymbol{x}'|\Phi_{n,\ell,m}(t)\rangle$$
$$\times\langle(m_e\boldsymbol{x}/M)+(m_p\boldsymbol{x}'/M)|\Phi_{\boldsymbol{K}}(t)\rangle. \tag{17.68}$$

I am emphasizing this to caution the reader. We frequently calculate scattering matrix elements and expectation values for many-particle states which are products of independent single-particle states like (17.67). However, we should keep in mind that these states may *temporarily* describe the state of a many-particle system, e.g. for $t \to \pm\infty$ or $t = 0$, but these tensor-product states never describe the full time evolution of many-particle systems with interactions. Two-particle systems with an interaction potential $V(\boldsymbol{x} - \boldsymbol{x}')$ (and without external fields affecting both particles differently, see Section 18.4 below) always allow for separation of the center of mass motion, and the states can be written in the form (17.68). However, the general form is (17.64). These remarks immediately generalize to N-particle states. Those states will be characterized by amplitudes $\langle \boldsymbol{x}_1, \sigma_1; \ldots; \boldsymbol{x}_N, \sigma_N | \Phi(t)\rangle$ with appropriate (anti-) symmetry properties for identical particles.

In spite of all those cautionary remarks about the limitations of tensor product states of the form (17.67) in the actual description of many-particle systems, we will now return to those states because they will help us to understand important aspects of expectation values in many-particle systems in Sections 17.6 and 17.7.

The state (17.67) is anti-symmetric or symmetric for fermions or bosons, respectively. In particular the state will vanish for identical fermions with identical spin and orbital quantum numbers $\sigma = \sigma'$, $n = n'$.

For the normalization of the state (17.67) we note that the following equations hold with upper signs for fermions,

$$\begin{aligned}\langle 0|\psi_{a',\sigma'}(\boldsymbol{y}')\psi_{a,\sigma}(\boldsymbol{y})\psi^+_{a,\sigma}(\boldsymbol{x})\psi^+_{a',\sigma'}(\boldsymbol{x}')|0\rangle &= \delta(\boldsymbol{y}-\boldsymbol{x})\delta(\boldsymbol{y}'-\boldsymbol{x}')\\ &\quad \mp \delta_{aa'}\delta_{\sigma\sigma'}\delta(\boldsymbol{y}-\boldsymbol{x}')\delta(\boldsymbol{y}'-\boldsymbol{x})\end{aligned}$$

and therefore

$$\begin{aligned}&\int d^3\boldsymbol{y}\int d^3\boldsymbol{y}'\int d^3\boldsymbol{x}\int d^3\boldsymbol{x}'\,\langle 0|\psi_{a',\sigma'}(\boldsymbol{y}')\psi_{a,\sigma}(\boldsymbol{y})\psi^+_{a,\sigma}(\boldsymbol{x})\psi^+_{a',\sigma'}(\boldsymbol{x}')|0\rangle\\ &\times\langle\Phi_{a',n'}(t)|\boldsymbol{y}'\rangle\langle\Phi_{a,n}(t)|\boldsymbol{y}\rangle\langle\boldsymbol{x}|\Phi_{a,n}(t)\rangle\langle\boldsymbol{x}'|\Phi_{a',n'}(t)\rangle\\ &=\int d^3\boldsymbol{x}\,|\langle\boldsymbol{x}|\Phi_{a,n}(t)\rangle|^2\int d^3\boldsymbol{x}'\,|\langle\boldsymbol{x}'|\Phi_{a',n'}(t)\rangle|^2\\ &\mp\delta_{aa'}\delta_{\sigma\sigma'}\int d^3\boldsymbol{x}\,\langle\Phi_{a',n'}(t)|\boldsymbol{x}\rangle\langle\boldsymbol{x}|\Phi_{a,n}(t)\rangle\int d^3\boldsymbol{x}'\,\langle\Phi_{a,n}(t)|\boldsymbol{x}'\rangle\langle\boldsymbol{x}'|\Phi_{a',n'}(t)\rangle\\ &=1\mp\delta_{aa'}\delta_{\sigma\sigma'}\delta_{nn'},\end{aligned}$$

i.e. the 2-particle state (17.67) is properly normalized to 1, except when it vanishes because it corresponds to two fermions with identical quantum numbers.

We can also form singlet states and triplet states for two spin 1/2 fermions with single-particle orbital quantum numbers n and n'. The triplet states are

$$|\Phi_{n,n';1,\pm 1}(t)\rangle = |\Phi_{n,\pm 1/2}(t), \Phi_{n',\pm 1/2}(t)\rangle = -|\Phi_{n',n;1,\pm 1}(t)\rangle, \qquad (17.69)$$

$$|\Phi_{n,n';1,0}(t)\rangle = \frac{|\Phi_{n,1/2}(t), \Phi_{n',-1/2}(t)\rangle + |\Phi_{n,-1/2}(t), \Phi_{n',1/2}(t)\rangle}{\sqrt{2}}$$
$$= -|\Phi_{n',n;1,0}(t)\rangle, \qquad (17.70)$$

and the singlet state is

$$|\Phi_{n,n';0,0}(t)\rangle = \frac{|\Phi_{n,1/2}(t), \Phi_{n',-1/2}(t)\rangle - |\Phi_{n,-1/2}(t), \Phi_{n',1/2}(t)\rangle}{\sqrt{2}}$$
$$= |\Phi_{n',n;0,0}(t)\rangle. \qquad (17.71)$$

17.6 Two-particle interaction potentials and equations of motion

It is now only a small step to describe particle interactions as exchange of virtual particles between particles. We will take this step in Section 19.7 for exchange of non-relativistic virtual particles, and in Chapters 22 for photon exchange between charged particles. However, a description of interactions through 2-particle interaction potentials V is often sufficient. Furthermore, interaction potentials also appear in quantum electrodynamics in the Coulomb gauge[6].

The Hamiltonian with time-independent particle-particle interaction potentials $V_{a,a'}(\boldsymbol{x})$ has the same form in the Schrödinger picture and in the Heisenberg picture,

$$H = \frac{1}{2}\int d^3\boldsymbol{x}\int d^3\boldsymbol{x}'\sum_{a,a'}\sum_{\sigma,\sigma'}\psi^+_{a,\sigma}(\boldsymbol{x})\psi^+_{a',\sigma'}(\boldsymbol{x}')V_{a,a'}(\boldsymbol{x}-\boldsymbol{x}')\psi_{a',\sigma'}(\boldsymbol{x}')\psi_{a,\sigma}(\boldsymbol{x})$$
$$+\int d^3\boldsymbol{x}\sum_{a,\sigma}\frac{\hbar^2}{2m_a}\nabla\psi^+_{a,\sigma}(\boldsymbol{x})\cdot\nabla\psi_{a,\sigma}(\boldsymbol{x}) \qquad (17.72)$$
$$= \frac{1}{2}\int d^3\boldsymbol{x}\int d^3\boldsymbol{x}'\sum_{a,a'}\sum_{\sigma,\sigma'}\Psi^+_{a,\sigma}(\boldsymbol{x},t)\Psi^+_{a',\sigma'}(\boldsymbol{x}',t)V_{a,a'}(\boldsymbol{x}-\boldsymbol{x}')$$
$$\times\Psi_{a',\sigma'}(\boldsymbol{x}',t)\Psi_{a,\sigma}(\boldsymbol{x},t)$$
$$+\int d^3\boldsymbol{x}\sum_{a,\sigma}\frac{\hbar^2}{2m_a}\nabla\Psi^+_{a,\sigma}(\boldsymbol{x},t)\cdot\nabla\Psi_{a,\sigma}(\boldsymbol{x},t). \qquad (17.73)$$

If the operators $\psi^+_{1,\sigma}(\boldsymbol{x})$ describe electrons, we would include e.g. the repulsive Coulomb potential $V_{11}(\boldsymbol{x}-\boldsymbol{x}') = e^2/(4\pi\epsilon_0|\boldsymbol{x}-\boldsymbol{x}'|)$ between pairs of electrons.

[6] We will derive this for non-relativistic charged particles in Section 18.5, see equation (18.66), and for relativistic charged particles in Section 21.4, see equation (21.94).

The ordering of annihilation and creation operators in the potential term in equation (17.72) is determined by the requirement that the expectation value of the interaction potential for the vacuum $|0\rangle$ and for single particle states $|\Phi_n(t)\rangle$ vanishes. The nested structure $\psi^+_{a,\sigma}(\boldsymbol{x})\psi^+_{a',\sigma'}(\boldsymbol{x}')\psi_{a',\sigma'}(\boldsymbol{x}')\psi_{a,\sigma}(\boldsymbol{x})$ of the operators ensures the correct sign for the interaction energy of two-fermion states.

It is also instructive to write the Hamiltonian (17.72) in wavevector space. We use the following conventions for the Fourier transformation of the potential:

$$V(\boldsymbol{x}) = \int d^3\boldsymbol{q}\, V(\boldsymbol{q}) \exp(\mathrm{i}\boldsymbol{q}\cdot\boldsymbol{x}),$$

$$V(\boldsymbol{q}) = V^+(-\boldsymbol{q}) = \frac{1}{(2\pi)^3}\int d^3\boldsymbol{x}\, V(\boldsymbol{x}) \exp(-\mathrm{i}\boldsymbol{q}\cdot\boldsymbol{x}),$$

to avoid extra factors of 2π in the particle interaction terms in wavevector space. This yields the representations

$$\begin{aligned} H &= \frac{1}{2}\int d^3\boldsymbol{k}\int d^3\boldsymbol{k}'\int d^3\boldsymbol{q}\sum_{\sigma,\sigma'} a^+_\sigma(\boldsymbol{k}+\boldsymbol{q})a^+_{\sigma'}(\boldsymbol{k}'-\boldsymbol{q})V(\boldsymbol{q})a_{\sigma'}(\boldsymbol{k}')a_\sigma(\boldsymbol{k}) \\ &\quad + \int d^3\boldsymbol{k}\sum_\sigma \frac{\hbar^2\boldsymbol{k}^2}{2m}a^+_\sigma(\boldsymbol{k})a_\sigma(\boldsymbol{k}) \\ &= \frac{1}{2}\int d^3\boldsymbol{k}\int d^3\boldsymbol{k}'\int d^3\boldsymbol{q}\sum_{\sigma,\sigma'} a^+_\sigma(\boldsymbol{k}+\boldsymbol{q},t)a^+_{\sigma'}(\boldsymbol{k}'-\boldsymbol{q},t)V(\boldsymbol{q}) \\ &\quad \times a_{\sigma'}(\boldsymbol{k}',t)a_\sigma(\boldsymbol{k},t) + \int d^3\boldsymbol{k}\sum_\sigma \frac{\hbar^2\boldsymbol{k}^2}{2m}a^+_\sigma(\boldsymbol{k},t)a_\sigma(\boldsymbol{k},t), \end{aligned}$$

where the labels a and a' for the particle species are suppressed. The representations in momentum space imply that the Fourier component $V(\boldsymbol{q})$ of the two-particle interaction potential describes exchange of momentum $\hbar\boldsymbol{q}$ between the two interacting particles. Note that symmetric interaction potentials $V(\boldsymbol{x}) = V(-\boldsymbol{x})$ are also symmetric in wavevector space $V(\boldsymbol{q}) = V(-\boldsymbol{q})$. The Coulomb potential e.g. is dominated by small momentum exchange, $V(\boldsymbol{q}) = (2\pi)^{-3}(e^2/\epsilon_0)\boldsymbol{q}^{-2}$.

The corresponding Hamiltonians in the Dirac picture are

$$\begin{aligned} H_0 &= \int d^3\boldsymbol{x}\sum_\sigma \frac{\hbar^2}{2m}\nabla\psi^+_\sigma(\boldsymbol{x})\cdot\nabla\psi_\sigma(\boldsymbol{x}) = \int d^3\boldsymbol{k}\sum_\sigma \frac{\hbar^2\boldsymbol{k}^2}{2m}a^+_\sigma(\boldsymbol{k})a_\sigma(\boldsymbol{k}) \\ &= \int d^3\boldsymbol{x}\sum_\sigma \frac{\hbar^2}{2m}\nabla\psi^+_\sigma(\boldsymbol{x},t)\cdot\nabla\psi_\sigma(\boldsymbol{x},t) \\ &= \int d^3\boldsymbol{k}\sum_\sigma \frac{\hbar^2\boldsymbol{k}^2}{2m}a^+_{D,\sigma}(\boldsymbol{k},t)a_{D,\sigma}(\boldsymbol{k},t) \end{aligned}$$

and

$$
\begin{aligned}
H_D(t) &= \exp\left(\frac{\mathrm{i}}{\hbar}H_0t\right)(H - H_0)\exp\left(-\frac{\mathrm{i}}{\hbar}H_0t\right) \\
&= \frac{1}{2}\int d^3\boldsymbol{x}\int d^3\boldsymbol{x}'\sum_{\sigma,\sigma'}\psi_\sigma^+(\boldsymbol{x},t)\psi_{\sigma'}^+(\boldsymbol{x}',t)V(\boldsymbol{x}-\boldsymbol{x}')\psi_{\sigma'}(\boldsymbol{x}',t)\psi_\sigma(\boldsymbol{x},t) \\
&= \frac{1}{2}\int d^3\boldsymbol{k}\int d^3\boldsymbol{k}'\int d^3\boldsymbol{q}\sum_{\sigma,\sigma'}a_{D,\sigma}^+(\boldsymbol{k}+\boldsymbol{q},t)a_{D,\sigma'}^+(\boldsymbol{k}'-\boldsymbol{q},t)V(\boldsymbol{q}) \\
&\quad\times a_{D,\sigma'}(\boldsymbol{k}',t)a_{D,\sigma}(\boldsymbol{k},t),
\end{aligned}
$$

with the time-dependent field operators in the Dirac picture. Recall that H_0 determines the time evolution of the operators, while $H_D(t)$ determines the time evolution of the states in the Dirac picture.

Contrary to H_0, we cannot simply replace the time-dependent field operators in the Dirac picture with the time-independent operators of the Schrödinger picture in $H_D(t)$, because $[H_0, V] \neq 0$. In applications within non-relativistic field theory one often uses the representation $H_D(t) = \exp(\mathrm{i}H_0t/\hbar)V\exp(-\mathrm{i}H_0t/\hbar)$, where the Schrödinger picture potential operator V is given in terms of the time-independent field operators.

Equation of motion

The derivation of the equation of motion for the Schrödinger picture state (17.64) with the Schrödinger picture Hamiltonian (17.72) is easily done with the relation

$$
\begin{aligned}
&\psi_{\rho'}(\boldsymbol{y}')\psi_\rho(\boldsymbol{y})\psi_\sigma^+(\boldsymbol{x})\psi_{\sigma'}^+(\boldsymbol{x}')|0\rangle = [\psi_{\rho'}(\boldsymbol{y}'),[\psi_\rho(\boldsymbol{y}),\psi_\sigma^+(\boldsymbol{x})\psi_{\sigma'}^+(\boldsymbol{x}')]_-]_\pm|0\rangle \\
&= \delta_{\rho\sigma}\delta(\boldsymbol{x}-\boldsymbol{y})\delta_{\rho'\sigma'}\delta(\boldsymbol{x}'-\boldsymbol{y}')|0\rangle \mp \delta_{\rho\sigma'}\delta(\boldsymbol{x}'-\boldsymbol{y})\delta_{\rho'\sigma}\delta(\boldsymbol{x}-\boldsymbol{y}')|0\rangle,
\end{aligned}
$$

with the upper signs for fermions. This yields both for bosons and fermions the equation

$$
\begin{aligned}
\mathrm{i}\hbar\frac{d}{dt}|\Phi_{a,a'}(t)\rangle &= H|\Phi_{a,a'}(t)\rangle \\
&= \frac{1}{\sqrt{1+\delta_{aa'}}}\int d^3\boldsymbol{x}\int d^3\boldsymbol{x}'\sum_{\sigma,\sigma'}\psi_{a,\sigma}^+(\boldsymbol{x})\psi_{a',\sigma'}^+(\boldsymbol{x}')|0\rangle \\
&\quad\times\left(-\frac{\hbar^2}{2m_a}\Delta - \frac{\hbar^2}{2m_{a'}}\Delta' + V_{a,a'}(\boldsymbol{x}-\boldsymbol{x}')\right)\langle\boldsymbol{x},\sigma;\boldsymbol{x}',\sigma'|\Phi_{a,a'}(t)\rangle. \quad (17.74)
\end{aligned}
$$

Here we used $\Delta \equiv \partial^2/\partial\boldsymbol{x}^2$, $\Delta' \equiv \partial^2/\partial\boldsymbol{x}'^2$ and symmetry of the potential: $V_{a,a'}(\boldsymbol{x} - \boldsymbol{x}') = V_{a',a}(\boldsymbol{x}' - \boldsymbol{x})$.

Linear independence of the states $\psi_\sigma^+(\boldsymbol{x})\psi_{\sigma'}^+(\boldsymbol{x}')|0\rangle$ (or equivalently application of the projector $\langle 0|\psi_{a',\rho'}(\boldsymbol{y}')\psi_{a,\rho}(\boldsymbol{y})$ and the symmetry property (17.65)) implies that equation (17.74) is equivalent to the two-particle Schrödinger equation

$$i\hbar\frac{\partial}{\partial t}\langle\boldsymbol{x},\sigma;\boldsymbol{x}',\sigma'|\Phi_{a,a'}(t)\rangle = \left(-\frac{\hbar^2}{2m_a}\Delta - \frac{\hbar^2}{2m_{a'}}\Delta' + V_{a,a'}(\boldsymbol{x}-\boldsymbol{x}')\right) \times\langle\boldsymbol{x},\sigma;\boldsymbol{x}',\sigma'|\Phi_{a,a'}(t)\rangle. \tag{17.75}$$

Time-independence of the Hamiltonian implies that we can also write this in the time-independent form

$$E\langle\boldsymbol{x},\sigma;\boldsymbol{x}',\sigma'|\Phi_{a,a'}\rangle = \left(-\frac{\hbar^2}{2m_a}\Delta - \frac{\hbar^2}{2m_{a'}}\Delta' + V_{a,a'}(\boldsymbol{x}-\boldsymbol{x}')\right) \times\langle\boldsymbol{x},\sigma;\boldsymbol{x}',\sigma'|\Phi_{a,a'}\rangle. \tag{17.76}$$

These are exactly the two-particle Schrödinger equations that we would have expected for a wave function which describes two particles interacting with a potential V. Indeed, we have used this expectation already in Chapter 7 to formulate the equation of motion for the electron-proton system that constitutes a hydrogen atom. The not entirely trivial observation at this point is that these two-particle Schrödinger equations also hold for identical particles. The only manifestation of statistics of the particles is the symmetry property (17.65) of the two-particle wave function[7].

The amplitude for a general N-particle state with particle species $a(i)$ for the i-th particle would satisfy

$$i\hbar\frac{\partial}{\partial t}\langle\boldsymbol{x}_1,\sigma_1;\ldots;\boldsymbol{x}_N,\sigma_N|\Phi_{a_1,\ldots,a_N}(t)\rangle = \left(-\sum_{i=1}^{N}\frac{\hbar^2}{2m_{a(i)}}\Delta_i + \sum_{i=1}^{N-1}\sum_{j=i+1}^{N} V_{a(i),aj}(\boldsymbol{x}_i-\boldsymbol{x}_j)\right)\langle\boldsymbol{x}_1,\sigma_1;\ldots;\boldsymbol{x}_N,\sigma_N|\Phi_{a_1,\ldots,a_N}(t)\rangle. \tag{17.78}$$

---

[7] *Formal* substitution e.g. of two-particle tensor product states of definite spin for two identical particles,

$$\langle\boldsymbol{x},\sigma;\boldsymbol{x}',\sigma'|\Phi_{n,n'}\rangle = \frac{\langle\boldsymbol{x}|\Phi_n\rangle\langle\boldsymbol{x}'|\Phi_{n'}\rangle \mp \delta_{\sigma\sigma'}\langle\boldsymbol{x}|\Phi_{n'}\rangle\langle\boldsymbol{x}'|\Phi_n\rangle}{\sqrt{2(1+\delta_{nn'}\delta_{\sigma\sigma'})}} \tag{17.77}$$

(cf. equation (17.66) for $a = a'$, $\rho = \sigma$, $\rho' = \sigma'$) and projection onto effective single particle equations using orthonormality of single particle wave functions yields exchange terms. However, the resulting equations are not identical with the Hartree-Fock equations from Problem 17.7, because E in equation (17.76) is the total energy of the system, whereas the Lagrange multipliers ϵ_n in Hartree-Fock equations do not add up to the total energy of a many particle system, see Problem 17.7b. The formal nature of the substitution (17.77) is emphasized because we know that solutions of equation (17.76) do not factorize in single particle tensor products.

The derivation of the two-particle equation (17.75) or the general N-particle equation (17.78) crucially relies on the fact that the Hamiltonian (17.73) preserves the number of particles for each species a. This is a simple consequence of the fact that each individual term does not have an excess of annihilation or creation operators for any particular species a. Otherwise we could not use linear independence of states within a particular subsector of $(\sum_a n_a = N)$-particle states to read off the first quantized evolution equations (17.78) for the time-dependent N-particle amplitudes $\langle \boldsymbol{x}_1, \sigma_1; \boldsymbol{x}_2, \sigma_2; \ldots |\Phi_{a_1,a_2,\ldots}(t)\rangle$. If the Hamiltonian would not preserve the numbers n_a of particles for each species, we could have derived coupled systems of equations for wave functions with different particle numbers n_a and possibly also different total numbers N of particles. However, the first quantized formalism becomes unwieldy if we have to include sets of coupled many particle Schrödinger equations with different particle numbers, and one rather calculates everything in the second quantized formalism then.

Relation to other equations of motion

In N-particle mechanics, the two-particle Schrödinger equations (17.75) and (17.76) and their extensions to $N > 2$ correspond to the equation for the total energy of the particle system. However, in mechanics we are used to deal with separate equations of motion for each particle in terms of the forces acting on the particle. Sometimes the experience from mechanics leads to the suspicion that there should be a separate Schrödinger type equation for each particle. We can infer that this naive expectation is not correct from the observation that the equations for individual particles in a classical particle system are equations for forces, not energies. We can easily recover the separate N-particle equations of classical mechanics from the N-particle equation (17.78), because this equation implies the N-particle version of the Ehrenfest theorem, e.g. for two particles

$$\begin{aligned}
\frac{d}{dt}\langle \boldsymbol{p}_a\rangle(t) &= \int d^3\boldsymbol{x}\int d^3\boldsymbol{x}' \left(\frac{\partial \Phi^+_{a,\sigma;a',\sigma'}(\boldsymbol{x},\boldsymbol{x}')}{\partial t}\frac{\hbar}{\mathrm{i}}\frac{\partial}{\partial \boldsymbol{x}}\Phi_{a,\sigma;a',\sigma'}(\boldsymbol{x},\boldsymbol{x}')\right. \\
&\quad \left. +\Phi^+_{a,\sigma;a',\sigma'}(\boldsymbol{x},\boldsymbol{x}')\cdot\frac{\hbar}{\mathrm{i}}\frac{\partial}{\partial \boldsymbol{x}}\frac{\partial \Phi_{a,\sigma;a',\sigma'}(\boldsymbol{x},\boldsymbol{x}')}{\partial t}\right) \\
&= -\int d^3\boldsymbol{x}\int d^3\boldsymbol{x}'\,\Phi^+_{a,\sigma;a',\sigma'}(\boldsymbol{x},\boldsymbol{x}')\frac{\partial V_{a,a'}(\boldsymbol{x}-\boldsymbol{x}')}{\partial \boldsymbol{x}}\Phi_{a,\sigma;a',\sigma'}(\boldsymbol{x},\boldsymbol{x}') \\
&= -\langle \nabla V_{a,a'}(\boldsymbol{x}-\boldsymbol{x}')\rangle(t), \qquad (17.79) \\
\frac{d}{dt}\langle \boldsymbol{p}_{a'}\rangle(t) &= -\langle \nabla' V_{a,a'}(\boldsymbol{x}-\boldsymbol{x}')\rangle(t).
\end{aligned}$$

Another reason why one might incorrectly suspect that there should be N separate Schrödinger type equations for an N-particle system is that each single particle field operator $\Psi_{a,\sigma}(\boldsymbol{x},t)$ in the Heisenberg picture still satisfies its own evolution equation,

$$i\hbar\frac{\partial}{\partial t}\Psi_{a,\sigma}(\boldsymbol{x},t) = -[H,\Psi_{a,\sigma}(\boldsymbol{x},t)]$$
$$= -\frac{\hbar^2}{2m_a}\Delta\Psi_{a,\sigma}(\boldsymbol{x},t) + V_a(\boldsymbol{x},t)\Psi_{a,\sigma}(\boldsymbol{x},t), \qquad (17.80)$$

with

$$V_a(\boldsymbol{x},t) = \sum_{a'}\sum_{\sigma'}\int d^3x'\,\Psi^+_{a',\sigma'}(\boldsymbol{x}',t)V_{a,a'}(\boldsymbol{x}-\boldsymbol{x}')\Psi_{a',\sigma'}(\boldsymbol{x}',t).$$

However, note that this *non-linear Schrödinger equation* is an *operator equation* which holds for the *operators* annihilating and creating the *particle species* a, but *not* separately for a wave function for each single particle of type a. Indeed, the second quantized many particle Schrödinger equation

$$i\hbar\frac{d}{dt}|\Phi_{\nu_1,\nu_2,\ldots}(t)\rangle = H|\Phi_{\nu_1,\nu_2,\ldots}(t)\rangle$$

and the corresponding first quantized many particle Schrödinger equation (17.78) can be derived from the single particle operator equations (17.80), see Problem 17.11.

17.7 Expectation values and exchange terms

It is not difficult to discuss expectation values for the general two-particle state (17.64). However, it is more instructive to do this for the tensor product of single-particle states (17.67) with $a = a'$.

The result for the kinetic Hamiltonian in the Schrödinger picture

$$H_0|\Phi_{n,\sigma}(t),\Phi_{n',\sigma'}(t)\rangle = \frac{1}{\sqrt{1+\delta_{\sigma\sigma'}\delta_{nn'}}}\int d^3x\int d^3x'\,\psi^+_\sigma(\boldsymbol{x})\psi^+_{\sigma'}(\boldsymbol{x}')|0\rangle$$
$$\times\left(-\frac{\hbar^2}{2m}\Delta\langle\boldsymbol{x}|\Phi_n(t)\rangle\langle\boldsymbol{x}'|\Phi_{n'}(t)\rangle - \frac{\hbar^2}{2m}\langle\boldsymbol{x}|\Phi_n(t)\rangle\Delta'\langle\boldsymbol{x}'|\Phi_{n'}(t)\rangle\right)$$

also immediately yields the expectation value of the kinetic energy operator in the two-particle state (17.67),

$$\langle H_0\rangle = \langle\Phi_{n,\sigma}(t),\Phi_{n',\sigma'}(t)|H_0|\Phi_{n,\sigma}(t),\Phi_{n',\sigma'}(t)\rangle = -\frac{\hbar^2}{2m(1+\delta_{\sigma\sigma'}\delta_{nn'})}$$
$$\times\int d^3x\int d^3x'\,\Big(\langle\Phi_{n'}(t)|\boldsymbol{x}'\rangle\langle\Phi_n(t)|\boldsymbol{x}\rangle\Delta\langle\boldsymbol{x}|\Phi_n(t)\rangle\langle\boldsymbol{x}'|\Phi_{n'}(t)\rangle$$
$$+\langle\Phi_{n'}(t)|\boldsymbol{x}'\rangle\langle\Phi_n(t)|\boldsymbol{x}\rangle\langle\boldsymbol{x}|\Phi_n(t)\rangle\Delta'\langle\boldsymbol{x}'|\Phi_{n'}(t)\rangle$$

$$\mp\delta_{\sigma\sigma'}\langle\Phi_{n'}(t)|\boldsymbol{x}\rangle\langle\Phi_n(t)|\boldsymbol{x}'\rangle\Delta\langle\boldsymbol{x}|\Phi_n(t)\rangle\langle\boldsymbol{x}'|\Phi_{n'}(t)\rangle$$
$$\mp\delta_{\sigma\sigma'}\langle\Phi_{n'}(t)|\boldsymbol{x}\rangle\langle\Phi_n(t)|\boldsymbol{x}'\rangle\langle\boldsymbol{x}|\Phi_n(t)\rangle\Delta'\langle\boldsymbol{x}'|\Phi_{n'}(t)\rangle\Big)$$
$$= \frac{1\mp\delta_{\sigma\sigma'}\delta_{nn'}}{1+\delta_{\sigma\sigma'}\delta_{nn'}}K_{nn'},$$

i.e. unless the two-particle state vanishes because it describes two fermions with identical quantum numbers, the kinetic energy is the sum of the kinetic energies in the orbital motion of the two particles,

$$K_{nn'} = \langle\Phi_{n,\sigma}(t),\Phi_{n',\sigma'}(t)|H_0|\Phi_{n,\sigma}(t),\Phi_{n',\sigma'}(t)\rangle\Big|_{(n,\sigma)\neq(n',\sigma')\text{ for fermions}}$$
$$= -\int d^3x\,\frac{\hbar^2}{2m}\Big(\langle\Phi_n(t)|\boldsymbol{x}\rangle\Delta\langle\boldsymbol{x}|\Phi_n(t)\rangle + \langle\Phi_{n'}(t)|\boldsymbol{x}\rangle\Delta\langle\boldsymbol{x}|\Phi_{n'}(t)\rangle\Big).$$

The potential operator in the Schrödinger picture

$$V = \frac{1}{2}\int d^3x\int d^3x'\sum_{\sigma,\sigma'}\psi_\sigma^+(\boldsymbol{x})\psi_{\sigma'}^+(\boldsymbol{x}')V(\boldsymbol{x}-\boldsymbol{x}')\psi_{\sigma'}(\boldsymbol{x}')\psi_\sigma(\boldsymbol{x})$$

satisfies with $V(\boldsymbol{x}-\boldsymbol{x}') = V(\boldsymbol{x}'-\boldsymbol{x})$ both for fermions and for bosons the equation

$$V|\Phi_{n,\sigma}(t),\Phi_{n',\sigma'}(t)\rangle = \frac{1}{\sqrt{1+\delta_{\sigma\sigma'}\delta_{nn'}}}\int d^3x\int d^3x'\,\psi_\sigma^+(\boldsymbol{x})\psi_{\sigma'}^+(\boldsymbol{x}')|0\rangle$$
$$\times V(\boldsymbol{x}-\boldsymbol{x}')\langle\boldsymbol{x}|\Phi_n(t)\rangle\langle\boldsymbol{x}'|\Phi_{n'}(t)\rangle\Big).$$

This yields again with upper signs for fermions the result

$$\langle V\rangle = \langle\Phi_{n,\sigma}(t),\Phi_{n',\sigma'}(t)|V|\Phi_{n,\sigma}(t),\Phi_{n',\sigma'}(t)\rangle$$
$$= \int d^3x\int d^3x'\,\frac{V(\boldsymbol{x}-\boldsymbol{x}')}{1+\delta_{nn'}\delta_{\sigma\sigma'}}\Big(\langle\Phi_n(t)|\boldsymbol{x}\rangle\langle\Phi_{n'}(t)|\boldsymbol{x}'\rangle\langle\boldsymbol{x}|\Phi_n(t)\rangle\langle\boldsymbol{x}'|\Phi_{n'}(t)\rangle$$
$$\mp\delta_{\sigma\sigma'}\langle\Phi_n(t)|\boldsymbol{x}\rangle\langle\Phi_{n'}(t)|\boldsymbol{x}'\rangle\langle\boldsymbol{x}'|\Phi_n(t)\rangle\langle\boldsymbol{x}|\Phi_{n'}(t)\rangle\Big),$$

i.e. the expectation value for the potential energy becomes

$$\langle V\rangle = \int d^3x\int d^3x'\,\Phi_n^+(\boldsymbol{x},t)\Phi_n(\boldsymbol{x},t)\frac{V(\boldsymbol{x}-\boldsymbol{x}')}{1+\delta_{nn'}\delta_{\sigma\sigma'}}\Phi_{n'}^+(\boldsymbol{x}',t)\Phi_{n'}(\boldsymbol{x}',t)$$
$$\mp\delta_{\sigma\sigma'}\int d^3x\int d^3x'\,\Phi_n^+(\boldsymbol{x},t)\Phi_{n'}(\boldsymbol{x},t)\frac{V(\boldsymbol{x}-\boldsymbol{x}')}{1+\delta_{nn'}}\Phi_{n'}^+(\boldsymbol{x}',t)\Phi_n(\boldsymbol{x}',t).$$

We can collect the results for the expectation values of kinetic and potential energy of the two-particle state (17.67) (with upper signs for fermions)

$$\langle H\rangle = \langle\Phi_{n,\sigma}(t),\Phi_{n',\sigma'}(t)|H|\Phi_{n,\sigma}(t),\Phi_{n',\sigma'}(t)\rangle$$
$$= \frac{1\mp\delta_{\sigma\sigma'}\delta_{nn'}}{1+\delta_{\sigma\sigma'}\delta_{nn'}}K_{nn'} + \frac{C_{nn'}\mp J_{nn'}\delta_{\sigma\sigma'}}{1+\delta_{\sigma\sigma'}\delta_{nn'}}, \tag{17.81}$$

with the Coulomb term[8]

$$C_{nn'} = \int d^3x \int d^3x'\, \Phi_n^+(x,t)\Phi_{n'}^+(x',t)V(x-x')\Phi_{n'}(x',t)\Phi_n(x,t) \tag{17.82}$$

and the *exchange integral*[9]

$$J_{nn'} = \int d^3x \int d^3x'\, \Phi_{n'}^+(x,t)\Phi_n^+(x',t)V(x-x')\Phi_{n'}(x',t)\Phi_n(x,t). \tag{17.83}$$

The Coulomb term is what we would have expected for the energy of the interaction of two particles with quantum numbers n and n'. The exchange interaction, on the other hand, is a pure quantum effect which only exists as a consequence of the canonical (anti-)commutation relations for bosonic or fermionic operators. In the first quantized theory it appears as a consequence of symmetrized boson wave functions and anti-symmetrized fermion wave functions.

For electrons with aligned spins (17.69) and also for the $m = 0$ triplet state (17.70) we must have $n \neq n'$, and the result (17.81) implies a shift of the ordinary Coulomb term $C_{nn'}$ by the exchange term $J_{nn'}$, $C_{nn'} \to C_{nn'} - J_{nn'}$. For the $m = 0$ triplet state the exchange integral arises from the cross multiplication terms in the evaluation of the expectation value. By the same token, the Coulomb term for the singlet state (17.71) gets shifted to $(C_{nn'} + J_{nn'})/(1+\delta_{nn'})$ due to the cross multiplication terms. The potential energy part of the Hamiltonian (17.72) can therefore be replaced by an effective spin interaction Hamiltonian[10] (with dimensionless spins: $S/\hbar \to S$)

$$H_{nn'} = \frac{1}{1+\delta_{nn'}}\left(C_{nn'} + J_{nn'} - J_{nn'}(S+S')^2\right)$$
$$= \frac{1}{1+\delta_{nn'}}\left(C_{nn'} - \frac{1}{2}J_{nn'}(1+4S\cdot S')\right), \tag{17.84}$$

where S and S' are the dimensionless spin operators for two electrons. Equation (17.84) gives the correct shifts by $\mp J_{nn'}$ because $(S+S')^2 = 2$, $S\cdot S' = 1/4$,

[8]As derived, this result applies to every 2-particle interaction potential. The most often studied case in atomic, molecular and condensed matter physics is the Coulomb interaction between electrons, and therefore the standard (non-exchange) interaction term is simply denoted as the Coulomb term.

[9]W. Heisenberg, Z. Phys. 38, 411 (1926); Z. Phys. 39, 499 (1926).

[10]P.A.M. Dirac, Proc. Roy. Soc. London A 123, 714 (1929).

in the triplet state and $(\boldsymbol{S}+\boldsymbol{S}')^2 = 0$, $\boldsymbol{S}\cdot\boldsymbol{S}' = -3/4$, in the singlet state. Note that $C_{nn} = J_{nn}$, and therefore

$$\frac{C_{nn'} - J_{nn'}}{1+\delta_{nn'}} = C_{nn'} - J_{nn'}.$$

The Hamiltonian (17.84) without the constant terms is the spin-spin coupling Hamiltonian which is also known as the Heisenberg Hamiltonian[11].

Equations (17.81) and (17.84) show that the Coulomb interaction between electrons, through the exchange integral, effectively generates an interaction of the same form as the magnetic spin-spin interaction. The exchange interaction usually dominates over the magnetic spin-spin interaction in materials. For example in atoms or molecules $J_{nn'}$ will be of order of a few eV, whereas the energy of the genuine magnetic dipole-dipole interaction will only be of order meV or smaller. Exchange interaction with $J_{nn'} > 0$ therefore can align electron spins to generate ferromagnetism[12], but the magnetic dipole interaction will certainly not accomplish this at room temperature.

17.8 From many particle theory to second quantization

Second quantization (or field quantization) of the Schrödinger field is relevant for condensed matter physics and statistical physics, but it is usually not introduced through quantization of the corresponding Lagrangian field theory. An alternative approach proceeds through the observation that field quantization yields the same matrix elements as symmetrized wave functions (for bosons) or anti-symmetrized wave functions (for fermions) in first quantized theory with a fixed number N of particles.

In short this reasoning goes as follows. We assume a finite volume $V = L^3$ of our system. Then we can restrict attention to discrete momenta

$$\boldsymbol{k} = \frac{2\pi}{L}\boldsymbol{n}, \quad \boldsymbol{n} \in \mathbb{N}^3,$$

and the N-particle momentum eigenstates are generated by states of the form

$$|\boldsymbol{k}_1, \ldots \boldsymbol{k}_N\rangle = |\boldsymbol{k}_1\rangle \ldots |\boldsymbol{k}_N\rangle.$$

This state needs to be symmetrized for indistinguishable bosons by summing over all $N!$ permutations P of the N momenta,

$$\sum_{P\in S_N} P|\boldsymbol{k}_1, \ldots \boldsymbol{k}_N\rangle$$

[11]Heisenberg had introduced exchange integrals in 1926, and he published an investigation of ferromagnetism based on the exchange interaction (17.81) in 1928 (Z. Phys. 49, 619 (1928)). However, the effective Hamiltonian (17.84) was introduced by Dirac in the previously mentioned reference in 1929. Therefore a better name for (17.84) would be Dirac-Heisenberg Hamiltonian.

[12]Ferromagnetism or anti-ferromagnetism in magnetic materials usually requires indirect exchange interactions, see e.g. [5, 12, 22, 40].

However, this state is not generically normalized. If the momentum $\boldsymbol{k}$ is realized $n_{\boldsymbol{k}}$ times in the state, then

$$\left|\sum_{P\in S_N} P|\boldsymbol{k}_1,\ldots \boldsymbol{k}_N\rangle\right|^2 = \frac{N!}{\prod_{\boldsymbol{k}} n_{\boldsymbol{k}}!}\left(\prod_{\boldsymbol{k}} n_{\boldsymbol{k}}!\right)^2 = N!\prod_{\boldsymbol{k}} n_{\boldsymbol{k}}!,$$

since there are $N!/\prod_{\boldsymbol{k}} n_{\boldsymbol{k}}!$ different distinguishable states in the symmetrized state, and each of these states occurs $\prod_{\boldsymbol{k}} n_{\boldsymbol{k}}!$ times. Therefore the correctly normalized Bose states are

$$|\{n_{\boldsymbol{k}}\}\rangle = \frac{1}{\sqrt{N!\prod_{\boldsymbol{k}} n_{\boldsymbol{k}}!}}\sum_{P\in S_N} P|\boldsymbol{k}_1,\ldots \boldsymbol{k}_N\rangle.$$

The action of the operator $|\boldsymbol{k}'\rangle\langle\boldsymbol{k}|$ on this state is for $\boldsymbol{k}\neq\boldsymbol{k}'$

$$\begin{aligned}
|\boldsymbol{k}'\rangle\langle\boldsymbol{k}|\ldots,n_{\boldsymbol{k}},\ldots,n_{\boldsymbol{k}'},\ldots\rangle &= n_{\boldsymbol{k}}\sqrt{\frac{n_{\boldsymbol{k}'}+1}{n_{\boldsymbol{k}}}}|\ldots,n_{\boldsymbol{k}}-1,\ldots,n_{\boldsymbol{k}'}+1,\ldots\rangle \\
&= \sqrt{n_{\boldsymbol{k}}(n_{\boldsymbol{k}'}+1)}|\ldots,n_{\boldsymbol{k}}-1,\ldots,n_{\boldsymbol{k}'}+1,\ldots\rangle \\
&= a^+(\boldsymbol{k}')a(\boldsymbol{k})|\ldots,n_{\boldsymbol{k}},\ldots,n_{\boldsymbol{k}'},\ldots\rangle,
\end{aligned}$$

and for $\boldsymbol{k}=\boldsymbol{k}'$,

$$|\boldsymbol{k}\rangle\langle\boldsymbol{k}|\ldots,n_{\boldsymbol{k}},\ldots\rangle = n_{\boldsymbol{k}}|\ldots,n_{\boldsymbol{k}},\ldots\rangle = a^+(\boldsymbol{k})a(\boldsymbol{k})|\ldots,n_{\boldsymbol{k}},\ldots\rangle.$$

i.e. we find that for 1-particle operators, the operator

$$K^{(1)} = \int d^3\boldsymbol{k}'\int d^3\boldsymbol{k}\,|\boldsymbol{k}'\rangle\langle\boldsymbol{k}'|K^{(1)}|\boldsymbol{k}\rangle\langle\boldsymbol{k}|$$

has the same effect in the first quantized theory as

$$K^{(2)} = \int d^3\boldsymbol{k}'\int d^3\boldsymbol{k}\,\langle\boldsymbol{k}'|K^{(1)}|\boldsymbol{k}\rangle a^+(\boldsymbol{k}')a(\boldsymbol{k})$$

has in the second quantized theory. E.g. the first quantized 1-particle Hamiltonian

$$H^{(1)} = \frac{\mathbf{p}^2}{2m} = \int d^3\boldsymbol{k}\,|\boldsymbol{k}\rangle\frac{\hbar^2\boldsymbol{k}^2}{2m}\langle\boldsymbol{k}|$$

becomes

$$H^{(2)} = \int d^3\boldsymbol{k}\,\frac{\hbar^2\boldsymbol{k}^2}{2m}a^+(\boldsymbol{k})a(\boldsymbol{k}).$$

Once the beasts $a^+(\boldsymbol{k})$ and $a(\boldsymbol{k})$ are let loose, it is easy to recognize from their commutation or anti-commutation relations that they create and annihilate particles, and the whole theory can be developed from there. The approach through quantization of Lagrangian field theories is preferred in this book because it also yields an elegant formalism for the identification of conservation laws and generalizes more naturally to the relativistic case.

17.9 Problems

17.1. Calculate the evolution equations $dN(t)/dt$ and $d\boldsymbol{P}(t)/dt$ for the number and momentum operators in the Heisenberg picture if the Hamiltonian is given by equation (17.2).

17.2. The relation $p_0 = -E/c$ in relativity motivates the identification of the Hamiltonian H with a timelike momentum operator $P_0 = -H/c$.

Show that the field operators in the Heisenberg picture satisfy the commutation relations

$$[P_\mu, \Psi(\boldsymbol{x}, t)] = \mathrm{i}\hbar\partial_\mu\Psi(\boldsymbol{x}, t).$$

17.3a. Calculate the expectation value of the operator

$$\mathbf{x} = \int d^3\boldsymbol{x} \sum_{a,\sigma} \psi^+_{a,\sigma}(\boldsymbol{x})\boldsymbol{x}\psi_{a,\sigma}(\boldsymbol{x})$$

for the two-particle state (17.67).

17.3b. Express the operator $\mathbf{x}$ in terms of $\boldsymbol{k}$ space operators $a_{a,\sigma}(\boldsymbol{k})$ and $a^+_{a,\sigma}(\boldsymbol{k})$.

17.4a. Suppose n is a set of orbital quantum numbers and σ is a spin quantum number such that $H^{(1)}|n, \sigma\rangle = E_n|n, \sigma\rangle$ and the completeness relations

$$\sum_{n,\sigma} |n, \sigma\rangle\langle n, \sigma| = 1, \quad \langle n, \sigma|n', \sigma'\rangle = \delta_{nn'}\delta_{\sigma\sigma'}$$

hold. The range of n may contain both discrete and continuous components such that the sum over n may also contain integrations over the continuous components and $\delta_{nn'} = \delta(n - n')$ for n and n' in the continuous components. Show that the creation and annihilation operators

$$\begin{aligned}\psi^+_{n,\sigma}(t) &= \sum_{\sigma'} \int d^3\boldsymbol{x}\, \psi^+_{\sigma'}(\boldsymbol{x})\langle \boldsymbol{x}, \sigma'|n, \sigma\rangle \exp(-\mathrm{i}E_n t/\hbar) \\ &= \int d^3\boldsymbol{x}\, \psi^+_\sigma(\boldsymbol{x})\langle \boldsymbol{x}|n\rangle \exp(-\mathrm{i}E_n t/\hbar)\end{aligned} \qquad (17.85)$$

and

$$\psi_{n,\sigma}(t) = \int d^3x\, \psi_\sigma(\boldsymbol{x})\langle n|\boldsymbol{x}\rangle \exp(\mathrm{i}E_n t/\hbar) \tag{17.86}$$

satisfy canonical (anti-)commutation relations

$$[\psi_{n,\sigma}(t), \psi_{n',\sigma'}(t)]_\pm = 0, \quad [\psi^+_{n,\sigma}(t), \psi^+_{n',\sigma'}(t)]_\pm = 0,$$
$$[\psi_{n,\sigma}(t), \psi^+_{n',\sigma'}(t)]_\pm = \delta_{nn'}\delta_{\sigma\sigma'}. \tag{17.87}$$

17.4b. Suppose the second quantized Hamiltonian is

$$H = \sum_\sigma \int d^3x \left(\frac{\hbar^2}{2m}\nabla\psi^+_\sigma(\boldsymbol{x})\cdot\nabla\psi_\sigma(\boldsymbol{x}) + \psi^+_\sigma(\boldsymbol{x})V(\boldsymbol{x})\psi_\sigma(\boldsymbol{x})\right).$$

Express H in terms of the operators $\psi^+_{n,\sigma}(t)$ and $\psi_{n,\sigma}(t)$. Show that

$$|\psi_{n,\sigma}(t)\rangle = \psi^+_{n,\sigma}(t)|0\rangle$$

is an eigenstate of H, and that

$$\psi_{n',\sigma'}(t)|\psi_{n,\sigma}(t)\rangle = \delta_{nn'}\delta_{\sigma\sigma'}|0\rangle.$$

17.5. Calculate the expectation values $\langle H_0\rangle$ and $\langle V\rangle$ for kinetic and potential energy in the two-particle state (17.64) with identical particles.

17.6. Show that for pairs of spin-1 bosons, the interaction energy for states (17.67)

$$\langle V\rangle = \langle\Psi_{n,\sigma}(t), \Psi_{n',\sigma'}(t)|V|\Psi_{n,\sigma}(t), \Psi_{n',\sigma'}(t)\rangle = \frac{C_{nn'} + J_{nn'}\delta_{\sigma\sigma'}}{1 + \delta_{\sigma\sigma'}\delta_{nn'}}$$

corresponds to the following values for the interaction energy in the singlet, triplet, and quintuplet states,

$$E^{(0)}_{n,n'} = E^{(2)}_{n,n'} = \frac{C_{nn'} + J_{nn'}}{1 + \delta_{nn'}}, \quad E^{(1)}_{nn'} = \frac{C_{nn'} - J_{nn'}}{1 + \delta_{nn'}} = C_{nn'} - J_{nn'}.$$

Show also that these energies can be reproduced with an effective spin-spin interaction Hamiltonian

$$H_{n,n'} = \frac{1}{1 + \delta_{nn'}}\left(C_{nn'} + \frac{1}{4}J_{nn'}\left(4 - 6(\boldsymbol{S} + \boldsymbol{S}')^2 + (\boldsymbol{S} + \boldsymbol{S}')^4\right)\right)$$
$$= \frac{1}{1 + \delta_{nn'}}\left(C_{nn'} - J_{nn'} + J_{nn'}\boldsymbol{S}\cdot\boldsymbol{S}' + J_{nn'}(\boldsymbol{S}\cdot\boldsymbol{S}')^2\right). \tag{17.88}$$

17.7. Hartree-Fock equations

17.7a. Calculate the expectation value of the Hamiltonian (17.72) for a three-particle state which is a tensor product of single particle factors for a helium nucleus and two electrons,

$$|\Phi\rangle = \int d^3x \int d^3x' \int d^3y\, \psi^+_{e,\sigma}(x)\psi^+_{e,\sigma'}(x')\psi^+_{\alpha,\nu}|0\rangle$$
$$\times \langle x|\phi_n(t)\rangle\langle x'|\phi_{n'}(t)\rangle\langle y|\phi_N(t)\rangle.$$

Show that the requirement of minimal expectation value $\langle\Phi|H|\Phi\rangle$ under the constraints of normalized single particle wave functions yields a set of three non-linear coupled equations for the single particle wave functions. You have to use Lagrange multipliers to include the normalization constraints, i.e. you have to calculate the variational derivatives of the functional

$$F[\phi_n(t), \phi_{n'}(t), \phi_N(t)] = \langle\Phi|H|\Phi\rangle - \epsilon_n\left(\int d^3x\,|\langle x|\phi_n(t)\rangle|^2 - 1\right)$$
$$- \epsilon_{n'}\left(\int d^3x\,|\langle x|\phi_{n'}(t)\rangle|^2 - 1\right) - \epsilon_N\left(\int d^3x\,|\langle x|\phi_N(t)\rangle|^2 - 1\right),$$

e.g.

$$\frac{\delta}{\delta\phi_n(x)}F[\phi_n(t), \phi_{n'}(t), \phi_N(t)] = 0.$$

This yields intuitive versions of non-linearly coupled equations which look like time-independent Schrödinger equations. These equations are examples of *Hartree-Fock* equations[13]. The equations for the electrons contain exchange terms due to the presence of identical particles, and Hartree-Fock type equations have been successfully applied to calculate electronic configurations in atoms, molecules and solids. However, the limitation of the variation of the N particle states to tensor product states is a principal limitation of the Hartree-Fock method. The lowly hydrogen atom already told us that translation invariant interaction potentials $V(x - x')$ entangle two-particle states in such a way that the energy eigenstates of the coupled system cannot be written as tensor products of single particle states.

17.7b. Show that the Lagrange multipliers ϵ_i add up to the sum of the kinetic energy plus *twice* the potential energy of the system.

17.8a. We consider field operators for spin-1/2 fermions,

$$\{\psi_\sigma(x), \psi^+_{\sigma'}(x')\} = \delta_{\sigma\sigma'}\delta(x - x'),$$
$$\{\psi_\sigma(x), \psi_{\sigma'}(x')\} = 0, \quad \{\psi^+_\sigma(x), \psi^+_{\sigma'}(x')\} = 0.$$

[13] Very good textbook discussions of Hartree-Fock equations can be found in [26, 35, 36], and a comprehensive discussion of the uses of Hartree-Fock type equations in chemistry and materials physics is contained in [11].

17.8a. The particle density operator is $n_\sigma(\boldsymbol{x}) = \psi_\sigma^+(\boldsymbol{x})\psi_\sigma(\boldsymbol{x})$. Show that the states $|\boldsymbol{x}_1,\sigma_1\rangle = \psi_{\sigma_1}^+(\boldsymbol{x}_1)|0\rangle$ and $|\boldsymbol{x}_1,\sigma_1;\boldsymbol{x}_2,\sigma_2\rangle = \psi_{\sigma_1}^+(\boldsymbol{x}_1)\psi_{\sigma_2}^+(\boldsymbol{x}_2)|0\rangle$ are eigenstates of $n_\sigma(\boldsymbol{x})$ in the sense that relations of the kind

$$n_\sigma(\boldsymbol{x})|\boldsymbol{x}_1,\sigma_1\rangle = \lambda_{\sigma,\sigma_1}(\boldsymbol{x},\boldsymbol{x}_1)|\boldsymbol{x}_1,\sigma_1\rangle,$$
$$n_\sigma(\boldsymbol{x})|\boldsymbol{x}_1,\sigma_1;\boldsymbol{x}_2,\sigma_2\rangle = \lambda_{\sigma,\sigma_1,\sigma_2}(\boldsymbol{x},\boldsymbol{x}_1,\boldsymbol{x}_2)|\boldsymbol{x}_1,\sigma_1;\boldsymbol{x}_2,\sigma_2\rangle$$

hold. Calculate the "eigenvalues" $\lambda_{\sigma,\sigma_1}(\boldsymbol{x},\boldsymbol{x}_1)$ and $\lambda_{\sigma,\sigma_1,\sigma_2}(\boldsymbol{x},\boldsymbol{x}_1,\boldsymbol{x}_2)$.

17.8b. We can define a density-density correlation operator

$$\mathcal{G}_{\sigma,\sigma'}(\boldsymbol{x},\boldsymbol{x}') = n_\sigma(\boldsymbol{x})n_{\sigma'}(\boldsymbol{x}').$$

Evaluate the matrix elements $\langle\boldsymbol{x}_1,\sigma_1|\mathcal{G}_{\sigma,\sigma'}(\boldsymbol{x},\boldsymbol{x}')|\boldsymbol{x}_2,\sigma_2\rangle$ of this operator in 1-particle states.

Solution.

17.8a. For the 1-particle state we can write

$$\begin{aligned} n_\sigma(\boldsymbol{x})|\boldsymbol{x}_1,\sigma_1\rangle &= \psi_\sigma^+(\boldsymbol{x})\{\psi_\sigma(\boldsymbol{x}),\psi_{\sigma_1}^+(\boldsymbol{x}_1)\}|0\rangle = \delta_{\sigma\sigma_1}\delta(\boldsymbol{x}-\boldsymbol{x}_1)\psi_\sigma^+(\boldsymbol{x})|0\rangle \\ &= \delta_{\sigma\sigma_1}\delta(\boldsymbol{x}-\boldsymbol{x}_1)|\boldsymbol{x}_1,\sigma_1\rangle. \end{aligned} \tag{17.89}$$

For the 2-particle state we use the relation $[A,BC] = \{A,B\}C - B\{C,A\}$ in

$$\begin{aligned} n_\sigma(\boldsymbol{x})|\boldsymbol{x}_1,\sigma_1;\boldsymbol{x}_2,\sigma_2\rangle &= \psi_\sigma^+(\boldsymbol{x})[\psi_\sigma(\boldsymbol{x}),\psi_{\sigma_1}^+(\boldsymbol{x}_1)\psi_{\sigma_2}^+(\boldsymbol{x}_2)]|0\rangle \\ &= \delta_{\sigma\sigma_1}\delta(\boldsymbol{x}-\boldsymbol{x}_1)\psi_\sigma^+(\boldsymbol{x})\psi_{\sigma_2}^+(\boldsymbol{x}_2)|0\rangle - \delta_{\sigma\sigma_2}\delta(\boldsymbol{x}-\boldsymbol{x}_2)\psi_\sigma^+(\boldsymbol{x})\psi_{\sigma_1}^+(\boldsymbol{x}_1)|0\rangle \\ &= \left(\delta_{\sigma\sigma_1}\delta(\boldsymbol{x}-\boldsymbol{x}_1) + \delta_{\sigma\sigma_2}\delta(\boldsymbol{x}-\boldsymbol{x}_2)\right)|\boldsymbol{x}_1,\sigma_1;\boldsymbol{x}_2,\sigma_2\rangle. \end{aligned} \tag{17.90}$$

17.8b. From the previous results and the orthogonality of the single-particle states (following from the anti-commutation relation between $\psi_{\sigma_1}(\boldsymbol{x}_1)$ and $\psi_{\sigma_2}^+(\boldsymbol{x}_2)$) we find

$$\begin{aligned} \langle\boldsymbol{x}_1,\sigma_1|\mathcal{G}_{\sigma,\sigma'}(\boldsymbol{x},\boldsymbol{x}')|\boldsymbol{x}_2,\sigma_2\rangle &= \langle\boldsymbol{x}_1,\sigma_1|n_\sigma(\boldsymbol{x})n_{\sigma'}(\boldsymbol{x}')|\boldsymbol{x}_2,\sigma_2\rangle \\ &= \delta_{\sigma\sigma_1}\delta(\boldsymbol{x}-\boldsymbol{x}_1)\delta_{\sigma'\sigma_2}\delta(\boldsymbol{x}'-\boldsymbol{x}_2)\delta_{\sigma_1\sigma_2}\delta(\boldsymbol{x}_1-\boldsymbol{x}_2). \end{aligned}$$

17.9a. Pair correlations in the Fermi gas

We know that the Pauli principle excludes two fermions from being in the same state, i.e. they cannot have the same quantum numbers. But what does that mean for continuous quantum numbers like the location $\boldsymbol{x}$ of a particle? What exactly does the statement mean: "Two electrons of equal spin cannot be in the same place"? How far apart do two electrons of equal spin have to be to satisfy this constraint? We will figure this out in this problem.

In the previous problem we have found that the density-density correlation operator $\mathcal{G}_{\sigma,\sigma'}(\boldsymbol{x},\boldsymbol{x}') = n_\sigma(\boldsymbol{x})n_{\sigma'}(\boldsymbol{x}')$ has non-vanishing matrix elements for 1-particle states. Therefore we define the pair correlation operator

$$\begin{aligned}\hat{g}_{\sigma,\sigma'}(\boldsymbol{x},\boldsymbol{x}') &= \mathcal{G}_{\sigma,\sigma'}(\boldsymbol{x},\boldsymbol{x}') - \delta_{\sigma\sigma'}\delta(\boldsymbol{x}-\boldsymbol{x}')n_\sigma(\boldsymbol{x}) \\ &= \psi_\sigma^+(\boldsymbol{x})\psi_{\sigma'}^+(\boldsymbol{x}')\psi_{\sigma'}(\boldsymbol{x}')\psi_\sigma(\boldsymbol{x})\end{aligned}$$

as a measure for the probability to find a fermion with spin orientation σ at the point $\boldsymbol{x}$, when we know that there is another fermion with spin orientation σ' at the point $\boldsymbol{x}'$. This ordering of operators eliminates the 1-particle matrix elements.

17.9a. Show that the corresponding combination of classical electron densities divided by 2,

$$\frac{1}{2}\tilde{g}_{\sigma,\sigma'}(\boldsymbol{x},\boldsymbol{x}') = \frac{1}{2}n_\sigma(\boldsymbol{x})n_{\sigma'}(\boldsymbol{x}') - \frac{1}{2}\delta_{\sigma\sigma'}\delta(\boldsymbol{x}-\boldsymbol{x}')n_\sigma(\boldsymbol{x}), \tag{17.91}$$

can be interpreted as a *probability density normalized to the number of fermion pairs* to find a fermion with spin projection $\hbar\sigma$ in $\boldsymbol{x}$ and a fermion with spin projection $\hbar\sigma'$ in $\boldsymbol{x}'$.

Hint: There are $N_\uparrow + N_\downarrow = N$ fermions in the volume V. What do you get from equation (17.91) by integrating over the volume?

17.9b. The ground state of a free fermion gas is

$$|\Omega\rangle = \prod_{\boldsymbol{k},|\boldsymbol{k}|\le k_F} a_{1/2}^+(\boldsymbol{k})a_{-1/2}^+(\boldsymbol{k})|0\rangle,$$

where k_F is the Fermi wave number (12.18). Calculate the pair distribution function

$$g_{\sigma,\sigma'}(\boldsymbol{x},\boldsymbol{x}') = \langle\Omega|\hat{g}_{\sigma,\sigma'}(\boldsymbol{x},\boldsymbol{x}')|\Omega\rangle$$

of the free fermion gas in the ground state.

Hints for the solution of 9b: With discrete momenta

$$\boldsymbol{k} = \frac{2\pi}{L}\boldsymbol{n}$$

the anti-commutation relation for fermionic creation and annihilation operators becomes

$$\{a_\sigma(\boldsymbol{k}), a_{\sigma'}^+(\boldsymbol{k}')\} = \delta_{\sigma\sigma'}\delta_{\boldsymbol{k},\boldsymbol{k}'}.$$

The corresponding mode expansion for the annihilation operator in $\boldsymbol{x}$-space becomes

$$\psi_\sigma(\boldsymbol{x}) = \frac{1}{\sqrt{V}}\sum_{\boldsymbol{k}} a_\sigma(\boldsymbol{k})\exp(\mathrm{i}\boldsymbol{k}\cdot\boldsymbol{x}),$$

and the inversion is

$$a_\sigma(\boldsymbol{k}) = \frac{1}{\sqrt{V}} \int_V d^3\boldsymbol{x}\, \psi_\sigma(\boldsymbol{x}) \exp(-\mathrm{i}\boldsymbol{k}\cdot\boldsymbol{x}).$$

Substitute the mode expansions for ψ and ψ^+ into the operator $\hat{g}_{\sigma,\sigma'}(\boldsymbol{x},\boldsymbol{x}')$. This yields a four-fold sum over momenta

$$\langle\Omega| \sum_{\boldsymbol{k},\boldsymbol{k}',\boldsymbol{q},\boldsymbol{q}'} \ldots |\Omega\rangle.$$

In the next step, you can use that

$$a_\sigma(\boldsymbol{k})|\Omega\rangle = \Theta(k_F - k)a_\sigma(\boldsymbol{k})|\Omega\rangle, \quad \langle\Omega|a_\sigma^+(\boldsymbol{q}) = \Theta(k_F - q)\langle\Omega|a_\sigma^+(\boldsymbol{q}), \tag{17.92}$$

because e.g. in the first equation, if $k > k_F$, $a_\sigma(\boldsymbol{k})$ would simply anti-commute through all the creation operators in $|\Omega\rangle$ and yield zero through action on the vacuum $|0\rangle$. This reduces the four sums to sums over momenta inside the Fermi sphere,

$$\langle\Omega| \sum_{\{\boldsymbol{k},\boldsymbol{k}',\boldsymbol{q},\boldsymbol{q}'\}\leq k_F} \ldots |\Omega\rangle.$$

For the following steps, you can use that for fermionic operators $(a_\sigma^+(\boldsymbol{k}))^2 = 0$ and therefore

$$k \leq k_F: \quad a_\sigma^+(\boldsymbol{k})|\Omega\rangle = 0. \tag{17.93}$$

This observation can be used to replace operator products with commutators or anti-commutators, e.g.

$$q' \leq k_F: \quad a_{\sigma'}^+(\boldsymbol{q}')a_{\sigma'}(\boldsymbol{k}')a_\sigma(\boldsymbol{k})|\Omega\rangle = [a_{\sigma'}^+(\boldsymbol{q}'), a_{\sigma'}(\boldsymbol{k}')a_\sigma(\boldsymbol{k})]|\Omega\rangle,$$
$$q \leq k_F: \quad a_\sigma^+(\boldsymbol{q})a_{\sigma'}(\boldsymbol{k}')|\Omega\rangle = \{a_\sigma^+(\boldsymbol{q}), a_{\sigma'}(\boldsymbol{k}')\}|\Omega\rangle.$$

This helps to get rid of all the operators in $g_{\sigma,\sigma'}(\boldsymbol{x},\boldsymbol{x}')$. For the last steps, you have to figure out what the sum over all momenta inside a Fermi sphere is, $\sum_{\boldsymbol{k},|\boldsymbol{k}|\leq k_F} 1$ (you know this sum because you know that there are N Fermions in the system). For another term, you have to use that for $N \gg 1$

$$\frac{1}{V} \sum_{\boldsymbol{k},|\boldsymbol{k}|\leq k_F} f(\boldsymbol{k}) \simeq \frac{1}{(2\pi)^3} \int_{k\leq k_F} d^3\boldsymbol{k}\, f(\boldsymbol{k}).$$

Solution.

17.9a. Integration of equation (17.91) over $\boldsymbol{x}$ and $\boldsymbol{x}'$ yields for identical spins e.g.

$$\frac{1}{2}\int d^3\boldsymbol{x} \int d^3\boldsymbol{x}'\, \tilde{g}_{\uparrow\uparrow}(\boldsymbol{x},\boldsymbol{x}') = \frac{1}{2}N_\uparrow(N_\uparrow - 1) = \mathcal{N}_{\uparrow\uparrow},$$

which is the number of independent fermion pairs with both fermions having spin up, and we also find

$$\frac{1}{2}\int d^3x \int d^3x'\, \tilde{g}_{\uparrow\downarrow}(x,x') = \frac{1}{2}N_\uparrow N_\downarrow = \frac{1}{2}\mathcal{N}_{\uparrow\downarrow},$$

which is the number of independent fermion pairs with opposite spins if we take into account that we want e.g. spin up in the location $\boldsymbol{x}$ and spin down in $\boldsymbol{x}'$. Note also that summation over spin polarizations then yields

$$\begin{aligned}\sum_{\sigma,\sigma'}\int d^3x \int d^3x'\, \frac{1}{2}\tilde{g}_{\sigma,\sigma'}(x,x') &= \frac{1}{2}N_\uparrow(N_\uparrow - 1) + \frac{1}{2}N_\downarrow(N_\downarrow - 1) + N_\uparrow N_\downarrow \\ &= \frac{1}{2}N(N-1) = \mathcal{N},\end{aligned}$$

i.e. the total number of independent fermions pairs.

17.9b. The discussion below is a modification of the discussion given by Schwabl [36]. Other derivations of the exchange hole of the effective charge density experienced by a Hartree-Fock electron in a metal can be found in [11, 17].

We have

$$\begin{aligned}g_{\sigma,\sigma'}(\boldsymbol{x},\boldsymbol{x}') = \frac{1}{V^2}\sum_{\boldsymbol{k},\boldsymbol{k}',\boldsymbol{q},\boldsymbol{q}'} &\exp\left[\mathrm{i}(\boldsymbol{k}\cdot\boldsymbol{x} + \boldsymbol{k}'\cdot\boldsymbol{x}' - \boldsymbol{q}\cdot\boldsymbol{x} - \boldsymbol{q}'\cdot\boldsymbol{x}')\right] \\ &\times\langle\Omega|a_\sigma^+(\boldsymbol{q})a_{\sigma'}^+(\boldsymbol{q}')a_{\sigma'}(\boldsymbol{k}')a_\sigma(\boldsymbol{k})|\Omega\rangle.\end{aligned}$$

The observation (17.92) limits the sums over wave numbers,

$$\begin{aligned}g_{\sigma,\sigma'}(\boldsymbol{x},\boldsymbol{x}') = \frac{1}{V^2}\sum_{\{\boldsymbol{k},\boldsymbol{k}',\boldsymbol{q},\boldsymbol{q}'\}\le k_F} &\exp\left[\mathrm{i}(\boldsymbol{k}\cdot\boldsymbol{x} + \boldsymbol{k}'\cdot\boldsymbol{x}' - \boldsymbol{q}\cdot\boldsymbol{x} - \boldsymbol{q}'\cdot\boldsymbol{x}')\right] \\ &\times\langle\Omega|a_\sigma^+(\boldsymbol{q})a_{\sigma'}^+(\boldsymbol{q}')a_{\sigma'}(\boldsymbol{k}')a_\sigma(\boldsymbol{k})|\Omega\rangle.\end{aligned}$$

For the next step we use (17.93), $\Theta(k_F - q')a_\sigma^+(\boldsymbol{q}')|\Omega\rangle = 0$, to replace operator products with commutators or anti-commutators:

$$\begin{aligned}g_{\sigma,\sigma'}(\boldsymbol{x},\boldsymbol{x}') &= \frac{1}{V^2}\sum_{\{\boldsymbol{k},\boldsymbol{k}',\boldsymbol{q},\boldsymbol{q}'\}\le k_F} \exp\left[\mathrm{i}(\boldsymbol{k}\cdot\boldsymbol{x} + \boldsymbol{k}'\cdot\boldsymbol{x}' - \boldsymbol{q}\cdot\boldsymbol{x} - \boldsymbol{q}'\cdot\boldsymbol{x}')\right] \\ &\quad\times\langle\Omega|a_\sigma^+(\boldsymbol{q})[a_{\sigma'}^+(\boldsymbol{q}'), a_{\sigma'}(\boldsymbol{k}')a_\sigma(\boldsymbol{k})]|\Omega\rangle \\ &= \frac{1}{V^2}\sum_{\{\boldsymbol{k},\boldsymbol{k}',\boldsymbol{q}\}\le k_F} \exp\left[\mathrm{i}(\boldsymbol{k}-\boldsymbol{q})\cdot\boldsymbol{x}\right]\langle\Omega|a_\sigma^+(\boldsymbol{q})a_\sigma(\boldsymbol{k})|\Omega\rangle\end{aligned}$$

$$- \delta_{\sigma\sigma'}\frac{1}{V^2}\sum_{\{k,k',q\}\leq k_F} \exp\left[\mathrm{i}(\boldsymbol{k}\cdot\boldsymbol{x}+\boldsymbol{k}'\cdot\boldsymbol{x}'-\boldsymbol{q}\cdot\boldsymbol{x}-\boldsymbol{k}\cdot\boldsymbol{x}')\right]$$

$$\times\langle\Omega|a^+_\sigma(\boldsymbol{q})a_{\sigma'}(\boldsymbol{k}')|\Omega\rangle$$

$$= \frac{1}{V^2}\sum_{\{k,k'\}\leq k_F} 1 - \delta_{\sigma\sigma'}\frac{1}{V^2}\sum_{\{k,k'\}\leq k_F} \exp\left[\mathrm{i}(\boldsymbol{k}-\boldsymbol{k}')\cdot(\boldsymbol{x}-\boldsymbol{x}')\right].$$

Here the notation $\leq k_F$ under the summation indicates that e.g. the summation over $\boldsymbol{k}$ is over all vectors $\boldsymbol{k}$ inside the Fermi sphere: $|\boldsymbol{k}| \leq k_F$.

For the further evaluation we note that there are two fermions per momentum inside the Fermi sphere, and therefore

$$\frac{1}{V}\sum_{k,|k|\leq k_F} 1 = \frac{N}{2V} = \frac{n}{2}.$$

This yields

$$g_{\sigma,\sigma'}(\boldsymbol{x},\boldsymbol{x}') = \frac{n^2}{4} - \delta_{\sigma\sigma'}\left|\frac{1}{V}\sum_{k,|k|\leq k_F} \exp\left[\mathrm{i}\boldsymbol{k}\cdot(\boldsymbol{x}-\boldsymbol{x}')\right]\right|^2$$

$$\approx \frac{n^2}{4} - \delta_{\sigma\sigma'}\left|\frac{1}{(2\pi)^3}\int_{k\leq k_F} d^3\boldsymbol{k}\, \exp\left[\mathrm{i}\boldsymbol{k}\cdot(\boldsymbol{x}-\boldsymbol{x}')\right]\right|^2$$

$$= \frac{n^2}{4} - \delta_{\sigma\sigma'}\left|\frac{1}{(2\pi)^2}\int_0^{k_F} dk\int_{-1}^1 d\xi\, k^2 \exp\left[\mathrm{i}k|\boldsymbol{x}-\boldsymbol{x}'|\xi\right]\right|^2$$

$$= \frac{n^2}{4} - \delta_{\sigma\sigma'}\frac{1}{4\pi^4|\boldsymbol{x}-\boldsymbol{x}'|^2}\left|\int_0^{k_F} dk\, k\sin\left(k|\boldsymbol{x}-\boldsymbol{x}'|\right)\right|^2$$

$$= \frac{n^2}{4} - \delta_{\sigma\sigma'}\frac{[\sin\left(k_F|\boldsymbol{x}-\boldsymbol{x}'|\right) - k_F|\boldsymbol{x}-\boldsymbol{x}'|\cos\left(k_F|\boldsymbol{x}-\boldsymbol{x}'|\right)]^2}{4\pi^4|\boldsymbol{x}-\boldsymbol{x}'|^6}.$$

In particular, the result for equal spin orientation can be written as

$$\frac{4}{n^2}g_{\sigma,\sigma}(\boldsymbol{x},\boldsymbol{x}') = 1 - 9\frac{[\sin\left(k_F|\boldsymbol{x}-\boldsymbol{x}'|\right) - k_F|\boldsymbol{x}-\boldsymbol{x}'|\cos\left(k_F|\boldsymbol{x}-\boldsymbol{x}'|\right)]^2}{\left(k_F|\boldsymbol{x}-\boldsymbol{x}'|\right)^6},$$

where $n = k_F^3/3\pi^2$ (12.18) was used. This means that up to a distance of order

$$\frac{\lambda_F}{2} = \frac{\pi}{k_F} = \left(\frac{\pi}{3n}\right)^{\frac{1}{3}}$$

the probability to find a fermion of like spin is significantly reduced: The Pauli principle prevents two fermions of like spin to be in the same place, even if there is no interaction between the fermions.

The function $\frac{4}{n^2}g_{\sigma,\sigma}(\boldsymbol{x},\boldsymbol{x}')$ is plotted for three ranges of the variable $x = k_F|\boldsymbol{x}-\boldsymbol{x}'|$ in Figures 17.1–17.3.

The first maximum $\frac{4}{n^2}g_{\sigma,\sigma}(\boldsymbol{x},\boldsymbol{x}') = 1$ is reached at

$$k_F|\boldsymbol{x}-\boldsymbol{x}'| \approx 4.4934,$$

i.e. depending on the maximal momentum in the fermion gas the minimal distance between two fermions of the *same* spin orientation is given by the minimal wavelength in the gas:

$$r_{hole} \approx 0.7\lambda_F. \tag{17.94}$$

On the other hand, the density of fermions with equal spin is $n/2$. Equal separation between those fermions would correspond to a distance

$$a = \left(\frac{2}{n}\right)^{\frac{1}{3}},$$

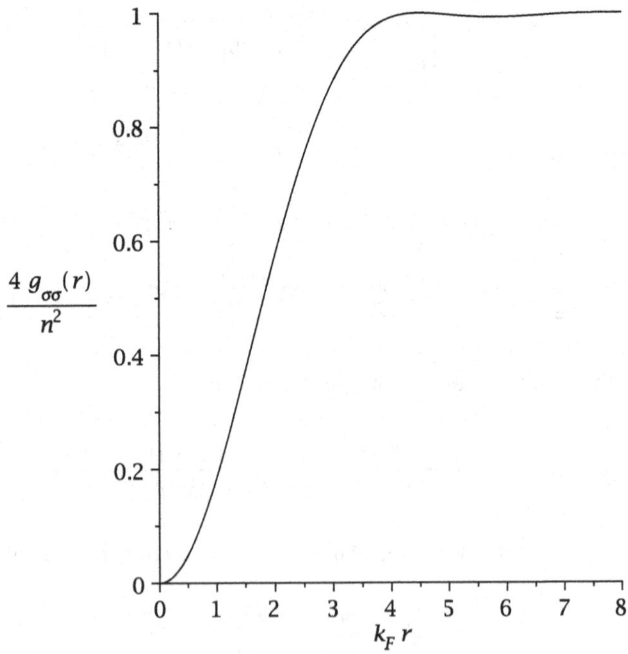

Fig. 17.1 The scaled pair correlation function $4g_{\sigma,\sigma}(r)/n^2$ for $0 \le k_F r \le 8$

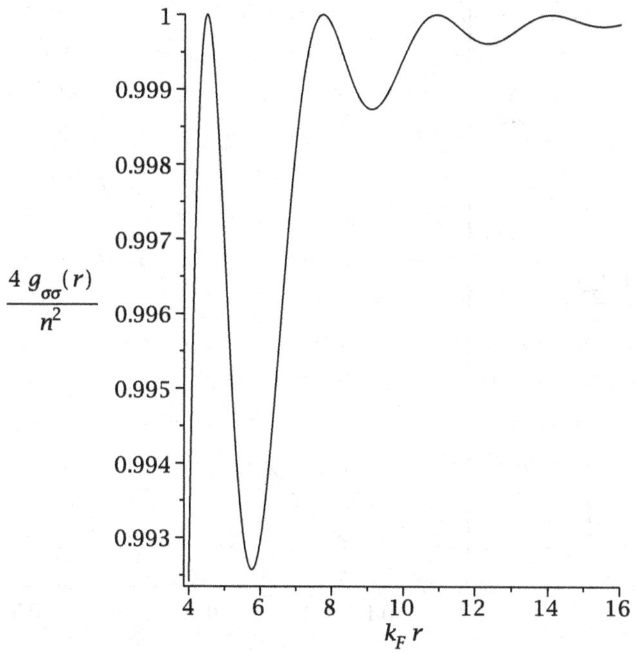

Fig. 17.2 The scaled pair correlation function $4g_{\sigma,\sigma}(r)/n^2$ for $4 \leq k_F r \leq 16$

and inserting the result for the Fermi momentum

$$k_F = \left(3\pi^2 n\right)^{\frac{1}{3}}, \quad \lambda_F = \left(\frac{8\pi}{3n}\right)^{\frac{1}{3}}$$

yields

$$a = \left(\frac{3}{4\pi}\right)^{\frac{1}{3}} \lambda_F \simeq 0.62\lambda_F. \tag{17.95}$$

Comparison with r_{hole} (17.94) shows that the Pauli principle effectively repels fermions of equal spin such that they try to fill the available volume uniformly. Note that the existence of this *exchange hole* in the pair correlation between identical fermions of like spin has nothing to do e.g. with any electromagnetic interaction between the fermions. It is only a consequence of avoidance due to the Pauli principle. Stated differently (and presumably in the simplest possible way): The Pauli principle implies that free fermions of the same spin orientation try to occupy a volume as uniformly as possible to avoid contact.

If we want to add free fermions to a fermion gas of constant volume we have to increase the energy in the gas, thereby increasing the maximal momentum in the gas,

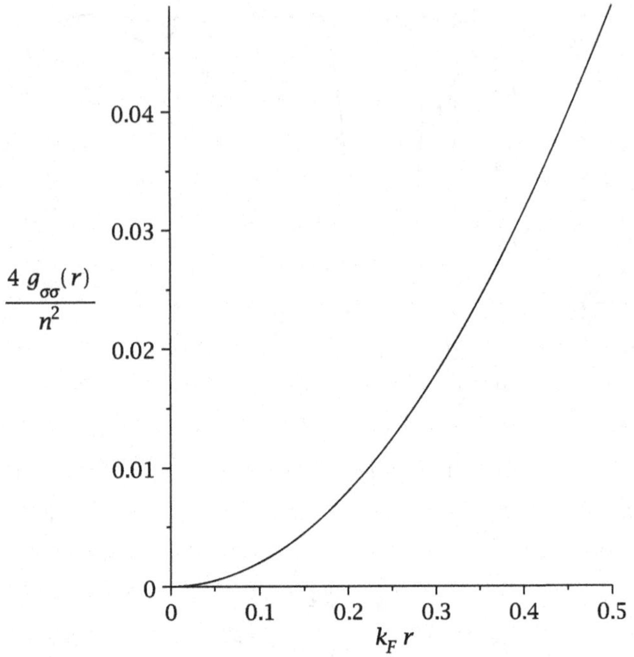

Fig. 17.3 The scaled pair correlation function $4g_{\sigma,\sigma}(r)/n^2$ for $0 \leq k_F r \leq 0.5$

to squeeze more fermions into the volume and reduce the mean distance between fermions of like spin.

The presence of the exchange hole implies a local reduction of the charge density of the other electrons seen by an electron in a metal (a "Hartree-Fock electron in a jellium model" [11, 17, 25]),

$$\begin{aligned}\rho_e(r) &= -e\frac{n}{2} - e\frac{n}{2}\left(1 - \frac{9}{(k_F r)^6}\left[\sin(k_F r) - k_F r\cos(k_F r)\right]^2\right) \\ &= -en\left(1 - \frac{9}{2\,(k_F r)^6}\left[\sin(k_F r) - k_F r\cos(k_F r)\right]^2\right).\end{aligned}$$

This effective electron charge density is plotted in Figure 17.4.

A fermion gas where the fermions fill the lowest possible energy states under the constraint of the Pauli principle is denoted as a degenerate fermion gas. Addition of a fermion to a degenerate free fermion gas or compression of the free fermion gas costs energy according to (12.18), and effectively this amounts to a repulsive force between the fermions. It is easy to calculate the corresponding degeneracy pressure for the non-relativistic degenerate fermion gas. The total energy of the degenerate free fermion gas is

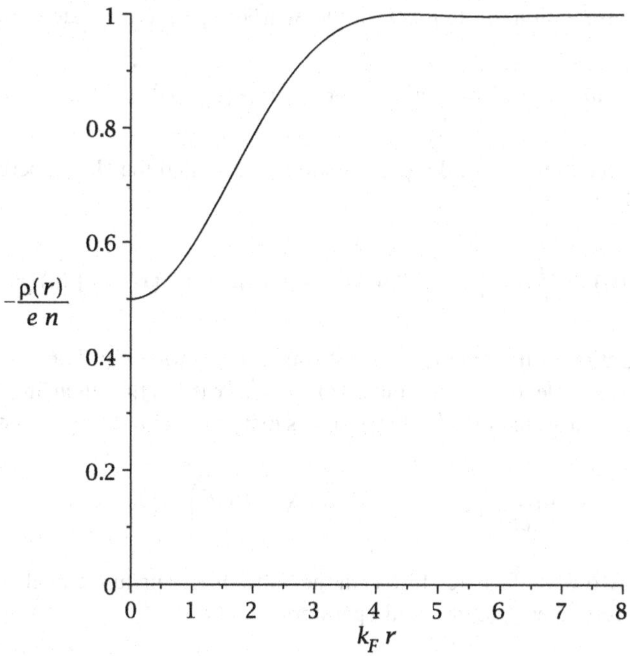

Fig. 17.4 The effective scaled electron charge density $\rho_e(r)/(-en)$ in a metal is plotted for $0 \leq k_F r \leq 8$

$$E = 2V \int_{|k|\leq k_F} \frac{d^3k}{(2\pi)^3} \frac{\hbar^2 k^2}{2m} = V \frac{\hbar^2 k_F^5}{10\pi^2 m} = \pi^{4/3} \frac{\hbar^2}{10m} \frac{(3N)^{5/3}}{V^{2/3}}.$$

The degeneracy pressure is therefore

$$p = -\frac{\partial E}{\partial V} = (3\pi^2)^{2/3} \frac{\hbar^2}{5m} n^{5/3}.$$

The corresponding chemical potential is the Fermi energy, of course,

$$\mu = \frac{\partial E}{\partial N} = \frac{\hbar^2}{2m}(3\pi^2 n)^{2/3} = \frac{\hbar^2}{2m} k_F^2,$$

and the average energy per particle is $E/N = 0.6\mu$.

17.10. Show that the time-dependence of the Heisenberg picture operator in (17.44) implies equivalence of the Heisenberg evolution equation (17.7) with the Schrödinger equation for the single particle wave function $\langle \boldsymbol{x}|\Phi(t)\rangle$.

Solution. We have found the following representations of a second quantized single particle state associated with a general wave function $\langle \boldsymbol{x}|\Phi(t)\rangle$,

$$|\Phi(t)\rangle = \int d^3\boldsymbol{x}\, \psi^+(\boldsymbol{x})|0\rangle\langle \boldsymbol{x}|\Phi(t)\rangle = \int d^3\boldsymbol{x}\, \Psi^+(\boldsymbol{x}, -t)|0\rangle\langle \boldsymbol{x}|\Phi(0)\rangle. \quad (17.96)$$

The first representation (cf. (17.42) for the Schrödinger picture state implies

$$\mathrm{i}\hbar\frac{d}{dt}|\Phi(t)\rangle = \int d^3x\,\psi^+(x)|0\rangle\mathrm{i}\hbar\frac{\partial}{\partial t}\langle x|\Phi(t)\rangle. \tag{17.97}$$

On the other hand, the second representation (17.44) and the Heisenberg evolution equation (17.7) imply

$$\mathrm{i}\hbar\frac{d}{dt}|\Phi(t)\rangle = \int d^3x\left(-\frac{\hbar^2}{2m}\Delta\Psi^+(x,-t) + V(x)\Psi^+(x,-t)\right)|0\rangle\langle x|\Phi(0)\rangle.$$

Partial integration of the kinetic term and the correspondence (17.96) together with the linear independence of the states $|x\rangle = \int d^3x\,\psi^+(x)|0\rangle$ then imply that the single particle wave function $\langle x|\Phi(t)\rangle$ must satisfy the Schrödinger equation

$$\mathrm{i}\hbar\frac{\partial}{\partial t}\langle x|\Phi(t)\rangle = \left(-\frac{\hbar^2}{2m}\Delta + V(x)\right)\langle x|\Phi(t)\rangle.$$

17.11. The two-particle state (17.64) in the Schrödinger picture can also be written in terms of Heisenberg picture field operators, cf. (17.44),

$$\begin{aligned}|\Phi_{a,a'}(t)\rangle &= \exp\left(-\frac{\mathrm{i}}{\hbar}Ht\right)|\Phi_{a,a'}(0)\rangle \\ &= \frac{1}{\sqrt{1+\delta_{a,a'}}}\sum_{\sigma,\sigma'}\int d^3x\int d^3x'\,\Psi^+_{a,\sigma}(x,-t)\Psi^+_{a',\sigma'}(x',-t)|0\rangle \\ &\quad\times\langle x,\sigma;x',\sigma'|\Phi_{a,a'}(0)\rangle.\end{aligned} \tag{17.98}$$

Show that the Heisenberg evolution equations (17.80) with the Hamiltonian (17.73) yield again the two-particle Schrödinger equations (17.74) and (17.75).

Show also that this derivation works for the general N-particle state and yields the N-particle equation (17.78). For the potential terms you can use the product rule for commutators,

$$\begin{aligned}[A,B_1B_2\ldots B_N] &= [A,B_1]B_2\ldots B_N + \sum_{I=1}^{N-2}B_1\ldots B_I[A,B_{I+1}]B_{I+2}\ldots B_N \\ &\quad + B_1\ldots B_{N-1}[A,B_N].\end{aligned}$$

Chapter 18
Quantization of the Maxwell Field: Photons

We will now start to quantize the Maxwell field $A_\mu(x) = \{-\Phi(x)/c, \boldsymbol{A}(x)\}$ similar to the quantization of the Schrödinger field. The fact that electromagnetism has a gauge invariance implies that there are more components than actual dynamical degrees of freedom in the Maxwell field. This will make quantization a little more challenging than for the Schrödinger field, but we will overcome those difficulties.

Electromagnetic field theory is implicitly relativistic, and quantized Maxwell theory therefore also provides us with a first example of a relativistic quantum field theory. Appendix B provides an introduction to 4-vector and tensor notation in electromagnetic theory.

18.1 Lagrange density and mode expansion for the Maxwell field

The equations of motion for the Maxwell field are the inhomogeneous Maxwell equations (recall from electrodynamics that the homogeneous equations were solved through the introduction of the potentials A_μ),

$$\partial_\mu F^{\mu\nu} = \partial_\mu \left(\partial^\mu A^\nu - \partial^\nu A^\mu\right) = -\mu_0 j^\nu.$$

These equations can be written as

$$j^\nu + \frac{1}{\mu_0}\partial_\mu \left(\partial^\mu A^\nu - \partial^\nu A^\mu\right) = \frac{\partial \mathcal{L}}{\partial A_\nu} - \partial_\mu \frac{\partial \mathcal{L}}{\partial(\partial_\mu A_\nu)} = 0$$

if we use the Lagrange density

$$\mathcal{L} = j^\nu A_\nu - \frac{1}{4\mu_0} F_{\mu\nu} F^{\mu\nu} = \frac{\epsilon_0}{2}\boldsymbol{E}^2 - \frac{1}{2\mu_0}\boldsymbol{B}^2 + \boldsymbol{j}\cdot\boldsymbol{A} - \varrho\Phi. \tag{18.1}$$

R. Dick, *Advanced Quantum Mechanics*, Graduate Texts in Physics,
DOI 10.1007/978-3-319-25675-7_18

This Lagrangian provides us with the canonically conjugate momentum for the vector potential $\boldsymbol{A}$:

$$\boldsymbol{\Pi}_A = \frac{\partial \mathcal{L}}{\partial \dot{\boldsymbol{A}}} = \epsilon_0(\dot{\boldsymbol{A}} + \nabla\Phi) = -\epsilon_0 \boldsymbol{E},$$

but

$$\Pi_\Phi = \frac{\partial \mathcal{L}}{\partial \dot{\Phi}} = 0$$

vanishes identically! Therefore we cannot simply impose canonical commutation relations between the four components A_μ of the 4-vector potential and four conjugate momenta Π_ν. To circumvent this problem we revisit Maxwell's equations,

$$\Delta\Phi + \nabla\cdot\dot{\boldsymbol{A}} = -\frac{1}{\epsilon_0}\varrho, \tag{18.2}$$

$$\nabla(\nabla\cdot\boldsymbol{A}) - \Delta\boldsymbol{A} + \frac{1}{c^2}\frac{\partial^2}{\partial t^2}\boldsymbol{A} + \frac{1}{c^2}\frac{\partial}{\partial t}\nabla\Phi = \mu_0\boldsymbol{j}. \tag{18.3}$$

One way to solve the problem with $\Pi_\Phi = 0$ is to eliminate $\nabla\cdot\boldsymbol{A}$ from the equations of motion through the gauge freedom

$$\Phi(\boldsymbol{x},t) \to \Phi_f(\boldsymbol{x},t) = \Phi(\boldsymbol{x},t) - \dot{f}(\boldsymbol{x},t), \tag{18.4}$$

$$\boldsymbol{A}(\boldsymbol{x},t) \to \boldsymbol{A}_f(\boldsymbol{x},t) = \boldsymbol{A}(\boldsymbol{x},t) + \nabla f(\boldsymbol{x},t), \tag{18.5}$$

i.e. we impose the gauge condition $\nabla\cdot\boldsymbol{A}_f = 0$. The equation

$$\Delta f(\boldsymbol{x},t) = -\nabla\cdot\boldsymbol{A}(\boldsymbol{x},t)$$

can be solved with the Green's function $G(r) = (4\pi r)^{-1}$ for the Laplace operator,

$$\Delta\frac{1}{4\pi|\boldsymbol{x}-\boldsymbol{x}'|} = -\delta(\boldsymbol{x}-\boldsymbol{x}'),$$

see equations (11.11) and (11.17) for $E = 0$,

$$f(\boldsymbol{x},t) = \frac{1}{4\pi}\int d^3x'\,\frac{1}{|\boldsymbol{x}-\boldsymbol{x}'|}\nabla\cdot\boldsymbol{A}(\boldsymbol{x}',t).$$

This gauge is denoted as *Coulomb gauge.*

We denote the gauge transformed fields again with Φ and $\boldsymbol{A}$, i.e. we have

$$\nabla\cdot\boldsymbol{A}(\boldsymbol{x},t) = 0. \tag{18.6}$$

and

$$\Delta\Phi = -\frac{1}{\epsilon_0}\varrho, \tag{18.7}$$

$$\frac{1}{c^2}\frac{\partial^2}{\partial t^2}\boldsymbol{A} - \Delta\boldsymbol{A} + \frac{1}{c^2}\frac{\partial}{\partial t}\nabla\Phi = \mu_0\boldsymbol{j}. \tag{18.8}$$

We can now get rid of Φ by solving (18.7) again with the Green's function for the Laplace operator,

$$\Phi(\boldsymbol{x},t) = \frac{1}{4\pi\epsilon_0}\int d^3x'\,\frac{\varrho(\boldsymbol{x}',t)}{|\boldsymbol{x}-\boldsymbol{x}'|}. \tag{18.9}$$

The resulting equation for $\boldsymbol{A}$ is

$$\begin{aligned}\left(\frac{1}{c^2}\frac{\partial^2}{\partial t^2} - \Delta\right)\boldsymbol{A}(\boldsymbol{x},t) &= \mu_0\boldsymbol{j}(\boldsymbol{x},t) + \frac{\mu_0}{4\pi}\int d^3x'\,\frac{\boldsymbol{x}-\boldsymbol{x}'}{|\boldsymbol{x}-\boldsymbol{x}'|^3}\frac{\partial}{\partial t}\varrho(\boldsymbol{x}',t)\\ &= \mu_0\boldsymbol{j}(\boldsymbol{x},t) + \frac{\mu_0}{4\pi}\int d^3x'\,\nabla\frac{1}{|\boldsymbol{x}-\boldsymbol{x}'|}\nabla'\cdot\boldsymbol{j}(\boldsymbol{x}',t)\\ &= \mu_0\boldsymbol{j}(\boldsymbol{x},t) + \frac{\mu_0}{4\pi}\int d^3x'\,\boldsymbol{j}(\boldsymbol{x}',t)\cdot\nabla\otimes\nabla\frac{1}{|\boldsymbol{x}-\boldsymbol{x}'|}\\ &= \mu_0\boldsymbol{J}(\boldsymbol{x},t).\end{aligned} \tag{18.10}$$

We can also evaluate the derivatives in the integral using

$$\nabla\otimes\nabla\frac{1}{r} = -\nabla\otimes\frac{\boldsymbol{e}_r}{r^2} = -\frac{4\pi}{3}\underline{1}\delta(\boldsymbol{x}) - \frac{\underline{1}-3\boldsymbol{e}_r\otimes\boldsymbol{e}_r}{r^3}. \tag{18.11}$$

This yields

$$\boldsymbol{J}(\boldsymbol{x},t) = \frac{2}{3}\boldsymbol{j}(\boldsymbol{x},t) + \int d^3x'\,\frac{3(\boldsymbol{x}-\boldsymbol{x}')\otimes(\boldsymbol{x}-\boldsymbol{x}') - |\boldsymbol{x}-\boldsymbol{x}'|^2\underline{1}}{4\pi|\boldsymbol{x}-\boldsymbol{x}'|^5}\cdot\boldsymbol{j}(\boldsymbol{x}',t).$$

The new current density $\boldsymbol{J}$ satisfies

$$\nabla\cdot\boldsymbol{J}(\boldsymbol{x},t) = 0. \tag{18.12}$$

This follows from the definition of $\boldsymbol{J}$ in the first line of (18.10) and charge conservation, or directly from the second or third line of (18.10), which can be considered as projections of the vector $\boldsymbol{j}$ onto its divergence-free part. We also require localization of charges and currents in the sense of

$$\lim_{|\boldsymbol{x}|\to\infty}|\boldsymbol{x}|\boldsymbol{j}(\boldsymbol{x},t) = 0. \tag{18.13}$$

Equation (18.10) can be solved e.g. with the retarded Green's function, cf. equation (J.60),

$$\left(\frac{1}{c^2}\frac{\partial^2}{\partial t^2}-\Delta\right)G(\boldsymbol{x},t)=\delta(\boldsymbol{x})\delta(t),\quad G(\boldsymbol{x},t)=\frac{1}{4\pi r}\delta\!\left(t-\frac{r}{c}\right), \tag{18.14}$$

in the form

$$\begin{aligned}\boldsymbol{A}_J(\boldsymbol{x},t)&=\mu_0\int d^3\boldsymbol{x}'\int dt'\,G(\boldsymbol{x}-\boldsymbol{x}',t-t')\boldsymbol{J}(\boldsymbol{x}',t')\\&=\frac{\mu_0}{4\pi}\int d^3\boldsymbol{x}'\,\frac{1}{|\boldsymbol{x}-\boldsymbol{x}'|}\boldsymbol{J}\!\left(\boldsymbol{x}',t-\frac{|\boldsymbol{x}-\boldsymbol{x}'|}{c}\right).\end{aligned} \tag{18.15}$$

This satisfies $\nabla\cdot\boldsymbol{A}_J(\boldsymbol{x},t)=0$ due to (18.12, 18.13).

The vector field is only a special solution of the inhomogeneous equation (18.10), and the general solution will be a superposition

$$\boldsymbol{A}(\boldsymbol{x},t)=\boldsymbol{A}_J(\boldsymbol{x},t)+\boldsymbol{A}_D(\boldsymbol{x},t)$$

of the special inhomogeneous solution with the general solution of the homogeneous equations

$$\left(\frac{1}{c^2}\frac{\partial^2}{\partial t^2}-\Delta\right)\boldsymbol{A}_D(\boldsymbol{x},t)=0. \tag{18.16}$$

The homogeneous solution still has to satisfy the gauge condition $\nabla\cdot\boldsymbol{A}_D=0$, because the total vector potential $\boldsymbol{A}$ has to satisfy this condition.

Fourier decomposition

$$\begin{aligned}\boldsymbol{A}_D(\boldsymbol{x},t)&=\frac{1}{4\pi^2}\int d^3\boldsymbol{k}\int d\omega\,\boldsymbol{A}_D(\boldsymbol{k},\omega)\exp[\mathrm{i}(\boldsymbol{k}\cdot\boldsymbol{x}-\omega t)],\\\boldsymbol{A}_D(\boldsymbol{k},\omega)&=\frac{1}{4\pi^2}\int d^3\boldsymbol{x}\int dt\,\boldsymbol{A}_D(\boldsymbol{x},t)\exp[-\mathrm{i}(\boldsymbol{k}\cdot\boldsymbol{x}-\omega t)]\end{aligned} \tag{18.17}$$

transforms the condition $\nabla\cdot\boldsymbol{A}_D(\boldsymbol{x},t)=0$ and the equation (18.16) into

$$\boldsymbol{k}\cdot\boldsymbol{A}_D(\boldsymbol{k},\omega)=0 \tag{18.18}$$

and

$$\left(\boldsymbol{k}^2-\frac{\omega^2}{c^2}\right)\boldsymbol{A}_D(\boldsymbol{k},\omega)=0. \tag{18.19}$$

Equation (18.18) is the statement that photons are transverse, whereas equation (18.19) implies that $\boldsymbol{A}_D(\boldsymbol{k},\omega)$ can be written as

$$\boldsymbol{A}_D(\boldsymbol{k},\omega) = \sqrt{\frac{\pi\hbar\mu_0 c}{k}}\sum_{\alpha=1}^{2}\boldsymbol{\epsilon}_\alpha(\boldsymbol{k})\left(a_\alpha(\boldsymbol{k})\delta(\omega-ck)+a_\alpha^+(-\boldsymbol{k})\delta(\omega+ck)\right), \tag{18.20}$$

where the prefactor so far is a matter of convention and the two vectors $\boldsymbol{\epsilon}_\alpha(\boldsymbol{k})$ are a Cartesian basis in the plane orthogonal to $\boldsymbol{k}$:

$$\boldsymbol{\epsilon}_\alpha(\boldsymbol{k})\cdot\boldsymbol{\epsilon}_\beta(\boldsymbol{k}) = \delta_{\alpha\beta}, \quad \boldsymbol{k}\cdot\boldsymbol{\epsilon}_\alpha(\boldsymbol{k}) = 0.$$

Inserting (18.20) into (18.17) yields

$$\begin{aligned}\boldsymbol{A}_D(\boldsymbol{x},t) = \sqrt{\frac{\hbar\mu_0 c}{(2\pi)^3}}\int\frac{d^3\boldsymbol{k}}{\sqrt{2k}}\sum_{\alpha=1}^{2}\boldsymbol{\epsilon}_\alpha(\boldsymbol{k})\Big(a_\alpha(\boldsymbol{k})\exp[\mathrm{i}(\boldsymbol{k}\cdot\boldsymbol{x}-ckt)]\\ +a_\alpha^+(\boldsymbol{k})\exp[-\mathrm{i}(\boldsymbol{k}\cdot\boldsymbol{x}-ckt)]\Big),\end{aligned} \tag{18.21}$$

and for the fields

$$\begin{aligned}\boldsymbol{E}_D(\boldsymbol{x},t) &= -\frac{\partial}{\partial t}\boldsymbol{A}_D(\boldsymbol{x},t)\\ &= \mathrm{i}\sqrt{\frac{\hbar\mu_0 c^3}{(2\pi)^3}}\int d^3\boldsymbol{k}\,\sqrt{\frac{k}{2}}\sum_{\alpha=1}^{2}\boldsymbol{\epsilon}_\alpha(\boldsymbol{k})\Big(a_\alpha(\boldsymbol{k})\exp[\mathrm{i}(\boldsymbol{k}\cdot\boldsymbol{x}-ckt)]\\ &\quad -a_\alpha^+(\boldsymbol{k})\exp[-\mathrm{i}(\boldsymbol{k}\cdot\boldsymbol{x}-ckt)]\Big),\end{aligned} \tag{18.22}$$

$$\begin{aligned}\boldsymbol{B}_D(\boldsymbol{x},t) &= \nabla\times\boldsymbol{A}_D(\boldsymbol{x},t)\\ &= \mathrm{i}\sqrt{\frac{\hbar\mu_0 c}{(2\pi)^3}}\int\frac{d^3\boldsymbol{k}}{\sqrt{2k}}\sum_{\alpha=1}^{2}\boldsymbol{k}\times\boldsymbol{\epsilon}_\alpha(\boldsymbol{k})\Big(a_\alpha(\boldsymbol{k})\exp[\mathrm{i}(\boldsymbol{k}\cdot\boldsymbol{x}-ckt)]\\ &\quad -a_\alpha^+(\boldsymbol{k})\exp[-\mathrm{i}(\boldsymbol{k}\cdot\boldsymbol{x}-ckt)]\Big).\end{aligned} \tag{18.23}$$

Inversion of equations (18.21, 18.22) yields

$$\begin{aligned}a_\alpha(\boldsymbol{k}) &= \int\frac{d^3\boldsymbol{x}}{\sqrt{(2\pi)^3 2\mu_0\hbar c}}\boldsymbol{\epsilon}_\alpha(\boldsymbol{k})\cdot\left(\sqrt{k}\boldsymbol{A}_D(\boldsymbol{x},t)+\frac{\mathrm{i}}{c\sqrt{k}}\dot{\boldsymbol{A}}_D(\boldsymbol{x},t)\right)\\ &\quad\times\exp[-\mathrm{i}(\boldsymbol{k}\cdot\boldsymbol{x}-ckt)],\end{aligned}$$

$$\begin{aligned}a_\alpha^+(\boldsymbol{k}) &= \int\frac{d^3\boldsymbol{x}}{\sqrt{(2\pi)^3 2\mu_0\hbar c}}\boldsymbol{\epsilon}_\alpha(\boldsymbol{k})\cdot\left(\sqrt{k}\boldsymbol{A}_D(\boldsymbol{x},t)-\frac{\mathrm{i}}{c\sqrt{k}}\dot{\boldsymbol{A}}_D(\boldsymbol{x},t)\right)\\ &\quad\times\exp[\mathrm{i}(\boldsymbol{k}\cdot\boldsymbol{x}-ckt)].\end{aligned}$$

We can think of the vector potential (18.21) as a state $|A_D(t)\rangle$ with components

$$\langle \boldsymbol{k},\alpha|A_D(t)\rangle = \sqrt{\frac{\hbar\mu_0 c}{2k}}\left(a_\alpha(\boldsymbol{k})\exp(-\mathrm{i}ckt) + a_\alpha^+(-\boldsymbol{k})\exp(\mathrm{i}ckt)\right) \tag{18.24}$$

in wave vector space, and

$$\langle \boldsymbol{x},i|A_D(t)\rangle = \sqrt{\frac{\hbar\mu_0 c}{(2\pi)^3}}\int\frac{d^3k}{\sqrt{2k}}\sum_{\alpha=1}^{2}\epsilon_\alpha^i(\boldsymbol{k})\Big(a_\alpha(\boldsymbol{k})\exp[\mathrm{i}(\boldsymbol{k}\cdot\boldsymbol{x}-ckt)]$$
$$+a_\alpha^+(\boldsymbol{k})\exp[-\mathrm{i}(\boldsymbol{k}\cdot\boldsymbol{x}-ckt)]\Big) \tag{18.25}$$

in $\boldsymbol{x}$ space. This corresponds to transformation matrices

$$\langle \boldsymbol{x},i|\boldsymbol{k},\alpha\rangle = \frac{1}{\sqrt{2\pi}^3}\epsilon_\alpha^i(\boldsymbol{k})\exp(\mathrm{i}\boldsymbol{k}\cdot\boldsymbol{x}),$$

and we can easily check the completeness relations

$$\langle \boldsymbol{k},\alpha|\boldsymbol{k}',\beta\rangle = \int d^3x\sum_i\langle \boldsymbol{k},\alpha|\boldsymbol{x},i\rangle\langle \boldsymbol{x},i|\boldsymbol{k}',\beta\rangle = \delta(\boldsymbol{k}-\boldsymbol{k}')\boldsymbol{\epsilon}_\alpha(\boldsymbol{k})\cdot\boldsymbol{\epsilon}_\beta(\boldsymbol{k})$$
$$= \delta(\boldsymbol{k}-\boldsymbol{k}')\delta_{\alpha\beta}, \tag{18.26}$$

$$\begin{aligned}\langle \boldsymbol{x},i|\boldsymbol{x}',j\rangle &= \int d^3k\sum_\alpha\langle \boldsymbol{x},i|\boldsymbol{k},\alpha\rangle\langle \boldsymbol{k},\alpha|\boldsymbol{x}',j\rangle\\ &= \frac{1}{(2\pi)^3}\int d^3k\exp[\mathrm{i}\boldsymbol{k}\cdot(\boldsymbol{x}-\boldsymbol{x}')]\sum_\alpha\epsilon_\alpha^i(\boldsymbol{k})\epsilon_\alpha^j(\boldsymbol{k})\\ &= \frac{1}{(2\pi)^3}\int d^3k\ \exp[\mathrm{i}\boldsymbol{k}\cdot(\boldsymbol{x}-\boldsymbol{x}')]P_\perp^{ij}(\boldsymbol{k})\\ &= \frac{1}{(2\pi)^3}\int d^3k\ \exp[\mathrm{i}\boldsymbol{k}\cdot(\boldsymbol{x}-\boldsymbol{x}')]\left(\delta^{ij}-\frac{k^ik^j}{k^2}\right)\\ &= \delta_\perp^{ij}(\boldsymbol{x}-\boldsymbol{x}').\end{aligned} \tag{18.27}$$

Here the equation

$$\sum_{\alpha=1}^{2}\boldsymbol{\epsilon}_\alpha(\boldsymbol{k})\otimes\boldsymbol{\epsilon}_\alpha(\boldsymbol{k}) = \underline{1}-\hat{\boldsymbol{k}}\otimes\hat{\boldsymbol{k}} \tag{18.28}$$

has been used, cf. the decomposition of unity (4.14). Equation (18.27) defines the *transverse δ function*.

Energy-momentum tensor for the free Maxwell field

The Lagrange density for the free Maxwell field,

$$\mathcal{L} = -\frac{1}{4\mu_0}F_{\mu\nu}F^{\mu\nu}, \tag{18.29}$$

yields a canonical energy-momentum tensor

$$\Theta_\mu{}^\nu = \eta_\mu{}^\nu\mathcal{L} - \partial_\mu A_\lambda\frac{\partial\mathcal{L}}{\partial(\partial_\nu A_\lambda)} = \frac{1}{\mu_0}\left(\partial_\mu A_\lambda \cdot F^{\nu\lambda} - \frac{1}{4}\eta_\mu{}^\nu F_{\kappa\lambda}F^{\kappa\lambda}\right) \tag{18.30}$$

which is not gauge invariant. However, the free equation $\partial_\nu F^{\nu\lambda} = 0$ implies a trivial conservation law

$$-\frac{1}{\mu_0}\partial_\nu\left(\partial_\lambda A_\mu \cdot F^{\nu\lambda}\right) = 0 \tag{18.31}$$

which can be added to the conservation law for the free fields, $\partial_\nu\Theta_\mu{}^\nu = 0$. In this way we can improve the energy-momentum tensor $\Theta_\mu{}^\nu$ to a gauge invariant energy-momentum tensor

$$T_\mu{}^\nu = \Theta_\mu{}^\nu - \frac{1}{\mu_0}\partial_\lambda A_\mu \cdot F^{\nu\lambda} = \frac{1}{\mu_0}\left(F_{\mu\lambda}F^{\nu\lambda} - \frac{1}{4}\eta_\mu{}^\nu F_{\kappa\lambda}F^{\kappa\lambda}\right). \tag{18.32}$$

The corresponding energy-momentum density vector $\mathcal{P}_\mu = T_\mu{}^0/c$ yields the well known expressions for the energy and momentum densities of electromagnetic fields,

$$\mathcal{H} = -c\mathcal{P}_0 = -T_0{}^0 = \frac{\epsilon_0}{2}\boldsymbol{E}^2 + \frac{1}{2\mu_0}\boldsymbol{B}^2, \tag{18.33}$$

$$\boldsymbol{\mathcal{P}} = \epsilon_0\boldsymbol{E}\times\boldsymbol{B}. \tag{18.34}$$

The components of the energy current density (the Poynting vector) are given by the components $-cT_0{}^i$ (because $-\partial_t T_0{}^0 = -c\partial_0 T_0{}^0 = c\partial_i T_0{}^i$):

$$\boldsymbol{\mathcal{S}} = \frac{1}{\mu_0}\boldsymbol{E}\times\boldsymbol{B} = c^2\boldsymbol{\mathcal{P}}. \tag{18.35}$$

We are interested in the energy and momentum densities for the free fields $\boldsymbol{A}_D(\boldsymbol{x},t)$, because those will become the freely evolving field operators in the Dirac picture.

18.2 Photons

In the previous section we got rid of Φ and even of the longitudinal component of $\boldsymbol{A}$. Now we might be tempted to impose canonical commutation relations $[A_i(\boldsymbol{x},t),\Pi_j(\boldsymbol{x}',t)] \sim i\hbar\delta_{ij}\delta(\boldsymbol{x}-\boldsymbol{x}')$. However, this would be inconsistent, since equation (18.6) implies that application of $\partial/\partial x_i$ and summation over i on the left hand side would yield zero, but on the right hand side would not yield zero! This problem arises irrespective of whether we wish to quantize the full vector potential $\boldsymbol{A}$ or only the free vector potential $\boldsymbol{A}_D$. Therefore we have to invoke the transverse δ-function (18.27) to formulate the canonical commutation relations for the Maxwell field. We will use these relations primarily for the Dirac picture operators, but we omit the index D from now on,

$$\begin{aligned}[A_i(\boldsymbol{x},t),\dot{A}_j(\boldsymbol{x}',t)] &= \frac{i\hbar}{\epsilon_0(2\pi)^3}\int d^3\boldsymbol{k}\left(\delta_{ij}-\frac{k_ik_j}{\boldsymbol{k}^2}\right)\exp[i\boldsymbol{k}\cdot(\boldsymbol{x}-\boldsymbol{x}')]\\ &= \frac{i\hbar}{\epsilon_0(2\pi)^3}\int d^3\boldsymbol{k}\sum_{\alpha=1}^{2}\epsilon_{\alpha,i}(\boldsymbol{k})\epsilon_{\alpha,j}(\boldsymbol{k})\exp[i\boldsymbol{k}\cdot(\boldsymbol{x}-\boldsymbol{x}')],\end{aligned}$$

or in short form

$$[A_i(\boldsymbol{x},t),\dot{A}_j(\boldsymbol{x}',t)] = \frac{i\hbar}{\epsilon_0}\delta_{ij}^{\perp}(\boldsymbol{x}-\boldsymbol{x}'). \tag{18.36}$$

This equation can also be written using the zero energy Green's function $G(\boldsymbol{x}-\boldsymbol{x}') = \langle\boldsymbol{x}|G(0)|\boldsymbol{x}'\rangle$ (cf. equations (J.18, J.19) and (J.24)),

$$[A_i(\boldsymbol{x},t),\dot{A}_j(\boldsymbol{x}',t)] = \frac{i\hbar}{\epsilon_0}\Big(\delta_{ij}\delta(\boldsymbol{x}-\boldsymbol{x}')+\partial_i\partial_jG(\boldsymbol{x}-\boldsymbol{x}')\Big), \tag{18.37}$$

and using (18.11) we find

$$\begin{aligned}[\boldsymbol{A}(\boldsymbol{x},t)\overset{\otimes}{,}\dot{\boldsymbol{A}}(\boldsymbol{x}',t)] &= \frac{i\hbar}{\epsilon_0}\left(\frac{2}{3}\underline{1}\,\delta(\boldsymbol{x}-\boldsymbol{x}')\right.\\ &\quad\left.+\frac{3(\boldsymbol{x}-\boldsymbol{x}')\otimes(\boldsymbol{x}-\boldsymbol{x}')-|\boldsymbol{x}-\boldsymbol{x}'|^2\underline{1}}{4\pi|\boldsymbol{x}-\boldsymbol{x}'|^5}\right).\end{aligned} \tag{18.38}$$

The remaining commutation relations are

$$[A_i(\boldsymbol{x},t),A_j(\boldsymbol{x}',t)] = 0, \quad [\dot{A}_i(\boldsymbol{x},t),\dot{A}_j(\boldsymbol{x}',t)] = 0. \tag{18.39}$$

The relations (18.36, 18.39) yield for the operators $a_\alpha(\boldsymbol{k})$, $a_\beta^+(\boldsymbol{k}')$ harmonic oscillator relations,

$$[a_\alpha(\boldsymbol{k}),a_\beta(\boldsymbol{k}')]=0, \quad [a_\alpha^+(\boldsymbol{k}),a_\beta^+(\boldsymbol{k}')]=0, \quad [a_\alpha(\boldsymbol{k}),a_\beta^+(\boldsymbol{k}')] = \delta_{\alpha\beta}\delta(\boldsymbol{k}-\boldsymbol{k}').$$

The prefactor in (18.20) was chosen such that no extra factor appears in the commutation relation of $a_\alpha(\boldsymbol{k})$ and $a^+_\beta(\boldsymbol{k}')$.

The energy and momentum densities (18.33) and (18.34) yield energy and momentum operators

$$H = \int d^3x \left(\frac{\epsilon_0}{2}\boldsymbol{E}^2 + \frac{1}{2\mu_0}\boldsymbol{B}^2\right) = \sum_\alpha \int d^3k\, \hbar ck\, a^+_\alpha(\boldsymbol{k})a_\alpha(\boldsymbol{k}), \qquad (18.40)$$

$$\boldsymbol{P} = \epsilon_0 \int d^3x\, \boldsymbol{E}\times\boldsymbol{B} = \sum_\alpha \int d^3k\, \hbar\boldsymbol{k}\, a^+_\alpha(\boldsymbol{k})a_\alpha(\boldsymbol{k}). \qquad (18.41)$$

From these expressions we can infer by the meanwhile standard methods that $a^+_\alpha(\boldsymbol{k})$ creates a photon of momentum $\hbar\boldsymbol{k}$, energy $\hbar ck$ and polarization $\boldsymbol{\epsilon}_\alpha(\boldsymbol{k})$, while $a_\alpha(\boldsymbol{k})$ annihilates such a photon. In particular,

$$|\boldsymbol{k},\alpha\rangle = a^+_\alpha(\boldsymbol{k})|0\rangle$$

is a single photon state with momentum $\hbar\boldsymbol{k}$, energy $\hbar ck$ and polarization $\boldsymbol{\epsilon}_\alpha(\boldsymbol{k})$.

In writing the $\boldsymbol{k}$ space integrals for H and $\boldsymbol{P}$ we have used the prescription of *normal ordering*, i.e. writing the creation operators on the left side of the annihilation operators. This ensures that vacuum expectation values of charges and currents vanish. Many authors like to explicitly indicate normal ordering for $\boldsymbol{x}$ space representations of charges or currents through double colons, e.g. for equation (18.40) this would read

$$H = : \int d^3x \left(\frac{\epsilon_0}{2}\boldsymbol{E}^2 + \frac{1}{2\mu_0}\boldsymbol{B}^2\right) : = \sum_\alpha \int d^3k\, \hbar ck\, a^+_\alpha(\boldsymbol{k})a_\alpha(\boldsymbol{k}).$$

We will not use the double colon notation and instead use the implicit convention that charges and currents have to be normal ordered in terms of creation and annihilation operators.

If we want to construct a creation operator $a^+_\alpha(\boldsymbol{x})$ in $\boldsymbol{x}$ space (corresponding to the operator $\psi^+(\boldsymbol{x})$ in Schrödinger theory) we find

$$\begin{aligned} a^+_\alpha(\boldsymbol{x}) &= \frac{1}{\sqrt{2\pi}^3}\int d^3k\, a^+_\alpha(\boldsymbol{k})\exp(-\mathrm{i}\boldsymbol{k}\cdot\boldsymbol{x}) \\ &= \frac{1}{(2\pi)^3}\int d^3k \int d^3x'\, \boldsymbol{\epsilon}_\alpha(\boldsymbol{k})\cdot\left(\sqrt{\frac{k}{2\mu_0 c\hbar}}\boldsymbol{A}(\boldsymbol{x}',t)\right. \\ &\quad \left. - \frac{\mathrm{i}}{\sqrt{2k\mu_0 c^3\hbar}}\dot{\boldsymbol{A}}(\boldsymbol{x}',t)\right)\exp[\mathrm{i}\boldsymbol{k}\cdot(\boldsymbol{x}'-\boldsymbol{x}) - \mathrm{i}ckt]. \end{aligned}$$

The expression on the right hand side is time-independent and can just as well be written in terms of the Schrödinger picture operators $\boldsymbol{A}(\boldsymbol{x}) = \boldsymbol{A}(\boldsymbol{x}, 0)$ and $\dot{\boldsymbol{A}}(\boldsymbol{x}) = \dot{\boldsymbol{A}}(\boldsymbol{x}, 0)$. However, the important observation is that contrary to Schrödinger theory, the original operator in $\boldsymbol{x}$ space, $\boldsymbol{A}(\boldsymbol{x}, t)$, is not a pure annihilation or creation operator any more, but instead is a superposition of annihilation and creation operators. This is a generic feature of relativistic field operators. The property $a^+(\boldsymbol{x}) = \psi^+(\boldsymbol{x})$ is a special feature of the non-relativistic Schrödinger field. It is because of this feature of the Schrödinger field that we did not have to use an explicit double colon notation or an implicit agreement to use normal ordering in Schrödinger field theory. Normal ordered expressions in $\boldsymbol{x}$ space were automatically normal ordered in $\boldsymbol{k}$ space.

The time evolution of the free photon operators in $\boldsymbol{k}$ space is given by the standard Heisenberg evolution equations,

$$a_{D\alpha}(\boldsymbol{k}, t) = a_\alpha(\boldsymbol{k}) \exp(-ickt) = \exp\!\left(\frac{\mathrm{i}}{\hbar} Ht\right) a_\alpha(\boldsymbol{k}) \exp\!\left(-\frac{\mathrm{i}}{\hbar} Ht\right), \quad (18.42)$$

$$\frac{\partial}{\partial t} a_{D\alpha}(\boldsymbol{k}, t) = \frac{\mathrm{i}}{\hbar}[H, a_{D\alpha}(\boldsymbol{k}, t)], \quad (18.43)$$

and therefore we also have for the field operators in $\boldsymbol{x}$ space

$$\boldsymbol{A}(\boldsymbol{x}, t) = \exp\!\left(\frac{\mathrm{i}}{\hbar} Ht\right) \boldsymbol{A}(\boldsymbol{x}) \exp\!\left(-\frac{\mathrm{i}}{\hbar} Ht\right), \quad (18.44)$$

but to recover the evolution equation (18.16) in $\boldsymbol{x}$ space we have to use iterated Heisenberg evolution equations,

$$\frac{\partial}{\partial t} \boldsymbol{A}(\boldsymbol{x}, t) = \frac{\mathrm{i}}{\hbar}[H, \boldsymbol{A}(\boldsymbol{x}, t)] = -\boldsymbol{E}(\boldsymbol{x}, t), \quad (18.45)$$

$$\frac{\partial^2}{\partial t^2} \boldsymbol{A}(\boldsymbol{x}, t) = -\frac{\mathrm{i}}{\hbar}[H, \boldsymbol{E}(\boldsymbol{x}, t)] = -\frac{1}{\hbar^2}[H, [H, \boldsymbol{A}(\boldsymbol{x}, t)]]. \quad (18.46)$$

This is a general property of bosonic relativistic fields.

18.3 Coherent states of the electromagnetic field

We can directly apply what we have learned about coherent oscillator states to construct a quantum state with the property that the operator $\boldsymbol{E}(\boldsymbol{x}, t)$ (18.22) yields a classical electromagnetic wave as expectation value,

$$\begin{aligned}\langle\mathcal{E}|\boldsymbol{E}(\boldsymbol{x},t)|\mathcal{E}\rangle &= \mathcal{E}(\boldsymbol{x},t)\\ &= \mathrm{i}\sqrt{\frac{\hbar\mu_0 c^3}{(2\pi)^3}}\int d^3\boldsymbol{k}\,\sqrt{\frac{k}{2}}\sum_{\alpha=1}^{2}\boldsymbol{\epsilon}_\alpha(\boldsymbol{k})\Big(\zeta_\alpha(\boldsymbol{k})\exp[\mathrm{i}(\boldsymbol{k}\cdot\boldsymbol{x}-ckt)]\\ &\quad -\zeta_\alpha^+(\boldsymbol{k})\exp[-\mathrm{i}(\boldsymbol{k}\cdot\boldsymbol{x}-ckt)]\Big).\end{aligned} \tag{18.47}$$

The results of Section 6.5 imply that the state $|\mathcal{E}\rangle$ can be unitarily generated out of the vacuum[1],

$$|\mathcal{E}\rangle = \exp\!\left(\int d^3\boldsymbol{k}\sum_{\alpha=1}^{2}\big(\zeta_\alpha(\boldsymbol{k})a_\alpha^+(\boldsymbol{k})-\zeta_\alpha^+(\boldsymbol{k})a_\alpha(\boldsymbol{k})\big)\right)|0\rangle.$$

The corresponding equations in the Schrödinger picture are

$$\langle\mathcal{E}(t)|\boldsymbol{E}(\boldsymbol{x})|\mathcal{E}(t)\rangle = \mathcal{E}(\boldsymbol{x},t),$$

$$|\mathcal{E}(t)\rangle = \exp\!\left(\int d^3\boldsymbol{k}\sum_{\alpha=1}^{2}\big(\zeta_\alpha(\boldsymbol{k},t)a_\alpha^+(\boldsymbol{k})-\zeta_\alpha^+(\boldsymbol{k},t)a_\alpha(\boldsymbol{k})\big)\right)|0\rangle, \tag{18.48}$$

with

$$\zeta_\alpha(\boldsymbol{k},t) = \zeta_\alpha(\boldsymbol{k})\exp(-\mathrm{i}ckt).$$

The average photon number in the electromagnetic wave is

$$\langle n\rangle = \langle\mathcal{E}|\int d^3\boldsymbol{k}\sum_{\alpha=1}^{2}a_\alpha^+(\boldsymbol{k})a_\alpha(\boldsymbol{k})|\mathcal{E}\rangle = \int d^3\boldsymbol{k}\sum_{\alpha=1}^{2}|\zeta_\alpha(\boldsymbol{k})|^2$$

and we find with

$$\begin{aligned}\left(\int d^3\boldsymbol{k}\sum_{\alpha=1}^{2}a_\alpha^+(\boldsymbol{k})a_\alpha(\boldsymbol{k})\right)^2 &= \int d^3\boldsymbol{k}\int d^3\boldsymbol{k}'\sum_{\alpha,\alpha'=1}^{2}a_\alpha^+(\boldsymbol{k})a_{\alpha'}^+(\boldsymbol{k}')a_{\alpha'}(\boldsymbol{k}')a_\alpha(\boldsymbol{k})\\ &\quad+\int d^3\boldsymbol{k}\sum_{\alpha=1}^{2}a_\alpha^+(\boldsymbol{k})a_\alpha(\boldsymbol{k})\end{aligned}$$

the relations

$$\langle n^2\rangle = \left(\int d^3\boldsymbol{k}\sum_{\alpha=1}^{2}|\zeta_\alpha(\boldsymbol{k})|^2\right)^2 + \int d^3\boldsymbol{k}\sum_{\alpha=1}^{2}|\zeta_\alpha(\boldsymbol{k})|^2,$$

[1] R.J. Glauber, Phys. Rev. 131, 2766 (1963).

$$\Delta n = \sqrt{\langle n\rangle} = \left(\int d^3k \sum_{\alpha=1}^{2} |\zeta_\alpha(\boldsymbol{k})|^2\right)^{1/2}, \quad \frac{\Delta n}{\langle n\rangle} = \frac{1}{\sqrt{\langle n\rangle}}.$$

Every free quantum field has an expansion in terms of oscillator operators, and therefore each quantum field has coherent states which yield classical expectation values for the field. They are particularly important for electromagnetic fields because classical electromagnetic waves are so abundant, readily available, and of technical relevance. This is a consequence of boson statistics and of the vanishing mass and charge of photons. Generating and packing together huge numbers of photons is very inexpensive in terms of energy.

18.4 Photon coupling to relative motion

The discussion of photon interactions with atoms or molecules usually does not involve discussions of photon interactions with the individual constituent electrons and nuclei, but assumes either an effective coupling to the quasiparticles which describe the relative motion between nuclei and electrons, or otherwise assumes coupling of the photons to only one kind of particle in a many particle system. A shortcut justification e.g. for assuming that photons should primarily couple to electrons rather than nuclei in atoms is that physical intuition would indicate that an electromagnetic wave should shake a lighter particle more easily than a heavier particle. Indeed, when we calculate the cross section for photon scattering off free charged particles in Section 22.3, we will find that scattering of low energy photons is suppressed with the mass of the scattering particle like m^{-2}, and for high energy photons like m^{-1}. Furthermore, atomic matrix elements for optical dipole transitions scale like m^{-1} which also indicates a preference for coupling to the lighter charged components in a composite system. However, the intuition can be misleading. A simple counterexample is provided e.g. by the absorption or emission of infrared photons by molecules. The dominant degrees of freedom which undergo transitions in these cases are molecular vibrations or rotations, but not electronic transitions, i.e. the dominant photon-matter interaction for infrared photons concerns coupling to clusters of ions or atomic nuclei. What this means is that we have to make judicious calls on which terms in a quantum electronics or quantum electrodynamics Hamiltonian will make the dominant contributions to photon interactions, depending on the photon energy range that we are interested in and the available atomic or molecular transitions. However, if we use standard atomic orbitals to model the states of the unperturbed matter system, then this does imply an approximation of photon coupling to the quasiparticle which describes relative motion, or to the electrons as the lightest charges. It is therefore instructive to revisit the problem of separation of center of mass motion and relative motion in the presence of electromagnetic fields.

The two-particle Hamiltonian (7.1) with the electromagnetic vector potentials included takes the form

$$H = \frac{1}{2m_1}\left(\mathbf{p}_1 - q_1\boldsymbol{A}(\mathbf{x}_1,t)\right)^2 + \frac{1}{2m_2}\left(\mathbf{p}_2 - q_2\boldsymbol{A}(\mathbf{x}_2,t)\right)^2 + V(|\mathbf{x}_1 - \mathbf{x}_2|). \tag{18.49}$$

Substitution of the single particle momenta with the total momentum and the effective momentum in the relative motion (7.6) yields

$$H = \frac{1}{2M}\left(\mathbf{P} - q_1\boldsymbol{A}(\mathbf{x}_1,t) - q_2\boldsymbol{A}(\mathbf{x}_2,t)\right)^2 + \frac{1}{2\mu}\left(\mathbf{p} - \frac{m_2 q_1\boldsymbol{A}(\mathbf{x}_1,t) - m_1 q_2\boldsymbol{A}(\mathbf{x}_2,t)}{M}\right)^2 + V(|\mathbf{x}_1 - \mathbf{x}_2|). \tag{18.50}$$

Now we assume that the electromagnetic potentials vary weakly over the extension of the two-particle system,

$$\boldsymbol{A}(\boldsymbol{x}_1,t) \simeq \boldsymbol{A}(\boldsymbol{x}_2,t) \simeq \boldsymbol{A}(\boldsymbol{R},t). \tag{18.51}$$

This yields an effective Hamiltonian

$$H = \frac{1}{2M}\left(\mathbf{P} - Q\boldsymbol{A}(\mathbf{R},t)\right)^2 + \frac{1}{2\mu}\left(\mathbf{p} - q\boldsymbol{A}(\mathbf{R},t)\right)^2 + V(|\mathbf{r}|), \tag{18.52}$$

with the total charge $Q = q_1 + q_2$ in the kinetic term for center of mass motion, and a reduced charge in the quasiparticle kinetic term,

$$q = \frac{m_2 q_1 - m_1 q_2}{m_1 + m_2}, \tag{18.53}$$

with inversions

$$q_1 = \frac{m_1}{M}Q + q, \quad q_2 = \frac{m_2}{M}Q - q.$$

The equations of motion of the classical two-particle system are in the approximations $\boldsymbol{B}(\boldsymbol{x}_1,t) \simeq \boldsymbol{B}(\boldsymbol{x}_2,t) \simeq \boldsymbol{B}(\boldsymbol{R},t)$, (same for $\boldsymbol{E}$) given by

$$M\ddot{\boldsymbol{R}} = (Q\dot{\boldsymbol{R}} + q\dot{\boldsymbol{r}}) \times \boldsymbol{B} + Q\boldsymbol{E}, \tag{18.54}$$

$$\mu\ddot{\boldsymbol{r}} = \left(q\dot{\boldsymbol{R}} + \frac{\mu Q + \sqrt{M(M-4\mu)}q}{M}\dot{\boldsymbol{r}}\right) \times \boldsymbol{B} + q\boldsymbol{E} - \frac{\partial V}{\partial \boldsymbol{r}}. \tag{18.55}$$

Even the simplifying assumption (18.51) does not allow for separation of the center of mass motion any more, and the equations do not separate in terms of center of mass and relative coordinates, nor in total and reduced masses or charges. However, equations (18.54, 18.55) show that the coupling of photons to the center of mass motion is suppressed with inverse total mass. Therefore the impact of the photons on the relative motion dominates over the impact on center of mass motion, and in leading order we are left with an effective single-particle Hamiltonian for the relative motion in the center of mass frame,

$$H = \frac{1}{2\mu}\left(\mathbf{p} - q\boldsymbol{A}(t)\right)^2 + V(|\mathbf{r}|). \tag{18.56}$$

The corresponding statement at the classical level (18.54, 18.55) is that in leading order of μ/M, the center of mass frame is preserved and we have an effective single particle problem for relative motion,

$$\mu\ddot{\boldsymbol{r}} = q\dot{\boldsymbol{r}} \times \boldsymbol{B} + q\boldsymbol{E} - \frac{\partial V}{\partial \boldsymbol{r}}.$$

Equation (18.53) for the effective charge q yields $q = q_1$ if $q_2 = -q_1$, and $m_2 \gg m_1$ also implies $q \simeq q_1$. This entails that in atoms or molecules, we can think of photons as effectively coupling to the electrons if the photon wavelengths hc/E_γ are large compared to the size of the atoms or molecules. An alternative justification of using effective single particle Hamiltonians like (18.56) for photon interactions with bound systems is therefore also to discard the contribution for the heavier particle with mass m_2 in the original two-particle Hamiltonian (18.49), but still assume the bound states $|n, \ell, m_\ell\rangle$ which were derived for the relative motion $\boldsymbol{r} = \boldsymbol{x}_1 - \boldsymbol{x}_2$ to hold for the coordinate $\boldsymbol{x}_1$ of the lighter particle. This is an equivalent approximation up to correction terms of the same order $m_1/m_2 \simeq \mu/M$. However, the approximation clearly becomes invalid if transitions for the lighter particles are prohibited either by selection rules or by absence of suitable energy levels.

The derivation of (18.56) required negligible spatial variation of the photon terms over the extension of the unperturbed wave functions for relative motion, to justify minimal photon coupling into the Hamiltonian for relative motion. Dipole approximation and minimal electromagnetic coupling to the Hamiltonian for relative motion in a bound system therefore use the same premise.

18.5 Energy-momentum densities and time evolution in quantum optics

Further discussions of photon-matter interactions and of time evolution of the quantized Maxwell field require the Hamiltonian and momentum operators for coupled electromagnetic and matter fields. The study of electromagnetic interactions with non-relativistic matter fields is the domain of quantum optics or quantum electronics.

A Lagrange density for coupled electromagnetic and non-relativistic matter fields is

$$\begin{aligned}\mathcal{L} = \sum_a \Bigg[& \frac{\mathrm{i}\hbar}{2}\left(\Psi_a^+ \cdot \frac{\partial \Psi_a}{\partial t} - \frac{\partial \Psi_a^+}{\partial t} \cdot \Psi_a\right) - q_a \Psi_a^+ \Phi \Psi_a - \frac{\hbar^2}{2m_a} \nabla \Psi_a^+ \cdot \nabla \Psi_a \\ & - \mathrm{i}\frac{q_a \hbar}{2m_a} \boldsymbol{A} \cdot \left(\Psi_a^+ \overset{\leftrightarrow}{\nabla} \Psi_a\right) - \frac{q_a^2}{2m_a} \Psi_a^+ \boldsymbol{A}^2 \Psi_a \Bigg] - \frac{1}{4\mu_0} F_{\mu\nu} F^{\mu\nu}.\end{aligned} \tag{18.57}$$

Here $\Phi = -cA_0$ is the electric potential, and we use the definition of an alternating derivative operator $\psi^+ \overleftrightarrow{\nabla} \psi \equiv \psi^+ \cdot \nabla\psi - \nabla\psi^+ \cdot \psi$.

The summation over a refers to different kinds of non-relativistic particles (e.g. electrons, protons etc.), and a summation over spin labels is implicitly understood.

Phase invariance yields the electric charge and current densities

$$\varrho = j^0/c = \sum_a q_a \Psi_a^+ \Psi_a, \tag{18.58}$$

$$\boldsymbol{j} = \sum_a \frac{q_a}{2\mathrm{i}m_a} \left(\hbar \Psi_a^+ \overleftrightarrow{\nabla} \Psi_a - 2\mathrm{i}q_a \Psi_a^+ \boldsymbol{A} \Psi_a \right). \tag{18.59}$$

Like the energy-momentum tensor (18.30), the canonical energy-momentum tensor following from the Lagrange density (18.57) according to the general result (16.16),

$$\Theta_\mu{}^\nu = \eta_\mu{}^\nu \mathcal{L} + \frac{1}{\mu_0} \partial_\mu A_\lambda \cdot F^{\nu\lambda} - \sum_a \left(\partial_\mu \Psi_a \frac{\partial \mathcal{L}}{\partial(\partial_\nu \Psi_a)} + \partial_\mu \Psi_a^+ \frac{\partial \mathcal{L}}{\partial(\partial_\nu \Psi_a^+)} \right),$$

is not gauge invariant. Just like in the case of the free Maxwell field, we can cure this by adding the trivially conserved tensor

$$\delta\Theta_\mu{}^\nu = -\frac{1}{\mu_0} \partial_\lambda (A_\mu F^{\nu\lambda}) = -A_\mu j^\nu - \frac{1}{\mu_0} \partial_\lambda A_\mu \cdot F^{\nu\lambda}, \quad \partial_\nu \delta\Theta_\mu{}^\nu \equiv 0.$$

The improved energy-momentum tensor $t_\mu{}^\nu = \Theta_\mu{}^\nu + \delta\Theta_\mu{}^\nu$ yields in particular the gauge invariant energy density for quantum optics,

$$\begin{aligned} \mathcal{H} = -t_0{}^0 = \frac{\epsilon_0}{2} \boldsymbol{E}^2 + \frac{1}{2\mu_0} \boldsymbol{B}^2 + \sum_a \frac{1}{2m_a} \Big[\hbar^2 \nabla \Psi_a^+ \cdot \nabla \Psi_a \\ + \mathrm{i}q_a \hbar \boldsymbol{A} \cdot \left(\Psi_a^+ \overleftrightarrow{\nabla} \Psi_a \right) + q_a^2 \Psi_a^+ \boldsymbol{A}^2 \Psi_a \Big], \end{aligned} \tag{18.60}$$

and the gauge invariant momentum density,

$$\mathcal{P} = \frac{1}{c} t_i^0 \boldsymbol{e}_i = \epsilon_0 \boldsymbol{E} \times \boldsymbol{B} + \frac{1}{2\mathrm{i}} \sum_a \left(\hbar \Psi_a^+ \overleftrightarrow{\nabla} \Psi_a - 2\mathrm{i}q_a \Psi_a^+ \boldsymbol{A} \Psi_a \right). \tag{18.61}$$

In materials science it is convenient to explicitly disentangle the contributions from Coulomb and photon terms in Coulomb gauge $\nabla \cdot \boldsymbol{A} = 0$. We split the electric field components in Coulomb gauge according to

$$\boldsymbol{E}_\parallel = -\nabla\Phi \tag{18.62}$$

and

$$\boldsymbol{E}_\perp = -\frac{\partial \boldsymbol{A}}{\partial t}. \tag{18.63}$$

The equation for the electrostatic potential decouples from the vector potential,

$$\Delta\Phi = -\frac{1}{\epsilon_0}\sum_a q_a \Psi_a^+ \Psi_a,$$

and is solved by

$$\Phi(\boldsymbol{x},t) = \frac{1}{4\pi\epsilon_0}\int d^3x' \sum_a \frac{q_a}{|\boldsymbol{x}-\boldsymbol{x}'|}\Psi_a^+(\boldsymbol{x}',t)\Psi_a(\boldsymbol{x}',t).$$

Furthermore, the two components of the electric field are orthogonal in the Coulomb gauge,

$$\begin{aligned}\int d^3x\, \boldsymbol{E}_\parallel(\boldsymbol{x},t)\cdot\boldsymbol{E}_\perp(\boldsymbol{x},t) &= \int d^3k\, \boldsymbol{E}_\parallel(\boldsymbol{k},t)\cdot\boldsymbol{E}_\perp(-\boldsymbol{k},t) \\ &= -\int d^3x\, \Phi(\boldsymbol{x},t)\frac{\partial}{\partial t}\nabla\cdot\boldsymbol{A}(\boldsymbol{x},t) = 0, \end{aligned} \tag{18.64}$$

and the contribution from $\boldsymbol{E}_\parallel$ to the Hamiltonian generates the Coulomb potentials

$$\begin{aligned} H_C &= \frac{\epsilon_0}{2}\int d^3x\, \boldsymbol{E}_\parallel^2(\boldsymbol{x},t) = -\frac{\epsilon_0}{2}\int d^3x\, \Phi(\boldsymbol{x},t)\Delta\Phi(\boldsymbol{x},t) \\ &= \frac{1}{2}\int d^3x\, \Phi(\boldsymbol{x},t)\varrho(\boldsymbol{x},t) \\ &= \sum_{aa'}\int d^3x \int d^3x'\, q_a q_{a'}\frac{\Psi_a^+(\boldsymbol{x},t)\Psi_{a'}^+(\boldsymbol{x}',t)\Psi_{a'}(\boldsymbol{x}',t)\Psi_a(\boldsymbol{x},t)}{8\pi\epsilon_0|\boldsymbol{x}-\boldsymbol{x}'|}, \end{aligned} \tag{18.65}$$

where the ordering of the field operators was performed to ensure correct expectation values for the interaction energy of 2-particle states after second quantization. The summation may also implicitly include spinor indices.

The resulting Hamiltonian in Coulomb gauge therefore has the form

$$\begin{aligned} H = \int d^3x \Bigg(&\sum_a \frac{1}{2m_a}\Big[\hbar^2\nabla\Psi_a^+(\boldsymbol{x},t)\cdot\nabla\Psi_a(\boldsymbol{x},t) \\ &+ \mathrm{i}q_a\hbar\boldsymbol{A}(\boldsymbol{x},t)\cdot\left(\Psi_a^+(\boldsymbol{x},t)\overset{\leftrightarrow}{\nabla}\Psi_a(\boldsymbol{x},t)\right) + q_a^2\Psi_a^+(\boldsymbol{x},t)\boldsymbol{A}^2(\boldsymbol{x},t)\Psi_a(\boldsymbol{x},t)\Big] \\ &+ \frac{\epsilon_0}{2}\boldsymbol{E}_\perp^2(\boldsymbol{x},t) + \frac{1}{2\mu_0}\boldsymbol{B}^2(\boldsymbol{x},t)\Bigg) \\ &+ \sum_{aa'}\int d^3x \int d^3x'\, q_a q_{a'}\frac{\Psi_a^+(\boldsymbol{x},t)\Psi_{a'}^+(\boldsymbol{x}',t)\Psi_{a'}(\boldsymbol{x}',t)\Psi_a(\boldsymbol{x},t)}{8\pi\epsilon_0|\boldsymbol{x}-\boldsymbol{x}'|}. \end{aligned} \tag{18.66}$$

The momentum operator in Coulomb gauge follows from (18.61) and

$$\int d^3x\,\epsilon_0\boldsymbol{E}_{\|}\times\boldsymbol{B} = -\int d^3x\,\epsilon_0\Phi\Delta\boldsymbol{A} = \int d^3x\,\varrho\boldsymbol{A} = \int d^3x\sum_a q_a\Psi_a^+\boldsymbol{A}\Psi_a$$

as

$$\boldsymbol{P} = \int d^3x\left(\frac{\hbar}{\mathrm{i}}\sum_a\Psi_a^+\nabla\Psi_a + \epsilon_0\boldsymbol{E}_\perp\times\boldsymbol{B}\right). \tag{18.67}$$

Recall that Heisenberg or Schrödinger picture field operators satisfy the same canonical commutation relations as the Dirac picture operators because the quantum pictures are related by unitary transformations. For the vector potential $\boldsymbol{A}(\boldsymbol{x},t)$ in Coulomb gauge this implies the same commutation relations (18.36, 18.39) as for the Dirac picture vector potential. The Hamiltonian (18.66) then yields the Schrödinger equations for the matter fields from

$$\mathrm{i}\hbar\frac{\partial}{\partial t}\Psi(\boldsymbol{x},t) = [\Psi(\boldsymbol{x},t),H], \tag{18.68}$$

and the electromagnetic wave equation in Coulomb gauge (18.10) from

$$\mathrm{i}\hbar\frac{\partial}{\partial t}\boldsymbol{A}(\boldsymbol{x},t) = [\boldsymbol{A}(\boldsymbol{x},t),H], \quad \frac{\partial^2}{\partial t^2}\boldsymbol{A}(\boldsymbol{x},t) = \frac{1}{\hbar^2}[H,[\boldsymbol{A}(\boldsymbol{x},t),H]]. \tag{18.69}$$

These relations imply that also after quantization of the Maxwell field, field operators in the Heisenberg and Schrödinger pictures are still related according to

$$\boldsymbol{A}(\boldsymbol{x},t) = \exp(\mathrm{i}Ht/\hbar)\boldsymbol{A}(\boldsymbol{x})\exp(-\mathrm{i}Ht/\hbar), \tag{18.70}$$

and the derivation of scattering matrix elements with the automatic emergence of the interaction picture proceeds exactly as in the previous cases of quantum mechanics and non-relativistic quantum field theory,

$$S_{fi} = \langle f|U_D(\infty,-\infty)|i\rangle, \tag{18.71}$$

$$\begin{aligned} U_D(t,t') &= \exp\!\left(\frac{\mathrm{i}}{\hbar}H_0t\right)\exp\!\left(-\frac{\mathrm{i}}{\hbar}H(t-t')\right)\exp\!\left(-\frac{\mathrm{i}}{\hbar}H_0t'\right) \\ &= \mathrm{T}\exp\!\left(-\frac{\mathrm{i}}{\hbar}\int_{t'}^{t}d\tau\,H_D(\tau)\right), \end{aligned} \tag{18.72}$$

$$H_D(t) = \exp\!\left(\frac{\mathrm{i}}{\hbar}H_0t\right)V\exp\!\left(-\frac{\mathrm{i}}{\hbar}H_0t\right), \tag{18.73}$$

where the identification of $H_0 = H - V$ depends on what part of H we can solve and what part we wish to take into account through perturbation theory. I.e. we find the same basic structure of time-dependent perturbation theory in terms of Hamilton operators also after introduction of the relativistic photon operators. We will see in Chapters 21 and 22 that this property persists in general in quantum field theory also after introduction of other relativistic field operators.

18.6 Photon emission rates

The calculation of transition probabilities between Fock states requires time-dependent perturbation theory in the second quantized formalism.

The relevant part of the Hamiltonian (18.66) for a coupled system of non-relativistic charged particles and photons is

$$\begin{aligned} H &= H_0 + H_I + H_{II} \\ &= \int d^3\boldsymbol{x}\left(\frac{\hbar^2}{2m}\sum_\sigma \nabla\psi_\sigma^+\cdot\nabla\psi_\sigma + \sum_\sigma \psi_\sigma^+ V\psi_\sigma + \frac{\epsilon_0}{2}\dot{\boldsymbol{A}}^2 + \frac{1}{2\mu_0}(\nabla\times\boldsymbol{A})^2\right. \\ &\quad \left. + \mathrm{i}\frac{q\hbar}{2m}\boldsymbol{A}\cdot\sum_\sigma\left(\psi_\sigma^+\overleftrightarrow{\nabla}\psi_\sigma\right) + \frac{q^2}{2m}\sum_\sigma \psi_\sigma^+\boldsymbol{A}^2\psi_\sigma\right), \end{aligned} \tag{18.74}$$

where V is an intra-atomic or intra-molecular potential and the interaction terms between the charged particles and the photons are

$$H_I = \int d^3\boldsymbol{x}\sum_\sigma \mathrm{i}\frac{q\hbar}{2m}\boldsymbol{A}\cdot\left(\psi_\sigma^+\overleftrightarrow{\nabla}\psi_\sigma\right), \quad H_{II} = \int d^3\boldsymbol{x}\,\frac{q^2}{2m}\sum_\sigma \psi_\sigma^+\boldsymbol{A}^2\psi_\sigma.$$

Here we explicitly included the spin summations and wrote the Hamiltonian in terms of the Schrödinger picture field operators ($\dot{\boldsymbol{A}}(\boldsymbol{x}) \equiv \dot{\boldsymbol{A}}(\boldsymbol{x}, 0)$). In principle there is also the electrostatic repulsion between the particles,

$$H_C = \frac{q^2}{8\pi\epsilon_0}\int d^3\boldsymbol{x}\int d^3\boldsymbol{x}'\sum_{\sigma,\sigma'}\psi_\sigma^+(\boldsymbol{x})\psi_{\sigma'}^+(\boldsymbol{x}')\frac{1}{|\boldsymbol{x}-\boldsymbol{x}'|}\psi_{\sigma'}(\boldsymbol{x}')\psi_\sigma(\boldsymbol{x}).$$

However, we will only study transitions with single matter particles in the initial state, where H_C will not contribute.

For the following calculations we use hydrogen states as an example to illustrate the method, and we use $\psi_\sigma(\boldsymbol{x})$ and $\psi_\sigma^+(\boldsymbol{x})$ as the Schrödinger picture field operators of the effective quasiparticle which describes relative motion of the proton and electron in the atom,

$$|n,\ell,m_\ell,\sigma;t'\rangle = |\Psi_{n,\ell,m_\ell,\sigma}(t')\rangle = \int d^3x\,\Psi_{n,\ell,m_\ell}(x,t')\psi_\sigma^+(x)|0\rangle$$
$$= \exp(-\mathrm{i}E_{n,\ell}t'/\hbar)\int d^3x\,\Psi_{n,\ell,m_\ell}(x)\psi_\sigma^+(x)|0\rangle$$
$$= \int d^3x\,\Psi_{n,\ell,m_\ell}(x)\exp(-\mathrm{i}H_0t'/\hbar)\psi_\sigma^+(x)|0\rangle, \qquad (18.75)$$

i.e. $\psi_\sigma(x)$ and $\psi_\sigma^+(x)$ are the Schrödinger picture field operators which arise from quantization of the wave function $\langle x,\sigma|\Psi(t)\rangle$ in Schrödinger's wave mechanics. See Problem 18.7 for the question why the state (18.75) is an eigenstate of H_0.

According to our results from Section 18.4, the Hamiltonian (18.74) includes an approximation if we use it for coupling the electromagnetic potential to the hydrogen atom, because we introduced the photon operators through minimal coupling into the effective single particle problem that resulted from separation of the center of mass motion. This is a good approximation if the electromagnetic potentials vary only weakly over the size of the atom, $A(x_p,t)\simeq A(x_e,t)$. Indeed, it is an excellent approximation for the study of transitions between bound hydrogen states, because in these cases $\lambda > hc/13.6\,\mathrm{eV} = 91\,\mathrm{nm}$.

We wish to calculate the photon emission rate, i.e. the transition rate from the initial state (18.75) into a final state with the electron in another atomic state and a photon with momentum $\hbar k$ and polarization $\epsilon_\alpha(k)$,

$$|n',\ell',m'_\ell,\sigma';k,\alpha;t\rangle = \int d^3x\,\Psi_{n',\ell',m'_\ell}(x)\exp(-\mathrm{i}H_0t/\hbar)\psi_{\sigma'}^+(x)a_\alpha^+(k)|0\rangle$$
$$= \exp\!\left[-\mathrm{i}\left(\frac{E_{n',\ell'}}{\hbar}+ck\right)t\right]\int d^3x\,\Psi_{n',\ell',m'_\ell}(x)\psi_{\sigma'}^+(x)a_\alpha^+(k)|0\rangle. \qquad (18.76)$$

The relevant transition matrix elements for photon emission between t' and t are

$$S_{fi}(t,t') \equiv S_{n',\ell',m'_\ell,\sigma';k,\alpha|n,\ell,m_\ell,\sigma}(t,t')$$
$$= \langle n',\ell',m'_\ell,\sigma';k,\alpha;t|\exp\!\left(-\frac{\mathrm{i}}{\hbar}H(t-t')\right)|n,\ell,m_\ell,\sigma;t'\rangle$$
$$= \langle n',\ell',m'_\ell,\sigma';k,\alpha|\mathrm{T}\exp\!\left(-\frac{\mathrm{i}}{\hbar}\int_{t'}^{t}d\tau H_D(\tau)\right)|n,\ell,m_\ell,\sigma\rangle,$$

where

$$H_D(\tau) = \exp\!\left(\frac{\mathrm{i}}{\hbar}H_0\tau\right)(H_I+H_{II})\exp\!\left(-\frac{\mathrm{i}}{\hbar}H_0\tau\right)$$

is the time evolution operator on the states in the interaction picture.

The scattering matrix element is in leading order

$$S_{fi} = S_{n',\ell',m'_\ell,\sigma';k,\alpha|n,\ell,m_\ell,\sigma} = \langle n',\ell',m'_\ell,\sigma';k,\alpha|U_D(\infty,-\infty)|n,\ell,m_\ell,\sigma;0\rangle$$

$$\simeq \frac{1}{\mathrm{i}\hbar}\int_{-\infty}^{\infty} dt\, \exp\left[\mathrm{i}(\omega_{n',\ell';n,\ell}+ck)t\right]\langle n',\ell',m'_\ell,\sigma';\boldsymbol{k},\alpha| \\ \times \int d^3\boldsymbol{x}\sum_\nu \frac{\mathrm{i}q\hbar}{2m}\boldsymbol{A}(\boldsymbol{x})\cdot\left(\psi_\nu^+(\boldsymbol{x})\overset{\leftrightarrow}{\nabla}\psi_\nu(\boldsymbol{x})\right)|n,\ell,m_\ell,\sigma;0\rangle \tag{18.77}$$

with the field operators in the Schrödinger picture. We also took into account that the energy levels are ℓ dependent through fine structure. At this stage we are still using $\boldsymbol{A}(\boldsymbol{x})$, although our reasoning in Section 18.4 already indicated that any $\boldsymbol{x}$ dependence in $\boldsymbol{A}(\boldsymbol{x})$ must be negligible to justify minimal photon coupling into the effective Hamiltonian for relative motion in the atom. We will return to this point below.

Substitution of the mode expansion (18.21) for the photon operator and evaluation of the second quantized matrix element transforms the transition matrix element for photon emission into a matrix element of first quantized theory,

$$S_{n',\ell',m'_\ell,\sigma';\boldsymbol{k},\alpha|n,\ell,m_\ell,\sigma} \simeq 2\pi\delta(\omega_{n',\ell';n,\ell}+ck)\frac{\mathrm{i}q}{m\hbar}\sqrt{\frac{\hbar\mu_0 c}{16\pi^3 k}}\delta_{\sigma\sigma'} \\ \times\langle n',\ell',m'_\ell|\boldsymbol{\epsilon}_\alpha(\boldsymbol{k})\cdot\mathbf{p}\exp(-\mathrm{i}\boldsymbol{k}\cdot\mathbf{x})|n,\ell,m_\ell\rangle. \tag{18.78}$$

The operators $\boldsymbol{\epsilon}_\alpha(\boldsymbol{k})\cdot\mathbf{p}$ and $\boldsymbol{k}\cdot\mathbf{x}$ commute, whence we do not encounter a normal ordering problem in the first quantized matrix element.

Equation (18.78) can be interpreted as a first quantized matrix element of the perturbation operator

$$V(t) = -\frac{q}{2m}\left(\mathbf{p}\cdot\boldsymbol{A}(\mathbf{x},t)+\boldsymbol{A}(\mathbf{x},t)\cdot\mathbf{p}\right), \tag{18.79}$$

which contains an operator corresponding to a classical transversely polarized plane wave

$$\boldsymbol{A}_\alpha^{(+)}(\boldsymbol{x},t) = \sqrt{\frac{\hbar\mu_0 c}{16\pi^3 k}}\boldsymbol{\epsilon}_\alpha(\boldsymbol{k})\exp[-\mathrm{i}(\boldsymbol{k}\cdot\boldsymbol{x}-ckt)]. \tag{18.80}$$

This classical plane wave apparently represents a single *emitted* photon of sharp energy $\hbar ck$ and momentum $\hbar\boldsymbol{k}$, and second quantization helped us to determine both the proper amplitude for the single photon wave and the $\boldsymbol{k}$-dependent term in the transition matrix element. The corresponding calculation for *absorption* of a photon yields a first quantized matrix element of the perturbation operator (18.79) with a single photon vector potential

$$\boldsymbol{A}_\alpha^{(-)}(\boldsymbol{x},t) = \sqrt{\frac{\hbar\mu_0 c}{16\pi^3 k}}\boldsymbol{\epsilon}_\alpha(\boldsymbol{k})\exp[\mathrm{i}(\boldsymbol{k}\cdot\boldsymbol{x}-ckt)], \tag{18.81}$$

see equation (18.93).

We can understand the amplitudes of the single photon wave functions (18.80) and (18.81) in the following way: The mode expansion (18.21) becomes in finite volume V

$$\boldsymbol{A}(\boldsymbol{x},t) = \sqrt{\frac{\hbar\mu_0 c}{V}}\sum_{\boldsymbol{k}}\sum_{\alpha=1}^{2}\frac{\boldsymbol{\epsilon}_\alpha(\boldsymbol{k})}{\sqrt{2k}}\Big(a_\alpha(\boldsymbol{k})\exp[\mathrm{i}(\boldsymbol{k}\cdot\boldsymbol{x}-ckt)]$$
$$+a_\alpha^+(\boldsymbol{k})\exp[-\mathrm{i}(\boldsymbol{k}\cdot\boldsymbol{x}-ckt)]\Big), \tag{18.82}$$

and the corresponding energy and momentum operators[2] are

$$H = \sum_{\boldsymbol{k}}\sum_\alpha \hbar ck\, a_\alpha^+(\boldsymbol{k})a_\alpha(\boldsymbol{k}), \quad \boldsymbol{P} = \sum_{\boldsymbol{k}}\sum_\alpha \hbar\boldsymbol{k}\, a_\alpha^+(\boldsymbol{k})a_\alpha(\boldsymbol{k}).$$

These equations tell us for a classical amplitude $a_\alpha(\boldsymbol{k})$ that this amplitude would (up to an arbitrary phase φ) have to be a Kronecker δ with respect to momentum and polarization to represent a single photon of momentum $\hbar\boldsymbol{k}$, energy $\hbar ck$ and polarization $\boldsymbol{\epsilon}_\alpha$, and therefore the classical vector potential for the single photon in the continuum limit $V \to 8\pi^3$ is

$$\boldsymbol{A}_{\gamma,\boldsymbol{k},\alpha}(\boldsymbol{x},t) = \sqrt{\frac{\hbar\mu_0 c}{16\pi^3 k}}\boldsymbol{\epsilon}_\alpha(\boldsymbol{k})\Big(\exp[\mathrm{i}(\boldsymbol{k}\cdot\boldsymbol{x}-ckt+\varphi)]$$
$$+\exp[-\mathrm{i}(\boldsymbol{k}\cdot\boldsymbol{x}-ckt+\varphi)]\Big)$$
$$= 2\sqrt{\frac{\hbar\mu_0 c}{16\pi^3 k}}\boldsymbol{\epsilon}_\alpha(\boldsymbol{k})\cos(\boldsymbol{k}\cdot\boldsymbol{x}-ckt+\varphi). \tag{18.83}$$

Note however that for emission only the plane wave with $\exp[-\mathrm{i}(\boldsymbol{k}\cdot\boldsymbol{x} - ckt + \varphi)]$ contributes to the transition matrix element, whereas for absorption only the other term contributes.

The vector potential in box normalization (18.82) does have the expected units Vs/m, whereas the continuum limit vector potentials (18.21, 18.83) come in units of $\mathrm{m}^{3/2}\mathrm{Vs/m}$. This is related to the fact that their transition matrix elements squared yield transition probability densities per volume unit $d^3\boldsymbol{k}$ in the photon state space, see e.g. equation (18.84) below. It is the same effect that we encountered in scattering theory for momentum eigenstates $\exp(\mathrm{i}\boldsymbol{k}\cdot\boldsymbol{x})/V^{1/2}$ in box normalization or $\exp(\mathrm{i}\boldsymbol{k}\cdot\boldsymbol{x})/(2\pi)^{3/2}$ in the continuum limit.

[2]Classically these equations would hold for time averages.

Evaluation of the transition matrix element in the dipole approximation

We have already emphasized that the coupling of the electromagnetic potentials to the effective single particle model for relative motion in atoms assumes a long wavelength approximation in the sense $\boldsymbol{A}(\boldsymbol{x}_p, t) \simeq \boldsymbol{A}(\boldsymbol{x}_e, t)$, see equations (18.51) and (18.56). Therefore the exponential factor $\exp(-\mathrm{i}\boldsymbol{k}\cdot\boldsymbol{x})$ must effectively be constant over the extension of the atomic wave functions and can be replaced by $\exp(-\mathrm{i}\boldsymbol{k}\cdot\boldsymbol{x}) \simeq 1$. For an estimate of the product $|\boldsymbol{k}\cdot\boldsymbol{x}|$, we recall that the energy of the emitted photon from an excited bound state cannot exceed the binding energy of hydrogen,

$$\frac{hc}{\lambda} < -E_1 = \frac{e^2}{8\pi\epsilon_0 a_0} = \frac{hc\alpha}{4\pi a_0},$$

and therefore

$$\lambda > \frac{4\pi}{\alpha}a_0 \simeq 1.72\times 10^3 a_0, \quad ka_0 < \frac{\alpha}{2} \simeq 3.65\times 10^{-3}.$$

This confirms that the exponential factor will be approximately constant over the extension of the wave functions,

$$\langle n', \ell', m'_\ell|\boldsymbol{\epsilon}_\alpha(\boldsymbol{k})\cdot\mathbf{p}\exp(-\mathrm{i}\boldsymbol{k}\cdot\mathbf{x})|n, \ell, m_\ell\rangle \approx \langle n', \ell', m'_\ell|\boldsymbol{\epsilon}_\alpha(\boldsymbol{k})\cdot\mathbf{p}|n, \ell, m_\ell\rangle.$$

The matrix element of the momentum operator between energy eigenstates is usually converted into matrix elements of the position operator $\mathbf{x}$ using the first quantized Hamiltonian $H_0 = (\mathbf{p}^2/2m) + V(\mathbf{x})$ and the relation

$$[H_0, \mathbf{x}] = \frac{\hbar}{\mathrm{i}m}\mathbf{p}.$$

This implies

$$\begin{aligned}\langle n', \ell', m'_\ell|\mathbf{p}|n, \ell, m_\ell\rangle &= \mathrm{i}\frac{m}{\hbar}\langle n', \ell', m'_\ell|[H_0, \mathbf{x}]|n, \ell, m_\ell\rangle \\ &= \mathrm{i}m\omega_{n',\ell';n,\ell}\langle n', \ell', m'_\ell|\mathbf{x}|n, \ell, m_\ell\rangle,\end{aligned}$$

where $\hbar\omega_{n',\ell';n,\ell} = E_{n',\ell'} - E_{n,\ell}$. In the case of emission we have $\omega_{n',\ell';n,\ell} < 0$. The transition matrix element (18.78) therefore becomes

$$\begin{aligned}S_{n',\ell',m'_\ell,\sigma';\boldsymbol{k},\alpha|n,\ell,m_\ell,\sigma} \simeq &-2\pi\delta(\omega_{n',\ell';n,\ell} + ck)q\sqrt{\frac{\mu_0 c}{16\pi^3\hbar k}}\delta_{\sigma\sigma'}\omega_{n',\ell';n,\ell} \\ &\times\langle n', \ell', m'_\ell|\boldsymbol{\epsilon}_\alpha(\boldsymbol{k})\cdot\mathbf{x}|n, \ell, m_\ell\rangle\end{aligned}$$

$$= \delta(\omega_{n',\ell';n,\ell} + ck) q \sqrt{\frac{\mu_0 c^3 k}{4\pi\hbar}} \delta_{\sigma\sigma'}$$
$$\times \langle n', \ell', m'_\ell | \boldsymbol{\epsilon}_\alpha(\boldsymbol{k}) \cdot \mathbf{x} | n, \ell, m_\ell \rangle.$$

The differential emission rate into a momentum volume element $d^3\boldsymbol{k}$ around $\boldsymbol{k}$ of a photon of polarization $\boldsymbol{\epsilon}_\alpha(\boldsymbol{k})$ is then with $q = -e$, $\delta(0) \to T/2\pi$,

$$d\Gamma^{(\alpha)}(\boldsymbol{k})_{n,\ell,m_\ell,\sigma\to n',\ell',m'_\ell,\sigma'} = d^3\boldsymbol{k} \frac{\left| S_{n',\ell',m'_\ell,\sigma';\boldsymbol{k},\alpha|n,\ell,m_\ell,\sigma} \right|^2}{T}$$
$$\simeq \frac{\mu_0 c^3 e^2}{8\pi^2\hbar} k \delta_{\sigma\sigma'} \left| \langle n', \ell', m'_\ell | \boldsymbol{\epsilon}_\alpha(\boldsymbol{k}) \cdot \mathbf{x} | n, \ell, m_\ell \rangle \right|^2 \delta(\omega_{n,\ell;n',\ell'} - ck) d^3\boldsymbol{k}, \quad (18.84)$$

or after integration over the wave number k of the emitted photon,

$$\frac{d\Gamma^{(\alpha)}(\hat{\boldsymbol{k}})_{n,\ell,m_\ell,\sigma\to n',\ell',m'_\ell,\sigma'}}{d\Omega} = \frac{\mu_0 e^2}{8\pi^2\hbar c} \omega^3_{n,\ell;n',\ell'} \delta_{\sigma\sigma'}$$
$$\times \left| \boldsymbol{\epsilon}_\alpha(\boldsymbol{k}) \cdot \langle n', \ell', m'_\ell | \mathbf{x} | n, \ell, m_\ell \rangle \right|^2. \quad (18.85)$$

Note that if we would have tried to calculate this only within a semi-classical first quantized theory for the monochromatic perturbation (18.79, 18.80), the δ function in energy and the units of the transition matrix element would have tempted us to introduce a density $\varrho(E_{n'})$ of final hydrogen states per energy, similar to the Golden Rule for transitions into a continuum. This factor would then have appeared instead of the factor[3] $\delta(E_n - E_{n'} - \hbar ck) d^3\boldsymbol{k}$ in (18.84). Indeed, we do have a transition into a continuum of final *photon* states, but the semi-classical approximation would have missed that and naive application of the Golden Rule would have tempted us to include a wrong factor with an unjustified interpretation, see also Problem 18.11.

As a consequence of the φ dependence of the spherical harmonics, the vector

$$\langle n', \ell', m'_\ell | \mathbf{x} | n, \ell, m_\ell \rangle = \langle n', \ell', m'_\ell | r \sin\vartheta \cos\varphi | n, \ell, m_\ell \rangle \boldsymbol{e}_x$$
$$+ \langle n', \ell', m'_\ell | r \sin\vartheta \sin\varphi | n, \ell, m_\ell \rangle \boldsymbol{e}_y + \langle n', \ell', m'_\ell | r \cos\vartheta | n, \ell, m_\ell \rangle \boldsymbol{e}_z$$

has real x and z components and an imaginary y component. We know already from the dipole selection rules from Section 15.1 that the z component $\langle n', \ell', m'_\ell | \mathrm{z} | n, \ell, m_\ell \rangle$ is only different from 0 if $\Delta m_\ell = m'_\ell - m_\ell = 0$, while the x and y components are only different from 0 if $\Delta m_\ell = \pm 1$.

The different conjugation properties and selection rules imply

[3]Recall that densities of states $\varrho(E) \sim k^2 dk/dE$ in the $V \to \infty$ limit have units of $\mathrm{cm}^{-3}\mathrm{eV}^{-1}$, see the remark after equation (12.8).

$$\begin{aligned}
\left|\boldsymbol{\epsilon}_\alpha(\boldsymbol{k})\cdot\langle n',\ell',m'_\ell|\mathbf{x}|n,\ell,m_\ell\rangle\right|^2 &= \Big(\boldsymbol{\epsilon}_\alpha(\boldsymbol{k})\cdot\big[\langle n',\ell',m'_\ell|\mathrm{x}|n,\ell,m_\ell\rangle\boldsymbol{e}_x \\
&\quad - \langle n',\ell',m'_\ell|\mathrm{y}|n,\ell,m_\ell\rangle\boldsymbol{e}_y + \langle n',\ell',m'_\ell|\mathrm{z}|n,\ell,m_\ell\rangle\boldsymbol{e}_z\big]\Big) \\
&\quad \times\Big(\boldsymbol{\epsilon}_\alpha(\boldsymbol{k})\cdot\big[\langle n',\ell',m'_\ell|\mathrm{x}|n,\ell,m_\ell\rangle\boldsymbol{e}_x + \langle n',\ell',m'_\ell|\mathrm{y}|n,\ell,m_\ell\rangle\boldsymbol{e}_y \\
&\quad + \langle n',\ell',m'_\ell|\mathrm{z}|n,\ell,m_\ell\rangle\boldsymbol{e}_z\big]\Big) \\
&= \big[\boldsymbol{\epsilon}_\alpha(\boldsymbol{k})\cdot\langle n',\ell',m'_\ell|\mathrm{x}|n,\ell,m_\ell\rangle\boldsymbol{e}_x\big]^2 + \big[\boldsymbol{\epsilon}_\alpha(\boldsymbol{k})\cdot\langle n',\ell',m'_\ell|\mathrm{z}|n,\ell,m_\ell\rangle\boldsymbol{e}_z\big]^2 \\
&\quad + \big[\mathrm{i}\boldsymbol{\epsilon}_\alpha(\boldsymbol{k})\cdot\langle n',\ell',m'_\ell|\mathrm{y}|n,\ell,m_\ell\rangle\boldsymbol{e}_y\big]^2.
\end{aligned}$$

This cannot be directly associated with an angle between the polarization $\boldsymbol{\epsilon}_\alpha(\boldsymbol{k})$ and one of the real vectors

$$\begin{aligned}
\langle n',\ell',m'_\ell|\mathbf{x}_\pm|n,\ell,m_\ell\rangle &= \langle n',\ell',m'_\ell|\mathrm{x}|n,\ell,m_\ell\rangle\boldsymbol{e}_x \pm \mathrm{i}\langle n',\ell',m'_\ell|\mathrm{y}|n,\ell,m_\ell\rangle\boldsymbol{e}_y \\
&\quad + \langle n',\ell',m'_\ell|\mathrm{z}|n,\ell,m_\ell\rangle\boldsymbol{e}_z
\end{aligned}$$

because of missing cross terms of the form

$$\pm 2\big[\boldsymbol{\epsilon}_\alpha(\boldsymbol{k})\cdot\langle n',\ell',m'_\ell|\mathrm{x}|n,\ell,m_\ell\rangle\boldsymbol{e}_x\big]\big[\mathrm{i}\boldsymbol{\epsilon}_\alpha(\boldsymbol{k})\cdot\langle n',\ell',m'_\ell|\mathrm{y}|n,\ell,m_\ell\rangle\boldsymbol{e}_y\big].$$

However, we can write

$$\begin{aligned}
\left|\boldsymbol{\epsilon}_\alpha(\boldsymbol{k})\cdot\langle n',\ell',m'_\ell|\mathbf{x}|n,\ell,m_\ell\rangle\right|^2 &= \frac{1}{2}\left|\langle n',\ell',m'_\ell|\mathbf{x}|n,\ell,m_\ell\rangle\right|^2 \\
&\quad \times\left(\cos^2\theta_{\alpha,-} + \cos^2\theta_{\alpha,+}\right)
\end{aligned}$$

where $\theta_{\alpha,-}$ and $\theta_{\alpha,+}$ are the angles between the polarization $\boldsymbol{\epsilon}_\alpha(\boldsymbol{k})$ and the real vectors $\langle n',\ell',m'_\ell|\mathbf{x}_-|n,\ell,m_\ell\rangle$ and $\langle n',\ell',m'_\ell|\mathbf{x}_+|n,\ell,m_\ell\rangle$, respectively.

This yields a differential emission rate

$$\begin{aligned}
d\Gamma^{(\alpha)}(\hat{\boldsymbol{k}})_{n,\ell,m_\ell,\sigma\to n',\ell',m'_\ell,\sigma'} &= \frac{\mu_0 e^2}{8\pi^2\hbar c}\omega^3_{n,\ell;n',\ell'}\delta_{\sigma\sigma'}\left|\langle n',\ell',m'_\ell|\mathbf{x}|n,\ell,m_\ell\rangle\right|^2 \\
&\quad \times\frac{\cos^2\theta_{\alpha,-} + \cos^2\theta_{\alpha,+}}{2}\,d\Omega. \qquad (18.86)
\end{aligned}$$

The solid angle element $d\Omega = \sin\vartheta\, d\vartheta\, d\varphi$ measures the direction of the emission vector $\boldsymbol{k}$ and the calculation of the total polarized emission rate $\Gamma^{(\alpha)}_{n,\ell,m_\ell\to n',\ell',m'_\ell}$ requires integration over $d\Omega$. We can do that e.g. by evaluating the angles $\theta_{\alpha,\pm}$ in terms of the angles $\{\vartheta_\alpha,\varphi_\alpha\}$ of the vector $\boldsymbol{\epsilon}_\alpha(\boldsymbol{k})$ and the angles $\{\vartheta_\pm,\varphi_\pm\}$ of the vectors $\langle n',\ell',m'_\ell|\mathbf{x}_\pm|n,\ell,m_\ell\rangle$. However, a faster way is to choose in each of the two terms the respective angle $\theta_{\alpha,\pm}$ and a corresponding azimuthal angle $\phi_{\alpha,\pm}$ as integration variables. This reduces the calculation of the angular integrals to

$$\int_0^{2\pi} d\phi_{\alpha,\pm}\int_0^{\pi} d\theta_{\alpha,\pm}\,\sin\theta_{\alpha,\pm}\cos^2\theta_{\alpha,\pm} = \frac{4\pi}{3}.$$

The total emission rate for *polarized* photons is therefore

$$\Gamma^{(\alpha)}_{n,\ell,m_\ell,\sigma\to n',\ell',m'_\ell,\sigma'} = \frac{\mu_0 e^2}{6\pi\hbar c}\omega^3_{n,\ell;n',\ell'}\delta_{\sigma\sigma'}\left|\langle n',\ell',m'_\ell|\mathbf{x}|n,\ell,m_\ell\rangle\right|^2, \tag{18.87}$$

and the total unpolarized emission rate is

$$\Gamma_{n,\ell,m_\ell,\sigma\to n',\ell',m'_\ell,\sigma'} = \frac{\mu_0 e^2}{3\pi\hbar c}\omega^3_{n,\ell;n',\ell'}\delta_{\sigma\sigma'}\left|\langle n',\ell',m'_\ell|\mathbf{x}|n,\ell,m_\ell\rangle\right|^2. \tag{18.88}$$

The relation $\Gamma_{n,\ell,m_\ell,\sigma\to n',\ell',m'_\ell,\sigma'} = 2\Gamma^{(\alpha)}_{n,\ell,m_\ell,\sigma\to n',\ell',m'_\ell,\sigma'}$ follows at a more formal level from the fact that

$$\sum_{\alpha=1}^{2}\boldsymbol{\epsilon}_\alpha(\boldsymbol{k})\otimes\boldsymbol{\epsilon}_\alpha(\boldsymbol{k}) = \underline{1} - \hat{\boldsymbol{k}}\otimes\hat{\boldsymbol{k}}$$

is the projector onto the plane orthogonal to $\boldsymbol{k}$, and therefore

$$\sum_\alpha\left|\boldsymbol{\epsilon}_\alpha(\boldsymbol{k})\cdot\langle n',\ell',m'_\ell|\mathbf{x}_\pm|n,\ell,m_\ell\rangle\right|^2 = \left|\langle n',\ell',m'_\ell|\mathbf{x}|n,\ell,m_\ell\rangle\right|^2\sin^2\theta_\pm,$$

where $\theta_\pm$ are the angles between the wave vector $\boldsymbol{k}$ and the two real vectors $\langle n',\ell',m'_\ell|\mathbf{x}_\pm|n,\ell,m_\ell\rangle$. Therefore we find for the unpolarized differential emission rate

$$\begin{aligned} d\Gamma(\hat{\boldsymbol{k}})_{n,\ell,m_\ell,\sigma\to n',\ell',m'_\ell,\sigma'} &= \frac{\mu_0 e^2}{8\pi^2\hbar c}\omega^3_{n,\ell;n',\ell'}\delta_{\sigma\sigma'}\left|\langle n',\ell',m'_\ell|\mathbf{x}|n,\ell,m_\ell\rangle\right|^2 \\ &\quad\times\frac{\sin^2\theta_- + \sin^2\theta_+}{2}\,d\Omega, \end{aligned} \tag{18.89}$$

and this time the angular integrals yield

$$\int_0^{2\pi}d\phi_\pm\int_0^\pi d\theta_\pm\,\sin^3\theta_\pm = \frac{8\pi}{3},$$

which implies the total emission rate (18.88).

We had to write the polarized and unpolarized differential emission rates (18.86) and (18.89) as averages over two real dipoles $-e\langle n',\ell',m'_\ell|\mathbf{x}_\pm|n,\ell,m_\ell\rangle$, where we used the dipole selection rules for hydrogen states. For general atomic or molecular states, all Cartesian components of $\langle f|\mathbf{x}|i\rangle$ may be complex, and we may have a sum of two dipoles of different magnitude,

$$\begin{aligned} |\boldsymbol{\epsilon}_\alpha\cdot\langle f|\mathbf{x}|i\rangle|^2 &= (\boldsymbol{\epsilon}_\alpha\cdot\Re\langle f|\mathbf{x}|i\rangle)^2 + (\boldsymbol{\epsilon}_\alpha\cdot\Im\langle f|\mathbf{x}|i\rangle)^2 \\ &= (\Re\langle f|\mathbf{x}|i\rangle)^2\cos^2\theta_{\alpha,1} + (\Im\langle f|\mathbf{x}|i\rangle)^2\cos^2\theta_{\alpha,2}, \end{aligned} \tag{18.90}$$

$$\sum_\alpha|\boldsymbol{\epsilon}_\alpha\cdot\langle f|\mathbf{x}|i\rangle|^2 = (\Re\langle f|\mathbf{x}|i\rangle)^2\sin^2\theta_1 + (\Im\langle f|\mathbf{x}|i\rangle)^2\sin^2\theta_2.$$

This yields the same results as (18.86) in a different parametrization. The difference between the construction in (18.86) and (18.90) is that we could construct two dipoles of the same magnitude $-e\left|\langle n',\ell',m'_\ell|\mathbf{x}|n,\ell,m_\ell\rangle\right|$ in (18.86) and express the result as an average, whereas the generic construction (18.90) yields a sum of two dipoles of different magnitude.

Since we are observing photons of certain frequency with no regard to the particular transition which generated those photons, it is customary to sum the emission rate over degenerate final states and average over degenerate initial states. The emission rate per excited atom for photons with angular frequency $\omega_{n,\ell;n',\ell'}$ follows from (18.88) as

$$\begin{aligned}\Gamma_{n,\ell\to n',\ell'} &= \frac{1}{2\ell+1}\sum_{m_\ell=-\ell}^{\ell}\sum_{m'_\ell=-\ell'}^{\ell'}\Gamma_{n,\ell,m_\ell\to n',\ell',m'_\ell}\\ &= \frac{\mu_0 e^2}{2\pi mc}\omega^2_{n,\ell;n',\ell'}\left|f_{n',\ell'|n,\ell}\right|. \end{aligned}\tag{18.91}$$

Here we have set $\sigma = \sigma'$ and omitted the spin indices, and we used the definition (15.29) of the averaged oscillator strength.

The quantity $\Gamma_{n,\ell\to n',\ell'} \equiv A_{n,\ell\to n',\ell'}$ provides a quantum mechanical expression for the Einstein A coefficient for spontaneous emission of photons. Einstein had introduced this coefficient in 1916 in his balance equations for the origin of the Planck spectrum.

We have seen that in leading order the relevant interaction Hamiltonian for photon emission or absorption is

$$H_I = \int d^3x \sum_\sigma \mathrm{i}\frac{q\hbar}{2m}\boldsymbol{A}\cdot\left(\psi^+_\sigma \overleftrightarrow{\nabla}\psi_\sigma\right),$$

and in the Schrödinger picture this operator contains only time-independent field operators $\boldsymbol{A}(\boldsymbol{x})$, $\psi_\sigma(\boldsymbol{x})$.

Substitution of the mode expansions in terms of the momentum space operators yields (note $q \neq |\boldsymbol{q}|$ in the following equation):

$$\begin{aligned}H_I = &-\frac{q\hbar}{m}\sqrt{\frac{\hbar\mu_0 c}{(2\pi)^3}}\int\frac{d^3q}{\sqrt{2|\boldsymbol{q}|}}\int d^3k\sum_\sigma\sum_\alpha \boldsymbol{k}\cdot\boldsymbol{\epsilon}_\alpha(\boldsymbol{q})\\ &\times\left(c^+_\sigma(\boldsymbol{k}+\boldsymbol{q})a_\alpha(\boldsymbol{q})c_\sigma(\boldsymbol{k}) + c^+_\sigma(\boldsymbol{k}-\boldsymbol{q})a^+_\alpha(\boldsymbol{q})c_\sigma(\boldsymbol{k})\right).\end{aligned}\tag{18.92}$$

The representation of interaction Hamiltonians in terms of $c_\sigma(\boldsymbol{k})$, $c^+_\sigma(\boldsymbol{k})$ is useful for processes involving (quasi)free electrons, e.g. for the Compton effect ("free-free scattering") or for the discussion of electron-photon interactions in metals (assuming e.g. a jellium model for the electrons). However, for the discussion of emission or absorption from atomic or molecular bound states the $\boldsymbol{x}$-representation is more convenient.

18.7 Photon absorption

We will continue to use energy labels n and n' such that $E_n > E_{n'}$. Therefore the previously discussed transition $n \to n'$ involved photon emission, while the process $n' \to n$ involves photon absorption. Later on we will also compare emission and absorption rates, and it is desirable to make the distinction between emission and absorption rates more visible in the notation. Therefore we will denote absorption rates with the symbol $\tilde{\Gamma}$.

The leading order scattering matrix element for photon absorption due to a transition from a state $|n', \ell', m'_\ell, \sigma'; \boldsymbol{k}, \alpha\rangle$ to a state $|n, \ell, m_\ell, \sigma; 0\rangle$,

$$S_{n,\ell,m_\ell,\sigma|n',\ell',m'_\ell,\sigma';\boldsymbol{k},\alpha} \simeq \frac{1}{\mathrm{i}\hbar}\int_{-\infty}^{\infty} dt\, \exp\left[\mathrm{i}(\omega_{n,\ell;n',\ell'} - ck)t\right]\mathrm{i}\frac{q\hbar}{2m}$$
$$\times\langle n, \ell, m_\ell, \sigma; 0|\int d^3\boldsymbol{x}\sum_\nu \boldsymbol{A}(\boldsymbol{x})\cdot\left(\psi_\nu^+(\boldsymbol{x})\overset{\leftrightarrow}{\nabla}\psi_\nu(\boldsymbol{x})\right)|n', \ell', m'_\ell, \sigma'; \boldsymbol{k}, \alpha\rangle$$

is just the negative complex conjugate of the emission matrix element (18.77). The resulting scattering matrix element after evaluation of the field operators,

$$S_{n,\ell,m_\ell,\sigma|n',\ell',m'_\ell,\sigma';\boldsymbol{k},\alpha} \simeq 2\pi\delta(\omega_{n,\ell;n',\ell'} - ck)\frac{\mathrm{i}q}{m\hbar}\sqrt{\frac{\hbar\mu_0 c}{16\pi^3 k}}\delta_{\sigma\sigma'}$$
$$\times\langle n, \ell, m_\ell|\boldsymbol{\epsilon}_\alpha(\boldsymbol{k})\cdot\mathbf{p}\exp(\mathrm{i}\boldsymbol{k}\cdot\mathbf{x})|n', \ell', m'_\ell\rangle, \tag{18.93}$$

therefore has the form of a first quantized scattering matrix element with perturbation (18.79) and vector potential (18.81).

The equality of the scattering matrix elements up to a phase factor also implies that the absorption rate per $\boldsymbol{k}$ space volume of the incoming photons has the same value as the corresponding emission rate (18.84) per $\boldsymbol{k}$ space volume of emitted photons,

$$\frac{d\tilde{\Gamma}^{(\alpha)}(\boldsymbol{k})_{n',\ell',m'_\ell,\sigma'\to n,\ell,m_\ell,\sigma}}{d^3k} = \frac{\left|S_{n,\ell,m_\ell,\sigma|n',\ell',m'_\ell,\sigma';\boldsymbol{k},\alpha}\right|^2}{T}$$
$$\simeq \frac{\mu_0 c^3 e^2}{8\pi^2\hbar}k\delta_{\sigma\sigma'}\left|\langle n, \ell, m_\ell|\boldsymbol{\epsilon}_\alpha(\boldsymbol{k})\cdot\mathbf{x}|n', \ell', m'_\ell\rangle\right|^2\delta(\omega_{n,\ell;n',\ell'} - ck), \tag{18.94}$$

where $q = -e$ was substituted.

This yields the differential absorption rate for polarized photons in terms of the angles $\theta_{\alpha,\pm}$ between the vectors $\langle n, \ell, m_\ell|\mathbf{x}_\pm|n', \ell', m'_\ell\rangle$ and the polarization $\boldsymbol{\epsilon}_\alpha(\boldsymbol{k})$,

$$\frac{d\tilde{\Gamma}^{(\alpha)}(\boldsymbol{k})_{n',\ell',m'_\ell,\sigma'\to n,\ell,m_\ell,\sigma}}{d^3\boldsymbol{k}} \simeq \frac{\mu_0 c^3 e^2}{8\pi^2\hbar}k\delta_{\sigma\sigma'}\left|\langle n,\ell,m_\ell|\mathbf{x}|n',\ell',m'_\ell\rangle\right|^2 \times\frac{\cos^2\theta_{\alpha,-}+\cos^2\theta_{\alpha,+}}{2}\delta(\omega_{n,\ell;n',\ell'}-ck).$$

The differential absorption rate for unpolarized photons, $d\tilde{\Gamma}(\boldsymbol{k}) = \sum_\alpha d\tilde{\Gamma}^{(\alpha)}(\boldsymbol{k})$, depends on the angles $\theta_\pm$ between the vectors $\langle n,\ell,m_\ell|\mathbf{x}_\pm|n',\ell',m'_\ell\rangle$ and the incident vector $\boldsymbol{k}$,

$$\frac{d\tilde{\Gamma}(\boldsymbol{k})_{n',\ell',m'_\ell,\sigma'\to n,\ell,m_\ell,\sigma}}{d^3\boldsymbol{k}} \simeq \frac{\mu_0 c^3 e^2}{8\pi^2\hbar}k\delta_{\sigma\sigma'}\left|\langle n,\ell,m_\ell|\mathbf{x}|n',\ell',m'_\ell\rangle\right|^2 \times\frac{\sin^2\theta_{-}+\sin^2\theta_{+}}{2}\delta(\omega_{n,\ell;n',\ell'}-ck). \quad (18.95)$$

The total absorption rate between the specified states follows as

$$\tilde{\Gamma}_{n',\ell',m'_\ell,\sigma'\to n,\ell,m_\ell,\sigma} = \frac{\mu_0 e^2}{3\pi\hbar c}\omega^3_{n,\ell;n',\ell'}\delta_{\sigma\sigma'}\left|\langle n,\ell,m_\ell|\mathbf{x}|n',\ell',m'_\ell\rangle\right|^2, \quad (18.96)$$

and the total absorption rate per atom for photons of angular frequency $\omega_{n,\ell;n',\ell'}$ is

$$\tilde{\Gamma}_{n',\ell'\to n,\ell} = \frac{1}{2\ell'+1}\sum_{m'_\ell=-\ell'}^{\ell'}\sum_{m_\ell=-\ell}^{\ell}\tilde{\Gamma}_{n',\ell',m'_\ell\to n,\ell,m_\ell} = \frac{\mu_0 e^2}{2\pi mc}\omega^2_{n,\ell;n',\ell'}f_{n,\ell|n',\ell'}. \quad (18.97)$$

This differs from the corresponding spontaneous emission rate (18.91) for photons of angular frequency $\omega_{n,\ell;n',\ell'}$ only through the different averaging factors for the respective initial states,

$$\tilde{\Gamma}_{n',\ell'\to n,\ell} = \frac{2\ell+1}{2\ell'+1}\Gamma_{n,\ell\to n',\ell'}. \quad (18.98)$$

The number of absorption events will be proportional to the flux of incoming photons, and therefore another observable of interest is the absorption rate per flux of incoming photons, i.e. the absorption cross section.

The *photon flux* or current density of monochromatic photons of momentum $\hbar\boldsymbol{k}$ can be calculated by dividing their energy current density $\mathcal{S}(\boldsymbol{k})$ by their energy $\hbar ck$. Equations (18.22, 18.23, 18.35) and (18.83) yield

$$\frac{\mathcal{S}(\boldsymbol{k})}{\hbar ck} = \frac{\boldsymbol{E}\times\boldsymbol{B}}{\mu_0\hbar ck} = \frac{c}{(2\pi)^3}\hat{\boldsymbol{k}}. \quad (18.99)$$

This is actually a photon flux $dj(\boldsymbol{k})/d^3\boldsymbol{k}$ per $\boldsymbol{k}$ space volume due to the use of the photon wave functions in the continuum limit[4].

Equations (18.94) and (18.99) yield the polarized photon absorption cross section

$$\sigma^{(\alpha)}(\boldsymbol{k})_{n',\ell',m'_\ell\to n,\ell,m_\ell} = \frac{d\tilde{\Gamma}^{(\alpha)}(\boldsymbol{k})_{n',\ell',m'_\ell\to n,\ell,m_\ell}}{dj(\boldsymbol{k})}$$

$$\simeq \frac{\pi\mu_0 ce^2}{\hbar}\omega_{n,\ell;n',\ell'}\left|\langle n,\ell,m_\ell|\boldsymbol{\epsilon}_\alpha(\boldsymbol{k})\cdot\mathbf{x}|n',\ell',m'_\ell\rangle\right|^2\delta(\omega_{n,\ell;n',\ell'}-ck)$$

$$= 4\pi^2\alpha\omega_{n,\ell;n',\ell'}\left|\langle n,\ell,m_\ell|\boldsymbol{\epsilon}_\alpha(\boldsymbol{k})\cdot\mathbf{x}|n',\ell',m'_\ell\rangle\right|^2\delta(\omega_{n,\ell;n',\ell'}-ck), \quad (18.100)$$

where we encounter again Sommerfeld's fine structure constant $\alpha=\mu_0 ce^2/4\pi\hbar$ (7.61) (not to be confused with the polarization index, of course).

To average (18.100) over the angles of the incident photons, we can use the same methods that we applied for the calculation of the total polarized emission rate (18.87), except for an extra factor of $(4\pi)^{-1}$ from the averaging over directions. This yields an isotropic cross section for polarized photons

$$\sigma^{(\alpha)}(k)_{n',\ell',m'_\ell\to n,\ell,m_\ell} \simeq \frac{4\pi^2}{3}\alpha\omega_{n,\ell;n',\ell'}\left|\langle n,\ell,m_\ell|\mathbf{x}|n',\ell',m'_\ell\rangle\right|^2$$
$$\times\delta(\omega_{n,\ell;n',\ell'}-ck), \quad (18.101)$$

and a total isotropic cross section

$$\sigma(k)_{n',\ell',m'_\ell\to n,\ell,m_\ell} \simeq \frac{8\pi^2}{3}\alpha\omega_{n,\ell;n',\ell'}\left|\langle n,\ell,m_\ell|\mathbf{x}|n',\ell',m'_\ell\rangle\right|^2$$
$$\times\delta(\omega_{n,\ell;n',\ell'}-ck). \quad (18.102)$$

The average absorption cross section per atom for photons of angular frequency $\omega_{n,\ell;n',\ell'}$ follows then again through averaging over initial states and summation over final states,

$$\sigma(k)_{n',\ell'\to n,\ell} = \frac{1}{2\ell'+1}\sum_{m'_\ell=-\ell'}^{\ell'}\sum_{m_\ell=-\ell}^{\ell}\sigma(k)_{n',\ell',m'_\ell\to n,\ell,m_\ell}$$
$$= \frac{4\pi^2\hbar}{m}\alpha f_{n,\ell|n',\ell'}\delta(\omega_{n,\ell;n',\ell'}-ck). \quad (18.103)$$

We get a more realistic representation for absorption cross sections if we take into account the representation (2.10) of the δ function,

---

[4]The result in box normalization is $\boldsymbol{j}(\boldsymbol{k}) = (c/V)\hat{\boldsymbol{k}}$.

$$\delta(\omega_{n,\ell;n',\ell'} - ck) = \lim_{\gamma\to 0}\frac{1}{2\pi}\int_{-\infty}^{\infty}dt\,\exp[\mathrm{i}(\omega_{n,\ell;n',\ell'} - ck)t - \gamma|t|]$$
$$= \lim_{\gamma\to 0}\frac{1}{\pi}\frac{\gamma}{(\omega_{n,\ell;n',\ell'} - ck)^2 + \gamma^2}. \quad (18.104)$$

Keeping a finite value of γ yields a Lorentzian absorption line shape of half width 2γ,

$$\sigma(k)_{n',\ell'\to n,\ell} = \frac{4\pi\hbar}{m}\alpha f_{n,\ell|n',\ell'}\frac{\gamma}{(\omega_{n,\ell;n',\ell'} - ck)^2 + \gamma^2}. \quad (18.105)$$

A finite width of line shapes arises from many sources. A certainly not exhaustive list of mechanisms includes adiabatic switching of perturbations, lifetime broadening, pressure broadening, Doppler broadening, and broadening through chemical shifts.

We have found $\Gamma \propto \delta_{\sigma\sigma'}$ both for photon emission and absorption, i.e. no spin-flips in either process. The same holds in arbitrary order with the Hamiltonian (18.74), since there are no spin flipping terms there. How then can a magnetic field flip spins even for non-relativistic electrons? There is actually a term missing in the Hamiltonian (18.74), the Pauli term:

$$H_B = -\frac{q}{m}\int d^3\boldsymbol{x}\sum_{\sigma,\sigma'}\psi_\sigma^+(\boldsymbol{x})\boldsymbol{S}_{\sigma,\sigma'}\cdot(\nabla\times\boldsymbol{A}(\boldsymbol{x}))\psi_{\sigma'}(\boldsymbol{x}). \quad (18.106)$$

This term induces spin flips through two of the three components of the vector of Pauli matrices $\underline{S} = \hbar\underline{\sigma}/2$, and it follows from the non-relativistic expansion of the relativistic wave equation for electrons, see Section 21.5. We could neglect the Pauli term in the present calculation because a derivative on the vector potential yields a factor k, whereas a derivative on the wave functions amounts approximately to a factor of order $1/a_0$. The Pauli term is therefore suppressed when dipole approximation $\lambda \gg a_0$ applies. E.g. for transition between bound states in hydrogen, $\hbar ck < -E_1$ implies that H_B is suppressed relative to H_I by approximately $ka_0 < \alpha/2$, which translates into a suppression of spin-flipping transitions between bound hydrogen states by about $\alpha^2/4 \simeq 1.3\times 10^{-5}$. An exception to negligibility of spin-flipping transitions with low energy photons concerns situations where spin-preserving electronic transitions do not exist in the same energy range. This is the case e.g. for the 21 cm transition in hydrogen.

Photon absorption into continuous states

The spin labels are omitted in the following discussion because H_I does not induce spin flips.

If we have photon absorption due to transition into continuous states, e.g. from $|n',\ell',m'_\ell\rangle$ to $|E,\ell,m_\ell\rangle$, we have to take into account the proper measure for the continuous states from the completeness relation. E.g. for hydrogen states we have (7.75)

$$\sum_{\ell=0}^{\infty}\sum_{m_\ell=-\ell}^{\ell}\left(\sum_{n=\ell+1}^{\infty}|n,\ell,m_\ell\rangle\langle n,\ell,m_\ell| + \int_0^\infty dK\,K^2|K,\ell,m_\ell\rangle\langle K,\ell,m_\ell|\right)$$
$$=\sum_{\ell=0}^{\infty}\sum_{m_\ell=-\ell}^{\ell}\left(\sum_{n=\ell+1}^{\infty}|n,\ell,m_\ell\rangle\langle n,\ell,m_\ell| + \int dE\,\varrho(E)\,|E,\ell,m_\ell\rangle\langle E,\ell,m_\ell|\right)$$
$$=1,$$

and if we directly use the Coulomb wave states for the continuous energy eigenstates without rescaling, $|E,\ell,m_\ell\rangle = |K,\ell,m_\ell\rangle$, we have in the continuous part of the spectrum

$$\varrho(E)=\Theta(E)K^2\frac{dK}{dE}=\Theta(E)\frac{1}{\hbar^3}\sqrt{2m^3E}.$$

The first order scattering matrix element with the interaction Hamiltonian H_I then yields a differential absorption rate for polarized photons

$$\frac{d\tilde{\Gamma}^{(\alpha)}(\boldsymbol{k})_{n',\ell',m'_\ell\to E,\ell,m_\ell}}{d^3\boldsymbol{k}dE}=\varrho(E)\frac{\left|S_{E,\ell,m_\ell|n',\ell',m'_\ell;\boldsymbol{k},\alpha}\right|^2}{T}$$
$$\simeq\varrho(E)\frac{e^2ck}{8\pi^2\epsilon_0}\left|\langle E,\ell,m_\ell|\boldsymbol{\epsilon}_\alpha(\boldsymbol{k})\cdot\mathbf{x}|n',\ell',m'_\ell\rangle\right|^2\delta(E-E_{n',\ell'}-\hbar ck),\quad(18.107)$$

and integration over the energy E of the ionized state yields

$$\frac{d\tilde{\Gamma}^{(\alpha)}(\boldsymbol{k})_{n',\ell',m'_\ell\to E,\ell,m_\ell}}{d^3\boldsymbol{k}}\simeq\frac{e^2c}{8\pi^2\epsilon_0}k\Big[\varrho(E)$$
$$\times\left|\langle E,\ell,m_\ell|\boldsymbol{\epsilon}_\alpha(\boldsymbol{k})\cdot\mathbf{x}|n',\ell',m'_\ell\rangle\right|^2\Big]_{E=E_{n',\ell'}+\hbar ck}.$$

The photons appear in the initial state and are therefore taken into account by dividing out their current density from the transition rate, thus yielding an absorption cross section, see the general discussion for initial continuous states in Sections 13.5 and 13.6.

However, for photon absorption due to transition from a discrete into a continuous atomic or electronic state we can also calculate a spectral absorption cross section $d\sigma^{(\alpha)}(\boldsymbol{k})/dE_\gamma$ since $E_\gamma=\hbar ck=E-E_{n',\ell'}$ implies $dE_\gamma=dE$. This allows us to define a spectral absorption cross section for polarized photons according to

$$\frac{d\sigma^{(\alpha)}(\boldsymbol{k})_{n',\ell',m'_\ell\to E,\ell,m_\ell}}{dE_\gamma} = \frac{d\tilde{\Gamma}^{(\alpha)}(\boldsymbol{k})_{n',\ell',m'_\ell\to E,\ell,m_\ell}}{dEdj(\boldsymbol{k})}$$

$$\simeq \varrho(E)\frac{\pi e^2}{\epsilon_0}k\left|\langle E,\ell,m_\ell|\boldsymbol{\epsilon}_\alpha(\boldsymbol{k})\cdot\mathbf{x}|n',\ell',m'_\ell\rangle\right|^2\delta(E-E_{n',\ell'}-\hbar ck),(18.108)$$

where (18.99) was used. In practical applications of (18.108) the energy preserving δ function could again be replaced by a Lorentzian line shape as in (18.104).

The absorption cross section for polarized photons with momentum $\hbar\boldsymbol{k}$ follows from $\sigma^{(\alpha)}(\boldsymbol{k}) = d\tilde{\Gamma}^{(\alpha)}(\boldsymbol{k})/dj(\boldsymbol{k}) = (8\pi^3/c)d\tilde{\Gamma}^{(\alpha)}(\boldsymbol{k})/d^3\boldsymbol{k}$ or from (18.108) as

$$\sigma^{(\alpha)}(\boldsymbol{k})_{n',\ell',m'_\ell\to E,\ell,m_\ell} \simeq \frac{\pi e^2}{\epsilon_0}k\Big[\varrho(E) \times\left|\langle E,\ell,m_\ell|\boldsymbol{\epsilon}_\alpha(\boldsymbol{k})\cdot\mathbf{x}|n',\ell',m'_\ell\rangle\right|^2\Big]_{E=E_{n',\ell'}+\hbar ck},$$

and averaging over the directions like in (18.101) yields

$$\sigma^{(\alpha)}(k)_{n',\ell',m'_\ell\to E,\ell,m_\ell} = \frac{1}{2}\sigma(k)_{n',\ell',m'_\ell\to E,\ell,m_\ell}$$
$$\simeq \frac{\pi e^2}{3\epsilon_0}k\Big[\varrho(E)\left|\langle E,\ell,m_\ell|\mathbf{x}|n',\ell',m'_\ell\rangle\right|^2\Big]_{E=E_{n',\ell'}+\hbar ck}.$$

18.8 Stimulated emission of photons

Here we use box normalization in a volume $V = L^3$, i.e. $\boldsymbol{k} = 2\pi\boldsymbol{n}/L$.

If we have already $n_{\boldsymbol{k},\alpha}$ photons of momentum $\hbar\boldsymbol{k}$ and polarization $\boldsymbol{\epsilon}_\alpha(\boldsymbol{k})$ in the initial state,

$$|n,\ell,m_\ell,\sigma;n_{\boldsymbol{k},\alpha}\rangle = \int d^3\boldsymbol{x}\,\psi_\sigma^+(\boldsymbol{x})\frac{(a_\alpha^+(\boldsymbol{k}))^{n_{\boldsymbol{k},\alpha}}}{\sqrt{n_{\boldsymbol{k},\alpha}!}}|0\rangle\langle\boldsymbol{x}|n,\ell,m_\ell\rangle,$$

the basic oscillator relation $\langle n+1|a^+|n\rangle = \sqrt{n+1}$ yields for the leading order scattering matrix elements the relation

$$S_{n',\ell',m'_\ell,\sigma';n_{\boldsymbol{k},\alpha}+1|n,\ell,m_\ell,\sigma;n_{\boldsymbol{k},\alpha}} = \sqrt{n_{\boldsymbol{k},\alpha}+1}S_{n',\ell',m'_\ell,\sigma';\boldsymbol{k},\alpha|n,\ell,m_\ell,\sigma},$$

i.e. the emission rate scales with the number of photons of momentum $\hbar\boldsymbol{k}$, energy $\hbar ck = \hbar\omega_{n,\ell;n',\ell'}$ and fixed polarization like

$$\Gamma^{(\alpha)}_{n,\ell;n_{\boldsymbol{k},\alpha}\to n',\ell';n_{\boldsymbol{k},\alpha}+1} = (n_{\boldsymbol{k},\alpha}+1)\Gamma^{(\alpha)}_{n,\ell;0\to n',\ell';1}$$

$$= \frac{n_{k,\alpha}+1}{2\ell+1}\sum_{m_\ell=-\ell}^{\ell}\sum_{m'_\ell=-\ell'}^{\ell'}\Gamma^{(\alpha)}_{n,\ell,m_\ell\to n',\ell',m'_\ell}$$

$$= (n_{k,\alpha}+1)\frac{\mu_0 e^2}{4\pi mc}\omega^2_{n,\ell;n',\ell'}\left|f_{n',\ell'|n,\ell}\right|.$$

The total polarized emission rate in the presence of the $n_{k,\alpha}$ photons therefore differs from the "spontaneous" emission rate $\Gamma^{(\alpha)}_{n,\ell;0\to n',\ell';1} \equiv \Gamma^{(\alpha)}_{n,\ell\to n',\ell'} = \Gamma_{n,\ell\to n',\ell'}/2$ (cf. equation (18.91)) by an additional "stimulated" emission rate

$$\Gamma^{(s,\alpha)}_{n,\ell;n_{k,\alpha}\to n',\ell';n_{k,\alpha}+1} = n_{k,\alpha}\Gamma^{(\alpha)}_{n,\ell;0\to n',\ell';1} = n_{k,\alpha}\frac{\mu_0 e^2}{4\pi mc}\omega^2_{n,\ell;n',\ell'}\left|f_{n',\ell'|n,\ell}\right|$$

which is proportional to the number of photons which are already present in the system. This is sometimes metaphorically explained as a consequence of one of the original photons stimulating the emission by shaking the excited state. However, in the end it is nothing but a combinatorial quantum effect of indistinguishable photon operators.

On the other hand, we find for the absorption of a photon in the initial state

$$|n',\ell',m'_\ell,\sigma';n_{k,\alpha}\rangle = \int d^3\boldsymbol{x}\,\psi^+_{\sigma'}(\boldsymbol{x})\frac{(a^+_\alpha(\boldsymbol{k}))^{n_{k,\alpha}}}{\sqrt{n_{k,\alpha}!}}|0\rangle\langle\boldsymbol{x}|n',\ell',m'_\ell\rangle,$$

from $\langle n-1|a|n\rangle = \sqrt{n}$ the relation

$$\begin{aligned}S_{n,\ell,m_\ell,\sigma;n_{k,\alpha}-1|n',\ell',m'_\ell,\sigma';n_{k,\alpha}} &= \sqrt{n_{k,\alpha}}S_{n,\ell,m_\ell,\sigma|n',\ell',m'_\ell,\sigma';\boldsymbol{k},\alpha}\\ &= -\sqrt{n_{k,\alpha}}S^*_{n',\ell',m'_\ell,\sigma';\boldsymbol{k},\alpha|n,\ell,m_\ell,\sigma}.\end{aligned}$$

Therefore the polarized absorption rate in the presence of $n_{k,\alpha}$ photons of momentum $\hbar\boldsymbol{k}$ and polarization $\boldsymbol{\epsilon}_\alpha(\boldsymbol{k})$ is

$$\tilde{\Gamma}^{(\alpha)}_{n',\ell';n_{k,\alpha}\to n,\ell;n_{k,\alpha}-1} = n_{k,\alpha}\tilde{\Gamma}^{(\alpha)}_{n',\ell';1\to n,\ell;0}$$

$$= \frac{n_{k,\alpha}}{2\ell'+1}\sum_{m'_\ell=-\ell'}^{\ell'}\sum_{m_\ell=-\ell}^{\ell}\tilde{\Gamma}^{(\alpha)}_{n',\ell',m'_\ell\to n,\ell,m_\ell} = n_{k,\alpha}\frac{\mu_0 e^2}{4\pi mc}\omega^2_{n,\ell;n',\ell'}f_{n,\ell|n',\ell'}.$$

This equals corresponding stimulated and total emission rates up to the different averaging factors for the different initial states which enter into the averaged and summed transition matrix elements,

$$\begin{aligned}\tilde{\Gamma}^{(\alpha)}_{n',\ell';n_{k,\alpha}\to n,\ell;n_{k,\alpha}-1} &= \frac{2\ell+1}{2\ell'+1}\Gamma^{(\alpha)}_{n,\ell;n_{k,\alpha}-1\to n',\ell';n_{k,\alpha}}\\ &= \frac{2\ell+1}{2\ell'+1}\Gamma^{(s,\alpha)}_{n,\ell;n_{k,\alpha}\to n',\ell';n_{k,\alpha}+1}.\end{aligned}$$

Note that it does not matter that we used the single photon absorption rate and current density in the calculation (18.100) of the polarized photon absorption cross section without explicitly taking into account the number $n_{k,\alpha}$ of available photons. The common factor $n_{k,\alpha}$ cancels in the ratio

$$\begin{aligned}\sigma^{(\alpha)}(\boldsymbol{k})_{n',\ell',m'_\ell\to n,\ell,m_\ell} &= \frac{d\tilde{\Gamma}^{(\alpha)}_{n',\ell',m'_\ell;n_{k,\alpha}\to n,\ell,m_\ell;n_{k,\alpha}-1}}{dJ^{(\alpha)}(\boldsymbol{k})}\\ &= \frac{8\pi^3}{n_{k,\alpha}c}\frac{d\tilde{\Gamma}^{(\alpha)}_{n',\ell',m'_\ell;n_{k,\alpha}\to n,\ell,m_\ell;n_{k,\alpha}-1}}{d^3k}.\end{aligned}$$

18.9 Photon scattering

For the following calculations we switch back to a generic notation $|n,\zeta\rangle$ for atomic or molecular states, where the energy levels E_n depend on the index set n and the index set ζ enumerates the degenerate states.

Scattering concerns transitions which involve a photon both in the initial and in the final state: $|n,\zeta;\boldsymbol{k},\alpha\rangle \to |n',\zeta';\boldsymbol{k}',\alpha'\rangle$. Here we consider scattering of photons by bound non-relativistic systems, i.e. the initial state $|n,\zeta\rangle$ and the final state $|n',\zeta'\rangle$ of the scattering system are discrete, and we use minimal coupling of the photon to effective single particle models for relative motion in the bound system. We have seen in Section 18.4 that photon coupling to the relative motion in materials effectively amounts to photon-electron coupling, and therefore we use photon scattering off bound electrons as the relevant paradigm for the following discussion.

To have a non-vanishing matrix element between different 1-photon states in lowest order requires two copies of the photon operator $\boldsymbol{A}$ – one to annihilate the initial photon and one to create the final photon. The relevant interaction Hamiltonian for photon interactions with non-relativistic electrons is

$$\begin{aligned}H_{int} &= \int d^3x\left(-\mathrm{i}\frac{e\hbar}{2m}\boldsymbol{A}\cdot\left(\psi^+\overleftrightarrow{\nabla}\psi\right)+\frac{e^2}{2m}\psi^+\boldsymbol{A}^2\psi+\frac{e\hbar}{2m}\psi^+\boldsymbol{\sigma}\cdot\boldsymbol{B}\psi\right)\\ &= H_I+H_{II}+H_B,\end{aligned} \tag{18.109}$$

where H_B is the Pauli term (18.106). Summations over spinor indices are tacitly understood. We have already substituted $q = -e$, because we have seen in Section 18.4 that the coupling of long wavelength photons to bound systems involving electrons can effectively be considered as coupling of the photons to a charge $-e$ if the charge binding the electron is $q_2 = e$ or if the mass m_2 of the binding charge is much larger than the electron mass, $m_2 \gg m_e$. The reduced mass m in the Hamiltonian (18.109) is usually also $m \simeq m_e$ in excellent approximation[5].

[5]An exception is positronium with $m = m_e/2$.

We can get two copies of $\boldsymbol{A}$ from H_I^2, H_IH_B, H_BH_I and H_B^2 in second order perturbation theory, and from H_{II} in first order perturbation theory. Among these terms, only those involving the Pauli term can induce spin flips. However, we will focus on photon energies in the soft X-ray regime, $E_\gamma \lesssim 1\,\mathrm{keV}$. Due to the suppression of the Pauli term by about a_0/λ the allowed transition matrix elements of H_I in the soft X-ray regime are typically at least an order of magnitude larger than the allowed matrix elements of H_B, see the discussion after (18.106). This implies that spin preserving scattering probabilities $|S_{fi}|^2$ of order H_I^4 will generically be at least two orders of magnitude larger than spin preserving scattering of order $H_I^2H_B^2$ or spin reversing scattering of order $(H_IH_B)^2$.

Therefore we neglect H_B in the following calculations. The relevant scattering matrix elements in order $\mathcal{O}(e^2)$ are then

$$\begin{aligned}
S_{n',\zeta';\boldsymbol{k}',\alpha'|n,\zeta;\boldsymbol{k},\alpha} &= \langle n',\zeta';\boldsymbol{k}',\alpha'|U_D(\infty,-\infty)|n,\zeta;\boldsymbol{k},\alpha\rangle|_{e^2}\\
&= \langle n',\zeta';\boldsymbol{k}',\alpha'|\mathrm{T}\exp\!\left(-\frac{\mathrm{i}}{\hbar}\int_{-\infty}^{\infty}dt\,H_D(t)\right)|n,\zeta;\boldsymbol{k},\alpha\rangle|_{e^2}\\
&= S^{(I)}_{n',\zeta';\boldsymbol{k}',\alpha'|n,\zeta;\boldsymbol{k},\alpha} + S^{(II)}_{n',\zeta';\boldsymbol{k}',\alpha'|n,\zeta;\boldsymbol{k},\alpha},
\end{aligned}$$

with contributions from H_I^2,

$$\begin{aligned}
S^{(I)}_{n',\zeta';\boldsymbol{k}',\alpha'|n,\zeta;\boldsymbol{k},\alpha} &= -\frac{1}{\hbar^2}\int_{-\infty}^{\infty}dt\int_{-\infty}^{t}dt'\,\exp[\mathrm{i}(\omega_{n'}+ck')t]\exp[-\mathrm{i}(\omega_n+ck)t']\\
&\quad\times\langle n',\zeta';\boldsymbol{k}',\alpha'|H_I\exp\!\left(-\frac{\mathrm{i}}{\hbar}H_0(t-t')\right)H_I|n,\zeta;\boldsymbol{k},\alpha\rangle,
\end{aligned}$$

and from H_{II},

$$S^{(II)}_{n',\zeta';\boldsymbol{k}',\alpha'|n,\zeta;\boldsymbol{k},\alpha} = \int_{-\infty}^{\infty}\frac{dt}{\mathrm{i}\hbar}\exp[\mathrm{i}(\omega_{n',n}+\omega_{k',k})t]\,\langle n',\zeta';\boldsymbol{k}',\alpha'|H_{II}|n,\zeta;\boldsymbol{k},\alpha\rangle.$$

The first order term $S^{(II)}$ is the easier one to evaluate. Insertion of the mode expansion (18.21) for the photon field yields

$$\begin{aligned}
S^{(II)}_{n',\zeta';\boldsymbol{k}',\alpha'|n,\zeta;\boldsymbol{k},\alpha} &= \frac{\mu_0ce^2}{8\pi^2\mathrm{i}m\sqrt{kk'}}\boldsymbol{\epsilon}_{\alpha'}(\boldsymbol{k}')\cdot\boldsymbol{\epsilon}_\alpha(\boldsymbol{k})\delta(\omega_{n',n}+\omega_{k',k})\\
&\quad\times\int d^3\boldsymbol{x}\,\exp[\mathrm{i}(\boldsymbol{k}-\boldsymbol{k}')\cdot\boldsymbol{x}]\,\Psi^+_{n',\zeta'}(\boldsymbol{x})\Psi_{n,\zeta}(\boldsymbol{x})\\
&= \frac{\mu_0ce^2}{8\pi^2\mathrm{i}m\sqrt{kk'}}\boldsymbol{\epsilon}_{\alpha'}(\boldsymbol{k}')\cdot\boldsymbol{\epsilon}_\alpha(\boldsymbol{k})\delta(\omega_{n',n}+\omega_{k',k})\\
&\quad\times\int d^3\boldsymbol{q}\,\Psi^+_{n',\zeta'}(\boldsymbol{q}+\boldsymbol{k}-\boldsymbol{k}')\Psi_{n,\zeta}(\boldsymbol{q}).
\end{aligned}$$

This leaves in dipole approximation $\exp\left[i(\boldsymbol{k}-\boldsymbol{k}')\cdot\boldsymbol{x}\right]\simeq 1$ the amplitude

$$S^{(II)}_{n',\zeta';\boldsymbol{k}',\alpha'|n,\zeta;\boldsymbol{k},\alpha}=\frac{\mu_0 e^2}{8\pi^2\mathrm{i}mk}\boldsymbol{\epsilon}_{\alpha'}(\boldsymbol{k}')\cdot\boldsymbol{\epsilon}_{\alpha}(\boldsymbol{k})\delta(k'-k)\delta_{n'n}\delta_{\zeta'\zeta},\tag{18.110}$$

i.e. only elastic photon scattering, but no Raman scattering from H_{II}.

The term $S^{(I)}$ splits into amplitudes with zero or two photons in virtual intermediate states,

$$S^{(I)}_{n',\zeta';\boldsymbol{k}',\alpha'|n,\zeta;\boldsymbol{k},\alpha}=S^{(I),0}_{n',\zeta';\boldsymbol{k}',\alpha'|n,\zeta;\boldsymbol{k},\alpha}+S^{(I),2}_{n',\zeta';\boldsymbol{k}',\alpha'|n,\zeta;\boldsymbol{k},\alpha}.$$

We omit the indices in the amplitudes $S^{(I),0}$ and $S^{(I),2}$ in the following calculations. The amplitude with no photons in the virtual intermediate state is

$$\begin{aligned}S^{(I),0}=&\frac{e^2}{4m^2}\sum\!\!\!\!\!\!\int_{n'',\zeta''}\int_{-\infty}^{\infty}dt\int_{-\infty}^{t}dt'\,\exp\left[\mathrm{i}(\omega_{n',n''}+ck')t\right]\exp\left[\mathrm{i}(\omega_{n'',n}-ck)t'\right]\\&\times\int d^3\boldsymbol{x}'\,\langle n',\zeta';\boldsymbol{k}',\alpha'|\boldsymbol{A}(\boldsymbol{x}')\cdot\left(\psi^+(\boldsymbol{x}')\overset{\leftrightarrow}{\nabla}\psi(\boldsymbol{x}')\right)|n'',\zeta'';0\rangle\\&\times\int d^3\boldsymbol{x}\,\langle n'',\zeta'';0|\boldsymbol{A}(\boldsymbol{x})\cdot\left(\psi^+(\boldsymbol{x})\overset{\leftrightarrow}{\nabla}\psi(\boldsymbol{x})\right)|n,\zeta;\boldsymbol{k},\alpha\rangle.\end{aligned}$$

The notation $\sum\!\!\!\!\!\int_{n'',\zeta''}$ takes into account that the intermediate states can also be part of the energy continuum of the scattering system.

We have already evaluated the time integrals in second order perturbation terms in (13.43),

$$\begin{aligned}&\int_{-\infty}^{\infty}dt\int_{-\infty}^{t}dt'\,\exp\left[\mathrm{i}(\omega_{n',n''}+ck')t\right]\exp\left[\mathrm{i}(\omega_{n'',n}-ck)t'+\epsilon t'\right]\\&=-2\pi\mathrm{i}\frac{\delta(\omega_{n',n}+\omega_{k',k})}{\omega_{n'',n}-ck-\mathrm{i}\epsilon}.\end{aligned}$$

Evaluation of the matrix elements of the field operators then yields again in dipole approximation $\exp(-\mathrm{i}\boldsymbol{k}'\cdot\boldsymbol{x}')\simeq 1$, $\exp(\mathrm{i}\boldsymbol{k}\cdot\boldsymbol{x})\simeq 1$ the result

$$\begin{aligned}S^{(I),0}=&\frac{\hbar\mu_0 ce^2}{32\pi^2\mathrm{i}m^2\sqrt{kk'}}\delta(\omega_{n',n}+\omega_{k',k})\sum\!\!\!\!\!\!\int_{n'',\zeta''}\frac{1}{\omega_{n'',n}-ck-\mathrm{i}\epsilon}\\&\times\int d^3\boldsymbol{x}'\,\boldsymbol{\epsilon}_{\alpha'}(\boldsymbol{k}')\cdot\left(\Psi^+_{n',\zeta'}(\boldsymbol{x}')\overset{\leftrightarrow}{\nabla}\Psi_{n'',\zeta''}(\boldsymbol{x}')\right)\\&\times\int d^3\boldsymbol{x}\,\boldsymbol{\epsilon}_{\alpha}(\boldsymbol{k})\cdot\left(\Psi^+_{n'',\zeta''}(\boldsymbol{x})\overset{\leftrightarrow}{\nabla}\Psi_{n,\zeta}(\boldsymbol{x})\right).\end{aligned}\tag{18.111}$$

We can transform this from velocity into length form using the by now standard trick $\hbar\mathbf{p}=\mathrm{i}m[H_0,\mathbf{x}]$ to find

$$S^{(I),0} = \frac{\mu_0 ce^2}{8\pi^2 \mathrm{i}\hbar\sqrt{kk'}}\delta(\omega_{n',n} + \omega_{k',k})\sum\!\!\!\!\!\!\int_{n'',\zeta''} \frac{\omega_{n',n''}\omega_{n'',n}}{\omega_{n'',n} - ck - \mathrm{i}\epsilon}$$
$$\times\langle n',\zeta'|\boldsymbol{\epsilon}_{\alpha'}(\boldsymbol{k}')\cdot\mathbf{x}|n'',\zeta''\rangle\langle n'',\zeta''|\boldsymbol{\epsilon}_{\alpha}(\boldsymbol{k})\cdot\mathbf{x}|n,\zeta\rangle. \qquad (18.112)$$

For the amplitude with two photons in the intermediate state we have to take into account that for two-photon states

$$\frac{1}{2}\int d^3\kappa'\int d^3\kappa\sum_{\beta',\beta}|\kappa',\beta';\kappa,\beta\rangle\langle\kappa',\beta';\kappa,\beta| = 1.$$

This yields

$$S^{(I),2} = \frac{e^2}{8m^2}\int d^3\kappa'\int d^3\kappa\sum\!\!\!\!\!\!\int_{n'',\zeta''}\sum_{\beta',\beta}\int_{-\infty}^{\infty}dt\int_{-\infty}^{t}dt'$$
$$\times\exp\!\big[\mathrm{i}(\omega_{n',n''}+ck'-c\kappa-c\kappa')t\big]\exp\!\big[\mathrm{i}(\omega_{n'',n}+c\kappa+c\kappa'-ck)t'\big]$$
$$\times\int d^3x'\langle n',\zeta';\boldsymbol{k}',\alpha'|\boldsymbol{A}(\boldsymbol{x}')\cdot\left(\psi^+(\boldsymbol{x}')\overleftrightarrow{\nabla}\psi(\boldsymbol{x}')\right)|n'',\zeta'';\boldsymbol{\kappa}',\beta';\boldsymbol{\kappa},\beta\rangle$$
$$\times\int d^3x\langle n'',\zeta'';\boldsymbol{\kappa}',\beta';\boldsymbol{\kappa},\beta|\boldsymbol{A}(\boldsymbol{x})\cdot\left(\psi^+(\boldsymbol{x})\overleftrightarrow{\nabla}\psi(\boldsymbol{x})\right)|n,\zeta;\boldsymbol{k},\alpha\rangle.$$

The matrix elements of the photon operators are given by

$$\langle\boldsymbol{\kappa}',\beta';\boldsymbol{\kappa},\beta|\boldsymbol{A}(\boldsymbol{x})|\boldsymbol{k},\alpha\rangle = \sqrt{\frac{\hbar\mu_0 c}{16\pi^3\kappa}}\boldsymbol{\epsilon}_\beta(\boldsymbol{\kappa})\exp(-\mathrm{i}\boldsymbol{\kappa}\cdot\boldsymbol{x})\delta(\boldsymbol{\kappa}'-\boldsymbol{k})\delta_{\beta'\alpha}$$
$$+\sqrt{\frac{\hbar\mu_0 c}{16\pi^3\kappa'}}\boldsymbol{\epsilon}_{\beta'}(\boldsymbol{\kappa}')\exp(-\mathrm{i}\boldsymbol{\kappa}'\cdot\boldsymbol{x})\delta(\boldsymbol{\kappa}-\boldsymbol{k})\delta_{\beta\alpha}$$

and a corresponding conjugate expression. This yields in dipole approximation

$$\int d^3\kappa'\int d^3\kappa\sum_{\beta',\beta}\exp\!\big[\mathrm{i}c(\kappa+\kappa')(t'-t)\big]$$
$$\times\langle\boldsymbol{k}',\alpha'|\boldsymbol{A}(\boldsymbol{x}')|\boldsymbol{\kappa}',\beta';\boldsymbol{\kappa},\beta\rangle\langle\boldsymbol{\kappa}',\beta';\boldsymbol{\kappa},\beta|\boldsymbol{A}(\boldsymbol{x})|\boldsymbol{k},\alpha\rangle$$
$$\simeq\frac{\hbar\mu_0 c}{8\pi^3}\delta_{\alpha\alpha'}\delta(\boldsymbol{k}-\boldsymbol{k}')\int d^3\kappa\sum_\beta\frac{\boldsymbol{\epsilon}_\beta(\boldsymbol{\kappa})\otimes\boldsymbol{\epsilon}_\beta(\boldsymbol{\kappa})}{\kappa}\exp\!\big[\mathrm{i}c(\kappa+k)(t'-t)\big]$$
$$+\frac{\hbar\mu_0 c}{8\pi^3}\frac{\boldsymbol{\epsilon}_\alpha(\boldsymbol{k})\otimes\boldsymbol{\epsilon}_{\alpha'}(\boldsymbol{k}')}{\sqrt{kk'}}\exp\!\big[\mathrm{i}c(k+k')(t'-t)\big]. \qquad (18.113)$$

The first term in (18.113) corresponds to an electron self-energy contribution where the external photon does not interact with the electron, but there are two

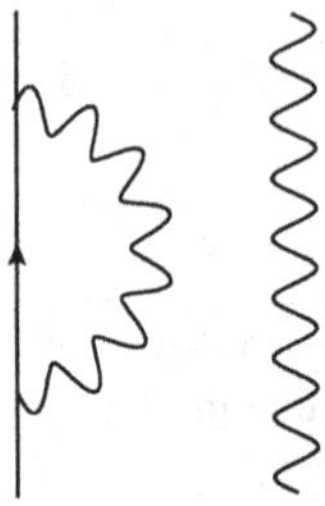

Fig. 18.1 A process with two photons in an intermediate state due to emission and re-absorption of a virtual photon. The straight line represents the electron and the wavy lines represent photons

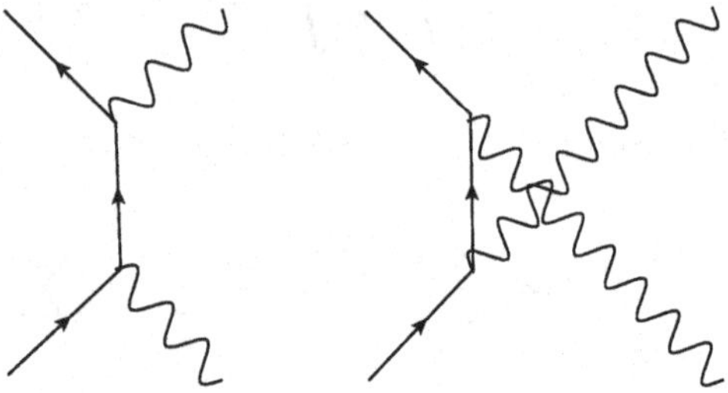

Fig. 18.2 The left diagram corresponds to absorption of the initial photon before emission of the final photon. The diagram on the right hand side corresponds to emission of the final photon before absorption of the initial photon

photons in the intermediate state due to emission and re-absorption of a virtual photon by the electron, see Figure 18.1.

This is an effect which leads to a renormalization of the electron mass in quantum field theory, but does not contribute to photon scattering.

The second term yields an expression for $S^{(I),2}$ which looks almost exactly like $S^{(I),0}$ (18.111), *except* that the polarization vectors are swapped $\boldsymbol{\epsilon}_{\alpha'}(\boldsymbol{k}') \leftrightarrow \boldsymbol{\epsilon}_{\alpha}(\boldsymbol{k})$, and $\omega_{n'',n} - ck - \mathrm{i}\epsilon$ is replaced by $\omega_{n'',n} + ck' - \mathrm{i}\epsilon$ in the denominator. After transformation into the length form, $S^{(I),0}$ and $S^{(I),2}$ yield the following expression,

$$\begin{aligned} S^{(I)}_{n',\zeta';k',\alpha'|n,\zeta;k,\alpha} &= \frac{\mu_0 ce^2}{8\pi^2 \mathrm{i}\hbar\sqrt{kk'}}\delta(\omega_{n',n} + \omega_{k',k}) \sum\!\!\!\!\!\!\!\int_{n'',\zeta''} \omega_{n',n''}\omega_{n'',n} \\ &\times \left(\frac{\langle n',\zeta'|\boldsymbol{\epsilon}_{\alpha'}(\boldsymbol{k}')\cdot\mathbf{x}|n'',\zeta''\rangle\langle n'',\zeta''|\boldsymbol{\epsilon}_{\alpha}(\boldsymbol{k})\cdot\mathbf{x}|n,\zeta\rangle}{\omega_{n'',n} - ck - \mathrm{i}\epsilon} \right. \\ &+ \left. \frac{\langle n',\zeta'|\boldsymbol{\epsilon}_{\alpha}(\boldsymbol{k})\cdot\mathbf{x}|n'',\zeta''\rangle\langle n'',\zeta''|\boldsymbol{\epsilon}_{\alpha'}(\boldsymbol{k}')\cdot\mathbf{x}|n,\zeta\rangle}{\omega_{n'',n} + ck' - \mathrm{i}\epsilon} \right). \end{aligned}$$

The first term corresponds to absorption of the initial photon before emission of the final photon, whereas the second term corresponds to emission of the final photon before absorption of the initial photon, see Figure 18.2.

The total scattering matrix element in order e^2 is

$$\begin{aligned} S_{n',\zeta';k',\alpha'|n,\zeta;k,\alpha} &= \frac{\mu_0 ce^2}{8\pi^2 \mathrm{i}\sqrt{kk'}}\delta(\omega_{n',n} + \omega_{k',k}) \left[\frac{1}{m}\delta_{n'n}\delta_{\zeta'\zeta}\boldsymbol{\epsilon}_{\alpha'}(\boldsymbol{k}')\cdot\boldsymbol{\epsilon}_{\alpha}(\boldsymbol{k}) \right. \\ &+ \sum\!\!\!\!\!\!\!\int_{n'',\zeta''} \omega_{n',n''}\omega_{n'',n} \left(\frac{\langle n',\zeta'|\boldsymbol{\epsilon}_{\alpha'}(\boldsymbol{k}')\cdot\mathbf{x}|n'',\zeta''\rangle\langle n'',\zeta''|\boldsymbol{\epsilon}_{\alpha}(\boldsymbol{k})\cdot\mathbf{x}|n,\zeta\rangle}{\hbar\omega_{n'',n} - \hbar ck - \mathrm{i}\epsilon} \right. \end{aligned}$$

$$+\frac{\langle n',\zeta'|\boldsymbol{\epsilon}_\alpha(\boldsymbol{k})\cdot\mathbf{x}|n'',\zeta''\rangle\langle n'',\zeta''|\boldsymbol{\epsilon}_{\alpha'}(\boldsymbol{k}')\cdot\mathbf{x}|n,\zeta\rangle}{\hbar\omega_{n'',n}+\hbar ck'-\mathrm{i}\epsilon}\Bigg)\Bigg].$$

We separate the energy conserving δ function for the calculation of the scattering cross section,

$$S_{n',\zeta';\boldsymbol{k}',\alpha'|n,\zeta;\boldsymbol{k},\alpha}=-\mathrm{i}\mathcal{M}_{n',\zeta';\boldsymbol{k}',\alpha'|n,\zeta;\boldsymbol{k},\alpha}\delta(\omega_{n',n}+\omega_{k',k}). \tag{18.114}$$

The differential scattering rate per $\boldsymbol{k}$ space volume of incident photons is then

$$\frac{d\Gamma_{n,\zeta;\boldsymbol{k},\alpha\to n',\zeta';\boldsymbol{k}',\alpha'}}{d^3\boldsymbol{k}}=d^3\boldsymbol{k}'\frac{\left|S_{n',\zeta';\boldsymbol{k}',\alpha'|n,\zeta;\boldsymbol{k},\alpha}\right|^2}{T}$$
$$=\frac{d^3\boldsymbol{k}'}{2\pi}\left|\mathcal{M}_{n',\zeta';\boldsymbol{k}',\alpha'|n,\zeta;\boldsymbol{k},\alpha}\right|^2\delta(\omega_{n',n}+\omega_{k',k}),$$

and the differential scattering cross section for polarized photons is with the incident photon current density per $\boldsymbol{k}$ space volume $dj/d^3\boldsymbol{k}=c\hat{\boldsymbol{k}}/(2\pi)^3$ (18.99),

$$d\sigma_{n,\zeta;\boldsymbol{k},\alpha\to n',\zeta';\boldsymbol{k}',\alpha'}=\frac{d\Gamma_{n,\zeta;\boldsymbol{k},\alpha\to n',\zeta';\boldsymbol{k}',\alpha'}}{dj(\boldsymbol{k})}$$
$$=\frac{4\pi^2}{c}\left|\mathcal{M}_{n',\zeta';\boldsymbol{k}',\alpha'|n,\zeta;\boldsymbol{k},\alpha}\right|^2\delta(\omega_{n',n}+\omega_{k',k})d^3\boldsymbol{k}'. \tag{18.115}$$

This yields after integration over k'

$$\frac{d\sigma_{n,\zeta;\boldsymbol{k},\alpha\to n',\zeta';\boldsymbol{k}',\alpha'}}{d\Omega}=\frac{4\pi^2}{c^2}k'^2\left|\mathcal{M}_{n',\zeta';\boldsymbol{k}',\alpha'|n,\zeta;\boldsymbol{k},\alpha}\right|^2\Big|_{k'=k-(\omega_{n',n}/c)}. \tag{18.116}$$

Substitution of our results for the scattering matrix element yields the result

$$\frac{d\sigma}{d\Omega}=\left(\frac{\mu_0e^2}{4\pi}\right)^2\frac{k'}{k}\Bigg|\frac{1}{m}\delta_{n'n}\delta_{\zeta'\zeta}\boldsymbol{\epsilon}_{\alpha'}(\boldsymbol{k}')\cdot\boldsymbol{\epsilon}_\alpha(\boldsymbol{k})+\sum\!\!\!\!\!\!\int_{n'',\zeta''}\omega_{n',n''}\omega_{n'',n}$$
$$\times\Bigg(\frac{\langle n',\zeta'|\boldsymbol{\epsilon}_{\alpha'}(\boldsymbol{k}')\cdot\mathbf{x}|n'',\zeta''\rangle\langle n'',\zeta''|\boldsymbol{\epsilon}_\alpha(\boldsymbol{k})\cdot\mathbf{x}|n,\zeta\rangle}{\hbar\omega_{n'',n}-\hbar ck-\mathrm{i}\epsilon}$$
$$+\frac{\langle n',\zeta'|\boldsymbol{\epsilon}_\alpha(\boldsymbol{k})\cdot\mathbf{x}|n'',\zeta''\rangle\langle n'',\zeta''|\boldsymbol{\epsilon}_{\alpha'}(\boldsymbol{k}')\cdot\mathbf{x}|n,\zeta\rangle}{\hbar\omega_{n'',n}+\hbar ck'-\mathrm{i}\epsilon}\Bigg)\Bigg|^2_{k'=k-(\omega_{n',n}/c)} \tag{18.117}$$

If there are non-vanishing transition matrix elements $\langle n',\zeta'|\boldsymbol{\epsilon}_{\alpha'}(\boldsymbol{k}')\cdot\mathbf{x}|n'',\zeta''\rangle$ and $\langle n'',\zeta''|\boldsymbol{\epsilon}_\alpha(\boldsymbol{k})\cdot\mathbf{x}|n,\zeta\rangle$ with the properties $\omega_{n'',n}\simeq ck$ and $\omega_{n',n''}\simeq -ck'$, or if there are any non-vanishing matrix elements $\langle n',\zeta'|\boldsymbol{\epsilon}_\alpha(\boldsymbol{k})\cdot\mathbf{x}|n'',\zeta''\rangle$ and

$\langle n'', \zeta''|\boldsymbol{\epsilon}_{\alpha'}(\boldsymbol{k}')\cdot\mathbf{x}|n,\zeta\rangle$ with the properties $\omega_{n'',n} \simeq -ck'$ and $\omega_{n',n''} \simeq ck$, then the differential scattering cross section will be dominated by the resonantly enhanced contributions from those matrix elements, and we will have $\omega_{n',n''}\omega_{n'',n} \simeq -c^2kk'$ for the dominant terms. In these cases we can approximate our result (18.117) by the equation

$$\frac{d\sigma}{d\Omega} \simeq \alpha^2c^2kk'^3\left|\sum_{n'',\zeta''}\left(\frac{\langle n',\zeta'|\boldsymbol{\epsilon}_{\alpha'}(\boldsymbol{k}')\cdot\mathbf{x}|n'',\zeta''\rangle\langle n'',\zeta''|\boldsymbol{\epsilon}_{\alpha}(\boldsymbol{k})\cdot\mathbf{x}|n,\zeta\rangle}{\omega_{n'',n}-ck-\mathrm{i}\epsilon}\right.\right.$$
$$\left.\left.+\frac{\langle n',\zeta'|\boldsymbol{\epsilon}_{\alpha}(\boldsymbol{k})\cdot\mathbf{x}|n'',\zeta''\rangle\langle n'',\zeta''|\boldsymbol{\epsilon}_{\alpha'}(\boldsymbol{k}')\cdot\mathbf{x}|n,\zeta\rangle}{\omega_{n'',n}+ck'-\mathrm{i}\epsilon}\right)\right|^2_{k'=k-(\omega_{n',n}/c)} \tag{18.118}$$

This is an equation for photon scattering which was proposed already in 1924 by Kramers and Heisenberg based on the correspondence principle[6]. However, note that this is only a suitable approximation to the actual cross section (18.117) if the near resonance conditions $\omega_{n'',n} \simeq ck$ and $\omega_{n',n''} \simeq -ck'$, or $\omega_{n'',n} \simeq -ck'$ and $\omega_{n',n''} \simeq ck$, can be fulfilled, and if there are allowed dipole transitions into the intermediate nearly resonant levels.

Thomson cross section

The contribution from the first term in (18.117) coincides with the classical Thomson cross section for elastic scattering of light which we will encounter again in Section 22.3 when we discuss photon scattering off free electrons. The first term yields for scattering of polarized photons

$$\left.\frac{d\sigma_T}{d\Omega}\right|_{\alpha\to\alpha'} = \left(\frac{\mu_0e^2}{4\pi m}\right)^2\left(\boldsymbol{\epsilon}_{\alpha'}(\boldsymbol{k}')\cdot\boldsymbol{\epsilon}_{\alpha}(\boldsymbol{k})\right)^2 = \left(\frac{\mu_0e^2}{4\pi m}\right)^2\cos^2\theta_{\alpha\alpha'},$$

The resulting cross section for unpolarized light involves a sum over final polarizations and an average over initial polarizations,

$$\frac{1}{2}\sum_{\alpha,\alpha'}\boldsymbol{\epsilon}_{\alpha'}(\boldsymbol{k}')\cdot\boldsymbol{\epsilon}_{\alpha}(\boldsymbol{k})\otimes\boldsymbol{\epsilon}_{\alpha}(\boldsymbol{k})\cdot\boldsymbol{\epsilon}_{\alpha'}(\boldsymbol{k}')$$
$$=\frac{1}{2}\sum_{\alpha'}\boldsymbol{\epsilon}_{\alpha'}(\boldsymbol{k}')\cdot\left(\underline{1}-\hat{\boldsymbol{k}}\otimes\hat{\boldsymbol{k}}\right)\cdot\boldsymbol{\epsilon}_{\alpha'}(\boldsymbol{k}')$$
$$=\frac{1}{2}\mathrm{tr}\left[\left(\underline{1}-\hat{\boldsymbol{k}}\otimes\hat{\boldsymbol{k}}\right)\cdot\left(\underline{1}-\hat{\boldsymbol{k}}'\otimes\hat{\boldsymbol{k}}'\right)\right]=\frac{1+\cos^2\theta}{2}, \tag{18.119}$$

[6]H.A. Kramers, W. Heisenberg, Z. Phys. 31, 681 (1925).

where $\hat{\boldsymbol{k}}\cdot\hat{\boldsymbol{k}}' = \cos\theta$, i.e. θ is the scattering angle. This yields[7]

$$\frac{d\sigma_T}{d\Omega} = \left(\frac{\mu_0 e^2}{4\pi m}\right)^2 \frac{1+\cos^2\theta}{2}, \tag{18.120}$$

and

$$\sigma_T = \frac{8\pi}{3}\left(\frac{\mu_0 e^2}{4\pi m}\right)^2. \tag{18.121}$$

The first term in equation (18.117) would hypothetically dominate the cross section $d\sigma/d\Omega$ if the photon energy is much larger than all the excitation energies of dipole allowed transitions, i.e. if $ck \gg |\omega_{n'',n}|$ for all $\langle n'',\zeta''|\mathbf{x}|n,\zeta\rangle \neq 0$. However, there will always be allowed transitions into intermediate continuum states. Therefore the condition $ck \gg |\omega_{n'',n}|$ for all dipole allowed transitions will not be fulfilled and the first term in (18.117) will never dominate light scattering by atoms or molecules[8]. However, the Thomson cross section plays an important role in the scattering of light by free electrons, which will be discussed in Section 22.3.

Rayleigh scattering

Molecules in a gas or a liquid have many dense lying rotational and vibrational levels, and the condition of dipole allowed resonant excitation of intermediate levels will practically always be fulfilled. The Kramers-Heisenberg formula (18.118) will therefore always be an excellent approximation to (18.117) for molecules in a fluid phase. In particular, the cross section for elastic photon scattering $|g;\boldsymbol{k},\alpha\rangle \to |g;\boldsymbol{k}',\alpha'\rangle$ from a ground state $|g\rangle$ or a state $|g\rangle$ near the ground state will be

$$\frac{d\sigma_R}{d\Omega} \simeq (\alpha c k^2)^2 \left|\sum_{n,\zeta,\omega_{n,g}\simeq ck} \frac{\langle g|\boldsymbol{\epsilon}_{\alpha'}(\boldsymbol{k}')\cdot\mathbf{x}|n,\zeta\rangle\langle n,\zeta|\boldsymbol{\epsilon}_\alpha(\boldsymbol{k})\cdot\mathbf{x}|g\rangle}{\omega_{n,g} - ck - \mathrm{i}\epsilon}\right|^2. \tag{18.122}$$

A formula for resonance fluorescence which is equivalent to (18.122) was given for the first time by Viktor Weisskopf in his Ph.D. thesis[9].

---

[7]The combination $r_e \equiv \mu_0 e^2/4\pi m = 2.82\,\mathrm{fm}$ is also denoted as the *classical radius of the electron.*

[8]A loophole in this argument concerns the remote possibility that all the matrix elements $\langle n'',\zeta''|\mathbf{x}|n,\zeta\rangle$ with $\omega_{n'',n} \gtrsim ck$ are extremely small.

[9]V. Weisskopf, Annalen Phys. 401, 23 (1931). He used a dipole operator $H = -e\boldsymbol{x}\cdot\dot{\boldsymbol{A}}(\boldsymbol{x},t)$ for atom-photon interactions throughout his calculations.

The reasoning with only one kind of resonantly enhanced terms is correct as long as the alternative resonance condition $\omega_{n,g} \simeq -ck$ cannot be fulfilled, i.e. as long as the energy E_g of the initial state $|g\rangle$ is less than $\hbar ck$ above the ground state energy. This applies e.g. to molecules at room temperature. These molecules will generically occupy states with energies less than 0.1 eV above their ground state energy. Scattering of optical photons by these molecules can be described by equation (18.122).

We can connect (18.122) to the polarizability properties of the scattering centers by noting that the dynamical polarizability tensor (15.26) for $\omega_{mn} \simeq \omega = ck$ has exactly the same form as the tensor multiplying the polarization vectors in (18.122). Therefore we can rewrite this equation also in the form

$$\left.\frac{d\sigma_R}{d\Omega}\right|_{\alpha\to\alpha'} = \left(\frac{\mu_0}{4\pi}\right)^2 \omega^4 \left(\boldsymbol{\epsilon}_{\alpha'}(\boldsymbol{k}')\cdot\underline{\alpha}_{(g)}\cdot\boldsymbol{\epsilon}_{\alpha}(\boldsymbol{k})\right)^2, \tag{18.123}$$

where it is understood that the sum over intermediate levels in (15.26) is dominated by terms which are almost resonant with the frequency ω of the elastically scattered photons.

Directional averaging over the orientation of the molecules will lead to an isotropic effective polarization tensor,

$$\boldsymbol{\epsilon}_{\alpha'}(\boldsymbol{k}')\cdot\underline{\alpha}_{(g)}\cdot\boldsymbol{\epsilon}_{\alpha}(\boldsymbol{k}) = \alpha_{(g)}\boldsymbol{\epsilon}_{\alpha'}(\boldsymbol{k}')\cdot\boldsymbol{\epsilon}_{\alpha}(\boldsymbol{k}),$$

$$\left.\frac{d\sigma_R}{d\Omega}\right|_{\alpha\to\alpha'} = \left(\frac{\mu_0}{4\pi}\alpha_{(g)}\right)^2 \omega^4 \cos^2\theta_{\alpha\alpha'},$$

and averaging and summation over the polarizations of the incoming and scattered photons (18.119) yields the same angular dependence on the scattering angle as for Thomson scattering (18.120),

$$\frac{d\sigma_R}{d\Omega} = \left(\frac{\mu_0}{4\pi}\alpha_{(g)}\right)^2 \omega^4 \frac{1+\cos^2\theta}{2} \tag{18.124}$$

and

$$\sigma_R = \frac{8\pi}{3}\left(\frac{\mu_0}{4\pi}\alpha_{(g)}\right)^2 \omega^4. \tag{18.125}$$

Equations (18.124, 18.125) are quantum mechanical versions of Lord Rayleigh's ω^4 law (Rayleigh 1871, 1899; see also Jackson [19] for a derivation of Rayleigh scattering in classical electrodynamics). It is sometimes stated (but neither in [19] nor in Weisskopf's thesis) that Rayleigh scattering is a small frequency approximation in the sense that $\hbar\omega = \hbar ck$ should be small compared to the internal excitations of the scattering system. This is not true. The quantum mechanical derivation (as well as Jackson's classical derivation) does not require this assumption. The only assumption that went into our derivation above was resonantly enhanced dipole scattering. Besides, energies of optical photons are not small compared to excitation energies for nitrogen or oxygen molecules. Indeed, the assumption of resonantly enhanced dipole scattering *implies* that the photon frequency $\omega = ck$ should be comparable to the transition frequencies of some dipole allowed transitions.

18.10 Problems

18.1. We consider a gauge invariant Lagrange density which contains matter fields $\Phi(x)$ besides the electromagnetic fields $A_\mu(x)$,

$$\mathcal{L} = \mathcal{L}_m(\Phi, \Phi^+, \partial\Phi - \mathrm{i}(q/\hbar)A\Phi, \partial\Phi^+ + \mathrm{i}(q/\hbar)\Phi^+ A) - \frac{1}{4\mu_0}F_{\mu\nu}F^{\mu\nu}.$$

Equation (16.13) yields for the conserved charged current density from phase invariance

$$\delta\Phi(x) = \frac{\mathrm{i}}{\hbar}q\varphi\Phi(x), \quad \delta\Phi^+(x) = -\frac{\mathrm{i}}{\hbar}q\varphi\Phi^+(x)$$

after division by the irrelevant constant factor φ the expression

$$\begin{aligned} j_q^\mu &= -\frac{1}{\varphi}\delta\Phi\cdot\frac{\partial\mathcal{L}}{\partial(\partial_\mu\Phi)} - \frac{1}{\varphi}\delta\Phi^+\cdot\frac{\partial\mathcal{L}}{\partial(\partial_\mu\Phi^+)} \\ &= -\frac{\mathrm{i}}{\hbar}q\Phi\cdot\frac{\partial\mathcal{L}}{\partial(\partial_\mu\Phi)} + \frac{\mathrm{i}}{\hbar}q\Phi^+\cdot\frac{\partial\mathcal{L}}{\partial(\partial_\mu\Phi^+)}. \end{aligned}$$

On the other hand, the current density that appears in Maxwell's equations

$$\partial_\mu F^{\mu\nu} = -\mu_0 j^\nu$$

is $j^\mu = \partial\mathcal{L}/\partial A_\mu$. Why are those two current densities the same, $j_q^\mu = j^\mu$?

18.2. Prove that the vector field (18.15) satisfies $\nabla\cdot\boldsymbol{A}_J(\boldsymbol{x},t) = 0$.

18.3. Show that in the gauge $\Phi = 0$ the conjugate momentum $\boldsymbol{\Pi}_A = \partial\mathcal{L}/\partial\dot{\boldsymbol{A}} = \epsilon_0\dot{\boldsymbol{A}}$ also yields the Hamiltonian density $\mathcal{H}$ through the standard Lagrangian expression

$$\mathcal{H} = \boldsymbol{\Pi}_A\cdot\dot{\boldsymbol{A}} - \mathcal{L} = \epsilon_0\dot{\boldsymbol{A}}^2 - \mathcal{L}.$$

18.4. We can solve the Coulomb equation (18.2) for the scalar potential Φ also without invoking any particular gauge. How does this generalize equations (18.9) and (18.10)?

Show that taking the divergence of the generalization of equation (18.10) yields a trivially fulfilled equation.

18.5a. The action (18.29) of electromagnetic fields is invariant under Lorentz transformations

$$\epsilon^{\mu} = -\delta x^{\mu} = -\varphi^{\mu\nu}x_{\nu}, \quad \varphi^{\mu\nu} = -\varphi^{\nu\mu},$$
$$\delta A_{\mu}(x) = A'_{\mu}(x') - A_{\mu}(x) = \varphi_{\mu\nu}A^{\nu}(x).$$

Use a procedure similar to the derivation of the energy-momentum tensor (18.32) to derive the densities and currents

$$\mathcal{M}_{\alpha\beta}{}^{\mu} = \frac{1}{c}\left(x_{\alpha}T_{\beta}{}^{\mu} - x_{\beta}T_{\alpha}{}^{\mu}\right) \tag{18.126}$$

of the corresponding conserved charges

$$M_{\alpha\beta} = \int d^3\boldsymbol{x}\,\mathcal{M}_{\alpha\beta}{}^{0}.$$

Hint: You have to add the improvement term $\partial_{\nu}(x_{\beta}A_{\alpha}F^{\mu\nu} - x_{\alpha}A_{\beta}F^{\mu\nu})/\mu_0$ to j^{μ} from equation (16.13) to get the gauge invariant expression (18.126) for the angular momentum densities and currents.

18.5b. The angular momentum of the electromagnetic fields is

$$\boldsymbol{M} = \frac{1}{2}\boldsymbol{e}_i\epsilon_{ijk}M_{jk} = \int d^3\boldsymbol{x}\,\epsilon_0\boldsymbol{x}\times(\boldsymbol{E}\times\boldsymbol{B}), \tag{18.127}$$

but what is the meaning of the conserved quantities M^{0i}?

We define an energy weighted location of the electromagnetic fields,

$$\langle\boldsymbol{x}\rangle = \frac{1}{E}\int d^3\boldsymbol{x}\,\boldsymbol{x}\mathcal{H}, \tag{18.128}$$

where $\mathcal{H}$ is the energy density (18.33) of the electromagnetic fields. Show that the conservation of M^{0i} implies a conservation law for "center of energy" motion for the freely evolving electromagnetic fields,

$$\langle\boldsymbol{x}\rangle(t) = \langle\boldsymbol{x}\rangle(0) + \frac{c^2\boldsymbol{P}}{E}t. \tag{18.129}$$

18.6. A helium-neon laser produces a light wave with a central wavelength of 632.8 nm and a power of 5 mW. Suppose the electric component is a sine oscillation $|\boldsymbol{E}(\boldsymbol{x},t)| \propto \sin(\boldsymbol{k}\cdot\boldsymbol{x} - ckt)$ and is polarized in x direction. We also assume that the frequency profile is Gaussian with a relative width $\Delta f/f = 3.16 \times 10^{-6}$. Which photon state describes this light wave? How many photons does the electromagnetic wave contain?

18.7. Show that the state

$$|n,\ell,m_\ell,\sigma;\boldsymbol{k},\alpha\rangle = \int d^3\boldsymbol{x}\,\Psi_{n,\ell,m_\ell}(\boldsymbol{x})\psi_\sigma^+(\boldsymbol{x})a_\alpha^+(\boldsymbol{k})|0\rangle$$

satisfies

$$H_0|n,\ell,m_\ell,\sigma;\boldsymbol{k},\alpha\rangle = (E_{n,\ell} + \hbar ck)|n,\ell,m_\ell,\sigma;\boldsymbol{k},\alpha\rangle,$$

where

$$H_0 = \int d^3\boldsymbol{x}\left(\frac{\hbar^2}{2m}\sum_\sigma \nabla\psi_\sigma^+\cdot\nabla\psi_\sigma + \sum_\sigma \psi_\sigma^+ V\psi_\sigma + \frac{\epsilon_0}{2}\dot{\boldsymbol{A}}^2 + \frac{(\nabla\times\boldsymbol{A})^2}{2\mu_0}\right).$$

You have to use that the atomic orbital satisfies

$$-\frac{\hbar^2}{2m}\Delta\Psi_{n,\ell,m_\ell}(\boldsymbol{x}) + V(\boldsymbol{x})\Psi_{n,\ell,m_\ell}(\boldsymbol{x}) = E_{n,\ell}\Psi_{n,\ell,m_\ell}(\boldsymbol{x}).$$

It is also useful to keep the $\boldsymbol{x}$ representation for the electronic part of H_0, but to use the $\boldsymbol{k}$ representation for the photon contributions in H_0.

18.8. Calculate the emission rate for unpolarized photons from the 2p state to the ground state of hydrogen in first order and dipole approximation.

Which estimate do you get from this for the lifetime of 2p states?

Which estimate do you get from this for the radiated power from decay of 2p states?

18.9. Calculate the integrated photon absorption cross section,

$$G_{1s\to 2p} = \int_0^\infty \frac{dk}{k}\,\sigma_{1,0\to 2,1}(k)$$

due to the transition from 1s to 2p states in hydrogen.

18.10a. Show that the first order scattering matrix elements (18.78) and (18.93) for emission and absorption can also be gotten in a semi-classical approximation from a perturbation operator

$$V(t) = -q\mathbf{x}\cdot\boldsymbol{E}(\mathbf{x},t) \tag{18.130}$$

with $\boldsymbol{E}(\mathbf{x},t)$ corresponding to a single photon electric field

$$\boldsymbol{E}_{\alpha}^{(+)}(\boldsymbol{x},t) = -\dot{\boldsymbol{A}}_{\alpha}^{(+)}(\boldsymbol{x},t) = -\mathrm{i}\sqrt{\frac{\hbar\mu_0 c^3 k}{16\pi^3}}\boldsymbol{\epsilon}_{\alpha}(\boldsymbol{k})\exp[-\mathrm{i}(\boldsymbol{k}\cdot\boldsymbol{x}-ckt)]$$

for emission, and to

$$\boldsymbol{E}_{\alpha}^{(-)}(\boldsymbol{x},t) = -\dot{\boldsymbol{A}}_{\alpha}^{(-)}(\boldsymbol{x},t) = \mathrm{i}\sqrt{\frac{\hbar\mu_0 c^3 k}{16\pi^3}}\boldsymbol{\epsilon}_{\alpha}(\boldsymbol{k})\exp[\mathrm{i}(\boldsymbol{k}\cdot\boldsymbol{x}-ckt)]$$

for absorption.

18.10b. If we would use the same substitution of semi-classical perturbation operators $V(t)$ from (18.79) to (18.130) for the calculation of scattering in dipole approximation $\exp(\pm\mathrm{i}\boldsymbol{k}\cdot\boldsymbol{x}) \simeq 1$, we would find the Kramers-Heisenberg formula (18.118) from (18.130), while (18.79) yields the correct result (18.117). Why does the substitution (18.79) → (18.130) not work beyond first order perturbation theory, except in the case of resonances?

Hint: The justification for the transition from the velocity form to the length form of matrix elements is based on

$$\frac{\mathbf{p}}{m} = \frac{\mathrm{i}}{\hbar}[H,\mathbf{x}] \;\Rightarrow\; \langle f|\frac{\mathbf{p}}{m}|i\rangle = \mathrm{i}\omega_{fi}\langle f|\mathbf{x}|i\rangle.$$

18.11. Show that the transition rate (18.85) can formally be derived by incorrectly assuming a Golden Rule for transition between the discrete states $|n,\ell,m_\ell\rangle \to |n',\ell',m'_\ell\rangle$ in a semi-classical approximation (18.130) for the monochromatic perturbation $V(t)$, if we use the density of final states

$$\varrho(E)dE = d^3\boldsymbol{k} = d\Omega k^2 dk = d\Omega E^2 dE/(\hbar c)^3. \tag{18.131}$$

This works because (18.131) is the *density of continuous final states of the emitted photon* in the infinite volume limit, but we would have missed that important piece of information if we would just have naively insisted on using the Golden Rule for calculating the transition rate between states $|n,\ell,m_\ell\rangle \to |n',\ell',m'_\ell\rangle$ due to the monochromatic perturbation $V(t)$. Instead, we would have tried to make sense of the energy preserving δ function by invoking a final electron density of states $\varrho(E_{n'})$, e.g. by using some finite energy width of the final electron state. Any such guess would certainly not have produced the correct factor E^2, and we would also have missed the factor $d\Omega$ because the final electron state $|n',\ell',m'_\ell\rangle$ uses angular momentum quantum numbers instead of angles.

18.12. Ultraviolet photons with an energy $E_\gamma = 10.15\,\text{eV}$ are nearly resonant with the $n = 1 \to n'' = 2$ transition in hydrogen. Use both the result (18.117) and the Kramers-Heisenberg formula (18.118) to estimate the differential scattering cross section for a photon scattering angle of $\pi/2$ if the incident photons are polarized in z direction and move in x direction. Assume that the scattered photons move in y direction with polarization $\boldsymbol{e}_z \cos\alpha + \boldsymbol{e}_x \sin\alpha$.

18.13. Express the photon absorption cross sections from Section 18.7 using the velocity form (instead of the length form) for the matrix elements.

Chapter 19
Quantum Aspects of Materials II

We have already seen in Chapter 10 that basic properties of electron states in materials are determined by quantum effects. This impacts all properties of materials, including their mechanical properties, electrical and thermal conductivities, and optical properties. An example of the inherently quantum mechanical nature of electrical properties is provided by the role of virtual intermediate states in the polarizability tensor in Section 15.3.

We will now continue to illustrate quantum effects in materials with a focus on effects that require the use of second quantization or Lagrangian field theory, or at least the knowledge of exchange interactions for a proper treatment. We will start at the molecular level and then discuss the second quantization of basic excitations in condensed materials.

The inception of the Schrödinger equation was accompanied by a large number of immediate successes, including atomic theory, the quantum theory of photon-atom interactions, and quantum tunneling. Another of these important successes was the development of the theory of covalent chemical bonding, which was initiated by Burrau[1], Heitler and London[2], and others. This is an extremely important and well studied subject in chemistry and molecular physics, and yet it never seemed to reach the level of popularity and recognition that other areas of applied quantum mechanics enjoy. One reason for this lack of popularity might be the lack of simple, beautiful model systems which can be solved analytically. Solvable model systems are of great instructive and illustrative value, and often provide a level of insight that is very hard to attain with systems which can only be analyzed by

[1]Ø. Burrau, Naturwissenschaften 15, 16 (1927); K. Danske Vidensk. Selsk., Mat.-fys. Medd. 7(14) (1927).

[2] W. Heitler, F. London, Z. Phys. 44, 455 (1927).

R. Dick, *Advanced Quantum Mechanics*, Graduate Texts in Physics,
DOI 10.1007/978-3-319-25675-7_19

approximation methods. However, the existence and stability of covalent bonds is clearly an important property of molecules and of materials in general, and a basic quantitative understanding of the covalent bond should be part of the toolbox of every chemist, physicist and materials scientist. Indeed, there is a model system which can be analyzed to some extent by analytic methods. If only basic qualitative features are required, the analytic formulation can then be used for numerical evaluations which do not require a huge amount of effort. This model system is the hydrogen molecule ion H_2^+, which is also known as the dihydrogen cation. The analysis of electron states for fixed locations of the two protons in this simplest molecular system have been investigated already in the early years of quantum mechanics[3], and have been a subject of research ever since, both in terms of the semi-analytic analysis in prolate spheroidal coordinates[4] used in Section 19.2, and in terms of high precision variational calculations[5]. Before specializing to H_2^+ we will discuss the interplay of nuclear and electronic coordinates and the role of the Born-Oppenheimer approximation in molecular physics.

19.1 The Born-Oppenheimer approximation

Molecules can be described by first quantized Hamiltonians of the form

$$H = \sum_i \frac{\boldsymbol{p}_i^2}{2m_e} + \sum_I \frac{\boldsymbol{P}_I^2}{2M_I} + \sum_{I<J} \frac{Z_I Z_J e^2}{4\pi\epsilon_0 |\boldsymbol{R}_I - \boldsymbol{R}_J|} + \sum_{i<j} \frac{e^2}{4\pi\epsilon_0 |\boldsymbol{r}_i - \boldsymbol{r}_j|} - \sum_{i,J} \frac{Z_J e^2}{4\pi\epsilon_0 |\boldsymbol{r}_i - \boldsymbol{R}_J|} \quad (19.1)$$

if we use properly anti-symmetrized wave functions for the electrons and symmetrized or anti-symmetrized wave functions for bosonic or fermionic nuclei of the same kind. Here lower case indices enumerate electrons while upper case indices refer to nuclei.

[3] A.H. Wilson, Proc. Roy. Soc. London A 118, 617, 635 (1928); E. Teller, Z. Phys. 61, 458 (1930); E.A. Hylleraas, Z. Phys. 71, 739 (1931); G. Jaffé, Z. Phys. 87, 535 (1934).

[4] See e.g. G. Hunter, H.O. Pritchard, J. Chem. Phys. 46, 2146 (1967); M. Aubert, N. Bessis, G. Bessis, Phys. Rev. A 10, 51 (1974); T.C. Scott, M. Aubert-Frécon, J. Grotendorf, Chem. Phys. 324, 323 (2006).

[5] B. Grémaud, D. Delande, N. Billy, J. Phys. B 31, 383 (1998); M.M. Cassar, G.W.F. Drake, J. Phys. B 37, 2485 (2004); H. Li, J. Wu, B.-L. Zhou, J.-M. Zhu, Z.-C. Yan, Phys. Rev. A 75, 012504 (2007).

Otherwise, we might just as well use the second quantized Schrödinger picture Hamiltonian

$$H = \int d^3x \left(\frac{\hbar^2}{2m_e} \nabla\psi_e^+(x) \cdot \nabla\psi_e(x) + \sum_A \frac{\hbar^2}{2M_A} \nabla\psi_A^+(x) \cdot \nabla\psi_A(x) \right)$$
$$+ \int d^3x \int d^3x' \frac{e^2}{4\pi\epsilon_0|x-x'|} \left(\sum_{A<B} Z_A Z_B \psi_A^+(x)\psi_B^+(x')\psi_B(x')\psi_A(x) \right.$$
$$+ \sum_A \frac{Z_A}{2} \psi_A^+(x)\psi_A^+(x')\psi_A(x')\psi_A(x) + \frac{1}{2}\psi_e^+(x)\psi_e^+(x')\psi_e(x')\psi_e(x)$$
$$\left. - \sum_A Z_A \psi_e^+(x)\psi_A^+(x')\psi_A(x')\psi_e(x) \right), \qquad (19.2)$$

where the labels A, B enumerate different kinds of nuclei. We assume that there are N_e electrons and $N_n = \sum_A N_A$ nuclei in our molecule. Realistically, we would restrict attention to valence electrons (rather than all electrons), and the numbers A would enumerate different kinds of ion cores. However, in the example of the hydrogen molecule ion below this distinction is void. The choice of kinetic terms also assumes that all the particles are non-relativistic. Indeed, this also informs the choice of interaction terms in the Born-Oppenheimer Hamiltonian. Electromagnetic interactions between non-relativistic charged particles are dominated by the Coulomb interaction, but if there are relativistic charged particles in the system, photon exchange between charged particles through their couplings to the vector potential $\boldsymbol{A}(x)$ becomes important. Domination of the Coulomb interaction in the case of non-relativistic electron-nucleus and electron-electron scattering is demonstrated in Sections 22.2 and 22.4, respectively. Equation (22.29) provides an estimate of the relative importance of photon exchange versus Coulomb interactions for non-relativistic electrons and nuclei.

Spin labels are suppressed in (19.2) and also in the corresponding states below, because they enter trivially in the equations of motion[6].

Note that even in the valence electrons plus ion cores approximation, the Hamiltonians (19.1, 19.2) describe an incredibly complicated quantum mechanical system, even in the case of a "simple" diatomic molecule. This is because the complete spectrum of energy levels and eigenstates of (19.1) does not only include bound molecular states (which is complicated enough), but also scattering states of electrons and of molecular fragments. The Hamiltonian for the hydrogen molecule H_2 describes not only bound states of two protons and two electrons, but also electron scattering off an H_2^+ ion, atomic hydrogen-hydrogen scattering, proton scattering off an H^- ion, and a plasma of free protons and electrons. However, our

[6]We would have to be more careful if we would discuss expectation values, because exchange integrals appear in the expectation values of potential terms, see Section 17.7.

primary interest concerns an understanding of the nature of covalent bonds and of ground state properties of molecules. In this case, we don't have to include the scattering states, and we can even neglect the motion of ion cores.

Born and Oppenheimer have pointed out that it makes intuitive sense to separate nuclear and electronic motion by first solving the electronic problem for fixed nuclear coordinates, and then substituting the electronic solution into a remnant nuclear Schrödinger equation[7]. In the framework of quantized Schrödinger theory this amounts to an electronic Hamiltonian

$$H_e = H - \int d^3x \sum_A \frac{\hbar^2}{2M_A} \nabla\psi_A^+(\boldsymbol{x}) \cdot \nabla\psi_A(\boldsymbol{x}) \tag{19.3}$$

with corresponding parameter dependent electronic states

$$\begin{aligned} |\boldsymbol{n}; \boldsymbol{X}_1, \ldots \boldsymbol{X}_{N_n}\rangle &= \prod_{i=1}^{N_e} \int d^3x_i\, \psi_e^+(\boldsymbol{x}_i) \prod_{I=1}^{N_n} \psi_{A(I)}^+(\boldsymbol{X}_I)|0\rangle \\ &\quad \times \langle \boldsymbol{x}_1, \ldots \boldsymbol{x}_{N_e}|\boldsymbol{n}; \boldsymbol{X}_1, \ldots \boldsymbol{X}_{N_n}\rangle. \end{aligned} \tag{19.4}$$

Here $\psi_e^+(\boldsymbol{x}_i)$ is an electronic creation operator and $\psi_{A(I)}^+(\boldsymbol{X}_I)$ is a creation operator for a nucleus of species A at the location $\boldsymbol{X}_I$. The set of quantum numbers $\boldsymbol{n}$ specifies the state (including the energy level), and the notation $|\boldsymbol{n}; \boldsymbol{X}_1, \ldots \boldsymbol{X}_{N_n}\rangle$ indicates that the electronic state also depends on the location of the nuclei.

The equation of motion for the electronic states (19.4) with the Hamiltonian (19.3) then follows as in Section 17.6, except that here we use a time-independent Schrödinger equation. The equation

$$E_{e,\boldsymbol{n}}(\boldsymbol{X}_1, \ldots \boldsymbol{X}_{N_n})|\boldsymbol{n}; \boldsymbol{X}_1, \ldots \boldsymbol{X}_{N_n}\rangle = H_e|\boldsymbol{n}; \boldsymbol{X}_1, \ldots \boldsymbol{X}_{N_n}\rangle$$

yields with the short hand notation $\langle \boldsymbol{x}|\boldsymbol{n}; \boldsymbol{X}\rangle \equiv \langle \boldsymbol{x}_1, \ldots \boldsymbol{x}_{N_e}|\boldsymbol{n}; \boldsymbol{X}_1, \ldots \boldsymbol{X}_{N_n}\rangle$ the equation

$$\begin{aligned} E_{e,\boldsymbol{n}}(\boldsymbol{X})\langle \boldsymbol{x}|\boldsymbol{n}; \boldsymbol{X}\rangle &= -\frac{\hbar^2}{2m_e} \sum_i \frac{\partial^2}{\partial \boldsymbol{x}_i^2} \langle \boldsymbol{x}|\boldsymbol{n}; \boldsymbol{X}\rangle + \frac{e^2}{4\pi\epsilon_0} \\ &\times \left(\sum_{i<j} \frac{1}{|\boldsymbol{x}_i - \boldsymbol{x}_j|} - \sum_{i,I} \frac{Z_{A(I)}}{|\boldsymbol{x}_i - \boldsymbol{X}_I|} + \sum_{I<J} \frac{Z_{A(I)}Z_{A(J)}}{|\boldsymbol{X}_I - \boldsymbol{X}_J|} \right) \langle \boldsymbol{x}|\boldsymbol{n}; \boldsymbol{X}\rangle. \end{aligned} \tag{19.5}$$

The N_e-electron wave functions $\langle \boldsymbol{x}|\boldsymbol{n}; \boldsymbol{X}\rangle$ are complete in the $3N_e$-dimensional configuration space of the electrons, and therefore the wave functions of the full $(N_e + N_n)$-particle problem can be expanded in the form

$$\langle \boldsymbol{x}, \boldsymbol{X}|E\rangle = \sum_{\boldsymbol{n}} c(\boldsymbol{n}; \boldsymbol{X})\langle \boldsymbol{x}|\boldsymbol{n}; \boldsymbol{X}\rangle. \tag{19.6}$$

[7] M. Born, J.R. Oppenheimer, Annalen Phys. 84, 457 (1927).

The sum over the quantum numbers $\boldsymbol{n}$ also involves at least one integration over a continuous quantum number for the scattering states.

On the level of the second quantized theory, the amplitude (19.6) corresponds to the $(N_e + N_n)$-particle state

$$\begin{aligned}|E\rangle &= \prod_{i=1}^{N_e}\int d^3\boldsymbol{x}_i\,\psi_e^+(\boldsymbol{x}_i)\prod_{I=1}^{N_n}\int d^3\boldsymbol{X}_I\,\psi_{A(I)}^+(\boldsymbol{X}_I)|0\rangle\langle\boldsymbol{x},\boldsymbol{X}|E\rangle\\ &= \prod_{I=1}^{N_n}\int d^3\boldsymbol{X}_I\sum_{\boldsymbol{n}} c(\boldsymbol{n};\boldsymbol{X})|\boldsymbol{n};\boldsymbol{X}\rangle,\end{aligned}$$

where the parameter-dependent electronic state $|\boldsymbol{n};\boldsymbol{X}\rangle$ is given in (19.4).

Substituting (19.6) into the full $(N_e + N_n)$-particle Schrödinger equation

$$H|E\rangle = E|E\rangle$$

yields the equation

$$\sum_{\boldsymbol{n}}\left(\sum_{I=1}^{N_n}\frac{\hbar^2}{2M_{A(I)}}\frac{\partial^2}{\partial\boldsymbol{X}_I^2} - E_{e,\boldsymbol{n}}(\boldsymbol{X}) + E\right)c(\boldsymbol{n};\boldsymbol{X})\langle\boldsymbol{x}|\boldsymbol{n};\boldsymbol{X}\rangle = 0. \tag{19.7}$$

This can be resolved into a set of coupled equations for the nuclear factors $c(\boldsymbol{n};\boldsymbol{X})$ through orthogonality of the electron factors $\langle\boldsymbol{x}|\boldsymbol{n};\boldsymbol{X}\rangle$. If this is done, no approximation has been made so far to the problem to solve the molecular Hamiltonian (19.2). However, if we are in the center of mass frame of the nuclei, and if both rotational and vibrational excitations are small, we can neglect the nuclear kinetic terms, and we find for these nuclear configurations $\boldsymbol{X}^{(0)}$ that their energy levels can be approximated by

$$E = E_{e,\boldsymbol{n}}(\boldsymbol{X}^{(0)}). \tag{19.8}$$

The corresponding full molecular eigenstate in this approximation has a wave function

$$\langle\boldsymbol{x},\boldsymbol{X}|E_{e,\boldsymbol{n}}(\boldsymbol{X}^{(0)})\rangle = \delta(\boldsymbol{X}-\boldsymbol{X}^{(0)})\langle\boldsymbol{x}|\boldsymbol{n};\boldsymbol{X}^{(0)}\rangle, \tag{19.9}$$

and a corresponding second quantized state

$$\begin{aligned}|E_{e,\boldsymbol{n}}(\boldsymbol{X}^{(0)})\rangle &= \prod_{i=1}^{N_e}\int d^3\boldsymbol{x}_i\,\psi_e^+(\boldsymbol{x}_i)\prod_{I=1}^{N_n}\int d^3\boldsymbol{X}_I\,\psi_{A(I)}^+(\boldsymbol{X}_I)|0\rangle\\ &\quad\times\langle\boldsymbol{x},\boldsymbol{X}|E_{e,\boldsymbol{n}}(\boldsymbol{X}^{(0)})\rangle = |\boldsymbol{n};\boldsymbol{X}^{(0)}\rangle.\end{aligned}$$

It might be tempting to conclude from (19.8) that the solution of the electronic equation (19.5) eventually allows us to calculate the nuclear equilibrium configuration $\boldsymbol{X}^{(0)}$ in the aftermath from a requirement $[\partial E_{e,\boldsymbol{n}}(\boldsymbol{X})/\partial\boldsymbol{X}]_{\boldsymbol{X}=\boldsymbol{X}^{(0)}} = 0$. However, *this is not true: The energy level $E_{e,\boldsymbol{n}}(\boldsymbol{X})$ for a general nuclear configuration $\boldsymbol{X}$ represents only the electronic energy plus the electrostatic nuclear potential energy for that configuration.* Equation (19.8) only states that within the Born-Oppenheimer approximation, the energy $E_{e,\boldsymbol{n}}(\boldsymbol{X})$ and the full molecular energy coincide in an equilibrium configuration, but that does *not* imply that the two energies coincide in a *neighborhood* of an equilibrium configuration. As a consequence the energy $E_{e,\boldsymbol{n}}(\boldsymbol{X})$ and the full molecular energy can (and generically will) have *different gradients* with respect to the nuclear configuration, even in a molecular equilibrium configuration. The function $E_{e,\boldsymbol{n}}(\boldsymbol{X})$ may have non-vanishing gradient in the molecular equilibrium configuration because it neglects the contributions from nuclear kinetic terms.

Therefore we have to use *a priori* knowledge of the equilibrium configuration $\boldsymbol{X}^{(0)}$, e.g. from scattering experiments, to calculate the molecular energy in the Born-Oppenheimer approximation. *We cannot calculate both the energy and the equilibrium configuration from* (19.5).

19.2 Covalent bonding: The dihydrogen cation

The stability of molecules is an issue in classical physics in the same sense as the stability of atoms is an issue. It is not surprising that sharing of electrons yields a net attractive force between positively charged nuclei or atomic cores. Consider e.g. two protons at separation b with an electron right in the middle between the protons. The net classical electrostatic energy of the system $\propto -3e^2/b$ is attractive, but the problem is again to prevent collapse of the system. The corresponding quantum mechanical system is again stabilized by wave particle duality. Squeezing the particles very tight together implies strongly peaked wave functions, hence too much curvature in the wave functions, and the ensuing increase in kinetic energy eventually cannot be compensated any more by gains in potential energy terms for normalizable wave functions.

We apply the basic tenet of the Born-Oppenheimer approximation to the hydrogen molecule ion H_2^+ and determine approximate molecular orbitals under the assumption that the two protons are fixed at their equilibrium separation b. The distances of the electron from the two protons are given by

$$r_\pm^2 = x^2 + y^2 + (z \pm (b/2))^2 \tag{19.10}$$

if we assume that the two protons are located on the z axis at $z = \pm b/2$. A suitable set of coordinates for the 2-center Coulomb problem are given by

$$\xi^+ = r_+ + r_-, \quad b \le \xi^+,$$
$$\xi^- = r_+ - r_-, \quad -b \le \xi^- \le b$$

and the azimuthal angle φ around the z axis. These coordinates are known as prolate spheroidal coordinates. They seem to have been used for the analysis of classical 2-center gravitational or electrostatic problems and for acoustic and electromagnetic radiation problems since the 19th century.

The surfaces $\xi^+ = const.$ are ellipsoids with the protons in the focal points, while the surfaces $\xi^- = const.$ are the corresponding hyperboloids. The ξ^- coordinate lines take us from one hyperboloid $\xi^- = const.$ to another hyperboloid $\xi^- = const.$ for constant ξ^+ and φ. For given value of ξ^+, going from $\xi^- = -b$ to $\xi^- = b$ takes us from the south pole of the ellipsoid $\xi^+ = const.$ to its north pole, i.e. ξ^-/b is similar to the ϑ coordinate on a sphere, except that we move from negative z to positive z for increasing ξ^-. The advantage of this is that $z > 0$ corresponds to $\xi^- > 0$, but the right handed prolate spheroidal coordinate system is then $\{\xi^-, \xi^+, \varphi\}$.

The ξ^+ coordinate lines are hyperbolas $\xi^- = const.$, $\varphi = const.$ with the protons in the focal points. $\xi^+ = b$ corresponds to the line $-b/2 \leq z \leq b/2$ on the z axis and $\xi^+ \to \infty$ takes us to infinite distance from the protons, i.e. ξ^+ plays a role similar to the radius r in spherical coordinates.

We apply the methods of Section 5.4 to determine tangent vectors to the coordinate lines and the relevant differential operators. We have

$$2r^2 + \frac{b^2}{2} = r_+^2 + r_-^2 = \frac{1}{2}(\xi^+)^2 + \frac{1}{2}(\xi^-)^2$$

and

$$z = \frac{\xi^+\xi^-}{2b},$$

and this implies also

$$x^2 + y^2 = \frac{b^2(\xi^+)^2 + b^2(\xi^-)^2 - (\xi^+\xi^-)^2 - b^4}{4b^2} = \frac{[(\xi^+)^2 - b^2][b^2 - (\xi^-)^2]}{4b^2},$$

$$x = \frac{1}{2b}\sqrt{[(\xi^+)^2 - b^2][b^2 - (\xi^-)^2]}\cos\varphi,$$

$$y = \frac{1}{2b}\sqrt{[(\xi^+)^2 - b^2][b^2 - (\xi^-)^2]}\sin\varphi.$$

The dual basis vectors (5.21) are in the present case

$$\nabla\xi^+ = \frac{1}{2r_+r_-}\left(2\xi^+\boldsymbol{r} - b\xi^-\boldsymbol{e}_z\right), \quad \nabla\xi^- = -\frac{1}{2r_+r_-}\left(2\xi^-\boldsymbol{r} - b\xi^+\boldsymbol{e}_z\right),$$

and

$$\nabla\varphi = \frac{x\boldsymbol{e}_y - y\boldsymbol{e}_x}{x^2 + y^2}.$$

This yields a diagonal inverse metric with components

$$g^{++}=4\frac{(\xi^+)^2-b^2}{(\xi^+)^2-(\xi^-)^2},\quad g^{--}=4\frac{b^2-(\xi^-)^2}{(\xi^+)^2-(\xi^-)^2},$$
$$g^{\varphi\varphi}=\frac{4b^2}{[(\xi^+)^2-b^2][b^2-(\xi^-)^2]},$$

and the volume measure (5.27) for $d\xi^- d\xi^+ d\varphi$ follows as

$$\sqrt{g}=\left(g^{++}g^{--}g^{\varphi\varphi}\right)^{-1/2}=\frac{1}{8b}[(\xi^+)^2-(\xi^-)^2]. \tag{19.11}$$

The Laplace operator (5.26) in spheroidal coordinates is therefore

$$\Delta=\frac{4}{(\xi^+)^2-(\xi^-)^2}\left[\partial_+\left((\xi^+)^2-b^2\right)\partial_+ +\partial_-\left(b^2-(\xi^-)^2\right)\partial_-\right] + \frac{4b^2}{[(\xi^+)^2-b^2][b^2-(\xi^-)^2]}\partial_\varphi^2. \tag{19.12}$$

On the other hand, the coordinate dependence of the electrostatic potential of the electron is

$$\frac{1}{r_+}+\frac{1}{r_-}=\frac{4\xi^+}{(\xi^+)^2-(\xi^-)^2},$$

and therefore the Hamiltonian in the $\{\xi^+,\xi^-,\varphi\}$ representation satisfies

$$\frac{m_e}{2\hbar^2}[(\xi^-)^2-(\xi^+)^2]H=\partial_+\left((\xi^+)^2-b^2\right)\partial_+ +\partial_-\left(b^2-(\xi^-)^2\right)\partial_- + \left(\frac{b^2}{(\xi^+)^2-b^2}+\frac{b^2}{b^2-(\xi^-)^2}\right)\partial_\varphi^2+\frac{m_e e^2}{2\pi\epsilon_0\hbar^2}\xi^+. \tag{19.13}$$

The Hamiltonian H commutes with the azimuthal angular momentum operator L_z, and therefore we can discuss the spectrum and eigenfunctions of H within the subspaces of L_z eigenvalues $m\hbar$,

$$\psi_m(\xi^+,\xi^-,\varphi)=\frac{1}{\sqrt{2\pi}}\psi(\xi^+,\xi^-)\exp(im\varphi).$$

Within these subspaces, the normalization condition on the bound electron states becomes with (19.11),

$$\int_b^\infty d\xi^+\int_{-b}^b d\xi^-\,[(\xi^+)^2-(\xi^-)^2]\left|\psi(\xi^+,\xi^-)\right|^2=8b, \tag{19.14}$$

and the Hamiltonian H_m acting within these subspaces satisfies

$$\frac{m_e}{2\hbar^2}[(\xi^-)^2 - (\xi^+)^2](H_m - E) = D_{+,m}(\xi^+) - D_{-,m}(\xi^-),$$

$$D_{+,m}(\xi^+) = \partial_+ \left((\xi^+)^2 - b^2\right) \partial_+ - \frac{m^2 b^2}{(\xi^+)^2 - b^2} + \frac{m_e}{2\hbar^2} E(\xi^+)^2 + \frac{m_e e^2}{2\pi\epsilon_0 \hbar^2}\xi^+,$$

$$D_{-,m}(\xi^-) = \partial_- \left((\xi^-)^2 - b^2\right) \partial_- - \frac{m^2 b^2}{(\xi^-)^2 - b^2} + \frac{m_e}{2\hbar^2} E(\xi^-)^2.$$

Here the energy E differs from the energy E_e (19.8) of the molecule in the Born-Oppenheimer approximation by the electrostatic energy of the nuclei,

$$E_e = E + \frac{e^2}{4\pi\epsilon_0 b}. \tag{19.15}$$

Since H_m is hermitian with respect to the scalar product appearing in (19.14), the differential operators $D_{+,m}$ and $D_{-,m}$ must be hermitian with respect to the scalar products

$$\langle \psi_+ | \phi_+ \rangle_+ = \int_b^\infty d\xi^+ \, \psi_+^+(\xi^+)\phi_+(\xi^+)$$

and

$$\langle \psi_- | \phi_- \rangle_- = \int_{-b}^b d\xi^+ \, \psi_-^+(\xi^-)\phi_-(\xi^-),$$

respectively. The corresponding Sturm-Liouville type boundary conditions can be read off from the differential operators. We must certainly have

$$\lim_{\xi^+ \to \infty} \psi_+(\xi^+) = 0. \tag{19.16}$$

For azimuthal quantum numbers $m \neq 0$ we must also require

$$\lim_{\xi^+ \to b} \psi_+(\xi^+) = 0, \quad \lim_{\xi^- \to \pm b} \psi_-(\xi^-) = 0. \tag{19.17}$$

Note that $\xi^+ = b$ corresponds to the interval $-b/2 \leq z \leq b/2$ on the z axis, while $\xi^- = -b$ and $\xi^- = b$ correspond to the half-lines $z \leq -b/2$ and $z \geq b/2$ on the z axis, respectively. The boundary conditions (19.17) therefore imply that the wave functions

$$\psi_m(\xi^+, \xi^-, \varphi) = \frac{1}{\sqrt{2\pi}} \psi_+(\xi^+)\psi_-(\xi^-) \exp(im\varphi)$$

must vanish on the z axis if $m \neq 0$, which apparently makes sense.

We certainly should not expect that the molecular orbitals with $m = 0$ vanish on the z axis, and the differential operators $D_{\pm,0}$ are actually hermitian on their respective domains without extra boundary conditions at $\xi^- = \pm b$ or $\xi^+ = b$ except that the wave functions should remain finite in those points.

The point of this discourse about hermiticity of the operators $D_{\pm,m}$ is that as a consequence, separation of the electronic Schrödinger equation for the hydrogen molecule ion H_2^+ in terms of prolate spheroidal coordinates will not only give us solutions, but a *complete set* of solutions in the form

$$\psi_{m,\lambda}(\xi^+,\xi^-,\varphi) = \frac{1}{\sqrt{2\pi}}\psi_{+,\lambda}(\xi^+)\psi_{-,\lambda}(\xi^-)\exp(im\varphi), \tag{19.18}$$

$$D_{-,m}(\xi^-)\psi_{-,\lambda}(\xi^-) = \lambda\psi_{-,\lambda}(\xi^-), \quad -b \le \xi^- \le b, \tag{19.19}$$

$$D_{+,m}(\xi^+)\psi_{+,\lambda}(\xi^+) = \lambda\psi_{+,\lambda}(\xi^+). \quad \xi^+ \ge b. \tag{19.20}$$

Energy is a third quantum number which is treated as implicit in the notation for the states.

The equation (19.19) and the equation (19.20) for $e^2 = 0$ are relevant for radiation problems and have been studied extensively, see [1] and references there. The solutions are known as angular spheroidal functions and radial spheroidal functions because of the angular and radial interpretation of the coordinates ξ^- and ξ^+, respectively.

The $\xi^+ \to \infty$ limit of equation (19.20) immediately tells us that we can satisfy the boundary condition (19.16) only for negative energy,

$$\frac{m_e}{2\hbar^2}E = -\kappa^2,$$

and the asymptotic form of the solution should be

$$\psi_{+,\lambda}(\xi^+) = f_{+,\lambda}(\xi^+)\exp(-\kappa\xi^+) \quad \kappa > 0. \tag{19.21}$$

with $\lim_{\xi\to\infty} f_{+,\lambda}(\xi)\exp(-\kappa\xi) = 0$.

We wish to analyze in particular the sector $m = 0$, which should contain the ground state of the H_2^+ ion. Equation (19.20) with $m = 0$ has the form

$$\partial_\xi \left(\xi^2 - b^2\right)\partial_\xi\psi_\lambda(\xi) - \kappa^2\xi^2\psi_\lambda(\xi) + \frac{2}{a_e}\xi\psi_\lambda(\xi) = \lambda\psi_\lambda(\xi), \tag{19.22}$$

where we substituted $\xi^+ \to \xi$, $\psi_+ \to \psi$ because in the following it will be clear from presence or absence of the Coulomb term $\propto 1/a_e$ whether we are considering the radial or the angular spheroidal coordinates and wave functions.

The length parameter

$$a_e = \frac{4\pi\epsilon_0\hbar^2}{m_e e^2} = \frac{\mu}{m_e}a$$

is closely related to the Bohr radius (7.62) of the hydrogen atom.

Since our solution should remain finite at $\xi = b$, we make an *ansatz*

$$\psi_{+,\lambda}(\xi) = \sum_{n\geq 0} c_n \left(\frac{\xi - b}{b}\right)^n \exp[\kappa(b - \xi)]. \tag{19.23}$$

Substitution into (19.22) yields a two-step recursion relation

$$2(n+1)^2 c_{n+1} = \left(\lambda + \kappa^2 b^2 + 2\kappa b - \frac{2b}{a_e} + 4\kappa bn - n(n+1)\right) c_n + 2b\left(\kappa n - \frac{1}{a_e}\right) c_{n-1}. \tag{19.24}$$

On the other hand, $\psi_{-,\lambda}(\xi^-)$ must satisfy the differential equation (19.22) without electrostatic term: $a_e \to \infty$,

$$\partial_\xi \left(\xi^2 - b^2\right) \partial_\xi \psi_\lambda(\xi) - \kappa^2 \xi^2 \psi_\lambda(\xi) = \lambda \psi_\lambda(\xi), \tag{19.25}$$

and on the interval $-b \leq \xi \leq b$. This equation allows for even and odd solutions under $\xi^- \to -\xi^-$, and we expect the ground state solution to be even. Therefore we try an *ansatz*

$$\psi_{-,\lambda}(\xi) = \sum_{n\geq 0} d_n \left(\frac{\xi}{b}\right)^{2n}, \tag{19.26}$$

where we can set e.g.

$$d_0 = 1 \tag{19.27}$$

because the product form $\psi_{+,\lambda}(\xi^+)\psi_{-,\lambda}(\xi^-)/\sqrt{2\pi}$ of the ground state implies a degeneracy between d_0 and the coefficient c_0 in the radial factor (19.23). The constant c_0 is then determined by the normalization condition (19.14).

Substitution of (19.26) into (19.25) yields the recursion relation

$$2(n+1)(2n+1)d_{n+1} = \left(4n^2 + 2n - \lambda\right) d_n - \kappa^2 b^2 d_{n-1}. \tag{19.28}$$

The expansions (19.23) and (19.26) are not the standard expansions. For the angular function (19.26) one rather uses an expansion in terms of Legendre polynomials $P_n(\xi/b)$ (or associated Legendre polynomials $P_n^m(\xi/b)$ for $m \neq 0$), which are orthogonal polynomials in $-b \leq \xi \leq b$ and satisfy (19.25) or (19.19) for $\kappa = 0$ and $\lambda = n(n+1)$. For the polynomial factors in the radial function (19.23) one rather uses Laguerre polynomials $L_n(2\kappa(\xi - b))$ or $L_n^m(2\kappa(\xi - b))$, because $L_n^m(2\kappa(\xi - b)) \exp[-\kappa(\xi - b)]$ are complete orthogonal functions in $b \leq \xi \leq \infty$. The corresponding two-step recursion relations for the coefficients in these expansions then follow from the differential equations and recursion relations of the orthogonal polynomials. However, for our purposes the simpler expansions (19.23) and (19.26) are sufficient for the illustration of basic solution techniques for the dihydrogen cation.

We cannot go ahead and simply solve the recursion relations (19.24) and (19.28) to some finite order to get approximate wave functions for the electron, because for generic values of λ and κ^2b^2 the resulting wave functions will not be regular and square integrable in the domains $-b \le \xi^- \le b$ and $1 \le \xi^+ \le \infty$. Therefore, one first has to determine which pairs of parameters λ and κ^2b^2 allow for regular and square integrable solutions.

A classical method for the approximate calculation of the allowed parameter pairs λ and κ^2b^2 in a two-step recursion relation like (19.28) uses the ratios $f_n = d_{n+1}/d_n$ with the initial condition from (19.28), $f_0 = -\lambda/2$. The recursion relation (19.28) can then be written as an upwards recursion $f_{n-1} \to f_n$,

$$f_n = \frac{n}{n+1} - \frac{\lambda}{2(n+1)(2n+1)} - \frac{\kappa^2b^2}{2(n+1)(2n+1)f_{n-1}}, \tag{19.29}$$

or as a downwards recursion $f_{n+1} \to f_n$,

$$f_n = \frac{\kappa^2b^2}{2(n+1)(2n+3) - \lambda - 2(n+2)(2n+3)f_{n+1}}. \tag{19.30}$$

The requirement of finite limits $\psi_{-,\lambda}(\pm b)$ of the angular wave function implies that the solution of (19.29, 19.30) should satisfy

$$\lim_{n\to\infty} f_n = 0.$$

One way to derive the resulting condition on λ and κ^2b^2 in approximate form is to use both relations (19.30) and (19.29) for f_n with the approximation $f_N = 0$ for some $N \gg n$. Iteration of equation (19.30) in $N - n - 1$ steps yields a relation of the form $f_n = f_n^{(-)}(\lambda, \kappa^2b^2, f_N) \simeq f_n^{(-)}(\lambda, \kappa^2b^2, 0)$, while on the other hand f_n is also determined in n steps from equation (19.29) and $f_0 = -\lambda/2$ to yield functions $f_n = f_n^{(+)}(\lambda, \kappa^2b^2)$. The condition

$$f_n^{(-)}(\lambda, \kappa^2b^2, 0) = f_n^{(+)}(\lambda, \kappa^2b^2)$$

then implicitly determines the relation between λ and κ^2b^2.

Another way to derive the relation between λ and κ^2b^2 writes the recursion relation (19.28) as a matrix relation

$$\underline{F}\cdot\boldsymbol{d} = \lambda\boldsymbol{d}$$

with matrix elements

$$F_{n\ge0,n'\ge0} = (4n^2 + 2n)\delta_{n,n'} - \kappa^2b^2\delta_{n,n'+1} - 2(n+1)(2n+1)\delta_{n,n'-1}.$$

The condition

$$\det(\underline{F} - \lambda\underline{1}) = 0 \tag{19.31}$$

is then cut off for an $(N+1) \times (N+1)$ submatrix $F_{0\leq n\leq N, 0\leq n'\leq N}$ to yield a relation between λ and $\kappa^2 b^2$.

Once the relation between λ and $\kappa^2 b^2$ is established, application of the same techniques to (19.24) implies a relation between the remaining parameter $\kappa^2 b^2$ and the parameter b/a_e. Since $\kappa^2 b^2 \propto -E$, this relation determines the quantized energies of the even states (due to the even *ansatz* (19.26)), with $m = 0$.

Application of the same techniques with an odd *ansatz* for $\psi_{-,\lambda}(\xi^-)$ or to the equations with general m yields the approximate energy levels and wave functions of the electron in the dihydrogen cation with fixed centers. The matrix and determinant condition for equation (19.24) are

$$\begin{aligned} C_{n\geq 0, n'\geq 0} &= \left(n(n+1) + 2\frac{b}{a_e} - 4n\kappa b - 2\kappa b - \kappa^2 b^2 \right) \delta_{n,n'} \\ &\quad + 2b\left(\frac{1}{a_e} - n\kappa\right)\delta_{n,n'+1} + 2(n+1)^2\delta_{n,n'-1}, \\ \det(\underline{C} - \lambda\underline{1}) &= 0 \end{aligned} \tag{19.32}$$

Using only 3×3 matrices $\underline{F}$ and $\underline{C}$ in the conditions (19.31) and (19.32) yields a ground state energy

$$E_e = \frac{e^2}{4\pi\epsilon_0 b} - \frac{2\hbar^2}{m_e}\kappa^2 = -14.2\,\text{eV}$$

with eigenvalues $\lambda = -0.490$ and $\kappa b = 1.42$ for a bond length $b = 105$ pm. Using the equivalent of a 4×4 matrix $\underline{F}$ and a 6×6 matrix $\underline{C}$ in the expansions with Legendre and Laguerre polynomials, Aubert *et al.*[8] found $E_e = -16.4\,\text{eV}$ with $\kappa b = 1.485$ for $b = 2a$. Either way, we find that the ground state energy E_e is smaller than the energy $E_1 = -13.6\,\text{eV}$ of a hydrogen atom and a proton at large distance, i.e. sharing the electron stabilizes the dihydrogen cation in spite of the electrostatic repulsion of the protons. The actual dissociation energy $D = E_1 - E_e$ for the dihydrogen cation is about 2.6 eV, i.e. the value of Aubert *et al.* from higher order approximation of the recursion relations is much better, as expected.

The coefficients which follow from the relations (19.24), (19.28), (19.27) and (19.14) for $\lambda = -0.490$ and $\kappa b = 1.42$ are

$$\begin{aligned} &d_0 = 1, \quad d_1 = 0.2451, \quad d_2 = -0.0357, \\ &c_0 = 1.869, \quad c_1 = 0.3760, \quad c_2 = -0.0712. \end{aligned} \tag{19.33}$$

[8] M. Aubert, N. Bessis, G. Bessis, Phys. Rev. A 10, 51 (1974).

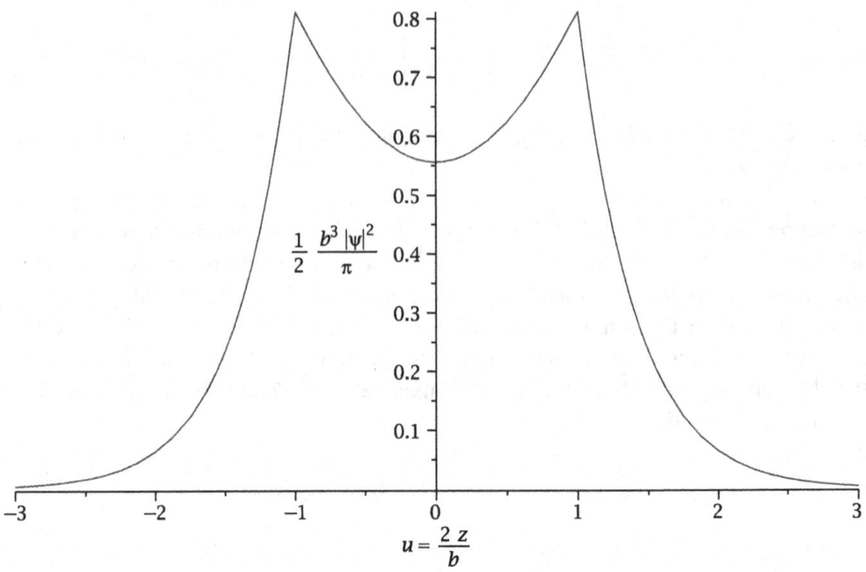

Fig. 19.1 The function $b^3|\psi(\xi^+, \xi^-)|^2/2\pi$ for the approximate ground state (19.33) is displayed along the symmetry axis of the dihydrogen cation. The protons are located at $u = \pm 1$. The abscissa $u = 2z/b$ is $u = \xi^-/b$ in the range $-1 < u < 1$, where $\xi^+ = b$. Outside of this range we have $u = -\xi^+/b$ for $u < -1$ ($\xi^- = -b$) and $u = \xi^+/b$ for $u > 1$ ($\xi^- = b$)

The resulting function $b^3|\psi(\xi^+, \xi^-)|^2/2\pi$ along the symmetry axis of the cation is displayed in Figure 19.1. The abscissa u is related to the z coordinate from equation (19.10) through $u = 2z/b$.

This low order approximation has already all the characteristic features of the real ground state as confirmed by higher order approximations. The electronic wave functions fall off with a linear exponential for large values of the radial coordinate ξ^+, and a double peak appears at the locations of the two protons. However, higher order approximations yield lower energies with a corresponding stronger exponential drop $\exp(-\kappa\xi^+)$, $\kappa b > 1.42$. This implies that the values of $b^3|\psi(\xi^+, \xi^-)|^2$ along the symmetry axis are actually underestimated in the approximation in Figure 19.1, and the cusps become more pronounced in higher order approximations.

Cusps are inevitable in many-particle wave functions for charged particles. Kato had demonstrated that these wave functions have cusps for coalescence of any two charged particles[9]. Specifically, if r_{12} is the separation between two particles with charges $Z_1 e$ and $Z_2 e$, and if the wave function does not vanish for $r_{12} \to 0$, the directional average of $\partial\psi/\partial r_{12}$ in the limit $r_{12} \to 0$ satisfies

$$\lim_{r_{12}\to 0} \frac{1}{4\pi} \int_0^{\pi} d\vartheta \int_0^{2\pi} d\varphi \, \sin\vartheta \frac{\partial\psi}{\partial r_{12}} = \gamma_{12}\psi\Big|_{r_{12}=0}.$$

[9]T. Kato, Commun. Pure Appl. Math. 10, 151 (1957). See also R.T. Pack, W.B. Brown, J. Chem. Phys. 45, 556 (1966) and Á. Nagy, C. Amovilli, Phys. Rev. A 82, 042510 (2010).

The constant γ_{12} is

$$\gamma_{12} = Z_1 Z_2 \alpha \frac{\mu_{12} c}{\hbar} = Z_1 Z_2 \frac{\mu_{12}}{m_e} \frac{1}{a_e},$$

where $\mu_{12} = m_1 m_2/(m_1 + m_2)$ is the reduced mass of the charged particles. In particular, coalescence of two electrons or of electrons and protons corresponds to

$$\gamma_{e^- e^-} = \frac{1}{2a_e} \quad \text{and} \quad \gamma_{e^- p^+} = -\frac{1}{a}.$$

19.3 Bloch and Wannier operators

The use of second quantized Hamiltonians is ubiquitous in condensed matter physics, and in the following sections we will introduce very common and useful examples for this, *viz.* the Hubbard Hamiltonian for electron-electron interactions, phonons, and a basic Hamiltonian for electron-phonon coupling. We will motivate the model Hamiltonians from basic Schrödinger field theory or the classical Hamiltonian for lattice vibrations, respectively, and refer the reader to more specialized monographs for alternative derivations of these Hamiltonians.

However, before we embark on this journey, we should generalize the results from Sections 10.1, 10.2 and 10.3 to three dimensions and combine them with what we had learned in Chapter 17 about quantization and Schrödinger field operators.

The basic Schrödinger picture Hamiltonian for an electron gas has the form

$$\begin{aligned} H &= \int d^3\boldsymbol{x} \int d^3\boldsymbol{x}' \sum_{\sigma,\sigma'} \psi_\sigma^+(\boldsymbol{x}) \psi_{\sigma'}^+(\boldsymbol{x}') \frac{e^2}{8\pi\epsilon_0 |\boldsymbol{x} - \boldsymbol{x}'|} \psi_{\sigma'}(\boldsymbol{x}') \psi_\sigma(\boldsymbol{x}) \\ &+ \int d^3\boldsymbol{x} \sum_\sigma \frac{\hbar^2}{2m} \nabla \psi_\sigma^+(\boldsymbol{x}) \cdot \nabla \psi_\sigma(\boldsymbol{x}) \\ &= \int d^3\boldsymbol{k} \int d^3\boldsymbol{k}' \int d^3\boldsymbol{q} \sum_{\sigma,\sigma'} a_\sigma^+(\boldsymbol{k} + \boldsymbol{q}) a_{\sigma'}^+(\boldsymbol{k}' - \boldsymbol{q}) \frac{e^2}{16\pi^3 \epsilon_0 \boldsymbol{q}^2} a_{\sigma'}(\boldsymbol{k}') a_\sigma(\boldsymbol{k}) \\ &+ \int d^3\boldsymbol{k} \sum_\sigma \frac{\hbar^2 \boldsymbol{k}^2}{2m} a_\sigma^+(\boldsymbol{k}) a_\sigma(\boldsymbol{k}). \end{aligned} \tag{19.34}$$

Suppose that this electron gas exists in a lattice with basis vectors $\boldsymbol{a}_i$ and dual basis vectors $\boldsymbol{a}^i$ (4.18). The lattice points are $\boldsymbol{\ell} = n^i \boldsymbol{a}_i$ with a triplet of integers n^i. However, we can also use the basis $\boldsymbol{a}_i$ as a basis in $\mathbb{R}^3$,

$$\boldsymbol{x} = x^i \boldsymbol{e}_i = \nu^i \boldsymbol{a}_i, \quad \nabla = \boldsymbol{e}^i \frac{\partial}{\partial x^i} = \boldsymbol{a}^i \frac{\partial}{\partial \nu^i}.$$

Note that the coordinates x^i and the lattice basis vectors $\boldsymbol{a}_i$ have the dimensions of length, while the dual basis vectors have dimension length^{-1}. The coordinates ν^i are dimensionless.

A Brillouin zone $\mathcal{B}$ is a unit cell in the dual lattice stretched by a factor 2π and then shifted such that the center of the Brillouin zone is a dual lattice point,

$$\boldsymbol{k} = \kappa_i \boldsymbol{a}^i, \quad -\pi < \kappa_i \leq \pi, \tag{19.35}$$

see also (10.10), where this notion was introduced for one-dimensional lattices.

The $\boldsymbol{k}$ vectors in a Brillouin zone have the following useful properties, which are easily derived from Fourier transformation on a one-dimensional lattice[10] $\kappa_i \equiv \kappa_i + 2\pi$,

$$\int_{\mathcal{B}} d^3\boldsymbol{k}\, \exp[i\boldsymbol{k}\cdot(\boldsymbol{\ell}-\boldsymbol{\ell}')] = (2\pi)^3 \tilde{V} \delta_{\boldsymbol{\ell},\boldsymbol{\ell}'}, \quad d^3\boldsymbol{k} = \tilde{V} d\kappa_1 d\kappa_2 d\kappa_3, \tag{19.36}$$

$$\sum_{\boldsymbol{\ell}} \exp[i(\boldsymbol{k}-\boldsymbol{k}')\cdot\boldsymbol{\ell}] = (2\pi)^3 \tilde{V} \delta(\boldsymbol{k}-\boldsymbol{k}'). \tag{19.37}$$

Recall that the volume of a unit cell $\tilde{V}$ in the dual lattice is related to the volume of a unit cell in the direct lattice through $\tilde{V} = 1/V$, (4.19).

If a unit cell in the lattice contains N ions, electrons in the lattice will also experience a lattice potential

$$H_V(\boldsymbol{x}) = -\sum_{\boldsymbol{\ell},A} \frac{n_A e^2}{4\pi\epsilon_0 |\boldsymbol{x} - \boldsymbol{r}_{\boldsymbol{\ell},A}|}, \tag{19.38}$$

where

$$\boldsymbol{r}_{\boldsymbol{\ell},A} = \boldsymbol{\ell} + \boldsymbol{r}_A \quad 1 \leq A \leq N,$$

enumerates the locations of the ions in the unit cell $\boldsymbol{\ell} = n^i \boldsymbol{a}_i$, and $n_A e$ is the effective charge of the A-th ion. On the level of the quantized Schrödinger field theory, the potential (19.38) adds the operator

$$H_V = -\sum_{\boldsymbol{\ell},A} \int d^3\boldsymbol{x} \sum_{\sigma} \psi_\sigma^+(\boldsymbol{x}) \frac{n_A e^2}{4\pi\epsilon_0 |\boldsymbol{x} - \boldsymbol{r}_{\boldsymbol{\ell},A}|} \psi_\sigma(\boldsymbol{x}), \tag{19.39}$$

to the Hamiltonian (19.34). We will focus on this potential term in the remainder of this section and neglect the electron-electron interaction term in (19.34). The corresponding first quantized Hamiltonian

$$H = \frac{\mathbf{p}^2}{2m} + H_V(\mathbf{x}),$$

[10] We have seen the corresponding one-dimensional equations in (10.1–10.4). However, when comparing equations (19.36) and (19.37) with (10.1–10.4) please keep in mind that the continuous variables κ_i play the role of x there, while the discrete lattice sites $\boldsymbol{\ell} = n^i \boldsymbol{a}_i$ compare to the discrete momenta $2\pi n/a$ in equations (10.1–10.4), see also (10.12).

is invariant under lattice translations,

$$\exp\left(\frac{\mathrm{i}}{\hbar}\boldsymbol{\ell}\cdot\mathbf{p}\right)H\exp\left(-\frac{\mathrm{i}}{\hbar}\boldsymbol{\ell}\cdot\mathbf{p}\right)=\frac{\mathbf{p}^2}{2m}+H_V(\mathbf{x}+\boldsymbol{\ell})=H,$$

and therefore admits a complete set of Bloch type eigenstates, see (10.14) for the one-dimensional case. We can decompose the Schrödinger picture field operators $\psi_\sigma(\boldsymbol{x})$ in terms of a complete set of Bloch type eigenstates

$$\psi_\sigma(\boldsymbol{x})=\sum_n\sqrt{\frac{V}{(2\pi)^3}}\int_{\mathcal{B}}d^3\boldsymbol{k}\,a_{n,\sigma}(\boldsymbol{k})\exp(\mathrm{i}\boldsymbol{k}\cdot\boldsymbol{x})u_n(\boldsymbol{k},\boldsymbol{x}), \tag{19.40}$$

$$a_{n,\sigma}(\boldsymbol{k})=\sqrt{\frac{V}{(2\pi)^3}}\int d^3\boldsymbol{x}\,\exp(-\mathrm{i}\boldsymbol{k}\cdot\boldsymbol{x})u_n^+(\boldsymbol{k},\boldsymbol{x})\psi_\sigma(\boldsymbol{x}), \tag{19.41}$$

with periodic Bloch factors

$$u_n(\boldsymbol{k},\boldsymbol{x}+\boldsymbol{\ell})=u_n(\boldsymbol{k},\boldsymbol{x}).$$

We denote integration over the unit cell of the lattice with $\int_V d^3\boldsymbol{x}$. Normalization of the Bloch energy eigenfunctions then yields

$$\begin{aligned}\delta_{mn}\delta(\boldsymbol{k}-\boldsymbol{k}')&=\frac{V}{(2\pi)^3}\int d^3\boldsymbol{x}\,\exp[\mathrm{i}(\boldsymbol{k}-\boldsymbol{k}')\cdot\boldsymbol{x}]u_m^+(\boldsymbol{k}',\boldsymbol{x})u_n(\boldsymbol{k},\boldsymbol{x})\\&=\frac{V}{(2\pi)^3}\sum_{\boldsymbol{\ell}}\exp[\mathrm{i}(\boldsymbol{k}-\boldsymbol{k}')\cdot\boldsymbol{\ell}]\\&\quad\times\int_V d^3\boldsymbol{x}\,\exp[\mathrm{i}(\boldsymbol{k}-\boldsymbol{k}')\cdot\boldsymbol{x}]u_m^+(\boldsymbol{k}',\boldsymbol{x})u_n(\boldsymbol{k},\boldsymbol{x}),\end{aligned} \tag{19.42}$$

and with (19.37) we find

$$\int_V d^3\boldsymbol{x}\,u_m^+(\boldsymbol{k},\boldsymbol{x})u_n(\boldsymbol{k},\boldsymbol{x})=\delta_{mn}.$$

Equation (19.42) also implies with the canonical anticommutation relations for the Schrödinger field operators $\psi_\sigma(\boldsymbol{x})$ and $\psi_\sigma^+(\boldsymbol{x})$ that the the operators $a_{n,\sigma}(\boldsymbol{k})$ satisfy the relations

$$\{a_{n,\sigma}(\boldsymbol{k}),a_{n',\sigma'}(\boldsymbol{k}')\}=0\quad\{a_{n,\sigma}(\boldsymbol{k}),a_{n',\sigma'}^+(\boldsymbol{k}')\}=\delta_{n,n'}\delta_{\sigma\sigma'}\delta(\boldsymbol{k}-\boldsymbol{k}').$$

The second quantized state

$$|n,\sigma,\boldsymbol{k}\rangle=a_{n,\sigma}^+(\boldsymbol{k})|0\rangle$$

is therefore a state with an electron in the first quantized orbital Bloch state

$$\phi_n(\boldsymbol{k},\boldsymbol{x}) = \sqrt{\frac{V}{(2\pi)^3}}\exp(\mathrm{i}\boldsymbol{k}\cdot\boldsymbol{x})u_n(\boldsymbol{k},\boldsymbol{x}) \tag{19.43}$$

and spin projection σ. Equation (19.41) and the conjugate equation for $a^+_{n,\sigma}(\boldsymbol{k})$ are a special case of our general observations (17.59) and (17.58) how annihilation and creation operators for particles in specific states relate to the generic operators $\psi_\sigma(\boldsymbol{x})$ and $\psi^+_\sigma(\boldsymbol{x})$.

Since the operators $a_{n,\sigma}(\boldsymbol{k})$ are restricted to the Brillouin zone, or equivalently are periodic in the rescaled dual lattice with the Brillouin zone as unit cell,

$$a_{n,\sigma}(\boldsymbol{k}) = a_{n,\sigma}(\boldsymbol{k}+2\pi\tilde{\boldsymbol{\ell}}),\quad \tilde{\boldsymbol{\ell}} = n_i\boldsymbol{a}^i,$$

we can expand them using equations (19.36, 19.37),

$$a^+_{n,\sigma}(\boldsymbol{k}) = \sqrt{\frac{V}{(2\pi)^3}}\sum_{\boldsymbol{\ell}}\psi^+_{n,\sigma}(\boldsymbol{\ell})\exp(\mathrm{i}\boldsymbol{k}\cdot\boldsymbol{\ell}), \tag{19.44}$$

$$\psi^+_{n,\sigma}(\boldsymbol{\ell}) = \sqrt{\frac{V}{(2\pi)^3}}\int_B d^3\boldsymbol{k}\, a^+_{n,\sigma}(\boldsymbol{k})\exp(-\mathrm{i}\boldsymbol{k}\cdot\boldsymbol{\ell}). \tag{19.45}$$

The operators $\psi_{n,\sigma}(\boldsymbol{\ell})$ in the direct lattice satisfy

$$\{\psi_{n,\sigma}(\boldsymbol{\ell}),\psi_{n',\sigma'}(\boldsymbol{\ell}')\} = 0 \quad \{\psi_{n,\sigma}(\boldsymbol{\ell}),\psi^+_{n',\sigma'}(\boldsymbol{\ell}')\} = \delta_{n,n'}\delta_{\sigma\sigma'}\delta(\boldsymbol{\ell}-\boldsymbol{\ell}').$$

Substitution of (19.41) into (19.45) yields

$$\psi^+_{n,\sigma}(\boldsymbol{\ell}) = \int d^3\boldsymbol{x}\, w_n(\boldsymbol{\ell},\boldsymbol{x})\psi^+_\sigma(\boldsymbol{x})$$

with the Wannier states

$$w_n(\boldsymbol{\ell},\boldsymbol{x}) = \frac{V}{(2\pi)^3}\int_B d^3\boldsymbol{k}\, u_n(\boldsymbol{k},\boldsymbol{x})\exp[\mathrm{i}\boldsymbol{k}\cdot(\boldsymbol{x}-\boldsymbol{\ell})] = w_n(\boldsymbol{x}-\boldsymbol{\ell}). \tag{19.46}$$

These states satisfy the usual completeness relations as a consequence of the completeness relations of the Bloch states $\phi_n(\boldsymbol{k},\boldsymbol{x})$,

$$\int d^3\boldsymbol{x}\, w^+_n(\boldsymbol{\ell},\boldsymbol{x})w_{n'}(\boldsymbol{\ell}',\boldsymbol{x}) = \delta_{n,n'}\delta_{\boldsymbol{\ell},\boldsymbol{\ell}'},$$

$$\sum_{n,\boldsymbol{\ell}} w_n(\boldsymbol{\ell},\boldsymbol{x})w^+_n(\boldsymbol{\ell},\boldsymbol{x}') = \delta(\boldsymbol{x}-\boldsymbol{x}').$$

The operator $\psi^+_{n,\sigma}(\boldsymbol{\ell})$ therefore generates an electron with spin projection σ in the Wannier state $w_n(\boldsymbol{\ell},\boldsymbol{x})$.

We denote the operators $a_{n,\sigma}(\boldsymbol{k})$ and $a^+_{n,\sigma}(\boldsymbol{k})$ as Bloch operators, and the operators $\psi_{n,\sigma}(\boldsymbol{\ell})$ and $\psi^+_{n,\sigma}(\boldsymbol{\ell})$ as Wannier operators.

19.4 The Hubbard model

The Hubbard model treats electron-electron interactions in a tight binding approximation. Therefore we wish to use the creation operators $\psi^+_{n,\sigma}(\boldsymbol{\ell})$ for electrons in Wannier states.

The kinetic electron operator transforms into Wannier type operators according to

$$
\begin{aligned}
H_0 &= \int d^3\boldsymbol{x} \sum_\sigma \frac{\hbar^2}{2m} \nabla\psi^+_\sigma(\boldsymbol{x})\cdot\nabla\psi_\sigma(\boldsymbol{x}) \\
&= \int d^3\boldsymbol{x} \sum_{\sigma,n,\boldsymbol{\ell},n',\boldsymbol{\ell}'} \psi^+_{n,\sigma}(\boldsymbol{\ell})\frac{\hbar^2}{2m}\nabla w^+_n(\boldsymbol{\ell},\boldsymbol{x})\cdot\nabla w_{n'}(\boldsymbol{\ell}',\boldsymbol{x})\psi_{n',\sigma}(\boldsymbol{\ell}'). \quad (19.47)
\end{aligned}
$$

This has the form of a hopping Hamiltonian for jumps $n',\boldsymbol{\ell}'\to n,\boldsymbol{\ell}$,

$$
H_0 = \sum_{\sigma,n,\boldsymbol{\ell},n',\boldsymbol{\ell}'} t_{n,\boldsymbol{\ell},n',\boldsymbol{\ell}'}\psi^+_{n,\sigma}(\boldsymbol{\ell})\psi_{n',\sigma}(\boldsymbol{\ell}') \quad (19.48)
$$

with a hopping parameter

$$
t_{n,\boldsymbol{\ell},n',\boldsymbol{\ell}'} = \int d^3\boldsymbol{x}\,\frac{\hbar^2}{2m}\nabla w^+_n(\boldsymbol{\ell},\boldsymbol{x})\cdot\nabla w_{n'}(\boldsymbol{\ell}',\boldsymbol{x}).
$$

On the other hand, the electron-electron interaction Hamiltonian becomes

$$
H_{ee} = \frac{1}{2}\sum_{\sigma,\sigma',m,\boldsymbol{l},m',\boldsymbol{l}',n,\boldsymbol{\ell},n',\boldsymbol{\ell}'} U_{m,\boldsymbol{l},m',\boldsymbol{l}',n',\boldsymbol{\ell}',n,\boldsymbol{\ell}}\psi^+_{m,\sigma}(\boldsymbol{l})\psi^+_{m',\sigma'}(\boldsymbol{l}')\psi_{n',\sigma'}(\boldsymbol{\ell}')\psi_{n,\sigma}(\boldsymbol{\ell})
$$

with the Coulomb matrix element

$$
\begin{aligned}
U_{m,\boldsymbol{l},m',\boldsymbol{l}',n',\boldsymbol{\ell}',n,\boldsymbol{\ell}} &= \int d^3\boldsymbol{x}\int d^3\boldsymbol{x}'\, w^+_m(\boldsymbol{l},\boldsymbol{x})w^+_{m'}(\boldsymbol{l}',\boldsymbol{x}') \\
&\quad\times\frac{e^2}{4\pi\epsilon_0|\boldsymbol{x}-\boldsymbol{x}'|}w_{n'}(\boldsymbol{\ell}',\boldsymbol{x}')w_n(\boldsymbol{\ell},\boldsymbol{x}).
\end{aligned}
$$

H_{ee} would certainly be dominated by terms on the same lattice site, and if we restrict the discussion to a single band index, the electron-electron interaction Hamiltonian $H = H_0 + H_{ee}$ assumes the simple form

$$H = \sum_{\ell,\ell',\sigma} t_{\ell,\ell'}\psi_\sigma^+(\ell)\psi_\sigma(\ell') + U\sum_\ell n_{\uparrow,\ell}n_{\downarrow,\ell}, \tag{19.49}$$

with the spin polarized occupation number operators for lattice site ℓ,

$$n_{\sigma,\ell} = \psi_\sigma^+(\ell)\psi_\sigma(\ell).$$

The Hamiltonian (19.49) is known as the *Hubbard Hamiltonian*[11]. This Hamiltonian was invented for the analysis of ferromagnetic behavior in transition metals, and soon became a very widely used model Hamiltonian in condensed matter theory not only for magnetic ordering, but also for the general investigation of electron correlations, conductivity properties and disorder effects in many different classes of materials[12]. However, the Hubbard model also provides basic insight into the relevance of delocalized Bloch states versus localized Wannier states, as we will now discuss.

We assume that the hopping term is invariant under translation and symmetric between sites, i.e.

$$t_{\ell,\ell'} = t_{\ell-\ell'} = t_{\Delta\ell} = t_{-\Delta\ell}.$$

If hopping is suppressed,

$$t_{\ell,\ell'} = t\delta_{\ell,\ell'},$$

the Hamiltonian involves only the number operators $n_{\sigma,\ell}$,

$$H = t\sum_{\sigma,\ell} n_{\sigma,\ell} + U\sum_\ell n_{\uparrow,\ell}n_{\downarrow,\ell}, \tag{19.50}$$

and the eigenstates and energy levels are given by $N = N_1 + 2N_2$ particle states

$$|\sigma_1,\ell_1;\ldots\sigma_N,\ell_N\rangle = \psi_{\sigma_1}^+(\ell_1)\ldots\psi_{\sigma_N}^+(\ell_N)|0\rangle$$

with energy

$$E(N_1,N_2) = t(N_1 + 2N_2) + UN_2.$$

---

[11] J. Hubbard, Proc. Roy. Soc. London A 276, 238 (1963), see also M.C. Gutzwiller, Phys. Rev. Lett. 10, 159 (1963).

[12] See e.g. J.E. Hirsch, Phys. Rev. B 31, 4403 (1985); I. Affleck, J.B. Marston, Phys. Rev. B 37, 3774 (1988); Y.M. Vilk, A.-M.S. Tremblay, J. Physique I 7, 1309 (1997). More comprehensive textbook discussions can be found in references [5, 11].

Here N_1 and N_2 are the numbers of single and double occupied lattice sites, respectively. This is also denoted as the atomic limit, since the electrons are fixed at the atoms and the total energy is a sum of atomic terms.

On the other hand, if we can neglect the electron-electron interaction term, $U = 0$, we end up with a quadratic Hamiltonian

$$H = \sum_{\boldsymbol{\ell},\Delta\boldsymbol{\ell},\sigma} t_{\Delta\boldsymbol{\ell}} \psi_\sigma^+(\boldsymbol{\ell} + \Delta\boldsymbol{\ell})\psi_\sigma(\boldsymbol{\ell}). \tag{19.51}$$

We can map the electron operators on lattice sites to electron operators (19.44) in the Brillouin zone,

$$a_\sigma(\boldsymbol{k}) = \frac{\sqrt{V}}{\sqrt{2\pi}^3} \sum_{\boldsymbol{\ell}} \psi_\sigma(\boldsymbol{\ell}) \exp(-\mathrm{i}\boldsymbol{k} \cdot \boldsymbol{\ell}), \tag{19.52}$$

This diagonalizes the Hamiltonian (19.51),

$$H = \int_{B} d^3k\, E(\boldsymbol{k}) \sum_\sigma a_\sigma^+(\boldsymbol{k}) a_\sigma(\boldsymbol{k}), \tag{19.53}$$

$$E(\boldsymbol{k}) = \sum_{\Delta\boldsymbol{\ell}} t_{\Delta\boldsymbol{\ell}} \exp(-\mathrm{i}\boldsymbol{k} \cdot \Delta\boldsymbol{\ell}) = \sum_{\Delta\boldsymbol{\ell}} t_{\Delta\boldsymbol{\ell}} \cos(\boldsymbol{k} \cdot \Delta\boldsymbol{\ell}). \tag{19.54}$$

The single particle eigenstate of the Hamiltonian (19.53) with energy $E(\boldsymbol{k})$,

$$a_\sigma^+(\boldsymbol{k})|0\rangle = \frac{\sqrt{V}}{\sqrt{2\pi}^3} \sum_{\boldsymbol{\ell}} \psi_\sigma^+(\boldsymbol{\ell}) \exp(\mathrm{i}\boldsymbol{k} \cdot \boldsymbol{\ell})|0\rangle,$$

is a Bloch state, while the single particle eigenstate $\psi_\sigma^+(\boldsymbol{\ell})|0\rangle$ of the Hamiltonian (19.50) is a Wannier state. The magnitude of the hopping terms $t_{\Delta\boldsymbol{\ell}\neq\boldsymbol{0}}$ relative to U will therefore determine the importance of itinerant (or delocalized) Bloch electron states versus localized Wannier electron states in the lattice.

19.5 Vibrations in molecules and lattices

Another basic excitation of lattices concerns oscillations of lattice ions or atoms around their equilibrium configurations. This kind of excitation is particularly amenable to description in classical mechanical terms, but at the quantum level lattice vibrations are very similar to quantum excitations of the vacuum like electrons or photons. In particular, elementary lattice vibrations can be spontaneously created and absorbed like photons, and therefore require a quantum field theory which is similar to the field theory for photons.

We will discuss the classical theory of small oscillations of N-particle systems in the present section as a preparation for the discussion of quantized lattice vibrations in Section 19.6. We suspend summation convention in this section, because we often encounter expressions with three identical indices in a multiplicative term, and also terms like $m_i\ddot{x}_i^j$ without summation over the repeated index.

Normal coordinates and normal oscillations

We consider an N particle system with potential $V(\boldsymbol{r}_1, \ldots \boldsymbol{r}_N)$. The equilibrium condition

$$\nabla_i V(\boldsymbol{r}_1, \ldots \boldsymbol{r}_N)\Big|_{\boldsymbol{r}_j=\boldsymbol{r}_j^{(0)}} = \boldsymbol{0} \tag{19.55}$$

implies for the second order expansion around an equilibrium configuration $\boldsymbol{r}_1^{(0)}, \ldots \boldsymbol{r}_N^{(0)}$,

$$V(\boldsymbol{r}_1, \ldots \boldsymbol{r}_N) = V(\boldsymbol{r}_1^{(0)}, \ldots \boldsymbol{r}_N^{(0)}) + \frac{1}{2}\sum_{ijkl} V_{ik,jl} x_i^k x_j^l,$$

where $\boldsymbol{x}_i = \boldsymbol{r}_i - \boldsymbol{r}_i^{(0)}$ parametrize the deviations from equilibrium and the coefficients $V_{ik,jl}$ are

$$V_{ik,jl} = \frac{\partial^2}{\partial y_i^k \partial y_j^l} V(\boldsymbol{r}_1^{(0)} + \boldsymbol{y}_1, \ldots \boldsymbol{r}_N^{(0)} + \boldsymbol{y}_N)\Big|_{\boldsymbol{y}_m=\boldsymbol{0}}.$$

The second order Lagrange function for small oscillations of the system,

$$L = \frac{1}{2}\sum_{ik} m_i \dot{x}_i^k \dot{x}_i^k - \frac{1}{2}\sum_{ijkl} V_{ik,jl} x_i^k x_j^l, \tag{19.56}$$

yields $3N$ coupled equations of motion

$$m_i \ddot{x}_i^k = -\sum_{jl} V_{ik,jl} x_j^l. \tag{19.57}$$

Fourier transformation

$$x_i^k(t) = \int d\omega\, a_i^k(\omega) \exp(-i\omega t), \quad [a_i^k(\omega)]^+ = a_i^k(-\omega), \tag{19.58}$$

yields the conditions

$$\sum_{jl} (V_{ik,jl} - m_i \omega^2 \delta_{ij} \delta_{kl}) a_j^l(\omega) = 0. \tag{19.59}$$

Writing this in the form

$$\sum_{jl}\left(\frac{V_{ik,jl}}{\sqrt{m_i m_j}} - \omega^2\delta_{ij}\delta_{kl}\right)\sqrt{m_j}a_j{}^l(\omega) = 0$$

tells us that the $3N$-dimensional vector

$$\boldsymbol{Q}(\omega) = \{\sqrt{m_1}a_1{}^1(\omega), \ldots \sqrt{m_N}a_N{}^3(\omega)\} = \boldsymbol{Q}^+(-\omega) \tag{19.60}$$

must have the form

$$\boldsymbol{Q}(\omega) = \sum_{I=1}^{3N}\left[\boldsymbol{Q}_I\delta(\omega - \omega_I) + \boldsymbol{Q}_{-I}\delta(\omega + \omega_I)\right], \tag{19.61}$$

where $\boldsymbol{Q}_I = \{\sqrt{m_1}a_{I,1}{}^1, \ldots \sqrt{m_N}a_{I,N}{}^3\} = \boldsymbol{Q}_{-I}^+$ is an eigenvector of the symmetric $3N \times 3N$ matrix

$$\Omega^2_{ik,jl} = \frac{V_{ik,jl}}{\sqrt{m_i m_j}} \tag{19.62}$$

with eigenvalue ω_I^2. We assume that $\boldsymbol{r}_1^{(0)}, \ldots \boldsymbol{r}_N^{(0)}$ is a stable equilibrium configuration such that all eigenvalues of $\Omega^2_{ik,jl}$ satisfy $\omega_I^2 \geq 0$, and we define $\omega_I = \sqrt{\omega_I^2} \geq 0$ as the positive semi-definite roots.

Since $\underline{\Omega}^2$ is a symmetric real $3N \times 3N$ matrix, we can find $3N$ orthogonal normalized real vectors

$$\hat{\boldsymbol{Q}}_I = \{\sqrt{m_1}\hat{a}_{I,1}{}^1, \ldots \sqrt{m_N}\hat{a}_{I,N}{}^3\}$$

which solve the eigenvalue problem

$$\underline{\Omega}^2 \cdot \hat{\boldsymbol{Q}}_I = \omega_I^2\hat{\boldsymbol{Q}}_I. \tag{19.63}$$

The general solution $\boldsymbol{Q}_I$ (19.60) of the eigenvalue problem with eigenvalue ω_I^2 will then have the form

$$\boldsymbol{Q}_I = q_I\hat{\boldsymbol{Q}}_I$$

with arbitrary complex factors $q_I = |q_I|\exp(\mathrm{i}\varphi_I)$. The mode expansion (19.58) will therefore take the form

$$\begin{aligned} x_i{}^k(t) &= \sum_{I=1}^{3N}\hat{a}_{I,i}{}^k\left[q_I\exp(-\mathrm{i}\omega_I t) + q_I^+\exp(\mathrm{i}\omega_I t)\right] \\ &= 2\sum_{I=1}^{3N}\hat{a}_{I,i}{}^k\,|q_I|\left[\cos(\varphi_I)\cos(\omega_I t) + \sin(\varphi_I)\sin(\omega_I t)\right]. \end{aligned} \tag{19.64}$$

Equation (19.63) and $V_{ik,jl} = V_{jl,ik} = V^{+}_{ik,jl}$ imply the orthogonality relations

$$0 = \sum_{ijkl} (\hat{a}_{I,i}{}^{k} V_{ik,jl} \hat{a}_{J,j}{}^{l} - \hat{a}_{I,i}{}^{k} V_{ik,jl} \hat{a}_{J,j}{}^{l}) = \sum_{ik} m_i \hat{a}_{\pm I,1}{}^{k} \hat{a}_{J,j}{}^{k} (\omega_I^2 - \omega_J^2).$$

This yields

$$\sum_{ik} m_i \hat{a}_{I,i}{}^{k} \hat{a}_{J,i}{}^{k} = \delta_{IJ}, \tag{19.65}$$

where we assume that eigenvectors $\hat{\boldsymbol{Q}}_I$ within degeneracy subspaces have been orthonormalized.

Note that the normalization changes the dimensions and the physical meaning of the coefficients. The amplitudes $a_{I,i}{}^{k}$ in equation (19.64) have the dimensions of a length, and the related eigenvectors $\boldsymbol{Q}_I$ and factors q_I have the dimension of $\text{mass}^{1/2} \times$ length. The normalized eigenvectors $\hat{\boldsymbol{Q}}_I$ are dimensionless, and therefore the related coefficients $\hat{a}_{I,i}{}^{k}$ have dimension $\text{mass}^{-1/2}$. We will denote the related $3N$ dimensional vector $\hat{\boldsymbol{a}}_I = \{\hat{a}_{I,1}{}^{1}, \ldots, \hat{a}_{I,N}{}^{3}\}$ as an *amplitude vector*.

The small oscillations of the system are then determined by the eigenmodes $\hat{\boldsymbol{a}}_I$ (or equivalently $\hat{\boldsymbol{Q}}_I$), and how strongly these eigenmodes of oscillation are excited,

$$x_i{}^{k}(t) = \sum_{I=1}^{3N} \hat{a}_{I,i}{}^{k} \left[q_I \exp(-\mathrm{i}\omega_I t) + q_I^{+} \exp(\mathrm{i}\omega_I t) \right], \tag{19.66}$$

$$\dot{x}_i{}^{k}(t) = -\mathrm{i} \sum_{I=1}^{3N} \omega_I \hat{a}_{I,i}{}^{k} \left[q_I \exp(-\mathrm{i}\omega_I t) - q_I^{+} \exp(\mathrm{i}\omega_I t) \right], \tag{19.67}$$

$$q_I = \frac{1}{2} \exp(\mathrm{i}\omega_I t) \sum_{ik} m_i \hat{a}_{I,i}{}^{k} \left(x_i{}^{k}(t) + \frac{\mathrm{i}}{\omega_I} \dot{x}_i{}^{k}(t) \right). \tag{19.68}$$

The $3N$ complex amplitudes q_I are denoted as *normal coordinates* of the oscillating N particle system, and the related eigenmodes of oscillation are also denoted as *normal modes*. Note from equations (19.66) or (19.68) that we can think of the coefficients $\hat{a}_{I,i}{}^{k}$ also as the components of a $3N \times 3N$ transformation matrix between the $3N$ Cartesian coordinates $x_i{}^{k}(t)$ and the $3N$ normal coordinates q_I of the oscillating system. These $3N \times 3N$ matrices satisfy the mass weighted orthogonality properties (19.65) and

$$\sum_{I} \hat{a}_{I,i}{}^{k} \hat{a}_{I,j}{}^{l} = \frac{1}{m_i} \delta_{ij} \delta^{kl}, \tag{19.69}$$

which follows from re-substitution of q_I (19.68) into $x_i{}^{k}(t)$ (19.66).

Appearance of the particular eigenvalue $\omega_I^2 = 0$ implies that the system is symmetric under rotations or translations. The corresponding amplitude vectors $\hat{\boldsymbol{a}}_I = \{\hat{a}_{I,i}{}^k\}$ denote the tangential directions to rotations or translations of the system.

We have learned that small oscillations of a system are always superpositions of the normal oscillation modes or eigenoscillations of the system. *A priori* this does not seem to be particularly helpful to determine the actual small oscillations of a system, because finding the eigenmodes is equivalent to the diagonalization of the $3N \times 3N$ matrix $\Omega^2_{ik,jl}$, which is anyhow the main task in the solution of the equations of motion (19.57) using the Fourier *ansatz* (19.64).

However, if the equilibrium configuration of the system has symmetries, then we can often guess the form of some of the eigenmodes which leaves us with a smaller diagonalization problem for the determination of the remaining eigenmodes.

Eigenmodes of three masses

A simple example for the identification of normal modes of a coupled particle system is given by three identical masses in a regular triangle, see Figure 19.2.

We will determine the eigenmodes in the plane of the triangle. The potential of the coupled system in the harmonic approximation is

$$\begin{aligned}
V &= \frac{K}{2}\Big((|\boldsymbol{r}_1 - \boldsymbol{r}_2| - d)^2 + (|\boldsymbol{r}_1 - \boldsymbol{r}_3| - d)^2 + (|\boldsymbol{r}_2 - \boldsymbol{r}_3| - d)^2\Big) \\
&\simeq \frac{K}{2}\Big((x_1{}^1 - x_2{}^1)^2 + \frac{1}{4}(x_1{}^1 - x_3{}^1)^2 + \frac{1}{4}(x_2{}^1 - x_3{}^1)^2 + \frac{3}{4}(x_1{}^2 - x_3{}^2)^2 \\
&\quad + \frac{3}{4}(x_2{}^2 - x_3{}^2)^2 + \frac{\sqrt{3}}{2}(x_1{}^1 x_1{}^2 - x_1{}^1 x_3{}^2 - x_2{}^1 x_2{}^2 + x_2{}^1 x_3{}^2 \\
&\quad - x_3{}^1 x_1{}^2 + x_3{}^1 x_2{}^2)\Big).
\end{aligned}$$

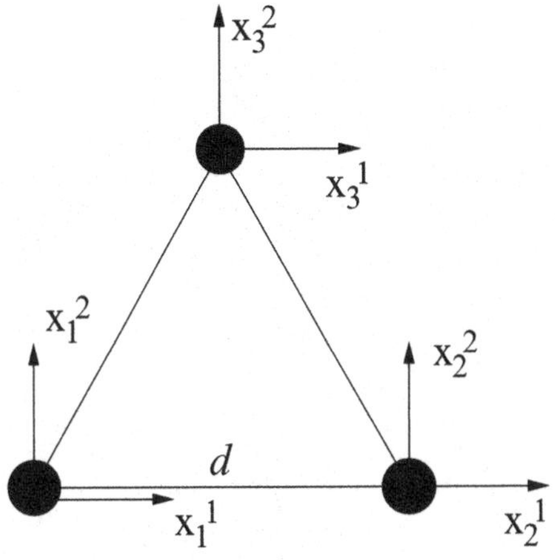

Fig. 19.2 Three elastically bound masses with equilibrium distance d

The matrix $V_{ik,jl}$ is

$$\underline{V} = m\underline{\Omega}^2 = K\begin{pmatrix} \frac{5}{4} & \frac{\sqrt{3}}{4} & -1 & 0 & -\frac{1}{4} & -\frac{\sqrt{3}}{4} \\ \frac{\sqrt{3}}{4} & \frac{3}{4} & 0 & 0 & -\frac{\sqrt{3}}{4} & -\frac{3}{4} \\ -1 & 0 & \frac{5}{4} & -\frac{\sqrt{3}}{4} & -\frac{1}{4} & \frac{\sqrt{3}}{4} \\ 0 & 0 & -\frac{\sqrt{3}}{4} & \frac{3}{4} & \frac{\sqrt{3}}{4} & -\frac{3}{4} \\ -\frac{1}{4} & -\frac{\sqrt{3}}{4} & -\frac{1}{4} & \frac{\sqrt{3}}{4} & \frac{1}{2} & 0 \\ -\frac{\sqrt{3}}{4} & -\frac{3}{4} & \frac{\sqrt{3}}{4} & -\frac{3}{4} & 0 & \frac{3}{2} \end{pmatrix},$$

and we must have

$$\mathrm{Det}(\underline{V} - m\omega^2\underline{1}) = 0.$$

Absence of external forces on the coupled system implies that there must be two translational and one rotational eigenmode, see Figures 19.2 and 19.3,

$$\hat{Q}_1 = \frac{1}{\sqrt{3}}\begin{pmatrix} 1 \\ 0 \\ 1 \\ 0 \\ 1 \\ 0 \end{pmatrix}, \quad \hat{Q}_2 = \frac{1}{\sqrt{3}}\begin{pmatrix} 0 \\ 1 \\ 0 \\ 1 \\ 0 \\ 1 \end{pmatrix}, \quad \hat{Q}_3 = \frac{1}{2\sqrt{3}}\begin{pmatrix} 1 \\ -\sqrt{3} \\ 1 \\ \sqrt{3} \\ -2 \\ 0 \end{pmatrix}.$$

The equations $\underline{V}\cdot\hat{Q}_I = \mathbf{0}$ for $I = 1, 2, 3$ are readily verified.
The symmetry reveals that another eigenmode can be read off from Figure 19.4. This yields the corresponding normalized eigenvector

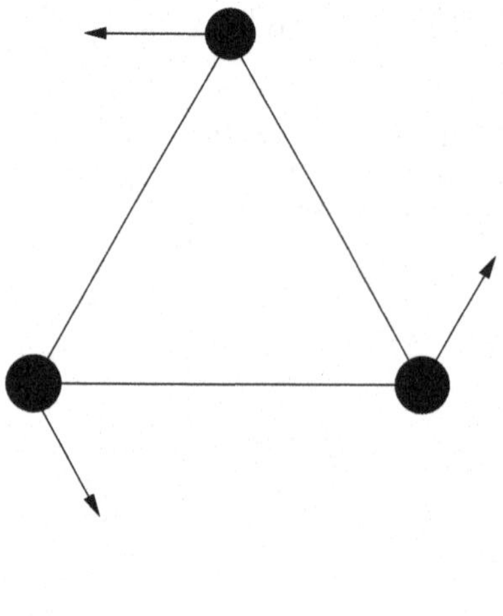

Fig. 19.3 The rotation mode $\hat{Q}_3$

Fig. 19.4 The eigenmode $\hat{Q}_4 = \sqrt{m}\hat{a}_4$

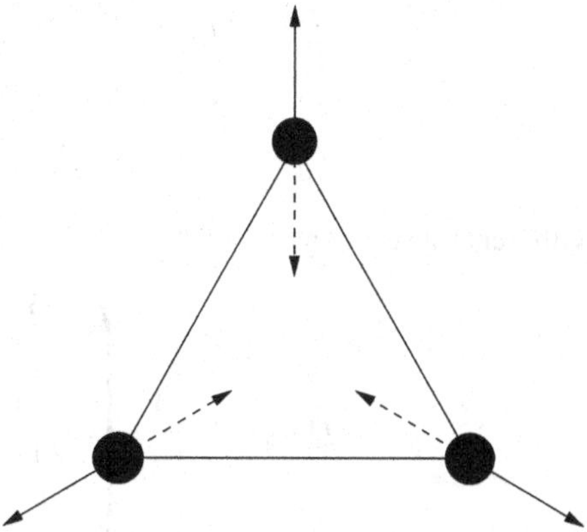

$$\hat{Q}_4 = \sqrt{m}\begin{pmatrix} a_{4,1}{}^1 \\ a_{4,1}{}^2 \\ a_{4,2}{}^1 \\ a_{4,2}{}^2 \\ a_{4,3}{}^1 \\ a_{4,3}{}^2 \end{pmatrix} = \frac{1}{2\sqrt{3}}\begin{pmatrix} \sqrt{3} \\ 1 \\ -\sqrt{3} \\ 1 \\ 0 \\ -2 \end{pmatrix},$$

and application of

$$\omega_4^2\hat{Q}_4 = \frac{1}{m}\underline{V}\cdot\hat{Q}_4,$$

yields for the corresponding frequency

$$\omega_4^2 = \frac{3K}{m}.$$

So far we have found four eigenmodes of the planar system, and there must still be two remaining eigenmodes, which must be orthogonal on the eigenmodes $\hat{Q}_1, \ldots \hat{Q}_4$. This yields for

$$\hat{Q}_I = \sqrt{m}\begin{pmatrix} a_{I,1}{}^1 \\ a_{I,1}{}^2 \\ a_{I,2}{}^1 \\ a_{I,2}{}^2 \\ a_{I,3}{}^1 \\ a_{I,3}{}^2 \end{pmatrix}, \quad I = 5, 6$$

the conditions

$$\sqrt{3}(a_{I,1}{}^1 - a_{I,2}{}^1) + a_{I,1}{}^2 + a_{I,2}{}^2 - 2a_{I,3}{}^2 = 0,$$

$$a_{I,1}{}^1 + a_{I,2}{}^1 + a_{I,3}{}^1 = 0,$$
$$a_{I,1}{}^2 + a_{I,2}{}^2 + a_{I,3}{}^2 = 0,$$
$$a_{I,1}{}^1 + a_{I,2}{}^1 - 2a_{I,3}{}^1 + \sqrt{3}(a_{I,2}{}^2 - a_{I,1}{}^2) = 0,$$

with general solutions

$$\hat{Q}_{I=5,6} \sim \frac{A}{2\sqrt{3}}\begin{pmatrix}\sqrt{3}\\-1\\-\sqrt{3}\\-1\\0\\2\end{pmatrix} + \frac{B}{2\sqrt{3}}\begin{pmatrix}1\\\sqrt{3}\\1\\-\sqrt{3}\\-2\\0\end{pmatrix}.$$

Application of $\underline{\Omega^2}$ reveals that these are degenerate eigenvectors with eigenvalue

$$\omega_5^2 = \omega_6^2 = \frac{3K}{2m},$$

and an orthonormal basis in the degeneracy subspace is provided by

$$\hat{Q}_5 = \frac{1}{2\sqrt{3}}\begin{pmatrix}\sqrt{3}\\-1\\-\sqrt{3}\\-1\\0\\2\end{pmatrix}, \quad \hat{Q}_6 = \frac{1}{2\sqrt{3}}\begin{pmatrix}1\\\sqrt{3}\\1\\-\sqrt{3}\\-2\\0\end{pmatrix}.$$

The corresponding eigenmodes are shown in Figure 19.5.

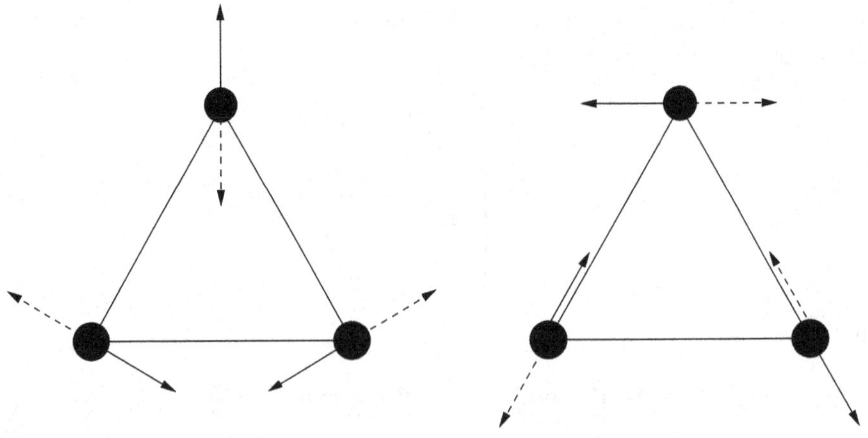

Fig. 19.5 The eigenmodes $\hat{Q}_5$ and $\hat{Q}_6$

The general small oscillation with $\omega > 0$ is then given by

$$\begin{pmatrix} x_1{}^1(t) \\ x_1{}^2(t) \\ x_2{}^1(t) \\ x_2{}^2(t) \\ x_3{}^1(t) \\ x_3{}^2(t) \end{pmatrix} = \sum_{I=4}^{6} \hat{Q}_I x_I(t)$$

with

$$x_I(t) = x_I(0)\cos(\omega_I t) + \frac{\dot{x}_I(0)}{\omega_I}\sin(\omega_I t).$$

The diatomic linear chain

Lines of harmonically bound atoms provide important model systems for oscillations in solid state physics. We consider in particular a diatomic chain of $2N$ atoms with masses m and M, respectively. This model is shown in Figure 19.6. The force constant between the atoms is K and their equilibrium distance is $a/2$. The number N of atom pairs is assumed to be even for simplicity.

We label the pairs of atoms with an index n, $1-(N/2) \leq n \leq N/2$, and we use periodic boundary conditions for the displadements x_n and X_n,

$$x_{n+N} = x_n, \quad X_{n+N} = X_n.$$

The Lagrange function

$$L = \sum_{n=1-(N/2)}^{N/2} \left(\frac{m}{2}\dot{x}_n^2 + \frac{M}{2}\dot{X}_n^2 - \frac{K}{2}(X_n - x_n)^2 - \frac{K}{2}(x_n - X_{n-1})^2\right)$$

yields equations of motion

$$m\ddot{x}_n = -K(2x_n - X_n - X_{n-1}), \quad M\ddot{X}_n = -K(2X_n - x_n - x_{n+1}), \tag{19.70}$$

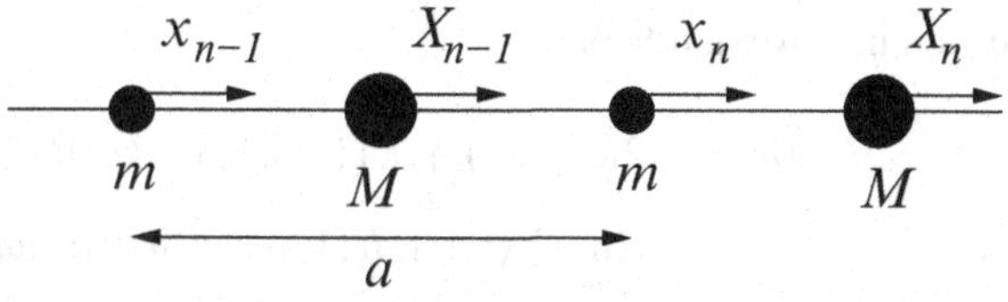

Fig. 19.6 A diatomic linear chain with masses m and M and lattice constant a

which can be solved using Fourier decomposition on a finite periodic chain,

$$x_n(t) = \frac{1}{\sqrt{N}} \sum_k \tilde{q}_k(t) \exp(inka), \tag{19.71}$$

$$X_n(t) = \frac{1}{\sqrt{N}} \sum_k \tilde{Q}_k(t) \exp(inka),$$

with

$$k = \frac{2\pi\tilde{n}}{Na}, \quad 1 - \frac{N}{2} \leq \tilde{n} \leq \frac{N}{2}.$$

The geometric series

$$\begin{aligned} \sum_{n=1-(N/2)}^{N/2} \exp\left(2\pi i n \frac{\tilde{n}-\tilde{m}}{N}\right) &= \exp\left[i\pi\left(\frac{2}{N}-1\right)(\tilde{n}-\tilde{m})\right] \\ \times \sum_{n=0}^{N-1} \exp\left(2\pi i n \frac{\tilde{n}-\tilde{m}}{N}\right) &= \exp\left[i\pi\left(\frac{2}{N}-1\right)(\tilde{n}-\tilde{m})\right] \\ \times \frac{1-\exp[2\pi i(\tilde{n}-\tilde{m})]}{1-\exp\left[\frac{2\pi i}{N}(\tilde{n}-\tilde{m})\right]} &= N\delta_{\tilde{n},\tilde{m}} \end{aligned} \tag{19.72}$$

implies that the inversion of (19.71) is

$$\tilde{q}_k(t) = \frac{1}{\sqrt{N}} \sum_{n=1-(N/2)}^{N/2} x_n(t) \exp(-inka) = \tilde{q}^{+}_{-k}(t).$$

Since the resulting system of ordinary differential equations for $\tilde{q}_k(t)$ and $\tilde{Q}_k(t)$ is linear with constant coefficients, we also use Fourier transformation to the frequency domain,

$$\tilde{q}_k(t) = \int d\omega\, \tilde{q}_k(\omega) \exp(-i\omega t),$$

and the coupled set of equations (19.70) separate into coupled pairs of equations for different wave numbers k,

$$(m\omega^2 - 2K)\tilde{q}_k(\omega) + K(1 + \exp(-ika))\tilde{Q}_k(\omega) = 0, \tag{19.73}$$

$$(M\omega^2 - 2K)\tilde{Q}_k(\omega) + K(1 + \exp(ika))\tilde{q}_k(\omega) = 0. \tag{19.74}$$

This implies that there is a unique set of frequencies $\omega = \omega_k$ for each wave number k which has to satisfy

$$mM\omega_k^4 - 2K(m+M)\omega_k^2 + 2K^2(1-\cos(ka)) = 0.$$

This condition has two solutions (up to irrelevant overall signs of $\omega_{k\pm}$),

$$\begin{aligned}\omega_{k\pm}^2 &= K\left(\frac{1}{M}+\frac{1}{m}\right) \pm K\sqrt{\frac{1}{M^2}+\frac{1}{m^2}+\frac{2}{mM}\cos(ka)} \\ &= K\left(\frac{1}{M}+\frac{1}{m}\right) \pm K\sqrt{\left(\frac{1}{M}+\frac{1}{m}\right)^2 - \frac{4}{mM}\sin^2\left(\frac{ka}{2}\right)},\end{aligned} \qquad (19.75)$$

and we have

$$\tilde{q}_k(\omega) = \tilde{q}_{k+}\delta(\omega-\omega_{k+}) + \tilde{q}_{k-}\delta(\omega-\omega_{k-}).$$

Equation (19.75) reads in terms of the reduced mass $\mu = mM/(m+M)$ of the atom pair in the unit cell

$$\omega_{k\pm}^2 = \frac{K}{\mu}\left(1 \pm \sqrt{1-\frac{4\mu}{m+M}\sin^2\left(\frac{ka}{2}\right)}\right). \qquad (19.76)$$

An example of these dispersion relations with $M = 1.5m$ is displayed in Figure 19.7.

Note that the Lagrange function for a single atom pair in the unit cell is

$$L = \frac{1}{2}(m+M)\dot{R}^2 + \frac{\mu}{2}\dot{r}^2 - \frac{K}{2}r^2, \quad r = x - X, \quad R = \frac{mx+MX}{m+M},$$

and therefore the oscillation frequency of the single pair is $\sqrt{K/\mu}$.

The frequencies at $k = 0$ are $\omega_{0-} = 0$ and $\omega_{0+} = \sqrt{2K/\mu}$.

The solution of (19.73, 19.74) for $\omega_{0-} = 0$: $\tilde{q}_{0-} = \tilde{Q}_{0-}$, is a uniform translation of the whole chain,

$$x_n(t) = X_n(t) = \tilde{q}_{0-}/\sqrt{N}.$$

The solution for ω_{0+}: $m\tilde{q}_{0+} = -M\tilde{Q}_{0+}$, is an oscillation

$$\begin{pmatrix} x_n(t) \\ X_n(t) \end{pmatrix} = A\begin{pmatrix} M \\ -m \end{pmatrix}\cos\left(\sqrt{\frac{2K}{\mu}}t + \varphi\right).$$

The acoustic solution for $ka = \pi$ is

$$\omega_{(\pi/a)-} = \sqrt{\frac{2K}{M}}, \quad \tilde{q}_{(\pi/a)-} = 0,$$

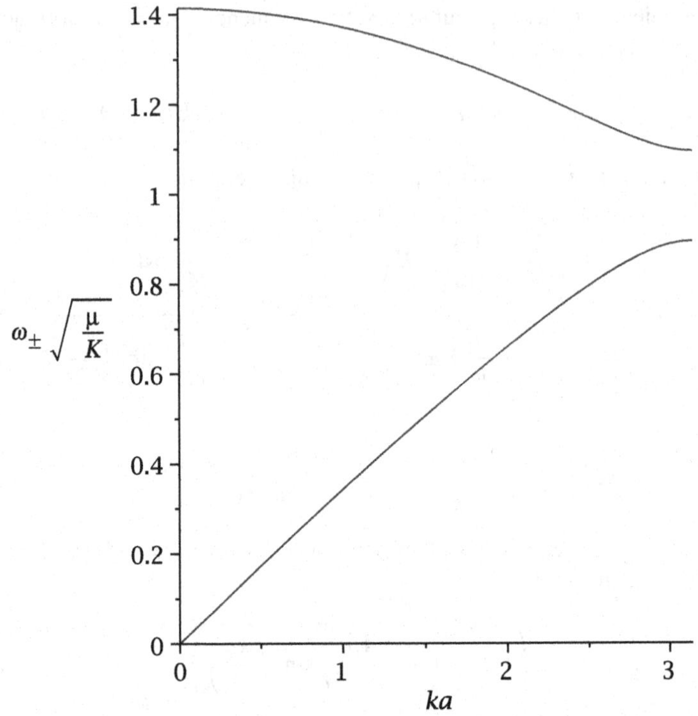

Fig. 19.7 The frequencies $\omega_{k\pm}$ from the dispersion relation (19.76) for $M = 1.5m$ and $0 \leq ka \leq \pi$. The frequencies $\omega_{k\pm}$ are displayed in units of $\sqrt{K/\mu}$, where μ is the reduced mass of the atom pair in a unit cell

i.e. only the heavy atoms oscillate,

$$\begin{pmatrix} x_n(t) \\ X_n(t) \end{pmatrix} = (-)^n A \begin{pmatrix} 0 \\ 1 \end{pmatrix} \cos\left(\sqrt{\frac{2K}{M}} t + \varphi\right).$$

On the other hand, the optical eigenmode with $ka = \pi$,

$$\omega_{(\pi/a)+} = \sqrt{\frac{2K}{m}}, \quad \tilde{Q}_{(\pi/a)+} = 0,$$

corresponds to an oscillation of the light atoms,

$$\begin{pmatrix} x_n(t) \\ X_n(t) \end{pmatrix} = (-)^n A \begin{pmatrix} 1 \\ 0 \end{pmatrix} \cos\left(\sqrt{\frac{2K}{m}} t + \varphi\right).$$

The general longitudinal oscillation will be a superposition of all longitudinal eigenvibrations.

Quantization of N-particle oscillations

The Lagrange function (19.56) implies canonical commutation relations

$$[x_i^k(t), \dot{x}_j^l(t)] = \frac{i\hbar}{m_i}\delta_{ij}\delta^{kl}, \quad [x_i^k(t), x_j^l(t)] = 0, \quad [\dot{x}_i^k(t), \dot{x}_j^l(t)] = 0.$$

This yields commutation relations for the normal coordinates

$$[q_I, q_J] = 0, \quad [q_I, q_J^+] = \frac{\hbar}{2\omega_I}\delta_{IJ}.$$

Therefore we find canonical annihilation and creation operators for the eigenvibrations in the form

$$a_I = \sqrt{\frac{2\omega_I}{\hbar}}q_I, \quad a_I^+ = \sqrt{\frac{2\omega_I}{\hbar}}q_I^+.$$

The discussion of the diatomic chain taught us that for lattice oscillations the eigenmodes also depend on wave vectors in a Brillouin zone, and the following section will show that there can be up to $3N$ branches if we have N atoms per unit cell. Therefore we will have annihilation and creation operators for lattice vibrations which are related to the corresponding normal modes through

$$a_I(\boldsymbol{k}) = \sqrt{\frac{2\omega_{I,\boldsymbol{k}}}{\hbar}}q_I(\boldsymbol{k}), \quad a_I^+(\boldsymbol{k}) = \sqrt{\frac{2\omega_{I,\boldsymbol{k}}}{\hbar}}q_I^+(\boldsymbol{k}).$$

The elementary excitations $a_I^+(\boldsymbol{k})|0\rangle$ of the lattice vibrations are denoted as *phonons*.

19.6 Quantized lattice vibrations: Phonons

We will first generalize the previous discussion of vibrations in N-particle systems to the case of three-dimensional lattices, and then quantize the lattice vibrations

We denote the three basis vectors of a three-dimensional lattice with $\boldsymbol{a}_i$, $1 \leq i \leq 3$. Each location $\boldsymbol{\ell} = n^i\boldsymbol{a}_i$ in the lattice denotes a particular location of a corresponding unit cell, and we can use $\boldsymbol{\ell}$ or equivalently the three integers n^i also to address the particular unit cell to which the point $\boldsymbol{\ell}$ belongs. Suppose we have N atoms (or ions) per unit cell in the lattice. We denote the displacement of the A-th atom from its equilibrium value in cell $\boldsymbol{\ell}$ by $\boldsymbol{x}_{\boldsymbol{\ell},A}(t)$, and in the harmonic approximation the displacements satisfy equations of motion

$$m_A\ddot{\boldsymbol{x}}_{\boldsymbol{\ell},A} + \sum_{\boldsymbol{\ell}',A'} \underline{V}_{\boldsymbol{\ell},A;\boldsymbol{\ell}',A'} \cdot \boldsymbol{x}_{\boldsymbol{\ell}',A'} = \boldsymbol{0}, \tag{19.77}$$

corresponding to a Lagrange function

$$L = \frac{1}{2}\sum_{\ell,A} m_A \dot{x}^2_{\ell,A} - \frac{1}{2}\sum_{\ell,A;\ell',A'} x_{\ell,A}\cdot \underline{V}_{\ell,A;\ell',A'}\cdot x_{\ell',A'}. \tag{19.78}$$

Substitution of Fourier transforms

$$x_{\ell,A}(t) = \frac{1}{\sqrt{m_A}}\int d\omega\, Q_{\ell,A}(\omega)\exp(-i\omega t)$$

into the equations of motion (19.77) yields the eigenvalue conditions

$$\sum_{\ell',A'} \underline{\Omega^2}_{\ell,A;\ell',A'}\cdot Q_{\ell',A'}(\omega) = \omega^2 Q_{\ell,A}(\omega) \tag{19.79}$$

with the symmetric matrices

$$\underline{\Omega^2}_{\ell,A;\ell',B} = \frac{1}{\sqrt{m_A m_B}}\underline{V}_{\ell,A;\ell',B} = \underline{\Omega^2}^T_{\ell',B;\ell,A}. \tag{19.80}$$

Translation invariance in the lattice implies that $\underline{\Omega^2}_{\ell,A;\ell',B}$ cannot depend on $\ell + \ell'$. Therefore we can write

$$\underline{\Omega^2}_{\ell,A;\ell',B} = \underline{\Omega^2}_{A,B}(\ell - \ell') = \frac{V}{(2\pi)^3}\int_B d^3k\, \underline{\tilde{\Omega}^2}_{A,B}(k)\exp[ik\cdot(\ell - \ell')], \tag{19.81}$$

with inversion

$$\underline{\tilde{\Omega}^2}_{A,B}(k) = \sum_{\ell}\underline{\Omega^2}_{A,B}(\ell)\exp(-ik\cdot\ell).$$

Symmetry of the real matrix $\underline{\Omega^2}_{\ell,A;\ell',B}$ under $i,\ell,A \leftrightarrow j,\ell',B$ implies

$$\begin{aligned}
\tilde{\Omega}^2_{iA,jB}(k) &= \sum_{\ell}\Omega^2_{iA,jB}(\ell)\exp(-ik\cdot\ell) = \sum_{\ell}\Omega^2_{jB,iA}(-\ell)\exp(-ik\cdot\ell)\\
&= \sum_{\ell}\Omega^2_{jB,iA}(\ell)\exp(ik\cdot\ell) = \tilde{\Omega}^{2,*}_{jB,iA}(k) = \tilde{\Omega}^{2,+}_{iA,jB}(k)\\
&= \tilde{\Omega}^2_{jB,iA}(-k),
\end{aligned}$$

i.e.

$$\underline{\tilde{\Omega}}^2(k) = \underline{\tilde{\Omega}}^{2^+}(k) = \underline{\tilde{\Omega}}^{2^T}(-k). \tag{19.82}$$

Substitution of (19.81) and

$$Q_{\ell,A}(\omega) = \frac{V}{(2\pi)^3}\int_{\mathcal{B}} d^3k\,\tilde{Q}_{k,A}(\omega)\exp(\mathrm{i}k\cdot\ell).$$

in (19.79) yields ,

$$\sum_B \underline{\tilde{\Omega}^2}_{A,B}(k)\cdot\tilde{Q}_{k,B}(\omega) = \omega^2\tilde{Q}_{k,A}(\omega). \tag{19.83}$$

For fixed value of k, this is a hermitian eigenvalue problem for the $3N$-dimensional complex vector

$$\tilde{Q}_k(\omega) = \{\tilde{Q}^i_{k,A}(\omega)\},\quad 1\le i\le 3,\quad 1\le A\le N.$$

Reality of the displacement vectors $x_{\ell,A}(t)$ implies $Q_{\ell,A}(\omega) = Q^+_{\ell,A}(-\omega)$ and

$$\tilde{Q}^+_k(\omega) = \tilde{Q}_{-k}(-\omega).$$

For each point k in the Brillouin zone, there will be $3N$ solutions $\omega_I^2(k)$ and $\hat{Q}_I(k)$ of (19.83) which satisfy the orthogonality property

$$\hat{Q}^+_I(k)\cdot\hat{Q}_J(k) \equiv \sum_A \hat{Q}^+_{I,A}(k)\cdot\hat{Q}_{J,A}(k) \equiv \sum_{i,A}\hat{Q}^{i+}_{I,A}(k)\hat{Q}^i_{J,A}(k)$$
$$= \delta_{IJ}. \tag{19.84}$$

The hermiticity and transposition properties imply that we have as a consequence of (19.83) for the normalized solutions,

$$\sum_B \underline{\tilde{\Omega}^2}_{A,B}(k)\cdot\hat{Q}_{I,B}(k) = \omega_I^2(k)\hat{Q}_{I,A}(k), \tag{19.85}$$

also the equations

$$\sum_B \underline{\tilde{\Omega}^2}_{A,B}(-k)\cdot\hat{Q}^+_{I,B}(k) = \omega_I^2(k)\hat{Q}^+_{I,A}(k) \tag{19.86}$$

and

$$\sum_A \hat{Q}_{I,A}(k)\cdot\underline{\tilde{\Omega}^2}_{A,B}(-k) = \omega_I^2(k)\hat{Q}_{I,B}(k). \tag{19.87}$$

Up to linear combinations within degeneracy subspaces, the general set of solutions of the conditions (19.83) will then have the form

$$\tilde{Q}_k(\omega) = \sum_I\Big(q_I(k)\hat{Q}_I(k)\delta(\omega-\omega_I(k)) + q_I^+(-k)\hat{Q}^+_I(-k)\delta(\omega+\omega_I(-k))\Big)$$

with complex factors $q_I(\boldsymbol{k})$. This yields the general lattice vibration in terms of the orthonormalized solutions of (19.83),

$$\boldsymbol{x}_{\boldsymbol{\ell},A}(t) = \frac{V}{(2\pi)^3\sqrt{m_A}}\int_{\mathcal{B}} d^3\boldsymbol{k}\sum_I \Big[q_I(\boldsymbol{k})\hat{\boldsymbol{Q}}_{I,A}(\boldsymbol{k})\exp\Big(\mathrm{i}[\boldsymbol{k}\cdot\boldsymbol{\ell} - \omega_I(\boldsymbol{k})t]\Big)$$
$$+ q_I^+(\boldsymbol{k})\hat{\boldsymbol{Q}}_{I,A}^+(\boldsymbol{k})\exp\Big(-\mathrm{i}[\boldsymbol{k}\cdot\boldsymbol{\ell} - \omega_I(\boldsymbol{k})t]\Big)\Big], \tag{19.88}$$

$$\dot{\boldsymbol{x}}_{\boldsymbol{\ell},A}(t) = \int_{\mathcal{B}} d^3\boldsymbol{k}\sum_I \frac{-\mathrm{i}\omega_I(\boldsymbol{k})V}{(2\pi)^3\sqrt{m_A}}\Big[q_I(\boldsymbol{k})\hat{\boldsymbol{Q}}_{I,A}(\boldsymbol{k})\exp\Big(\mathrm{i}[\boldsymbol{k}\cdot\boldsymbol{\ell} - \omega_I(\boldsymbol{k})t]\Big)$$
$$- q_I^+(\boldsymbol{k})\hat{\boldsymbol{Q}}_{I,A}^+(\boldsymbol{k})\exp\Big(-\mathrm{i}[\boldsymbol{k}\cdot\boldsymbol{\ell} - \omega_I(\boldsymbol{k})t]\Big)\Big], \tag{19.89}$$

$$q_I(\boldsymbol{k}) = \frac{1}{2}\sum_{\boldsymbol{\ell},A}\exp\Big(-\mathrm{i}[\boldsymbol{k}\cdot\boldsymbol{\ell} - \omega_I(\boldsymbol{k})t]\Big)\sqrt{m_A}$$
$$\times\hat{\boldsymbol{Q}}_{I,A}^+(\boldsymbol{k})\cdot\left(\boldsymbol{x}_{\boldsymbol{\ell},A}(t) + \frac{\mathrm{i}}{\omega_I(\boldsymbol{k})}\dot{\boldsymbol{x}}_{\boldsymbol{\ell},A}(t)\right). \tag{19.90}$$

The dual orthogonality relation to (19.84) follows from re-substitution of $q_I(\boldsymbol{k})$ into (19.88),

$$\frac{V}{(2\pi)^3}\int_{\mathcal{B}} d^3\boldsymbol{k}\sum_I \hat{\boldsymbol{Q}}_{I,A}(\boldsymbol{k})\otimes\hat{\boldsymbol{Q}}_{I,B}^+(\boldsymbol{k})\exp[\mathrm{i}\boldsymbol{k}\cdot(\boldsymbol{\ell} - \boldsymbol{\ell}')] = \delta_{AB}\delta_{\boldsymbol{\ell},\boldsymbol{\ell}'}\underline{1}. \tag{19.91}$$

This is actually fulfilled due to two more fundamental completeness relations. The first relation is completeness of $3N$ orthonormal unit vectors $\hat{\boldsymbol{Q}}_I(\boldsymbol{k}) \equiv \{\hat{\boldsymbol{Q}}_{I,A}(\boldsymbol{k})\}_{1\le A\le N}$ in a $3N$-dimensional vector space,

$$\sum_I \hat{\boldsymbol{Q}}_I(\boldsymbol{k})\otimes\hat{\boldsymbol{Q}}_I^+(\boldsymbol{k}) = \underline{1},$$

where $\underline{1}$ is the $3N \times 3N$ unit matrix, or if the atomic indices are spelled out,

$$\sum_I \hat{\boldsymbol{Q}}_{I,A}(\boldsymbol{k})\otimes\hat{\boldsymbol{Q}}_{I,B}^+(\boldsymbol{k}) = \delta_{AB}\underline{1}, \tag{19.92}$$

where now $\underline{1}$ is the 3×3 unit matrix referring to the spatial indices. The second relation is the completeness relation (19.36).

The canonical quantization relations

$$[x^i_{\boldsymbol{\ell},A}(t), \dot{x}^j_{\boldsymbol{\ell}',B}(t)] = \frac{\mathrm{i}\hbar}{m_A}\delta_{AB}\delta_{\boldsymbol{\ell},\boldsymbol{\ell}'}\delta^{ij},$$
$$[x^i_{\boldsymbol{\ell},A}(t), x^j_{\boldsymbol{\ell}',B}(t)] = 0, \quad [\dot{x}^i_{\boldsymbol{\ell},A}(t), \dot{x}^j_{\boldsymbol{\ell}',B}(t)] = 0,$$

imply

$$[q_I(\boldsymbol{k}), q_J(\boldsymbol{k}')] = 0, \quad [q_I(\boldsymbol{k}), q_J^+(\boldsymbol{k}')] = \frac{\hbar}{2\omega_I(\boldsymbol{k})}\frac{(2\pi)^3}{V}\delta_{IJ}\delta(\boldsymbol{k}-\boldsymbol{k}'),$$

i.e. the phonon annihilation operator for the Ith mode with wave vector $\boldsymbol{k}$ in the lattice is

$$a_I(\boldsymbol{k}) = \frac{1}{2}\sqrt{\frac{\omega_I(\boldsymbol{k})V}{\pi^3\hbar}}q_I(\boldsymbol{k}), \tag{19.93}$$

and the displacement operators in terms of the phonon operators are given by

$$\boldsymbol{x}_{\ell,A}(t) = \sum_I \boldsymbol{x}_{I,\ell,A}(t), \tag{19.94}$$

with

$$\boldsymbol{x}_{I,\ell,A}(t) = \sqrt{\frac{\hbar V}{(2\pi)^3 m_A}}\int_{\mathcal{B}}\frac{d^3k}{\sqrt{2\omega_I(\boldsymbol{k})}}\Big[a_I(\boldsymbol{k})\hat{\boldsymbol{Q}}_{I,A}(\boldsymbol{k})\exp\Big(\mathrm{i}[\boldsymbol{k}\cdot\boldsymbol{\ell}-\omega_I(\boldsymbol{k})t]\Big)$$
$$+\,a_I^+(\boldsymbol{k})\hat{\boldsymbol{Q}}_{I,A}^+(\boldsymbol{k})\exp\Big(-\mathrm{i}[\boldsymbol{k}\cdot\boldsymbol{\ell}-\omega_I(\boldsymbol{k})t]\Big)\Big]. \tag{19.95}$$

The Lagrange function (19.78) implies a Hamiltonian for the lattice vibrations,

$$H = \frac{1}{2}\sum_{\ell,A} m_A\dot{\boldsymbol{x}}_{\ell,A}^2 + \frac{1}{2}\sum_{\ell,A;\ell',A'}\sqrt{m_A m_{A'}}\boldsymbol{x}_{\ell,A}\cdot\underline{\Omega}^2{}_{\ell,A;\ell',A'}\cdot\boldsymbol{x}_{\ell',A'}.$$

This yields after substitution of equations (19.94, 19.95) and use of the eigenvalue, hermiticity and orthogonality conditions for the eigenvalue problem (19.85–19.87) the result[13]

$$H = \int_{\mathcal{B}} d^3k \sum_I \hbar\omega_I(\boldsymbol{k})a_I^+(\boldsymbol{k})a_I(\boldsymbol{k}). \tag{19.96}$$

It is uncommon but helpful for a better understanding of Bloch and Wannier states of electrons to point out an analogy with lattice vibrations at this point.

---

[13]You also have to use that the matrix $\underline{\tilde{\Omega}}^2(\boldsymbol{k})$ has a positive semi-definite square root $\underline{\tilde{\Omega}}(\boldsymbol{k})$, see Problem 19.2. Therefore we also have e.g.

$$\sum_{A,B}\hat{\boldsymbol{Q}}_{I,A}(\boldsymbol{k})\cdot\underline{\tilde{\Omega}}^2{}_{A,B}(-\boldsymbol{k})\cdot\hat{\boldsymbol{Q}}_{J,B}(-\boldsymbol{k}) = \omega_I(\boldsymbol{k})\omega_J(-\boldsymbol{k})\sum_A\hat{\boldsymbol{Q}}_{I,A}(\boldsymbol{k})\hat{\boldsymbol{Q}}_{J,A}(-\boldsymbol{k}).$$

We have seen in Sections 10.1, 10.2 and 10.3 that electrons in lattices can be described in terms of delocalized Bloch states $\psi_n(k,x,t) = \psi_n(k,x)\exp[-i\omega_n(k)t]$ or corresponding Wannier states $w_{n,\nu}(x)$, $w_{n,\nu}(x,t)$. Here ν labelled the different cells in the lattice and n labelled the different electron energy bands in the periodic potential of the crystal. We have encountered the corresponding states in three-dimensional lattices in equations (19.43, 19.46). To make the connection to lattice vibrations, we re-express the result (19.88) for the particular phonon energy band I in the form

$$\boldsymbol{x}_{I,\boldsymbol{\ell},A}(t) = \frac{V}{(2\pi)^3}\int d^3\boldsymbol{k}\,\tilde{\boldsymbol{x}}_{I,\boldsymbol{k},A}(t)\exp(\mathrm{i}\boldsymbol{k}\cdot\boldsymbol{\ell}),$$

$$\begin{aligned}\tilde{\boldsymbol{x}}_{I,\boldsymbol{k},A}(t) &= \sum_{\boldsymbol{\ell}}\boldsymbol{x}_{I,\boldsymbol{\ell},A}(t)\exp(-\mathrm{i}\boldsymbol{k}\cdot\boldsymbol{\ell}) = \tilde{\boldsymbol{x}}^{+}_{I,-\boldsymbol{k},A}(t)\\ &= \frac{\boldsymbol{q}_I(\boldsymbol{k})}{\sqrt{m_A}}\hat{Q}_{I,A}(\boldsymbol{k})\exp[-\mathrm{i}\omega_I(\boldsymbol{k})t]\\ &\quad + \frac{\boldsymbol{q}_I^{+}(-\boldsymbol{k})}{\sqrt{m_A}}\hat{Q}^{+}_{I,A}(-\boldsymbol{k})\exp[\mathrm{i}\omega_I(-\boldsymbol{k})t].\end{aligned}$$

Instead of the continuous dependence of the Bloch or Wannier type wave functions $\phi_n(\boldsymbol{k},\boldsymbol{x},t) \sim \exp(\mathrm{i}\boldsymbol{k}\cdot\boldsymbol{x})u_n(\boldsymbol{k},\boldsymbol{x},t)$ and $w_n(\boldsymbol{\ell},\boldsymbol{x},t)$ on location $\boldsymbol{x}$, we have displacement variables at the discrete locations $\{\boldsymbol{\ell},A\}$ in the lattice. However with the correspondence of band indices $n \leftrightarrow I$, the Brillouin zone representation $\tilde{\boldsymbol{x}}_{I,\boldsymbol{k},A}(t)$ of the displacements corresponds to the Bloch waves (19.43) for electron states, while the set of displacements $\{\boldsymbol{x}_{I,\boldsymbol{\ell},A}(t)\}_{1\le A\le N}$ in the unit cell at $\boldsymbol{\ell}$ corresponds to the Wannier states (19.46).

19.7 Electron-phonon interactions

Phonons in the lattice of a solid material naturally couple to electrons through the electrostatic interaction between the electrons and the ion cores. If we neglect electron-electron interactions, the basic Schrödinger picture Hamiltonian for quantized electrons in a lattice of ion cores with N atoms in the unit cell has the form

$$H = -\int d^3\boldsymbol{x}\sum_\sigma \psi_\sigma^{+}(\boldsymbol{x})\left(\frac{\hbar^2}{2m}\Delta + \sum_{\boldsymbol{\ell},A}\frac{n_A e^2}{4\pi\epsilon_0\,|\boldsymbol{x}-\boldsymbol{r}_{\boldsymbol{\ell},A}|}\right)\psi_\sigma(\boldsymbol{x}).$$

We assume that the A-th atom or ion in the unit cell couples to the electron with an effective charge $n_A e$, and we treat the atoms or ions as classical sources of electrostatic fields. However, we treat the lattice vibrations on the quantum level, which according to Sections 19.5 and 19.6 amounts to canonical quantization of the lattice displacements

$$\boldsymbol{x}_{\boldsymbol{\ell},A} = \boldsymbol{r}_{\boldsymbol{\ell},A} - \boldsymbol{x}^{(0)}_{\boldsymbol{\ell},A}.$$

The leading order expansion of the Coulomb term

$$\frac{n_A e}{|\boldsymbol{x}-\boldsymbol{r}_{\ell,A}|} \simeq \frac{n_A e}{|\boldsymbol{x}-\boldsymbol{x}^{(0)}_{\ell,A}|} + n_A e \frac{(\boldsymbol{x}-\boldsymbol{x}^{(0)}_{\ell,A})\cdot\boldsymbol{x}_{\ell,A}}{|\boldsymbol{x}-\boldsymbol{x}^{(0)}_{\ell,A}|^3} \tag{19.97}$$

corresponds to a dipole approximation in the language of Chapter 15, except that here the dipole operator $\boldsymbol{d}_{\ell,A} = n_A e \boldsymbol{x}_{\ell,A}$ is quantized according to (19.94, 19.95). This yields an electron-phonon interaction Hamiltonian of the form

$$H_{e-q} = -\sqrt{\frac{\hbar V}{(2\pi)^3}} \int d^3\boldsymbol{x} \sum_{\sigma,I,\ell,A} \psi^+_\sigma(\boldsymbol{x})\psi_\sigma(\boldsymbol{x}) \int_B \frac{d^3\boldsymbol{q}}{\sqrt{2\omega_I(\boldsymbol{q})}} \frac{e}{\sqrt{m_A}}$$
$$\times \boldsymbol{E}_{\ell,A}(\boldsymbol{x})\cdot\left[a_I(\boldsymbol{q})\hat{\boldsymbol{Q}}_{I,A}(\boldsymbol{q})\exp(\mathrm{i}\boldsymbol{q}\cdot\boldsymbol{\ell}) + a^+_I(\boldsymbol{q})\hat{\boldsymbol{Q}}^+_{I,A}(\boldsymbol{q})\exp(-\mathrm{i}\boldsymbol{q}\cdot\boldsymbol{\ell})\right],$$

where we substituted the time-independent phonon operators $\boldsymbol{x}_{\ell,A}(0)$ for the Hamiltonian in the Schrödinger picture. For the electron operators, we could substitute Bloch or Wannier type operators. However, Bloch operators make much more sense, because the dipole approximation (19.97) is a small oscillation approximation in the sense $|\boldsymbol{x}_{\ell,A}| \ll |\boldsymbol{x}-\boldsymbol{x}^{(0)}_{\ell,A}|$, or otherwise we should include quadrupole and higher order terms. This implies that matrix elements of electron states with the lattice electric fields $\boldsymbol{E}_{\ell,A}(\boldsymbol{x})$ must not be dominated by large terms from the ion cores. The linear phonon coupling Hamiltonian H_{e-q} should therefore not be a good approximation for the localized electrons in Wannier states. Evaluation of the substitution of the free electron operators through Bloch operators (19.41) in H_{e-q} uses the fact that integration over $\boldsymbol{x}$ can be split into summation over the lattice $\boldsymbol{l}$ and integration over the unit lattice cell V,

$$\int d^3\boldsymbol{x}\, f(\boldsymbol{x}) = \sum_{\boldsymbol{l}} \int_V d^3\boldsymbol{x}\, f(\boldsymbol{l}+\boldsymbol{x}),$$

and that the lattice electric fields satisfy

$$\boldsymbol{E}_{\ell,A}(\boldsymbol{x}) = \boldsymbol{E}_{0,A}(\boldsymbol{x}-\boldsymbol{\ell}).$$

We denote the Bloch operators for the electrons by $c_{n,\sigma}(\boldsymbol{k})$ to avoid confusion with the phonon operators. This yields the following form for the electron-phonon interaction operator,

$$H_{e-q} = -\sqrt{\frac{\hbar V}{(2\pi)^3}} \sum_{\sigma,I,\ell,A,n,n'} \int_B \frac{d^3\boldsymbol{q}}{\sqrt{2\omega_I(\boldsymbol{q})}} \int_B d^3\boldsymbol{k} \int_V d^3\boldsymbol{x} \frac{e}{\sqrt{m_A}}$$
$$\boldsymbol{E}_{\ell,A}(\boldsymbol{x})\cdot\Big[u^+_n(\boldsymbol{k}+\boldsymbol{q},\boldsymbol{x})c^+_{n,\sigma}(\boldsymbol{k}+\boldsymbol{q})a_I(\boldsymbol{q})\hat{\boldsymbol{Q}}_{I,A}(\boldsymbol{q})\exp[\mathrm{i}\boldsymbol{q}\cdot(\boldsymbol{\ell}-\boldsymbol{x})]$$
$$+ u^+_n(\boldsymbol{k}-\boldsymbol{q},\boldsymbol{x})c^+_{n,\sigma}(\boldsymbol{k}-\boldsymbol{q})a^+_I(\boldsymbol{q})\hat{\boldsymbol{Q}}^+_{I,A}(\boldsymbol{q})\exp[\mathrm{i}\boldsymbol{q}\cdot(\boldsymbol{x}-\boldsymbol{\ell})]\Big]$$
$$\times c_{n',\sigma}(\boldsymbol{k})u_{n'}(\boldsymbol{k},\boldsymbol{x}). \tag{19.98}$$

We can also write this as

$$H_{e-q} = \sum_{\sigma,I} \int_{\mathcal{B}} \frac{d^3q}{\sqrt{2\omega_I(\boldsymbol{q})}} \int_{\mathcal{B}} d^3k \left[c_\sigma^+(\boldsymbol{k}+\boldsymbol{q}) \cdot \underline{U}_I(\boldsymbol{k},\boldsymbol{q}) \cdot c_\sigma(\boldsymbol{k}) a_I(\boldsymbol{q}) \right.$$
$$\left. + a_I^+(\boldsymbol{q}) c_\sigma^+(\boldsymbol{k}) \cdot \underline{U}_I^+(\boldsymbol{k},\boldsymbol{q}) \cdot c_\sigma(\boldsymbol{k}+\boldsymbol{q}) \right], \tag{19.99}$$

with coupling matrices between the phonons and the Bloch electrons,

$$U_{I,n,n'}(\boldsymbol{k},\boldsymbol{q}) = -\sqrt{\frac{\hbar V}{(2\pi)^3}} \int_V d^3x \sum_{\boldsymbol{\ell},A} \frac{e}{\sqrt{m_A}} \exp[\mathrm{i}\boldsymbol{q}\cdot(\boldsymbol{\ell}-\boldsymbol{x})]$$
$$\times u_n^+(\boldsymbol{k}+\boldsymbol{q},\boldsymbol{x}) \boldsymbol{E}_{\boldsymbol{\ell},A}(\boldsymbol{x}) \cdot \hat{\boldsymbol{Q}}_{I,A}(\boldsymbol{q}) u_{n'}(\boldsymbol{k},\boldsymbol{x}). \tag{19.100}$$

The products in (19.99) contain summations over the electron energy band indices n, n'.

Below we will need the following property of the electron-phonon coupling functions,

$$U_{I,n,n'}(\boldsymbol{k}+\boldsymbol{q},-\boldsymbol{q}) = U^+_{I,n',n}(\boldsymbol{k},\boldsymbol{q}). \tag{19.101}$$

The full Hamiltonian also contains the free Hamiltonian for the phonons and the Bloch electrons

$$H_0 = \int_{\mathcal{B}} d^3k \left(\sum_I \hbar\omega_I(\boldsymbol{k}) a_I^+(\boldsymbol{k}) a_I(\boldsymbol{k}) + \sum_\sigma c_\sigma^+(\boldsymbol{k}) \cdot \underline{E}(\boldsymbol{k}) \cdot c_\sigma(\boldsymbol{k}) \right)$$

with

$$E_{n,n'}(\boldsymbol{k}) = \frac{\hbar^2}{2m} \left(\int_V d^3x\, \nabla u_n^+(\boldsymbol{k},\boldsymbol{x}) \cdot \nabla u_{n'}(\boldsymbol{k},\boldsymbol{x}) \right.$$
$$\left. - \mathrm{i}\boldsymbol{k} \cdot \int_V d^3x\, u_n^+(\boldsymbol{k},\boldsymbol{x}) \overleftrightarrow{\nabla} u_{n'}(\boldsymbol{k},\boldsymbol{x}) + \boldsymbol{k}^2 \delta_{n,n'} \right). \tag{19.102}$$

The two interaction terms in (19.99) describe absorption and emission of a phonon of wave number $\boldsymbol{q}$ by a Bloch electron. The resulting exchange of virtual phonons between electron pairs will generate an effective interaction between the electrons. If interband couplings can be neglected, $U_{I,n,n'}(\boldsymbol{k},\boldsymbol{q}) \propto \delta_{n,n'}$ and $E_{n,n'}(\boldsymbol{k}) \propto \delta_{n,n'}$, a simple method to estimate this phonon mediated electron-electron interaction eliminates the first order phonon coupling through the Lemma 1 (6.22) for exponentials of operators. A unitary transformation $|\Phi\rangle \to |\Phi'\rangle = \exp(A)|\Phi\rangle$ with

$$A = \sum_{\sigma,I} \int_{\mathcal{B}} \frac{d^3\boldsymbol{q}}{\sqrt{2\omega_I(\boldsymbol{q})}} \int_{\mathcal{B}} d^3\boldsymbol{k}\, \frac{1}{E(\boldsymbol{k}+\boldsymbol{q}) - E(\boldsymbol{k}) - \hbar\omega_I(\boldsymbol{q})}$$
$$\times \left[a_I^+(\boldsymbol{q})c_\sigma^+(\boldsymbol{k})U_I^+(\boldsymbol{k},\boldsymbol{q})c_\sigma(\boldsymbol{k}+\boldsymbol{q}) - c_\sigma^+(\boldsymbol{k}+\boldsymbol{q})U_I(\boldsymbol{k},\boldsymbol{q})c_\sigma(\boldsymbol{k})a_I(\boldsymbol{q})\right]$$

eliminates the leading order electron-phonon coupling term due to

$$[A, H_0] + H_{e-q} = 0,$$

and generates a direct electron-electron coupling term

$$H_{e-e}^{(q)} = \left[\frac{1}{2}{}^{2}[A, H_0] + [A, H_{e-q}]\right]_{c^+c^+cc} = \frac{1}{2}[A, H_{e-q}]\Big|_{c^+c^+cc}$$
$$= \sum_{\sigma,\sigma',I} \int_{\mathcal{B}} \frac{d^3\boldsymbol{q}}{4\omega_I(\boldsymbol{q})} \int_{\mathcal{B}} d^3\boldsymbol{k} \int_{\mathcal{B}} d^3\boldsymbol{k}'\, \frac{1}{E(\boldsymbol{k}+\boldsymbol{q}) - E(\boldsymbol{k}) - \hbar\omega_I(\boldsymbol{q})}$$
$$\times \left[c_\sigma^+(\boldsymbol{k}+\boldsymbol{q})c_{\sigma'}^+(\boldsymbol{k}')U_I^+(\boldsymbol{k}',\boldsymbol{q})U_I(\boldsymbol{k},\boldsymbol{q})c_{\sigma'}(\boldsymbol{k}'+\boldsymbol{q})c_\sigma(\boldsymbol{k})\right.$$
$$\left. + \, c_\sigma^+(\boldsymbol{k})c_{\sigma'}^+(\boldsymbol{k}'+\boldsymbol{q})U_I^+(\boldsymbol{k},\boldsymbol{q})U_I(\boldsymbol{k}',\boldsymbol{q})c_{\sigma'}(\boldsymbol{k}')c_\sigma(\boldsymbol{k}+\boldsymbol{q})\right].$$

In the next step we substitute

$$\boldsymbol{k} \to \boldsymbol{k}+\boldsymbol{q}, \quad \boldsymbol{k}' \to \boldsymbol{k}'+\boldsymbol{q}, \quad \boldsymbol{q} \to -\boldsymbol{q},$$

in the second term in $H_{e-e}^{(q)}$ and use the properties (19.101) and $\omega_I(\boldsymbol{q}) = \omega_I(-\boldsymbol{q})$. This yields

$$H_{e-e}^{(q)} = \sum_{\sigma,\sigma',I} \int_{\mathcal{B}} \frac{d^3\boldsymbol{q}}{4\omega_I(\boldsymbol{q})} \int_{\mathcal{B}} d^3\boldsymbol{k} \int_{\mathcal{B}} d^3\boldsymbol{k}'\, c_\sigma^+(\boldsymbol{k}+\boldsymbol{q})c_{\sigma'}^+(\boldsymbol{k}')U_I^+(\boldsymbol{k}',\boldsymbol{q})$$
$$\times U_I(\boldsymbol{k},\boldsymbol{q})c_{\sigma'}(\boldsymbol{k}'+\boldsymbol{q})c_\sigma(\boldsymbol{k})$$
$$\times \left[\frac{1}{E(\boldsymbol{k}+\boldsymbol{q}) - E(\boldsymbol{k}) - \hbar\omega_I(\boldsymbol{q})} - \frac{1}{E(\boldsymbol{k}+\boldsymbol{q}) - E(\boldsymbol{k}) + \hbar\omega_I(\boldsymbol{q})}\right]$$
$$= \frac{\hbar}{2}\sum_{\sigma,\sigma',I} \int_{\mathcal{B}} d^3\boldsymbol{q} \int_{\mathcal{B}} d^3\boldsymbol{k} \int_{\mathcal{B}} d^3\boldsymbol{k}'\, \frac{1}{[E(\boldsymbol{k}+\boldsymbol{q}) - E(\boldsymbol{k})]^2 - \hbar^2\omega_I^2(\boldsymbol{q})}.$$
$$\times c_\sigma^+(\boldsymbol{k}+\boldsymbol{q})c_{\sigma'}^+(\boldsymbol{k}')U_I^+(\boldsymbol{k}',\boldsymbol{q})U_I(\boldsymbol{k},\boldsymbol{q})c_{\sigma'}(\boldsymbol{k}'+\boldsymbol{q})c_\sigma(\boldsymbol{k}). \qquad (19.103)$$

Phonons with frequencies which are large compared to the electron energy difference,

$$\hbar\omega_I(\boldsymbol{q}) > |E(\boldsymbol{k}+\boldsymbol{q}) - E(\boldsymbol{k})|,$$

lower the energy of a two-electron state, thus implying an energetically favorable correlation between electrons. Effectively, a negative coefficient of $c_\sigma^+(\boldsymbol{k}+\boldsymbol{q})c_{\sigma'}^+(\boldsymbol{k}')c_{\sigma'}(\boldsymbol{k}'+\boldsymbol{q})c_\sigma(\boldsymbol{k})$ also amounts to an electron-electron attraction. Compare (19.103) with the simplified expression for free fermion operators,

$$H' = \Lambda\sum_{\sigma,\sigma'}\int d^3\boldsymbol{q}\int d^3\boldsymbol{k}\int d^3\boldsymbol{k}'\, c_\sigma^+(\boldsymbol{k}+\boldsymbol{q})c_{\sigma'}^+(\boldsymbol{k}')c_{\sigma'}(\boldsymbol{k}'+\boldsymbol{q})c_\sigma(\boldsymbol{k}).$$

In $\boldsymbol{x}$ space this becomes

$$H' = (2\pi)^3\Lambda\sum_{\sigma,\sigma'}\int d^3\boldsymbol{x}\,\psi_\sigma^+(\boldsymbol{x})\psi_{\sigma'}^+(\boldsymbol{x})\psi_{\sigma'}(\boldsymbol{x})\psi_\sigma(\boldsymbol{x}),$$

which is an attractive interaction for $\Lambda < 0$ and repulsive otherwise.

The possible instability of Fermi surfaces against phonon-induced energetically favored correlations between electrons, and the ensuing suppression of electron scattering, had been identified in the 1950s as the mechanism for low temperature superconductivity[14]. Please consult [5, 17, 22, 25] for textbook discussions of low temperature superconductivity.

19.8 Problems

19.1. Suppose we are using the Born-Oppenheimer approximation for the hydrogen atom, i.e. we treat the proton as fixed at location $\boldsymbol{X}_p = \boldsymbol{0}$. This would yield the same energy levels and energy eigenfunctions that we had found in the exact solution in Chapter 7, *except* that the reduced mass $\mu = m_e m_p/(m_e + m_p)$ would be replaced by the electron mass m_e in the result for the Bohr radius a, and therefore also in the energy eigenvalues and the wave functions.

Show that the corresponding change in the mass value $\delta\mu = m_e - \mu$ satisfies $\delta\mu/\mu = m_e/m_p$. Show also that in the center of mass frame, the neglected kinetic energy of the proton is related to the kinetic energies of the electron and of the relative motion according to

$$K_p = \frac{m_e}{m_p}K_e = \frac{m_e}{m_e + m_p}K_r.$$

Expand the ground state wave function in the Born-Oppenheimer approximation in first order in m_e/m_p in terms of the exact energy eigenstates from Chapter 7.

[14] J. Bardeen, L.N. Cooper, J.R. Schrieffer, Phys. Rev. 108, 1175 (1957); see also H. Fröhlich, Phys. Rev. 79, 845 (1950) and J. Bardeen, D. Pines, Phys. Rev. 99, 1140 (1955).

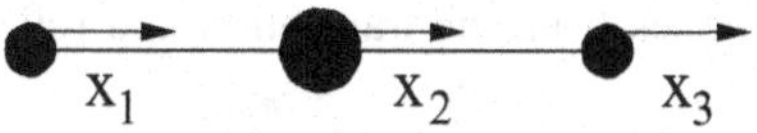

Fig. 19.8 Three particles with masses m and M. It is supposed that the particles can only move along the line connecting them

19.2. Show that the hermitian symmetric matrix $\underline{\tilde{\Omega}}^2(\boldsymbol{k})$ (19.80) with eigenvalues $\omega_I^2(\boldsymbol{k}) \geq 0$ and corresponding normalized eigenvectors $\hat{\boldsymbol{Q}}_I(\boldsymbol{k})$ has square roots $\underline{\tilde{\Omega}}(\boldsymbol{k})$,

$$\underline{\tilde{\Omega}}^2(\boldsymbol{k}) = \underline{\tilde{\Omega}}^2(\boldsymbol{k}).$$

Hint: The column vectors $\hat{\boldsymbol{Q}}_I(\boldsymbol{k})$ can be used to form a unitary matrix $\underline{Q}(\boldsymbol{k})$. The matrix $\underline{Q}(\boldsymbol{k})$ transforms $\underline{\tilde{\Omega}}^2(\boldsymbol{k})$ into diagonal form, or in turn can be used to generate $\underline{\tilde{\Omega}}^2(\boldsymbol{k})$ from its diagonal form $\text{diag}(\omega_1^2(\boldsymbol{k}), \ldots, \omega_{3N}^2(\boldsymbol{k}))$. Use this observation to construct all the possible square roots $\underline{\tilde{\Omega}}(\boldsymbol{k})$ in terms of $\underline{Q}(\boldsymbol{k})$ and $\text{diag}(\pm\omega_1(\boldsymbol{k}), \ldots, \pm\omega_{3N}(\boldsymbol{k}))$.

19.3. Suppose the three particles with masses m and M in Figure 19.8 can only move in one dimension.

The potential energy of the system is

$$V = \frac{K}{2}(x_1 - x_2)^2 + \frac{K}{2}(x_2 - x_3)^2.$$

Calculate the eigenvibrations and the eigenfrequencies of the system.

Solution. The potential in matrix notation is

$$V = \frac{K}{2}(x_1, x_2, x_3)\begin{pmatrix} 1 & -1 & 0 \\ -1 & 2 & -1 \\ 0 & -1 & 1 \end{pmatrix}\begin{pmatrix} x_1 \\ x_2 \\ x_3 \end{pmatrix},$$

and we have to find the eigenvectors of the corresponding matrix

$$\underline{\Omega}^2 = K\begin{pmatrix} \frac{1}{m} & -\frac{1}{\sqrt{mM}} & 0 \\ -\frac{1}{\sqrt{mM}} & \frac{2}{M} & -\frac{1}{\sqrt{mM}} \\ 0 & -\frac{1}{\sqrt{mM}} & \frac{1}{m} \end{pmatrix}, \tag{19.104}$$

cf. 19.62.

Rather than trying to solve

$$\det(\underline{\Omega}^2 - \omega^2\underline{1}) = 0,$$

we can infer two eigenmodes from the translation and reflection symmetry of the system.

Invariance of the potential under translations $x_1 = x_2 = x_3$ implies that one eigenvector of $\underline{\Omega}^2$ has the form

$$\hat{Q}_{\omega_1=0} = \frac{1}{\sqrt{2m+M}}\begin{pmatrix}\sqrt{m}\\ \sqrt{M}\\ \sqrt{m}\end{pmatrix}.$$

Reflection symmetry also suggests an eigenmode $x_1 = -x_3$, $x_2 = 0$,

$$\hat{Q}_{\omega_2} = \frac{1}{\sqrt{2}}\begin{pmatrix}1\\ 0\\ -1\end{pmatrix},$$

and application of $\underline{\Omega}^2$ yields the corresponding eigenvalue

$$\omega_2^2 = \frac{K}{m}.$$

The remaining eigenvector follows from orthogonality on $\hat{Q}_{\omega_1}$ and $\hat{Q}_{\omega_2}$,

$$\hat{Q}_{\omega_3} = \frac{1}{\sqrt{2(2m+M)}}\begin{pmatrix}\sqrt{M}\\ -2\sqrt{m}\\ \sqrt{M}\end{pmatrix},$$

and application of $\underline{\Omega}^2$ confirms that this is an eigenmode with frequency

$$\omega_3^2 = \frac{K}{m} + \frac{2K}{M}.$$

For the actual eigenvibration we have to go back to the amplitude vector $\boldsymbol{a}_{\omega_2}$ (19.60), because different masses participate in the oscillation. The normalized amplitude vector (19.65) is

$$\hat{\boldsymbol{a}}_{\omega_3} = \frac{1}{\sqrt{2mM(2m+M)}}\begin{pmatrix}M\\ -2m\\ M\end{pmatrix}.$$

The eigenvibrations $\boldsymbol{a}_{\omega_2}$ and $\boldsymbol{a}_{\omega_3}$ are shown in Figure 19.9.

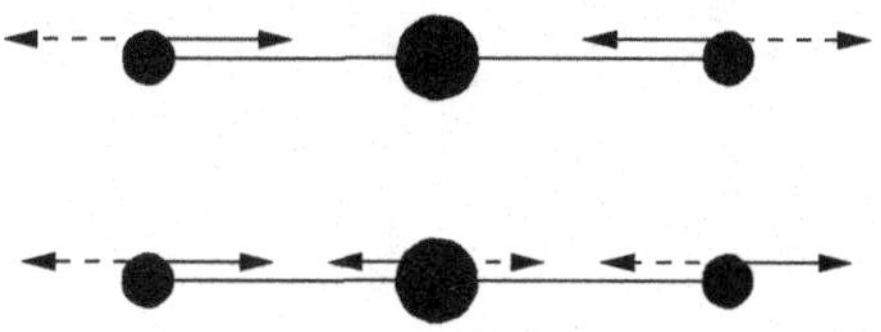

Fig. 19.9 The eigenvibrations $\boldsymbol{a}_{\omega_2}$ and $\boldsymbol{a}_{\omega_3}$

19.4. Calculate the positive semi-definite square root of the matrix $\underline{\Omega}^2$ in equation (19.104). Use the hint from Problem 2.

Answer.

$$\underline{\Omega} = \frac{\sqrt{K}}{2\sqrt{mM(2m+M)}} \times \begin{pmatrix} M + \sqrt{M(2m+M)} & -2\sqrt{mM} & M - \sqrt{M(2m+M)} \\ -2\sqrt{mM} & 4m & -2\sqrt{mM} \\ M - \sqrt{M(2m+M)} & -2\sqrt{mM} & M + \sqrt{M(2m+M)} \end{pmatrix}.$$

19.5. The electron-phonon interaction Hamiltonian (19.99) is very similar to the electron-photon interaction Hamiltonian in the representation (18.92),

$$H_{e-\gamma} = \frac{e\hbar c}{m_e}\sqrt{\frac{\hbar\mu_0}{(2\pi)^3}} \sum_{\sigma,\alpha} \int \frac{d^3\boldsymbol{q}}{\sqrt{2\omega(\boldsymbol{q})}} \int d^3\boldsymbol{k}\,\boldsymbol{k}\cdot\boldsymbol{\epsilon}_\alpha(\boldsymbol{q}) \times \left[c_\sigma^+(\boldsymbol{k}+\boldsymbol{q})a_\alpha(\boldsymbol{q})c_\sigma(\boldsymbol{k}) + c_\sigma^+(\boldsymbol{k})a_\alpha^+(\boldsymbol{q})c_\sigma(\boldsymbol{k}+\boldsymbol{q})\right]. \quad (19.105)$$

Which effective electron-electron interaction Hamiltonian $H_{e-e}^{(\gamma)}$ would you get if you eliminate the photon operators through a unitary transformation $|\Phi\rangle \to |\Phi'\rangle = \exp(A)|\Phi\rangle$ similar to the transformation that we performed to transform H_{e-q} into $H_{e-e}^{(q)}$ (19.103)?

Chapter 20
Dimensional Effects in Low-dimensional Systems

Surfaces, interfaces, thin films, and quantum wires provide abundant examples of quasi two-dimensional or one-dimensional systems in science and technology. Quantum mechanics in low dimensions has become an important tool for modeling properties of these systems. Here we wish to go beyond the simple low-dimensional potential models of Chapter 3 and discuss in particular implications of the dependence of energy-dependent Green's functions on the number d of spatial dimensions. However, if it is true that the behavior of electrons in certain systems and parameter ranges can be described by low-dimensional quantum mechanics, then there must also exist ranges of parameters for quasi low-dimensional systems where the behavior of electrons exhibits *inter-dimensional* behavior in the sense that there must exist continuous interpolations e.g. between two-dimensional and three-dimensional behavior. We will see that inter-dimensional (or "dimensionally hybrid") Green's functions provide a possible avenue to the identification and discussion of inter-dimensional behavior in physical systems.

20.1 Quantum mechanics in d dimensions

Suppose an electron is strictly confined to a two-dimensional quantum well. Unless the material in the quantum well has special dielectric properties, that electron would still "know" that it exists in three spatial dimensions, because it feels the $1/r$ Coulomb interaction with other charged particles, and this $1/r$ distance law is characteristic for three dimensions. If the electron would not just be confined to two dimensions, but exist in a genuine two-dimensional world, it would experience a logarithmic distance law for the Coulomb potential. The reason for the $1/r$ Coulomb law in three dimensions is that the solution of the equation

$$\Delta G(r) = -\delta(\boldsymbol{x}) \qquad (20.1)$$

R. Dick, *Advanced Quantum Mechanics*, Graduate Texts in Physics,
DOI 10.1007/978-3-319-25675-7_20

in three dimensions is given by

$$G(r)=\frac{1}{4\pi r},$$

but in general the solution of equation (20.1) depends on the number d of spatial dimensions. Appendix J explains in detail the derivation of the d-dimensional version of $G(r)$ with the result

$$G_d(r)=\begin{cases}(a-r)/2, & d=1,\\ -(2\pi)^{-1}\ln(r/a), & d=2,\\ \Gamma\left(\frac{d-2}{2}\right)\left(4\sqrt{\pi}^d r^{d-2}\right)^{-1}, & d\geq 3.\end{cases} \tag{20.2}$$

The most direct application of these results are electrostatic potentials. Equation (20.1) implies that the electrostatic potential of a point charge q in d dimensions is given by

$$\Phi_d(r)=\frac{q}{\epsilon_0}G_d(r). \tag{20.3}$$

However, from the point of view of non-relativistic quantum mechanics, the Green's functions (20.2) are the special zero energy values of the energy-dependent free Schrödinger Green's functions, $G_d(r) = G_d(\boldsymbol{x}, E = 0)$. These energy dependent Green's functions satisfy

$$\Delta G_d(\boldsymbol{x},E)+\frac{2m}{\hbar^2}EG_d(\boldsymbol{x},E)=-\delta(\boldsymbol{x}). \tag{20.4}$$

We have solved this condition in three dimensions in Section 11.1, see equation (11.17). The solution in d dimensions by two different methods is described in Appendix J, with the result (J.27)

$$\begin{aligned}G_d(\boldsymbol{x},E)=&\frac{\Theta(-E)}{\sqrt{2\pi}^d}\left(\frac{\sqrt{-2mE}}{\hbar r}\right)^{\frac{d-2}{2}}K_{\frac{d-2}{2}}\left(\sqrt{-2mE}\frac{r}{\hbar}\right)\\ &+\mathrm{i}\frac{\pi}{2}\frac{\Theta(E)}{\sqrt{2\pi}^d}\left(\frac{\sqrt{2mE}}{\hbar r}\right)^{\frac{d-2}{2}}H^{(1)}_{\frac{d-2}{2}}\left(\sqrt{2mE}\frac{r}{\hbar}\right).\end{aligned} \tag{20.5}$$

The functions K_ν and $H^{(1)}_\nu$ are modified Bessel functions and Hankel functions of the first kind, respectively. These are exponential functions for $d = 1$ or $d = 3$,

$$K_{-\frac{1}{2}}(x)=K_{\frac{1}{2}}(x)=\sqrt{\frac{\pi}{2x}}\exp(-x),\quad H^{(1)}_{-\frac{1}{2}}(x)=\mathrm{i}H^{(1)}_{\frac{1}{2}}(x)=\sqrt{\frac{2}{\pi x}}\exp(\mathrm{i}x),$$

and therefore

$$G_1(x,E) = \frac{\hbar\Theta(-E)}{2\sqrt{-2mE}}\exp\!\left(-\sqrt{-2mE}\frac{|x|}{\hbar}\right) + \mathrm{i}\frac{\hbar\Theta(E)}{2\sqrt{2mE}}\exp\!\left(\mathrm{i}\sqrt{2mE}\frac{|x|}{\hbar}\right), \tag{20.6}$$

and we recover the already known three-dimensional result from Section 11.1,

$$G_3(\boldsymbol{x},E) = \frac{\Theta(-E)}{4\pi r}\exp\!\left(-\sqrt{-2mE}\frac{r}{\hbar}\right) + \frac{\Theta(E)}{4\pi r}\exp\!\left(\mathrm{i}\sqrt{2mE}\frac{r}{\hbar}\right). \tag{20.7}$$

These results also give us the screened or Yukawa potentials in d dimensions if we substitute $\hbar/\sqrt{-2mE} \to \lambda$, see Figure 20.2.

The Figures 20.1 and 20.2 illustrate that long distance effects of interactions become more prominent in lower dimensions, while short distance effects become stronger with increasing number of dimensions.

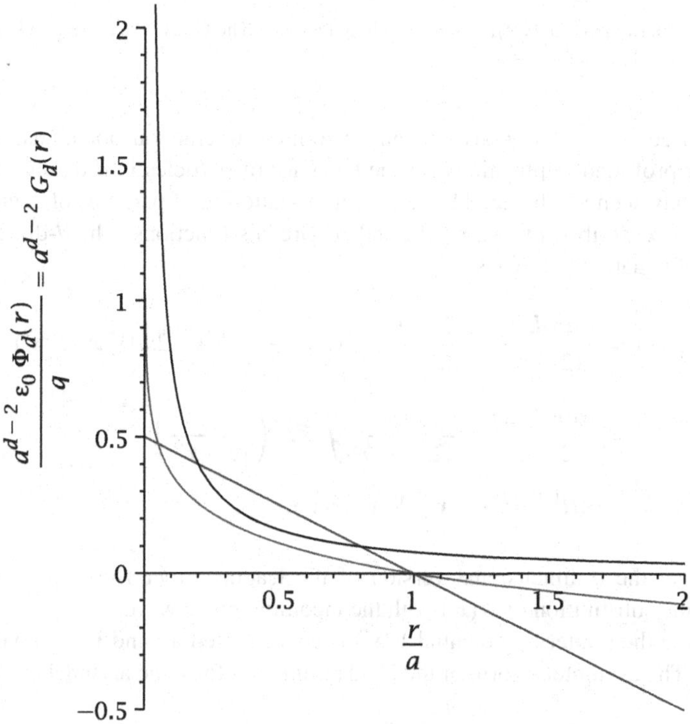

Fig. 20.1 The Green's functions (20.2) and the related specific electrostatic potentials $\Phi_d(r)/q$ in 1, 2 and 3 dimensions. The blue curve is for $d = 1$, red is for $d = 2$ and the black curve is the 3-dimensional Coulomb potential

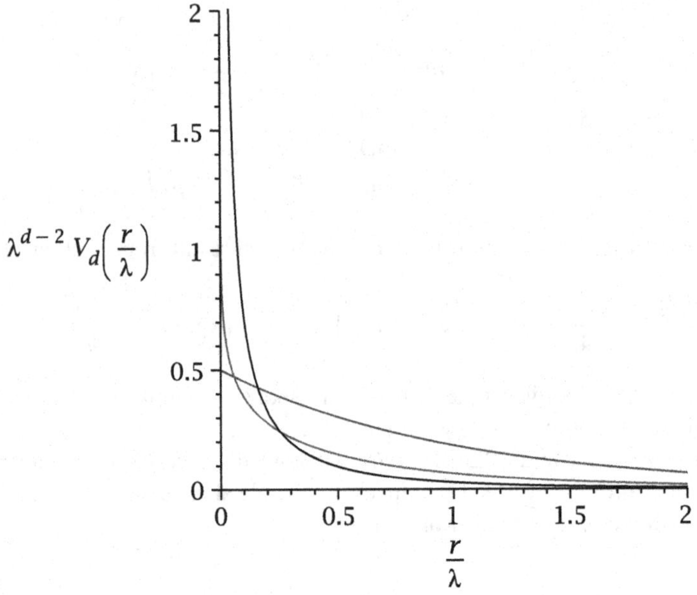

Fig. 20.2 Yukawa potentials $V_d(r)$ with screening length λ. The blue curve corresponds to $d = 1$, red to $d = 2$ and black to $d = 3$

The Green's function does not only determine interaction potentials, but also has other profound implications for the behavior of particles in d dimensions. We have already seen in Chapter 11 that potential scattering of particles of energy $E = \hbar^2 k^2/2m$ is described by energy-dependent Green's functions. The d-dimensional version of equation (11.20) is

$$\begin{aligned}\psi(\boldsymbol{x}) &= \frac{\exp(\mathrm{i}\boldsymbol{k}\cdot\boldsymbol{x})}{(2\pi)^{d/2}} - \frac{2m}{\hbar^2}\int d^d x'\, G_d(\boldsymbol{x}-\boldsymbol{x}', \hbar^2 k^2/2m) V(\boldsymbol{x}')\psi(\boldsymbol{x}') \\ &= \frac{\exp(\mathrm{i}\boldsymbol{k}\cdot\boldsymbol{x})}{(2\pi)^{d/2}} - \frac{\mathrm{i}\pi m}{(2\pi)^{d/2}\hbar^2}\int d^d x' \left(\frac{k}{|\boldsymbol{x}-\boldsymbol{x}'|}\right)^{\frac{d-2}{2}} \\ &\quad \times H^{(1)}_{\frac{d-2}{2}}\left(k|\boldsymbol{x}-\boldsymbol{x}'|\right) V(\boldsymbol{x}')\psi(\boldsymbol{x}'),\end{aligned} \tag{20.8}$$

and we get the d-dimensional version of the leading order Born approximation through the substitution of $\psi(\boldsymbol{x}')$ with the incoming plane wave.

Suppose the scattering potential $V(\boldsymbol{x}')$ is concentrated around $\boldsymbol{x}' = 0$ with finite range R. The asymptotic form of the Hankel function for large argument $x \gg 1$,

$$H^{(1)}_{\frac{d-2}{2}}(x) \to \sqrt{\frac{2}{\pi x}}\exp\left(\mathrm{i}x - \mathrm{i}\pi\frac{d-1}{4}\right),$$

yields the asymptotic form for the scattered wave function in the limit $|\boldsymbol{x}| \gg R$ (here we neglect again the normalization factors $(2\pi)^{-d/2}$ because they cancel in the cross section),

$$\begin{aligned}\psi(\boldsymbol{x}) &= \exp(\mathrm{i}\boldsymbol{k}\cdot\boldsymbol{x}) - \frac{m}{(2\pi)^{(d-1)/2}\hbar^2}\frac{k^{(d-3)/2}}{r^{(d-1)/2}}\exp\!\left(\mathrm{i}kr - \mathrm{i}\pi\frac{d-3}{4}\right)\\ &\quad\times\int d^d x'\,\exp[\mathrm{i}(\boldsymbol{k}-k\hat{\boldsymbol{x}})\cdot\boldsymbol{x}']V(\boldsymbol{x}')\\ &= \exp(\mathrm{i}\boldsymbol{k}\cdot\boldsymbol{x}) + f(k\hat{\boldsymbol{x}}-\boldsymbol{k})\frac{1}{r^{(d-1)/2}}\exp(\mathrm{i}kr)\\ &= \psi^{(\mathrm{in})}(\boldsymbol{x}) + \psi^{(\mathrm{out})}(\boldsymbol{x}),\end{aligned} \tag{20.9}$$

with the scattering amplitude

$$\begin{aligned}f(\Delta\boldsymbol{k}) &= -\frac{mk^{(d-3)/2}}{(2\pi)^{(d-1)/2}\hbar^2}\exp\!\left(-\mathrm{i}\pi\frac{d-3}{4}\right)\int d^d x\,\exp(-\mathrm{i}\Delta\boldsymbol{k}\cdot\boldsymbol{x})V(\boldsymbol{x})\\ &= -\sqrt{2\pi}\frac{mk^{(d-3)/2}}{\hbar^2}\exp\!\left(-\mathrm{i}\pi\frac{d-3}{4}\right)V(\Delta\boldsymbol{k}).\end{aligned} \tag{20.10}$$

The d-dimensional version of (11.2),

$$\frac{d\sigma}{d\Omega} = \frac{1}{j_{in}(\hat{\boldsymbol{k}})}\frac{dn(\Omega)}{d\Omega dt} = \lim_{r\to\infty} r^{d-1}\frac{j_{out}(\hat{\boldsymbol{k}}')}{j_{in}(\hat{\boldsymbol{k}})}, \tag{20.11}$$

then yields the same relation between the scattering cross section and the scattering amplitude as in three dimensions,

$$\frac{d\sigma_k}{d\Omega} = |f(k\hat{\boldsymbol{x}}-\boldsymbol{k})|^2.$$

Note that the relative prominence of long distance effects in low dimensions should be amplified in scattering effects, because the Green's function generically determines *both* the scattering potentials $V(\boldsymbol{x}')$ *and* the kernels $G_d(\boldsymbol{x}-\boldsymbol{x}', E)$ which are convoluted into the potentials to calculate the scattered wave functions.

Yet another instance where the number of dimensions plays a prominent role is the density of states. We have seen this already in equations (12.12, 12.13). However, this result is also closely linked to the energy-dependent Green's functions.

First we note that we can write the generalization of equation (20.4) for the Hamilton operator $H = H_0 + V$,

$$\Delta G_{d,V}(\boldsymbol{x},\boldsymbol{x}';E) + \frac{2m}{\hbar^2}[E - V(\boldsymbol{x})]G_{d,V}(\boldsymbol{x},\boldsymbol{x}';E) = -\delta(\boldsymbol{x}-\boldsymbol{x}'). \tag{20.12}$$

also in representation free operator notation,

$$(E-H)\mathcal{G}_{d,V}(E) = 1, \tag{20.13}$$

where the connection to (20.12) is recovered through the matrix elements[1]

$$\langle \boldsymbol{x}|\mathcal{G}_{d,V}(E)|\boldsymbol{x}'\rangle = -\frac{2m}{\hbar^2}G_{d,V}(\boldsymbol{x},\boldsymbol{x}';E)$$

The solution of (20.13) requires a small complex shift in E to avoid the singularities on the real axis which arise from the real spectrum of H,

$$\mathcal{G}_{d,V}(E) = \frac{1}{E - H + \mathrm{i}\epsilon}.$$

The positive imaginary shift yields the retarded Green's function with only outgoing spherical waves from scattering centers in scattering theory, and only forward evolution in time, see Section 11.1.

The Hamilton operator in the denominator of $\mathcal{G}(E)$ can be replaced by energy eigenvalues if we use energy eigenstates

$$H|E',\nu(E')\rangle = E'|E',\nu(E')\rangle, \quad \sum\!\!\!\!\!\!\int dE'\,d\nu(E')\,|E',\nu(E')\rangle\langle E',\nu(E')| = 1,$$

where $\nu(E')$ is a set of degeneracy indices for energy level E'. This yields

$$\mathcal{G}_{d,V}(E) = \sum\!\!\!\!\!\!\int dE'\,d\nu(E')\,\frac{|E',\nu(E')\rangle\langle E',\nu(E')|}{E - E' + \mathrm{i}\epsilon}. \tag{20.14}$$

The connection to the density of states follows if we rewrite this with the Sokhotsky-Plemelj relation (2.11),

$$\mathcal{G}_{d,V}(E) = \mathcal{P}\sum\!\!\!\!\!\!\int dE'\,d\nu(E')\,\frac{|E',\nu(E')\rangle\langle E',\nu(E')|}{E - E'} \\ - \mathrm{i}\pi\sum\!\!\!\!\!\!\int dE'\,d\nu(E')\,\delta(E-E')|E',\nu(E')\rangle\langle E',\nu(E')|. \tag{20.15}$$

Comparison with equation (12.17) shows that

$$\varrho_d(E,\boldsymbol{x}) = -\frac{1}{\pi}\Im\langle \boldsymbol{x}|\mathcal{G}_{d,V}(E)|\boldsymbol{x}\rangle = \frac{2m}{\pi\hbar^2}\Im\langle \boldsymbol{x}|G_{d,V}(E)|\boldsymbol{x}\rangle. \tag{20.16}$$

In particular, substitution of the free Green's functions (20.5) yields again the result (12.13) which we had initially derived from equation (12.12)

$$\varrho_d(E) = g\Theta(E)\sqrt{\frac{m}{2\pi}}^{d}\,\frac{\sqrt{E}^{d-2}}{\Gamma(d/2)\hbar^d}. \tag{20.17}$$

[1]These concepts are further discussed in Appendix J. However, it is not necessary to read Appendix J before reading this section.

For the derivation of (20.17) from (20.16) and (20.5), we recall that

$$\langle \boldsymbol{x}|G_{d,V}(E)|\boldsymbol{x}'\rangle\big|_{V=0} \equiv G_{d,V}(\boldsymbol{x},\boldsymbol{x}';E)\big|_{V=0} = G_d(\boldsymbol{x}-\boldsymbol{x}',E)$$

is translation invariant for free particles, and use the property

$$\Re H^{(1)}_{\frac{d-2}{2}}\left(\sqrt{2mE}\frac{r}{\hbar}\right)\Big|_{r\to 0} = J_{\frac{d-2}{2}}\left(\sqrt{2mE}\frac{r}{\hbar}\right)\Big|_{r\to 0} \sim \frac{1}{\Gamma(d/2)}\left(\sqrt{\frac{mE}{2}}\frac{r}{\hbar}\right)^{\frac{d-2}{2}}$$

of the Hankel functions. The spin or helicity factor g arises from the summation over spin states included in the summation over degeneracy indices ν in equation (20.15) if we take into account that the Green's functions (20.5) in the presence of spin multiply $g \times g$ unit matrices in spin space[2].

20.2 Inter-dimensional effects in interfaces and thin layers

The dependence of energy-dependent Green's functions on the number of dimensions begs the question whether this can have observable consequences in (quasi)two-dimensional or one-dimensional systems like interfaces, layers, thin films, or nanowires. Indeed, the density of states $\varrho_d(E)$ (20.16) is often used to estimate densities of electron states in low-dimensional systems in nanotechnology. However, is this really justified? After all, we are still dealing with electrons with non-vanishing extensions of their wave functions in every direction, including directions perpendicular to any confining potential barriers. Wave functions can be squeezed, but they will will never be genuine two-dimensional or one-dimensional. Furthermore, if the behavior of low-energy particles in confining structures can be approximated by the laws of one-dimensional or two-dimensional quantum mechanics, there must exist a transition regime at higher energy levels, where inter-dimensional effects between low-dimensional and three-dimensional behavior should be observable.

To examine these questions, we consider a model system of electrons moving in a bulk material which also contains a layer of thickness $2a$ located at $z = z_0$. The potential energy of the electrons inside the layer is shifted by an amount V_0,

$$V(\boldsymbol{x}) \equiv V(z) = V_0\Theta(z_0 + a - z)\Theta(z - z_0 + a), \tag{20.18}$$

and we also assume that electrons in the bulk move with (effective) mass m, while the effective mass inside the layer is m_*. This yields a Hamiltonian which in the first

[2]For spin or helicity, there is actually a transition from a tensor product to a trace operation in making the connection between (20.15) and (20.16): $1 = \sum_s |s\rangle\langle s| \to \sum_s \langle s|s\rangle = g$. Otherwise equation (20.16) would yield the density of states per spin state.

quantized formalism has the form

$$H = \frac{\mathbf{p}^2}{2m}\left[1 - \Theta(z_0 + a - z)\Theta(z - z_0 + a)\right]$$
$$+ \Theta(z_0 + a - z)\Theta(z - z_0 + a)\left(\frac{\mathbf{p}^2}{2m_*} + V_0\right). \qquad (20.19)$$

We might expect two-dimensional behavior in the limit $a \to 0$ both from the difference of effective mass in the interface and from the interface potential. Indeed, the different effective mass m_* in the interface implies different propagation properties inside the interface and yields quasi two-dimensional behavior both in terms of propagators and in the density of states[3], even for vanishing interface potential. The corresponding second quantized Hamiltonian is

$$H = \int d^2\boldsymbol{x}_\parallel \int dz\, \frac{\hbar^2}{2m} \nabla\psi^+(\boldsymbol{x}_\parallel, z) \cdot \nabla\psi(\boldsymbol{x}_\parallel, z)$$
$$+ \int d^2\boldsymbol{x}_\parallel\, \frac{\hbar^2}{2\mu} \nabla_\parallel \psi^+(\boldsymbol{x}, z_0) \cdot \nabla_\parallel \psi(\boldsymbol{x}_\parallel, z_0), \qquad (20.20)$$

where the index $\parallel$ is used for two-dimensional vectors parallel to the interface at z_0. The parameter μ has dimensions of mass per length.

In the following we will investigate the emergence of quasi two-dimensional behavior from an attractive interface potential.

Two-dimensional behavior from a thin quantum well

We wish to examine the appearance of quasi two-dimensional behavior from a quantum well potential, i.e. we assume $m_* = m$. An infinitely thin attractive quantum well arises from the potential (20.18) if we set $V_0 = -\mathcal{W}/2a$ and take the limit $a \to 0$,

$$H = \frac{\mathbf{p}^2}{2m} - \mathcal{W}\delta(z - z_0).$$

The corresponding Schrödinger equation separates, and the z component is the Schrödinger equation with the attractive δ potential that we had solved in Section 3.3. This implies three kinds of energy eigenstates. First we have eigenstates which are moving along the interface,

$$\langle \boldsymbol{x}|\boldsymbol{k}_\parallel, \kappa\rangle = \frac{\sqrt{\kappa}}{2\pi} \exp\left(\mathrm{i}\boldsymbol{k}_\parallel \cdot \boldsymbol{x}_\parallel - \kappa|z - z_0|\right), \quad \kappa = \frac{m}{\hbar^2}\mathcal{W}, \qquad (20.21)$$
$$E(\boldsymbol{k}_\parallel, \kappa) = \frac{\hbar^2}{2m}\boldsymbol{k}_\parallel^2 - \frac{m}{2\hbar^2}\mathcal{W}^2.$$

[3] R. Dick, Physica E 40, 2973 (2008); Nanoscale Res. Lett. 5, 1546 (2010).

We also have free states with odd or even parity under $z \to 2z_0 - z$, cf. (3.21, 3.22),

$$\langle \boldsymbol{x}|\boldsymbol{k}_\|, k_\perp, -\rangle = \frac{1}{2\sqrt{\pi}^3} \exp(\mathrm{i}\boldsymbol{k}_\| \cdot \boldsymbol{x}_\|) \sin[k_\perp(z - z_0)], \tag{20.22}$$

$$\langle \boldsymbol{x}|\boldsymbol{k}_\|, k_\perp, +\rangle = \exp(\mathrm{i}\boldsymbol{k}_\| \cdot \boldsymbol{x}_\|) \frac{k_\perp \cos[k_\perp(z - z_0)] - \kappa \sin[k_\perp|z - z_0|]}{2\sqrt{\pi^3(\kappa^2 + k_\perp^2)}}. \tag{20.23}$$

The wave number $k_\perp$ in (20.22) and (20.23) is constrained to the positive half-line $k_\perp > 0$, and the energy levels of the free states are

$$E(\boldsymbol{k}_\|, k_\perp) = \frac{\hbar^2}{2m}\left(\boldsymbol{k}_\|^2 + k_\perp^2\right).$$

The energy-dependent Green's function

$$\langle \boldsymbol{x}_\|, z|G(E)|\boldsymbol{x}'_\|, z'\rangle \equiv \langle z|G(\boldsymbol{x}_\| - \boldsymbol{x}'_\|, E)|z'\rangle \equiv -\frac{\hbar^2}{2m}\langle z|\mathcal{G}(\boldsymbol{x}_\| - \boldsymbol{x}'_\|, E)|z'\rangle$$

of this system must satisfy

$$\left(\Delta + \frac{2m}{\hbar^2}[E + \mathcal{W}\delta(z - z_0)]\right)\langle z|G(\boldsymbol{x}_\|, E)|z'\rangle = -\delta(\boldsymbol{x}_\|)\delta(z - z'). \tag{20.24}$$

We would not have to solve this equation explicitly, since we know the complete set of energy eigenstates of the system. However, there is a neat way to solve these kinds of problems which also works for interfaces in which particles move with different effective mass[4].

We can solve equation (20.24) in a mixed representation using

$$\langle \boldsymbol{k}_\|, k_\perp|G(E)|\boldsymbol{k}'_\|, z'\rangle = \frac{1}{\sqrt{2\pi}^5}\int d^2\boldsymbol{x}_\| \int d^2\boldsymbol{x}'_\| \int dz\, \langle \boldsymbol{x}_\|, z|G(E)|\boldsymbol{x}'_\|, z'\rangle$$
$$\times \exp\left[\mathrm{i}\left(\boldsymbol{k}'_\| \cdot \boldsymbol{x}'_\| - \boldsymbol{k}_\| \cdot \boldsymbol{x}_\| - k_\perp z\right)\right] \tag{20.25}$$
$$= \langle k_\perp|G(\boldsymbol{k}_\|, E)|z'\rangle \delta\left(\boldsymbol{k}_\| - \boldsymbol{k}'_\|\right), \tag{20.26}$$

$$\langle k_\perp|G(\boldsymbol{k}_\|, E)|z'\rangle = \frac{1}{\sqrt{2\pi}}\int d^2\boldsymbol{x}_\| \int dz\, \langle z|G(\boldsymbol{x}_\|, E)|z'\rangle$$
$$\times \exp[-\mathrm{i}(\boldsymbol{k}_\| \cdot \boldsymbol{x}_\| + k_\perp z)]. \tag{20.27}$$

[4] R. Dick, Int. J. Theor. Phys. 42, 569 (2003). See also the previous references.

Substitution into equation (20.24) yields with $\kappa = m\mathcal{W}/\hbar^2$

$$\frac{\exp[\mathrm{i}k_\perp(z_0 - z')]}{\sqrt{2\pi}} = \left(\boldsymbol{k}_\parallel^2 + k_\perp^2 - \frac{2mE}{\hbar^2}\right)\exp(\mathrm{i}k_\perp z_0)\,\langle k_\perp|G(\boldsymbol{k}_\parallel, E)|z'\rangle$$
$$- \frac{\kappa}{\pi}\int dq_\perp\, \exp(\mathrm{i}q_\perp z_0)\,\langle q_\perp|G(\boldsymbol{k}_\parallel, E)|z'\rangle. \qquad (20.28)$$

This result implies that the retarded Green's function $\langle k_\perp|G(\boldsymbol{k}_\parallel, E)|z'\rangle$ must have the form

$$\exp(\mathrm{i}k_\perp z_0)\,\langle k_\perp|G(\boldsymbol{k}_\parallel, E)|z'\rangle = \left(\frac{\exp[\mathrm{i}k_\perp(z_0 - z')]}{\sqrt{2\pi}} + f(\boldsymbol{k}_\parallel, E, z')\right)$$
$$\times \frac{1}{k_\perp^2 + \boldsymbol{k}_\parallel^2 - (2mE/\hbar^2) - \mathrm{i}\epsilon}, \qquad (20.29)$$

with the yet to be determined function $f(\boldsymbol{k}_\parallel, E, z')$ satisfying

$$f(\boldsymbol{k}_\parallel, E, z') - \frac{\kappa}{\pi}\int dk_\perp\, \frac{\left(\exp[\mathrm{i}k_\perp(z_0 - z')]\,/\sqrt{2\pi}\right) + f(\boldsymbol{k}_\parallel, E, z')}{k_\perp^2 + \boldsymbol{k}_\parallel^2 - (2mE/\hbar^2) - \mathrm{i}\epsilon} = 0,$$

which follows from substituting (20.29) back into (20.28).

The integral is readily evaluated with the residue theorem,

$$\int \frac{dk_\perp}{\pi}\,\frac{\exp(\mathrm{i}k_\perp z)}{k_\perp^2 + \boldsymbol{k}_\parallel^2 - (2mE/\hbar^2) - \mathrm{i}\epsilon} = \frac{\Theta(\hbar^2\boldsymbol{k}_\parallel^2 - 2mE)}{\sqrt{\boldsymbol{k}_\parallel^2 - (2mE/\hbar^2)}}$$
$$\times \exp\!\left(-\sqrt{\boldsymbol{k}_\parallel^2 - \frac{2mE}{\hbar^2}}|z|\right) + \mathrm{i}\frac{\Theta(2mE - \hbar^2\boldsymbol{k}_\parallel^2)}{\sqrt{(2mE/\hbar^2) - \boldsymbol{k}_\parallel^2}}\exp\!\left(\mathrm{i}\sqrt{\frac{2mE}{\hbar^2} - \boldsymbol{k}_\parallel^2}|z|\right).$$

This yields the condition for $f(\boldsymbol{k}_\parallel, E, z')$ in the form

$$\left[1 - \hbar\kappa\left(\frac{\Theta(\hbar^2\boldsymbol{k}_\parallel^2 - 2mE)}{\sqrt{\hbar^2\boldsymbol{k}_\parallel^2 - 2mE}} + \mathrm{i}\frac{\Theta(2mE - \hbar^2\boldsymbol{k}_\parallel^2)}{\sqrt{2mE - \hbar^2\boldsymbol{k}_\parallel^2}}\right)\right] f(\boldsymbol{k}_\parallel, E, z')$$
$$= \frac{\hbar\kappa}{\sqrt{2\pi}}\left[\frac{\Theta(\hbar^2\boldsymbol{k}_\parallel^2 - 2mE)}{\sqrt{\hbar^2\boldsymbol{k}_\parallel^2 - 2mE}}\exp\!\left(-\sqrt{\hbar^2\boldsymbol{k}_\parallel^2 - 2mE}\frac{|z' - z_0|}{\hbar}\right)\right.$$
$$\left. + \mathrm{i}\frac{\Theta(2mE - \hbar^2\boldsymbol{k}_\parallel^2)}{\sqrt{2mE - \hbar^2\boldsymbol{k}_\parallel^2}}\exp\!\left(\mathrm{i}\sqrt{2mE - \hbar^2\boldsymbol{k}_\parallel^2}\frac{|z' - z_0|}{\hbar}\right)\right],$$

and therefore we find with the proper treatment of poles for retarded Green's functions the result

$$\langle k_\perp|G(\boldsymbol{k}_\parallel,E)|z'\rangle = \frac{1}{\sqrt{2\pi}}\frac{1}{k_\perp^2+\boldsymbol{k}_\parallel^2-(2mE/\hbar^2)-\mathrm{i}\epsilon}\left[\exp(-\mathrm{i}k_\perp z')\right.$$
$$+\frac{\hbar\kappa\Theta(\hbar^2\boldsymbol{k}_\parallel^2-2mE)}{\sqrt{\hbar^2\boldsymbol{k}_\parallel^2-2mE}-\hbar\kappa-\mathrm{i}\epsilon}\exp\!\left(-\mathrm{i}k_\perp z_0-\sqrt{\hbar^2\boldsymbol{k}_\parallel^2-2mE}\,\frac{|z'-z_0|}{\hbar}\right)$$
$$\left.+\frac{\mathrm{i}\hbar\kappa\Theta(2mE-\hbar^2\boldsymbol{k}_\parallel^2)}{\sqrt{2mE-\hbar^2\boldsymbol{k}_\parallel^2}-\mathrm{i}\hbar\kappa}\exp\!\left(-\mathrm{i}k_\perp z_0+\mathrm{i}\sqrt{2mE-\hbar^2\boldsymbol{k}_\parallel^2}\,\frac{|z'-z_0|}{\hbar}\right)\right]. \tag{20.30}$$

Fourier transformation of equation (20.30) with respect to $k_\perp$ yields finally

$$\langle z|G(\boldsymbol{k}_\parallel,E)|z'\rangle = \frac{\hbar\Theta(\hbar^2\boldsymbol{k}_\parallel^2-2mE)}{2\sqrt{\hbar^2\boldsymbol{k}_\parallel^2-2mE}}\left[\exp\!\left(-\sqrt{\hbar^2\boldsymbol{k}_\parallel^2-2mE}\,\frac{|z-z'|}{\hbar}\right)\right.$$
$$\left.+\frac{\hbar\kappa}{\sqrt{\hbar^2\boldsymbol{k}_\parallel^2-2mE}-\hbar\kappa-\mathrm{i}\epsilon}\exp\!\left(-\sqrt{\hbar^2\boldsymbol{k}_\parallel^2-2mE}\,\frac{|z-z_0|+|z'-z_0|}{\hbar}\right)\right]$$
$$+\mathrm{i}\frac{\hbar\Theta(2mE-\hbar^2\boldsymbol{k}_\parallel^2)}{2\sqrt{2mE-\hbar^2\boldsymbol{k}_\parallel^2}}\left[\exp\!\left(\mathrm{i}\sqrt{2mE-\hbar^2\boldsymbol{k}_\parallel^2}\,\frac{|z-z'|}{\hbar}\right)\right.$$
$$\left.+\frac{\mathrm{i}\hbar\kappa}{\sqrt{2mE-\hbar^2\boldsymbol{k}_\parallel^2}-\mathrm{i}\hbar\kappa}\exp\!\left(\mathrm{i}\sqrt{2mE-\hbar^2\boldsymbol{k}_\parallel^2}\,\frac{|z-z_0|+|z'-z_0|}{\hbar}\right)\right]. \tag{20.31}$$

The limit $\kappa\to 0$ in equations (20.30, 20.31), as well as in equation (20.34) below reproduces the corresponding representations of the free retarded Green's function in three dimensions.

Our results describe the Green's function for a particle in the presence of the thin quantum well, but for arbitrary energy and both near and far from the quantum well. Therefore we cannot easily identify any two-dimensional limit from the Green's function. To explore this question further, we will look at the density of electron states in the presence of the quantum well.

The quantum well at z_0 breaks translational invariance in z direction, and we have with equation (20.16)

$$\varrho(E,z) = \frac{4m}{\pi\hbar^2}\Im\langle\boldsymbol{x}_\parallel,z|G(E)|\boldsymbol{x}_\parallel,z\rangle = \frac{m}{\pi^3\hbar^2}\Im\int d^2\boldsymbol{k}_\parallel\,\langle z|G(\boldsymbol{k}_\parallel,E)|z\rangle,$$

where a factor $g=2$ was taken into account for spin 1/2 states.

If there is any quasi two-dimensional behavior in this system, we would expect it in the quantum well region. Therefore we use the result (20.31) to calculate the density of states $\varrho(E, z_0)$ in the quantum well. Substitution yields

$$\begin{aligned}\varrho(E, z_0) &= \frac{m}{\pi^3\hbar^2}\Im\int d^2\boldsymbol{k}_\parallel\,\langle z_0|G(\boldsymbol{k}_\parallel, E)|z_0\rangle \\ &= \frac{m}{\pi\hbar}\int_0^\infty dk\,k\,\delta\Big(\sqrt{\hbar^2k^2-2mE}-\hbar\kappa\Big) \\ &+ \frac{m}{\pi^2\hbar}\Theta(E)\int_0^{\sqrt{2mE}/\hbar} dk\,\frac{k\sqrt{2mE-\hbar^2k^2}}{2mE-\hbar^2k^2+\hbar^2\kappa^2},\end{aligned}$$

and after evaluation of the integrals,

$$\begin{aligned}\varrho(E, z_0) &= \Theta(2mE+\hbar^2\kappa^2)\kappa\frac{m}{\pi\hbar^2} \\ &+ \Theta(E)\frac{m}{\pi^2\hbar^3}\left[\sqrt{2mE}-\hbar\kappa\arctan\left(\frac{\sqrt{2mE}}{\hbar\kappa}\right)\right].\end{aligned} \tag{20.32}$$

We can also express this in terms of the free two-dimensional and three-dimensional densities of electron states (cf. (20.17)),

$$\begin{aligned}\varrho(E, z_0) &= \kappa\varrho_{d=2}\big(E+(\hbar^2\kappa^2/2m)\big) \\ &+ \varrho_{d=3}(E)\left[1-\frac{\hbar\kappa}{\sqrt{2mE}}\arctan\left(\frac{\sqrt{2mE}}{\hbar\kappa}\right)\right].\end{aligned} \tag{20.33}$$

We note that the states which are exponentially suppressed perpendicular to the quantum well indeed contribute a term proportional to the two-dimensional density of states $\varrho_{d=2}(E')$ with the kinetic energy E' of motion of particles along the quantum well, but with a dimensional proportionality constant κ which is the inverse penetration depth of those states. Such a dimensional factor has to be there, because densities of states in three dimensions enumerate states per energy and per volume, while $\varrho_{d=2}(E')$ counts states per energy and per area. Furthermore, the unbound states yield a contribution which approaches the free three-dimensional density of states $\varrho_{d=3}(E)$ in the limit $\kappa \to 0$. The result can also be derived directly from the energy eigenstates (20.21–20.23) and the definition (12.15) of the local density of states, see Problem 20.7. However, the derivation from the Green's function, while more lengthy for the pure quantum well, has the advantage to also work in the case of an interface in which the electrons move with different effective mass.

The density of states in the quantum well region is displayed for binding energy $B = \hbar^2\kappa^2/2m = 1\,\mathrm{eV}$, mass $m = m_e = 511\,\mathrm{keV}/c^2$, and different energy ranges in Figures 20.3 and 20.4.

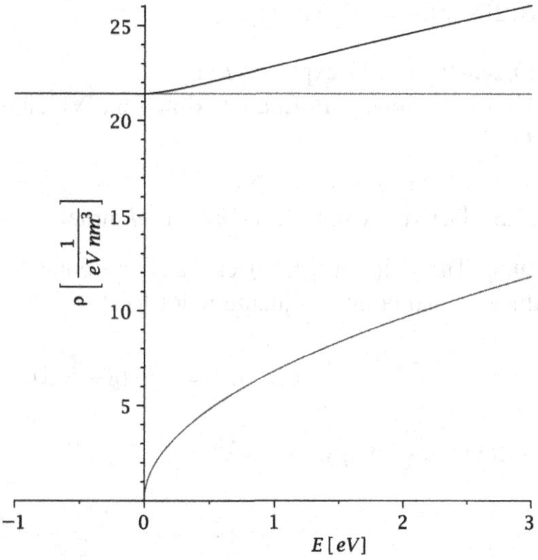

Fig. 20.3 The density of states in the quantum well location $z = z_0$ for binding energy $B = 1\,\text{eV}$, mass $m = m_e = 511\,\text{keV}/c^2$, and energies $-B \leq E \leq 3\,\text{eV}$. The red curve is the contribution from states bound inside the quantum well, the blue curve is the pure three-dimensional density of states in absence of a quantum well, and the black curve is the density of states according to equation (20.32)

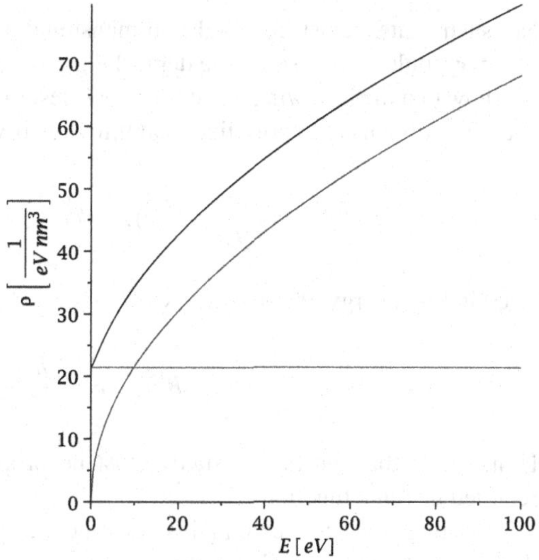

Fig. 20.4 The density of states (20.32) in the quantum well location $z = z_0$ for higher energies $0 \leq E \leq 100\,\text{eV}$. The binding energy, mass and color coding are the same as in Figure 20.3. The full density of states (20.32) approximates the three-dimensional $\sqrt{E}$ behavior for energies $E \gg B$, but there remains a finite offset compared to $\varrho_{d=3}$ due to the presence of the quantum well

20.3 Problems

20.1. Derive the d-dimensional version of equation (11.28) for scattering off spherically symmetric potentials.

20.2. Calculate the differential scattering cross sections for the potentials

20.2a. $V(r) = V_0\Theta(R - r)$,

20.2b. $V(r) = V_0 \exp(-r/R)$,

20.2c. $V(r) = V_0 \exp(-r^2/R^2)$,
in d dimensions in Born approximation. Which results do you find in particular for $d = 2$?

20.3. Derive the optical theorem in d dimensions.

20.4. The solution (3.16) can also be considered as the bound state in a one-dimensional pointlike quantum dot $V(x) = -\mathcal{W}\delta(x)$,

$$\psi_{d=1}(x) = \kappa \exp(-\kappa|x|), \quad \kappa = \frac{m}{\hbar^2}\mathcal{W},$$

with binding energy

$$B = -E = \frac{\hbar^2\kappa^2}{2m} = \frac{m}{2\hbar^2}\mathcal{W}^2.$$

No such states exist for higher-dimensional pointlike quantum dots $V(\boldsymbol{x}) = -\mathcal{W}\delta(\boldsymbol{x})$, unless we also let the depth $\mathcal{W}$ go to zero in a judicious way.

Show that the following wave functions describe bound states in two-dimensional and three-dimensional pointlike quantum dots if we let $\mathcal{W}$ go to zero,

$$\psi_{d=2}(r) = \frac{\kappa}{\sqrt{\pi}}K_0(\kappa r), \quad \psi_{d=3}(r) = \sqrt{\frac{\kappa}{2\pi}}\frac{\exp(-\kappa r)}{r}.$$

The binding energy of the states is

$$B = -E = \frac{\hbar^2\kappa^2}{2m}.$$

Hint: Show that the bound states must be proportional to the energy-dependent retarded Green's functions.

Note that we cannot extend this construction to four or more dimensions because the corresponding Green's functions are not square integrable any more.

20.5. Suppose we consider a proton and an electron in $d \geq 3$ spatial dimensions. The electromagnetic interaction potential of these particles is

$$V_d(r) = -\frac{e^2}{\epsilon_0}G_d(r).$$

Suppose that there are normalizable bound energy eigenstates in this system. Which relation between the expectation values $\langle K \rangle$ and $\langle V \rangle$ of kinetic and potential energy would then be implied by the virial theorem (4.41)? Can atoms exist in $d \geq 4$ dimensions?

20.6. Show that substitution of the Fourier transform

$$\langle \boldsymbol{k}_{\|}, k_{\perp}|G(E)|\boldsymbol{k}'_{\|}, k'_{\perp}\rangle = \int d^2\boldsymbol{x}_{\|}\int d^2\boldsymbol{x}'_{\|}\int dz\int dz'\,\frac{\langle \boldsymbol{x}_{\|}, z|G(E)|\boldsymbol{x}'_{\|}, z'\rangle}{(2\pi)^3}$$
$$\times\exp\!\left[\mathrm{i}\left(\boldsymbol{k}'_{\|}\cdot\boldsymbol{x}'_{\|}+k'_{\perp}z'-\boldsymbol{k}_{\|}\cdot\boldsymbol{x}_{\|}-k_{\perp}z\right)\right]$$
$$=\langle k_{\perp}|G(\boldsymbol{k}_{\|}, E)|k'_{\perp}\rangle\delta\!\left(\boldsymbol{k}_{\|}-\boldsymbol{k}'_{\|}\right),$$

with

$$\langle k_{\perp}|G(\boldsymbol{k}_{\|}, E)|k'_{\perp}\rangle = \frac{1}{2\pi}\int d^2\boldsymbol{x}_{\|}\int dz\,\langle z|G(\boldsymbol{x}_{\|}, E)|z'\rangle$$
$$\times\exp[-\mathrm{i}(\boldsymbol{k}_{\|}\cdot\boldsymbol{x}_{\|}+k_{\perp}z-k'_{\perp}z')]$$

in equation (20.24) yields with the same technique that we used to solve (20.28) the result

$$\langle k_{\perp}|G(E,\boldsymbol{k}_{\|})|k'_{\perp}\rangle = \frac{1}{k_{\perp}^2+\boldsymbol{k}_{\|}^2-(2mE/\hbar^2)-\mathrm{i}\epsilon}\Bigg[\delta(k_{\perp}-k'_{\perp})$$
$$+\frac{\kappa}{\pi}\frac{\exp[\mathrm{i}(k'_{\perp}-k_{\perp})z_0]}{k'^2_{\perp}+\boldsymbol{k}_{\|}^2-(2mE/\hbar^2)-\mathrm{i}\epsilon}\left(\frac{\sqrt{\hbar^2\boldsymbol{k}_{\|}^2-2mE}\,\Theta(\hbar^2\boldsymbol{k}_{\|}^2-2mE)}{\sqrt{\hbar^2\boldsymbol{k}_{\|}^2-2mE}-\hbar\kappa-\mathrm{i}\epsilon}\right.$$
$$\left.+\frac{\sqrt{2mE-\hbar^2\boldsymbol{k}_{\|}^2}\,\Theta(2mE-\hbar^2\boldsymbol{k}_{\|}^2)}{\sqrt{2mE-\hbar^2\boldsymbol{k}_{\|}^2}-\mathrm{i}\hbar\kappa}\right)\Bigg]. \qquad (20.34)$$

Show also that Fourier transformation yields again the result (20.30).

20.7. Derive the result (20.32) directly from the energy eigenstates (20.21), (20.22) and (20.23) for particles in the presence of the quantum well.

Solution.
The decomposition of unity in terms of the eigenstates is

$$\int d^2\boldsymbol{k}_{\|}\,|\boldsymbol{k}_{\|},\kappa\rangle\langle\boldsymbol{k}_{\|},\kappa| + \sum_{\pm}\int d^2\boldsymbol{k}_{\|}\int_0^{\infty}dk_{\perp}\,|\boldsymbol{k}_{\|},k_{\perp},\pm\rangle\langle\boldsymbol{k}_{\|},k_{\perp},\pm| = 1.$$

For the application of the definition (12.15) we have to take into account that

$$d^2\boldsymbol{k}_{\|} = k_{\|}dk_{\|}d\varphi \to \frac{m}{\hbar^2}dEd\varphi$$

holds both for the two dimensional integration measure $d^2\boldsymbol{k}_\parallel$ from $E = \hbar^2(\boldsymbol{k}_\parallel^2 - \kappa^2)/2m$, and also in the three-dimensional integration measure $d^2\boldsymbol{k}_\parallel \wedge dk_\perp$, where $E = \hbar^2(\boldsymbol{k}_\parallel^2 + k_\perp^2)/2m$. This yields with a factor of 4π from $g = 2$ for electrons and from integration over φ the result

$$\varrho(E, z_0) = \frac{4\pi m}{\hbar^2}\left(\Theta(2mE + \hbar^2\kappa^2)\frac{\kappa}{4\pi^2} + \Theta(E)\int_0^{\sqrt{2mE}/\hbar}\frac{k_\perp^2\, dk_\perp}{4\pi^3(\kappa^2 + k_\perp^2)}\right).$$

Evaluation of the integral yields again the result (20.32).

20.8. Generalize the derivation of the relation (20.16) to the relativistic case.

Solution.
The relativistic scalar Green's function is

$$G = \frac{\hbar^2}{p^2 + m^2c^2 - \mathrm{i}\epsilon}, \qquad \langle k|G|k'\rangle = \frac{\delta(k - k')}{k^2 + (mc/\hbar)^2 - \mathrm{i}\epsilon},$$

see also (J.66–J.70). We can write this with $H = c\sqrt{\boldsymbol{p}^2 + (mc)^2}$ in the form

$$G = -\frac{\hbar^2c^2}{E^2 - H^2 + \mathrm{i}\epsilon} = -\frac{\hbar^2c^2}{2E}\left(\frac{1}{E - H + \mathrm{i}\epsilon} + \frac{1}{E + H - \mathrm{i}\epsilon}\right).$$

Here $E = cp^0$ is still an operator, but we can make the transition to the energy-dependent Green's operator $G(E)$ with classical variable $E = \hbar ck^0$ through $|k\rangle = |\boldsymbol{k}\rangle \otimes |k^0\rangle$ and

$$\langle k^0|G|k'^0\rangle = G(E)\delta(k^0 - k'^0). \tag{20.35}$$

Use of the Sokhotsky-Plemelj relation (2.11) yields

$$\begin{aligned}\Im G(E) &= \frac{\pi\hbar^2c^2}{2E}\left[\delta(E - H) - \delta(E + H)\right]\\ &= \frac{\pi\hbar^2c^2}{2E}\sum\!\!\!\!\!\!\int_{n,\nu}\left[\delta(E - E_n) - \delta(E + E_n)\right]|n,\nu\rangle\langle n,\nu|,\end{aligned} \tag{20.36}$$

and therefore[5]

$$\Im\langle\boldsymbol{x}|G(E)|\boldsymbol{x}\rangle = \frac{\pi\hbar^2c^2}{2E}\left[\varrho(E) - \overline{\varrho}(\overline{E})\right]. \tag{20.37}$$

[5]A.C. Zulkoskey, R. Dick, K. Tanaka, Phys. Rev. A 89, 052103 (2014).

Here $\varrho(E)$ and $\overline{\varrho}(\overline{E})$ denote the densities of states of particles of energy E, and of anti-particles (or holes) of energy $\overline{E} = -E$, respectively.

We can test our result in the free (anti-)particle case where the density of states per helicity state is

$$\hat{\varrho}(E) = \varrho(E) + \overline{\varrho}(\overline{E}) = \frac{2\Theta(E^2 - m^2c^4)}{(2\sqrt{\pi}\hbar c)^d\Gamma(d/2)}|E|\sqrt{E^2 - m^2c^4}^{\,d-2}, \tag{20.38}$$

see equation (12.23).

The $\boldsymbol{x}$-representation $\langle\boldsymbol{x}|G(E)|\boldsymbol{x}'\rangle = G(\boldsymbol{x} - \boldsymbol{x}', \omega)$ of the energy-dependent free Green's function has been calculated in Appendix J, equation (J.42). The modified Bessel function $K_\nu(z)$ with real argument is real, and the imaginary part of the Hankel function $\mathrm{i}H_\nu^{(1)}(z)$ for real z satisfies [1]

$$\lim_{z\to 0}\Re H_\nu^{(1)}(z) = \frac{(z/2)^\nu}{\Gamma(\nu + 1)}.$$

Substitution into (J.42) for $r = |\boldsymbol{x} - \boldsymbol{x}'| \to 0$ yields

$$\Im\langle\boldsymbol{x}|G(E)|\boldsymbol{x}\rangle = \frac{\pi\hbar^2c^2}{\Gamma(d/2)}\frac{\Theta(E^2 - m^2c^4)}{(2\sqrt{\pi}\hbar c)^d}\sqrt{E^2 - m^2c^4}^{\,d-2},$$

in agreement with equations (20.37) and (20.38).

Chapter 21
Relativistic Quantum Fields

The quantized Maxwell field provided us already with an example of a relativistic quantum field theory. On the other hand, the description of relativistic charged particles requires Klein-Gordon fields for scalar particles and Dirac fields for fermions. Relativistic fields are apparently relevant for high energy physics. However, relativistic effects are also important in photon-matter interactions, spectroscopy, spin dynamics, and for the generation of brilliant photon beams from ultra-relativistic electrons in synchrotrons. Quasirelativistic effects from linear dispersion relations $E \propto \boldsymbol{p}$ in materials like Graphene and in Dirac semimetals have also reinvigorated the need to reconsider the role of Dirac and Weyl equations in materials science. In applications to materials with quasirelativistic dispersion relations c and m become effective velocity and mass parameters to describe cones or hyperboloids in regions of $(E, \boldsymbol{k})$ space.

We start our discussion of relativistic matter fields with the simpler Klein-Gordon equation and then move on to the more widely applicable Dirac equation. We will also discuss covariant quantization of photons, since this is more convenient for the calculation of basic scattering events than quantization in Coulomb gauge.

21.1 The Klein-Gordon equation

A limitation of the Schrödinger equation in the framework of ordinary quantum mechanics is its lack of covariance under Lorentz transformations[1]. On the other hand, we have encountered an example of a relativistic wave equation in Chapter 18,

[1]However, we will see that in the second quantized formalism in the Heisenberg and Dirac pictures, the time evolution of the field operators is given by Heisenberg equations of motion, and the corresponding time evolution of states in the Schrödinger and Dirac pictures is given by corresponding Schrödinger equations with relativistic Hamiltonians.

R. Dick, *Advanced Quantum Mechanics*, Graduate Texts in Physics,
DOI 10.1007/978-3-319-25675-7_21

viz. the inhomogeneous Maxwell equation

$$\partial_\mu \left(\partial^\mu A^\nu - \partial^\nu A^\mu\right) = -\mu_0 j^\nu.$$

This equation is manifestly covariant (or rather, form invariant) under Lorentz transformations because it is composed of quantities with simple tensorial transformation behavior under Lorentz transformations, and it relates a 4-vector $\partial_\mu F^{\mu\nu}$ to a 4-vector j^ν, such that the equation holds in this form in every inertial reference frame.

Another, simple reasoning to come up with a relativistic wave equation goes as follows. We know that the standard Schrödinger equation for a free massive particle arises from the non-relativistic energy-momentum dispersion relation $E = -cp_0 = \boldsymbol{p}^2/2m$ upon substitution of the classical energy-momentum vector through differential operators, $p_\mu \to -\mathrm{i}\hbar\partial_\mu$. Following the same procedure in the relativistic dispersion relation

$$-\frac{E^2}{c^2} + \boldsymbol{p}^2 + m^2c^2 = p^2 + m^2c^2 = 0$$

yields the free Klein-Gordon equation[2]

$$\left(\partial^2 - \frac{m^2c^2}{\hbar^2}\right)\phi(x) = \left(\Delta - \frac{1}{c^2}\frac{\partial^2}{\partial t^2} - \frac{m^2c^2}{\hbar^2}\right)\phi(x) = 0. \tag{21.1}$$

Furthermore, the gauge principle or minimal coupling prescription $\partial_\mu \to D_\mu = \partial_\mu - \mathrm{i}(q/\hbar)A_\mu$ yields the coupling of the charged Klein-Gordon field to electromagnetic potentials,

$$\begin{aligned}\left[\left(\partial - \mathrm{i}\frac{q}{\hbar}A(x)\right)^2 - \frac{m^2c^2}{\hbar^2}\right]\phi(x) = \Bigg[\left(\nabla - \mathrm{i}\frac{q}{\hbar}\boldsymbol{A}(x)\right)^2 \\ - \frac{1}{c^2}\left(\frac{\partial}{\partial t} + \mathrm{i}\frac{q}{\hbar}\Phi(x)\right)^2 - \frac{m^2c^2}{\hbar^2}\Bigg]\phi(x) = 0.\end{aligned} \tag{21.2}$$

Complex conjugation of equation (21.2) leads to the Klein-Gordon equation for a scalar field with charge $-q$. Therefore the *charge conjugate* Klein-Gordon field is simply gotten by complex conjugation,

$$\phi^c(x) = \phi^*(x). \tag{21.3}$$

The Klein-Gordon field is relevant in particle physics. E.g. π-mesons are described by Klein-Gordon fields as soon as their kinetic energy becomes

[2] E. Schrödinger, Annalen Phys. 386, 109 (1926); W. Gordon, Z. Phys. 40, 117 (1926); O. Klein, Z. Phys. 41, 407 (1927).

comparable to their mass $mc^2 \simeq 140\,\mathrm{MeV}$, when relativistic effects have to be taken into account. Another important application of the Klein-Gordon field is the Higgs field for electroweak symmetry breaking in the Standard Model of particle physics.

The Klein-Gordon field also provides a simple introduction into the relativistic quantum mechanics of charged particles. Therefore it is also useful as a preparation for the study of the Dirac field. We will focus in particular on the canonical quantization of freely evolving Klein-Gordon fields, since this describes Klein-Gordon operators in the practically relevant interaction picture representation. Conservation laws for full scalar quantum electrodynamics are discussed in Problems 21.6a and 21.7.

Mode expansion and quantization of the Klein-Gordon field

Fourier transformation of equation (21.1) yields the general solution of the free Klein-Gordon equation in $k = (\omega/c, \boldsymbol{k})$ space,

$$\begin{aligned}\phi(k) &= \phi(\boldsymbol{k},\omega) = \langle \boldsymbol{k}|\phi(\omega)\rangle \\ &= \sqrt{\frac{\pi}{\omega_k}}\left[a(\boldsymbol{k})\delta(\omega-\omega_k) + b^+(-\boldsymbol{k})\delta(\omega+\omega_k)\right],\end{aligned} \tag{21.4}$$

where ω_k is just the k space expression for the relativistic dispersion relation,

$$\omega_k = c\sqrt{\boldsymbol{k}^2 + (m^2c^2/\hbar^2)}.$$

Frequency-time Fourier transformation (5.12) yields

$$\langle \boldsymbol{k}|\phi(t)\rangle = \frac{1}{\sqrt{2\omega_k}}\left[a(\boldsymbol{k})\exp(-\mathrm{i}\omega_k t) + b^+(-\boldsymbol{k})\exp(\mathrm{i}\omega_k t)\right]$$

and the general free Klein-Gordon wave function in $x = (ct, \boldsymbol{x})$ space is

$$\begin{aligned}\phi(x) = \langle \boldsymbol{x}|\phi(t)\rangle = \frac{1}{\sqrt{2\pi}^3}\int \frac{d^3\boldsymbol{k}}{\sqrt{2\omega_k}}\Big(&a(\boldsymbol{k})\exp[\mathrm{i}(\boldsymbol{k}\cdot\boldsymbol{x}-\omega_k t)] \\ &+ b^+(\boldsymbol{k})\exp[-\mathrm{i}(\boldsymbol{k}\cdot\boldsymbol{x}-\omega_k t)]\Big).\end{aligned} \tag{21.5}$$

For the inversion of the Fourier transformation in the sense of solving for $a(\boldsymbol{k})$ and $b(\boldsymbol{k})$ we need equation (21.5) and

$$\begin{aligned}\dot{\phi}(\boldsymbol{x},t) = \frac{\mathrm{i}}{\sqrt{2\pi}^3}\int d^3\boldsymbol{k}\,\sqrt{\frac{\omega_k}{2}}\Big(&-a(\boldsymbol{k})\exp[\mathrm{i}(\boldsymbol{k}\cdot\boldsymbol{x}-\omega_k t)] \\ &+ b^+(\boldsymbol{k})\exp[-\mathrm{i}(\boldsymbol{k}\cdot\boldsymbol{x}-\omega_k t)]\Big).\end{aligned}$$

Inversion of both equations yields

$$a(\boldsymbol{k}) = \frac{1}{\sqrt{2\pi}^3}\int\frac{d^3\boldsymbol{x}}{\sqrt{2\omega_k}}\big(\omega_k\phi(\boldsymbol{x},t)+\mathrm{i}\dot{\phi}(\boldsymbol{x},t)\big)\exp[-\mathrm{i}(\boldsymbol{k}\cdot\boldsymbol{x}-\omega_k t)]$$
$$= \frac{1}{\sqrt{2\pi}^3}\int\frac{d^3\boldsymbol{x}}{\sqrt{2\omega_k}}\exp[-\mathrm{i}(\boldsymbol{k}\cdot\boldsymbol{x}-\omega_k t)]\,\mathrm{i}\overleftrightarrow{\partial_t}\phi(\boldsymbol{x},t), \tag{21.6}$$

$$b(\boldsymbol{k}) = \frac{1}{\sqrt{2\pi}^3}\int\frac{d^3\boldsymbol{x}}{\sqrt{2\omega_k}}\big(\omega_k\phi^+(\boldsymbol{x},t)+\mathrm{i}\dot{\phi}^+(\boldsymbol{x},t)\big)\exp[-\mathrm{i}(\boldsymbol{k}\cdot\boldsymbol{x}-\omega_k t)]$$
$$= \frac{1}{\sqrt{2\pi}^3}\int\frac{d^3\boldsymbol{x}}{\sqrt{2\omega_k}}\exp[-\mathrm{i}(\boldsymbol{k}\cdot\boldsymbol{x}-\omega_k t)]\,\mathrm{i}\overleftrightarrow{\partial_t}\phi^+(\boldsymbol{x},t). \tag{21.7}$$

Here the alternating derivative is defined as

$$f\overleftrightarrow{\partial_t}g = f\frac{\partial g}{\partial t}-\frac{\partial f}{\partial t}g.$$

Substituting (21.6, 21.7) back into (21.5) and formal exchange of integrations yields

$$\phi(\boldsymbol{x},t) = \int d^3\boldsymbol{x}'\,\mathcal{K}(\boldsymbol{x}-\boldsymbol{x}',t-t')\,\overleftrightarrow{\partial_{t'}}\,\phi(\boldsymbol{x}',t') \tag{21.8}$$

with the time evolution kernel for free scalar fields,

$$\mathcal{K}(\boldsymbol{x},t) = \frac{1}{(2\pi)^3}\int\frac{d^3\boldsymbol{k}}{\omega_k}\exp(\mathrm{i}\boldsymbol{k}\cdot\boldsymbol{x})\sin(\omega_k t). \tag{21.9}$$

This distribution satisfies the initial value problem

$$\left(\partial^2-\frac{m^2c^2}{\hbar^2}\right)\mathcal{K}(\boldsymbol{x},t)=0,\quad \mathcal{K}(\boldsymbol{x},0)=0,\quad \frac{\partial}{\partial t}\mathcal{K}(\boldsymbol{x},t)\bigg|_{t=0}=\delta(\boldsymbol{x}).$$

For canonical quantization we need the Lagrange density for the complex Klein-Gordon field

$$\mathcal{L} = \hbar\dot{\phi}^+\cdot\dot{\phi}-\hbar c^2\nabla\phi^+\cdot\nabla\phi-\frac{m^2c^4}{\hbar}\phi^+\cdot\phi$$
$$= -\hbar c^2\partial\phi^+\cdot\partial\phi-\frac{m^2c^4}{\hbar}\phi^+\cdot\phi, \tag{21.10}$$

or the real Klein-Gordon field

$$\mathcal{L} = \frac{\hbar}{2}\dot{\phi}\cdot\dot{\phi}-\frac{\hbar c^2}{2}\nabla\phi\cdot\nabla\phi-\frac{m^2c^4}{2\hbar}\phi^2 = -\frac{\hbar c^2}{2}(\partial\phi)^2-\frac{m^2c^4}{2\hbar}\phi^2. \tag{21.11}$$

In the following we will continue with the discussion of the complex Klein-Gordon field.

Canonical quantization proceeds from (21.10) without any problems. The conjugate momenta

$$\Pi_\phi = \frac{\partial\mathcal{L}}{\partial\dot{\phi}} = \hbar\dot{\phi}^+, \quad \Pi_{\phi^+} = \frac{\partial\mathcal{L}}{\partial\dot{\phi}^+} = \hbar\dot{\phi},$$

yield the canonical commutation relations in $\boldsymbol{x}$ space,

$$[\phi(\boldsymbol{x},t), \dot{\phi}^+(\boldsymbol{x}',t)] = \mathrm{i}\delta(\boldsymbol{x}-\boldsymbol{x}'), \quad [\phi^+(\boldsymbol{x},t), \dot{\phi}(\boldsymbol{x}',t)] = \mathrm{i}\delta(\boldsymbol{x}-\boldsymbol{x}'),$$

$$[\phi(\boldsymbol{x},t), \dot{\phi}(\boldsymbol{x}',t)] = 0, \quad [\phi(\boldsymbol{x},t), \phi(\boldsymbol{x}',t)] = 0, \quad [\phi(\boldsymbol{x},t), \phi^+(\boldsymbol{x}',t)] = 0,$$

$$[\dot{\phi}(\boldsymbol{x},t), \dot{\phi}(\boldsymbol{x}',t)] = 0, \quad [\dot{\phi}(\boldsymbol{x},t), \dot{\phi}^+(\boldsymbol{x}',t)] = 0,$$

and in $\boldsymbol{k}$ space,

$$[a(\boldsymbol{k}), a^+(\boldsymbol{k}')] = \delta(\boldsymbol{k}-\boldsymbol{k}'), \quad [a(\boldsymbol{k}), a(\boldsymbol{k}')] = 0, \quad [b(\boldsymbol{k}), b^+(\boldsymbol{k}')] = \delta(\boldsymbol{k}-\boldsymbol{k}'),$$

$$[b(\boldsymbol{k}), b(\boldsymbol{k}')] = 0, \quad [a(\boldsymbol{k}), b(\boldsymbol{k}')] = 0, \quad [a(\boldsymbol{k}), b^+(\boldsymbol{k}')] = 0.$$

The Lagrangian for interacting Klein-Gordon and Maxwell fields is

$$\mathcal{L} = -\frac{c^2}{\hbar}\left(\hbar\partial_\mu\phi^+ + \mathrm{i}q\phi^+\cdot A_\mu\right)\cdot\left(\hbar\partial^\mu\phi - \mathrm{i}qA^\mu\cdot\phi\right) - \frac{m^2c^4}{\hbar}\phi^+\cdot\phi$$
$$-\frac{1}{4\mu_0}F_{\mu\nu}F^{\mu\nu}. \tag{21.12}$$

The charge operator of the Klein-Gordon field

The Klein-Gordon Lagrangian (21.10) is invariant under phase transformations

$$\phi(x) \to \phi'(x) = \exp\!\left(\mathrm{i}\frac{q}{\hbar}\alpha\right)\phi(x), \quad \delta\phi(x) = \mathrm{i}\frac{q}{\hbar}\alpha\phi(x).$$

According to Section 16.2 this implies a local conservation law (16.13) for a conserved charge Q. After cancelling the superfluous factor α, the charge following from (16.14) is (after normal ordering of the integrand in $\boldsymbol{k}$ space, see the remarks following equations (18.40, 18.41))

$$Q = -\mathrm{i}\frac{q}{\hbar}\int d^3\boldsymbol{x}\left(\frac{\partial\mathcal{L}}{\partial\dot{\phi}}\cdot\phi - \phi^+\cdot\frac{\partial\mathcal{L}}{\partial\dot{\phi}^+}\right)$$
$$= -\mathrm{i}q\int d^3\boldsymbol{x}\left(\dot{\phi}^+(\boldsymbol{x},t)\cdot\phi(\boldsymbol{x},t) - \phi^+(\boldsymbol{x},t)\cdot\dot{\phi}(\boldsymbol{x},t)\right)$$
$$= q\int d^3\boldsymbol{k}\left(a^+(\boldsymbol{k})a(\boldsymbol{k}) - b^+(\boldsymbol{k})b(\boldsymbol{k})\right). \tag{21.13}$$

The charge density $\mathrm{i}q\phi^+ \overset{\leftrightarrow}{\partial_t} \phi$ is not positive definite, and therefore division of the charge density by q does not yield a probability density for the location of a particle, contrary to the Schrödinger field. Lack of a single particle interpretation is a generic property of relativistic fields which we had also encountered for the Maxwell field.

Hamiltonian and momentum operators for the Klein-Gordon field

The invariance of the Klein-Gordon Lagrangian (21.10) under constant translations

$$x^\mu \to x'^\mu = x^\mu + \delta x^\mu$$

implies a local conservation law (16.15) with corresponding conserved Hamilton and momentum operators (16.17). This yields the following expressions for energy and momentum of Klein-Gordon fields,

$$\mathcal{H} = -\Theta_0{}^0 = \hbar\dot{\phi}^+ \cdot \dot{\phi} + \hbar c^2 \nabla\phi^+ \cdot \nabla\phi + \frac{m^2c^4}{\hbar}\phi^+ \cdot \phi, \tag{21.14}$$

$$H = \int d^3x\, \mathcal{H} = \int d^3k\, \hbar\omega_k \big(a^+(k)a(k) + b^+(k)b(k)\big), \tag{21.15}$$

$$\begin{aligned}\mathcal{P} &= \frac{1}{c} e_i \Theta_i{}^0 = -\frac{\partial\mathcal{L}}{\partial\dot{\phi}} \cdot \nabla\phi - \nabla\phi^+ \cdot \frac{\partial\mathcal{L}}{\partial\dot{\phi}^+} \\ &= -\hbar\dot{\phi}^+ \cdot \nabla\phi - \hbar\nabla\phi^+ \cdot \dot{\phi},\end{aligned} \tag{21.16}$$

$$\boldsymbol{P} = \int d^3x\, \mathcal{P} = \int d^3k\, \hbar\boldsymbol{k} \big(a^+(\boldsymbol{k})a(\boldsymbol{k}) + b^+(\boldsymbol{k})b(\boldsymbol{k})\big). \tag{21.17}$$

The commutation relations and the charge operator (21.13), the Hamilton operator (21.15), and the momentum operators (21.17) imply that the operator $a^+(\boldsymbol{k})$ creates a particle of momentum $\hbar\boldsymbol{k}$, energy $\hbar\omega_k$ and charge q, while $b^+(\boldsymbol{k})$ creates a particle of momentum $\hbar\boldsymbol{k}$, energy $\hbar\omega_k$ and charge $-q$.

The operators (21.5) and $a(\boldsymbol{k},t) = a(\boldsymbol{k})\exp(-\mathrm{i}\omega_k t)$, $a^+(\boldsymbol{k},t) = a^+(\boldsymbol{k})\exp(\mathrm{i}\omega_k t)$ are the field operators in the Dirac picture, or the free field operators in the Heisenberg picture. They satisfy the Heisenberg evolution equations

$$\frac{\partial}{\partial t}a(\boldsymbol{k},t) = \frac{\mathrm{i}}{\hbar}[H, a(\boldsymbol{k},t)], \quad \frac{\partial}{\partial t}\phi(\boldsymbol{x},t) = \frac{\mathrm{i}}{\hbar}[H, \phi(\boldsymbol{x},t)],$$

with the free Hamiltonian (21.15). The corresponding integrals follow in the standard way,

$$a(\boldsymbol{k},t) = \exp\!\left(\frac{\mathrm{i}}{\hbar}Ht\right) a(\boldsymbol{k}) \exp\!\left(-\frac{\mathrm{i}}{\hbar}Ht\right),$$

$$\phi(\boldsymbol{x},t) = \exp\!\left(\frac{\mathrm{i}}{\hbar}Ht\right) \phi(\boldsymbol{x}) \exp\!\left(-\frac{\mathrm{i}}{\hbar}Ht\right),$$

etc. In the Schrödinger picture theory, this amounts to operators $a(\boldsymbol{k})$, $\phi(\boldsymbol{x})$, and time evolution of the states

$$\mathrm{i}\hbar\frac{d}{dt}|\Psi(t)\rangle = H|\Psi(t)\rangle$$

with the free Hamiltonian (21.15) for free states or a corresponding minimally coupled Hamiltonian which follows from (21.12) for interacting states, see Problem 21.5. This is the statement that we have Heisenberg and Schrödinger type evolution equations also in relativistic quantum field theory.

The Klein-Gordon equation also follows from the iterated Heisenberg equation,

$$\frac{\partial^2}{\partial t^2}\phi(\boldsymbol{x},t) = -\frac{1}{\hbar^2}[H,[H,\phi(\boldsymbol{x},t)]], \tag{21.18}$$

cf. (18.46) for photons.

Non-relativistic limit of the Klein-Gordon field

We have in the non-relativistic limit

$$\omega_k = c\sqrt{\boldsymbol{k}^2 + \frac{m^2c^2}{\hbar^2}} \simeq \frac{mc^2}{\hbar} + \frac{\hbar\boldsymbol{k}^2}{2m},$$

and therefore in leading order also $1/\sqrt{2\omega_k} \simeq \sqrt{\hbar/2mc^2}$.

Suppose that the $\boldsymbol{k}$-space amplitudes $a(\boldsymbol{k})$ and $b^+(\boldsymbol{k})$ are negligibly small unless $\hbar|\boldsymbol{k}| \ll mc$. In this case we can approximate equation (21.5) by

$$\phi(\boldsymbol{x},t) \simeq \frac{1}{\sqrt{2\pi}^3}\sqrt{\frac{\hbar}{2mc^2}}\int d^3\boldsymbol{k}\left[a(\boldsymbol{k})\exp\!\left(\mathrm{i}\boldsymbol{k}\cdot\boldsymbol{x} - \mathrm{i}\frac{\hbar\boldsymbol{k}^2}{2m}t\right)\exp\!\left(-\mathrm{i}\frac{mc^2}{\hbar}t\right)\right.$$
$$\left. + b^+(\boldsymbol{k})\exp\!\left(-\mathrm{i}\boldsymbol{k}\cdot\boldsymbol{x} + \mathrm{i}\frac{\hbar\boldsymbol{k}^2}{2m}t\right)\exp\!\left(\mathrm{i}\frac{mc^2}{\hbar}t\right)\right].$$

However, this expression automatically contains two fields

$$\psi(\boldsymbol{x},t) = \frac{1}{\sqrt{2\pi}^3}\int d^3\boldsymbol{k}\,a(\boldsymbol{k})\exp\!\left[\mathrm{i}\left(\boldsymbol{k}\cdot\boldsymbol{x} - \frac{\hbar\boldsymbol{k}^2}{2m}t\right)\right]$$

and

$$\varphi(\boldsymbol{x},t) = \frac{1}{\sqrt{2\pi}^3}\int d^3k\, b(\boldsymbol{k}) \exp\left[\mathrm{i}\left(\boldsymbol{k}\cdot\boldsymbol{x} - \frac{\hbar k^2}{2m}t\right)\right],$$

which satisfy the free Schrödinger equation, i.e. the complex Klein-Gordon field will reduce to a Schrödinger field $\psi(\boldsymbol{x},t)$ if the $\boldsymbol{k}$-space amplitudes also satisfy $|\langle a(\boldsymbol{k})\rangle| \gg |\langle b^+(\boldsymbol{k})\rangle|$.

Substitution of the remaining approximation

$$\phi(\boldsymbol{x},t) \simeq \sqrt{\frac{\hbar}{2mc^2}}\psi(\boldsymbol{x},t)\exp\left(-\mathrm{i}\frac{mc^2}{\hbar}t\right) \tag{21.19}$$

into the charge, current, energy and momentum densities of the Klein-Gordon field yields the corresponding expressions for the Schrödinger field,

$$\begin{aligned}
\varrho &= -\mathrm{i}q\left(\dot{\phi}^+\cdot\phi - \phi^+\cdot\dot{\phi}\right) \simeq q\psi^+\psi, \\
\boldsymbol{j} &= \mathrm{i}qc^2\left(\nabla\phi^+\cdot\phi - \phi^+\cdot\nabla\phi\right) \simeq q\frac{\hbar}{2\mathrm{i}m}\left(\psi^+\cdot\nabla\psi - \psi\cdot\nabla\psi^+\right), \\
\mathcal{H} &= \hbar\dot{\phi}^+\cdot\dot{\phi} + \hbar c^2\nabla\phi^+\cdot\nabla\phi + \frac{m^2c^4}{\hbar}\phi^+\cdot\phi \simeq \frac{\hbar^2}{2m}\nabla\psi^+\cdot\nabla\psi + mc^2\psi^+\cdot\psi, \\
\mathcal{P} &= -\hbar\dot{\phi}^+\cdot\nabla\phi - \hbar\nabla\phi^+\cdot\dot{\phi} \simeq \frac{\hbar}{2\mathrm{i}}\left(\psi^+\cdot\nabla\psi - \psi\cdot\nabla\psi^+\right) = \frac{m}{q}\boldsymbol{j}.
\end{aligned} \tag{21.20}$$

Furthermore, the free Klein-Gordon equation (21.1) becomes with

$$\frac{1}{c^2}\frac{\partial^2}{\partial t^2}\phi(\boldsymbol{x},t) \simeq \sqrt{\frac{\hbar}{2mc^2}}\exp\left(-\mathrm{i}\frac{mc^2}{\hbar}t\right)\left(-\frac{m^2c^2}{\hbar^2}\psi(\boldsymbol{x},t) - \mathrm{i}\frac{2m}{\hbar}\frac{\partial}{\partial t}\psi(\boldsymbol{x},t)\right)$$

the free Schrödinger equation

$$\mathrm{i}\hbar\frac{\partial}{\partial t}\psi(\boldsymbol{x},t) = -\frac{\hbar^2}{2m}\Delta\psi(\boldsymbol{x},t),$$

as it should, because we have already observed in the derivation of (21.19) that $\psi(\boldsymbol{x},t)$ satisfies the free Schrödinger equation.

For the non-relativistic limit of the real Klein-Gordon field we find

$$\phi(\boldsymbol{x},t) \simeq \sqrt{\frac{\hbar}{2mc^2}}\left[\psi(\boldsymbol{x},t)\exp\left(-\mathrm{i}\frac{mc^2}{\hbar}t\right) + \psi^+(\boldsymbol{x},t)\exp\left(\mathrm{i}\frac{mc^2}{\hbar}t\right)\right],$$

but we have to include first order time derivatives of $\psi(\boldsymbol{x},t)$ and $\psi^+(\boldsymbol{x},t)$ in the evaluation of $\mathcal{H}$ and $\boldsymbol{P}$, and then use the Schrödinger equation to find that remnant fast oscillation terms proportional to $\exp(\pm 2\mathrm{i}mc^2t/\hbar)$ reduce to boundary terms.

21.2 Klein's paradox

The commutation relations for the field operators $a(\boldsymbol{k})$ and $b^+(\boldsymbol{k})$ imply that the operator $\phi(\boldsymbol{x},t)$ (21.5) describes both particles and anti-particles simultaneously, and therefore the Klein-Gordon equation cannot support a single particle interpretation. This is also obvious from the charge operator (21.13) and the corresponding lack of a conserved probability density for Klein-Gordon particles. Klein's paradox provides a particularly neat illustration of the failure of single particle interpretations of relativistic wave equations.

Klein observed that using relativistic quantum fields to describe a relativistic particle running against a potential step yields results for the transmission and reflection probabilities which are incompatible with a single particle interpretation[3]. This observation can be explained by pair creation in strong fields and the fact that relativistic fields describe both particles and anti-particles simultaneously. We will explain Klein's paradox for the Klein-Gordon field.

In the following we can neglect the y and z coordinates and deal only with the x and t coordinates. We are interested in a scalar particle of charge q scattered off a potential step of height $V > 0$. The step is located at $x = 0$, and can be implemented through an electrostatic potential $\Phi(x)$,

$$V(x) = q\Phi(x) = -cqA_0(x) = V\Theta(x). \tag{21.21}$$

Minimal coupling then yields the free Klein-Gordon equation for $x < 0$, and[4]

$$(\hbar\partial_t + \mathrm{i}V)^2\phi - \hbar^2c^2\partial_x^2\phi + m^2c^4\phi = 0 \tag{21.22}$$

for $x > 0$.

A monochromatic solution without any apparent left moving component for $x > 0$ is (after omission of an irrelevant constant prefactor)

$$\phi(x,t) = \begin{cases} [\exp(\mathrm{i}kx) + \beta\exp(-\mathrm{i}kx)]\exp(-\mathrm{i}\omega t), & x < 0 \\ \theta\exp[\mathrm{i}(\kappa x - \omega t)], & x > 0. \end{cases} \tag{21.23}$$

The frequency follows from the solution of the Klein-Gordon equation in the two domains,

$$\omega = c\sqrt{k^2 + \frac{m^2c^2}{\hbar^2}} = \frac{V}{\hbar} \pm c\sqrt{\kappa^2 + \frac{m^2c^2}{\hbar^2}}. \tag{21.24}$$

It has to be the same in both regions for continuity of the wave function at $x = 0$.

[3]O. Klein, Z. Phys. 53, 157 (1929). Klein actually discussed reflection and transmission of relativistic spin 1/2 fermions which are described by the Dirac equation (21.38).

[4]We cannot try to discuss motion of particles of mass m in the presence of a potential by simply including a scalar potential term in the form $\left(\hbar^2\partial_t^2 - \hbar^2c^2\partial_x^2 + m^2c^4\right)\phi = \Theta(x)V^2\phi$ in the Klein-Gordon equation. This would correspond to a local mass $M(x)c^2 = \sqrt{m^2c^4 - \Theta(x)V^2}$ rather than to a local potential, and yield tachyons in $x > 0$ for $V^2 > m^2c^4$.

The sign in the last equation of (21.24) depends on the sign of $\hbar\omega - V$. We apparently have to use the minus sign if and only if $\hbar\omega - V < 0$. Note that in our solution we always have $\hbar\omega \geq mc^2$.

Solving for κ yields

$$\kappa = \pm\frac{1}{\hbar}\sqrt{\frac{(\hbar\omega - V)^2}{c^2} - m^2c^2} \in \mathbb{R}, \quad (\hbar\omega - V)^2 > m^2c^4, \tag{21.25}$$

$$\kappa = \frac{\mathrm{i}}{\hbar}\sqrt{m^2c^2 - \frac{(\hbar\omega - V)^2}{c^2}} \in \mathrm{i}\mathbb{R}_+, \quad (\hbar\omega - V)^2 < m^2c^4. \tag{21.26}$$

However, we have to be careful with the sign in (21.25). The group velocity in $x > 0$ for $\hbar\omega + mc^2 < V$ (i.e. for the negative sign in (21.24)) is

$$\frac{d\omega}{d\kappa} = -c\frac{\hbar\kappa}{\sqrt{\hbar^2\kappa^2 + m^2c^2}},$$

i.e. we have to take the *negative* root for κ for $V > \hbar\omega + mc^2$ to ensure positive group velocity in the region $x > 0$. We can collect the results for κ in the equations

$$\begin{aligned}
V < \hbar\omega - mc^2: \quad & \kappa = \tfrac{1}{\hbar}\sqrt{\tfrac{(\hbar\omega - V)^2}{c^2} - m^2c^2} \in \mathbb{R}_+,\\
\hbar\omega - mc^2 < V < \hbar\omega + mc^2: \quad & \kappa = \tfrac{\mathrm{i}}{\hbar}\sqrt{m^2c^2 - \tfrac{(\hbar\omega - V)^2}{c^2}} \in \mathrm{i}\mathbb{R}_+,\\
V > \hbar\omega + mc^2: \quad & \kappa = -\tfrac{1}{\hbar}\sqrt{\tfrac{(\hbar\omega - V)^2}{c^2} - m^2c^2} \in \mathbb{R}_-.
\end{aligned}$$

The current density $j = \mathrm{i}qc^2(\partial_x\phi^+ \cdot \phi - \phi^+ \cdot \partial_x\phi)$ is

$$\begin{aligned}
j &= 2qc^2k\left(1 - |\beta|^2\right), \quad x < 0,\\
j &= 2qc^2\kappa|\theta|^2, \quad x > 0,\ \kappa \in \mathbb{R},\\
j &= 0, \quad x > 0,\ \kappa \in \mathrm{i}\mathbb{R}.
\end{aligned} \tag{21.27}$$

Note that in $x > 0$ we have $j/q < 0$ if $V > \hbar\omega + mc^2$, in spite of the fact of positive group velocity in the region. Since charges q cannot move to the left in $x > 0$, this means that the negative value of j/q in $x > 0$ for $V > \hbar\omega + mc^2$ must correspond to right moving charges $-q$. We will see that this arises as a consequence of the generation of anti-particles near the potential step for $V > \hbar\omega + mc^2$.

The junction conditions

$$1 + \beta = \theta, \quad k(1 - \beta) = \kappa\theta \tag{21.28}$$

yield

$$\beta = \frac{k - \kappa}{k + \kappa}, \quad \theta = \frac{2k}{k + \kappa},$$

Table 21.1 Reflection and transmission for different relations between height V of the potential step and energy $\hbar\omega$ of the incident particle

$-\infty < V \leq 0$	$\infty > \kappa \geq k$	$1 > R \geq 0$	$0 < T \leq 1$
$0 \leq V \leq \hbar\omega - mc^2$	$k \geq \kappa \geq 0$	$0 \leq R \leq 1$	$1 \geq T \geq 0$
$\hbar\omega - mc^2 < V < \hbar\omega + mc^2$	$\kappa \in \mathrm{i}\mathbb{R}_+$	$R = 1$	$T = 0$
$\hbar\omega + mc^2 \leq V \leq 2\hbar\omega$	$0 \geq \kappa \geq -k$	$1 \leq R \leq \infty$	$0 \geq T \geq -\infty$
$2\hbar\omega \leq V < \infty$	$-k \geq \kappa > -\infty$	$\infty \geq R > 1$	$-\infty \leq T < 0$

and the corresponding reflection and transmission coefficients are

$$R = |\beta|^2 = \frac{k^2 + |\kappa|^2 - 2k\Re\kappa}{k^2 + |\kappa|^2 + 2k\Re\kappa}, \tag{21.29}$$

$$T = \frac{\Re\kappa}{k}|\theta|^2 = \frac{4k\Re\kappa}{k^2 + |\kappa|^2 + 2k\Re\kappa} = 1 - R. \tag{21.30}$$

The resulting behavior of the reflection coefficient is summarized in Table 21.1.

For an explanation of the unexpected result $R > 1$ for $V > \hbar\omega + mc^2 \geq 2mc^2$, recall that the solution for $V > \hbar\omega + mc^2$ in $x > 0$ has $\kappa < 0$. If we write the solution as

$$\phi(x, t) = \theta \exp[-\mathrm{i}(-\kappa x + \omega t)], \quad x > 0, \tag{21.31}$$

and compare with the anti-particle contribution to the free solution (21.5), we recognize the solution in the region $x > 0$ as an anti-particle solution with momentum $\hbar\kappa' = -\hbar\kappa > 0$ and energy

$$\overline{E}_p = -\hbar\omega < 0, \quad mc^2 - V \leq \overline{E}_p \leq -mc^2. \tag{21.32}$$

This is acceptable, because the anti-particle has charge $-q$ and therefore experiences a potential $U = -V$ in the region $x > 0$. Further support for this energy assignment for the anti-particles comes from the equality for the kinetic+rest energy of the anti-particles,

$$K_{\overline{p}} = c\sqrt{\hbar^2\kappa^2 + m^2c^2} = V - \hbar\omega, \quad mc^2 \leq K_{\overline{p}} \leq V - mc^2. \tag{21.33}$$

We expect $E_{\overline{p}} = K_{\overline{p}} - V$ at least in the non-relativistic limit for the anti-particles.

The anti-particles move to the right, $d(-\omega)/d(-\kappa) > 0$, and yield a negative particle current density $j/q \propto -q\kappa'/q = -\kappa' < 0$ due to the opposite charge. We therefore get $R > 1$ and $T < 0$ for $V - \hbar\omega > mc^2$ due to pair creation. The generated particles move to the left because they are repelled by the potential $V > mc^2 + \hbar\omega$. They add to the reflected particle in $x < 0$ to generate a formal reflection coefficient $R > 1$. The anti-particles move to the right because they can only move in the

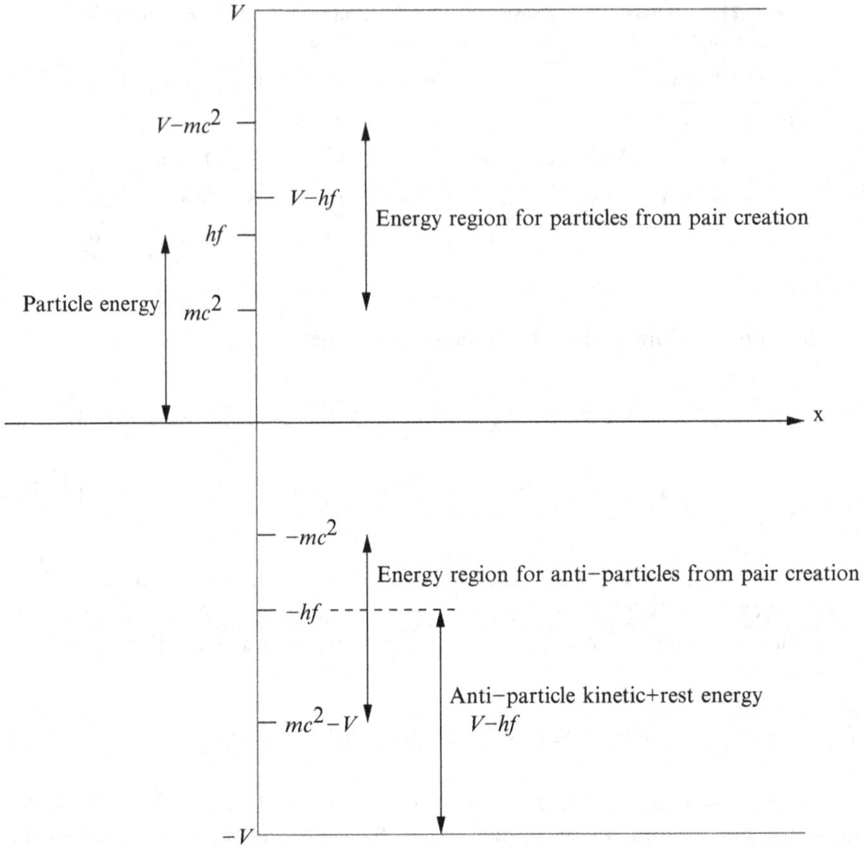

Fig. 21.1 Particles of charge q experience the potential V for $x > 0$, while anti-particles with charge $-q$ experience the potential $-V$. If the potential satisfies $V > 2mc^2$, it can produce particles with energy E_p, $mc^2 \leq E_p = hf \leq V - mc^2$, in the region $x < 0$ and anti-particles with energy $E_{\overline{p}} = -hf$, $mc^2 - V \leq E_{\overline{p}} \leq -mc^2$, in the region $x > 0$. This corresponds to a kinetic+rest energy $K_{\overline{p}} = V - hf$, $mc^2 \leq K_{\overline{p}} \leq V - mc^2$, see equation (21.33). Pair creation is most efficient for $hf = E_p = K_{\overline{p}} = -E_{\overline{p}} = V/2$

attractive potential $-V$ in $x > 0$. The movement of charges $-q$ to the right generates a negative apparent transmission coefficient $T = j_{x>0}/j_{in} < 0$.

Please note that the last two lines in Table 21.1 do *not* state that extremely large potentials $V \gg 2mc^2$ are less efficient for pair creation. They only state that a potential $V > 2mc^2$ is particularly efficient for generation of particle–anti-particle pairs with energies $E_p = K_{\overline{p}} = -E_{\overline{p}} = \hbar\omega = V/2$.

The conclusion in a nutshell is that if we wish to calculate scattering in the potential $V > 2mc^2$ for incident particles with energies in the pair creation region $mc^2 \leq \hbar\omega \leq V - mc^2$, then the ongoing pair creation will yield the seemingly paradoxical results $R > 1$ and $T = 1 - R < 0$, see Figure 21.1.

Please note that a more satisfactory discussion of energetics of the problem would also have to take into account the dynamics of the electromagnetic field $\Phi = V/q$, and then use the Hamiltonian density (21.132) of quantum electrodynamics with scalar matter. This would also imply an additional energy cost for separating the oppositely charged particles and anti-particles. The potential V would therefore decay due to pair creation until it satisfies the condition $V \leq 2mc^2$, when pair creation would seize and the standard single particle results $0 \leq T = 1 - R \leq 1$ apply for incident particles with any energy, or the potential would have to be maintained through an external energy source.

21.3 The Dirac equation

We have seen in equation (21.13) that the conserved charge of the complex Klein-Gordon field does not yield a conserved probability, and therefore has no single particle interpretation. This had motivated Paul Dirac in 1928 to propose a relativistic wave equation which is linear in the derivatives[5],

$$i\hbar\gamma^{\mu}\partial_{\mu}\Psi(x) - mc\Psi(x) = 0. \tag{21.34}$$

Since the relativistic dispersion relation $p^2 + m^2c^2 = 0$ implies that the field Ψ should still satisfy the Klein-Gordon equation, equation (21.34) should imply the Klein-Gordon equation. Applying the operator $i\hbar\gamma^{\mu}\partial_{\mu} + mc$ yields

$$-\hbar^2\gamma^{\mu}\gamma^{\nu}\partial_{\mu}\partial_{\nu}\Psi(x) - m^2c^2\Psi(x) = 0.$$

This is the Klein-Gordon equation if the coefficients γ^{μ} can be chosen to satisfy

$$\{\gamma^{\mu}, \gamma^{\nu}\} = -2\eta^{\mu\nu}. \tag{21.35}$$

In four dimensions, equation (21.35) has an up to equivalence transformations unique solution in terms of (4×4)-matrices (see Appendix G for the relevant proofs and for the construction of γ matrices in d spacetime dimensions).

The Dirac basis for γ matrices is

$$\gamma_0 = \begin{pmatrix} -\underline{1} & \underline{0} \\ \underline{0} & \underline{1} \end{pmatrix}, \quad \gamma_i = \begin{pmatrix} \underline{0} & \underline{\sigma}_i \\ -\underline{\sigma}_i & \underline{0} \end{pmatrix}, \tag{21.36}$$

where the (4×4)-matrices are expressed in terms of (2×2)-matrices. Another often used basis is the Weyl basis:

$$\gamma_0 = \begin{pmatrix} \underline{0} & \underline{1} \\ \underline{1} & \underline{0} \end{pmatrix}, \quad \gamma_i = \begin{pmatrix} \underline{0} & \underline{\sigma}_i \\ -\underline{\sigma}_i & \underline{0} \end{pmatrix}. \tag{21.37}$$

[5]P.A.M. Dirac, Proc. Roy. Soc. London A 117, 610 (1928). Dirac's relativistic wave equation was a great success, but like every relativistic wave equation, it also does not yield a single particle interpretation. It immediately proved itself by explaining the anomalous magnetic moment of the electron and the fine structure of spectral lines, and by predicting positrons.

The two bases are related by the orthogonal transformation

$$\gamma_W^\mu = \frac{1}{2}\begin{pmatrix} \underline{1} & \underline{1} \\ -\underline{1} & \underline{1} \end{pmatrix} \cdot \gamma_D^\mu \cdot \begin{pmatrix} \underline{1} & -\underline{1} \\ \underline{1} & \underline{1} \end{pmatrix},$$

$$\gamma_D^\mu = \frac{1}{2}\begin{pmatrix} \underline{1} & -\underline{1} \\ \underline{1} & \underline{1} \end{pmatrix} \cdot \gamma_W^\mu \cdot \begin{pmatrix} \underline{1} & \underline{1} \\ -\underline{1} & \underline{1} \end{pmatrix}.$$

The Dirac equation with minimal photon coupling

$$\gamma^\mu(\mathrm{i}\hbar\partial_\mu + qA_\mu)\Psi(x) - mc\Psi(x) = 0 \tag{21.38}$$

follows from the Lagrange density of quantum electrodynamics,

$$\mathcal{L} = c\overline{\Psi}\left[\gamma^\mu(\mathrm{i}\hbar\partial_\mu + qA_\mu) - mc\right]\Psi - \frac{1}{4\mu_0}F_{\mu\nu}F^{\mu\nu}, \quad \overline{\Psi} = \Psi^+\gamma^0. \tag{21.39}$$

The conserved current density for the phase invariance

$$\Psi' = \exp\!\left(\mathrm{i}\frac{q}{\hbar}\alpha\right)\Psi$$

is

$$j^\mu = cq\overline{\Psi}\gamma^\mu\Psi, \quad \varrho = j^0/c = q\Psi^+\Psi, \quad \boldsymbol{j} = cq\overline{\Psi}\boldsymbol{\gamma}\Psi. \tag{21.40}$$

Variation of (21.39) with respect to the vector potential shows that j^μ appears as the source term in Maxwell's equations,

$$\partial_\mu F^{\mu\nu} = -\mu_0 j^\nu.$$

Solutions of the free Dirac equation

We temporarily set $\hbar = 1$ and $c = 1$ for the construction of the general solution of the free Dirac equation.

Substitution of the Fourier *ansatz*

$$\Psi(x) = \int \frac{d^4p}{(2\pi)^2}\,\Psi(p)\exp(\mathrm{i}p\cdot x)$$

into (21.34) yields the equation

$$(\gamma^\mu p_\mu + m)\Psi(p) = 0. \tag{21.41}$$

We can use any representation of the γ matrices to find

$$\det(\gamma^\mu p_\mu + m) = (m^2 + p^2)^2 = (m^2 + \boldsymbol{p}^2 - E^2)^2 = (E^2(\boldsymbol{p}) - E^2)^2,$$

i.e. the solutions of (21.41) must have the form

$$\Psi(p) = \sqrt{\frac{\pi}{E(\boldsymbol{p})}}u(\boldsymbol{p})\delta(E - E(\boldsymbol{p})) + \sqrt{\frac{\pi}{E(\boldsymbol{p})}}v(-\boldsymbol{p})\delta(E + E(\boldsymbol{p})) \tag{21.42}$$

with $E(\boldsymbol{p}) = \sqrt{\boldsymbol{p}^2 + m^2}$ and

$$\left[\boldsymbol{\gamma}\cdot\boldsymbol{p} - \gamma^0 E(\boldsymbol{p}) + m\right]\cdot u(\boldsymbol{p}) = 0, \tag{21.43}$$

$$\left[\boldsymbol{\gamma}\cdot\boldsymbol{p} + \gamma^0 E(\boldsymbol{p}) + m\right]\cdot v(-\boldsymbol{p}) = 0. \tag{21.44}$$

The normalization factors in (21.42) are included for later convenience when we quantize the Dirac field.

To find the eigenspinors $u(\boldsymbol{p})$, $v(-\boldsymbol{p})$, we observe

$$(\gamma^\mu p_\mu + m)(m - \gamma^\mu p_\mu) = m^2 + p^2,$$

i.e. the columns $\zeta_{i+}(\boldsymbol{p})$ of the matrix $(m - \gamma^\mu p_\mu)_{E=E(\boldsymbol{p})}$ solve equation (21.43) while the columns $\zeta_{i-}(\boldsymbol{p})$ of the matrix $(m - \gamma^\mu p_\mu)_{E=-E(\boldsymbol{p})}$ solve equation (21.44). However, only two columns of each of the two matrices $\zeta_{i\pm}(\boldsymbol{p})$ are linearly independent.

We initially use a Dirac basis (21.36) for the γ matrices. A suitable basis for the general solution of the free Dirac equation is then given by the spin basis in the Dirac representation,

$$\begin{aligned} u(\boldsymbol{p}, \tfrac{1}{2}) = u_\uparrow(\boldsymbol{p}) &= \frac{1}{\sqrt{E(\boldsymbol{p}) + m}}\zeta_{1+}(\boldsymbol{p}) \\ &= \frac{1}{\sqrt{E(\boldsymbol{p}) + m}}\begin{pmatrix} E(\boldsymbol{p}) + m \\ 0 \\ p_3 \\ p_+ \end{pmatrix}, \end{aligned} \tag{21.45}$$

$$\begin{aligned} u(\boldsymbol{p}, -\tfrac{1}{2}) = u_\downarrow(\boldsymbol{p}) &= \frac{1}{\sqrt{E(\boldsymbol{p}) + m}}\zeta_{2+}(\boldsymbol{p}) \\ &= \frac{1}{\sqrt{E(\boldsymbol{p}) + m}}\begin{pmatrix} 0 \\ E(\boldsymbol{p}) + m \\ p_- \\ -p_3 \end{pmatrix}, \end{aligned} \tag{21.46}$$

$$v(-\boldsymbol{p}, -\tfrac{1}{2}) = v_{\downarrow}(-\boldsymbol{p}) = \frac{1}{\sqrt{E(\boldsymbol{p}) + m}} \zeta_{3-}(\boldsymbol{p})$$

$$= \frac{1}{\sqrt{E(\boldsymbol{p}) + m}} \begin{pmatrix} -p_3 \\ -p_+ \\ E(\boldsymbol{p}) + m \\ 0 \end{pmatrix}, \tag{21.47}$$

$$v(-\boldsymbol{p}, \tfrac{1}{2}) = v_{\uparrow}(-\boldsymbol{p}) = \frac{1}{\sqrt{E(\boldsymbol{p}) + m}} \zeta_{4-}(\boldsymbol{p})$$

$$= \frac{1}{\sqrt{E(\boldsymbol{p}) + m}} \begin{pmatrix} -p_- \\ p_3 \\ 0 \\ E(\boldsymbol{p}) + m \end{pmatrix}, \tag{21.48}$$

where $p_{\pm} = p_1 \pm \mathrm{i}p_2$ was used. The spin labels indicate that $u(\boldsymbol{p}, \pm\frac{1}{2})$ describes spin up or down particles, while $v(\boldsymbol{p}, \pm\frac{1}{2})$ describes spin up or down anti-particles.

It is also convenient to express the 4-spinors (21.45–21.48) in terms of the 2-spinors

$$\chi_{\uparrow} = \begin{pmatrix} 1 \\ 0 \end{pmatrix}, \quad \chi_{\downarrow} = \begin{pmatrix} 0 \\ 1 \end{pmatrix},$$

in the form

$$u_{\uparrow}(\boldsymbol{p}) = \frac{1}{\sqrt{E(\boldsymbol{p}) + m}} \begin{pmatrix} (E(\boldsymbol{p}) + m)\chi_{\uparrow} \\ (\boldsymbol{p} \cdot \underline{\sigma}) \cdot \chi_{\uparrow} \end{pmatrix},$$

$$u_{\downarrow}(\boldsymbol{p}) = \frac{1}{\sqrt{E(\boldsymbol{p}) + m}} \begin{pmatrix} (E(\boldsymbol{p}) + m)\chi_{\downarrow} \\ (\boldsymbol{p} \cdot \underline{\sigma}) \cdot \chi_{\downarrow} \end{pmatrix},$$

$$v_{\downarrow}(\boldsymbol{p}) = \frac{1}{\sqrt{E(\boldsymbol{p}) + m}} \begin{pmatrix} (\boldsymbol{p} \cdot \underline{\sigma}) \cdot \chi_{\uparrow} \\ (E(\boldsymbol{p}) + m)\chi_{\uparrow} \end{pmatrix},$$

$$v_{\uparrow}(\boldsymbol{p}) = \frac{1}{\sqrt{E(\boldsymbol{p}) + m}} \begin{pmatrix} (\boldsymbol{p} \cdot \underline{\sigma}) \cdot \chi_{\downarrow} \\ (E(\boldsymbol{p}) + m)\chi_{\downarrow} \end{pmatrix}.$$

The general solution of the free Dirac equation then has the form

$$\Psi(x) = \frac{1}{\sqrt{2\pi}^3} \int \frac{d^3\boldsymbol{p}}{\sqrt{2E(\boldsymbol{p})}} \sum_{s \in \{\downarrow, \uparrow\}} \big[b_s(\boldsymbol{p}) u(\boldsymbol{p}, s) \exp(\mathrm{i}p \cdot x)$$

$$+ d_s^+(\boldsymbol{p}) v(\boldsymbol{p}, s) \exp(-\mathrm{i}p \cdot x) \big], \tag{21.49}$$

where $p^0 = E(\boldsymbol{p})$ is understood: $p \cdot x = \boldsymbol{p} \cdot \boldsymbol{x} - E(\boldsymbol{p})t$.

Calculations involving 4-spinors are often conveniently carried out with $\hbar = 1$ and $c = 1$, and restoration of the constants is usually only done in the final results from the requirement of correct units. For completeness I would also like to give the general solution of the free Dirac equation with the constants $\hbar$ and c restored. We can choose the basic spinors (21.45–21.48) to have units of square roots of energy, e.g.

$$u_{\uparrow}(\boldsymbol{k}) = \frac{1}{\sqrt{E(\boldsymbol{k}) + mc^2}} \begin{pmatrix} E(\boldsymbol{k}) + mc^2 \\ 0 \\ \hbar c k_3 \\ \hbar c k_+ \end{pmatrix}, \tag{21.50}$$

and the solution (21.49) is

$$\Psi(x) = \frac{1}{\sqrt{2\pi}^3} \int \frac{d^3\boldsymbol{k}}{\sqrt{2E(\boldsymbol{k})}} \sum_{s\in\{\downarrow,\uparrow\}} \big[b_s(\boldsymbol{k})u(\boldsymbol{k},s)\exp(\mathrm{i}k\cdot x) + d_s^+(\boldsymbol{k})v(\boldsymbol{k},s)\exp(-\mathrm{i}k\cdot x)\big] \tag{21.51}$$

with $k\cdot x \equiv \boldsymbol{k}\cdot\boldsymbol{x} - \omega(\boldsymbol{k})t$. In these conventions the Dirac field has the same dimensions length$^{-3/2}$ as the Schrödinger field. The free field $\Psi(x)$ also describes the freely evolving field operator $\Psi_D(x)$ in the interaction picture.

Some useful algebraic properties of the spinors (21.45–21.48) are frequently used in the calculations of cross sections and other observables,

$$u^+(\boldsymbol{k},s)\cdot u(\boldsymbol{k},s') = 2E(\boldsymbol{k})\delta_{ss'}, \quad v^+(\boldsymbol{k},s)\cdot v(\boldsymbol{k},s') = 2E(\boldsymbol{k})\delta_{ss'}, \tag{21.52}$$

$$u^+(\boldsymbol{k},s)\cdot v(-\boldsymbol{k},s') = 0, \quad \overline{u}(\boldsymbol{k},s)\cdot v(\boldsymbol{k},s') = 0, \tag{21.53}$$

$$\overline{u}(\boldsymbol{k},s)\cdot u(\boldsymbol{k},s') = 2mc^2\delta_{ss'}, \quad \overline{v}(\boldsymbol{k},s)\cdot v(\boldsymbol{k},s') = -2mc^2\delta_{ss'}, \tag{21.54}$$

$$\overline{u}(\boldsymbol{k},+)\cdot v(-\boldsymbol{k},-) = -2cp_3, \quad \overline{u}(\boldsymbol{k},+)\cdot v(-\boldsymbol{k},+) = -2cp_-, \tag{21.55}$$

$$\overline{u}(\boldsymbol{k},-)\cdot v(-\boldsymbol{k},-) = -2cp_+, \quad \overline{u}(\boldsymbol{k},-)\cdot v(-\boldsymbol{k},+) = 2cp_3. \tag{21.56}$$

The following equations contain 4×4 unit matrices $\underline{1}$ on the right hand sides,

$$\sum_s u(\boldsymbol{k},s)u^+(\boldsymbol{k},s) + \sum_s v(-\boldsymbol{k},s)v^+(-\boldsymbol{k},s) = 2E(\boldsymbol{k})\underline{1}, \tag{21.57}$$

$$\sum_s u(\boldsymbol{k},s)\overline{u}(\boldsymbol{k},s) = mc^2\underline{1} - c\gamma^\mu p_\mu\Big|_{cp^0=E(\boldsymbol{k})}, \tag{21.58}$$

$$\sum_s u(-\boldsymbol{k},s)\overline{u}(-\boldsymbol{k},s) = mc^2\underline{1} + c\gamma^\mu p_\mu\Big|_{cp^0=-E(\boldsymbol{k})}, \tag{21.59}$$

$$\sum_s v(\boldsymbol{k},s)\overline{v}(\boldsymbol{k},s) = -mc^2\underline{1} - c\gamma^\mu p_\mu\Big|_{cp^0=E(\boldsymbol{k})}, \tag{21.60}$$

$$\sum_s v(-\boldsymbol{k},s)\overline{v}(-\boldsymbol{k},s) = -mc^2\underline{1} + c\gamma^\mu p_\mu\Big|_{cp^0=-E(\boldsymbol{k})}. \tag{21.61}$$

It is actually clumsy to write down unit matrices when their presence is clear from the context, and the action e.g. of the scalar mc^2 on a 4-spinor Ψ has the same effect as the matrix $mc^2\underline{1}$. Therefore we will usually adopt the practice of not writing down 4×4 unit matrices explicitly.

Equations (21.52) and (21.53) are used e.g. in the inversion of the Fourier representation (21.51),

$$b_s(\boldsymbol{k}) = \frac{1}{\sqrt{2\pi}^3}\int\frac{d^3x}{\sqrt{2E(\boldsymbol{k})}}\exp(-\mathrm{i}k\cdot x)u^+(\boldsymbol{k},s)\cdot\Psi(x), \tag{21.62}$$

$$d_s(\boldsymbol{k}) = \frac{1}{\sqrt{2\pi}^3}\int\frac{d^3x}{\sqrt{2E(\boldsymbol{k})}}\exp(-\mathrm{i}k\cdot x)\Psi^+(x)\cdot v(\boldsymbol{k},s). \tag{21.63}$$

Substituting these equations back into (21.51) yields

$$\Psi(\boldsymbol{x},t) = \int d^3x'\,\mathcal{W}(\boldsymbol{x}-\boldsymbol{x}',t-t')\cdot\Psi(\boldsymbol{x}',t') \tag{21.64}$$

with the time evolution kernel

$$\begin{aligned}\mathcal{W}(\boldsymbol{x},t) &= \frac{1}{(2\pi)^3}\int\frac{d^3k}{2E(\boldsymbol{k})}\exp(\mathrm{i}\boldsymbol{k}\cdot\boldsymbol{x})\sum_s[u(\boldsymbol{k},s)u^+(\boldsymbol{k},s)\exp(-\mathrm{i}\omega(\boldsymbol{k})t)\\ &\quad + v(-\boldsymbol{k},s)v^+(-\boldsymbol{k},s)\exp(\mathrm{i}\omega(\boldsymbol{k})t)]\\ &= \frac{1}{(2\pi)^3}\int\frac{d^3k}{E(\boldsymbol{k})}\exp(\mathrm{i}\boldsymbol{k}\cdot\boldsymbol{x})[E(\boldsymbol{k})\cos(\omega(\boldsymbol{k})t)\\ &\quad + \mathrm{i}c(\hbar\boldsymbol{\gamma}\cdot\boldsymbol{k} - mc)\gamma^0\sin(\omega(\boldsymbol{k})t)].\end{aligned} \tag{21.65}$$

This satisfies the initial value problem

$$(\mathrm{i}\hbar\gamma^\mu\partial_\mu - mc)\mathcal{W}(\boldsymbol{x},t) = 0, \quad \mathcal{W}(\boldsymbol{x},0) = \delta(\boldsymbol{x}). \tag{21.66}$$

It is related to the time evolution kernel (21.9) of the Klein-Gordon field through

$$\mathrm{i}\mathcal{W}(\boldsymbol{x},t)\gamma^0 = c\left(\mathrm{i}\gamma\cdot\partial + \frac{mc}{\hbar}\right)\mathcal{K}(\boldsymbol{x},t). \tag{21.67}$$

It is sometimes useful to express equation (21.49) and the corresponding equation in $\boldsymbol{k}$ space in bra-ket notation, similar to equations (18.24, 18.25) for the Maxwell field. With the definitions

$$b_{+,s}(\boldsymbol{k}) = b_s(\boldsymbol{k}), \quad b_{-,s}(\boldsymbol{k}) = d_s^+(-\boldsymbol{k}),$$

$$u_{+,s}(\boldsymbol{k}) = u_s(\boldsymbol{k}), \quad u_{-,s}(\boldsymbol{k}) = v_s(-\boldsymbol{k}),$$

we can write the free Dirac field in the forms

$$\langle \boldsymbol{k}, \sigma, s|\Psi(t)\rangle = b_{\sigma,s}(\boldsymbol{k}) \exp[-\mathrm{i}\sigma\omega(\boldsymbol{k})t] \tag{21.68}$$

and

$$\langle \boldsymbol{x}, a|\Psi(t)\rangle = \int \frac{d^3\boldsymbol{k}}{\sqrt{2\pi}^3} \sum_{\sigma\in\{+,-\}} \sum_{s\in\{\uparrow,\downarrow\}} \frac{b_{\sigma,s}(\boldsymbol{k})u^a_{\sigma,s}(\boldsymbol{k})}{\sqrt{2\hbar\omega(\boldsymbol{k})}} \exp[\mathrm{i}(\boldsymbol{k}\cdot\boldsymbol{x} - \sigma\omega(\boldsymbol{k})t)],$$

where $a \in \{1,\ldots 4\}$ is a Dirac spinor index, $\sigma \in \{+,-\}$ labels particles $(+)$ or anti-particles $(-)$, and s is the spin label. The equations (21.52) and the first equation in (21.53) are

$$u^+_{\sigma,s}(\boldsymbol{k}) \cdot u_{\sigma',s'}(\boldsymbol{k}) = 2\hbar\omega(\boldsymbol{k})\delta_{\sigma,\sigma'}\delta_{s,s'}. \tag{21.69}$$

Equation (21.54) is

$$\overline{u}_{\sigma,s}(\boldsymbol{k}) \cdot u_{\sigma,s'}(\boldsymbol{k}) = 2mc^2\sigma\delta_{ss'}, \tag{21.70}$$

and equation (21.57) is

$$\sum_{\sigma,s} u_{\sigma,s}(\boldsymbol{k})u^+_{\sigma,s}(\boldsymbol{k}) = 2\hbar\omega(\boldsymbol{k})\underline{1}. \tag{21.71}$$

The $\boldsymbol{x}$ representations of the spinor momentum eigenstates are

$$\langle \boldsymbol{x}, a|\boldsymbol{k}, \sigma, s\rangle = \frac{\exp(\mathrm{i}\boldsymbol{k}\cdot\boldsymbol{x})}{4\pi\sqrt{\pi\hbar\omega(\boldsymbol{k})}} u^a_{\sigma,s}(\boldsymbol{k}), \tag{21.72}$$

and using equations (21.69, 21.71) we can easily verify the relations

$$\langle \boldsymbol{k}, \sigma, s|\boldsymbol{k}', \sigma', s'\rangle = \delta_{\sigma,\sigma'}\delta_{s,s'}\delta(\boldsymbol{k}-\boldsymbol{k}'), \quad \langle \boldsymbol{x}, a|\boldsymbol{x}', a'\rangle = \delta_{a,a'}\delta(\boldsymbol{x}-\boldsymbol{x}'). \tag{21.73}$$

Charge operators and quantization of the Dirac field

We can apply the results from Section 16.2 to calculate the energy and momentum operator for the Dirac field. The free Dirac Lagrangian

$$\mathcal{L} = c\overline{\Psi}\left(\mathrm{i}\hbar\gamma^\mu\partial_\mu - mc\right)\Psi \tag{21.74}$$

yields the positive definite normal ordered Hamiltonian

$$H = \int d^3\boldsymbol{x}\, c\overline{\Psi}(\boldsymbol{x}, t)\,(mc - \mathrm{i}\hbar\boldsymbol{\gamma}\cdot\nabla)\,\Psi(\boldsymbol{x}, t)$$

$$= \int d^3k\, \hbar\omega(\boldsymbol{k}) \sum_{s\in\{\downarrow,\uparrow\}} \left[b_s^+(\boldsymbol{k})b_s(\boldsymbol{k}) + d_s^+(\boldsymbol{k})d_s(\boldsymbol{k})\right], \tag{21.75}$$

but only if we assume anti-commutation properties of the d_s and d_s^+ operators.

The normal ordered momentum operator is then

$$\boldsymbol{P} = \int d^3x\, \Psi^+(\boldsymbol{x},t)\frac{\hbar}{\mathrm{i}}\nabla\Psi(\boldsymbol{x},t)$$
$$= \int d^3k\, \hbar\boldsymbol{k} \sum_{s\in\{\downarrow,\uparrow\}} \left[b_s^+(\boldsymbol{k})b_s(\boldsymbol{k}) + d_s^+(\boldsymbol{k})d_s(\boldsymbol{k})\right]. \tag{21.76}$$

The electromagnetic current density (21.40) yields the charge operator

$$Q = q\int d^3x\, \Psi^+(\boldsymbol{x},t)\Psi(\boldsymbol{x},t)$$
$$= q\int d^3k \sum_{s\in\{\downarrow,\uparrow\}} \left[b_s^+(\boldsymbol{k})b_s(\boldsymbol{k}) - d_s^+(\boldsymbol{k})d_s(\boldsymbol{k})\right]. \tag{21.77}$$

The normalization in equation (21.51) has been chosen such that the quantization condition

$$\{\Psi_\alpha(\boldsymbol{x},t), \Psi_\beta{}^+(\boldsymbol{x}',t)\} = \delta_{\alpha\beta}\delta(\boldsymbol{x}-\boldsymbol{x}')$$

for the components of $\Psi(x)$ yields

$$\{b(\boldsymbol{k},s), b^+(\boldsymbol{k}',s')\} = \delta_{ss'}\delta(\boldsymbol{k}-\boldsymbol{k}'), \quad \{d(\boldsymbol{k},s), d^+(\boldsymbol{k}',s')\} = \delta_{ss'}\delta(\boldsymbol{k}-\boldsymbol{k}'),$$

with the other anti-commutators vanishing. The equations (21.75–21.77) then imply that the operator $b^+(\boldsymbol{k},s)$ creates a fermion of mass m, momentum $\hbar\boldsymbol{k}$ and charge q, while $d^+(\boldsymbol{k},s)$ creates a particle with the same mass and momentum, but opposite charge $-q$.

For an explanation of the spin labels of the spinors $u(\boldsymbol{k},\pm\frac{1}{2})$, we notice that the spin operators corresponding to the rotation generators

$$M_i = -\mathrm{i}L_i = \frac{1}{2}\epsilon_{ijk}M_{jk}$$

are both in the Dirac and in the Weyl representation given by

$$S_i = \frac{\hbar}{2}\epsilon_{ijk}S_{jk} = \frac{\mathrm{i}\hbar}{4}\epsilon_{ijk}\gamma_j\gamma_k = \frac{\hbar}{2}\begin{pmatrix} \underline{\sigma}_i & 0 \\ 0 & \underline{\sigma}_i \end{pmatrix}, \tag{21.78}$$

see Appendix H for an explanation of generators of Lorentz boosts and rotations for Dirac spinors.

Equation (21.78) implies that the rest frame spinors $u(\mathbf{0}, \pm\frac{1}{2})$ transform under rotations around the z axis as spinors with z-component of spin $\hbar s = \pm\hbar/2$.

For an explanation of the spin labels of the spinors $v(\boldsymbol{p}, \pm\frac{1}{2})$, we have to look at charge conjugation. Both in the Dirac and the Weyl representation of γ matrices we have

$$\gamma_\mu^* = \gamma_2\gamma_\mu\gamma_2.$$

Therefore complex conjugation of the Dirac equation

$$[\mathrm{i}\gamma^\mu\partial_\mu + q\gamma^\mu A_\mu(x) - m]\Psi(x) = 0,$$

followed by multiplication with $\mathrm{i}\gamma_2$ from the left yields

$$[\mathrm{i}\gamma^\mu\partial_\mu - q\gamma^\mu A_\mu(x) - m]\Psi^c(x) = 0$$

with the charge conjugate field

$$\Psi^c(x) = \mathrm{i}\gamma_2\Psi^*(x). \tag{21.79}$$

In particular, we have

$$v^c(\boldsymbol{k}, \tfrac{1}{2}) = \mathrm{i}\gamma_2 v^*(\boldsymbol{k}, \tfrac{1}{2}) = u(\boldsymbol{k}, \tfrac{1}{2})$$

and

$$v^c(\boldsymbol{k}, -\tfrac{1}{2}) = \mathrm{i}\gamma_2 v^*(\boldsymbol{k}, -\tfrac{1}{2}) = -u(\boldsymbol{k}, -\tfrac{1}{2}),$$

i.e. the negative energy spinors for charge q, momentum $\hbar\boldsymbol{k}$ and spin projection $\hbar s$ correspond to positive energy spinors for charge $-q$, momentum $\hbar\boldsymbol{k}$ and spin projection $\hbar s$.

21.4 The energy-momentum tensor for quantum electrodynamics

We use the symmetrized form of the QED Lagrangian (21.39),

$$\mathcal{L} = c\overline{\Psi}\left[\gamma^\mu\left(\frac{\mathrm{i}\hbar}{2}\overleftrightarrow{\partial}_\mu + qA_\mu\right) - mc\right]\Psi - \frac{1}{4\mu_0}F_{\mu\nu}F^{\mu\nu}. \tag{21.80}$$

This yields according to (16.16) a conserved energy-momentum tensor

$$\Theta_\mu{}^\nu = \eta_\mu{}^\nu \mathcal{L} - \partial_\mu \overline{\Psi} \frac{\partial \mathcal{L}}{\partial(\partial_\nu \overline{\Psi})} - \partial_\mu \Psi \frac{\partial \mathcal{L}}{\partial(\partial_\nu \Psi)} - \partial_\mu A_\lambda \frac{\partial \mathcal{L}}{\partial(\partial_\nu A_\lambda)}$$
$$= \eta_\mu{}^\nu \left(c\overline{\Psi} \left[\gamma^\lambda \left(\frac{\mathrm{i}\hbar}{2} \overleftrightarrow{\partial}_\lambda + qA_\lambda \right) - mc \right] \Psi - \frac{1}{4\mu_0} F_{\kappa\lambda} F^{\kappa\lambda} \right)$$
$$- \frac{\mathrm{i}\hbar}{2} c\overline{\Psi}\gamma^\nu \overleftrightarrow{\partial}_\mu \Psi + \frac{1}{\mu_0} \partial_\mu A_\lambda F^{\nu\lambda}.$$

According to the results of Section 16.2, this yields on-shell conserved charges, i.e. we can use the equations of motion to simplify this expression. The Dirac equation then implies

$$\Theta_\mu{}^\nu = -\frac{\mathrm{i}\hbar}{2} c\overline{\Psi}\gamma^\nu \overleftrightarrow{\partial}_\mu \Psi + \frac{1}{\mu_0} \partial_\mu A_\lambda F^{\nu\lambda} - \frac{1}{4\mu_0} \eta_\mu{}^\nu F_{\kappa\lambda} F^{\kappa\lambda}.$$

We can also add the identically conserved improvement term

$$-\frac{1}{\mu_0} \partial_\lambda \left(A_\mu F^{\nu\lambda} \right) = -\frac{1}{\mu_0} \partial_\lambda A_\mu F^{\nu\lambda} - \frac{1}{\mu_0} A_\mu \partial_\lambda F^{\nu\lambda}$$
$$= -\frac{1}{\mu_0} \partial_\lambda A_\mu F^{\nu\lambda} - qcA_\mu \overline{\Psi}\gamma^\nu \Psi,$$

where Maxwell's equations $\partial_\mu F^{\mu\nu} = -\mu_0 qc\overline{\Psi}\gamma^\nu\Psi$ have been used. This yields the gauge invariant tensor

$$t_\mu{}^\nu = \Theta_\mu{}^\nu - \frac{1}{\mu_0} \partial_\lambda \left(A_\mu F^{\nu\lambda} \right)$$
$$= -\frac{\mathrm{i}\hbar}{2} c\overline{\Psi}\gamma^\nu \overleftrightarrow{\partial}_\mu \Psi - qc\overline{\Psi}\gamma^\nu A_\mu \Psi + \frac{1}{\mu_0} F_{\mu\lambda} F^{\nu\lambda} - \eta_\mu{}^\nu \frac{1}{4\mu_0} F_{\kappa\lambda} F^{\kappa\lambda}. \quad (21.81)$$

However, we can go one step further and replace $t_\mu{}^\nu$ with a symmetric energy-momentum tensor. The divergence of the spinor term in $t_\mu{}^\nu$ is

$$\partial_\nu \left(\frac{\mathrm{i}\hbar}{2} \overline{\Psi}\gamma^\nu \overleftrightarrow{\partial}_\mu \Psi + q\overline{\Psi}\gamma^\nu A_\mu \Psi \right) = -q\overline{\Psi} F_{\mu\nu} \gamma^\nu \Psi, \quad (21.82)$$

where again the Dirac equation was used.

The symmetrization of $t_\mu{}^\nu$ also involves the commutators of γ matrices,

$$S_{\mu\nu} = \frac{\mathrm{i}}{4} [\gamma_\mu, \gamma_\nu] = \gamma_0 \cdot S_{\mu\nu}^+ \cdot \gamma_0. \quad (21.83)$$

Since we can write a product always as a sum of an anti-commutator and a commutator, we have

$$\gamma_\mu \cdot \gamma_\nu = -\eta_{\mu\nu} - 2\mathrm{i} S_{\mu\nu}, \quad (21.84)$$

and the commutators also satisfy[6]

$$\eta_{\mu\alpha}\gamma_\beta - \eta_{\mu\beta}\gamma_\alpha + \mathrm{i}[S_{\alpha\beta}, \gamma_\mu] = 0. \tag{21.85}$$

Equations (21.83–21.85) together with

$$\begin{aligned}\hbar^2\partial^2\Psi &= \mathrm{i}\hbar\gamma^\mu\partial_\mu\left(mc\Psi - q\gamma^\nu A_\nu\Psi\right)\\ &= \mathrm{i}\hbar q\left[\partial_\mu\left(A^\mu\Psi\right) + 2\mathrm{i}S^{\mu\nu}\partial_\mu\left(A_\nu\Psi\right)\right] + mc\left[mc\Psi - q\gamma^\nu A_\nu\Psi\right]\end{aligned}$$

imply also

$$\partial_\nu\left(\frac{\mathrm{i}\hbar}{2}\overline{\Psi}\gamma_\mu\overset{\leftrightarrow}{\partial^\nu}\Psi + q\overline{\Psi}\gamma_\mu A^\nu\Psi\right) = -q\overline{\Psi}F_{\mu\nu}\gamma^\nu\Psi. \tag{21.86}$$

Therefore the local conservation law $\partial_\nu T_\mu{}^\nu = 0$ also holds for the symmetrized energy-momentum tensor

$$\begin{aligned}T_\mu{}^\nu = &-\frac{c}{2}\overline{\Psi}\left[\frac{\mathrm{i}\hbar}{2}\gamma^\nu\overset{\leftrightarrow}{\partial_\mu} + \frac{\mathrm{i}\hbar}{2}\gamma_\mu\overset{\leftrightarrow}{\partial^\nu} + q\gamma^\nu A_\mu + q\gamma_\mu A^\nu\right]\Psi\\ &+\frac{1}{\mu_0}F_{\mu\lambda}F^{\nu\lambda} - \eta_\mu{}^\nu\frac{1}{4\mu_0}F_{\kappa\lambda}F^{\kappa\lambda}.\end{aligned} \tag{21.87}$$

This yields in particular the Hamiltonian density

$$\begin{aligned}\mathcal{H} = -T_0{}^0 &= c\Psi^+\left[\frac{\mathrm{i}\hbar}{2}\overset{\leftrightarrow}{\partial_0} + qA_0\right]\Psi + \frac{\epsilon_0}{2}\boldsymbol{E}^2 + \frac{1}{2\mu_0}\boldsymbol{B}^2\\ &= c\overline{\Psi}\left[mc - \frac{\mathrm{i}\hbar}{2}\boldsymbol{\gamma}\cdot\overset{\leftrightarrow}{\nabla} - q\boldsymbol{\gamma}\cdot\boldsymbol{A}\right]\Psi + \frac{\epsilon_0}{2}\boldsymbol{E}^2 + \frac{1}{2\mu_0}\boldsymbol{B}^2,\end{aligned} \tag{21.88}$$

and the momentum density with components $\mathcal{P}_i = T_i{}^0/c$,

$$\mathcal{P} = \frac{1}{2}\Psi^+\left[\frac{\hbar}{2\mathrm{i}}\overset{\leftrightarrow}{\nabla} - q\boldsymbol{A}\right]\Psi + \frac{1}{2}\overline{\Psi}\boldsymbol{\gamma}\left[\frac{\mathrm{i}\hbar}{2}\overset{\leftrightarrow}{\partial_0} + qA_0\right]\Psi + \epsilon_0\boldsymbol{E}\times\boldsymbol{B}. \tag{21.89}$$

Elimination of the time derivatives using the Dirac equation yields

$$\mathcal{P} = \Psi^+\left[\frac{\hbar}{2\mathrm{i}}\overset{\leftrightarrow}{\nabla} - q\boldsymbol{A}\right]\Psi + \epsilon_0\boldsymbol{E}\times\boldsymbol{B} + \frac{1}{2}\nabla\times(\Psi^+\cdot\boldsymbol{S}\cdot\Psi). \tag{21.90}$$

[6]The commutators $S_{\mu\nu}$ provide the spinor representation of the generators of Lorentz transformations. Furthermore, equation (21.85) is the invariance of the γ matrices under Lorentz transformations, see Appendix H.

The spin contribution $\mathcal{P}_S = \nabla \times (\Psi^+ \cdot \boldsymbol{S} \cdot \Psi)/2$ with the vector of 4×4 spin matrices $\boldsymbol{S} = i\hbar\boldsymbol{\gamma}\times\boldsymbol{\gamma}/4$ (21.78) appears here as an additional contribution compared to the orbital momentum density $\mathcal{P}_O = \mathcal{P} - \mathcal{P}_S$ that follows directly from the tensor (21.81). The spin term in the momentum density (21.90) generates the spin contribution in the total angular momentum density $\mathcal{J} = \boldsymbol{x} \times \mathcal{P} = \mathcal{M} + \mathcal{S}$ from $\mathcal{S} = \boldsymbol{x} \times \mathcal{P}_S \to \Psi^+ \cdot \boldsymbol{S} \cdot \Psi$ if the symmetric energy-momentum tensor is used in the calculation of angular momentum. This is explained in Problem 21.16c, see in particular equations (21.147–21.150).

Energy and momentum in QED in Coulomb gauge

In materials science it is convenient to explicitly disentangle the contributions from Coulomb and photon terms in Coulomb gauge $\nabla \cdot \boldsymbol{A} = 0$. We split the electric field components in Coulomb gauge according to

$$\boldsymbol{E}_\parallel = -\nabla\Phi, \quad \boldsymbol{E}_\perp = -\frac{\partial \boldsymbol{A}}{\partial t}. \tag{21.91}$$

The equation for the electrostatic potential decouples from the vector potential in Coulomb gauge,

$$\Delta\Phi = -\frac{q}{\epsilon_0}\Psi^+\Psi,$$

and is solved by

$$\Phi(\boldsymbol{x}, t) = \frac{q}{4\pi\epsilon_0}\int d^3x' \, \frac{1}{|\boldsymbol{x} - \boldsymbol{x}'|}\Psi^+(\boldsymbol{x}', t)\Psi(\boldsymbol{x}', t).$$

Furthermore, the two components (21.91) of the electric field are orthogonal in Coulomb gauge,

$$\begin{aligned}\int d^3x \, \boldsymbol{E}_\parallel(\boldsymbol{x}, t) \cdot \boldsymbol{E}_\perp(\boldsymbol{x}, t) &= \int d^3k \, \boldsymbol{E}_\parallel(\boldsymbol{k}, t) \cdot \boldsymbol{E}_\perp(-\boldsymbol{k}, t) \\ &= -\int d^3x \, \Phi(\boldsymbol{x}, t)\frac{\partial}{\partial t}\nabla \cdot \boldsymbol{A}(\boldsymbol{x}, t) = 0,\end{aligned} \tag{21.92}$$

and the contribution from $\boldsymbol{E}_\parallel$ to the Hamiltonian is

$$\begin{aligned}H_C &= \frac{\epsilon_0}{2}\int d^3x \, \boldsymbol{E}_\parallel^2(\boldsymbol{x}, t) = -\frac{\epsilon_0}{2}\int d^3x \, \Phi(\boldsymbol{x}, t)\Delta\Phi(\boldsymbol{x}, t) \\ &= \frac{1}{2}\int d^3x \, \Phi(\boldsymbol{x}, t)\varrho(\boldsymbol{x}, t)\end{aligned}$$

$$= q^2 \sum_{ss'} \int d^3x \int d^3x' \, \frac{\Psi_s^+(x,t)\Psi_{s'}^+(x',t)\Psi_{s'}(x',t)\Psi_s(x,t)}{8\pi\epsilon_0|x-x'|}, \quad (21.93)$$

where the summation is over 4-spinor indices. The presentation of the ordering of the field operators was conventionally chosen as the correct ordering in the non-relativistic limit, cf. (18.65), but (21.93) must actually be normal ordered such that the particle and anti-particle creation operators $b_s^+(k)$ and $d_s^+(k)$ appear leftmost in the Coulomb term in the forms b^+d^+db, d^+d^+dd, etc. Substituting the mode expansions $\Psi \sim b + d^+$ and normal ordering therefore leads to the attractive Coulomb terms between particles and their anti-particles.

The resulting Hamiltonian in Coulomb gauge therefore has the form

$$\begin{aligned} H = &\int d^3x \Big(c\overline{\Psi}(x,t)\,[mc - \gamma\cdot(\mathrm{i}\hbar\nabla + qA(x,t))]\,\Psi(x,t) \\ &+ \frac{\epsilon_0}{2}E_\perp^2(x,t) + \frac{1}{2\mu_0}B^2(x,t)\Big) \\ &+ q^2 \sum_{ss'} \int d^3x \int d^3x' \, \frac{\Psi_s^+(x,t)\Psi_{s'}^+(x',t)\Psi_{s'}(x',t)\Psi_s(x,t)}{8\pi\epsilon_0|x-x'|}. \end{aligned} \quad (21.94)$$

This Hamiltonian yields the corresponding Dirac equation in the Heisenberg form

$$\mathrm{i}\hbar\frac{\partial}{\partial t}\Psi(x,t) = [\Psi(x,t), H]$$

if canonical anti-commutation relations are used for the spinor field. The Coulomb gauge wave equation (18.10) with the relativistic current density j (21.40) follows in the form

$$\mathrm{i}\hbar\frac{\partial}{\partial t}A(x,t) = [A(x,t), H], \quad \frac{\partial^2}{\partial t^2}A(x,t) = \frac{1}{\hbar^2}[H, [A(x,t), H]]. \quad (21.95)$$

if the commutation relations (18.36, 18.39) are used. This confirms the canonical relations between Heisenberg, Schrödinger and Dirac pictures, and the consistency of Coulomb gauge quantization with the transverse δ function (18.27) also in the fully relativistic theory. It also implies appearance of the Dirac picture time evolution operator in the scattering matrix in the now familiar form.

The momentum operator in Coulomb gauge follows from (21.90) and

$$\int d^3x\,\epsilon_0 E_\parallel \times B = -\int d^3x\,\epsilon_0 \Phi\Delta A = \int d^3x\,\varrho A = q\int d^3x\,\Psi^+ A\Psi$$

as

$$P = \int d^3x \left(\Psi^+\frac{\hbar}{\mathrm{i}}\nabla\Psi + \epsilon_0 E_\perp \times B\right), \quad (21.96)$$

where boundary terms at infinity were discarded.

21.5 The non-relativistic limit of the Dirac equation

The Dirac basis (21.36) for the γ-matrices is convenient for the non-relativistic limit. Splitting off the time dependence due to the rest mass term

$$\Psi(\boldsymbol{x},t)=\Upsilon(\boldsymbol{x},t)\exp\!\left(-\mathrm{i}\frac{mc^2}{\hbar}t\right)=\begin{pmatrix}\psi(\boldsymbol{x},t)\\ \phi(\boldsymbol{x},t)\end{pmatrix}\exp\!\left(-\mathrm{i}\frac{mc^2}{\hbar}t\right) \tag{21.97}$$

in the Dirac equation (21.38) yields the equations

$$(\mathrm{i}\hbar\partial_t-q\Phi)\psi+c\underline{\sigma}\cdot(\mathrm{i}\hbar\nabla+q\boldsymbol{A})\phi=0, \tag{21.98}$$

$$(\mathrm{i}\hbar\partial_t-q\Phi+2mc^2)\phi+c\underline{\sigma}\cdot(\mathrm{i}\hbar\nabla+q\boldsymbol{A})\psi=0. \tag{21.99}$$

This yields in the non-relativistic regime

$$\phi\simeq-\frac{1}{2mc}\underline{\sigma}\cdot(\mathrm{i}\hbar\nabla+q\boldsymbol{A})\psi \tag{21.100}$$

and substitution into the equation for ψ yields Pauli's equation[7]

$$\mathrm{i}\hbar\partial_t\psi=-\frac{1}{2m}(\hbar\nabla-\mathrm{i}q\boldsymbol{A})^2\psi-\frac{q\hbar}{2m}\underline{\sigma}\cdot\boldsymbol{B}\psi+q\Phi\psi. \tag{21.101}$$

The spin matrices for spin-1/2 Schrödinger fields are the upper block matrices in the spin matrices (21.78) for the full Dirac fields, $\underline{S}=\hbar\underline{\sigma}/2$, see also Section 8.1 and in particular equation (8.12).

If the external magnetic field $\boldsymbol{B}$ is approximately constant over the extension of the wave function $\psi(\boldsymbol{x},t)$ we can use

$$\boldsymbol{A}(\boldsymbol{x},t)=\frac{1}{2}\boldsymbol{B}(t)\times\boldsymbol{x}.$$

Substitution of the vector potential in equation (21.101) then yields the following linear terms in $\boldsymbol{B}$ in the Hamiltonian on the right hand side,

$$\begin{aligned}\mathrm{i}\frac{q\hbar}{2m}(\boldsymbol{B}\times\boldsymbol{x})\cdot\nabla-\frac{q}{m}\boldsymbol{B}\cdot\underline{S}&=-\frac{q}{2m}\boldsymbol{B}\cdot(\boldsymbol{M}+2\underline{S})\\ &=-\frac{q}{e}\frac{\mu_B}{\hbar}\boldsymbol{B}\cdot(\boldsymbol{M}+2\underline{S}).\end{aligned} \tag{21.102}$$

[7]W. Pauli, Z. Phys. 43, 601 (1927). Pauli actually only studied the time-independent Schrödinger equation with the Pauli term in the Hamiltonian, and although he mentions Schrödinger in the beginning, he seems to be more comfortable with Heisenberg's matrix mechanics in the paper.

Here $\mu_B = e\hbar/2m$ is the Bohr magneton, and we used the short hand notation $-i\hbar\boldsymbol{x}\times\nabla \to \boldsymbol{M}$ for the $\boldsymbol{x}$ representation of the angular momentum operator. Recall that this operator is actually given by

$$\boldsymbol{M} = \mathbf{x}\times\mathbf{p} = -\,\mathrm{i}\hbar\int d^3\boldsymbol{x}\,|\boldsymbol{x}\rangle\boldsymbol{x}\times\nabla\langle\boldsymbol{x}|.$$

Equation (21.102) shows that the Dirac equation explains the double strength magnetic coupling of spin as compared to orbital angular momentum (often denoted as the *magneto-mechanical anomaly of the electron* or the *anomalous magnetic moment of the electron*). The corresponding electromagnetic currents in the non-relativistic regime are

$$\begin{aligned}
\varrho &= q\psi^+\psi,\\
\boldsymbol{j} &= cq\left(\psi^+\underline{\sigma}\phi + \phi^+\underline{\sigma}\psi\right)\\
&= -\frac{q}{2m}\left(\psi^+\underline{\sigma}\otimes\underline{\sigma}\cdot(\mathrm{i}\hbar\nabla + q\boldsymbol{A})\psi - (\mathrm{i}\hbar\nabla\psi^+ - q\psi^+\boldsymbol{A})\cdot\underline{\sigma}\otimes\underline{\sigma}\psi\right),
\end{aligned}$$

where $\underline{\sigma}\otimes\underline{\sigma}$ is the three-dimensional tensor with the (2×2)-matrix entries $\underline{\sigma}_i\cdot\underline{\sigma}_j$ (we can think of it as a (3×3)-matrix containing (2×2)-matrices as entries). Substitution of

$$\underline{\sigma}\otimes\underline{\sigma} = \underline{1} + \mathrm{i}\boldsymbol{e}_i\otimes\boldsymbol{e}_j\epsilon_{ijk}\underline{\sigma}_k$$

yields

$$\boldsymbol{j} = \frac{q}{2\mathrm{i}m}\left(\psi^+\cdot\hbar\nabla\psi - \hbar\nabla\psi^+\cdot\psi - 2\mathrm{i}q\psi^+\boldsymbol{A}\psi\right) + \boldsymbol{j}_s, \tag{21.103}$$

with a spin term

$$\boldsymbol{j}_s = \frac{q\hbar}{2m}\nabla\times\left(\psi^+\underline{\sigma}\psi\right).$$

However, this term does not accumulate or diminish charges in any volume, $\nabla\cdot\boldsymbol{j}_s = 0$, and can therefore be neglected in the calculation of electric currents.

The non-relativistic approximations for the Lagrange density $\mathcal{L}$, the energy density $\mathcal{H}$ and the momentum density $\mathcal{P}$ are

$$\begin{aligned}
\mathcal{L} &= \frac{\mathrm{i}\hbar}{2}\left(\psi^+\cdot\frac{\partial}{\partial t}\psi - \frac{\partial}{\partial t}\psi^+\cdot\psi\right) - q\psi^+\Phi\psi + \frac{q\hbar}{2m}\psi^+\underline{\sigma}\cdot\boldsymbol{B}\psi\\
&\quad + \frac{1}{2m}(\mathrm{i}\hbar\nabla\psi^+ - q\psi^+\boldsymbol{A})\cdot(\mathrm{i}\hbar\nabla\psi + q\boldsymbol{A}\psi) - \frac{1}{4\mu_0}F_{\mu\nu}F^{\mu\nu},
\end{aligned} \tag{21.104}$$

$$\begin{aligned}
\mathcal{H} &= \frac{1}{2m}(\hbar\nabla\psi^+ + \mathrm{i}q\psi^+\boldsymbol{A})\cdot(\hbar\nabla\psi - \mathrm{i}q\boldsymbol{A}\psi) - \frac{q\hbar}{2m}\psi^+\underline{\sigma}\cdot\boldsymbol{B}\psi\\
&\quad + \frac{\epsilon_0}{2}\boldsymbol{E}^2 + \frac{1}{2\mu_0}\boldsymbol{B}^2,
\end{aligned} \tag{21.105}$$

$$\mathcal{P} = \frac{\hbar}{2\mathrm{i}}\left(\psi^+\cdot\nabla\psi - \nabla\psi^+\cdot\psi\right) - q\psi^+\boldsymbol{A}\psi + \epsilon_0\boldsymbol{E}\times\boldsymbol{B}. \tag{21.106}$$

The Hamiltonian and momentum operators in Coulomb gauge are

$$\begin{aligned} H = &\int d^3x\left(-\frac{1}{2m}\psi^+(x,t)[\hbar\nabla - \mathrm{i}q\boldsymbol{A}(x,t)]^2\psi(x,t)\right.\\ &\left.+\frac{\epsilon_0}{2}\boldsymbol{E}_\perp^2(x,t) + \frac{1}{2\mu_0}\boldsymbol{B}^2(x,t) - \frac{q\hbar}{2m}\psi^+(x,t)\underline{\sigma}\cdot\boldsymbol{B}(x,t)\psi(x,t)\right)\\ &+q^2\sum_{ss'=1}^{2}\int d^3x\int d^3x'\,\frac{\psi_s^+(x,t)\psi_{s'}^+(x',t)\psi_{s'}(x',t)\psi_s(x,t)}{8\pi\epsilon_0|x-x'|}. \end{aligned} \tag{21.107}$$

and (cf. equation (21.96))

$$\boldsymbol{P} = \int d^3x\left(\psi^+\frac{\hbar}{\mathrm{i}}\nabla\psi + \epsilon_0\boldsymbol{E}_\perp\times\boldsymbol{B}\right). \tag{21.108}$$

It is interesting to note that if we write the current density (21.103) as

$$\boldsymbol{j} = \boldsymbol{J} + \frac{q\hbar}{2m}\nabla\times\left(\psi^+\underline{\sigma}\psi\right) = \boldsymbol{J} + \frac{q}{e}\mu_B\nabla\times\left(\psi^+\underline{\sigma}\psi\right)$$

we can write Ampère's law with Maxwell's correction term as

$$\nabla\times\left(\boldsymbol{B} - \frac{q}{e}\mu_0\mu_B\psi^+\underline{\sigma}\psi\right) = \nabla\times\boldsymbol{B}_{class} = \mu_0\boldsymbol{J} + \mu_0\epsilon_0\frac{\partial}{\partial t}\boldsymbol{E},$$

i.e. the "spin density"

$$\boldsymbol{S}(x,t) = \frac{\hbar}{2}\psi^+(x,t)\underline{\sigma}\psi(x,t)$$

adds a spin magnetic field to the magnetic field $\boldsymbol{B}_{class}$ which is generated by orbital currents $\boldsymbol{J}$ and time-dependent electric fields $\boldsymbol{E}$,

$$\boldsymbol{B}(x,t) = \boldsymbol{B}_{class}(x,t) + \frac{2q}{e\hbar}\mu_0\mu_B\boldsymbol{S}(x,t) = \boldsymbol{B}_{class}(x,t) + \mu_0\frac{q}{m}\boldsymbol{S}(x,t).$$

Higher order terms and spin-orbit coupling

We will discuss higher order terms in the framework of relativistic quantum mechanics, i.e. our basic quantum operators are **x** and **p** etc., but not quantum fields. This also entails a semi-classical approximation for the electromagnetic fields and potentials.

For the discussion of higher order terms, we write the Dirac equation in Schrödinger form,

$$\mathrm{i}\hbar\frac{d}{dt}|\Upsilon(t)\rangle = H(t)|\Upsilon(t)\rangle,$$

with the Hamilton operator

$$H(t) = (\gamma^0 - 1)mc^2 + q\Phi(\mathbf{x},t) + c\,\boldsymbol{\alpha}\cdot[\mathbf{p} - q\mathbf{A}(\mathbf{x},t)]. \tag{21.109}$$

The operator $\boldsymbol{\alpha}$ is

$$\boldsymbol{\alpha} = \gamma^0\boldsymbol{\gamma}, \quad \alpha^i{}_{ab} = \langle a|\alpha^i|b\rangle = \gamma^0{}_{ac}\gamma^i{}_{cb},$$

and $\langle \boldsymbol{x}, a|\Upsilon(t)\rangle = \Upsilon_a(\boldsymbol{x},t)$ is the a-th component of the 4-spinor Υ (21.97) in $\boldsymbol{x}$ representation.

We continue to use the Dirac basis (21.36) of γ matrices in this section, such that as a matrix valued vector $\boldsymbol{\alpha}$ is given by

$$\boldsymbol{\alpha} = \begin{pmatrix} \underline{0} & \underline{\boldsymbol{\sigma}} \\ \underline{\boldsymbol{\sigma}} & \underline{0} \end{pmatrix}.$$

The part of the Hamiltonian (21.109) which mixes the upper and lower components of the 4-spinor Υ is

$$K(t) = c\,\boldsymbol{\alpha}\cdot[\mathbf{p} - q\mathbf{A}(\mathbf{x},t)].$$

Operators which mix upper and lower 2-spinors in 4-spinors are also denoted as *odd* terms in the Hamiltonian.

We can remove the odd contribution $K(t)$ by using the anti-hermitian operator

$$T(t) = \frac{\gamma^0}{2mc^2}K(t) = \frac{1}{2mc}\boldsymbol{\gamma}\cdot[\mathbf{p} - q\mathbf{A}(\mathbf{x},t)],$$

$$[T(t), \gamma^0 mc^2] = -K(t), \tag{21.110}$$

which implies subtraction of $K(t)$ from the new transformed Hamiltonian $\exp[T(t)]H(t)\exp[-T(t)]$. However, we also have to take into account that the transformed state $|\Upsilon_T(t)\rangle = \exp[T(t)]|\Upsilon(t)\rangle$ satisfies the equation

$$\mathrm{i}\hbar\frac{d}{dt}|\Upsilon_T(t)\rangle = \exp[T(t)]H(t)\exp[-T(t)]|\Upsilon(t)\rangle + \mathrm{i}\hbar\frac{d\exp[T(t)]}{dt}\exp[-T(t)]|\Upsilon(t)\rangle.$$

Therefore the transformed Hamiltonian is actually

$$\begin{aligned}
H_T(t) &= \exp[T(t)]H(t)\exp[-T(t)] - \mathrm{i}\hbar\exp[T(t)]\frac{d}{dt}\exp[-T(t)] \\
&= \sum_{n=0}^{\infty}\frac{1}{n!}\overset{n}{[T(t),H(t)]} - \mathrm{i}\hbar\sum_{n=0}^{\infty}\frac{1}{n!}\overset{n}{[T(t),d/dt]} \\
&= \sum_{n=0}^{\infty}\frac{1}{n!}\overset{n}{[T(t),H(t)]} + \mathrm{i}\hbar\sum_{n=1}^{\infty}\frac{1}{n!}\overset{n-1}{[\,T(t),dT(t)/dt]} \\
&= \sum_{n=0}^{\infty}\frac{1}{n!}\overset{n}{[T(t),H(t)]} - \frac{\mathrm{i}q\hbar}{2mc}\sum_{n=1}^{\infty}\frac{1}{n!}\overset{n-1}{[\,T(t),\boldsymbol{\gamma}\cdot\dot{\boldsymbol{A}}(t)]}.
\end{aligned}$$

We also wish to expand the Hamiltonian up to terms of order $(\mathcal{E}/mc^2)^3$, where $\mathcal{E}$ contains contributions from the kinetic energy of the particle and from its interactions with the electromagnetic fields.

Equation (21.110) implies

$$\begin{aligned}
H_T(t) &= (\gamma^0-1)mc^2 + q\Phi(t) + mc^2\sum_{n=2}^{4}\frac{1}{n!}\overset{n}{[T(t),\gamma^0]} \\
&\quad + \sum_{n=1}^{3}\frac{1}{n!}\overset{n}{[T(t),q\Phi(t)+K(t)]} - \frac{\mathrm{i}q\hbar}{2mc}\boldsymbol{\gamma}\cdot\dot{\boldsymbol{A}}(t) \\
&\quad - \frac{\mathrm{i}q\hbar}{2mc}\sum_{n=1}^{2}\frac{1}{(n+1)!}\overset{n}{[T(t),\boldsymbol{\gamma}\cdot\dot{\boldsymbol{A}}(t)]} + \mathcal{O}\left(\frac{\mathcal{E}}{mc^2}\right)^4.
\end{aligned}$$

The relevant commutators are

$$[T(t),q\Phi(t)] - \frac{\mathrm{i}q\hbar}{2mc}\boldsymbol{\gamma}\cdot\dot{\boldsymbol{A}}(t) = \frac{\mathrm{i}q\hbar}{2mc}\boldsymbol{\gamma}\cdot\boldsymbol{E}(\mathbf{x},t),$$

$$\begin{aligned}
\overset{2}{[T(t),q\Phi(t)]} - \frac{\mathrm{i}q\hbar}{2mc}[T(t),\boldsymbol{\gamma}\cdot\dot{\boldsymbol{A}}(t)] &= \frac{\mathrm{i}q\hbar}{2mc}[T(t),\boldsymbol{\gamma}\cdot\boldsymbol{E}(t)] \\
= -\frac{q\hbar^2}{4m^2c^2}\nabla\cdot\boldsymbol{E}(\mathbf{x},t) - \mathrm{i}\frac{q\hbar^2}{4m^2c^2}(\nabla\times\boldsymbol{E}(\mathbf{x},t))\cdot&\begin{pmatrix}\underline{\sigma} & \underline{0}\\ \underline{0} & \underline{\sigma}\end{pmatrix} \\
-\frac{q\hbar}{2m^2c^2}(\boldsymbol{E}(\mathbf{x},t)\times[\mathbf{p}-q\boldsymbol{A}(\mathbf{x},t)])\cdot&\begin{pmatrix}\underline{\sigma} & \underline{0}\\ \underline{0} & \underline{\sigma}\end{pmatrix},
\end{aligned}$$

$$[T(t),K(t)] = \frac{\gamma^0}{mc^2}K^2(t) = \frac{\gamma^0}{m}[\mathbf{p}-q\boldsymbol{A}(\mathbf{x},t)]^2 - \frac{q\hbar}{m}\boldsymbol{B}(\mathbf{x},t)\cdot\begin{pmatrix}\underline{\sigma} & \underline{0}\\ \underline{0} & -\underline{\sigma}\end{pmatrix},$$

$$\overset{2}{[T(t),K(t)]} = -\frac{1}{m^2c^4}K^3(t), \quad \overset{3}{[T(t),K(t)]} = -\frac{\gamma^0}{m^3c^6}K^4(t).$$

and

$$mc^2\,\overset{2}{[T(t),\gamma^0]} = [K(t),T(t)] = -\frac{\gamma^0}{mc^2}K^2(t),$$

$$mc^2\,\overset{3}{[T(t),\gamma^0]} = \frac{1}{m^2c^4}K^3(t), \quad mc^2\,\overset{4}{[T(t),\gamma^0]} = \frac{\gamma^0}{m^3c^6}K^4(t).$$

We don't need to evaluate the final higher order commutator

$$C^{(3)}_{odd}(t) = \overset{3}{[T(t),q\Phi(t)]} - \frac{\mathrm{i}q\hbar}{2mc}\,\overset{2}{[T(t),\boldsymbol{\gamma}\cdot\dot{\boldsymbol{A}}(t)]},$$

because this is an odd term of order $(\mathcal{E}/mc^2)^3$, which is eliminated in the next step through a unitary transformation, to which it contributes in order $(\mathcal{E}/mc^2)^4$. We only need to observe that $C^{(3)}_{odd}(t)$ contains one term proportional to $\boldsymbol{\gamma}$, and other terms proportional to

$$\gamma^i\cdot\begin{pmatrix}\underline{\sigma}^j & \underline{0}\\ \underline{0} & \underline{\sigma}^j\end{pmatrix} = \delta^{ij}\begin{pmatrix}\underline{0} & \underline{1}\\ -\underline{1} & \underline{0}\end{pmatrix} + \mathrm{i}\epsilon^{ijk}\gamma_k,$$

such that $\{\gamma^0, C^{(3)}_{odd}(t)\} = 0$. This will become relevant for the elimination of $C^{(3)}_{odd}(t)$ in the next step.

However, for now our transformed Hamiltonian is

$$\begin{aligned}
H_T(t) = {} & (\gamma^0-1)mc^2 + q\Phi(t) + \frac{\gamma^0}{2mc^2}K^2(t) - \frac{\gamma^0}{8m^3c^6}K^4(t)\\
& - \frac{q\hbar^2}{8m^2c^2}\nabla\cdot\boldsymbol{E}(t) - \mathrm{i}\frac{q\hbar^2}{8m^2c^2}(\nabla\times\boldsymbol{E}(t))\cdot\begin{pmatrix}\underline{\boldsymbol{\sigma}} & \underline{0}\\ \underline{0} & \underline{\boldsymbol{\sigma}}\end{pmatrix}\\
& - \frac{q\hbar}{4m^2c^2}(\boldsymbol{E}(t)\times[\mathbf{p}-q\boldsymbol{A}(\mathbf{x},t)])\cdot\begin{pmatrix}\underline{\boldsymbol{\sigma}} & \underline{0}\\ \underline{0} & \underline{\boldsymbol{\sigma}}\end{pmatrix}\\
& + \frac{\mathrm{i}q\hbar}{2mc}\boldsymbol{\gamma}\cdot\boldsymbol{E}(t) - \frac{1}{3m^2c^4}K^3(t) + \frac{1}{6}C^{(3)}_{odd}(t) + \mathcal{O}\!\left(\frac{\mathcal{E}}{mc^2}\right)^4.
\end{aligned}$$

The last line contains three odd contributions

$$L(t) = \frac{\mathrm{i}q\hbar}{2mc}\boldsymbol{\gamma}\cdot\boldsymbol{E}(t) - \frac{1}{3m^2c^4}K^3(t) + \frac{1}{6}C^{(3)}_{odd}(t),$$

which we can eliminate exactly as in the previous step by using a unitary transformation $|\Upsilon_{WT}(t)\rangle = \exp[W(t)]|\Upsilon_T(t)\rangle$ with

$$W(t) = \frac{\gamma^0}{2mc^2}L(t) = \frac{\mathrm{i}q\hbar}{4m^2c^3}\boldsymbol{\alpha}\cdot\boldsymbol{E}(t) - \frac{\gamma^0}{6m^3c^6}K^3(t) + \mathcal{O}\left(\frac{\mathcal{E}}{mc^2}\right)^4.$$

This yields a new Hamiltonian

$$\begin{aligned} H_{WT}(t) &= \exp[W(t)]H_T(t)\exp[-W(t)] - \mathrm{i}\hbar\exp[W(t)]\frac{d}{dt}\exp[-W(t)] \\ &= \sum_{n=0}^{\infty}\frac{1}{n!}\,{}^{n}[W(t), H_T(t)] + \mathrm{i}\hbar\sum_{n=1}^{\infty}\frac{1}{n!}\,{}^{n-1}[\,W(t), dW(t)/dt], \end{aligned}$$

which is in the required order

$$\begin{aligned} H_{WT}(t) &= (\gamma^0 - 1)mc^2 + q\Phi(t) + \frac{\gamma^0}{2mc^2}K^2(t) - \frac{\gamma^0}{8m^3c^6}K^4(t) \\ &\quad - \frac{q\hbar^2}{8m^2c^2}\nabla\cdot\boldsymbol{E}(t) - \mathrm{i}\frac{q\hbar^2}{8m^2c^2}(\nabla\times\boldsymbol{E}(t))\cdot\begin{pmatrix}\underline{\sigma} & \underline{0}\\ \underline{0} & \underline{\sigma}\end{pmatrix} \\ &\quad - \frac{q\hbar}{4m^2c^2}(\boldsymbol{E}(t)\times[\mathbf{p} - q\boldsymbol{A}(\mathbf{x},t)])\cdot\begin{pmatrix}\underline{\sigma} & \underline{0}\\ \underline{0} & \underline{\sigma}\end{pmatrix} \\ &\quad - \frac{q\gamma^0}{6m^3c^6}[K^3(t), \Phi(t)] + \frac{\mathrm{i}q\hbar}{8m^3c^5}[\boldsymbol{\alpha}\cdot\boldsymbol{E}(t), \gamma^0K^2(t)] \\ &\quad + \mathrm{i}\hbar\frac{dW(t)}{dt} + \mathcal{O}\left(\frac{\mathcal{E}}{mc^2}\right)^4. \end{aligned}$$

This contains again an odd piece

$$M(t) = \mathrm{i}\hbar\frac{dW(t)}{dt} - \frac{q\gamma^0}{6m^3c^6}[K^3(t), \Phi(t)] + \frac{\mathrm{i}q\hbar}{8m^3c^5}[\boldsymbol{\alpha}\cdot\boldsymbol{E}(t), \gamma^0K^2(t)]$$

which is eliminated by another unitary transformation of the form $|\Upsilon_{FWT}(t)\rangle = \exp[F(t)]|\Upsilon_{WT}(t)\rangle$ with

$$F(t) = \frac{\gamma^0}{2mc^2}M(t) = -\frac{q\hbar^2}{8m^3c^5}\boldsymbol{\gamma}\cdot\dot{\boldsymbol{E}}(t) + \mathcal{O}\left(\frac{\mathcal{E}}{mc^2}\right)^4.$$

The resulting Hamiltonian after this transformation still contains an odd piece

$$N(t) = -\mathrm{i}\frac{q\hbar^3}{8m^3c^5}\boldsymbol{\gamma}\cdot\ddot{\boldsymbol{E}}(t)$$

which is eliminated in a final transformation

$$G(t) = \frac{\gamma^0}{2mc^2} N(t) = \mathcal{O}\left(\frac{\mathcal{E}}{mc^2}\right)^4 .$$

Therefore up to terms of order $\mathcal{O}(\mathcal{E}/mc^2)^4$, we finally find an equation which is diagonal in upper and lower 2-spinors

$$|\Upsilon_{FWT}(t)\rangle = \exp[F(t)]\exp[W(t)]\exp[T(t)]|\Upsilon(t)\rangle, \tag{21.111}$$

$$\mathrm{i}\hbar \frac{d}{dt}|\Upsilon_{FWT}(t)\rangle = H_{FWT}(t)|\Upsilon_{FWT}(t)\rangle \tag{21.112}$$

with

$$\begin{aligned} H_{FWT}(t) = {} & (\gamma^0 - 1)mc^2 + q\Phi(t) + \frac{\gamma^0}{2mc^2}K^2(t) - \frac{\gamma^0}{8m^3c^6}K^4(t) \\ & - \frac{q\hbar^2}{8m^2c^2}\nabla\cdot\boldsymbol{E}(t) - \mathrm{i}\frac{q\hbar^2}{8m^2c^2}\left(\nabla\times\boldsymbol{E}(t)\right)\cdot\begin{pmatrix}\underline{\boldsymbol{\sigma}} & \underline{0} \\ \underline{0} & \underline{\boldsymbol{\sigma}}\end{pmatrix} \\ & - \frac{q\hbar}{4m^2c^2}\left(\boldsymbol{E}(t)\times[\mathbf{p} - q\boldsymbol{A}(\mathbf{x},t)]\right)\cdot\begin{pmatrix}\underline{\boldsymbol{\sigma}} & \underline{0} \\ \underline{0} & \underline{\boldsymbol{\sigma}}\end{pmatrix}. \end{aligned} \tag{21.113}$$

The transformation (21.111, 21.113) is known as a Foldy-Wouthuysen transformation[8].

The Hamiltonian acting on the upper 2-spinor is

$$\begin{aligned} H(t) = {} & \frac{[\mathbf{p} - q\boldsymbol{A}(\mathbf{x},t)]^2}{2m} + q\Phi(\mathbf{x},t) - \frac{q\hbar}{2m}\boldsymbol{B}(\mathbf{x},t)\cdot\underline{\boldsymbol{\sigma}} - \frac{q\hbar^2}{8m^2c^2}\nabla\cdot\boldsymbol{E}(\mathbf{x},t) \\ & - \frac{q\hbar}{8m^2c^2}\left(\mathrm{i}\hbar\nabla\times\boldsymbol{E}(\mathbf{x},t) + 2\boldsymbol{E}(\mathbf{x},t)\times[\mathbf{p} - q\boldsymbol{A}(\mathbf{x},t)]\right)\cdot\underline{\boldsymbol{\sigma}} \\ & - \frac{1}{8m^3c^2}\left([\mathbf{p} - q\boldsymbol{A}(\mathbf{x},t)]^2 - q\hbar\boldsymbol{B}(\mathbf{x},t)\cdot\underline{\boldsymbol{\sigma}}\right)^2 . \end{aligned} \tag{21.114}$$

The first three terms are again the Pauli Hamiltonian from (21.101).

It is of interest to write some of the higher order terms in the Hamiltonian (21.114) also in terms of the charge density $\varrho(\boldsymbol{x},t)$ which generates the electromagnetic fields.

The term

$$-\frac{q\hbar^2}{8m^2c^2}\nabla\cdot\boldsymbol{E}(\mathbf{x},t) = -\frac{q\hbar^2}{8m^2c^2\epsilon_0}\varrho(\mathbf{x},t) \tag{21.115}$$

[8] L.L. Foldy, S.A. Wouthuysen, Phys. Rev. 78, 29 (1950).

amounts to a contact interaction between the particles described by equation (21.112) (e.g. electrons) and the particles which generate the electromagnetic fields. This term is known as the Darwin term. The contact interaction has the counter-intuitive property to lower the interaction energy between like charges, but recall that it emerged from eliminating the anti-particle components up to terms of order $\mathcal{O}(\mathcal{E}/mc^2)^4$. It should not surprise us that a positronic component in electron wave functions contributes an attractive term to the electron-electron interaction. The Hamiltonian (21.114) is in excellent agreement with spectroscopy if radiative corrections are also taken into account, see e.g. [18].

The term

$$-\mathrm{i}\frac{q\hbar^2}{8m^2c^2}\big(\nabla\times\boldsymbol{E}(\mathbf{x},t)\big)\cdot\underline{\sigma}=\mathrm{i}\frac{q\hbar}{4m^2c^2}\dot{\boldsymbol{B}}(\mathbf{x},t)\cdot\underline{S} \tag{21.116}$$

is apparently a coupling between spin $\underline{S}=\hbar\underline{\sigma}/2$ and induced potentials from time-dependent charge-current distributions.

In the static case we can write the $\boldsymbol{E}(\mathbf{x})\times\mathbf{p}$ term in (21.114) in the form

$$-\frac{q}{2m^2c^2}\big(\boldsymbol{E}(\mathbf{x})\times\mathbf{p}\big)\cdot\underline{S}=-\frac{\mu_0 q}{8\pi m^2}\int d^3x'\,\frac{\varrho(\boldsymbol{x}')}{|\mathbf{x}-\boldsymbol{x}'|^3}\boldsymbol{M}(\mathbf{x}-\boldsymbol{x}',\mathbf{p})\cdot\underline{S}.$$

Here

$$\boldsymbol{M}(\mathbf{x}-\boldsymbol{x}',\mathbf{p})=(\mathbf{x}-\boldsymbol{x}')\times\mathbf{p}$$

is the orbital angular momentum operator with respect to the point $\boldsymbol{x}'$, and the $\boldsymbol{E}(\mathbf{x})\times\mathbf{p}$ term apparently contains a charge weighted sum over angular momentum operators. The $\boldsymbol{E}(\mathbf{x})\times\mathbf{p}$ term is therefore the origin of spin-orbit coupling. In particular, for a radially symmetric charge distribution

$$\boldsymbol{E}(\boldsymbol{x})=-\frac{\boldsymbol{x}}{r}\frac{d\Phi(r)}{dr}$$

one finds

$$-\frac{q}{2m^2c^2}\big(\boldsymbol{E}(\mathbf{x})\times\mathbf{p}\big)\cdot\underline{S}=\frac{q}{2m^2c^2r}\frac{d\Phi(r)}{dr}\boldsymbol{M}\cdot\underline{S}. \tag{21.117}$$

This implies equation (8.20) for spin-orbit coupling in hydrogen atoms.

So far we have emphasized the emergence of $\boldsymbol{M}\cdot\boldsymbol{S}$ terms from the $\boldsymbol{E}(\mathbf{x})\times\mathbf{p}$ term, and historically the coupling of spin and orbital angular momentum had provided the initial motivation for the designation as spin-orbit coupling term. However, the direct coupling of orbital momentum $\boldsymbol{p}$ and spin provides just as good a reason for the name spin-orbit coupling, and another important special case of the $\boldsymbol{E}(\mathbf{x})\times\mathbf{p}$ term arises for a uni-directional electric field e.g. in z direction. In this case the term takes the form

$$-\frac{q}{2m^2c^2}\big(\boldsymbol{E}(\mathbf{x})\times\mathbf{p}\big)\cdot\boldsymbol{S}=-\frac{q}{2m^2c^2}E_z(\mathrm{z})\left(\mathrm{p}_xS_y-\mathrm{p}_yS_x\right). \tag{21.118}$$

For homogeneous electric field this yields a spin-orbit coupling term of the form $\alpha_R(p_xS_y - p_yS_x)$ with constant α_R. This particular form of a spin-orbit coupling term is known as a Rashba term[9]. Spin-orbit coupling was always relevant not only for atomic and molecular spectroscopy, but also for electronic energy band structure in materials where they are often significantly enhanced e.g. due to low effective masses. In recent years spin-orbit coupling terms in low-dimensional systems, and Rashba terms in particular, have also attracted a lot of interest because of their relevance for spintronics[10].

21.6 Covariant quantization of the Maxwell field

We have seen in Section 18.2 how to quantize the Maxwell field and describe photons in Coulomb gauge. This is useful if our problem contains non-relativistic charged particles, since the Hamiltonian in Coulomb gauge conveniently describes the electromagnetic interaction between the charged particles through Coulomb terms. The free interaction picture photon operators $\boldsymbol{A}(\boldsymbol{x},t)$ or the corresponding Schrödinger picture operators $\boldsymbol{A}(\boldsymbol{x})$ are then only needed for the calculation of absorption, emission or scattering of external photons. Exchange of virtual photons provides only small corrections to Coulomb interactions for non-relativistic charged particles. The relevant Hamiltonian is (21.107) with Schrödinger fields and Coulomb terms for all the different kinds of charged particles.

Coulomb gauge can also be used for problems involving relativistic fermions. These can be described by the Hamiltonian (21.94) including Dirac fields and Coulomb interaction terms for all the different kinds of spin-$1/2$ particles in the problem. Indeed, we will calculate basic scattering processes involving relativistic charged particles in Sections 22.2 and 22.4 in Coulomb gauge, and the calculations will explicitly show how the Coulomb interaction terms between charged particles dominate over photon exchange terms if the kinetic energies of the charged particles are small compared to their rest energies, see in particular equation (22.29).

However, if the problem indeed contains relativistic charged particles, then the interaction of those particles with other charged particles is more conveniently described through a covariant quantization of photons in Lorentz gauge,

$$\partial_\mu A^\mu(x) = 0. \tag{21.119}$$

[9]E.I. Rashba, Sov. Phys. Solid State 2, 1109 (1960); Yu.A. Bychkov, E.I. Rashba, JETP Lett. 39, 78 (1984); J. Phys. C 17, 6039 (1984).

[10]See e.g. J. Nitta, T. Akazaki, H. Takayanagi, T. Enoki, Phys. Rev. Lett. 78, 1335 (1997); D. Grundler, Phys. Rev. Lett. 84, 6074 (2000); J. Sinova *et al.*, Phys. Rev. Lett. 92, 126603 (2004); E.Y. Sherman, D.J. Lockwood, Phys. Rev. B 72, 125340 (2005); K.C. Hall *et al.*, Appl. Phys. Lett. 86, 202114 (2005); P. Pietiläinen, T. Chakraborty, Phys. Rev. B 73, 155315 (2006); E. Cappelluti, C. Grimaldi, F. Marsiglio, Phys. Rev. Lett. 98, 167002 (2007).

Suppose the potential $\mathcal{A}_\mu(x)$ does not satisfy the Lorentz gauge condition. We can construct the Lorentz gauge vector potential $A^\mu(x)$ by performing a gauge transformation

$$A_\mu(x) = \mathcal{A}_\mu(x) + \partial_\mu f(x) \tag{21.120}$$

with

$$\begin{aligned} f(x) &= \int d^4x'\, G_d^{(m=0)}(x-x')\partial'_\mu \mathcal{A}^\mu(x') \\ &= \int d^3\boldsymbol{x}' \frac{1}{4\pi|\boldsymbol{x}-\boldsymbol{x}'|}\partial'_\mu \mathcal{A}^\mu(x')\Big|_{ct'=ct-|\boldsymbol{x}-\boldsymbol{x}'|}. \end{aligned} \tag{21.121}$$

Here

$$G_d^{(m=0)}(x) = \frac{1}{c}G_d^{(r,m=0)}(\boldsymbol{x},t) = \frac{1}{4\pi r}\delta(r-ct)$$

is the retarded massless scalar Green's function, cf. (J.37, J.60). This also helps us to solve Maxwell's equations in Lorentz gauge,

$$\partial_\mu\partial^\mu A^\nu(x) = -\mu_0 j^\nu(x) \tag{21.122}$$

in the form

$$A^\mu(x) = A^\mu_{LW}(x) + A^\mu_D(x), \tag{21.123}$$

where the Liénard-Wiechert potentials

$$\begin{aligned} A^\mu_{LW}(x) &= \mu_0 \int d^4x'\, G_d^{(m=0)}(x-x')j^\mu(x') \\ &= \frac{\mu_0}{4\pi}\int d^3\boldsymbol{x}'\, \frac{1}{|\boldsymbol{x}-\boldsymbol{x}'|}j^\mu\big(\boldsymbol{x}', ct-|\boldsymbol{x}-\boldsymbol{x}'|\big) \end{aligned} \tag{21.124}$$

solve the inhomogeneous Maxwell equations (21.122) and satisfy the Lorentz gauge condition due to charge conservation. The remainder $A^\mu_D(x)$ must therefore satisfy

$$\partial_\mu\partial^\mu A^\nu_D(x) = 0, \quad \partial_\mu A^\mu_D(x) = 0. \tag{21.125}$$

To quantize this, we observe that Maxwell's equations in Lorentz gauge follow from the Lagrange density of electromagnetic fields (18.1) if we take into account the Lorentz gauge condition,

$$\mathcal{L} = A_\mu j^\mu - \frac{1}{2\mu_0}\partial_\nu A_\mu \cdot \partial^\nu A^\mu. \tag{21.126}$$

This yields canonically conjugate momentum fields for all components of the vector potential,

$$\Pi_\mu = \frac{\partial\mathcal{L}}{\partial\dot{A}^\mu} = \epsilon_0\dot{A}_\mu.$$

The principles of canonical quantization and the Lorentz gauge condition then motivate the following quantization condition for electromagnetic potentials in Lorentz gauge (with $k^0 = |\boldsymbol{k}|$),

$$[A_\mu(x), \dot{A}_\nu(x')]_{t=t'} = \frac{\mathrm{i}\hbar}{\epsilon_0}\int d^3\boldsymbol{k}\left(\eta_{\mu\nu} - \frac{k_\mu k_\nu}{k^2 - \mathrm{i}\epsilon}\right)\frac{\exp[\mathrm{i}\boldsymbol{k}\cdot(\boldsymbol{x}-\boldsymbol{x}')]}{(2\pi)^3}.$$

The general solution of (21.125)

$$A_D^\mu(x) = \langle x, \mu|A_D\rangle = \sqrt{\frac{\hbar\mu_0 c}{(2\pi)^3}}$$

$$\times\int\frac{d^3\boldsymbol{k}}{\sqrt{2|\boldsymbol{k}|}}\sum_{\alpha=1}^{3}\epsilon_\alpha^\mu(\boldsymbol{k})\Big(a_\alpha(\boldsymbol{k})\exp(\mathrm{i}k\cdot x) + a_\alpha^+(\boldsymbol{k})\exp(-\mathrm{i}k\cdot x)\Big), \quad (21.127)$$

$$k\cdot\epsilon_\alpha(\boldsymbol{k}) = 0, \quad \sum_{\alpha=1}^{3}\epsilon_\alpha^\mu(\boldsymbol{k})\epsilon_\alpha^\nu(\boldsymbol{k}) = \eta^{\mu\nu} - \frac{k^\mu k^\nu}{k^2 - \mathrm{i}\epsilon},$$

satisfies the quantization condition if

$$[a_\alpha(\boldsymbol{k}), a_\beta^+(\boldsymbol{k}')] = \delta_{\alpha\beta}\delta(\boldsymbol{k}-\boldsymbol{k}') \quad (21.128)$$

and the other commutators vanish.

A possible choice for the polarization vectors $\epsilon_\alpha^\mu(\boldsymbol{k})$ is e.g. to choose $\epsilon_1^\mu(\boldsymbol{k})$ and $\epsilon_2^\mu(\boldsymbol{k})$ as spatial orthonormal vectors without time-like components and perpendicular to $\boldsymbol{k}$ such that

$$\sum_{\alpha=1}^{2}\boldsymbol{\epsilon}_\alpha(\boldsymbol{k})\otimes\boldsymbol{\epsilon}_\alpha(\boldsymbol{k}) = \underline{1} - \hat{\boldsymbol{k}}\otimes\hat{\boldsymbol{k}},$$

and choose

$$\epsilon_3(\boldsymbol{k}) = \frac{(|\boldsymbol{k}|, k^0\hat{\boldsymbol{k}})}{\sqrt{-k^2 + \mathrm{i}\epsilon}}.$$

This formalism can be motivated as a limiting case of the quantization of massive vector fields, and it has the advantage of faster and easier calculation of scattering amplitudes involving electromagnetic interactions of relativistic charged particles, because there are no separate amplitudes for photon exchange and Coulomb interactions, which need to be added to give the full scattering amplitude. The spatially longitudinal photons generated by $a_3^+(\boldsymbol{k})$, $\boldsymbol{k}\cdot\boldsymbol{\epsilon}_3(\boldsymbol{k}) = k^0\epsilon_3^0(\boldsymbol{k}) \neq 0$, incorporate the contributions from the Coulomb interactions[11]. Why then don't we see photon states $a_3^+(\boldsymbol{k})|0\rangle$? These photon states are actually spurious gauge degrees of freedom. We could perform another gauge transformation

$$A_\mu(x) \to \tilde{A}_\mu(x) = A_\mu(x) + \partial_\mu g(x) \tag{21.129}$$

with

$$\begin{aligned} g(\boldsymbol{x},t) &= \int_{-\infty}^{t} dt' \left(cA^0(\boldsymbol{x},t') - \int d^3x' \frac{\varrho(\boldsymbol{x}',t')}{4\pi\epsilon_0|\boldsymbol{x}-\boldsymbol{x}'|} \right) \\ &+ \int d^3x' \frac{\nabla'\cdot\boldsymbol{A}(\boldsymbol{x}',-\infty)}{4\pi|\boldsymbol{x}-\boldsymbol{x}'|} = \int d^3x' \frac{\nabla'\cdot\boldsymbol{A}(\boldsymbol{x}',t)}{4\pi|\boldsymbol{x}-\boldsymbol{x}'|}, \end{aligned} \tag{21.130}$$

which takes us right back to Coulomb gauge,

$$\tilde{A}^0(x) = \int d^3x' \frac{\varrho(\boldsymbol{x}',t)c}{4\pi\epsilon_0|\boldsymbol{x}-\boldsymbol{x}'|}, \quad \nabla\cdot\tilde{\boldsymbol{A}}(x) = 0,$$

without any freely oscillating time-like component. Since $a_3^+(\boldsymbol{k})|0\rangle$ was the only photon state with a time-like component, (21.129) has removed that photon state. We can think of the photons with longitudinal spatial components and corresponding time-like components as virtual place holders for the Coulomb interaction.

21.7 Problems

21.1. Show that for an appropriate class of integration contours C in the complex k^0 plane the scalar propagator (21.9) can be written in the form

$$\mathcal{K}(x) = -\frac{1}{(2\pi)^4 c} \oint_C dk^0 \int d^3\boldsymbol{k} \frac{\exp(ik\cdot x)}{k^2 + (mc/\hbar)^2}.$$

[11]Of course, this implies that one cannot naively invoke Hamiltonians with Coulomb interaction terms if we describe photons in Lorentz gauge. Otherwise we would overcount interactions. Remember that the Coulomb interaction terms came from the contributions to Hamiltonians from electromagnetic fields *in Coulomb gauge*, see Section 21.4.

21.2. We have discussed the non-relativistic limit of the Klein-Gordon field in the case $|\langle b^+(\boldsymbol{k})\rangle| \ll |\langle a(\boldsymbol{k})\rangle|$. However, there must also exist a non-relativistic limit for the anti-particles. How does the non-relativistic limit work in the case of negligible particle amplitude $|\langle a(\boldsymbol{k})\rangle| \ll |\langle b^+(\boldsymbol{k})\rangle|$?

21.3. Derive the energy density $\mathcal{H}$ and the momentum density $\mathcal{P}$ for the real Klein-Gordon field.

21.4. Calculate the non-relativistic limits for the Hamilton operator H and the momentum operator $\boldsymbol{P}$ of the real Klein-Gordon field.

21.5. Derive the energy-momentum tensor for QED with scalar matter (21.2),

$$\begin{aligned} T_\nu{}^\mu = &-\eta_\nu{}^\mu\hbar c^2\left(\partial_\rho\phi^+ + \mathrm{i}\frac{q}{\hbar}\phi^+A_\rho\right)\left(\partial^\rho\phi - \mathrm{i}\frac{q}{\hbar}A^\rho\phi\right) - \eta_\nu{}^\mu\frac{m^2c^4}{\hbar}\phi^+\phi \\ &-\eta_\nu{}^\mu\frac{1}{4\mu_0}F_{\rho\sigma}F^{\rho\sigma} + \hbar c^2\left(\partial_\nu\phi^+ + \mathrm{i}\frac{q}{\hbar}\phi^+A_\nu\right)\left(\partial^\mu\phi - \mathrm{i}\frac{q}{\hbar}A^\mu\phi\right) \\ &+\hbar c^2\left(\partial^\mu\phi^+ + \mathrm{i}\frac{q}{\hbar}\phi^+A^\mu\right)\left(\partial_\nu\phi - \mathrm{i}\frac{q}{\hbar}A_\nu\phi\right) + \frac{1}{\mu_0}F_{\nu\rho}F^{\mu\rho}. \end{aligned} \tag{21.131}$$

The corresponding densities of energy, momentum, and energy current are

$$\begin{aligned} \mathcal{H} = T^{00} = &\frac{\epsilon_0}{2}\boldsymbol{E}^2 + \frac{1}{2\mu_0}\boldsymbol{B}^2 + \frac{m^2c^4}{\hbar}\phi^+\phi + \hbar\left(\dot{\phi}^+ - \mathrm{i}\frac{q}{\hbar}\phi^+\Phi\right)\left(\dot{\phi} + \mathrm{i}\frac{q}{\hbar}\Phi\phi\right) \\ &+\hbar c^2\left(\nabla\phi^+ + \mathrm{i}\frac{q}{\hbar}\phi^+\boldsymbol{A}\right)\cdot\left(\nabla\phi - \mathrm{i}\frac{q}{\hbar}\boldsymbol{A}\phi\right), \end{aligned} \tag{21.132}$$

$$\begin{aligned} \mathcal{P} = \frac{1}{c}\boldsymbol{e}_iT^{i0} = &\,\epsilon_0\boldsymbol{E}\times\boldsymbol{B} - \hbar\left(\dot{\phi}^+ - \mathrm{i}\frac{q}{\hbar}\phi^+\Phi\right)\left(\nabla\phi - \mathrm{i}\frac{q}{\hbar}\boldsymbol{A}\phi\right) \\ &-\hbar\left(\nabla\phi^+ + \mathrm{i}\frac{q}{\hbar}\phi^+\boldsymbol{A}\right)\left(\dot{\phi} + \mathrm{i}\frac{q}{\hbar}\Phi\phi\right), \end{aligned} \tag{21.133}$$

$$\mathcal{S} = c\boldsymbol{e}_iT^{0i} = c^2\mathcal{P}.$$

Solution. The Lagrange density for quantum electrodynamics with scalar matter is

$$\mathcal{L} = -\hbar c^2\left(\partial\phi^+ + \mathrm{i}\frac{q}{\hbar}\phi^+A\right)\cdot\left(\partial\phi - \mathrm{i}\frac{q}{\hbar}A\phi\right) - \frac{m^2c^4}{\hbar}\phi^+\cdot\phi - \frac{1}{4\mu_0}F_{\mu\nu}F^{\mu\nu}.$$

This yields according to (16.16) a conserved energy-momentum tensor

$$\Theta_\mu{}^\nu = \eta_\mu{}^\nu\mathcal{L} - \partial_\mu\phi^+\frac{\partial\mathcal{L}}{\partial(\partial_\nu\phi^+)} - \partial_\mu\phi\frac{\partial\mathcal{L}}{\partial(\partial_\nu\phi)} - \partial_\mu A_\lambda\frac{\partial\mathcal{L}}{\partial(\partial_\nu A_\lambda)}$$

$$= \eta_\mu{}^\nu \mathcal{L} + \hbar c^2 \partial_\mu \phi^+ \left(\partial^\nu \phi - \mathrm{i}\frac{q}{\hbar} A^\nu \phi\right) + \hbar c^2 \left(\partial^\nu \phi^+ + \mathrm{i}\frac{q}{\hbar} \phi^+ A^\nu\right) \partial_\mu \phi$$
$$+ \frac{1}{\mu_0} \partial_\mu A_\lambda F^{\nu\lambda}.$$

To find a gauge invariant energy-momentum tensor we add the identically conserved improvement term

$$-\frac{1}{\mu_0} \partial_\lambda \left(A_\mu F^{\nu\lambda}\right) = -\frac{1}{\mu_0} \partial_\lambda A_\mu F^{\nu\lambda} - \frac{1}{\mu_0} A_\mu \partial_\lambda F^{\nu\lambda}$$
$$= -\frac{1}{\mu_0} \partial_\lambda A_\mu F^{\nu\lambda} + \mathrm{i}qc^2 \left(\phi^+ A_\mu \cdot \partial^\nu \phi - \partial^\nu \phi^+ \cdot A_\mu \phi\right) + 2\frac{q^2 c^2}{\hbar} \phi^+ A_\mu A^\nu \phi,$$

where Maxwell's equations

$$\partial_\mu F^{\mu\nu} = -\mu_0 \frac{\partial \mathcal{L}}{\partial A_\nu} = \mathrm{i}\frac{q}{\epsilon_0} \left(\phi^+ \cdot \partial^\nu \phi - \partial^\nu \phi^+ \cdot \phi\right) + 2\frac{q^2}{\epsilon_0 \hbar} \phi^+ A^\nu \phi$$

were used. This yields the gauge invariant tensor (21.131) from

$$T_\mu{}^\nu = \Theta_\mu{}^\nu - \frac{1}{\mu_0} \partial_\lambda \left(A_\mu F^{\nu\lambda}\right).$$

21.6. If we write the solution (21.5) of the free Klein-Gordon equation as the sum of the positive and negative energy components,

$$\phi(x) = \phi_+(x) + \phi_-(x),$$
$$\phi_+(x) = \frac{1}{\sqrt{2\pi}^3} \int \frac{d^3\boldsymbol{k}}{\sqrt{2\omega_k}} a(\boldsymbol{k}) \exp[\mathrm{i}(\boldsymbol{k} \cdot \boldsymbol{x} - \omega_{\boldsymbol{k}} t)],$$
$$\phi_-(x) = \frac{1}{\sqrt{2\pi}^3} \int \frac{d^3\boldsymbol{k}}{\sqrt{2\omega_k}} b^+(\boldsymbol{k}) \exp[-\mathrm{i}(\boldsymbol{k} \cdot \boldsymbol{x} - \omega_{\boldsymbol{k}} t)],$$

the charge densities $\varrho_\pm = -\mathrm{i}q(\dot{\phi}_\pm^+ \cdot \phi_\pm - \phi_\pm^+ \cdot \dot{\phi}_\pm)$ are separately conserved, and therefore we can also identify conserved particle and anti-particle numbers

$$N_\pm = \pm \frac{Q_\pm}{q} = \pm \frac{1}{q} \int d^3\boldsymbol{x}\, \varrho_\pm(\boldsymbol{x}, t).$$

21.6a. Show that the conserved current density for QED with scalar matter (21.2) is

$$j_\mu = \mathrm{i}qc^2 \left(\partial_\mu \phi^+ \cdot \phi - \phi^+ \cdot \partial_\mu \phi + 2\mathrm{i}\frac{q}{\hbar} \phi^+ \cdot A_\mu \cdot \phi\right).$$

21.6b. Why is it not possible to derive separately conserved (anti-)particle numbers $N_\pm$ for the scalar particles in the interacting theory (21.2)?

21.7. Show that contrary to the case of spinor electrodynamics, it is not possible in relativistic scalar electrodynamics to derive a Coulomb gauge Hamiltonian (although Coulomb gauge can still be imposed on the Maxwell field, of course).

Why can we nevertheless find a Coulomb gauge Hamiltonian in the non-relativistic limit for the scalar fields?

Hint: The Gauss law of scalar electrodynamics in Coulomb gauge takes the form

$$\Delta\Phi = -\mathrm{i}\frac{q}{\epsilon_0}\left(\phi^+ \cdot \partial_t\phi - \partial_t\phi^+ \cdot \phi + 2\mathrm{i}\frac{q}{\hbar}\phi^+\Phi\phi\right).$$

21.8. Show that the junction conditions (21.28) are necessary and sufficient to ensure that the Klein-Gordon equation holds at the step of the potential.

21.9. Generalize the reasoning from Section 21.2 to the case of oblique incidence against the potential step, e.g. by considering a scalar boson running against the potential (21.21) with initial momentum components $\hbar k_x > 0$ and $\hbar k_y > 0$.

Remarks on the Solution. The *ansatz* for the Klein-Gordon wave function which complies with the boundary conditions on the incoming particle and the requirement of smoothness for all times t and values of y along the interface $x = 0$ is

$$\phi(x,y,t) = \begin{cases} [\exp(\mathrm{i}k_x x) + \beta\exp(-\mathrm{i}k_x x)]\exp[\mathrm{i}(k_y y - \omega t)], & x < 0, \\ \theta\exp[\mathrm{i}(\kappa_x x + k_y y - \omega t)], & x > 0. \end{cases}$$

The frequency follows again from the solution of the Klein-Gordon equation in the two domains,

$$\omega = c\sqrt{k_x^2 + k_y^2 + \frac{m^2c^2}{\hbar^2}} = \frac{V}{\hbar} \pm c\sqrt{\kappa_x^2 + k_y^2 + \frac{m^2c^2}{\hbar^2}}. \tag{21.134}$$

All other pertinent results follow also exactly as in Section 21.2 if we make the substitutions $k \to k_x$, $\kappa \to \kappa_x$ and $mc \to \sqrt{m^2c^2 + \hbar^2k_y^2}$. This applies in particular also to Table 21.1 and Figure 21.1. In particular, we have generation of pairs of particles and anti-particles in the energy range

$$c\sqrt{\hbar^2k_y^2 + m^2c^2} < \hbar\omega = c\sqrt{\hbar^2k_x^2 + \hbar^2k_y^2 + m^2c^2} < V - c\sqrt{\hbar^2k_y^2 + m^2c^2}$$

if the height of the potential step satisfies

$$V > 2c\sqrt{\hbar^2k_y^2 + m^2c^2}.$$

The wave number κ_x in this energy range is

$$\kappa_x = -\frac{1}{\hbar}\sqrt{\frac{(\hbar\omega - V)^2}{c^2} - m^2c^2 - \hbar^2k_y^2},$$

and writing the solution for $x > 0$ as

$$\phi(x, y, t) = \theta \exp[-\mathrm{i}(-\kappa_x x - k_y y + \omega t)], \quad x > 0,$$

shows that it is an anti-particle solution with energy $\overline{E}_p = -\hbar\omega$ and momentum components $-\hbar\kappa_x > 0$, $-\hbar k_y < 0$. The kinetic+rest energy of the generated anti-particles in the region $x > 0$ is

$$K_{\overline{p}} = c\sqrt{\hbar^2\kappa_x^2 + \hbar^2k_y^2 + m^2c^2} = V - \hbar\omega, \tag{21.135}$$

and the energy of the anti-particles is just the sum of their kinetic+rest energy and their potential energy, $\overline{E}_p = K_{\overline{p}} - V$.

21.10. Calculate the boson number operator

$$N_b = \int d^3\boldsymbol{k}\left(a^+(\boldsymbol{k})a(\boldsymbol{k}) + b^+(\boldsymbol{k})b(\boldsymbol{k})\right)$$

for the free Klein-Gordon field in $\boldsymbol{x}$ representation.

21.11. Show that scattering of a Klein-Gordon field off a hard sphere yields the same result (11.36) as the non-relativistic Schrödinger theory, except that the definition $k = \sqrt{2mE}/\hbar$ (where E is the kinetic energy of the scattered particle) is replaced by

$$k = \frac{1}{\hbar c}\sqrt{\hbar^2\omega^2 - m^2c^4}.$$

The hard sphere is taken into account through a boundary condition of vanishing Klein-Gordon field on the surface of the sphere, like the condition on the Schrödinger wave function in Section 11.3, i.e. we do not model it as a potential. We could think of the hard sphere in this case as arising from a hypothetical interaction which repels particles and anti-particles alike (just like gravity is equally attractive for particles and anti-particles).

21.12. You could also model an impenetrable wall for a Klein-Gordon field in the manner of the hard sphere of Problem 11. Which wave function for the Klein-Gordon field do you get if the impenetrable wall prevents the field from entering the region $x > 0$? Why does this result not contradict the Klein paradox?

21.13. Calculate the fermion number operator

$$N_f = \int d^3\boldsymbol{k} \sum_{s\in\{\downarrow,\uparrow\}} \left[b_s^+(\boldsymbol{k})b_s(\boldsymbol{k}) + d_s^+(\boldsymbol{k})d_s(\boldsymbol{k})\right]$$

for the free Dirac field in $\boldsymbol{x}$ representation.

21.14. Calculate the reflection and transmission coefficients for a Dirac field of charge q in the presence of a potential step $q\Phi(x) = V(x) = V\Theta(x)$. How do your results compare with the results for the Klein-Gordon field in Section 21.2?

21.15a. What are the non-relativistic limits of the spinor plane waves (21.72)?

21.15b. Verify the relations (21.73) for the relativistic spinor plane wave states.

21.16. Angular momentum in relativistic field theory

21.16a. Show that if $T_{\mu\nu}$ is a symmetric conserved energy momentum tensor, then the currents

$$\mathcal{M}_{\alpha\beta}{}^{\mu} = \frac{1}{c}\left(x_\alpha T_\beta{}^\mu - x_\beta T_\alpha{}^\mu\right) \tag{21.136}$$

are also conserved:

$$\partial_\mu \mathcal{M}_{\alpha\beta}{}^{\mu} = 0. \tag{21.137}$$

21.16b. The quantities $\mathcal{M}_{\alpha\beta}{}^{\mu}$ have the properties $\mathcal{M}_{\alpha\beta}{}^{0} = x_\alpha \mathcal{P}_\beta - x_\beta \mathcal{P}_\alpha$ and are therefore associated with angular momentum conservation and conservation of the center of energy motion (18.128, 18.129) in relativistic field theories. Show that invariance of the relativistic field theory

$$\begin{aligned}\mathcal{L} = &-\hbar c^2\left(\partial_\mu\phi^+ + \mathrm{i}\frac{Q}{\hbar}\phi^+ A_\mu\right)\cdot\left(\partial^\mu\phi - \mathrm{i}\frac{Q}{\hbar}A^\mu\phi\right) - \frac{m^2c^4}{\hbar}\phi^+\cdot\phi\\ &+ c\overline{\Psi}\left[\gamma^\mu\left(\frac{\mathrm{i}\hbar}{2}\overleftrightarrow{\partial}_\mu + qA_\mu\right) - mc\right]\Psi - \frac{1}{4\mu_0}F_{\mu\nu}F^{\mu\nu}.\end{aligned} \tag{21.138}$$

under rotations and Lorentz boosts

$$\epsilon^\mu = -\delta x^\mu = -\varphi^{\mu\nu}x_\nu, \quad \varphi^{\mu\nu} = -\varphi^{\nu\mu}, \tag{21.139}$$

$$\delta\phi(x) \equiv \phi'(x') - \phi(x) = 0, \quad \delta A^\mu(x) = \varphi^{\mu\nu}A_\nu(x), \tag{21.140}$$

$$\delta\Psi(x) = \frac{\mathrm{i}}{2}\varphi^{\alpha\beta}S_{\alpha\beta}\cdot\Psi(x), \quad \delta\overline{\Psi}(x) = -\frac{\mathrm{i}}{2}\varphi^{\alpha\beta}\,\overline{\Psi}(x)\cdot S_{\alpha\beta}, \tag{21.141}$$

yields the conservations laws (21.137) from the results of Section 16.2 if proper improvement terms are added. The Lorentz generators $S_{\alpha\beta}$ in the spinor representation are defined in equation (H.12).

21.16c. We have seen in the previous problem that invariance of the relativistic theory (21.138) under the rotations $\delta x^i = \varphi^{ij}x_j = \epsilon^{ijk}x_j\varphi_k$ yields densities of conserved charges $\mathcal{M}_{ij}{}^0$ which we can express in vector form through $\mathcal{M}_{ij}{}^0 = \epsilon_{ijk}\mathcal{J}_k$, $\mathcal{J}_i = \epsilon_{ijk}\mathcal{M}_{jk}{}^0/2$, i.e. $\mathcal{J} = \boldsymbol{x}\times\mathcal{P}$. On the other hand, we have seen in Problem 16.6 that the total angular momentum density $\mathcal{J}$ of non-relativistic fermions contains a spin term which is *not* proportional to any space-time coordinates x_α, and yet we have also seen in Problem 16.7 that only the *combination* of both terms in (16.26) yields the density of a conserved quantity in the presence of spin-orbit coupling. How can that be?

Replace time derivatives on spinor fields in the momentum density using the Dirac equation. This yields spin contributions to the momentum density. Show that partial integration of the resulting spin contributions to the angular momentum density yields spin terms which reduce to the spin term in equation (16.26) in the non-relativistic limit.

Solution for 16b. The electric current density for the Lagrange density (21.138) is

$$j_q^\mu = \frac{\partial\mathcal{L}}{\partial A_\mu} = -\mathrm{i}Qc^2\phi^+\left(\overleftrightarrow{\partial^\mu} - 2\mathrm{i}\frac{Q}{\hbar}A^\mu\right)\phi + qc\overline{\Psi}\gamma^\mu\Psi. \tag{21.142}$$

Addition of the identically conserved improvement term

$$-\frac{1}{\mu_0}\partial_\nu\left(x_\alpha A_\beta F^{\mu\nu}\right) = -\frac{1}{\mu_0}A_\beta F^\mu{}_\alpha - \frac{1}{\mu_0}x_\alpha\partial_\nu A_\beta\cdot F^{\mu\nu} - x_\alpha A_\beta j_q^\mu$$

to the conserved current (16.13) for the transformation (21.139–21.141) yields the gauge invariant conserved current

$$\begin{aligned}
j^\mu = & -\varphi^{\mu\nu}x_\nu\mathcal{L} + \varphi^{\alpha\beta}x_\alpha\left[\hbar c^2\left(\partial^\mu\phi^+ + \mathrm{i}\frac{Q}{\hbar}\phi^+A^\mu\right)\left(\partial_\beta\phi - \mathrm{i}\frac{Q}{\hbar}A_\beta\phi\right)\right. \\
& + \hbar c^2\left(\partial_\beta\phi^+ + \mathrm{i}\frac{Q}{\hbar}\phi^+A_\beta\right)\left(\partial^\mu\phi - \mathrm{i}\frac{Q}{\hbar}A^\mu\phi\right) \\
& \left. - \mathrm{i}\frac{\hbar c}{2}\overline{\Psi}\gamma^\mu\left(\overleftrightarrow{\partial_\beta} - 2\mathrm{i}\frac{q}{\hbar}A_\beta\right)\Psi + \frac{1}{\mu_0}F_{\beta\nu}F^{\mu\nu}\right] \\
& + \varphi^{\alpha\beta}\frac{\hbar c}{4}\overline{\Psi}\left(\gamma^\mu S_{\alpha\beta} + S_{\alpha\beta}\gamma^\mu\right)\Psi.
\end{aligned} \tag{21.143}$$

The divergence of the spinor contributions to this current density is

$$\partial_\mu j^\mu_\Psi = \varphi^{\alpha\beta}\frac{\hbar c}{4}\,\partial_\mu\!\left[\overline{\Psi}\left(\gamma^\mu S_{\alpha\beta}+S_{\alpha\beta}\gamma^\mu\right)\Psi\right]-\frac{c}{2}\varphi^{\alpha\beta}\overline{\Psi}\gamma_\alpha\left(\mathrm{i}\hbar\,\overleftrightarrow{\partial}_\beta+2qA_\beta\right)\Psi$$
$$-\frac{c}{2}\varphi^{\alpha\beta}x_\alpha\partial_\mu\left[\overline{\Psi}\gamma^\mu\left(\mathrm{i}\hbar\,\overleftrightarrow{\partial}_\beta+2qA_\beta\right)\Psi\right]. \tag{21.144}$$

The relation (21.85) implies that on-shell

$$\varphi^{\alpha\beta}\frac{\hbar c}{4}\,\partial_\mu\!\left[\overline{\Psi}\left(\gamma^\mu S_{\alpha\beta}+S_{\alpha\beta}\gamma^\mu\right)\Psi\right]-\frac{c}{2}\varphi^{\alpha\beta}\overline{\Psi}\gamma_\alpha\left(\mathrm{i}\hbar\,\overleftrightarrow{\partial}_\beta+2qA_\beta\right)\Psi=0,$$

and comparison of (21.82) and (21.86) implies that we can write the remaining part of $\partial_\mu j^\mu_\Psi$ in the form

$$\partial_\mu j^\mu_\Psi = -\frac{c}{4}\varphi^{\alpha\beta}x_\alpha\partial_\mu\left[\overline{\Psi}\gamma^\mu\left(\mathrm{i}\hbar\,\overleftrightarrow{\partial}_\beta+2qA_\beta\right)\Psi+\overline{\Psi}\gamma_\beta\left(\mathrm{i}\hbar\,\overleftrightarrow{\partial}^\mu+2qA^\mu\right)\Psi\right]$$
$$= -\frac{c}{4}\varphi^{\alpha\beta}\partial_\mu\left[x_\alpha\overline{\Psi}\gamma^\mu\left(\mathrm{i}\hbar\,\overleftrightarrow{\partial}_\beta+2qA_\beta\right)\Psi+x_\alpha\overline{\Psi}\gamma_\beta\left(\mathrm{i}\hbar\,\overleftrightarrow{\partial}^\mu+2qA^\mu\right)\Psi\right].$$

The conserved current (21.144) is therefore equivalent to the conserved current

$$j^\mu=\frac{1}{2}\varphi^{\alpha\beta}\left(x_\alpha T_\beta{}^\mu-x_\beta T_\alpha{}^\mu\right)=\frac{c}{2}\varphi^{\alpha\beta}\mathcal{M}_{\alpha\beta}{}^\mu, \tag{21.145}$$

with the symmetric stress-energy tensor for the Lagrange density (21.138) (cf. (21.87, 21.131))

$$T_\mu{}^\nu=\eta_\mu{}^\nu\mathcal{L}-\frac{c}{2}\overline{\Psi}\left[\frac{\mathrm{i}\hbar}{2}\gamma^\nu\,\overleftrightarrow{\partial}_\mu+\frac{\mathrm{i}\hbar}{2}\gamma_\mu\,\overleftrightarrow{\partial}^\nu+q\gamma^\nu A_\mu+q\gamma_\mu A^\nu\right]\Psi+\frac{1}{\mu_0}F_{\mu\lambda}F^{\nu\lambda}$$
$$+\hbar c^2\left(\partial_\mu\phi^++\mathrm{i}\frac{Q}{\hbar}\phi^+A_\mu\right)\left(\partial^\nu\phi-\mathrm{i}\frac{Q}{\hbar}A^\nu\phi\right)$$
$$+\hbar c^2\left(\partial^\nu\phi^++\mathrm{i}\frac{Q}{\hbar}\phi^+A^\nu\right)\left(\partial_\mu\phi-\mathrm{i}\frac{Q}{\hbar}A_\mu\phi\right). \tag{21.146}$$

Solution for 16c. We discuss the angular momentum densities in vector form, $\mathcal{J}_i=\epsilon_{ijk}\mathcal{M}_{jk}{}^0/2=\epsilon_{ijk}x_j\mathcal{P}_k$, with the momentum densities $\mathcal{P}_k=T_k{}^0/c$ (cf. (21.89, 21.133)),

$$\mathcal{P}=\frac{1}{2}\Psi^+\left[\frac{\hbar}{2\mathrm{i}}\overleftrightarrow{\nabla}-q\boldsymbol{A}\right]\Psi+\frac{1}{2}\overline{\Psi}\boldsymbol{\gamma}\left[\frac{\mathrm{i}\hbar}{2}\overleftrightarrow{\partial}_0-\frac{q}{c}\Phi\right]\Psi+\epsilon_0\boldsymbol{E}\times\boldsymbol{B}$$
$$-\hbar\left(\dot{\phi}^+-\mathrm{i}\frac{Q}{\hbar}\phi^+\Phi\right)\left(\nabla\phi-\mathrm{i}\frac{Q}{\hbar}\boldsymbol{A}\phi\right)$$
$$-\hbar\left(\nabla\phi^++\mathrm{i}\frac{Q}{\hbar}\phi^+\boldsymbol{A}\right)\left(\dot{\phi}+\mathrm{i}\frac{Q}{\hbar}\Phi\phi\right). \tag{21.147}$$

The γ matrices satisfy

$$\gamma_i \cdot \gamma_j = -\delta_{ij}\begin{pmatrix} \underline{1} & 0 \\ 0 & \underline{1} \end{pmatrix} + \frac{1}{2}\epsilon_{ijk}\epsilon_{kmn}\gamma_m \cdot \gamma_n = -\delta_{ij}\begin{pmatrix} \underline{1} & 0 \\ 0 & \underline{1} \end{pmatrix} + \frac{2}{\mathrm{i}\hbar}\epsilon_{ijk}S_k$$

with the vector of 4×4 spin matrices $\boldsymbol{S} = \mathrm{i}\hbar\boldsymbol{\gamma}\times\boldsymbol{\gamma}/4$, cf. (21.78). The Dirac equation then implies

$$\overline{\Psi}\boldsymbol{\gamma}\left[\frac{\mathrm{i}\hbar}{2}\overset{\leftrightarrow}{\partial}_0 - \frac{q}{c}\Phi\right]\Psi = \Psi^+\left[\frac{\hbar}{2\mathrm{i}}\overset{\leftrightarrow}{\nabla} - q\boldsymbol{A}\right]\Psi + \nabla\times(\Psi^+\cdot\boldsymbol{S}\cdot\Psi),$$

and we can write the momentum density in the form

$$\begin{aligned}\mathcal{P} = {}& \Psi^+\left[\frac{\hbar}{2\mathrm{i}}\overset{\leftrightarrow}{\nabla} - q\boldsymbol{A}\right]\Psi + \frac{1}{2}\nabla\times(\Psi^+\cdot\boldsymbol{S}\cdot\Psi) + \epsilon_0\boldsymbol{E}\times\boldsymbol{B} \\ & - \hbar\left(\dot{\phi}^+ - \mathrm{i}\frac{Q}{\hbar}\phi^+\Phi\right)\left(\nabla\phi - \mathrm{i}\frac{Q}{\hbar}\boldsymbol{A}\phi\right) \\ & - \hbar\left(\nabla\phi^+ + \mathrm{i}\frac{Q}{\hbar}\phi^+\boldsymbol{A}\right)\left(\dot{\phi} + \mathrm{i}\frac{Q}{\hbar}\Phi\phi\right).\end{aligned} \tag{21.148}$$

The spin term $\mathcal{P}_S = \nabla\times(\Psi^+\cdot\boldsymbol{S}\cdot\Psi)/2$ in the momentum density contributes a term to the angular momentum of the form

$$\boldsymbol{J}_S = \int d^3x\,\boldsymbol{x}\times\frac{1}{2}[\nabla\times(\Psi^+\cdot\boldsymbol{S}\cdot\Psi)] = \int d^3x\,\Psi^+\cdot\boldsymbol{S}\cdot\Psi, \tag{21.149}$$

such that we can write the total angular momentum density also in the form

$$\mathcal{J} = \mathcal{M} + \mathcal{S}, \tag{21.150}$$

with a spin contribution $\mathcal{S} = \Psi^+\cdot\boldsymbol{S}\cdot\Psi$, and an orbital angular momentum $\mathcal{M} = \boldsymbol{x}\times\mathcal{P}_O$ with the orbital momentum density:

$$\begin{aligned}\mathcal{P}_O = {}& \Psi^+\left[\frac{\hbar}{2\mathrm{i}}\overset{\leftrightarrow}{\nabla} - q\boldsymbol{A}\right]\Psi - \hbar\left(\dot{\phi}^+ - \mathrm{i}\frac{Q}{\hbar}\phi^+\Phi\right)\left(\nabla\phi - \mathrm{i}\frac{Q}{\hbar}\boldsymbol{A}\phi\right) \\ & - \hbar\left(\nabla\phi^+ + \mathrm{i}\frac{Q}{\hbar}\phi^+\boldsymbol{A}\right)\left(\dot{\phi} + \mathrm{i}\frac{Q}{\hbar}\Phi\phi\right) + \epsilon_0\boldsymbol{E}\times\boldsymbol{B}.\end{aligned} \tag{21.151}$$

Substitution of the non-relativistic approximations in the Dirac representation of the γ matrices (cf. (21.19, 21.97)),

$$\phi(\boldsymbol{x},t) \to \sqrt{\frac{\hbar}{2mc^2}}\phi(\boldsymbol{x},t)\exp\!\left(-\mathrm{i}\frac{mc^2}{\hbar}t\right),$$

$$\Psi(\boldsymbol{x},t) = \begin{pmatrix}\psi(\boldsymbol{x},t) \\ \chi(\boldsymbol{x},t)\end{pmatrix}\exp\!\left(-\mathrm{i}\frac{mc^2}{\hbar}t\right)$$

into (21.148) yields after neglecting the subleading χ components (cf. (21.100))

$$\mathcal{P}_O = \psi^+ \left[\frac{\hbar}{2\mathrm{i}}\overleftrightarrow{\nabla} - qA\right]\psi + \phi^+ \left[\frac{\hbar}{2\mathrm{i}}\overleftrightarrow{\nabla} - QA\right]\phi + \epsilon_0 E \times B \tag{21.152}$$

and

$$\mathcal{S} = \frac{\hbar}{2}\psi^+ \cdot \underline{\sigma} \cdot \psi. \tag{21.153}$$

If we would not have used the equations and motion and equation (21.85) to write the conserved current in the form (21.145), the spin term $\mathcal{S} = \Psi^+ \cdot \boldsymbol{S} \cdot \Psi$ would have come from the last line in equation (21.143).

21.17. New charges from local phase invariance?

We have derived expressions for charge and current densities from phase invariance

$$\delta\Psi(x) = \mathrm{i}\frac{q}{\hbar}\varphi\Psi(x), \quad \delta\Psi^+(x) = -\mathrm{i}\frac{q}{\hbar}\varphi\Psi^+(x),$$

of Lagrange densities, see e.g. (16.31, 16.32) for the charge and current densities of non-relativistic charged matter fields, and (21.142) for relativistic charged matter fields. In the final expressions we always divided out the irrelevant constant parameter φ. However, introduction of the electromagnetic potentials rendered the Lagrange densities invariant under *local* phase transformations

$$\delta\Psi(x) = \mathrm{i}\frac{q}{\hbar}\varphi(x)\Psi(x), \quad \delta\Psi^+(x) = -\mathrm{i}\frac{q}{\hbar}\varphi(x)\Psi^+(x), \quad \delta A_\mu(x) = \partial_\mu\varphi(x).$$

In this case we cannot discard the phase parameter $\varphi(x)$ from the current densities for the local symmetry. Does this provide us with additional useful notions of conserved charges for quantum electronics and quantum electrodynamics?

21.17a. Show that application of the result (16.13) to local phase transformations yields current densities

$$J^\mu = \varphi j^\mu + \frac{1}{\mu_0}F^{\mu\nu}\partial_\nu\varphi, \tag{21.154}$$

where j^μ are the current densities which were derived for constant phase parameter φ, e.g. (16.31, 16.32) or (21.142).

Show that the currents J^μ can also be written in the *strong form*[12]

$$J^\mu = \frac{1}{\mu_0}\partial_\nu\left(\varphi F^{\mu\nu}\right). \tag{21.155}$$

[12] A current J^μ is sometimes denoted as *strongly conserved* if the local conservation law $\partial_\mu J^\mu = 0$ is an identity.

21.17b. Show in particular that the charge density $\mathcal{Q}_\varphi = J^0/c$ can be written in the form

$$\mathcal{Q}_\varphi = \epsilon_0 \nabla \cdot (\varphi \boldsymbol{E}), \tag{21.156}$$

and that the current density is

$$\boldsymbol{J} = \frac{1}{\mu_0} \nabla \times (\varphi \boldsymbol{B}) - \epsilon_0 \partial_t (\varphi \boldsymbol{E}). \tag{21.157}$$

Apply these results to a static charge distribution $\varrho(\boldsymbol{x}) = j^0(\boldsymbol{x})/c$. The charge

$$Q_\varphi = \int d^3x\, \mathcal{Q}_\varphi(\boldsymbol{x}, t)$$

is only conserved if Q_φ charges do not escape or enter at $|\boldsymbol{x}| \to \infty$:

$$\lim_{|\boldsymbol{x}|\to\infty} \int d^2\Omega\, |\boldsymbol{x}|\boldsymbol{x} \cdot \boldsymbol{J}(\boldsymbol{x}, t) = 0.$$

Show that this implies

$$\lim_{|\boldsymbol{x}|\to\infty} \dot{\varphi}(\boldsymbol{x}, t) = 0.$$

Show also that Q_φ differs from the standard electric charge

$$Q = \int d^3x\, \varrho(\boldsymbol{x})$$

only by a constant factor

$$Q_\varphi = \langle\varphi\rangle Q, \tag{21.158}$$

where

$$\langle\varphi\rangle = \frac{1}{4\pi} \int d^2\Omega \lim_{|\boldsymbol{x}|\to\infty} \varphi(\boldsymbol{x}, t)$$

is the angular average of the phase parameter $\varphi(\boldsymbol{x}, t)$ at $|\boldsymbol{x}| \to \infty$.

21.18. Show for a free electron that equation (21.100) implies that a positron component ϕ in the wave function is not negligible any more relative to the electron wave function ψ at a distance of order

$$d \simeq \frac{4mc}{\hbar} \Delta x^2 \simeq 10^4\,\mathrm{nm}^{-1} \Delta x^2 = 10^{11}\,\mathrm{cm}^{-1} \Delta x^2. \tag{21.159}$$

This implies that we cannot use the wave packet for a strongly localized free electron with $\Delta x = 1\,\text{Å}$ beyond a distance of about $0.1\,\mu$m from the center. However, for a free electron wave packet with $\Delta x = 1\,$mm the limit (21.159) is much larger than the confines of any physics or chemistry lab and therefore of no concern.

Show also that at time t, the estimate for the usable range of the wave packet is

$$d(t) \lesssim \frac{4mc}{\hbar}\Delta x(0)\Delta x(t). \tag{21.160}$$

21.19. Show that the Foldy-Wouthuysen transformation (21.111) can also be written as

$$|\Upsilon_{FWT}(t)\rangle = \exp[F(t) + W(t) + T(t) + C(t) + \mathcal{O}(\mathcal{E}/mc^2)^4]|\Upsilon(t)\rangle,$$

with

$$C(t) = \frac{q\hbar}{16m^3c^4}\gamma^0\big(\hbar\nabla\cdot\boldsymbol{E}(\mathbf{x},t) + 2\mathrm{i}\boldsymbol{E}(\mathbf{x},t)\cdot[\mathbf{p} - q\boldsymbol{A}(\mathbf{x},t)]\big)$$
$$+\mathrm{i}\frac{q\hbar^2}{16m^3c^4}\big(\nabla\times\boldsymbol{E}(\mathbf{x},t)\big)\cdot\begin{pmatrix}\underline{\sigma} & \underline{0}\\ \underline{0} & -\underline{\sigma}\end{pmatrix}.$$

21.20. We have derived the equations (12.21, 12.23) for particles which satisfy the relativistic dispersion equation. In the meantime, we have seen that at the quantum level these particles are described by scalar fields ϕ, spinors ψ, or vector fields A_μ. The factor g counts spin and internal symmetry degrees of freedom and has the form $g = g_s \times d_{\mathrm{rep}(G)}$, where $d_{\mathrm{rep}(G)}$ is the dimension of the representation of the internal symmetry group G under which the fields transform.

21.20a. Scalar fields have $g_s = 1$. Show that in $d+1$ space-time dimensions, $g_s = 2^{\lfloor (d-1)/2\rfloor}$ for Dirac fields and $g_s = d-1$ for vector fields.

Hint: A Dirac spinor in $d+1$ space-time dimensions has $2^{\lfloor (d+1)/2\rfloor}$ components, see Appendix G (note that there d denotes the number of space-time dimensions).

21.20b. We have seen that scalar fields can be either real or complex (and similar remarks apply to spinor and vector fields if we go beyond quantum electrodynamics into the standard model of particle physics). However, a complex field has twice as many degrees of freedom as a real field. Should g therefore not include an additional factor g_c with $g_c = 2$ for complex fields and $g_c = 1$ for real fields?

21.21. Formulate the basic relations for basis kets $|x,\mu\rangle$, $|k,\alpha\rangle$ for the potentials $|A_D\rangle$ in Lorentz gauge in analogy to the corresponding relations (18.24–18.27) in Coulomb gauge.

21.22. Show that the two representations given in equation (21.130) for the gauge transformation function $g(\boldsymbol{x}, t)$ are indeed equivalent (hint: use the fact that $A^\mu(x)$ satisfies the Gauss law $\boldsymbol{\nabla} \cdot \boldsymbol{E} = \varrho/\epsilon_0$).

Show also that the gauge transformation (21.129, 21.130) takes us from any vector potential $A_\mu(x)$ which satisfies Maxwell's equations into the vector potential in Coulomb gauge.

Chapter 22
Applications of Spinor QED

We have seen in Chapter 18 that inclusion of the quantized Maxwell field did not change the basic formalism of time-dependent perturbation theory, see equations (18.71–18.73), and this property also persists after promotion of the matter fields in the Hamiltonian to relativistic Klein-Gordon or Dirac fields. In the following sections we will use the Hamiltonian (21.94) of spinor quantum electrodynamics for the calculation of scattering processes. However, we first should generalize our previous results for the scattering matrix to the case of two free particles in the initial and final state.

22.1 Two-particle scattering cross sections

We have discussed events with one free particle in the initial or final state of a scattering event in the framework of potential scattering theory in Chapters 11 and 13, or in photon emission, absorption or scattering off bound electrons in Sections 18.6–18.9. The techniques that we have discussed so far cover many applications of scattering theory, but eventually we also wish to understand scattering involving two (quasi-)free particles in the initial and final states. Electron scattering off atomic nuclei, electron-electron scattering, electron-photon scattering, or electron-phonon scattering provide examples of these kinds of scattering events which happen all the time in materials. In these cases we are discussing scattering events with two particles in the initial or final states. We should therefore address the question how to generalize the equations from Sections 13.6 and 18.9, which dealt with the case of one free particle in the initial and final state.

Let us recall from Section 13.6 that with a free particle with wave vectors $\boldsymbol{k}$ and $\boldsymbol{k}'$ in the initial and final state, the scattering matrix element for a monochromatic perturbation $W(t) \sim \exp(-\mathrm{i}\omega t)$

$$S_{\boldsymbol{k}',\boldsymbol{k}} = \langle \boldsymbol{k}'|U_D(\infty, -\infty)|\boldsymbol{k}\rangle = -\mathrm{i}\mathcal{M}_{\boldsymbol{k}',\boldsymbol{k}}\delta(\omega(\boldsymbol{k}') - \omega(\boldsymbol{k}) - \omega)$$

R. Dick, *Advanced Quantum Mechanics*, Graduate Texts in Physics,
DOI 10.1007/978-3-319-25675-7_22

has dimension length^3 due to the length dimensions of the external states, and yields a differential scattering cross section

$$d\sigma_{k\to k'} = d^3k'\frac{|S_{k',k}|^2}{Tdj(\boldsymbol{k})/d^3\boldsymbol{k}} = d^3k'\frac{|\mathcal{M}_{k',k}|^2}{2\pi dj(\boldsymbol{k})/d^3\boldsymbol{k}}\delta(\omega(k') - \omega(k) - \omega).$$

Here we substituted the more precise notation $dj(\boldsymbol{k})/d^3\boldsymbol{k}$ for the incoming current density j_{in} per $\boldsymbol{k}$ space volume. Substitution of the current density for a free particle of momentum $\hbar\boldsymbol{k}$

$$\frac{dj(\boldsymbol{k})}{d^3\boldsymbol{k}} = \frac{\boldsymbol{v}}{(2\pi)^3}$$

yields

$$v d\sigma_{k\to k'} = 4\pi^2 |\mathcal{M}_{k',k}|^2 \delta(\omega(k') - \omega(k) - \omega) d^3\boldsymbol{k}', \tag{22.1}$$

see also equation (18.115), where we found this equation for photon scattering off atoms or molecules.

Now suppose that we have two free particles with momenta $\boldsymbol{k}$ and $\boldsymbol{q}$ in the initial state, and they scatter into free particles with momenta $\boldsymbol{k}'$ and $\boldsymbol{q}'$ in the final state. We also assume that the scattering preserves total energy and momentum. The corresponding scattering matrix element

$$S_{k',q';k,q} = \langle \boldsymbol{k}', \boldsymbol{q}'|U_D(\infty, -\infty)|\boldsymbol{k}, \boldsymbol{q}\rangle = -\mathrm{i}\mathcal{M}_{k',q';k,q}\delta(k' + q' - k - q) \tag{22.2}$$

has dimension length^6. This is consistent with the fact that $|S_{k',q';k,q}|^2$ is a transition probability density per volume units $d^3\boldsymbol{k}'d^3\boldsymbol{q}'d^3\boldsymbol{k}d^3\boldsymbol{q}$ in wave vector space.

For ease of the present discussion, we also assume that the scattering particles are different, like in electron-photon or electron-phonon scattering, and we will use markers e and γ to label quantities referring to the different particles. The notation is motivated from electron-photon scattering, but we will develop the formalism in this section with general pairs of particles of masses m_e and m_γ in mind.

Suppose the two particles have momentum 4-vectors

$$p_e = \hbar k = \hbar(\omega_e/c, \boldsymbol{k}), \quad p_\gamma = \hbar q = \hbar(\omega_\gamma/c, \boldsymbol{q})$$

relative to the laboratory frame in which we observe the collisions. The scattering rate will be proportional to the product

$$\frac{d\varrho_e(\boldsymbol{k})}{d^3\boldsymbol{k}}\frac{dj_\gamma(\boldsymbol{q})}{d^3\boldsymbol{q}} = \frac{d\varrho_e(\boldsymbol{k})}{d^3\boldsymbol{k}}\frac{d\varrho_\gamma(\boldsymbol{q})}{d^3\boldsymbol{q}}\tilde{v}_{e\gamma},$$

where

$$\tilde{v}_{e\gamma} = c^2\left|\frac{\boldsymbol{k}}{\omega_e} - \frac{\boldsymbol{q}}{\omega_\gamma}\right| \tag{22.3}$$

is the relative speed between the two particles that *we* assign from the point of view of our laboratory frame. The speed $\tilde{v}_{e\gamma}$ is usually replaced with another measure for relative speed between the two particles,

$$v_{e\gamma}=\frac{c^3}{\omega_e\omega_\gamma}\sqrt{(k\cdot q)^2-\frac{m_e^2m_\gamma^2c^4}{\hbar^4}}, \tag{22.4}$$

which agrees with $\tilde{v}_{e\gamma}$ in laboratory frames in which the two momentum vectors $\boldsymbol{p}_e$ and $\boldsymbol{p}_\gamma$ are parallel or anti-parallel, or where the laboratory frame coincides with the rest frame of one of the two particles:

$$\tilde{v}_{e\gamma}^2-v_{e\gamma}^2=\frac{c^6}{\omega_e^2\omega_\gamma^2}\,|\boldsymbol{k}|^2\,|\boldsymbol{q}|^2\left(1-\cos^2\vartheta\right).$$

Here ϑ is the angle between $\boldsymbol{p}_e$ and $\boldsymbol{p}_\gamma$. E.g. in the rest frame of the e-particle, $\hbar k_e=\hbar(\omega_e/c,\boldsymbol{0})=(m_e c,\boldsymbol{0})$, we find

$$v_{e\gamma}=\frac{c}{\omega_\gamma}\sqrt{\omega_\gamma^2-\frac{m_\gamma^2c^4}{\hbar^2}}=c^2\frac{|\boldsymbol{p}_\gamma|}{E_\gamma}=v_\gamma,$$

and in the center of mass frame of the two particles, $\boldsymbol{k}=-\boldsymbol{q}$, we also find the difference of particle velocities,

$$\begin{aligned}v_{e\gamma}^2&=c^2\frac{\hbar^4(\omega_e\omega_\gamma+c^2\boldsymbol{k}^2)^2-m_e^2m_\gamma^2c^8}{\hbar^4\omega_e^2\omega_\gamma^2}\\&=c^4\frac{2\hbar^2c^2\boldsymbol{k}^4+2\hbar^2\omega_e\omega_\gamma\boldsymbol{k}^2+c^4\boldsymbol{k}^2(m_e^2+m_\gamma^2)}{\hbar^2\omega_e^2\omega_\gamma^2}=c^4\boldsymbol{k}^2\frac{\omega_e^2+\omega_\gamma^2+2\omega_e\omega_\gamma}{\omega_e^2\omega_\gamma^2}\\&=\left(c^2\frac{\boldsymbol{k}}{\omega_e}-c^2\frac{\boldsymbol{q}}{\omega_\gamma}\right)^2.\end{aligned} \tag{22.5}$$

As a byproduct we also find another useful formula for the relative speed in the center of mass frame,

$$v_{e\gamma}=c^2|\boldsymbol{k}|\frac{\omega_e+\omega_\gamma}{\omega_e\omega_\gamma}. \tag{22.6}$$

Please keep in mind that (22.6) is the relative particle speed assigned to the two colliding particles by an observer at rest in the center of mass frame, but not the speed of one particle relative to the other particle as measured in the rest frame of one of the particles.

The differential cross section for two-particle scattering can then be defined through the equation

$$v d\sigma_{k,q\to k',q'} = v d^3k' d^3q' \frac{\left|S_{k',q';k,q}\right|^2}{VT(d\varrho_e/d^3k)(dj_\gamma/d^3q)}$$

$$= d^3k' d^3q' \frac{\left|S_{k',q';k,q}\right|^2}{VT(d\varrho_e/d^3k)(d\varrho_\gamma/d^3q)}. \tag{22.7}$$

In words, we divide the scattering rate $d^3k'd^3q'\left|S_{k',q';k,q}\right|^2 d^3kd^3q/T$ between wave vector volumes $d^3kd^3q \to d^3k'd^3q'$ by the number of scattering centers $Vd\varrho_e$ in the phase space volume Vd^3k and the incoming particle flux dj_γ in the wave vector volume d^3q to calculate $d\sigma_{k,q\to k',q'}$.

If we substitute the scattering amplitude $\mathcal{M}_{k',q';k,q}$ for the scattering matrix element and use for the 4-dimensional δ function in momentum space the equation

$$\delta^4(0) = \lim_{k\to 0} \frac{1}{(2\pi)^4} \int d^4x \, \exp(ik\cdot x) = \frac{cVT}{(2\pi)^4},$$

we find

$$v d\sigma_{k,q\to k',q'} = d^3k' d^3q' \frac{c\left|\mathcal{M}_{k',q';k,q}\right|^2}{(2\pi)^4(d\varrho_e/d^3k)(d\varrho_\gamma/d^3q)} \delta^4(k'+q'-k-q). \tag{22.8}$$

The density per $\boldsymbol{x}$ space volume and per unit d^3k of $\boldsymbol{k}$ space volume for momentum eigenstates is

$$\frac{d\varrho}{d^3k} = \frac{1}{(2\pi)^3}.$$

This yields

$$v d\sigma_{k,q\to k',q'} = 4\pi^2 c \left|\mathcal{M}_{k',q';k,q}\right|^2 \delta^4(k'+q'-k-q) d^3k' d^3q'. \tag{22.9}$$

Note from equation (22.2) that the two-particle scattering amplitude $\mathcal{M}_{k',q';k,q}$ has the dimension length^2 while the single-particle scattering amplitude $\mathcal{M}_{k',k}$ has dimension length^3/time due to the use of a δ function in frequencies rather than wave numbers in the single-particle case.

We can derive a single-particle scattering cross section from (22.9) by integrating over the final wave number of one of the two particles, e.g. $\boldsymbol{q}'$, while considering its initial wave number fixed, e.g. $\boldsymbol{q} = \boldsymbol{0}$. This yields

$$v d\sigma_{k\to k'} = v \int d^3q' \frac{d\sigma_{k,0\to k',q'}}{d^3q'}$$

$$= 4\pi^2 c^2 \left|\mathcal{M}_{k',k-k';k,0}\right|^2 \delta(\omega(\boldsymbol{k}') - \omega(\boldsymbol{k}) - \omega_q) d^3k',$$

with

$$\omega_q = \omega(\boldsymbol{q} = \boldsymbol{0}) - \omega(\boldsymbol{q}' = \boldsymbol{k} - \boldsymbol{k}')$$

and a resulting single particle scattering amplitude $\mathcal{M}_{k',k} = c\mathcal{M}_{k',k-k';k,0}$.

Measures for final states with two identical particles

To explain the necessary modifications of the previous results if we have two identical particles in the final state, we first consider decay of a normalizable state $|i\rangle$ into two identical particles with momenta $\hbar\mathbf{k}_1$ and $\hbar\mathbf{k}_2$.

The initial state belongs to a set of orthonormal states, $\langle i|j\rangle = \delta_{ij}$. For the final states with two identical particles, we have to take into account that the decomposition of unity on identical 2-particle states is

$$\begin{aligned} 1_{\text{identical 2-particle states}} &= \frac{1}{2}\int d^3\mathbf{k}_1 \int d^3\mathbf{k}_2\, |\mathbf{k}_1,\mathbf{k}_2\rangle\langle\mathbf{k}_1,\mathbf{k}_2| \\ &= \frac{1}{2}\int d^3\mathbf{k}_1 \int d^3\mathbf{k}_2\, a^+(\mathbf{k}_1)a^+(\mathbf{k}_2)|0\rangle\langle 0|a(\mathbf{k}_2)a(\mathbf{k}_1). \end{aligned}$$

If the scattering matrix allows only for decay into the two-particle states, unitarity $U_D^+U_D = 1$ (or equivalently $S^+S = 1$) implies

$$\frac{1}{2}\int d^3\mathbf{k}_1 \int d^3\mathbf{k}_2\, |\langle\mathbf{k}_1,\mathbf{k}_2|U_D(\infty,-\infty)|i\rangle|^2 = 1.$$

The proper probability density for the transition $|i\rangle \to |\mathbf{k}_1,\mathbf{k}_2\rangle$ is therefore

$$w_{i\to\mathbf{k}_1,\mathbf{k}_2} = \frac{1}{2}d^3\mathbf{k}_1 d^3\mathbf{k}_2\, |\langle\mathbf{k}_1,\mathbf{k}_2|U_D(\infty,-\infty)|i\rangle|^2.$$

Equation (22.9) for the two-particle scattering cross section must therefore be modified if the final state contains two identical particles,

$$\begin{aligned} v\,d\sigma_{\mathbf{k}_1,\mathbf{k}_2\to\mathbf{k}_1',\mathbf{k}_2'} &= 4\pi^2 c\left|\mathcal{M}_{\mathbf{k}_1',\mathbf{k}_2';\mathbf{k}_1,\mathbf{k}_2}\right|^2 \delta^4(k_1'+k_2'-k_1-k_2) \\ &\quad\times\frac{1}{2}d^3\mathbf{k}_1' d^3\mathbf{k}_2', \end{aligned} \tag{22.10}$$

and the total two-particle cross section is

$$\begin{aligned} \sigma &= \int d^3\mathbf{k}_1' \int d^3\mathbf{k}_2' \frac{d\sigma_{\mathbf{k}_1,\mathbf{k}_2\to\mathbf{k}_1',\mathbf{k}_2'}}{d^3\mathbf{k}_1' d^3\mathbf{k}_2'} \\ &= \frac{1}{2}\int d^3\mathbf{k}_1' \int d^3\mathbf{k}_2'\, 4\pi^2\frac{c}{v}\left|\mathcal{M}_{\mathbf{k}_1',\mathbf{k}_2';\mathbf{k}_1,\mathbf{k}_2}\right|^2 \delta^4(k_1'+k_2'-k_1-k_2). \end{aligned} \tag{22.11}$$

However, if we want to derive an effective *single-particle* differential scattering cross section $d\sigma/d\Omega$ from $d\sigma_{\mathbf{k}_1,\mathbf{k}_2\to\mathbf{k}_1',\mathbf{k}_2'}$ by integrating over the momentum of one particle and the magnitude of momentum of the second particle using the energy-momentum conserving δ function, we have to take into account that the particle observed in direction $d\Omega$ can be either one of the two scattered particles:

$$\frac{d\sigma}{d\Omega} = \int d^3k'_1 \int_0^\infty d|k'_2|\,|k'_2|^2 \frac{d\sigma_{k_1,k_2\to k'_1,k'_2}}{d^3k'_1 d^3k'_2}$$
$$+ \int d^3k'_2 \int_0^\infty d|k'_1|\,|k'_1|^2 \frac{d\sigma_{k_1,k_2\to k'_1,k'_2}}{d^3k'_1 d^3k'_2}.$$

In the center of mass frame this reduces to a factor of 2,

$$\frac{d\sigma}{d\Omega} = 2\int d^3k'_2 \int_0^\infty d|k'_1|\,|k'_1|^2 \frac{d\sigma_{k_1,k_2\to k'_1,k'_2}}{d^3k'_1 d^3k'_2}. \tag{22.12}$$

If we then wish to calculate the total two-particle scattering cross section (22.11) from the single-particle differential cross section (22.12), we have to compensate with a factor 1/2,

$$\sigma = \frac{1}{2}\int d\Omega\,\frac{d\sigma}{d\Omega}. \tag{22.13}$$

In practice one is often only interested in the effective single-particle differential cross section (22.12) and the total two-particle scattering cross section σ. The factor 1/2 is then usually neglected in the differential two-particle cross section $d\sigma_{k_1,k_2\to k'_1,k'_2}/(d^3k'_1 d^3k'_2)$, so that the factor of 2 is not needed in the calculation of $d\sigma/d\Omega$ (22.12) because it has been absorbed in $d\sigma_{k_1,k_2\to k'_1,k'_2}/(d^3k'_1 d^3k'_2)$. However, the factor 1/2 is still needed in the calculation of the total two-particle cross section σ from $d\sigma/d\Omega$ according to equation (22.13).

22.2 Electron scattering off an atomic nucleus

As a first application of two-particle scattering, we discuss scattering of an electron off an atomic nucleus. We assume that the nucleus is also a fermion and that the electrons are not energetic enough to resolve the internal structure of the nucleus. In that case we can use an effective description of the nucleus through Dirac field operators for a particle of charge Ze and mass M for the nucleus.

The Coulomb gauge Hamiltonian (21.94) has the form

$$H = H_0 + H_{e\gamma} + H_{N\gamma} + H_C, \tag{22.14}$$

where the free part H_0 contains the kinetic and mass terms and we have separated the different interaction terms.

The electron-photon and nucleus-photon interaction terms are

$$H_{e\gamma} = ec\int d^3x\,\overline{\psi}(x,t)\boldsymbol{\gamma}\cdot\boldsymbol{A}(x,t)\psi(x,t)$$

and

$$H_{N\gamma} = -Zec\int d^3x\,\overline{\Psi}(x,t)\boldsymbol{\gamma}\cdot\boldsymbol{A}(x,t)\Psi(x,t),$$

respectively. The relevant part of the Coulomb interaction term is the term describing the interaction of the electron and the nucleus,

$$H_{eN} = -Z\frac{e^2}{4\pi\epsilon_0}\sum_{cc'}\int d^3x\int d^3X\,\frac{\psi_c^+(x,t)\Psi_{c'}^+(X,t)\Psi_{c'}(X,t)\psi_c(x,t)}{|x-X|},$$

where the sum is over 4-spinor indices. The relevant leading order matrix element contains two terms,

$$S_{fi} = -\mathrm{i}\mathcal{M}_{fi}\delta(k+K-k'-K') = S_{fi}^{(\gamma)} + S_{fi}^{(C)},$$

which correspond to photon exchange,

$$\begin{aligned} S_{fi}^{(\gamma)} &= \langle K',S';k',s'|\frac{Ze^2}{\hbar^2}\mathrm{T}\int d^4x\int d^4X\,\overline{\psi}(x)\boldsymbol{\gamma}\cdot\boldsymbol{A}(x)\psi(x) \\ &\quad\times\overline{\Psi}(X)\boldsymbol{\gamma}\cdot\boldsymbol{A}(X)\Psi(X)|K,S;k,s\rangle, \end{aligned} \tag{22.15}$$

or Coulomb scattering,

$$\begin{aligned} S_{fi}^{(C)} &= \langle K',S';k',s'|\frac{\mathrm{i}Ze^2\mu_0c}{4\pi\hbar}\int d^4x\int d^3X \\ &\quad\times\sum_{cc'}\frac{\psi_c^+(x,t)\Psi_{c'}^+(X,t)\Psi_{c'}(X,t)\psi_c(x,t)}{|x-X|}|K,S;k,s\rangle. \end{aligned} \tag{22.16}$$

We first calculate the Coulomb contribution to the scattering amplitude. Evaluation of the operators yields

$$\begin{aligned} S_{fi}^{(C)} &= \mathrm{i}\frac{Ze^2\mu_0c}{8(2\pi)^7\hbar}\frac{u_{s'}^+(k')\cdot u_s(k)u_{S'}^+(K')\cdot u_S(K)}{\sqrt{E_e(k')E_e(k)E_N(K')E_N(K)}} \\ &\quad\times\int d^4x\int d^3X\,\frac{\exp[\mathrm{i}(K-K')\cdot X+\mathrm{i}(k-k')\cdot x]}{|x-X|} \\ &\quad\times\exp[-\mathrm{i}(\omega_e(k)+\omega_N(K)-\omega_e(k')-\omega_N(K'))t]. \end{aligned}$$

In the next step we use the Fourier decomposition of the Coulomb potential

$$\int d^3x\,\frac{\exp(-\mathrm{i}q\cdot x)}{|x|} = \frac{4\pi}{q^2} \tag{22.17}$$

to find

$$\begin{aligned} S_{fi}^{(C)} &= \mathrm{i}\frac{Ze^2\mu_0c}{4(2\pi)^2\hbar}\frac{u_{s'}^+(k')\cdot u_s(k)u_{S'}^+(K')\cdot u_S(K)}{\sqrt{E_e(k')E_e(k)E_N(K')E_N(K)}} \\ &\quad\times\frac{\delta(k+K-k'-K')}{|k-k'|^2}. \end{aligned} \tag{22.18}$$

For the evaluation of the photon exchange contribution (22.15), we first note that the photon operators between the photon vacuum states yield

$$\langle 0|\mathrm{T}\boldsymbol{A}(x)\otimes\boldsymbol{A}(X)|0\rangle = \frac{\hbar\mu_0 c}{(2\pi)^3}\int\frac{d^3\boldsymbol{q}}{2|\boldsymbol{q}|}\sum_\alpha \boldsymbol{\epsilon}_\alpha(\boldsymbol{q})\otimes\boldsymbol{\epsilon}_\alpha(\boldsymbol{q})$$
$$\times\Big[\Theta(t-T)\exp[\mathrm{i}q\cdot(x-X)]+\Theta(T-t)\exp[-\mathrm{i}q\cdot(x-X)]\Big]_{q^0=\omega_\gamma(\boldsymbol{q})/c=|\boldsymbol{q}|}$$
$$=\frac{\hbar\mu_0 c}{(2\pi)^3}\frac{1}{2\pi\mathrm{i}}\int d^4q\,\frac{\exp[\mathrm{i}q\cdot(x-X)]}{q^2-\mathrm{i}\epsilon}\left(\underline{1}-\frac{\boldsymbol{q}\otimes\boldsymbol{q}}{\boldsymbol{q}^2}\right).$$

Evaluation of the fermion operators yields

$$\langle \boldsymbol{K}',S';\boldsymbol{k}',s'|\overline{\psi}(x)\boldsymbol{\gamma}\psi(x)\otimes\overline{\Psi}(X)\boldsymbol{\gamma}\Psi(X)|\boldsymbol{K},S;\boldsymbol{k},s\rangle$$
$$=\frac{\overline{u}_{s'}(\boldsymbol{k}')\boldsymbol{\gamma}u_s(\boldsymbol{k})\otimes\overline{u}_{S'}(\boldsymbol{K}')\boldsymbol{\gamma}u_S(\boldsymbol{K})}{4(2\pi)^6\sqrt{E_e(\boldsymbol{k}')E_e(\boldsymbol{k})E_N(\boldsymbol{K}')E_N(\boldsymbol{K})}}\exp[\mathrm{i}(k-k')\cdot x+\mathrm{i}(K-K')\cdot X].$$

Assembling the pieces yields

$$S_{fi}^{(\gamma)}=\frac{Ze^2\mu_0 c}{4\mathrm{i}(2\pi)^2\hbar}\frac{\delta(k+K-k'-K')}{\sqrt{E_e(\boldsymbol{k}')E_e(\boldsymbol{k})E_N(\boldsymbol{K}')E_N(\boldsymbol{K})}}\frac{1}{(k-k')^2-\mathrm{i}\epsilon}$$
$$\times\overline{u}_{s'}(\boldsymbol{k}')\boldsymbol{\gamma}u_s(\boldsymbol{k})\left(\underline{1}+\frac{(\boldsymbol{k}-\boldsymbol{k}')\otimes(\boldsymbol{K}-\boldsymbol{K}')}{|\boldsymbol{k}-\boldsymbol{k}'|^2}\right)\overline{u}_{S'}(\boldsymbol{K}')\boldsymbol{\gamma}u_S(\boldsymbol{K}),$$

where $\boldsymbol{k}-\boldsymbol{k}'=-(\boldsymbol{K}-\boldsymbol{K}')$ from momentum conservation was used in the projection term. Substitution of the free Dirac equation for the external fermion states yields

$$S_{fi}^{(\gamma)}=\frac{Ze^2\mu_0 c}{4\mathrm{i}(2\pi)^2\hbar}\frac{\delta(k+K-k'-K')}{\sqrt{E_e(\boldsymbol{k}')E_e(\boldsymbol{k})E_N(\boldsymbol{K}')E_N(\boldsymbol{K})}}\frac{1}{(k-k')^2-\mathrm{i}\epsilon}$$
$$\times\Big(\overline{u}_{s'}(\boldsymbol{k}')\boldsymbol{\gamma}u_s(\boldsymbol{k})\cdot\overline{u}_{S'}(\boldsymbol{K}')\boldsymbol{\gamma}u_S(\boldsymbol{K})$$
$$+u_{s'}^+(\boldsymbol{k}')u_s(\boldsymbol{k})\frac{[E_e(\boldsymbol{k})-E_e(\boldsymbol{k}')][E_N(\boldsymbol{K})-E_N(\boldsymbol{K}')]}{\hbar^2c^2|\boldsymbol{k}-\boldsymbol{k}'|^2}u_{S'}^+(\boldsymbol{K}')u_S(\boldsymbol{K})\Big)$$

and energy conservation yields finally

$$S_{fi}^{(\gamma)}=\frac{Ze^2\mu_0 c}{4\mathrm{i}(2\pi)^2\hbar}\frac{\delta(k+K-k'-K')}{\sqrt{E_e(\boldsymbol{k}')E_e(\boldsymbol{k})E_N(\boldsymbol{K}')E_N(\boldsymbol{K})}}\frac{1}{(k-k')^2-\mathrm{i}\epsilon}$$
$$\times\Big(\overline{u}_{s'}(\boldsymbol{k}')\boldsymbol{\gamma}u_s(\boldsymbol{k})\cdot\overline{u}_{S'}(\boldsymbol{K}')\boldsymbol{\gamma}u_S(\boldsymbol{K})$$
$$-\overline{u}_{s'}(\boldsymbol{k}')\gamma^0u_s(\boldsymbol{k})\frac{(E_e(\boldsymbol{k})-E_e(\boldsymbol{k}'))^2}{\hbar^2c^2|\boldsymbol{k}-\boldsymbol{k}'|^2}\overline{u}_{S'}(\boldsymbol{K}')\gamma^0u_S(\boldsymbol{K})\Big). \tag{22.19}$$

The sum $S_{fi} = S_{fi}^{(\gamma)} + S_{fi}^{(C)}$ contains a term

$$\frac{1}{|\boldsymbol{k} - \boldsymbol{k}'|^2}\left(1 + \frac{(E_e(\boldsymbol{k}) - E_e(\boldsymbol{k}'))^2}{\hbar^2 c^2 (k - k')^2 - \mathrm{i}\epsilon}\right) = \frac{1}{(k - k')^2 - \mathrm{i}\epsilon}.$$

This yields finally the scattering matrix element

$$\begin{aligned} S_{fi} &= -\mathrm{i}\mathcal{M}_{fi}\delta(k + K - k' - K') \\ &= \frac{Z\alpha}{4\pi\mathrm{i}} \frac{\overline{u}_{s'}(\boldsymbol{k}')\gamma^\mu u_s(\boldsymbol{k})\overline{u}_{S'}(\boldsymbol{K}')\gamma_\mu u_S(\boldsymbol{K})}{\sqrt{E_e(\boldsymbol{k}')E_e(\boldsymbol{k})E_N(\boldsymbol{K}')E_N(\boldsymbol{K})}} \frac{\delta(k + K - k' - K')}{(k - k')^2 - \mathrm{i}\epsilon}, \end{aligned} \tag{22.20}$$

where the fine structure constant $\alpha = \mu_0 c e^2/4\pi\hbar$ (7.61) was substituted.

The differential scattering cross section for electron-nucleus scattering is then given by (22.9)

$$v_{eN} d\sigma = 4\pi^2 c |\mathcal{M}_{fi}|^2 \delta(k + K - k' - K') d^3\boldsymbol{K}' d^3\boldsymbol{k}'$$

with the relativistic expression for the relative velocity of the electron and the nucleus,

$$v_{eN} = \frac{c^3}{E_e(\boldsymbol{k})E_N(\boldsymbol{K})}\sqrt{\hbar^4 (K \cdot k)^2 - m^2 M^2 c^4}, \tag{22.21}$$

where m is the electron mass. Integration over the momentum $\boldsymbol{K}'$ of the scattered nucleus and the magnitude k' of the scattered electron momentum yields

$$v_{eN}\frac{d\sigma}{d\Omega_{k'}} = 4\pi^2 c |\mathcal{M}_{fi}|^2 \left.\frac{k'^2}{|\partial f(k')/\partial k'|}\right|_{f(k')=0}, \tag{22.22}$$

where k' has to satisfy the condition

$$\begin{aligned} f(k') = &\sqrt{(\boldsymbol{k}' - \boldsymbol{K} - \boldsymbol{k})^2 + (Mc/\hbar)^2} + \sqrt{k'^2 + (mc/\hbar)^2} \\ &- \sqrt{K^2 + (Mc/\hbar)^2} - \sqrt{k^2 + (mc/\hbar)^2} = 0. \end{aligned} \tag{22.23}$$

Usually we are not interested in the scattering with fixed initial and final spin polarizations. Therefore we average over initial spins and sum over final spins to calculate the unpolarized scattering cross section,

$$|\mathcal{M}_{fi}|^2 \to \frac{1}{4}\sum_{s,s',S,S'} |\mathcal{M}_{fi}|^2. \tag{22.24}$$

We will further evaluate these expressions for the case of negligible momenta and momentum transfers compared to Mc. We can take this into account through the limit $M \to \infty$. The condition (22.23) then reduces to the elastic electron scattering condition $k = k'$ which yields

$$\frac{k'^2}{|\partial f(\boldsymbol{k}')/\partial k'|} = k\sqrt{k^2 + (mc/\hbar)^2}.$$

Furthermore, using the Dirac representation (21.36) of the γ matrices yields with equations (21.45, 21.46, 21.50) for the 4-spinors the limit

$$\lim_{M\to\infty} \frac{\overline{u}_{S'}(\boldsymbol{K}')\gamma^\mu u_S(\boldsymbol{K})}{\sqrt{E_N(\boldsymbol{K}')E_N(\boldsymbol{K})}} = 2\delta_{SS'}\eta^\mu{}_0. \tag{22.25}$$

Note that this equation is invariant under similarity transformations of the γ matrices and therefore holds in every representation. Furthermore, the speed v_{eN} (22.21) becomes the electron speed in the rest frame of the heavy nucleus, $v_{eN} = c^2\hbar k/E_e(\boldsymbol{k})$.

The spin polarized scattering amplitude following from (22.20) in the heavy nucleus limit is

$$\mathcal{M}_{fi} = -\frac{Z\alpha}{2\pi}\frac{\overline{u}_{s'}(\boldsymbol{k}')\gamma^0 u_s(\boldsymbol{k})}{\sqrt{E_e(\boldsymbol{k}')E_e(\boldsymbol{k})}}\frac{\delta_{SS'}}{(\boldsymbol{k}-\boldsymbol{k}')^2 - \mathrm{i}\epsilon}. \tag{22.26}$$

The remaining electron spin averaging is easily accomplished using equation (21.58),

$$\frac{1}{2}\sum_{ss'}\left|\overline{u}_{s'}(\boldsymbol{k}')\gamma^0 u_s(\boldsymbol{k})\right|^2 = \frac{c^2}{2}\mathrm{tr}\big[(mc - \hbar k\cdot\gamma)\gamma^0(mc - \hbar k'\cdot\gamma)\gamma^0\big].$$

The trace theorems for products of γ matrices in Appendix G, in particular equation (G.19) and equation (G.20) in the form

$$\mathrm{tr}\big(\gamma_\kappa\gamma^0\gamma_\mu\gamma^0\big) = 8\eta_\kappa{}^0\eta_\mu{}^0 + 4\eta_{\mu\kappa}, \tag{22.27}$$

and the vanishing traces of odd numbers of γ matrices yield

$$\begin{aligned}\frac{1}{2}\sum_{ss'}\left|\overline{u}_{s'}(\boldsymbol{k}')\gamma^0 u_s(\boldsymbol{k})\right|^2 &= 2m^2c^4 + 4\hbar^2c^2k^0k'^0 + 2\hbar^2c^2k\cdot k' \\ &= 2c^2(\hbar^2k^0k'^0 + \hbar^2\boldsymbol{k}\cdot\boldsymbol{k}' + m^2c^2).\end{aligned}$$

This yields the unpolarized differential scattering cross section for electrons in the field of a heavy nucleus,

$$\frac{d\sigma}{d\Omega_{k'}} = \frac{Z^2\alpha^2}{2\hbar^2}\frac{2m^2c^2 + \hbar^2k^2(1+\cos\theta)}{k^4(1-\cos\theta)^2}, \tag{22.28}$$

where $k \equiv |\boldsymbol{k}|$. The Rutherford scattering formula (11.45) follows for $\hbar k \ll mc$. Electron scattering off a heavy nucleus is equivalent to scattering in an external Coulomb field $Ze/4\pi\epsilon_0 r$. This is known as *Mott scattering*[1]. From our calculation, we can easily understand why scattering off heavy nuclei is equivalent to scattering in an external Coulomb field. The Coulomb scattering matrix element (22.18) yields already the amplitude (22.26) in the heavy nucleus limit, while the photon exchange matrix element (22.19) vanishes in that limit. If we take into account that terms of the form $\overline{u}_{s'}(\boldsymbol{k}')\boldsymbol{\gamma}u_s(\boldsymbol{k})/\sqrt{E_e(\boldsymbol{k}')E_e(\boldsymbol{k})}$ are of order $\hbar^2|\boldsymbol{k}||\boldsymbol{k}'|/m^2c^2 \ll 1$ in the non-relativistic limit, we find that the ratio between the photon exchange amplitude and the Coulomb amplitude in the non-relativistic limit is of order

$$\left|\frac{S_{fi}^{(\gamma)}}{S_{fi}^{(C)}}\right| \simeq \frac{(E_e(\boldsymbol{k}) - E_e(\boldsymbol{k}'))^2}{\hbar^2c^2(\boldsymbol{k}-\boldsymbol{k}')^2} \simeq \frac{\hbar^2(k^2-k'^2)^2}{4m^2c^2(\boldsymbol{k}-\boldsymbol{k}')^2}$$

$$= \frac{\boldsymbol{k}-\boldsymbol{k}'}{|\boldsymbol{k}-\boldsymbol{k}'|}\cdot\frac{\hbar^2(\boldsymbol{k}+\boldsymbol{k}')\otimes(\boldsymbol{k}+\boldsymbol{k}')}{4m^2c^2}\cdot\frac{\boldsymbol{k}-\boldsymbol{k}'}{|\boldsymbol{k}-\boldsymbol{k}'|} \ll 1, \qquad (22.29)$$

If we denote the average velocity of the incoming and the scattered electron with $\boldsymbol{v}_e$, equation (22.29) tells us that photon exchange is suppressed by about $\boldsymbol{p}_e^2/m^2c^2 = \boldsymbol{v}_e^2/c^2$ compared to the Coulomb interaction in the non-relativistic limit. That is the reason why Coulomb gauge is convenient for the description of systems with non-relativistic charged particles. We can use Coulomb potentials in the calculation of scattering events and bound states of the non-relativistic particles without worrying about photon exchange. The photon terms are only needed for photon absorption and emission, and for photon scattering. On the other hand, if we are primarily concerned with interactions of relativistic charged particles, then use of a Hamiltonian like (21.88) with covariantly gauged photons as in Section 21.6 is more efficient.

22.3 Photon scattering by free electrons

Photon scattering by free or quasifree electrons is also known as Compton scattering. The cross section for this process had been calculated in leading order by Klein and Nishina[2].

The electron-photon interaction term from (21.94) is

$$\mathcal{H}_{e\gamma} = ec\overline{\Psi}\boldsymbol{\gamma}\cdot\boldsymbol{A}\Psi. \qquad (22.30)$$

We denote the wave vectors of the incoming photon and electron with $\boldsymbol{q}$ and $\boldsymbol{k}$, respectively. The relevant second order matrix element for scattering of photons by

[1]N.F. Mott, Proc. Roy. Soc. London A 124, 425 (1929).

[2]O. Klein, Y. Nishina, Z. Phys. 52, 853 (1929).

free electrons is

$$\begin{aligned}
S_{fi} &= S_{k',s';q',\alpha'|k,s;q,\alpha}\\
&= -\frac{e^2c^2}{\hbar^2}\int d^3\boldsymbol{x}\int d^3\boldsymbol{x}'\int_{-\infty}^{\infty}dt\int_{-\infty}^{t}dt'\,\langle \boldsymbol{k}',s';\boldsymbol{q}',\alpha'|\exp\!\left(\frac{\mathrm{i}}{\hbar}H_0 t\right)\\
&\quad\times\overline{\Psi}(\boldsymbol{x})\boldsymbol{\gamma}\cdot\boldsymbol{A}(\boldsymbol{x})\Psi(\boldsymbol{x})\exp\!\left(-\frac{\mathrm{i}}{\hbar}H_0(t-t')\right)\overline{\Psi}(\boldsymbol{x}')\boldsymbol{\gamma}\cdot\boldsymbol{A}(\boldsymbol{x}')\Psi(\boldsymbol{x}')\\
&\quad\times\exp\!\left(-\frac{\mathrm{i}}{\hbar}H_0 t'\right)|\boldsymbol{k},s;\boldsymbol{q},\alpha\rangle\\
&= -\frac{e^2}{\hbar^2}\int d^4x\int d^4x'\,\Theta(t-t')\langle \boldsymbol{k}',s';\boldsymbol{q}',\alpha'|\overline{\Psi}(x)\boldsymbol{\gamma}\cdot\boldsymbol{A}(x)\Psi(x)\\
&\quad\times\overline{\Psi}(x')\boldsymbol{\gamma}\cdot\boldsymbol{A}(x')\Psi(x')|\boldsymbol{k},s;\boldsymbol{q},\alpha\rangle.
\end{aligned}$$

Here $\boldsymbol{A}(x) \equiv \boldsymbol{A}_D(\boldsymbol{x},t)$ and $\Psi(x) \equiv \Psi_D(\boldsymbol{x},t)$ are the freely evolving field operators (18.21, 21.51) in the interaction picture.

We can insert a decomposition of unity between the two vertex operators $\overline{\Psi}\boldsymbol{\gamma}\cdot\boldsymbol{A}\Psi$ with a fermionic and a photon factor,

$$1 = 1_f \otimes 1_\gamma.$$

The relevant parts in the photon factor have zero or two intermediate photons,

$$1_\gamma \Rightarrow |0\rangle\langle 0| + \frac{1}{2}\sum_{\beta,\beta'}\int d^3\boldsymbol{K}\int d^3\boldsymbol{K}'\,|\boldsymbol{K},\beta;\boldsymbol{K}',\beta'\rangle\langle \boldsymbol{K},\beta;\boldsymbol{K}',\beta'|,$$

while for the intermediate fermion states only states with one intermediate electron or with two intermediate electrons and a positron contribute,

$$\begin{aligned}
1_f &\Rightarrow \sum_\sigma\int d^3\boldsymbol{\kappa}\,b_\sigma^+(\boldsymbol{\kappa})|0\rangle\langle 0|b_\sigma(\boldsymbol{\kappa})\\
&\quad+\frac{1}{2}\sum_{\sigma,\sigma',\nu}\int d^3\boldsymbol{\kappa}\int d^3\boldsymbol{\kappa}'\int d^3\boldsymbol{\lambda}\,b_\sigma^+(\boldsymbol{\kappa})b_{\sigma'}^+(\boldsymbol{\kappa}')d_\nu^+(\boldsymbol{\lambda})|0\rangle\\
&\quad\times\langle 0|d_\nu(\boldsymbol{\lambda})b_{\sigma'}(\boldsymbol{\kappa}')b_\sigma(\boldsymbol{\kappa}).
\end{aligned}$$

The full photon matrix element is

$$\begin{aligned}
\langle \boldsymbol{q}',\alpha'|\boldsymbol{A}(x)\otimes\boldsymbol{A}(x')|\boldsymbol{q},\alpha\rangle &= \frac{\hbar\mu_0 c}{16\pi^3\sqrt{|\boldsymbol{q}||\boldsymbol{q}'|}}\Big(\boldsymbol{\epsilon}_{\alpha'}(\boldsymbol{q}')\otimes\boldsymbol{\epsilon}_\alpha(\boldsymbol{q})\\
&\quad\times\exp[\mathrm{i}(q\cdot x'-q'\cdot x)]+\boldsymbol{\epsilon}_\alpha(\boldsymbol{q})\otimes\boldsymbol{\epsilon}_{\alpha'}(\boldsymbol{q}')\exp[\mathrm{i}(q\cdot x-q'\cdot x')]\Big),
\end{aligned}$$

where the first term arises from the term without intermediate photons and the second term arises from the term with two intermediate photons after integrating the intermediate photon momenta $\boldsymbol{K}$ and $\boldsymbol{K}'$.

Evaluation of the fermion matrix element with an electron in the intermediate state yields

$$\sum_\sigma \langle \boldsymbol{k}', s'|\overline{\Psi}(x)\boldsymbol{\gamma}\Psi(x)b_\sigma^+(\boldsymbol{\kappa})|0\rangle \otimes \langle 0|b_\sigma(\boldsymbol{\kappa})\overline{\Psi}(x')\boldsymbol{\gamma}\Psi(x')|\boldsymbol{k}, s\rangle$$
$$= \sum_\sigma \frac{\exp[\mathrm{i}\kappa\cdot(x-x') - \mathrm{i}k'\cdot x + \mathrm{i}k\cdot x']}{(2\pi)^6 4E(\boldsymbol{\kappa})\sqrt{E(\boldsymbol{k}')E(\boldsymbol{k})}}\overline{u}_{s'}(\boldsymbol{k}')\boldsymbol{\gamma}u_\sigma(\boldsymbol{\kappa}) \otimes \overline{u}_\sigma(\boldsymbol{\kappa})\boldsymbol{\gamma}u_s(\boldsymbol{k}).$$

We can substitute the sum over intermediate u spinors using equation (21.58),

$$\sum_\sigma \langle \boldsymbol{k}', s'|\overline{\Psi}(x)\boldsymbol{\gamma}\Psi(x)b_\sigma^+(\boldsymbol{\kappa})|0\rangle \otimes \langle 0|b_\sigma(\boldsymbol{\kappa})\overline{\Psi}(x')\boldsymbol{\gamma}\Psi(x')|\boldsymbol{k}, s\rangle$$
$$= \frac{\exp[\mathrm{i}\kappa\cdot(x-x') - \mathrm{i}k'\cdot x + \mathrm{i}k\cdot x']}{(2\pi)^6 4E(\boldsymbol{\kappa})\sqrt{E(\boldsymbol{k}')E(\boldsymbol{k})}}\boldsymbol{e}_i \otimes \boldsymbol{e}_j$$
$$\times \overline{u}_{s'}(\boldsymbol{k}')\gamma^i\left(mc^2 - \hbar c\boldsymbol{\gamma}\cdot\boldsymbol{\kappa} - \gamma^0 E(\boldsymbol{\kappa})\right)\gamma^j u_s(\boldsymbol{k}).$$

Assembling the pieces so far then yields the amplitude with a single intermediate fermion,

$$\begin{aligned} S_{fi}^{(1)} = &-\int\frac{d^3\kappa}{E(\boldsymbol{\kappa})}\int d^4x\int d^4x'\,\frac{e^2\Theta(t-t')}{8(2\pi)^9\epsilon_0\hbar c\sqrt{|\boldsymbol{q}||\boldsymbol{q}'|E(\boldsymbol{k})E(\boldsymbol{k}')}} \\ &\times\overline{u}_{s'}(\boldsymbol{k}')\Big[\left(\boldsymbol{\epsilon}_{\alpha'}(\boldsymbol{q}')\cdot\boldsymbol{\gamma}\right)\left(mc^2 - \hbar c\boldsymbol{\kappa}\cdot\boldsymbol{\gamma} - \gamma^0 E(\boldsymbol{\kappa})\right)\left(\boldsymbol{\epsilon}_\alpha(\boldsymbol{q})\cdot\boldsymbol{\gamma}\right) \\ &\times\exp\left[\mathrm{i}(\kappa - k' - q')\cdot x + \mathrm{i}(k + q - \kappa)\cdot x'\right] \\ &+\left(\boldsymbol{\epsilon}_\alpha(\boldsymbol{q})\cdot\boldsymbol{\gamma}\right)\left(mc^2 - \hbar c\boldsymbol{\kappa}\cdot\boldsymbol{\gamma} - \gamma^0 E(\boldsymbol{\kappa})\right)\left(\boldsymbol{\epsilon}_{\alpha'}(\boldsymbol{q}')\cdot\boldsymbol{\gamma}\right) \\ &\times\exp\left[\mathrm{i}(\kappa - k' + q)\cdot x + \mathrm{i}(k - q' - \kappa)\cdot x'\right]\Big]u_s(\boldsymbol{k}). \end{aligned} \tag{22.31}$$

The fermion matrix element with three intermediate fermions is

$$\frac{1}{2}\sum_{\sigma,\sigma',\nu}\langle \boldsymbol{k}', s'|\overline{\Psi}(x)\boldsymbol{\gamma}\Psi(x)b_\sigma^+(\boldsymbol{\kappa})b_{\sigma'}^+(\boldsymbol{\kappa}')d_\nu^+(\boldsymbol{\lambda})|0\rangle$$
$$\otimes\langle 0|d_\nu(\boldsymbol{\lambda})b_{\sigma'}(\boldsymbol{\kappa}')b_\sigma(\boldsymbol{\kappa})\overline{\Psi}(x')\boldsymbol{\gamma}\Psi(x')|\boldsymbol{k}, s\rangle$$
$$= -\sum_\nu \delta(\boldsymbol{\kappa}' - \boldsymbol{k}')\delta(\boldsymbol{\kappa} - \boldsymbol{k})\frac{\exp[\mathrm{i}\lambda\cdot(x-x') + \mathrm{i}k\cdot x - \mathrm{i}k'\cdot x']}{(2\pi)^6 4E(\boldsymbol{\lambda})\sqrt{E(\boldsymbol{k}')E(\boldsymbol{k})}}$$
$$\times\overline{v}_\nu(\boldsymbol{\lambda})\boldsymbol{\gamma}u_s(\boldsymbol{k}) \otimes \overline{u}_{s'}(\boldsymbol{k}')\boldsymbol{\gamma}v_\nu(\boldsymbol{\lambda}).$$

The last line has been simplified from an expression which is symmetric in the intermediate momenta $\boldsymbol{\kappa}$ and $\boldsymbol{\kappa}'$ by taking into account that those momenta will be integrated.

We can substitute the sum over intermediate v spinors using equation (21.60),

$$\frac{1}{2}\sum_{\sigma,\sigma',\nu}\langle \boldsymbol{k}',s'|\overline{\Psi}(x)\boldsymbol{\gamma}\Psi(x)b^+_\sigma(\boldsymbol{\kappa})b^+_{\sigma'}(\boldsymbol{\kappa}')d^+_\nu(\boldsymbol{\lambda})|0\rangle$$
$$\otimes\langle 0|d_\nu(\boldsymbol{\lambda})b_{\sigma'}(\boldsymbol{\kappa}')b_\sigma(\boldsymbol{\kappa})\overline{\Psi}(x')\boldsymbol{\gamma}\Psi(x')|\boldsymbol{k},s\rangle$$
$$=\delta(\boldsymbol{\kappa}'-\boldsymbol{k}')\delta(\boldsymbol{\kappa}-\boldsymbol{k})\frac{\exp[\mathrm{i}\lambda\cdot(x-x')+\mathrm{i}k\cdot x-\mathrm{i}k'\cdot x']}{(2\pi)^6 4E(\boldsymbol{\lambda})\sqrt{E(\boldsymbol{k}')E(\boldsymbol{k})}}$$
$$\times\boldsymbol{e}_i\otimes\boldsymbol{e}_j\overline{u}_{s'}(\boldsymbol{k}')\gamma^j\left(mc^2+\hbar c\boldsymbol{\gamma}\cdot\boldsymbol{\lambda}+\gamma^0E(\boldsymbol{\lambda})\right)\gamma^i u_s(\boldsymbol{k}).$$

If we substitute $\boldsymbol{\lambda}\to\boldsymbol{\kappa}$ (after integration over the intermediate electron momenta) for the wave vector of the intermediate positron, the contribution from three intermediate fermions to the scattering matrix element is

$$S^{(3)}_{fi}=-\int\frac{d^3\boldsymbol{\kappa}}{E(\boldsymbol{\kappa})}\int d^4x\int d^4x'\,\frac{e^2\Theta(t-t')}{8(2\pi)^9\epsilon_0\hbar c\sqrt{|\boldsymbol{q}||\boldsymbol{q}'|E(\boldsymbol{k})E(\boldsymbol{k}')}}$$
$$\times\overline{u}_{s'}(\boldsymbol{k}')\Big[\left(\boldsymbol{\epsilon}_{\alpha'}(\boldsymbol{q}')\cdot\boldsymbol{\gamma}\right)\left(mc^2+\hbar c\boldsymbol{\kappa}\cdot\boldsymbol{\gamma}+\gamma^0E(\boldsymbol{\kappa})\right)\left(\boldsymbol{\epsilon}_\alpha(\boldsymbol{q})\cdot\boldsymbol{\gamma}\right)$$
$$\times\exp\left[\mathrm{i}(\kappa+k+q)\cdot x-\mathrm{i}(\kappa+k'+q')\cdot x'\right]$$
$$+\left(\boldsymbol{\epsilon}_\alpha(\boldsymbol{q})\cdot\boldsymbol{\gamma}\right)\left(mc^2+\hbar c\boldsymbol{\kappa}\cdot\boldsymbol{\gamma}+\gamma^0E(\boldsymbol{\kappa})\right)\left(\boldsymbol{\epsilon}_{\alpha'}(\boldsymbol{q}')\cdot\boldsymbol{\gamma}\right)$$
$$\times\exp\left[\mathrm{i}(\kappa+k-q')\cdot x-\mathrm{i}(\kappa+k'-q)\cdot x'\right]\Big]u_s(\boldsymbol{k}).\tag{22.32}$$

We can simplify $S_{fi}=S^{(1)}_{fi}+S^{(3)}_{fi}$ by swapping $x\leftrightarrow x'$ in $S^{(3)}_{fi}$ and taking into account that

$$\int d^4\kappa\,\frac{(mc^2-\hbar c\gamma\cdot\kappa)f(\boldsymbol{\kappa})}{\kappa^2+(m^2c^2/\hbar^2)-\mathrm{i}\epsilon}\exp(\mathrm{i}\kappa\cdot x)$$
$$=-\int d^3\boldsymbol{\kappa}\int d\kappa^0\,\frac{\exp(\mathrm{i}\boldsymbol{\kappa}\cdot\boldsymbol{x})(mc^2-\hbar c\gamma\cdot\kappa)f(\boldsymbol{\kappa})\exp(-\mathrm{i}\kappa^0ct)}{[\kappa^0-(\omega(\boldsymbol{\kappa})/c)+\mathrm{i}\epsilon][\kappa^0+(\omega(\boldsymbol{\kappa})/c)-\mathrm{i}\epsilon]}$$
$$=2\pi\mathrm{i}c\Theta(t)\int d^3\boldsymbol{\kappa}\,\exp(\mathrm{i}\boldsymbol{\kappa}\cdot\boldsymbol{x})\frac{mc^2-\hbar c\boldsymbol{\kappa}\cdot\boldsymbol{\gamma}+\gamma^0E(\boldsymbol{\kappa})}{2\omega(\boldsymbol{\kappa})}f(\boldsymbol{\kappa})\exp(-\mathrm{i}\omega(\boldsymbol{\kappa})t)$$
$$-2\pi\mathrm{i}c\Theta(-t)\int d^3\boldsymbol{\kappa}\,\exp(\mathrm{i}\boldsymbol{\kappa}\cdot\boldsymbol{x})\frac{mc^2-\hbar c\boldsymbol{\kappa}\cdot\boldsymbol{\gamma}-\gamma^0E(\boldsymbol{\kappa})}{-2\omega(\boldsymbol{\kappa})}f(\boldsymbol{\kappa})\exp(\mathrm{i}\omega(\boldsymbol{\kappa})t)$$
$$=\mathrm{i}\pi c\Theta(t)\int\frac{d^3\boldsymbol{\kappa}}{\omega(\boldsymbol{\kappa})}(mc^2-\hbar c\gamma\cdot\kappa)f(\boldsymbol{\kappa})\exp(\mathrm{i}\kappa\cdot x)\Big|_{\kappa^0=\omega(\boldsymbol{\kappa})/c}$$
$$+\mathrm{i}\pi c\Theta(-t)\int\frac{d^3\boldsymbol{\kappa}}{\omega(\boldsymbol{\kappa})}(mc^2+\hbar c\gamma\cdot\kappa)f(\boldsymbol{\kappa})\exp(-\mathrm{i}\kappa\cdot x)\Big|_{\kappa^0=\omega(\boldsymbol{\kappa})/c}.\tag{22.33}$$

This yields the total scattering matrix element in the form

$$S_{fi} = \int d^4\kappa \int d^4x \int d^4x' \, \frac{ie^2}{4(2\pi)^{10}\epsilon_0\hbar^2 c\sqrt{|\boldsymbol{q}||\boldsymbol{q}'|E(\boldsymbol{k})E(\boldsymbol{k}')}}$$

$$\times \overline{u}_{s'}(\boldsymbol{k}')\Big[\left(\boldsymbol{\epsilon}_{\alpha'}(\boldsymbol{q}')\cdot\boldsymbol{\gamma}\right)\frac{mc - \hbar\gamma\cdot\kappa}{\kappa^2 + (m^2c^2/\hbar^2) - i\epsilon}\left(\boldsymbol{\epsilon}_{\alpha}(\boldsymbol{q})\cdot\boldsymbol{\gamma}\right)$$

$$\times \exp\left[i(\kappa - k' - q')\cdot x + i(k + q - \kappa)\cdot x'\right]$$

$$+ \left(\boldsymbol{\epsilon}_{\alpha}(\boldsymbol{q})\cdot\boldsymbol{\gamma}\right)\frac{mc - \hbar\gamma\cdot\kappa}{\kappa^2 + (m^2c^2/\hbar^2) - i\epsilon}\left(\boldsymbol{\epsilon}_{\alpha'}(\boldsymbol{q}')\cdot\boldsymbol{\gamma}\right)$$

$$\times \exp\left[i(\kappa - k' + q)\cdot x + i(k - q' - \kappa)\cdot x'\right]\Big]u_s(\boldsymbol{k}).$$

After performing the trivial integrations, we find

$$S_{fi} = \delta(k' + q' - k - q)\frac{ie^2}{16\pi^2\epsilon_0\hbar^2 c\sqrt{|\boldsymbol{q}||\boldsymbol{q}'|E(\boldsymbol{k})E(\boldsymbol{k}')}}$$

$$\times \overline{u}_{s'}(\boldsymbol{k}')\Big[\left(\boldsymbol{\epsilon}_{\alpha'}(\boldsymbol{q}')\cdot\boldsymbol{\gamma}\right)\frac{mc - \hbar\gamma\cdot(k+q)}{(k+q)^2 + (m^2c^2/\hbar^2) - i\epsilon}\left(\boldsymbol{\epsilon}_{\alpha}(\boldsymbol{q})\cdot\boldsymbol{\gamma}\right)$$

$$+ \left(\boldsymbol{\epsilon}_{\alpha}(\boldsymbol{q})\cdot\boldsymbol{\gamma}\right)\frac{mc - \hbar\gamma\cdot(k-q')}{(k-q')^2 + (m^2c^2/\hbar^2) - i\epsilon}\left(\boldsymbol{\epsilon}_{\alpha'}(\boldsymbol{q}')\cdot\boldsymbol{\gamma}\right)\Big]u_s(\boldsymbol{k}). \quad (22.34)$$

The first contribution to the amplitude corresponds to absorption of a photon with wave vector q followed by emission of a photon with wave vector q', see Figure 22.1, while the second contribution to the amplitude corresponds to emission of the photon with wave vector q' before absorption of the photon with wave vector q as shown in Figure 22.2.

The denominators in (22.34) can be simplified by noting that $k^2+(m^2c^2/\hbar^2) = 0$, $q^2 = q'^2 = 0$, and

$$k\cdot q = \boldsymbol{k}\cdot\boldsymbol{q} - |\boldsymbol{q}|\sqrt{\boldsymbol{k}^2 + (m^2c^2/\hbar^2)} < 0.$$

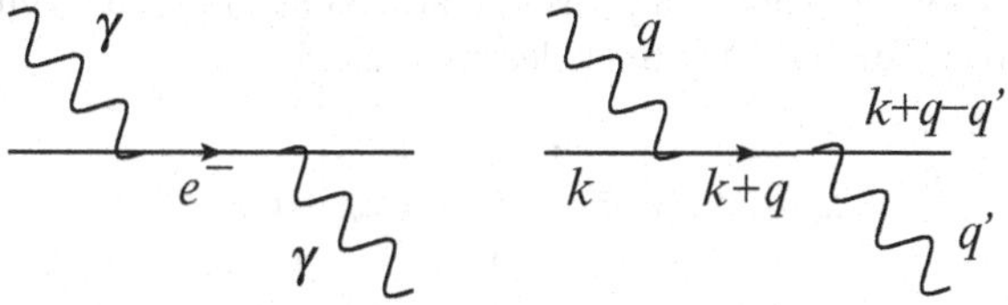

Fig. 22.1 Absorption of the incoming photon with momentum $\hbar\boldsymbol{q}$ before emission of the outgoing photon with momentum $\hbar\boldsymbol{q}'$. The virtual intermediate electron has 4-momentum $\hbar(k+q)$. The left panel uses particle labels and the right panel uses momentum labels

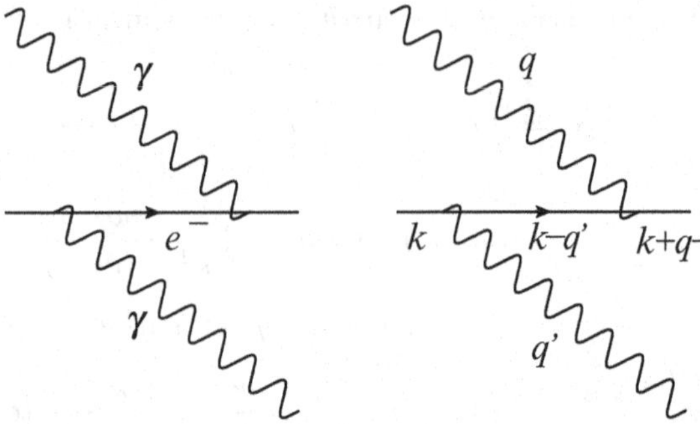

Fig. 22.2 Emission of the outgoing photon with momentum $\hbar\boldsymbol{q}'$ before absorption of the incoming photon with momentum $\hbar\boldsymbol{q}$. The virtual intermediate electron has 4-momentum $\hbar(k-q')$

This yields with the definition $\alpha = e^2/(4\pi\epsilon_0\hbar c)$ (7.61) of Sommerfeld's fine structure constant the result

$$\begin{aligned} S_{fi} &= \delta(k'+q'-k-q)\frac{\mathrm{i}\alpha}{8\pi\hbar\sqrt{|\boldsymbol{q}||\boldsymbol{q}'|E(\boldsymbol{k})E(\boldsymbol{k}')}} \\ &\times \overline{u}_{s'}(\boldsymbol{k}')\Big[\big(\boldsymbol{\epsilon}_{\alpha'}(\boldsymbol{q}')\cdot\boldsymbol{\gamma}\big)\frac{mc-\hbar\gamma\cdot(k+q)}{k\cdot q}\big(\boldsymbol{\epsilon}_{\alpha}(\boldsymbol{q})\cdot\boldsymbol{\gamma}\big) \\ &- \big(\boldsymbol{\epsilon}_{\alpha}(\boldsymbol{q})\cdot\boldsymbol{\gamma}\big)\frac{mc-\hbar\gamma\cdot(k-q')}{k\cdot q'}\big(\boldsymbol{\epsilon}_{\alpha'}(\boldsymbol{q}')\cdot\boldsymbol{\gamma}\big)\Big]u_s(\boldsymbol{k}). \end{aligned} \tag{22.35}$$

The spin and helicity polarized differential scattering cross section then follows from (22.9),

$$\begin{aligned} v d\sigma_{\boldsymbol{k},s;\boldsymbol{q},\alpha\to\boldsymbol{k}',s';\boldsymbol{q}',\alpha'} &= c d^3\boldsymbol{k}' d^3\boldsymbol{q}'\frac{\alpha^2\delta(k'+q'-k-q)}{16\hbar^2|\boldsymbol{q}||\boldsymbol{q}'|E(\boldsymbol{k})E(\boldsymbol{k}')} \\ &\times \Big|\overline{u}_{s'}(\boldsymbol{k}')\Big[\big(\boldsymbol{\epsilon}_{\alpha'}(\boldsymbol{q}')\cdot\boldsymbol{\gamma}\big)\frac{mc-\hbar\gamma\cdot(k+q)}{k\cdot q}\big(\boldsymbol{\epsilon}_{\alpha}(\boldsymbol{q})\cdot\boldsymbol{\gamma}\big) \\ &- \big(\boldsymbol{\epsilon}_{\alpha}(\boldsymbol{q})\cdot\boldsymbol{\gamma}\big)\frac{mc-\hbar\gamma\cdot(k-q')}{k\cdot q'}\big(\boldsymbol{\epsilon}_{\alpha'}(\boldsymbol{q}')\cdot\boldsymbol{\gamma}\big)\Big]u_s(\boldsymbol{k})\Big|^2. \end{aligned} \tag{22.36}$$

Spin-polarized cross sections are usually of less physical interest than electron-photon cross sections which average over polarizations of initial electron states and sum over the polarizations of the final electron states[3],

$$d\sigma_{\boldsymbol{k};\boldsymbol{q},\alpha\to\boldsymbol{k}';\boldsymbol{q}',\alpha'} = \frac{1}{2}\sum_{s,s'} d\sigma_{\boldsymbol{k},s;\boldsymbol{q},\alpha\to\boldsymbol{k}',s';\boldsymbol{q}',\alpha'}.$$

[3]However, spin polarized cross sections for electron scattering will likely become important in the framework of spintronics and spin based quantum computing.

Use of the property (21.58)

$$\sum_s u_s(\boldsymbol{k})\overline{u}_s(\boldsymbol{k}) = mc^2 - \hbar c\gamma^\mu k_\mu\Big|_{k^0=\omega(\boldsymbol{k})/c}$$

and of the relations $(\overline{u})^+ = \gamma^0 u$, $\gamma_\mu^+ = \gamma^0\gamma_\mu\gamma^0$, yields

$$\begin{aligned}
v d\sigma_{\boldsymbol{k};\boldsymbol{q},\alpha\to\boldsymbol{k}';\boldsymbol{q}',\alpha'} &= d^3\boldsymbol{k}'d^3\boldsymbol{q}'\frac{\alpha^2c^3\delta(k'+q'-k-q)}{32\hbar^2|\boldsymbol{q}||\boldsymbol{q}'|E(\boldsymbol{k})E(\boldsymbol{k}')}\,\mathrm{tr}\Bigg[\left(mc-\hbar\gamma\cdot k'\right)\\
&\times\Bigg((\boldsymbol{\epsilon}_{\alpha'}(\boldsymbol{q}')\cdot\boldsymbol{\gamma})\frac{mc-\hbar\gamma\cdot(k+q)}{k\cdot q}(\boldsymbol{\epsilon}_\alpha(\boldsymbol{q})\cdot\boldsymbol{\gamma})\\
&-(\boldsymbol{\epsilon}_\alpha(\boldsymbol{q})\cdot\boldsymbol{\gamma})\frac{mc-\hbar\gamma\cdot(k-q')}{k\cdot q'}(\boldsymbol{\epsilon}_{\alpha'}(\boldsymbol{q}')\cdot\boldsymbol{\gamma})\Bigg)(mc-\hbar\gamma\cdot k)\\
&\times\Bigg((\boldsymbol{\epsilon}_\alpha(\boldsymbol{q})\cdot\boldsymbol{\gamma})\frac{mc-\hbar\gamma\cdot(k+q)}{k\cdot q}(\boldsymbol{\epsilon}_{\alpha'}(\boldsymbol{q}')\cdot\boldsymbol{\gamma})\\
&-(\boldsymbol{\epsilon}_{\alpha'}(\boldsymbol{q}')\cdot\boldsymbol{\gamma})\frac{mc-\hbar\gamma\cdot(k-q')}{k\cdot q'}(\boldsymbol{\epsilon}_\alpha(\boldsymbol{q})\cdot\boldsymbol{\gamma})\Bigg)\Bigg].
\end{aligned}\tag{22.37}$$

This can be evaluated using the trace theorems for γ matrices from Appendix G. The full evaluation of $d\sigma_{\boldsymbol{k};\boldsymbol{q},\alpha\to\boldsymbol{k}';\boldsymbol{q}',\alpha'}$ needs in particular the trace theorems (G.20–G.22) for products of 4, 6 and 8 γ matrices.

We can simplify the evaluation in the rest frame of the initial electron,

$$k = \frac{1}{\hbar c}\begin{pmatrix}\sqrt{m^2c^4+\hbar^2\boldsymbol{k}^2}\\ \hbar c\boldsymbol{k}\end{pmatrix} \Rightarrow \frac{mc}{\hbar}\begin{pmatrix}1\\ \boldsymbol{0}\end{pmatrix}.$$

We can also use that $\boldsymbol{\epsilon}_\alpha(\boldsymbol{q})\cdot\boldsymbol{q} = \epsilon_\alpha(\boldsymbol{q})\cdot q = 0$ implies

$$(\gamma\cdot q)\,(\boldsymbol{\epsilon}_\alpha(\boldsymbol{q})\cdot\boldsymbol{\gamma}) = -\,(\boldsymbol{\epsilon}_\alpha(\boldsymbol{q})\cdot\boldsymbol{\gamma})\,(\gamma\cdot q).$$

This reduces products according to

$$\begin{aligned}
&(mc-\hbar\gamma\cdot(k+q))\,(\boldsymbol{\epsilon}_\alpha(\boldsymbol{q})\cdot\boldsymbol{\gamma})\,(mc-\hbar\gamma\cdot k)\\
&= \left(mc+mc\gamma^0-\hbar\gamma\cdot q\right)(\boldsymbol{\epsilon}_\alpha(\boldsymbol{q})\cdot\boldsymbol{\gamma})\,mc\left(1+\gamma^0\right)\\
&= (\boldsymbol{\epsilon}_\alpha(\boldsymbol{q})\cdot\boldsymbol{\gamma})\left(mc-mc\gamma^0+\hbar\gamma\cdot q\right)mc\left(1+\gamma^0\right)\\
&= m\hbar c\,(\boldsymbol{\epsilon}_\alpha(\boldsymbol{q})\cdot\boldsymbol{\gamma})\,(\gamma\cdot q)\left(1+\gamma^0\right).
\end{aligned}$$

The resulting cross section in the rest frame of the electron before scattering is

$$d\sigma_{0;q,\alpha\to k';q',\alpha'} = d^3k'd^3q'\frac{\alpha^2\hbar^2\delta(k'+q'-k-q)}{32m^2c|\boldsymbol{q}||\boldsymbol{q}'|E(\boldsymbol{k}')}\operatorname{tr}\Bigg[(mc-\hbar\gamma\cdot k')$$
$$\times\left((\boldsymbol{\epsilon}_{\alpha'}(\boldsymbol{q}')\cdot\boldsymbol{\gamma})(\boldsymbol{\epsilon}_{\alpha}(\boldsymbol{q})\cdot\boldsymbol{\gamma})\frac{\gamma\cdot q}{|\boldsymbol{q}|}+(\boldsymbol{\epsilon}_{\alpha}(\boldsymbol{q})\cdot\boldsymbol{\gamma})(\boldsymbol{\epsilon}_{\alpha'}(\boldsymbol{q}')\cdot\boldsymbol{\gamma})\frac{\gamma\cdot q'}{|\boldsymbol{q}'|}\right)(1+\gamma^0)$$
$$\times\left(\frac{\gamma\cdot q}{|\boldsymbol{q}|}(\boldsymbol{\epsilon}_{\alpha}(\boldsymbol{q})\cdot\boldsymbol{\gamma})(\boldsymbol{\epsilon}_{\alpha'}(\boldsymbol{q}')\cdot\boldsymbol{\gamma})+\frac{\gamma\cdot q'}{|\boldsymbol{q}'|}(\boldsymbol{\epsilon}_{\alpha'}(\boldsymbol{q}')\cdot\boldsymbol{\gamma})(\boldsymbol{\epsilon}_{\alpha}(\boldsymbol{q})\cdot\boldsymbol{\gamma})\right)\Bigg]. \tag{22.38}$$

Traces over products of an odd number of γ matrices vanish. The terms under the trace proportional to mc^2 contain products of six γ matrices, but two of these products vanish due to $(\gamma\cdot q)^2 = -q^2 = 0$ and $(\gamma\cdot q')^2 = -q'^2 = 0$. The remaining two terms involving six γ matrices turn out to yield the same result, such that the contribution to the trace term from products of six γ matrices is

$$tr_6 = \frac{8mc}{|\boldsymbol{q}||\boldsymbol{q}'|}\Big(q\cdot q' - 2(\boldsymbol{\epsilon}_{\alpha}(\boldsymbol{q})\cdot\boldsymbol{\epsilon}_{\alpha'}(\boldsymbol{q}'))^2 q\cdot q'$$
$$+2(\boldsymbol{\epsilon}_{\alpha}(\boldsymbol{q})\cdot\boldsymbol{\epsilon}_{\alpha'}(\boldsymbol{q}'))(\boldsymbol{\epsilon}_{\alpha}(\boldsymbol{q})\cdot\boldsymbol{q}')(\boldsymbol{\epsilon}_{\alpha'}(\boldsymbol{q}')\cdot\boldsymbol{q})\Big). \tag{22.39}$$

For the traces over products of eight γ matrices, we observe that those which contain the products $(\gamma\cdot q)\gamma^0(\gamma\cdot q)$ or $(\gamma\cdot q')\gamma^0(\gamma\cdot q')$ can be simplified to products of six γ matrices due to

$$\gamma^\mu\gamma^0\gamma^\nu q_\mu q_\nu = \left(-2\eta^{\mu 0}\gamma^\nu - \gamma^0\gamma^\mu\gamma^\nu\right)q_\mu q_\nu = -2|\boldsymbol{q}|\gamma^\nu q_\nu.$$

This yields for the sum of those terms which contain the products $(\gamma\cdot q)\gamma^0(\gamma\cdot q)$ or $(\gamma\cdot q')\gamma^0(\gamma\cdot q')$ the result

$$tr_{8a} = \frac{8\hbar}{|\boldsymbol{q}|}\Big(2(\boldsymbol{\epsilon}_{\alpha'}(\boldsymbol{q}')\cdot\boldsymbol{k}')(\boldsymbol{\epsilon}_{\alpha'}(\boldsymbol{q}')\cdot\boldsymbol{q}) - k'\cdot q\Big)$$
$$+\frac{8\hbar}{|\boldsymbol{q}'|}\Big(2(\boldsymbol{\epsilon}_{\alpha}(\boldsymbol{q})\cdot\boldsymbol{k}')(\boldsymbol{\epsilon}_{\alpha}(\boldsymbol{q})\cdot\boldsymbol{q}') - k'\cdot q'\Big),$$

and after substitution of $k' = k + q - q'$,

$$tr_{8a} = 16mc + \frac{8\hbar}{|\boldsymbol{q}|}\Big(q\cdot q' + 2(\boldsymbol{\epsilon}_{\alpha'}(\boldsymbol{q}')\cdot\boldsymbol{q})^2\Big)$$
$$-\frac{8\hbar}{|\boldsymbol{q}'|}\Big(q\cdot q' + 2(\boldsymbol{\epsilon}_{\alpha}(\boldsymbol{q})\cdot\boldsymbol{q}')^2\Big). \tag{22.40}$$

The traces over products of eight γ matrices which contain terms $(\gamma\cdot q)\gamma^0(\gamma\cdot q')$ or $(\gamma\cdot q')\gamma^0(\gamma\cdot q)$ can also be reduced to traces over products of six γ matrices by using the fact that γ^0 can only by contracted with one of the three γ matrices in products with 4-vectors. This yields after a bit of calculation and after substitution of $k' = k + q - q'$,

$$\begin{aligned}tr_{8b} &= 16mc\Big(2(\boldsymbol{\epsilon}_\alpha(\boldsymbol{q})\cdot\boldsymbol{\epsilon}_{\alpha'}(\boldsymbol{q}'))^2 - 1\Big) - \frac{8mc}{|\boldsymbol{q}||\boldsymbol{q}'|}\Big(q\cdot q' \\ &\quad - 2(\boldsymbol{\epsilon}_\alpha(\boldsymbol{q})\cdot\boldsymbol{\epsilon}_{\alpha'}(\boldsymbol{q}'))^2 q\cdot q' + 2(\boldsymbol{\epsilon}_\alpha(\boldsymbol{q})\cdot\boldsymbol{\epsilon}_{\alpha'}(\boldsymbol{q}'))(\boldsymbol{\epsilon}_\alpha(\boldsymbol{q})\cdot\boldsymbol{q}')(\boldsymbol{\epsilon}_{\alpha'}(\boldsymbol{q}')\cdot\boldsymbol{q})\Big) \\ &\quad - 16\frac{\hbar}{|\boldsymbol{q}|}(\boldsymbol{\epsilon}_{\alpha'}(\boldsymbol{q}')\cdot\boldsymbol{q})^2 + 16\frac{\hbar}{|\boldsymbol{q}'|}(\boldsymbol{\epsilon}_\alpha(\boldsymbol{q})\cdot\boldsymbol{q}')^2. \end{aligned} \tag{22.41}$$

The total trace term is therefore

$$\begin{aligned} tr &= tr_6 + tr_{8a} + tr_{8b} \\ &= 32mc(\boldsymbol{\epsilon}_\alpha(\boldsymbol{q})\cdot\boldsymbol{\epsilon}_{\alpha'}(\boldsymbol{q}'))^2 + 8\hbar\left(\frac{1}{|\boldsymbol{q}|} - \frac{1}{|\boldsymbol{q}'|}\right) q\cdot q', \end{aligned}$$

and combining all the terms yields

$$\begin{aligned} d\sigma_{0;\boldsymbol{q},\alpha\to\boldsymbol{k}';\boldsymbol{q}',\alpha'} &= d^3\boldsymbol{k}'d^3\boldsymbol{q}'\frac{\alpha^2\hbar^2\delta(k'+q'-k-q)}{4m^2c|\boldsymbol{q}||\boldsymbol{q}'|E(\boldsymbol{k}')} \\ &\quad\times\left[4mc(\boldsymbol{\epsilon}_\alpha(\boldsymbol{q})\cdot\boldsymbol{\epsilon}_{\alpha'}(\boldsymbol{q}'))^2 + \hbar\left(\frac{1}{|\boldsymbol{q}|} - \frac{1}{|\boldsymbol{q}'|}\right) q\cdot q'\right]. \end{aligned} \tag{22.42}$$

The product $q\cdot q'$ is directly related to the photon scattering angle,

$$q\cdot q' = -|\boldsymbol{q}||\boldsymbol{q}'|(1-\cos\theta).$$

However, energy and momentum conservation also imply

$$q\cdot q' = -\frac{1}{2}(q-q')^2 = -\frac{1}{2}(k'-k)^2 = \frac{m^2c^2}{\hbar^2} - \frac{mc}{\hbar}k'^0 = \frac{mc}{\hbar}\left(|\boldsymbol{q}'| - |\boldsymbol{q}|\right).$$

The relation between scattering angle and scattered photon wave number is therefore

$$\cos\theta = 1 - \frac{mc}{\hbar}\left(\frac{1}{|\boldsymbol{q}'|} - \frac{1}{|\boldsymbol{q}|}\right), \quad |\boldsymbol{q}'| = \frac{mc|\boldsymbol{q}|}{mc + \hbar|\boldsymbol{q}|(1-\cos\theta)}. \tag{22.43}$$

This is of course nothing but the Compton relation (1.11) for the wavelength of the scattered photon in terms of the scattering angle,

$$\lambda' = \lambda + \frac{h}{mc}(1-\cos\theta).$$

The four-dimensional δ function

$$\delta(k'+q'-k-q)=\delta\Big(\sqrt{\boldsymbol{k}'^2+(m^2c^2/\hbar^2)}+|\boldsymbol{q}'|-(mc/\hbar)-|\boldsymbol{q}|\Big)$$
$$\times\delta(\boldsymbol{k}'+\boldsymbol{q}'-\boldsymbol{q})$$

reduces the six-dimensional final state measure $d^3\boldsymbol{k}'d^3\boldsymbol{q}'$ to the two-dimensional measure $d\Omega(\hat{\boldsymbol{q}}')\equiv d\Omega$ over direction of the scattered photon after integration over $d^3\boldsymbol{k}'$ and $d|\boldsymbol{q}'|$. We include already the factor $|\boldsymbol{q}'|E(\boldsymbol{k}')$ in the denominator in equation (22.42) in the calculation:

$$\int d^3\boldsymbol{k}'\int_0^\infty d|\boldsymbol{q}'|\,\frac{|\boldsymbol{q}'|}{E(\boldsymbol{k}')}f(|\boldsymbol{q}'|)\delta(k'+q'-k-q)$$
$$=\int_0^\infty d|\boldsymbol{q}'|\,\frac{|\boldsymbol{q}'|f(|\boldsymbol{q}'|)}{c\sqrt{\hbar^2|\boldsymbol{q}'|^2+\hbar^2|\boldsymbol{q}|^2-2\hbar^2|\boldsymbol{q}'||\boldsymbol{q}|\cos\theta+m^2c^2}}$$
$$\times\delta\Big(\sqrt{|\boldsymbol{q}'|^2+|\boldsymbol{q}|^2-2|\boldsymbol{q}'||\boldsymbol{q}|\cos\theta+(m^2c^2/\hbar^2)}+|\boldsymbol{q}'|-(mc/\hbar)-|\boldsymbol{q}|\Big)$$
$$=\frac{1}{c}\left[\frac{|\boldsymbol{q}'|f(|\boldsymbol{q}'|)}{mc+\hbar|\boldsymbol{q}|(1-\cos\theta)}\right]_{|\boldsymbol{q}'|=mc|\boldsymbol{q}|/[mc+\hbar|\boldsymbol{q}|(1-\cos\theta)]}$$
$$=\frac{m|\boldsymbol{q}|}{[mc+\hbar|\boldsymbol{q}|\,(1-\cos\theta)]^2}f\Big(\frac{mc|\boldsymbol{q}|}{mc+\hbar|\boldsymbol{q}|\,(1-\cos\theta)}\Big).$$

This yields the Klein-Nishina cross section

$$d\sigma_{0;q,\alpha\to q-q';q',\alpha'}=d\Omega\frac{\alpha^2\hbar^2}{4mc\,[mc+\hbar|\boldsymbol{q}|\,(1-\cos\theta)]^2}$$
$$\times\left(4mc(\boldsymbol{\epsilon}_\alpha(\boldsymbol{q})\cdot\boldsymbol{\epsilon}_{\alpha'}(\boldsymbol{q}'))^2+\frac{\hbar^2|\boldsymbol{q}|^2\,(1-\cos\theta)^2}{mc+\hbar|\boldsymbol{q}|\,(1-\cos\theta)}\right). \tag{22.44}$$

Averaging over the initial photon polarization and summing over the final polarization (18.119) yields the unpolarized differential cross section

$$d\sigma_{0;q\to q-q';q'}=\frac{1}{2}\sum_{\alpha,\alpha'}d\sigma_{0;q,\alpha\to q-q';q',\alpha'}=d\Omega\frac{\alpha^2\hbar^2}{2mc\,[mc+\hbar|\boldsymbol{q}|\,(1-\cos\theta)]^2}$$
$$\times\left(mc(1+\cos^2\theta)+\frac{\hbar^2|\boldsymbol{q}|^2\,(1-\cos\theta)^2}{mc+\hbar|\boldsymbol{q}|\,(1-\cos\theta)}\right). \tag{22.45}$$

The resulting total cross section is

$$\sigma_{0;q\to q-q';q'} = \frac{\pi\alpha^2}{mc\hbar|\boldsymbol{q}|^3}\left[2\hbar|\boldsymbol{q}|\frac{2(mc)^3+8(mc)^2\hbar|\boldsymbol{q}|+9mc(\hbar|\boldsymbol{q}|)^2+(\hbar|\boldsymbol{q}|)^3}{(mc+2\hbar|\boldsymbol{q}|)^2}\right.$$
$$\left.-\left[2(mc)^2+2mc\hbar|\boldsymbol{q}|-(\hbar|\boldsymbol{q}|)^2\right]\ln\left(1+\frac{2\hbar|\boldsymbol{q}|}{mc}\right)\right]. \tag{22.46}$$

Photons below the hard X-ray regime satisfy $\hbar|\boldsymbol{q}| \ll mc$. This limit is also often denoted as the non-relativistic limit of Compton scattering because the kinetic energy imparted on the recoiling electron is small in this case,

$$\hbar^2(\boldsymbol{q}-\boldsymbol{q}')^2 \simeq 2\hbar^2\boldsymbol{q}^2(1-\cos\theta) \ll m^2c^2.$$

The cross section in the non-relativistic limit yields the Thomson cross section (18.120, 18.121) for photon scattering,

$$\frac{d\sigma_{0;q,\alpha\to q-q';q',\alpha'}}{d\Omega} = \left(\frac{\alpha\hbar}{mc}\right)^2(\boldsymbol{\epsilon}_\alpha(\boldsymbol{q})\cdot\boldsymbol{\epsilon}_{\alpha'}(\boldsymbol{q}'))^2,$$
$$\frac{d\sigma_{0;q\to q-q';q'}}{d\Omega} = \left(\frac{\alpha\hbar}{mc}\right)^2\frac{1+\cos^2\theta}{2},$$
$$\sigma_{0;q\to q-q';q'} = \frac{8\pi}{3}\left(\frac{\alpha\hbar}{mc}\right)^2 \equiv \sigma_T = 6.652\times 10^{-9}\,\text{Å}^2 = 0.6652\,\text{barn}. \tag{22.47}$$

The unpolarized differential scattering cross section (22.45) for Compton scattering is displayed for various photon energies in Figure 22.3. Forward scattering is energy independent, but scattering in other directions is suppressed with energy.

The energy dependence of the total Compton scattering cross section (22.46) is displayed in Figure 22.4.

22.4 Møller scattering

The leading order scattering cross section for electron-electron scattering was calculated in the framework of quantum electrodynamics by C. Møller[4].

The Hamiltonian (21.94) in Coulomb gauge $\nabla\cdot\boldsymbol{A}=0$ for the photon field is

$$H = H_0 + H_I + H_C, \tag{22.48}$$

[4]C. Møller, Annalen Phys. 406, 531 (1932).

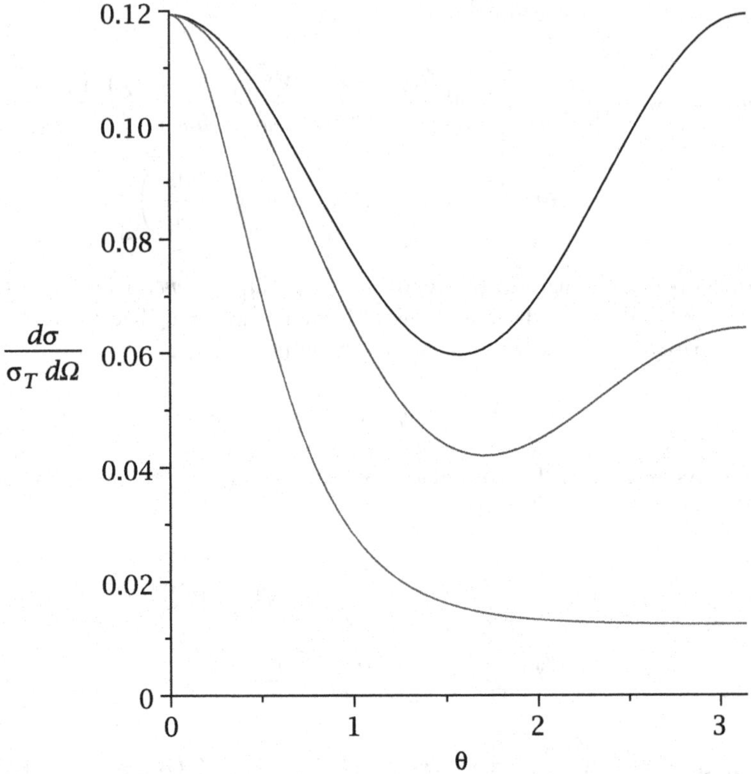

Fig. 22.3 The differential scattering cross section (22.45) for scattering angle $0 \leq \theta \leq \pi$. The energy of the incident photon is $E_\gamma = 0$ (top black curve), $E_\gamma = 0.2mc^2$ (center blue curve) and $E_\gamma = 2mc^2$ (lower red curve)

with the electron-photon interaction term

$$H_I \equiv H_{e\gamma} = ec \int d^3\boldsymbol{x}\, \overline{\Psi}(\boldsymbol{x}, t)\boldsymbol{\gamma} \cdot \boldsymbol{A}(\boldsymbol{x}, t)\Psi(\boldsymbol{x}, t)$$

and the Coulomb interaction term

$$H_C = \frac{e^2}{8\pi\epsilon_0} \sum_{ss'} \int d^3\boldsymbol{x} \int d^3\boldsymbol{x}'\, \Psi_s^+(\boldsymbol{x}, t)\Psi_{s'}^+(\boldsymbol{x}', t)\frac{1}{|\boldsymbol{x} - \boldsymbol{x}'|}\Psi_{s'}(\boldsymbol{x}', t)\Psi_s(\boldsymbol{x}, t).$$

Note that the summation is over Dirac indices, which are related to spin projections through the corresponding u or v spinors.

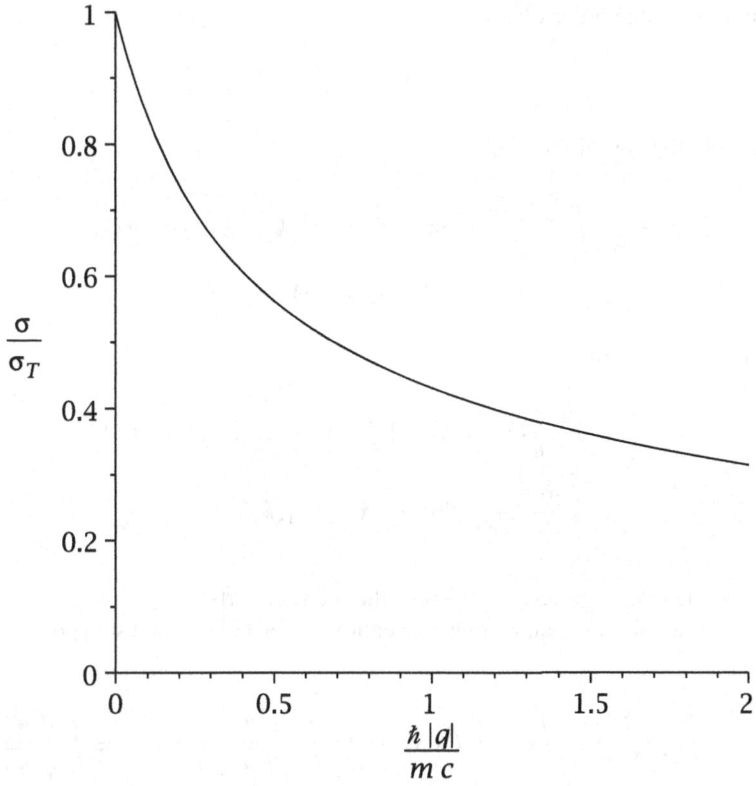

Fig. 22.4 The total Compton scattering cross section (22.46) in units of the Thomson cross section (22.47) for incident photon energy $0 < E_\gamma < 2mc^2$

The corresponding Hamiltonian on the states in the interaction picture is

$$H_D(t) = ec\sum_s \int d^3\boldsymbol{x}\,\overline{\Psi}_s(\boldsymbol{x},t)\boldsymbol{\gamma}\cdot\boldsymbol{A}(\boldsymbol{x},t)\Psi_s(\boldsymbol{x},t)$$

$$+\frac{e^2}{8\pi\epsilon_0}\sum_{ss'}\int d^3\boldsymbol{x}\int d^3\boldsymbol{x}'\,\Psi_s^+(\boldsymbol{x},t)\Psi_{s'}^+(\boldsymbol{x}',t)\frac{1}{|\boldsymbol{x}-\boldsymbol{x}'|}\Psi_{s'}(\boldsymbol{x}',t)\Psi_s(\boldsymbol{x},t)$$

with the freely evolving field operators $\boldsymbol{A}(\boldsymbol{x},t)$ (18.21) and $\Psi(\boldsymbol{x},t)$ (21.51) of the interaction picture.

The scattering matrix element for electron-electron scattering

$$S_{fi} \equiv S_{\boldsymbol{k}_1',s_1';\boldsymbol{k}_2',s_2'|\boldsymbol{k}_1,s_1;\boldsymbol{k}_2,s_2}$$

$$= \langle \boldsymbol{k}_1', s_1'; \boldsymbol{k}_2', s_2'|\mathrm{T}\exp\left(-\frac{\mathrm{i}}{\hbar}\int_{-\infty}^{\infty} dt\, H_D(t)\right)|\boldsymbol{k}_1, s_1; \boldsymbol{k}_2, s_2\rangle$$

becomes in leading order $\mathcal{O}(e^2)$

$$S_{fi} = S_{fi}^{(\gamma)} + S_{fi}^{(C)},$$

with the photon contribution

$$\begin{aligned}S_{fi}^{(\gamma)} &= -\frac{e^2}{\hbar^2}\int d^4x\int d^4x'\,\Theta(t-t')\langle \boldsymbol{k}_1', s_1'; \boldsymbol{k}_2', s_2'|\overline{\Psi}(x)\boldsymbol{\gamma}\cdot\boldsymbol{A}(x)\Psi(x)\\ &\quad\times\overline{\Psi}(x')\boldsymbol{\gamma}\cdot\boldsymbol{A}(x')\Psi(x')|\boldsymbol{k}_1, s_1; \boldsymbol{k}_2, s_2\rangle\end{aligned}$$

and the Coulomb term

$$\begin{aligned}S_{fi}^{(C)} &= \frac{\mu_0 e^2}{8\pi\mathrm{i}\hbar}\langle \boldsymbol{k}_1', s_1'; \boldsymbol{k}_2', s_2'|\int d^4x\int d^4x'\,\Psi^+(x)\Psi^+(x')\\ &\quad\times\frac{\delta(ct-ct')}{|\boldsymbol{x}-\boldsymbol{x}'|}\Psi(x')\Psi(x)|\boldsymbol{k}_1, s_1; \boldsymbol{k}_2, s_2\rangle.\end{aligned}$$

We evaluate $S_{fi}^{(\gamma)}$ first. Substitution of the relevant parts of the mode expansions yields (here we also use summation convention for the helicity and spin polarization indices)

$$\begin{aligned}S_{fi}^{(\gamma)} &= -\frac{e^2}{\hbar^2}\frac{\hbar\mu_0 c}{8(2\pi)^9}\int d^4x\int d^4x'\,\Theta(t-t')\int\frac{d^3\boldsymbol{q}_1'}{\sqrt{E(\boldsymbol{q}_1')}}\int\frac{d^3\boldsymbol{q}_2'}{\sqrt{E(\boldsymbol{q}_2')}}\int\frac{d^3\boldsymbol{q}'}{\sqrt{|\boldsymbol{q}'|}}\\ &\quad\times\int\frac{d^3\boldsymbol{q}_1}{\sqrt{E(\boldsymbol{q}_1)}}\int\frac{d^3\boldsymbol{q}_2}{\sqrt{E(\boldsymbol{q}_2)}}\int\frac{d^3\boldsymbol{q}}{\sqrt{|\boldsymbol{q}|}}\exp[\mathrm{i}(q'+q_2'-q_1')\cdot x-\mathrm{i}(q+q_1-q_2)\cdot x']\\ &\quad\times\overline{u}(\boldsymbol{q}_1',\sigma)\boldsymbol{\gamma}u(\boldsymbol{q}_2',\sigma')\cdot\boldsymbol{\epsilon}_\beta(\boldsymbol{q}')\boldsymbol{\epsilon}_\alpha(\boldsymbol{q})\cdot\overline{u}(\boldsymbol{q}_1,s)\boldsymbol{\gamma}u(\boldsymbol{q}_2,s')\langle 0|b(\boldsymbol{k}_1',s_1')b(\boldsymbol{k}_2',s_2')\\ &\quad\times b^+(\boldsymbol{q}_1',\sigma)b(\boldsymbol{q}_2',\sigma')a_\beta(\boldsymbol{q}')a_\alpha^+(\boldsymbol{q})b^+(\boldsymbol{q}_1,s)b(\boldsymbol{q}_2,s')b^+(\boldsymbol{k}_2,s_2)b^+(\boldsymbol{k}_1,s_1)|0\rangle.\end{aligned}$$

Elimination of the photon operators yields

$$\begin{aligned}S_{fi}^{(\gamma)} &= \frac{\mu_0 e^2 c}{8(2\pi)^9\hbar}\int d^4x\int d^4x'\,\Theta(t-t')\int\frac{d^3\boldsymbol{q}_1'}{\sqrt{E(\boldsymbol{q}_1')}}\int\frac{d^3\boldsymbol{q}_2'}{\sqrt{E(\boldsymbol{q}_2')}}\int\frac{d^3\boldsymbol{q}}{|\boldsymbol{q}|}\\ &\quad\times\int\frac{d^3\boldsymbol{q}_1}{\sqrt{E(\boldsymbol{q}_1)}}\int\frac{d^3\boldsymbol{q}_2}{\sqrt{E(\boldsymbol{q}_2)}}\exp[\mathrm{i}(q+q_2'-q_1')\cdot x-\mathrm{i}(q+q_1-q_2)\cdot x']\\ &\quad\times\overline{u}(\boldsymbol{q}_1',\sigma)\boldsymbol{\gamma}u(\boldsymbol{q}_2',\sigma')\cdot\boldsymbol{\epsilon}_\alpha(\boldsymbol{q})\boldsymbol{\epsilon}_\alpha(\boldsymbol{q})\cdot\overline{u}(\boldsymbol{q}_1,s)\boldsymbol{\gamma}u(\boldsymbol{q}_2,s')\langle 0|b(\boldsymbol{k}_1',s_1')b(\boldsymbol{k}_2',s_2')\\ &\quad\times b^+(\boldsymbol{q}_1',\sigma)b^+(\boldsymbol{q}_1,s)b(\boldsymbol{q}_2',\sigma')b(\boldsymbol{q}_2,s')b^+(\boldsymbol{k}_2,s_2)b^+(\boldsymbol{k}_1,s_1)|0\rangle,\end{aligned}$$

where fermionic operators were also re-ordered such that only the connected amplitude contributes. Evaluation of the fermionic operators yields

$$S_{fi}^{(\gamma)} = \frac{e^2}{8\hbar}\frac{\mu_0 c}{(2\pi)^9}\int d^4x \int d^4x' \int \frac{d^3\boldsymbol{q}}{|\boldsymbol{q}|}\frac{\Theta(t-t')\exp[\mathrm{i}q\cdot(x-x')]}{\sqrt{E(\boldsymbol{k}_1')E(\boldsymbol{k}_2')E(\boldsymbol{k}_1)E(\boldsymbol{k}_2)}}$$

$$\times\Big[\overline{u}(\boldsymbol{k}_2',s_2')\boldsymbol{\gamma}u(\boldsymbol{k}_1,s_1)\cdot\boldsymbol{\epsilon}_\alpha(\boldsymbol{q})\boldsymbol{\epsilon}_\alpha(\boldsymbol{q})\cdot\overline{u}(\boldsymbol{k}_1',s_1')\boldsymbol{\gamma}u(\boldsymbol{k}_2,s_2)$$

$$\times\Big(\exp[\mathrm{i}(k_1-k_2')\cdot x - \mathrm{i}(k_1'-k_2)\cdot x'] + \exp[\mathrm{i}(k_2-k_1')\cdot x - \mathrm{i}(k_2'-k_1)\cdot x']\Big)$$

$$-\overline{u}(\boldsymbol{k}_1',s_1')\boldsymbol{\gamma}u(\boldsymbol{k}_1,s_1)\cdot\boldsymbol{\epsilon}_\alpha(\boldsymbol{q})\boldsymbol{\epsilon}_\alpha(\boldsymbol{q})\cdot\overline{u}(\boldsymbol{k}_2',s_2')\boldsymbol{\gamma}u(\boldsymbol{k}_2,s_2)$$

$$\times\Big(\exp[\mathrm{i}(k_1-k_1')\cdot x - \mathrm{i}(k_2'-k_2)\cdot x']$$

$$+\exp[\mathrm{i}(k_2-k_2')\cdot x - \mathrm{i}(k_1'-k_1)\cdot x']\Big)\Big].$$

This yields after changing the integration variables $x \leftrightarrow x'$ in the second and fourth term

$$S_{fi}^{(\gamma)} = \frac{e^2}{8\hbar}\frac{\mu_0 c}{(2\pi)^9}\frac{1}{\sqrt{E(\boldsymbol{k}_1')E(\boldsymbol{k}_2')E(\boldsymbol{k}_1)E(\boldsymbol{k}_2)}}\int d^4x \int d^4x' \int \frac{d^3\boldsymbol{q}}{|\boldsymbol{q}|}$$

$$\times\Big(\overline{u}(\boldsymbol{k}_2',s_2')\boldsymbol{\gamma}u(\boldsymbol{k}_1,s_1)\cdot\boldsymbol{\epsilon}_\alpha(\boldsymbol{q})\boldsymbol{\epsilon}_\alpha(\boldsymbol{q})\cdot\overline{u}(\boldsymbol{k}_1',s_1')\boldsymbol{\gamma}u(\boldsymbol{k}_2,s_2)$$

$$\times\exp[\mathrm{i}(k_1-k_2')\cdot x + \mathrm{i}(k_2-k_1')\cdot x']$$

$$-\overline{u}(\boldsymbol{k}_1',s_1')\boldsymbol{\gamma}u(\boldsymbol{k}_1,s_1)\cdot\boldsymbol{\epsilon}_\alpha(\boldsymbol{q})\boldsymbol{\epsilon}_\alpha(\boldsymbol{q})\cdot\overline{u}(\boldsymbol{k}_2',s_2')\boldsymbol{\gamma}u(\boldsymbol{k}_2,s_2)$$

$$\times\exp[\mathrm{i}(k_1-k_1')\cdot x + \mathrm{i}(k_2-k_2')\cdot x']\Big)$$

$$\times\Big(\Theta(t-t')\exp[\mathrm{i}q\cdot(x-x')] + \Theta(t'-t)\exp[\mathrm{i}q\cdot(x'-x)]\Big).$$

We can use the following equation with $\omega(\boldsymbol{q}) = c\sqrt{\boldsymbol{q}^2 + (m^2c^2/\hbar^2)}$

$$\int d^4q\,\frac{f(\boldsymbol{q})\exp(\mathrm{i}q\cdot x)}{q^2+(m^2c^2/\hbar^2)-\mathrm{i}\epsilon} = -\int d^3\boldsymbol{q}\int d\omega\,\frac{cf(\boldsymbol{q})\exp(\mathrm{i}\boldsymbol{q}\cdot\boldsymbol{x})\exp(-\mathrm{i}\omega t)}{[\omega-\omega(\boldsymbol{q})+\mathrm{i}\epsilon][\omega+\omega(\boldsymbol{q})-\mathrm{i}\epsilon]}$$

$$= 2\pi c\mathrm{i}\Theta(t)\int d^3\boldsymbol{q}f(\boldsymbol{q})\exp(\mathrm{i}\boldsymbol{q}\cdot\boldsymbol{x})\frac{\exp(-\mathrm{i}\omega(\boldsymbol{q})t)}{2\omega(\boldsymbol{q})}$$

$$-2\pi c\mathrm{i}\Theta(-t)\int d^3\boldsymbol{q}f(\boldsymbol{q})\exp(\mathrm{i}\boldsymbol{q}\cdot\boldsymbol{x})\frac{\exp(\mathrm{i}\omega(\boldsymbol{q})t)}{-2\omega(\boldsymbol{q})}$$

$$= \mathrm{i}\pi c\,\Theta(t)\int\frac{d^3\boldsymbol{q}}{\omega(\boldsymbol{q})}f(\boldsymbol{q})\exp(\mathrm{i}q\cdot x)\Big|_{\omega=\omega(\boldsymbol{q})}$$

$$+\mathrm{i}\pi c\,\Theta(-t)\int\frac{d^3\boldsymbol{q}}{\omega(\boldsymbol{q})}f(\boldsymbol{q})\exp(-\mathrm{i}q\cdot x)\Big|_{\omega=\omega(\boldsymbol{q})} \tag{22.49}$$

to find

$$
\begin{aligned}
S_{fi}^{(\gamma)} &= \frac{\mu_0 e^2 c}{4\mathrm{i}\hbar(2\pi)^{10}} \frac{1}{\sqrt{E(\boldsymbol{k}_1')E(\boldsymbol{k}_2')E(\boldsymbol{k}_1)E(\boldsymbol{k}_2)}} \int d^4x \int d^4x' \int \frac{d^4q}{q^2 - \mathrm{i}\epsilon} \\
&\times \Big(\overline{u}(\boldsymbol{k}_2', s_2')\boldsymbol{\gamma} u(\boldsymbol{k}_1, s_1) \cdot \boldsymbol{\epsilon}_\alpha(\boldsymbol{q})\boldsymbol{\epsilon}_\alpha(\boldsymbol{q}) \cdot \overline{u}(\boldsymbol{k}_1', s_1')\boldsymbol{\gamma} u(\boldsymbol{k}_2, s_2) \\
&\times \exp[\mathrm{i}(k_1 - k_2' + q)\cdot x + \mathrm{i}(k_2 - k_1' - q)\cdot x'] \\
&- \overline{u}(\boldsymbol{k}_1', s_1')\boldsymbol{\gamma} u(\boldsymbol{k}_1, s_1) \cdot \boldsymbol{\epsilon}_\alpha(\boldsymbol{q})\boldsymbol{\epsilon}_\alpha(\boldsymbol{q}) \cdot \overline{u}(\boldsymbol{k}_2', s_2')\boldsymbol{\gamma} u(\boldsymbol{k}_2, s_2) \\
&\times \exp[\mathrm{i}(k_1 - k_1' + q)\cdot x + \mathrm{i}(k_2 - k_2' - q)\cdot x'] \Big).
\end{aligned}
$$

The integrations then yield

$$
\begin{aligned}
S_{fi}^{(\gamma)} &= \frac{\mu_0 e^2 c}{16\mathrm{i}\pi^2\hbar} \frac{\delta(k_1 + k_2 - k_1' - k_2')}{\sqrt{E(\boldsymbol{k}_1')E(\boldsymbol{k}_2')E(\boldsymbol{k}_1)E(\boldsymbol{k}_2)}} \\
&\times \left(\frac{\overline{u}(\boldsymbol{k}_2', s_2')\boldsymbol{\gamma} u(\boldsymbol{k}_1, s_1) \cdot \boldsymbol{\epsilon}_\alpha(\boldsymbol{k}_2' - \boldsymbol{k}_1)\boldsymbol{\epsilon}_\alpha(\boldsymbol{k}_2' - \boldsymbol{k}_1) \cdot \overline{u}(\boldsymbol{k}_1', s_1')\boldsymbol{\gamma} u(\boldsymbol{k}_2, s_2)}{(k_2' - k_1)^2 - \mathrm{i}\epsilon} \right. \\
&- \left. \frac{\overline{u}(\boldsymbol{k}_1', s_1')\boldsymbol{\gamma} u(\boldsymbol{k}_1, s_1) \cdot \boldsymbol{\epsilon}_\alpha(\boldsymbol{k}_1' - \boldsymbol{k}_1)\boldsymbol{\epsilon}_\alpha(\boldsymbol{k}_1' - \boldsymbol{k}_1) \cdot \overline{u}(\boldsymbol{k}_2', s_2')\boldsymbol{\gamma} u(\boldsymbol{k}_2, s_2)}{(k_1' - k_1)^2 - \mathrm{i}\epsilon} \right).
\end{aligned}
$$

Taking into account the energy-momentum conserving δ function, the transversal projectors can e.g. be written as

$$
\boldsymbol{\epsilon}_\alpha(\boldsymbol{k}_1' - \boldsymbol{k}_1) \otimes \boldsymbol{\epsilon}_\alpha(\boldsymbol{k}_1' - \boldsymbol{k}_1) = \underline{1} + \frac{(\boldsymbol{k}_1' - \boldsymbol{k}_1) \otimes (\boldsymbol{k}_2' - \boldsymbol{k}_2)}{(\boldsymbol{k}_1' - \boldsymbol{k}_1)^2}.
$$

The Dirac equation implies

$$
\overline{u}(\boldsymbol{k}', s')\boldsymbol{\gamma} \cdot (\boldsymbol{k}' - \boldsymbol{k}) u(\boldsymbol{k}, s) = \frac{E(\boldsymbol{k}') - E(\boldsymbol{k})}{\hbar c} \overline{u}(\boldsymbol{k}', s')\gamma^0 u(\boldsymbol{k}, s).
$$

This yields the photon exchange contribution to the electron-electron scattering matrix element,

$$
\begin{aligned}
S_{fi}^{(\gamma)} &= \frac{\mu_0 c e^2}{16\mathrm{i}\pi^2\hbar} \frac{\delta(k_1 + k_2 - k_1' - k_2')}{\sqrt{E(\boldsymbol{k}_1')E(\boldsymbol{k}_2')E(\boldsymbol{k}_1)E(\boldsymbol{k}_2)}} \\
&\times \left(\frac{\overline{u}(\boldsymbol{k}_2', s_2')\boldsymbol{\gamma} u(\boldsymbol{k}_1, s_1) \cdot \overline{u}(\boldsymbol{k}_1', s_1')\boldsymbol{\gamma} u(\boldsymbol{k}_2, s_2)}{(k_2' - k_1)^2 - \mathrm{i}\epsilon} \right. \\
&- \frac{\overline{u}(\boldsymbol{k}_1', s_1')\boldsymbol{\gamma} u(\boldsymbol{k}_1, s_1) \cdot \overline{u}(\boldsymbol{k}_2', s_2')\boldsymbol{\gamma} u(\boldsymbol{k}_2, s_2)}{(k_1' - k_1)^2 - \mathrm{i}\epsilon}
\end{aligned}
$$

$$
-\frac{\overline{u}(\boldsymbol{k}_2',s_2')\gamma^0u(\boldsymbol{k}_1,s_1)\overline{u}(\boldsymbol{k}_1',s_1')\gamma^0u(\boldsymbol{k}_2,s_2)}{(k_2'-k_1)^2-\mathrm{i}\epsilon}\frac{[E(\boldsymbol{k}_2')-E(\boldsymbol{k}_1)]^2}{\hbar^2c^2(\boldsymbol{k}_2'-\boldsymbol{k}_1)^2}
$$

$$
+\frac{\overline{u}(\boldsymbol{k}_1',s_1')\gamma^0u(\boldsymbol{k}_1,s_1)\overline{u}(\boldsymbol{k}_2',s_2')\gamma^0u(\boldsymbol{k}_2,s_2)}{(k_1'-k_1)^2-\mathrm{i}\epsilon}\frac{[E(\boldsymbol{k}_1')-E(\boldsymbol{k}_1)]^2}{\hbar^2c^2(\boldsymbol{k}_1'-\boldsymbol{k}_1)^2}\Bigg). \tag{22.50}
$$

For the evaluation of the Coulomb term, substitution of the mode expansions and evaluation of the operators in $S_{fi}^{(C)}$ yields

$$
\begin{aligned}
S_{fi}^{(C)} =& \frac{\mu_0ce^2}{8\pi\mathrm{i}\hbar}\frac{2}{4(2\pi)^6\sqrt{E(\boldsymbol{k}_1')E(\boldsymbol{k}_2')E(\boldsymbol{k}_1)E(\boldsymbol{k}_2)}}\int d^4x\int d^4x'\,\frac{\delta(ct-ct')}{|\boldsymbol{x}-\boldsymbol{x}'|}\\
&\times\Big(u^+(\boldsymbol{k}_1',s_1')u(\boldsymbol{k}_1,s_1)u^+(\boldsymbol{k}_2',s_2')u(\boldsymbol{k}_2,s_2)\exp\!\big[\mathrm{i}(k_2'-k_2)\cdot x'+\mathrm{i}(k_1'-k_1)\cdot x\big]\\
&-u^+(\boldsymbol{k}_2',s_2')u(\boldsymbol{k}_1,s_1)u^+(\boldsymbol{k}_1',s_1')u(\boldsymbol{k}_2,s_2)\exp\!\big[\mathrm{i}(k_1'-k_2)\cdot x'+\mathrm{i}(k_2'-k_1)\cdot x\big]\Big)\\
=&\frac{\mu_0ce^2}{16\pi\mathrm{i}\hbar}\frac{1}{(2\pi)^6\sqrt{E(\boldsymbol{k}_1')E(\boldsymbol{k}_2')E(\boldsymbol{k}_1)E(\boldsymbol{k}_2)}}\\
&\times\int d^4x\int d^3\boldsymbol{x}'\,\frac{\exp\!\big[\mathrm{i}(k_1'+k_2'-k_1-k_2)\cdot x\big]}{|\boldsymbol{x}-\boldsymbol{x}'|}\\
&\times\Big(\overline{u}(\boldsymbol{k}_1',s_1')\gamma^0u(\boldsymbol{k}_1,s_1)\overline{u}(\boldsymbol{k}_2',s_2')\gamma^0u(\boldsymbol{k}_2,s_2)\exp\!\big[\mathrm{i}(\boldsymbol{k}_2'-\boldsymbol{k}_2)\cdot(\boldsymbol{x}'-\boldsymbol{x})\big]\\
&-\overline{u}(\boldsymbol{k}_2',s_2')\gamma^0u(\boldsymbol{k}_1,s_1)\overline{u}(\boldsymbol{k}_1',s_1')\gamma^0u(\boldsymbol{k}_2,s_2)\exp\!\big[\mathrm{i}(\boldsymbol{k}_1'-\boldsymbol{k}_2)\cdot(\boldsymbol{x}'-\boldsymbol{x})\big]\Big).
\end{aligned}
$$

In the next step we use the Fourier decomposition (22.17) of the Coulomb potential to find

$$
\begin{aligned}
S_{fi}^{(C)} =& \frac{\mu_0ce^2}{16\pi^2\mathrm{i}\hbar}\frac{\delta(k_1'+k_2'-k_1-k_2)}{\sqrt{E(\boldsymbol{k}_1')E(\boldsymbol{k}_2')E(\boldsymbol{k}_1)E(\boldsymbol{k}_2)}}\\
&\times\left(\frac{\overline{u}(\boldsymbol{k}_1',s_1')\gamma^0u(\boldsymbol{k}_1,s_1)\overline{u}(\boldsymbol{k}_2',s_2')\gamma^0u(\boldsymbol{k}_2,s_2)}{(\boldsymbol{k}_2'-\boldsymbol{k}_2)^2}\right.\\
&\left.-\frac{\overline{u}(\boldsymbol{k}_2',s_2')\gamma^0u(\boldsymbol{k}_1,s_1)\overline{u}(\boldsymbol{k}_1',s_1')\gamma^0u(\boldsymbol{k}_2,s_2)}{(\boldsymbol{k}_2'-\boldsymbol{k}_1)^2}\right).
\end{aligned} \tag{22.51}
$$

For the addition of $S_{fi}^{(\gamma)}$ and $S_{fi}^{(C)}$, we observe

$$
\frac{1}{(\boldsymbol{k}'-\boldsymbol{k})^2}\left(\frac{[\omega(\boldsymbol{k}')-\omega(\boldsymbol{k})]^2}{c^2(k'-k)^2}+1\right)=\frac{1}{(k'-k)^2} \tag{22.52}
$$

to find

$$S_{fi} = i\frac{\mu_0 ce^2}{16\pi^2\hbar}\frac{\delta(k_1' + k_2' - k_1 - k_2)}{\sqrt{E(\boldsymbol{k}_1')E(\boldsymbol{k}_2')E(\boldsymbol{k}_1)E(\boldsymbol{k}_2)}}$$
$$\times\left(\frac{\overline{u}(\boldsymbol{k}_1', s_1')\gamma^\mu u(\boldsymbol{k}_1, s_1)\overline{u}(\boldsymbol{k}_2', s_2')\gamma_\mu u(\boldsymbol{k}_2, s_2)}{(k_1' - k_1)^2}\right.$$
$$\left.-\frac{\overline{u}(\boldsymbol{k}_2', s_2')\gamma^\mu u(\boldsymbol{k}_1, s_1)\overline{u}(\boldsymbol{k}_1', s_1')\gamma_\mu u(\boldsymbol{k}_2, s_2)}{(k_2' - k_1)^2}\right)$$
$$= -i\mathcal{M}_{fi}\delta(k_1 + k_2 - k_1' - k_2'), \tag{22.53}$$

where the last equation defines the scattering amplitude

$$\mathcal{M}_{fi} \equiv \mathcal{M}_{\boldsymbol{k}_1', s_1'; \boldsymbol{k}_2', s_2' | \boldsymbol{k}_1, s_1; \boldsymbol{k}_2, s_2}$$

for Møller scattering.

The two contributions to the scattering amplitude can be interpreted as virtual photon exchange with virtual photon 4-momentum $k_1 - k_1'$ or $k_1 - k_2'$, respectively. This is shown in Figure 22.5.

The scattering amplitude (22.53) yields the spin polarized differential cross section (22.10)

$$v d\sigma_{\boldsymbol{k}_1', s_1'; \boldsymbol{k}_2', s_2' | \boldsymbol{k}_1, s_1; \boldsymbol{k}_2, s_2} = 4\pi^2 c \left|\mathcal{M}_{\boldsymbol{k}_1', s_1'; \boldsymbol{k}_2', s_2' | \boldsymbol{k}_1, s_1; \boldsymbol{k}_2, s_2}\right|^2$$
$$\times\delta(k_1 + k_2 - k_1' - k_2')\frac{1}{2}d^3\boldsymbol{k}_1' d^3\boldsymbol{k}_2',$$

where

$$v = \frac{c^3}{E(\boldsymbol{k}_1)E(\boldsymbol{k}_2)}\sqrt{(\hbar^2 k_1 \cdot k_2)^2 - m^4 c^4} \tag{22.54}$$

is the relative speed (22.4) between the two electrons with momentum 4-vectors $\hbar k_1$ and $\hbar k_2$.

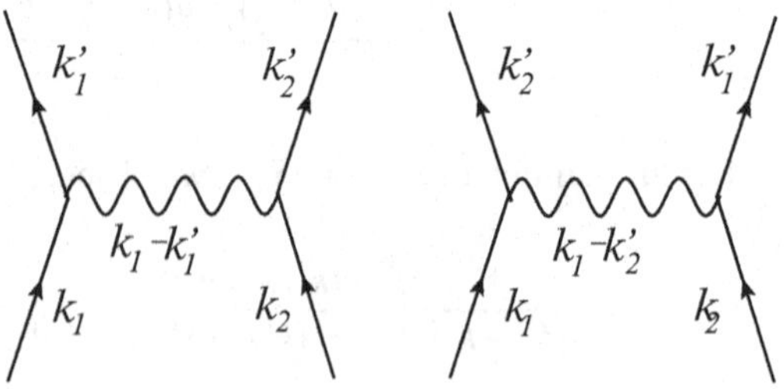

Fig. 22.5 Contributions to the Møller scattering amplitude (22.53)

The differential cross section is often averaged over initial spin states and summed over final spin states,

$$d\sigma_{k'_1;k'_2|k_1;k_2} = \frac{1}{4}\sum_{s_1,s_2,s'_1,s'_2} d\sigma_{k'_1,s'_1;k'_2,s'_2|k_1,s_1;k_2,s_2}.$$

The property (21.58)

$$\sum_s u(\boldsymbol{k},s)\overline{u}(\boldsymbol{k},s) = mc^2 - \hbar c\gamma^\mu k_\mu\Big|_{k^0=\omega(\boldsymbol{k})/c}$$

yields

$$\begin{aligned}
v d\sigma = &\frac{c}{2} d^3\boldsymbol{k}'_1 d^3\boldsymbol{k}'_2 \left(\frac{\mu_0 c e^2}{16\pi\hbar}\right)^2 \frac{\delta(k_1+k_2-k'_1-k'_2)}{E(\boldsymbol{k}'_1)E(\boldsymbol{k}'_2)E(\boldsymbol{k}_1)E(\boldsymbol{k}_2)} \\
&\times\Bigg(\frac{1}{(k'_1-k_1)^4}\mathrm{tr}\big[\left(mc^2-\hbar c\gamma\cdot k'_1\right)\gamma^\mu\left(mc^2-\hbar c\gamma\cdot k_1\right)\gamma^\nu\big] \\
&\times\mathrm{tr}\big[\left(mc^2-\hbar c\gamma\cdot k'_2\right)\gamma_\mu\left(mc^2-\hbar c\gamma\cdot k_2\right)\gamma_\nu\big] \\
&+\frac{1}{(k'_2-k_1)^4}\mathrm{tr}\big[\left(mc^2-\hbar c\gamma\cdot k'_2\right)\gamma^\mu\left(mc^2-\hbar c\gamma\cdot k_1\right)\gamma^\nu\big] \\
&\times\mathrm{tr}\big[\left(mc^2-\hbar c\gamma\cdot k'_1\right)\gamma_\mu\left(mc^2-\hbar c\gamma\cdot k_2\right)\gamma_\nu\big] \\
&-\frac{2}{(k'_1-k_1)^2(k'_2-k_1)^2}\mathrm{tr}\big[\left(mc^2-\hbar c\gamma\cdot k_2\right)\gamma^\mu\left(mc^2-\hbar c\gamma\cdot k'_2\right) \\
&\times\gamma^\nu\left(mc^2-\hbar c\gamma\cdot k_1\right)\gamma_\mu\left(mc^2-\hbar c\gamma\cdot k'_1\right)\gamma_\nu\big]\Bigg),
\end{aligned} \tag{22.55}$$

where it is understood that all 4-momenta of electrons are on shell. The 4-momenta of the intermediate photons are then automatically off shell with dominant spacelike components, $(k'-k)^2 > 0$ (except in the zero momentum transfer limit $\boldsymbol{k}' = \boldsymbol{k}$).

The traces in equation (22.55) are readily evaluated using the contraction and trace theorems for γ matrices from Appendix G. This yields together with 4-momentum conservation $k'_1 + k'_2 = k_1 + k_2$ the result

$$\begin{aligned}
v d\sigma = &c d^3\boldsymbol{k}'_1 d^3\boldsymbol{k}'_2 \left(\frac{e^2 c}{4\pi\epsilon_0\hbar}\right)^2 \frac{\delta(k_1+k_2-k'_1-k'_2)}{E(\boldsymbol{k}'_1)E(\boldsymbol{k}'_2)E(\boldsymbol{k}_1)E(\boldsymbol{k}_2)} \\
&\times\left(\frac{\hbar^4(k_1\cdot k_2)^2+\hbar^4(k_1\cdot k'_2)^2+2m^2c^2\hbar^2 k_1\cdot k'_1+2m^4c^4}{(k'_1-k_1)^4}\right.
\end{aligned}$$

$$+\frac{\hbar^4(k_1\cdot k_2)^2+\hbar^4(k_1\cdot k_1')^2+2m^2c^2\hbar^2 k_1\cdot k_2'+2m^4c^4}{(k_2'-k_1)^4}$$
$$+2\frac{\hbar^4(k_1\cdot k_2)^2+2m^2c^2\hbar^2 k_1\cdot k_2}{(k_1'-k_1)^2(k_2'-k_1)^2}\Bigg). \tag{22.56}$$

We further evaluate the cross section through integration over $d^3\boldsymbol{k}_2'$ and $d|\boldsymbol{k}_1'|$ in the center of mass frame $\boldsymbol{k}_1+\boldsymbol{k}_2=\boldsymbol{0}$ of the colliding electrons. If we integrate over the final states $d^3\boldsymbol{k}_2'$ of one of the electrons to get a single-electron differential cross section $d\sigma/d\Omega$, we have to include a factor of 2 because we could just as well observe the electron with momentum $\boldsymbol{k}_2'$ being scattered into the direction $d\Omega$, see equation (22.12).

It is convenient to define $\boldsymbol{k}=\boldsymbol{k}_1$, $\boldsymbol{k}'=\boldsymbol{k}_1'$. The integration with the energy-momentum δ function then yields

$$\int d^3\boldsymbol{k}_2'\int_0^\infty d|\boldsymbol{k}'|\,|\boldsymbol{k}'|^2 f(\boldsymbol{k},\boldsymbol{k}',\boldsymbol{k}_2')\delta(\boldsymbol{k}_2'+\boldsymbol{k}')$$
$$\times\delta\left(2\sqrt{|\boldsymbol{k}'|^2+(mc/\hbar)^2}-2\sqrt{|\boldsymbol{k}|^2+(mc/\hbar)^2}\right)$$
$$=\frac{|\boldsymbol{k}|}{2}\sqrt{|\boldsymbol{k}|^2+(mc/\hbar)^2}f(\boldsymbol{k},|\boldsymbol{k}|\hat{\boldsymbol{k}}',-|\boldsymbol{k}|\hat{\boldsymbol{k}}').$$

The scalar products in the center of mass frame are

$$k_1\cdot k_2=-2|\boldsymbol{k}|^2-(mc/\hbar)^2,$$
$$k_1\cdot k_1'=-|\boldsymbol{k}|^2(1-\cos\theta)-(mc/\hbar)^2,$$
$$k_1\cdot k_2'=-|\boldsymbol{k}|^2(1+\cos\theta)-(mc/\hbar)^2,$$

where θ is the angle between $\boldsymbol{k}$ and $\boldsymbol{k}'$. The relative speed (22.54) of the electrons in the center of mass frame is

$$v=\frac{2c\hbar|\boldsymbol{k}|}{\sqrt{\hbar^2|\boldsymbol{k}|^2+m^2c^2}}.$$

The differential scattering cross section is then with the factor of 2 from equation (22.12), and using the fine structure constant $\alpha=e^2/(4\pi\epsilon_0\hbar c)$,

$$\frac{d\sigma}{d\Omega}=\alpha^2\frac{\hbar^4k^4(3+\cos^2\theta)^2+m^2c^2(4\hbar^2k^2+m^2c^2)(1+3\cos^2\theta)}{4\hbar^2k^4(\hbar^2k^2+m^2c^2)\sin^4\theta}. \tag{22.57}$$

This is symmetric under $\theta\to(\pi/2)-\theta$ with a minimum for scattering angle $\theta=\pi/2$ and divergences in forward and backward direction. This divergence in the zero momentum transfer limit is due to the vanishing photon mass, or in other words due to the infinite range of electromagnetic interactions. It is the same divergence which rendered the Rutherford cross section non-integrable.

Equation (22.57) looks fairly complicated, but in terms of energy it essentially entails that Møller scattering is suppressed with kinetic energy K like K^{-2}: The low energy result for $K \simeq \hbar^2 k^2/2m \ll mc^2$ is

$$\frac{d\sigma}{d\Omega} = \left(\frac{\alpha\hbar c}{4K\sin^2\theta}\right)^2 (1 + 3\cos^2\theta),$$

and the high energy result for $K \simeq \hbar ck \gg mc^2$ is

$$\frac{d\sigma}{d\Omega} = \left(\frac{\alpha\hbar c}{2K}\frac{3+\cos^2\theta}{\sin^2\theta}\right)^2.$$

With respect to low-energy electron-electron scattering, we also note that the scattering matrix element (22.53) is dominated by the Coulomb contribution (22.51) if the electrons are non-relativistic, $\hbar|\boldsymbol{k}| \ll mc$, $\hbar|\boldsymbol{k}'| \ll mc$. The estimate (22.29) for the ratio of scattering amplitudes in the low energy limit applies here too, and this confirms again the domination of Coulomb interactions between non-relativistic charged particles.

22.5 Problems

22.1. Derive the relation (22.9) between the differential scattering cross section and the scattering amplitude in box normalization.

22.2. Calculate the differential scattering cross section for scattering of a relativistic charged scalar particle off a heavy nucleus. Assume that the heavy nucleus is non-relativistic and that the scalar particle cannot resolve its substructure, such that you can describe the nucleus with Schrödinger field operators.

22.3. Calculate the differential scattering cross section $d\sigma_{0;\boldsymbol{q},\alpha\to\boldsymbol{k}';\boldsymbol{q}'}/d\Omega$ for electron-photon scattering with polarized initial photons, i.e. sum over the polarizations of the scattered photons but do not average over the initial polarization.

22.4. Show that the differential cross sections (22.42, 22.44) for Compton scattering can also be written in the form

$$d\sigma_{0;\boldsymbol{q},\alpha\to\boldsymbol{k}';\boldsymbol{q}',\alpha'} = d^3k'd^3q'\frac{\alpha^2\hbar^2\delta(k'+q'-k-q)}{4m|\boldsymbol{q}||\boldsymbol{q}'|E(\boldsymbol{k}')} \times\left(4(\boldsymbol{\epsilon}_\alpha(\boldsymbol{q})\cdot\boldsymbol{\epsilon}_{\alpha'}(\boldsymbol{q}'))^2 + \frac{|\boldsymbol{q}'|}{|\boldsymbol{q}|} + \frac{|\boldsymbol{q}|}{|\boldsymbol{q}'|} - 2\right), \tag{22.58}$$

$$\frac{d\sigma_{0;\boldsymbol{q},\alpha\to\boldsymbol{q}-\boldsymbol{q}';\boldsymbol{q}',\alpha'}}{d\Omega} = \left(\frac{\alpha\hbar|\boldsymbol{q}'|}{2mc|\boldsymbol{q}|}\right)^2\left(4(\boldsymbol{\epsilon}_\alpha(\boldsymbol{q})\cdot\boldsymbol{\epsilon}_{\alpha'}(\boldsymbol{q}'))^2 + \frac{|\boldsymbol{q}'|}{|\boldsymbol{q}|} + \frac{|\boldsymbol{q}|}{|\boldsymbol{q}'|} - 2\right).$$

22.5. Calculate the kinetic energy imparted on the recoiling electron in Compton scattering as a function of $|\boldsymbol{q}|$ and θ.

22.6. Derive the scattering amplitude for electron-nucleus scattering using covariant quantization for the photon.

22.7. Derive the scattering amplitude for Compton scattering using covariant quantization for the photons.

22.8. Derive the scattering amplitude for Møller scattering using covariant quantization for the photon.

Appendix A: Lagrangian Mechanics

Lagrangian mechanics is not only a very beautiful and powerful formulation of mechanics, but it is also needed as a preparation for a deeper understanding of all fundamental interactions in physics. All fundamental equations of motion in physics are encoded in Lagrangian field theory, which is a generalization of Lagrangian mechanics for fields. Furthermore, the connection between symmetries and conservation laws of physical systems is best explored in the framework of the Lagrangian formulation of dynamics, and we also need Lagrangian field theory as a basis for field quantization.

Suppose we consider a particle with coordinates $\boldsymbol{x}(t)$ moving in a potential $V(\boldsymbol{x})$. Then Newton's equation of motion

$$m\ddot{\boldsymbol{x}} = -\nabla V(\boldsymbol{x})$$

is equivalent to the following statement (Hamilton's principle, 1834): The action integral

$$S[x] = \int_{t_0}^{t_1} dt\, L(\boldsymbol{x}, \dot{\boldsymbol{x}}) = \int_{t_0}^{t_1} dt \left(\frac{m}{2}\dot{\boldsymbol{x}}^2 - V(\boldsymbol{x})\right)$$

is in first order stationary under arbitrary perturbations $\boldsymbol{x}(t) \to \boldsymbol{x}(t) + \delta\boldsymbol{x}(t)$ of the path of the particle between fixed endpoints $\boldsymbol{x}(t_0)$ and $\boldsymbol{x}(t_1)$ (i.e. the perturbation is only restricted by the requirement of fixed endpoints: $\delta\boldsymbol{x}(t_0) = 0$ and $\delta\boldsymbol{x}(t_1) = 0$). This is demonstrated by straightforward calculation of the first order variation of S,

$$\begin{aligned}\delta S[x] &= S[x + \delta x] - S[x] = \int_{t_0}^{t_1} dt\, [m\dot{\boldsymbol{x}} \cdot \delta\dot{\boldsymbol{x}} - \delta\boldsymbol{x} \cdot \nabla V(\boldsymbol{x})] \\ &= -\int_{t_0}^{t_1} dt\, \delta\boldsymbol{x} \cdot (m\ddot{\boldsymbol{x}} + \nabla V(\boldsymbol{x}))\,. \end{aligned} \tag{A.1}$$

R. Dick, *Advanced Quantum Mechanics*, Graduate Texts in Physics,
DOI 10.1007/978-3-319-25675-7

Partial integration and $\delta\boldsymbol{x}(t_0) = 0$, $\delta\boldsymbol{x}(t_1) = 0$ were used in the last step. Equation (A.1) tells us that $\delta S[\boldsymbol{x}] = 0$ holds for arbitrary path variation with fixed endpoints if and only if the path $\boldsymbol{x}(t)$ satisfies Newton's equations,

$$m\ddot{\boldsymbol{x}} + \nabla V(\boldsymbol{x}) = 0.$$

This generalizes to arbitrary numbers of particles ($\boldsymbol{x}(t) \to \boldsymbol{x}_I(t)$, $1 \leq I \leq N$), and to the case that the motion of the particles is restricted through constraints, like e.g. a particle that can only move on a sphere[1]. In the case of constraints one substitutes generalized coordinates $q_i(t)$ which correspond to the actual degrees of freedom of the particle or system of particles (e.g. polar angles for the particle on the sphere), and one ends up with an action integral of the form

$$S[q] = \int_{t_0}^{t_1} dt\, L(q, \dot{q}).$$

The function $L(q, \dot{q})$ is the *Lagrange function* of the mechanical system with generalized coordinates $q_i(t)$, and a shorthand notation is used for a mechanical system with N degrees of freedom,

$$(q, \dot{q}) = (q_1(t), q_2(t), \ldots, q_N(t), \dot{q}_1(t), \dot{q}_2(t), \ldots, \dot{q}_N(t)).$$

First order variation of the action with fixed endpoints (i.e. $\delta q(t_0) = 0$, $\delta q(t_1) = 0$) yields after partial integration

$$\begin{aligned}\delta S[q] = S[q + \delta q] - S[q] &= \int_{t_0}^{t_1} dt \left(\sum_i \delta q_i \frac{\partial L}{\partial q_i} + \sum_i \delta \dot{q}_i \frac{\partial L}{\partial \dot{q}_i} \right) \\ &= \int_{t_0}^{t_1} dt \sum_i \delta q_i \left(\frac{\partial L}{\partial q_i} - \frac{d}{dt}\frac{\partial L}{\partial \dot{q}_i} \right),\end{aligned} \tag{A.2}$$

where again fixation of the endpoints was used.

$\delta S[q] = 0$ for arbitrary path variation $q_i(t) \to q_i(t) + \delta q_i(t)$ with fixed endpoints then immediately tells us the equations of motion in terms of the generalized coordinates,

$$\frac{\partial L}{\partial q_i} - \frac{d}{dt}\frac{\partial L}{\partial \dot{q}_i} = 0. \tag{A.3}$$

[1] And it also applies to relativistic particles, see Appendix B.

These equations of motion are called Lagrange equations of the second kind or Euler-Lagrange equations or simply Lagrange equations. The quantity

$$p_i = \frac{\partial L}{\partial \dot{q}_i}$$

is denoted as the *conjugate momentum* to the coordinate q_i.

The conjugate momentum is conserved if the Lagrange function depends only on the generalized velocity component $\dot{q}_i$ but not on q_i, $dp_i/dt = 0$.

Furthermore, if the Lagrange function does not explicitly depend on time, we have

$$\frac{dL}{dt} = p_i\ddot{q}_i + \frac{\partial L}{\partial q_i}\dot{q}_i.$$

The Euler-Lagrange equation then implies that the Hamilton function

$$H = p_i\dot{q}_i - L$$

is conserved, $dH/dt = 0$.

For a simple example, consider a particle of mass m in a gravitational field $\boldsymbol{g} = -g\boldsymbol{e}_z$. The particle is constrained so that it can only move on a sphere of radius r. An example of generalized coordinates are angles ϑ, φ on the sphere, and the Cartesian coordinates $\{X, Y, Z\}$ of the particle are related to the generalized coordinates through

$$\begin{aligned} X(t) &= r\sin\vartheta(t)\cdot\cos\varphi(t), \\ Y(t) &= r\sin\vartheta(t)\cdot\sin\varphi(t), \\ Z(t) &= r\cos\vartheta(t). \end{aligned}$$

The kinetic energy of the particle can be expressed in terms of the generalized coordinates,

$$K = \frac{m}{2}\dot{\boldsymbol{r}}^2 = \frac{m}{2}\left(\dot{X}^2 + \dot{Y}^2 + \dot{Z}^2\right) = \frac{m}{2}r^2\left(\dot{\vartheta}^2 + \dot{\varphi}^2\sin^2\vartheta\right),$$

and the potential energy is

$$V = mgZ = mgr\cos\vartheta.$$

This yields the Lagrange function in the generalized coordinates,

$$L = \frac{m}{2}\dot{\boldsymbol{r}}^2 - mgZ = \frac{m}{2}r^2\left(\dot{\vartheta}^2 + \dot{\varphi}^2\sin^2\vartheta\right) - mgr\cos\vartheta,$$

and the Euler-Lagrange equations yield the equations of motion of the particle,

$$\ddot{\vartheta} = \dot{\varphi}^2 \sin\vartheta \cos\vartheta + \frac{g}{r}\sin\vartheta, \tag{A.4}$$

$$\frac{d}{dt}\left(\dot{\varphi}\sin^2\vartheta\right) = 0. \tag{A.5}$$

The conjugate momenta

$$p_\vartheta = \frac{\partial L}{\partial \dot{\vartheta}} = mr^2\dot{\vartheta}$$

and

$$p_\varphi = \frac{\partial L}{\partial \dot{\varphi}} = mr^2\dot{\varphi}\sin^2\vartheta$$

are just the angular momenta for rotation in ϑ or φ direction. The Hamilton function is the conserved energy

$$H = p_\vartheta\dot{\vartheta} + p_\varphi\dot{\varphi} - L = \frac{p_\vartheta^2}{2mr^2} + \frac{p_\varphi^2}{2mr^2\sin^2\vartheta} + mgr\cos\vartheta = K + U.$$

The immediately apparent advantage of this formalism is that it directly yields the correct equations of motion (A.4, A.5) for the system without ever having to worry about finding the force that keeps the particle on the sphere. Beyond that the formalism also provides a systematic way to identify conservation laws in mechanical systems, and if one actually wants to know the force that keeps the particle on the sphere (which is actually trivial here, but more complicated e.g. for a system of two particles which have to maintain constant distance), a simple extension of the formalism to the Lagrange equations of the first kind can yield that, too.

The Lagrange function is not simply the difference between kinetic and potential energy if the forces are velocity dependent. This is the case for the Lorentz force. The Lagrange function for a non-relativistic charged particle in electromagnetic fields is

$$L = \frac{m}{2}\dot{\boldsymbol{x}}^2 + q\dot{\boldsymbol{x}}\cdot\boldsymbol{A} - q\Phi.$$

This yields the correct Lorentz force law $m\ddot{\boldsymbol{x}} = q(\boldsymbol{E} + \boldsymbol{v}\times\boldsymbol{B})$ for the particle, cf. Section 15.1. The relativistic versions of the Lagrange function for the particle can be found in equations (B.24, B.25).

Direct derivation of the Euler-Lagrange equations for the generalized coordinates q_a from Newton's equation in Cartesian coordinates

We can derive the Euler-Lagrange equations for the generalized coordinates of a constrained N-particle system directly from Newton's equations. This works in the following way:

Suppose we have N particles with coordinates x_i^j, $1 \leq i \leq N$, $1 \leq j \leq 3$, moving in a potential $V(\boldsymbol{x}_{1\ldots N})$. The Newton equations

$$\frac{d}{dt}(m_i \dot{x}_i^j) + \frac{\partial}{\partial x_i^j} V(\boldsymbol{x}_{1\ldots N}) = 0$$

can be written as

$$\left(\frac{d}{dt}\frac{\partial}{\partial \dot{x}_i^j} - \frac{\partial}{\partial x_i^j}\right)\left(\frac{1}{2}\sum_k m_k \dot{\boldsymbol{x}}_k^2 - V(\boldsymbol{x}_{1\ldots N})\right) = 0, \tag{A.6}$$

or equivalently

$$\left(\frac{d}{dt}\frac{\partial}{\partial \dot{x}_i^j} - \frac{\partial}{\partial x_i^j}\right) L(\boldsymbol{x}_{1\ldots N}, \dot{\boldsymbol{x}}_{1\ldots N}) = 0, \tag{A.7}$$

with the Lagrange function

$$L(\boldsymbol{x}_{1\ldots N}, \dot{\boldsymbol{x}}_{1\ldots N}) = \frac{1}{2}\sum_i m_i \dot{\boldsymbol{x}}_i^2 - V(\boldsymbol{x}_{1\ldots N}). \tag{A.8}$$

If there are C holonomic constraints on the motion of the N-particle system, we can describe its trajectories through $3N - C$ generalized coordinates q_a, $1 \leq a \leq 3N - C$:

$$x_i^j = x_i^j(q, t). \tag{A.9}$$

Note that in general $x_i^j(q, t)$ will implicitly depend on time t through the time dependence of the generalized coordinates $q_a(t)$, but it may also explicitly depend on t because there may be a time dependence in the C constraints[2].

The velocity components of the system are

$$\dot{x}_i^j = \frac{dx_i^j}{dt} = \sum_a \dot{q}_a \frac{\partial x_i^j}{\partial q_a} + \frac{\partial x_i^j}{\partial t}. \tag{A.10}$$

[2]A simple example for the latter would e.g. be a particle that is bound to a sphere with radius $R(t)$, where $R(t)$ is a given time-dependent function.

This implies in particular the equations

$$\frac{\partial \dot{x}_i^j}{\partial \dot{q}_a} = \frac{\partial x_i^j}{\partial q_a} \tag{A.11}$$

and

$$\frac{\partial \dot{x}_i^j}{\partial q_a} = \sum_b \dot{q}_b \frac{\partial^2 x_i^j}{\partial q_a \partial q_b} + \frac{\partial^2 x_i^j}{\partial t \partial q_a}. \tag{A.12}$$

Substitution of (A.11) into (A.12) yields

$$\frac{\partial \dot{x}_i^j}{\partial q_a} = \sum_b \dot{q}_b \frac{\partial^2 \dot{x}_i^j}{\partial q_b \partial \dot{q}_a} + \frac{\partial^2 \dot{x}_i^j}{\partial t \partial \dot{q}_a}. \tag{A.13}$$

Equation (A.11) also yields

$$\frac{\partial^2 \dot{x}_i^j}{\partial \dot{q}_a \partial \dot{q}_b} = 0,$$

and this implies with (A.13)

$$\frac{d}{dt}\frac{\partial \dot{x}_i^j}{\partial \dot{q}_a} = \sum_b \dot{q}_b \frac{\partial^2 \dot{x}_i^j}{\partial q_b \partial \dot{q}_a} + \frac{\partial^2 \dot{x}_i^j}{\partial t \partial \dot{q}_a} = \frac{\partial \dot{x}_i^j}{\partial q_a}. \tag{A.14}$$

With these preliminaries we can now look at the following linear combinations of the Newton equations (A.7):

$$\sum_{i,j} \frac{\partial x_i^j}{\partial q_a} \cdot \left(\frac{d}{dt}\frac{\partial L}{\partial \dot{x}_i^j} - \frac{\partial L}{\partial x_i^j} \right) = 0.$$

Insertion of equations (A.11, A.14) yields

$$\sum_{i,j} \left[\frac{\partial \dot{x}_i^j}{\partial \dot{q}_a}\frac{d}{dt}\frac{\partial L}{\partial \dot{x}_i^j} - \frac{\partial x_i^j}{\partial q_a}\frac{\partial L}{\partial x_i^j} + \left(\frac{d}{dt}\frac{\partial \dot{x}_i^j}{\partial \dot{q}_a} - \frac{\partial \dot{x}_i^j}{\partial q_a} \right) \frac{\partial L}{\partial \dot{x}_i^j} \right] = 0,$$

or after combining terms:

$$\frac{d}{dt}\sum_{i,j} \left(\frac{\partial \dot{x}_i^j}{\partial \dot{q}_a}\frac{\partial L}{\partial \dot{x}_i^j} \right) - \frac{\partial L}{\partial q_a} = 0. \tag{A.15}$$

However, the coordinates x_i^j are independent of the generalized velocities $\dot{q}_a$, and therefore equation (A.15) is just the Lagrange equation (A.3):

$$\frac{d}{dt}\frac{\partial L(q,\dot{q})}{\partial \dot{q}_a} - \frac{\partial L(q,\dot{q})}{\partial q_a} = 0. \tag{A.16}$$

Symmetries and conservation laws in classical mechanics

We call a set of first order transformations

$$t \to t' = t - \epsilon(t), \quad q_a(t) \to q'_a(t') = q_a(t) + \delta q_a(t) \tag{A.17}$$

a symmetry of a mechanical system with action

$$S[q] = \int_{t_o}^{t_1} dt\, L(q(t), \dot{q}(t), t)$$

if it changes the form $dt\, L(q(t), \dot{q}(t), t)$ in first order of $\epsilon(t)$ and $\delta q_a(t)$ at most by a term of the form $dB = dt(dB/dt)$:

$$\begin{aligned}\delta(dt\, L(q, \dot{q}, t)) &\equiv dt'\, L(q'(t'), \dot{q}'(t'), t') - dt\, L(q(t), \dot{q}(t), t)\\ &= dt\frac{d}{dt}B_{\delta q,\epsilon}(q(t), t).\end{aligned} \tag{A.18}$$

To see how this implies conservation laws in the mechanical system, we have to evaluate $\delta(dt\, L(q, \dot{q}, t))$ for the transformations (A.17). We have to take into account that (A.17) implies

$$dt' = dt(1 - \dot{\epsilon}(t)), \quad \frac{d}{dt'} = (1 + \dot{\epsilon}(t))\frac{d}{dt},$$

and therefore also

$$\delta\dot{q}_a(t) = \frac{d}{dt'}q'_a(t') - \frac{d}{dt}q_a(t) = \dot{\epsilon}(t)\frac{d}{dt}q_a(t) + \frac{d}{dt}\delta q_a(t).$$

The first order change in $dt\, L$ is therefore

$$\begin{aligned}\delta(dt\, L) &= dt'\, L(q'(t'), \dot{q}'(t'), t') - dt\, L(q(t), \dot{q}(t), t)\\ &= dt\left[\delta q_a\frac{\partial L}{\partial q_a} + \left(\dot{\epsilon}\frac{d}{dt}q_a + \frac{d}{dt}\delta q_a\right)\frac{\partial L}{\partial \dot{q}_a} - \dot{\epsilon}L - \epsilon\frac{\partial L}{\partial t}\right].\end{aligned}$$

Now we substitute

$$\delta\dot{q}_a\frac{\partial L}{\partial\dot{q}_a}=\frac{d}{dt}\left(\delta q_a\frac{\partial L}{\partial\dot{q}_a}\right)-\delta q_a\frac{d}{dt}\frac{\partial L}{\partial\dot{q}_a}$$

and

$$\begin{aligned}\dot{\epsilon}\left(\dot{q}_a\frac{\partial L}{\partial\dot{q}_a}-L\right)&=\frac{d}{dt}\left(\epsilon\dot{q}_a\frac{\partial L}{\partial\dot{q}_a}-\epsilon L\right)-\epsilon\ddot{q}_a\frac{\partial L}{\partial\dot{q}_a}-\epsilon\dot{q}_a\frac{d}{dt}\frac{\partial L}{\partial\dot{q}_a}+\epsilon\frac{dL}{dt}\\&=\frac{d}{dt}\left(\epsilon\dot{q}_a\frac{\partial L}{\partial\dot{q}_a}-\epsilon L\right)+\epsilon\dot{q}_a\left(\frac{\partial L}{\partial q_a}-\frac{d}{dt}\frac{\partial L}{\partial\dot{q}_a}\right)+\epsilon\frac{\partial L}{\partial t}.\end{aligned}$$

This yields

$$\begin{aligned}\delta(dt\,L)&=dt\,(\delta q_a+\epsilon\dot{q}_a)\left(\frac{\partial L}{\partial q_a}-\frac{d}{dt}\frac{\partial L}{\partial\dot{q}_a}\right)\\&\quad+dt\frac{d}{dt}\left((\delta q_a+\epsilon\dot{q}_a)\frac{\partial L}{\partial\dot{q}_a}-\epsilon L\right).\end{aligned}\tag{A.19}$$

Comparison of equations (A.18) and (A.19) implies an on-shell conservation law

$$\frac{d}{dt}Q_{\delta q,\epsilon}=0$$

with the conserved charge

$$Q_{\delta q,\epsilon}=\epsilon\left(L-\dot{q}_a\frac{\partial L}{\partial\dot{q}_a}\right)-\delta q_a\frac{\partial L}{\partial\dot{q}_a}+B_{\delta q,\epsilon}.\tag{A.20}$$

$B_{\delta q,\epsilon}$ is the one-dimensional version of the current K^μ in Lagrangian field theory and $Q_{\delta q,\epsilon}$ is the one-dimensional version of the conserved current J^μ, see the paragraph after equation (16.14). $B_{\delta q,\epsilon}=0$ in most cases. However, a noticeable exception are Galilei boosts in nonrelativistic N-particle mechanics (where I enumerates the particles). The Lagrange function

$$L=\frac{1}{2}m_I\dot{\boldsymbol{x}}_I^2-V_{I<J}(|\boldsymbol{x}_I-\boldsymbol{x}_J|)\tag{A.21}$$

satisfies (A.18) for Galilei transformations $\epsilon(t)=0$, $\delta\boldsymbol{x}_I(t)=-\boldsymbol{v}t$. In this case $B=-m_I\boldsymbol{x}_I(t)\cdot\boldsymbol{v}$ and the conservation law

$$Q=m_I\boldsymbol{v}\cdot[\dot{\boldsymbol{x}}_I(t)t-\boldsymbol{x}_I(t)]=\boldsymbol{v}\cdot[\boldsymbol{P}t-M\boldsymbol{X}(t)]$$

assures uniform center of mass motion $\boldsymbol{X}(t)$ with velocity $\boldsymbol{P}/M$. Other familiar symmetry transformations of (A.21) include time translations $\epsilon(t) = \text{const.}$, which yields energy conservation,

$$H = -Q/\epsilon = \dot{\boldsymbol{x}}_I \cdot \frac{\partial L}{\partial \dot{\boldsymbol{x}}_I} - L = \frac{1}{2} m_I \dot{\boldsymbol{x}}_I^2 + V_{I<J}(|\boldsymbol{x}_I - \boldsymbol{x}_J|),$$

spatial translations $\delta \boldsymbol{x}_I(t) = \boldsymbol{\epsilon} = \text{const.}$, which yields conservation of total momentum,

$$\boldsymbol{P} = -\frac{\partial Q}{\partial \boldsymbol{\epsilon}} = m_I \dot{\boldsymbol{x}}_I,$$

and rotations $\delta \boldsymbol{x}_I(t) = \boldsymbol{\varphi} \times \boldsymbol{x}_I(t)$, which yields conservation of total angular momentum,

$$\boldsymbol{M} = -\frac{\partial Q}{\partial \boldsymbol{\varphi}} = m_I \boldsymbol{x}_I \times \dot{\boldsymbol{x}}_I.$$

The corresponding conserved charges from the symmetries of the Lagrange function $L = -mc\sqrt{c^2 - \dot{\boldsymbol{x}}^2(t)}$ of a free relativistic particle are

$$\boldsymbol{p} = \frac{\partial L}{\partial \dot{\boldsymbol{x}}} = \frac{m\dot{\boldsymbol{x}}}{\sqrt{1 - (\dot{\boldsymbol{x}}/c)^2}},$$

$$H = cp^0 = \dot{\boldsymbol{x}} \cdot \boldsymbol{p} - L = \frac{mc^2}{\sqrt{1 - (\dot{\boldsymbol{x}}/c)^2}} = c\sqrt{\boldsymbol{p}^2 + m^2c^2},$$

$$\boldsymbol{M} = \boldsymbol{x} \times \boldsymbol{p},$$

and the conserved charges which follow from Lorentz invariance $\delta x^\mu = \omega^{\mu\nu} x_\nu$, $\omega^{\mu\nu} = -\omega^{\nu\mu}$, are

$$M^{\mu\nu} = x^\mu p^\nu - x^\nu p^\mu,$$

which of course includes the charges $L_i = \epsilon_{ijk} M^{jk}/2$ from rotations. See also Appendix B.

Appendix B: The Covariant Formulation of Electrodynamics

Electrodynamics is a relativistic field theory for every frequency or energy of electromagnetic waves because photons are massless. Understanding of electromagnetism and of photon-matter interactions therefore requires an understanding of special relativity. Furthermore, we also want to understand the quantum mechanics of relativistic electrons and other relativistic particles, and the covariant formulation of electrodynamics is also very helpful as a preparation for relativistic wave equations like the Klein-Gordon and Dirac equations.

Lorentz transformations

The scientific community faced several puzzling problems around 1900. Some of these problems led to the development of quantum mechanics, but two of the problems motivated Einstein's Special Theory of Relativity:

- In 1881 and 1887 Michelson had demonstrated that light from a terrestrial light source always moves with the same speed c in each direction, irrespective of Earth's motion.
- The basic equation of Newtonian mechanics, $\boldsymbol{F} = d(m\boldsymbol{u})/dt$, is invariant under Galilei transformations of the coordinates:

$$t' = t, \quad \boldsymbol{x}' = \boldsymbol{x} - \boldsymbol{v}t. \tag{B.1}$$

  Therefore any two observers who use coordinates related through a Galilei transformation are physically equivalent in Newtonian mechanics.

However, in 1887 (at the latest) it was realized that Galilei transformations do not leave Maxwell's equations invariant, i.e. if Maxwell's equations describe electromagnetic phenomena for one observer, they would not hold for another observer

R. Dick, *Advanced Quantum Mechanics*, Graduate Texts in Physics,
DOI 10.1007/978-3-319-25675-7

moving with constant velocity $\boldsymbol{v}$ relative to the first observer (because it was assumed that the coordinates of these two observers are related through the Galilei transformation (B.1)). Instead, Voigt (1887) and Lorentz (1892–1904) realized that Maxwell's equations would hold for the two observers if their coordinates would be related e.g. through a transformation of the form

$$ct' = \frac{ct - (v/c)x}{\sqrt{1-(v^2/c^2)}}, \quad x' = \frac{x - vt}{\sqrt{1-(v^2/c^2)}}, \quad y' = y, \quad z' = z, \tag{B.2}$$

and they also realized that coordinate transformations of this kind imply that light would move with the same speed c in both coordinate systems,

$$\Delta x^2 + \Delta y^2 + \Delta z^2 - c^2\Delta t^2 = \Delta x'^2 + \Delta y'^2 + \Delta z'^2 - c^2\Delta t'^2. \tag{B.3}$$

Voigt was interested in the most general symmetry transformation of the wave equations for electromagnetic fields, while Lorentz tried to explain the results of the Michelson experiment.

In 1905 Einstein took the bold step to propose that then the coordinates measured by two observers with constant relative speed $\boldsymbol{v}$ must be described by transformations like (B.2), but *not* by Galilei transformations[1]. This was a radical step, because it implies that two observers with non-vanishing relative speed assign different time coordinates to one and the same event, and they also have different notions of simultaneity of events. The same statement in another formulation: Two different observers with non-vanishing relative speed slice the four-dimensional universe differently into three-dimensional regions of simultaneity, or into three-dimensional universes. Einstein abandoned the common prejudice that everybody always assigns the same time coordinate to one and the same event. Time is not universal. The speed of light in vacuum is universal.

In the following we use the abbreviations

$$\beta = \frac{v}{c}, \quad \gamma = \frac{1}{\sqrt{1-\beta^2}}.$$

The transformation (B.2) and its inversion then read

$$ct' = \gamma(ct - \beta x), \quad x' = \gamma(x - \beta ct), \quad y' = y, \quad z' = z, \tag{B.4}$$

$$ct = \gamma(ct' + \beta x'), \quad x = \gamma(x' + \beta ct'). \tag{B.5}$$

The spatial origin $x' = y' = z' = 0$ of the (ct', x', y', z') system satisfies $x = \beta ct = vt$ (use $x' = 0$ in $x' = \gamma(x - \beta ct)$), and therefore moves with velocity $v\boldsymbol{e}_x$

[1] This idea was also enunciated by Poincaré in 1904, but Einstein went beyond the statement of the idea and also worked out the consequences.

relative to the (ct, x, y, z) system. In the same way one finds that the spatial origin of the (ct, x, y, z) system moves with velocity $-v\boldsymbol{e}'_x$ through the (ct', x', y', z') system. Therefore this is the special Lorentz transformation between two coordinate frames with a relative motion with speed v in x-direction as seen from the unprimed frame, or in $(-x')$-direction as seen from the primed frame.

Equation (B.2) tells us that for motion in a certain direction (x-direction in (B.2)), the coordinate in that direction is affected non-trivially by the transformation, while any orthogonal coordinate does not change its value. This immediately allows for a generalization of (B.2) in the case that the relative velocity $\boldsymbol{v}$ points in an arbitrary direction.

It is convenient to introduce a rescaled velocity vector $\boldsymbol{\beta} = \boldsymbol{v}/c$ and the corresponding unit vector $\hat{\boldsymbol{\beta}} = \boldsymbol{\beta}/\beta = \boldsymbol{v}/v$. The (3×3)-matrix $\hat{\boldsymbol{\beta}} \otimes \hat{\boldsymbol{\beta}}^T$ projects any vector $\boldsymbol{x}$ onto its component parallel to $\boldsymbol{\beta}$,

$$\boldsymbol{x}_{\|\beta} = \hat{\boldsymbol{\beta}} \otimes \hat{\boldsymbol{\beta}}^T \cdot \boldsymbol{x},$$

while the component orthogonal to $\boldsymbol{\beta}$ is

$$\boldsymbol{x}_{\perp\beta} = \boldsymbol{x} - \boldsymbol{x}_{\|\beta} = (\underline{1} - \hat{\boldsymbol{\beta}} \otimes \hat{\boldsymbol{\beta}}^T) \cdot \boldsymbol{x}.$$

From the form of the special Lorentz transformation (B.2) we know that the coordinate $|\boldsymbol{x}_{\|\beta}|$ parallel to $\boldsymbol{v}$ will be rescaled by a factor

$$\gamma = \frac{1}{\sqrt{1 - (v^2/c^2)}} = \frac{1}{\sqrt{1 - \beta^2}},$$

and be shifted by an amount $-\gamma vt = -\gamma\beta ct$. Similarly, the time coordinate ct will be rescaled by the factor γ and be shifted by an amount $-\gamma\beta|\boldsymbol{x}_{\|\beta}|$. Finally, nothing will happen to the transverse component $\boldsymbol{x}_{\perp\beta}$. We can collect these observations in a (4×4)-matrix equation relating the two four-dimensional coordinate vectors,

$$\begin{pmatrix} ct' \\ \boldsymbol{x}' \end{pmatrix} = \begin{pmatrix} \gamma & -\gamma\boldsymbol{\beta}^T \\ -\gamma\boldsymbol{\beta} & \underline{1} - \hat{\boldsymbol{\beta}} \otimes \hat{\boldsymbol{\beta}}^T + \gamma\hat{\boldsymbol{\beta}} \otimes \hat{\boldsymbol{\beta}}^T \end{pmatrix} \cdot \begin{pmatrix} ct \\ \boldsymbol{x} \end{pmatrix}$$

This is the general Lorentz transformation between two observers if the spatial sections of their coordinate frames were coincident at $t = 0$. The most general transformation of this kind also allows for constant shifts of the coordinates and for a rotation of the spatial axes,

$$\begin{pmatrix} ct' \\ \boldsymbol{x}' \end{pmatrix} = \begin{pmatrix} \gamma & -\gamma\boldsymbol{\beta}^T \\ -\gamma\boldsymbol{\beta} & \underline{1} - \hat{\boldsymbol{\beta}} \otimes \hat{\boldsymbol{\beta}}^T + \gamma\hat{\boldsymbol{\beta}} \otimes \hat{\boldsymbol{\beta}}^T \end{pmatrix} \cdot \begin{pmatrix} 1 & \boldsymbol{0}^T \\ \boldsymbol{0} & \underline{R} \end{pmatrix} \cdot \begin{pmatrix} ct - cT \\ \boldsymbol{x} - \boldsymbol{X} \end{pmatrix}, \tag{B.6}$$

where $\underline{R}$ is a 3×3 rotation matrix. Without the coordinate shifts this is the most general orthochronous Lorentz transformation, where *orthochronous* refers to the fact that we did not include a possible reversal of the time axis. With the coordinate shifts included, (B.6) is denoted as an inhomogeneous Lorentz transformation or a Poincaré transformation. The Poincaré transformations (B.6) and the subset of Lorentz transformations ($T = 0$, $\boldsymbol{X} = \mathbf{0}$) form the *Poincaré group* and the *Lorentz group*, respectively. The Lorentz group is apparently a subgroup of the Poincaré group, and the rotation group is a subgroup of the Lorentz group.

In four-dimensional notation the 4-vector of coordinates is $x^\mu = (ct, \boldsymbol{x})$, and the 4-vector short hand for equation (B.6) is

$$x'^\mu = \Lambda^\mu{}_\nu (x^\nu - X^\nu). \tag{B.7}$$

where the (4×4) transformation matrix Λ is

$$\Lambda = \{\Lambda^\mu{}_\nu\} = \begin{pmatrix} \gamma & -\gamma\boldsymbol{\beta}^T \\ -\gamma\boldsymbol{\beta} & \underline{1} - \hat{\boldsymbol{\beta}} \otimes \hat{\boldsymbol{\beta}}^T + \gamma\hat{\boldsymbol{\beta}} \otimes \hat{\boldsymbol{\beta}}^T \end{pmatrix} \cdot \begin{pmatrix} 1 & \mathbf{0}^T \\ \mathbf{0} & \underline{R} \end{pmatrix}. \tag{B.8}$$

A homogeneous Lorentz transformation of this form is denoted as a *proper orthochronous Lorentz transformation* if we also exclude inversions of an odd number of spatial axis[2]. This is equivalent to the requirement $\det\underline{R} = 1$.

We will see below that it plays a role where we attach the indices for the explicit numerical representation of the matrix Λ in terms of matrix elements. Usually, if a matrix is given for Λ without explicitly defining index positions, the default convention is that it refers to a superscript row index and a subscript column index, as above, $\Lambda = \{\Lambda^\mu{}_\nu\}$. This is important, because as soon as a boost is involved (i.e. $\boldsymbol{\beta} \neq \mathbf{0}$), we will find that e.g.

$$\Lambda^\mu{}_\nu(\boldsymbol{\beta}) = \Lambda_\nu{}^\mu(-\boldsymbol{\beta}) \neq \Lambda_\nu{}^\mu(\boldsymbol{\beta}).$$

The transformation equation (B.6) is the general solution to the following problem: Find the most general coordinate transformation $\{ct, \boldsymbol{x}\} = \{ct, x, y, z\} \to \{ct', \boldsymbol{x}'\} = \{ct', x', y', z'\}$ which leaves the expression $\Delta\boldsymbol{x}^2 - c^2\Delta t^2$ invariant, i.e. such that for arbitrary coordinate differentials $c\Delta t$, $\Delta\boldsymbol{x}$ we have

$$\Delta\boldsymbol{x}^2 - c^2\Delta t^2 = \Delta\boldsymbol{x}'^2 - c^2\Delta t'^2. \tag{B.9}$$

This equation implies in particular that if one of our observers sees a light wave moving at speed c, then this light wave will also move with speed c for the second observer,

$$\Delta\boldsymbol{x}^2 - c^2\Delta t^2 = 0 \;\Leftrightarrow\; \Delta\boldsymbol{x}'^2 - c^2\Delta t'^2 = 0.$$

[2]The two factors of a proper orthochronous Lorentz transformation can be written as exponentials. This is discussed in Appendix H.

In fact it suffices to require *only* that anything moving with speed c will also have speed c in the new coordinates, and that the spatial coordinates are Cartesian in both frames. Up to rescalings of the coordinates the most general coordinate transformation is then the general inhomogeneous Lorentz transformation (B.6).

Any constant offset X^μ between coordinate systems vanishes for differences of coordinates. Equation (B.7) therefore implies the following equation for Lorentz transformation of coordinate differentials,

$$dx'^\mu = \Lambda^\mu{}_\alpha dx^\alpha. \tag{B.10}$$

The condition (B.9),

$$d\boldsymbol{x}^2 - c^2dt^2 = d\boldsymbol{x}'^2 - c^2dt'^2$$

can also be written as

$$\eta_{\mu\nu}dx'^\mu dx'^\nu = \eta_{\alpha\beta}dx^\alpha dx^\beta \tag{B.11}$$

with the special (4×4)-matrix ($\underline{1}$ is the 3×3 unit matrix)

$$\eta_{\mu\nu} = \begin{pmatrix} -1 & \mathbf{0}^T \\ \mathbf{0} & \underline{1} \end{pmatrix},$$

Equation (B.9) also implies

$$\eta_{\mu\nu}dx'^\mu dy'^\nu = \eta_{\alpha\beta}dx^\alpha dy^\beta$$

for any pair of Lorentz transformed 4-vectors dx and dy (simply insert the 4-vector $dx + dy$ into (B.11)). This implies that Lorentz transformations leave the *Minkowski metric* $\eta_{\mu\nu}$ invariant:

$$\eta_{\mu\nu}\Lambda^\mu{}_\alpha\Lambda^\nu{}_\beta = \eta_{\alpha\beta}. \tag{B.12}$$

If we multiply this equation with the components $\eta^{\beta\gamma}$ of the inverse Minkowski tensor, we find

$$\eta_{\mu\nu}\Lambda^\mu{}_\alpha\Lambda^\nu{}_\beta\eta^{\beta\gamma} = \delta_\alpha{}^\gamma \equiv \eta_\alpha{}^\gamma.$$

This tells us a relation between the (4×4)-matrix Λ with "pulled indices" and its inverse Λ^{-1}:

$$\Lambda_\mu{}^\gamma \equiv \eta_{\mu\nu}\Lambda^\nu{}_\beta\eta^{\beta\gamma} = (\Lambda^{-1})^\gamma{}_\mu. \tag{B.13}$$

Explicitly, if

$$\{\Lambda^\mu{}_\nu\} = \begin{pmatrix} \gamma & -\gamma\boldsymbol{\beta}^T \\ -\gamma\boldsymbol{\beta} & \underline{1} - \hat{\boldsymbol{\beta}} \otimes \hat{\boldsymbol{\beta}}^T + \gamma\hat{\boldsymbol{\beta}} \otimes \hat{\boldsymbol{\beta}}^T \end{pmatrix} \cdot \begin{pmatrix} 1 & \mathbf{0}^T \\ \mathbf{0} & \underline{R} \end{pmatrix}$$

then

$$\{\Lambda_\mu{}^\nu\} \equiv \{\eta_{\mu\rho}\Lambda^\rho{}_\sigma\eta^{\sigma\nu}\} = \begin{pmatrix} \gamma & \gamma\boldsymbol{\beta}^T \\ \gamma\boldsymbol{\beta} & \underline{1} - \hat{\boldsymbol{\beta}} \otimes \hat{\boldsymbol{\beta}}^T + \gamma\hat{\boldsymbol{\beta}} \otimes \hat{\boldsymbol{\beta}}^T \end{pmatrix} \cdot \begin{pmatrix} 1 & \mathbf{0}^T \\ \mathbf{0} & \underline{R} \end{pmatrix}.$$

We can also "pull" or "draw" indices on 4-vectors, e.g. $dx_\alpha \equiv \eta_{\alpha\beta}dx^\beta = (-cdt, d\boldsymbol{x})$. Let us figure out how this 4-vector transforms under the Lorentz transformation (B.10):

$$dx'_\mu = \eta_{\mu\nu}dx'^\nu = \eta_{\mu\nu}\Lambda^\nu{}_\alpha dx^\alpha = \eta_{\mu\nu}\Lambda^\nu{}_\alpha\eta^{\alpha\beta}dx_\beta = \Lambda_\mu{}^\beta dx_\beta = dx_\beta(\Lambda^{-1})^\beta{}_\mu.$$

4-vectors with this kind of transformation behavior are denoted as *covariant 4-vectors*, while dx^μ is an example of a *contravariant 4-vector*. Another example of a covariant 4-vector is the vector of partial derivatives

$$\partial_\mu \equiv \frac{\partial}{\partial x^\mu} = \left(\frac{1}{c}\frac{\partial}{\partial t}, \nabla\right).$$

We can check that this is really a covariant 4-vector by calculating how it transforms under Lorentz transformations. According to the chain rule of differentiation we find

$$\partial'_\mu \equiv \frac{\partial}{\partial x'^\mu} = \frac{\partial x^\alpha}{\partial x'^\mu}\frac{\partial}{\partial x^\alpha}.$$

However, we have

$$dx^\alpha = (\Lambda^{-1})^\alpha{}_\nu dx'^\nu \quad \Rightarrow \quad \frac{\partial x^\alpha}{\partial x'^\mu} = (\Lambda^{-1})^\alpha{}_\mu$$

and therefore

$$\partial'_\mu = (\Lambda^{-1})^\alpha{}_\mu\partial_\alpha = \Lambda_\mu{}^\alpha\partial_\alpha.$$

Pairs of co- and contravariant indices do not transform if they are summed over. Assume e.g. that $F^{\alpha\beta}$ are the components of a 4×4 matrix which transform according to

$$F^{\alpha\beta} \to F'^{\mu\nu} = \Lambda^\mu{}_\alpha\Lambda^\nu{}_\beta F^{\alpha\beta}.$$

The combination $\partial_\alpha F^{\alpha\beta}$ then transforms under Lorentz transformations according to

$$\partial'_\mu F'^{\mu\nu} = \Lambda_\mu{}^\alpha\partial_\alpha\Lambda^\mu{}_\beta\Lambda^\nu{}_\gamma F^{\beta\gamma} = \Lambda^\nu{}_\gamma\eta^\alpha{}_\beta\partial_\alpha F^{\beta\gamma} = \Lambda^\nu{}_\gamma\partial_\alpha F^{\alpha\gamma},$$

i.e. the summed index pair does not contribute to the transformation. Only "free" indices (i.e. indices which are not paired and summed from 0 to 3) transform under Lorentz transformations.

The manifestly covariant formulation of electrodynamics

Electrodynamics is a Lorentz invariant theory, i.e. all equations have the same form in all coordinate systems which are related by Poincaré transformations. However, this property is hardly recognizable if one looks at Maxwell's equations in traditional notation,

$$\nabla\cdot\boldsymbol{E}=\frac{1}{\epsilon_0}\varrho,\quad \nabla\times\boldsymbol{E}+\frac{\partial}{\partial t}\boldsymbol{B}=\boldsymbol{0},$$

$$\nabla\cdot\boldsymbol{B}=0,\quad \nabla\times\boldsymbol{B}-\frac{1}{c^2}\frac{\partial}{\partial t}\boldsymbol{E}=\mu_0\boldsymbol{j}.$$

"Lorentz invariance" seems far from obvious: How, e.g. would the electric and magnetic fields transform under a Lorentz transformation of the coordinates? Apparently we seem to have three 3-dimensional vectors and one scalar in the equations. We can combine the current density $\boldsymbol{j}$ and the charge density ϱ into a current 4-vector

$$j^\nu=(\varrho c,\boldsymbol{j}).$$

For the field strengths it helps to recall that the homogeneous Maxwell's equations are solved through potentials Φ, $\boldsymbol{A}$,

$$\boldsymbol{B}=\nabla\times\boldsymbol{A},\quad \boldsymbol{E}=-\frac{\partial}{\partial t}\boldsymbol{A}-\nabla\Phi.$$

If one combines the potentials into a 4-vector,

$$A_\mu=(-\Phi/c,\boldsymbol{A}),$$

it is possible to realize that the electromagnetic field strengths E_i, B_i are related to antisymmetric combinations of the 4-vectors ∂_μ, A_ν,

$$F_{\mu\nu}=\partial_\mu A_\nu-\partial_\nu A_\mu=\begin{pmatrix}0 & -E_1/c & -E_2/c & -E_3/c\\ E_1/c & 0 & B_3 & -B_2\\ E_2/c & -B_3 & 0 & B_1\\ E_3/c & B_2 & -B_1 & 0\end{pmatrix}. \tag{B.14}$$

This electromagnetic field strength tensor $\underline{F}$ was introduced by Minkowski in 1907[3]. The matrix elements $F_{\mu\nu}$ are its covariant components. The contravariant components of $\underline{F}$ are

$$F^{\mu\nu} = \eta^{\mu\alpha}\eta^{\nu\beta}F_{\alpha\beta} = \partial^{\mu}A^{\nu} - \partial^{\nu}A^{\mu} = \begin{pmatrix} 0 & E_1/c & E_2/c & E_3/c \\ -E_1/c & 0 & B_3 & -B_2 \\ -E_2/c & -B_3 & 0 & B_1 \\ -E_3/c & B_2 & -B_1 & 0 \end{pmatrix}.$$

From this one can easily read off the transformation behavior of the fields under Lorentz transformations,

$$\begin{aligned} x^{\mu} &\to x'^{\mu} = \Lambda^{\mu}{}_{\alpha}x^{\alpha}, \\ \partial_{\mu} &\to \partial'_{\mu} = \Lambda_{\mu}{}^{\alpha}\partial_{\alpha}, \\ A_{\mu}(x) &\to A'_{\mu}(x') = \Lambda_{\mu}{}^{\alpha}A_{\alpha}(x) \end{aligned}$$

$$\begin{aligned} F_{\mu\nu}(x) \to F'_{\mu\nu}(x') &= \partial'_{\mu}A'_{\nu}(x') - \partial'_{\nu}A'_{\mu}(x') = \Lambda_{\mu}{}^{\alpha}\Lambda_{\nu}{}^{\beta}(\partial_{\alpha}A_{\beta}(x) - \partial_{\beta}A_{\alpha}(x)) \\ &= \Lambda_{\mu}{}^{\alpha}\Lambda_{\nu}{}^{\beta}F_{\alpha\beta}(x). \end{aligned}$$

Evaluation of $F'_{\mu\nu}(x')$ for a boost

$$\{\Lambda^{\mu}{}_{\nu}\} = \begin{pmatrix} \gamma & -\gamma\boldsymbol{\beta}^T \\ -\gamma\boldsymbol{\beta} & \underline{1} - \hat{\boldsymbol{\beta}} \otimes \hat{\boldsymbol{\beta}}^T + \gamma\hat{\boldsymbol{\beta}} \otimes \hat{\boldsymbol{\beta}}^T \end{pmatrix}$$

yields with $\boldsymbol{\beta} = \boldsymbol{v}/c$

$$\begin{aligned} \boldsymbol{E}'(\boldsymbol{x}',t') &= \gamma\Big(\boldsymbol{E}(\boldsymbol{x},t) + \boldsymbol{v} \times \boldsymbol{B}(\boldsymbol{x},t)\Big) - (\gamma - 1)\hat{\boldsymbol{\beta}}\Big(\hat{\boldsymbol{\beta}} \cdot \boldsymbol{E}(\boldsymbol{x},t)\Big) \\ &= \gamma\Big(\boldsymbol{E}(\boldsymbol{x},t) + \boldsymbol{v} \times \boldsymbol{B}(\boldsymbol{x},t)\Big) - \frac{\gamma^2}{(\gamma+1)c^2}\boldsymbol{v}\Big(\boldsymbol{v} \cdot \boldsymbol{E}(\boldsymbol{x},t)\Big), \\ \boldsymbol{B}'(\boldsymbol{x}',t') &= \gamma\left(\boldsymbol{B}(\boldsymbol{x},t) - \frac{1}{c^2}\boldsymbol{v} \times \boldsymbol{E}(\boldsymbol{x},t)\right) - (\gamma - 1)\hat{\boldsymbol{\beta}}\Big(\hat{\boldsymbol{\beta}} \cdot \boldsymbol{B}(\boldsymbol{x},t)\Big) \\ &= \gamma\left(\boldsymbol{B}(\boldsymbol{x},t) - \frac{1}{c^2}\boldsymbol{v} \times \boldsymbol{E}(\boldsymbol{x},t)\right) - \frac{\gamma^2}{(\gamma+1)c^2}\boldsymbol{v}\Big(\boldsymbol{v} \cdot \boldsymbol{B}(\boldsymbol{x},t)\Big), \end{aligned}$$

[3] H. Minkowski, Math. Ann. 68, 472 (1910). A translation of his results for dielectric materials into contemporary tensor notation can be found in R. Dick, Annalen Phys. (Berlin) 18, 174 (2009).

or expressed in terms of the field strength components parallel and perpendicular to the relative velocity $\boldsymbol{v}$ between the two observers,

$$\boldsymbol{E}'_{\|}(\boldsymbol{x}',t') = \boldsymbol{E}_{\|}(\boldsymbol{x},t), \quad \boldsymbol{B}'_{\|}(\boldsymbol{x}',t') = \boldsymbol{B}_{\|}(\boldsymbol{x},t),$$

$$\boldsymbol{E}'_{\perp}(\boldsymbol{x}',t') = \gamma\Big(\boldsymbol{E}_{\perp}(\boldsymbol{x},t) + \boldsymbol{v}\times\boldsymbol{B}_{\perp}(\boldsymbol{x},t)\Big),$$

$$\boldsymbol{B}'_{\perp}(\boldsymbol{x}',t') = \gamma\left(\boldsymbol{B}_{\perp}(\boldsymbol{x},t) - \frac{1}{c^2}\boldsymbol{v}\times\boldsymbol{E}_{\perp}(\boldsymbol{x},t)\right).$$

Electric and magnetic fields mix under Lorentz transformations, i.e. the distinction between electric and magnetic fields depends on the observer.

The equations

$$\partial_\mu F^{\mu\nu} = -\mu_0 j^\nu$$

are the inhomogeneous Maxwell's equations

$$\nabla\cdot\boldsymbol{E} = \frac{1}{\epsilon_0}\varrho, \quad \nabla\times\boldsymbol{B} - \frac{1}{c^2}\frac{\partial}{\partial t}\boldsymbol{E} = \mu_0\boldsymbol{j},$$

while the identities (with the 4-dimensional ϵ-tensor, $\epsilon^{0123} = -1$)

$$\epsilon^{\kappa\lambda\mu\nu}\partial_\lambda F_{\mu\nu} = 2\epsilon^{\kappa\lambda\mu\nu}\partial_\lambda\partial_\mu A_\nu \equiv 0$$

are the homogeneous Maxwell's equations

$$\nabla\cdot\boldsymbol{B} = 0, \quad \nabla\times\boldsymbol{E} + \frac{\partial}{\partial t}\boldsymbol{B} = \boldsymbol{0}.$$

These identities can also written in terms of the dual field strength tensor

$$\tilde{F}^{\mu\nu} = \frac{1}{2}\epsilon^{\mu\nu\alpha\beta}F_{\alpha\beta} = \begin{pmatrix} 0 & -B_1 & -B_2 & -B_3 \\ B_1 & 0 & E_3/c & -E_2/c \\ B_2 & -E_3/c & 0 & E_1/c \\ B_3 & E_2/c & -E_1/c & 0 \end{pmatrix}$$

as

$$\partial_\mu \tilde{F}^{\mu\nu} = 0.$$

The gauge freedom $A_\mu(x) \to A'_\mu(x) = A_\mu(x) + \partial_\mu f(x)$ apparently leaves the field strength tensor $F_{\mu\nu}$ invariant. In conventional terms this is

$$\Phi'(x) = \Phi(x) - \dot{f}(x), \quad \boldsymbol{A}'(x) = \boldsymbol{A}(x) + \nabla f(x).$$

We have written Maxwell's equations explicitly as equations between 4-vectors,

$$\partial_\mu F^{\mu\nu} = -\mu_0 j^\nu, \quad \partial_\mu \tilde{F}^{\mu\nu} = 0,$$

and this ensures that they hold in this form for every inertial observer. This is the *form invariance* (or simply "invariance") of Maxwell's equations under Lorentz transformations.

With the identification of the current 4-vector $j^\mu = (\varrho c, \boldsymbol{j})$, the local conservation law for charges can also be written in manifestly Lorentz invariant form,

$$\frac{\partial}{\partial t}\varrho + \nabla\cdot\boldsymbol{j} = \partial_\mu j^\mu = 0.$$

Relativistic mechanics

In special relativity it is better to express everything in quantities which transform linearly with combinations of the matrices Λ and Λ^{-1}. As a consequence of the transformation law

$$dx'^\mu = \Lambda^\mu{}_\nu dx^\nu, \tag{B.15}$$

ordinary velocities $d\boldsymbol{x}/dt$ and accelerations $d^2\boldsymbol{x}/dt^2$ transform nonlinearly under Lorentz boosts, due to the transformation of the time coordinates in the denominators. Therefore it is convenient to substitute the physical velocities and accelerations with "proper" velocities and accelerations, which do not require division by a transforming time parameter t.

Suppose the x'-frame is the frame of a moving object. In its own frame the trajectory of the object is $\boldsymbol{x}' = 0$. However, we know that the Lorentz transformation (B.15) leaves the product $dx^\mu dx_\mu$ invariant,

$$dx'^\mu dx'_\mu = d\boldsymbol{x}'^2 - c^2 dt'^2 = dx^\mu dx_\mu = d\boldsymbol{x}^2 - c^2 dt^2.$$

Therefore we have in particular for the time $dt' \equiv d\tau$ measured by the moving object along its own path $\boldsymbol{x}' = 0$

$$d\tau^2 = dt^2 - \frac{1}{c^2}d\boldsymbol{x}^2 = \left(1 - \frac{v^2}{c^2}\right)dt^2,$$

i.e. up to a constant

$$\tau = \int dt\,\sqrt{1-(v^2/c^2)} = \int \frac{dt}{\gamma}.$$

This is an *invariant*, i.e. it has the same value for each observer. Every observer will measure their own specific time interval Δt between any two events happening to the moving object, but all observers agree on the same value

$$\Delta\tau = \int_0^{\Delta t} dt\,\sqrt{1-(v^2/c^2)}$$

which elapsed on a clock moving with the object.

The time $\Delta\tau$ measured by an object between any two events happening to the object is denoted as the *proper time* or *eigentime* of the object.

The definition of eigentime entails a corresponding definition of the *proper velocity* or *eigenvelocity* of an object in an observer's frame: Divide the change in the object's coordinates $d\boldsymbol{x}$ *in the observer's frame* by the time interval $d\tau$ elapsed *for the object itself* while it was moving by $d\boldsymbol{x}$:

$$\boldsymbol{U} = \frac{d\boldsymbol{x}}{d\tau} = \gamma\frac{d\boldsymbol{x}}{dt} = \gamma\boldsymbol{v}.$$

This is a hybrid construction in the sense that a set of coordinate intervals $d\boldsymbol{x}$ measured in the observer's frame is divided by a coordinate interval $d\tau$ measured in the object's frame[4].

The notion of proper velocity may seem a little artificial, but it is useful because it can be extended to a 4-vector using the fact that $\{dx^\mu\} = (dx^0, d\boldsymbol{x}) = (cdt, d\boldsymbol{x})$ is a 4-vector under Lorentz transformations. If we define

$$U^0 = \frac{dx^0}{d\tau} = \frac{cdt}{d\tau} = \gamma c,$$

then

$$U^\mu = dx^\mu/d\tau = (U^0, \boldsymbol{U}) = \gamma(c, \boldsymbol{v})$$

is a 4-vector which transforms according to $U^\mu \to U'^\mu = \Lambda^\mu{}_\alpha U^\alpha$ under Lorentz transformations. This *4-velocity* vector satisfies

$$U^2 \equiv U^\mu U_\mu \equiv \eta_{\mu\nu}U^\mu U^\nu = \boldsymbol{U}^2 - (U^0)^2 = -c^2.$$

The conservation laws

$$\sum_i \boldsymbol{p}_i^{(in)} = \sum_i \boldsymbol{p}_i^{(out)}$$

$$\sum_i E_i^{(in)} = \sum_i E_i^{(out)}$$

[4]There is a limit $v \le c$ on the *physical speed* $v = |\boldsymbol{v}|$ of moving objects. No such limit holds for the "eigenspeed" $|\boldsymbol{U}|$, but the speed of signal transmission relative to an observer is v, not $|\boldsymbol{U}|$.

for momentum and energy in a collision would not be preserved under Lorentz transformations if the nonrelativistic definitions for momentum and energy would be employed, due to the nonlinear transformations of the particle velocities. This would mean that if momentum and energy conservation would hold for one observer, they would not hold for another observer with different velocity!

However, the conservation laws are preserved if energy and momentum transform linearly, like a 4-vector, under Lorentz transformations. We have already identified 4-velocities $\{U^{\mu}\} = \gamma(c, \boldsymbol{v})$ with the property $\lim_{\beta\to 0} \boldsymbol{U} = \boldsymbol{v}$.

This motivates the definition of the *4-momentum*

$$p^0 = mU^0, \quad \boldsymbol{p} = m\boldsymbol{U},$$

i.e. the relativistic definition of the spatial momentum of a particle of mass m and physical velocity $\boldsymbol{v}$ is

$$\boldsymbol{p} = m\boldsymbol{U} = \gamma m\boldsymbol{v} = \frac{m\boldsymbol{v}}{\sqrt{1-(v^2/c^2)}}. \tag{B.16}$$

The physical meaning of the fourth component

$$p^0 = mU^0 = \gamma mc = \frac{mc}{\sqrt{1-(v^2/c^2)}}$$

can be inferred from the nonrelativistic limit: $v \ll c$ yields

$$p^0 \simeq mc\left(1 + \frac{v^2}{2c^2}\right).$$

This motivates the identification of cp^0 with the energy of a particle of mass m and speed v:

$$E = cp^0 = \gamma mc^2 = \frac{mc^2}{\sqrt{1-(v^2/c^2)}}. \tag{B.17}$$

Division of the two equations (B.16) and (B.17) yields

$$\boldsymbol{v} = c^2\frac{\boldsymbol{p}}{E}, \tag{B.18}$$

and subtracting squares yields the relativistic dispersion relation

$$E^2 - c^2\boldsymbol{p}^2 = m^2c^4. \tag{B.19}$$

This is usually written as $p_{\mu}p^{\mu} = -m^2c^2$.

Equations (B.19) and (B.18) imply in particular for massless particles the relations $E = cp$ and $v = c$.

For the formulation of the relativistic version of Newton's law, we observe that the rate of change of 4-momentum with eigentime defines a 4-vector with the units of force,

$$f^\mu = \frac{d}{d\tau} m \frac{dx^\mu}{d\tau} = \frac{dp^\mu}{d\tau} = \gamma \frac{d}{dt} p^\mu.$$

It transforms linearly under Lorentz transformations because we divided a 4-vector dx^μ or dp^μ by invariants $d\tau^2$ or $d\tau$, respectively.

By convention one still defines three-dimensional forces according to

$$\boldsymbol{F} = \frac{d}{dt}\boldsymbol{p} = \frac{1}{\gamma}\boldsymbol{f},$$

i.e. $\boldsymbol{F}$ is *not* the spatial component of a 4-vector, but $\boldsymbol{f} = \gamma \boldsymbol{F}$ is.

For the 0-component f^0 we find with the relativistic dispersion relation $E = c\sqrt{\boldsymbol{p}^2 + m^2c^2}$,

$$\begin{aligned} f^0 &= \frac{d}{d\tau} m \frac{dx^0}{d\tau} = \frac{d}{d\tau}\left(m\gamma \frac{dx^0}{dt}\right) = \frac{d}{d\tau}(\gamma mc) = \frac{d}{d\tau}\frac{E}{c} = \frac{d}{d\tau}\sqrt{\boldsymbol{p}^2 + m^2c^2} \\ &= \frac{\boldsymbol{p}}{\sqrt{\boldsymbol{p}^2 + m^2c^2}} \cdot \frac{d}{d\tau}\boldsymbol{p} = \frac{\boldsymbol{v}}{c} \cdot \frac{d}{d\tau}\boldsymbol{p} = \frac{\boldsymbol{v}}{c} \cdot \boldsymbol{f}. \end{aligned} \quad \text{(B.20)}$$

The 4-vector of the force is therefore

$$(f^0, \boldsymbol{f}) = (\boldsymbol{\beta} \cdot \boldsymbol{f}, \boldsymbol{f}) = (\gamma \boldsymbol{\beta} \cdot \boldsymbol{F}, \gamma \boldsymbol{F}).$$

Multiplication of (cf. (B.20))

$$\frac{d}{d\tau}\frac{E}{c} = \frac{\boldsymbol{v}}{c} \cdot \boldsymbol{f}$$

with c/γ gives energy balance in conventional form,

$$\frac{d}{dt}E = \frac{c}{\sqrt{\boldsymbol{p}^2 + m^2c^2}}\boldsymbol{p} \cdot \frac{d}{dt}\boldsymbol{p} = \boldsymbol{v} \cdot \frac{d}{dt}\boldsymbol{p} = \boldsymbol{v} \cdot \boldsymbol{F}.$$

The nonrelativistic Newton equation for motion of a charged particle in electromagnetic fields contains the Lorentz force

$$\boldsymbol{F} = q\boldsymbol{E} + q\boldsymbol{v} \times \boldsymbol{B}.$$

We can get a hint at how the relativistic equation has to look like by expressing this combination of fields in terms of the field strength tensor (B.14),

$$E^i = cF^i{}_0 = F^i{}_0\frac{dx^0}{dt}, \quad \varepsilon^i{}_{jk}B^k = F^i{}_j.$$

The latter equation implies

$$(\boldsymbol{v}\times\boldsymbol{B})^i = \varepsilon^i{}_{jk}v^jB^k = F^i{}_jv^j,$$

and therefore

$$F^i = qE^i + q\varepsilon^i{}_{jk}v^jB^k = qF^i{}_0\frac{dx^0}{dt} + qF^i{}_j\frac{dx^j}{dt} = qF^i{}_\nu\frac{dx^\nu}{dt}.$$

This would be a spatial part of a 4-vector if we would not derive with respect to the laboratory time t, but with respect to the eigentime τ of the charged particle:

$$f^i = \gamma F^i = qF^i{}_\nu\frac{dx^\nu}{d\tau}.$$

The time component is then

$$f^0 = qF^0{}_i\frac{dx^i}{d\tau} = q\gamma\frac{1}{c}E_i\frac{dx^i}{dt} = \gamma q\boldsymbol{\beta}\cdot\boldsymbol{E}$$

and the electromagnetic force 4-vector is

$$f^\mu = qF^\mu{}_\nu\frac{dx^\nu}{d\tau} = (\gamma q\boldsymbol{\beta}\cdot\boldsymbol{E}, \gamma q(\boldsymbol{E}+\boldsymbol{v}\times\boldsymbol{B})).$$

The equation of motion of the charged particle in 4-vector notation is therefore

$$m\frac{d^2x^\mu}{d\tau^2} = qF^\mu{}_\nu\frac{dx^\nu}{d\tau},$$

or

$$m\ddot{x}^\mu(\tau) = qF^\mu{}_\nu(x(\tau))\dot{x}^\nu(\tau). \tag{B.21}$$

The time component yields after rescaling with c/γ again the energy balance equation

$$\frac{dE}{dt} = q\boldsymbol{v}\cdot\boldsymbol{E}. \tag{B.22}$$

The spatial part is after rescaling with γ^{-1}:

$$\frac{d}{dt}\boldsymbol{p} = q(\boldsymbol{E}+\boldsymbol{v}\times\boldsymbol{B}). \tag{B.23}$$

The only changes in (B.22, B.23) with respect to the nonrelativistic equations are the velocity dependences of E and $\boldsymbol{p}$:

$$\boldsymbol{p} = \frac{m\boldsymbol{v}}{\sqrt{1-(v^2/c^2)}}, \quad E = \frac{mc^2}{\sqrt{1-(v^2/c^2)}}.$$

The equations (B.21) are completely equivalent to equations (B.23) and (B.22). Note that equation (B.22) is a consequence of (B.23) just like the equation (B.21) with $\mu = 0$ is also a consequence of the other three equations with spatial values for μ.

The virtue of equations (B.21) is the *manifest covariance* of these equations, since linearly transforming equations between 4-vectors must hold in every inertial frame. Contrary to this, covariance is *not apparent* in the equations (B.22, B.23), but since they are equivalent to the manifestly covariant equations (B.21), they also must hold in every inertial frame. Covariance is only *hidden* in the nonlinear transformation behavior of equations (B.22, B.23). However, for practical purposes the equations (B.22, B.23) are often more useful.

The relativistic Lagrange function for a charged particle in terms of the laboratory time t is

$$L_{(t)} = -mc\sqrt{c^2 - \dot{\boldsymbol{x}}^2(t)} + q\dot{\boldsymbol{x}}(t)\cdot\boldsymbol{A}(\boldsymbol{x}(t),t) - q\Phi(\boldsymbol{x}(t),t). \tag{B.24}$$

This yields the canonical momentum

$$\boldsymbol{p}_{can} = \frac{\partial L_{(t)}}{\partial\dot{\boldsymbol{x}}} = \frac{mc\boldsymbol{v}}{\sqrt{c^2-\boldsymbol{v}^2}} + q\boldsymbol{A} = \boldsymbol{p} + q\boldsymbol{A},$$

and the equations of motion in the form (B.23). The relativistic action is

$$\begin{aligned} S &= \int dt\, L_{(t)} = \int\left(-mc\sqrt{c^2dt^2 - d\boldsymbol{x}^2} + qd\boldsymbol{x}\cdot\boldsymbol{A} - qdt\Phi\right) \\ &= \int d\tau\left(-mc\sqrt{-\eta_{\mu\nu}\frac{dx^\mu}{d\tau}\frac{dx^\nu}{d\tau}} + qA_\mu\frac{dx^\mu}{d\tau}\right) = \int d\tau\, L_{(\tau)}. \end{aligned} \tag{B.25}$$

The formulation in terms of the eigentime τ of the particle yields the canonical momentum (use $\eta_{\mu\nu}(dx^\mu/d\tau)(dx^\nu/d\tau) = -c^2$ from the equation $c^2d\tau^2 = -\eta_{\mu\nu}$ $dx^\mu dx^\nu$ after the derivative)

$$p_{can,\mu} = \frac{\partial L_{(\tau)}}{\partial\dot{x}^\mu} = m\eta_{\mu\nu}\frac{dx^\nu}{d\tau} + qA_\mu = p_\mu + qA_\mu,$$

and the Lagrange equation is the manifestly covariant formulation (B.21) of the equations of motion.

The gauge-dependent contributions qA_μ to the conserved momenta disappear in the fully covariant energy-momentum tensor of a classical charged particle of mass m and charge q coupled to electromagnetic fields,

$$\begin{aligned} T_\nu{}^\mu &= \frac{1}{\mu_0}\left(F_{\nu\rho}F^{\mu\rho} - \frac{1}{4}\eta_\nu{}^\mu F_{\rho\sigma}F^{\rho\sigma}\right) + \int d\tau\, mcU_\nu(\tau)U^\mu(\tau)\delta(x - x(\tau)) \\ &= \frac{1}{\mu_0}\left(F_{\nu\rho}F^{\mu\rho} - \frac{1}{4}\eta_\nu{}^\mu F_{\rho\sigma}F^{\rho\sigma}\right) + mc\frac{v_\nu v^\mu}{\sqrt{c^2 - v^2}}\delta(\boldsymbol{x} - \boldsymbol{x}(t)). \end{aligned} \tag{B.26}$$

Contrary to the 4-velocity U^μ, the four quantities $v^\mu = dx^\mu/dt = U^\mu/\gamma = (c, \boldsymbol{v})$ are not components of a 4-vector, but still convenient for the representation of the classical energy-momentum tensor after integration over the eigentime of the particle.

The corresponding results for the energy density, energy current density and momentum density of the classical particle plus fields system are

$$\mathcal{H} = c\mathcal{P}^0 = T^{00} = \frac{\epsilon_0}{2}\boldsymbol{E}^2 + \frac{1}{2\mu_0}\boldsymbol{B}^2 + \frac{mc^3}{\sqrt{c^2 - v^2}}\delta(\boldsymbol{x} - \boldsymbol{x}(t)), \tag{B.27}$$

$$\mathcal{S} = c\boldsymbol{e}_i T^{0i} = \frac{1}{\mu_0}\boldsymbol{E}\times\boldsymbol{B} + \frac{mc^3\boldsymbol{v}}{\sqrt{c^2 - v^2}}\delta(\boldsymbol{x} - \boldsymbol{x}(t)), \tag{B.28}$$

$$\mathcal{P} = \frac{1}{c}\boldsymbol{e}_i T^{i0} = \epsilon_0\boldsymbol{E}\times\boldsymbol{B} + \frac{mc\boldsymbol{v}}{\sqrt{c^2 - v^2}}\delta(\boldsymbol{x} - \boldsymbol{x}(t)) = \frac{1}{c^2}\mathcal{S}, \tag{B.29}$$

and the stress tensor is

$$\begin{aligned} \underline{T} = {} & \left(\frac{\epsilon_0}{2}\boldsymbol{E}^2 + \frac{1}{2\mu_0}\boldsymbol{B}^2\right)\underline{1} - \epsilon_0\boldsymbol{E}\otimes\boldsymbol{E} - \frac{1}{\mu_0}\boldsymbol{B}\otimes\boldsymbol{B} \\ & + mc\frac{\boldsymbol{v}\otimes\boldsymbol{v}}{\sqrt{c^2 - v^2}}\delta(\boldsymbol{x} - \boldsymbol{x}(t)). \end{aligned} \tag{B.30}$$

Relativistic center of mass frame

Considerations of two-particle systems in relativistic quantum mechanics also require the relativistic notion of center of mass frames. If two particles have momenta $\boldsymbol{p}_1$ and $\boldsymbol{p}_2$ in an inertial frame, every inertial frame with the property that the total momentum $\boldsymbol{P}' = \boldsymbol{p}_1' + \boldsymbol{p}_2'$ of the two-particle system in that frame vanishes is traditionally denoted as a center of mass frame for the two-particle system, although "zero total momentum frame" would be a more appropriate name. We will nevertheless continue to use the traditional name "center of mass frame". Two center of mass frames for the system then differ at most by a combination of a translation and a rotation and possibly an inversion of the time axis. According to equation (B.6) the task to actually transform into a center of mass frame then amounts to find a Lorentz boost into a frame moving with velocity $\boldsymbol{v} = c\boldsymbol{\beta}$ such that with $E = E_1 + E_2$

$$\boldsymbol{P}'_{\parallel} = \gamma\boldsymbol{P}_{\parallel} - \gamma\boldsymbol{\beta}\frac{E}{c} = 0 \tag{B.31}$$

and

$$\boldsymbol{P}'_{\perp} = \boldsymbol{P}_{\perp} = 0. \tag{B.32}$$

The condition (B.32) implies that we have to choose $\boldsymbol{\beta}\|\boldsymbol{P}$, and that $\boldsymbol{P}_{\parallel} = \boldsymbol{P}, \boldsymbol{P}'_{\parallel} = \boldsymbol{P}'$. The condition (B.31) is then solved by

$$\begin{aligned}\boldsymbol{\beta} &= c\frac{\boldsymbol{P}}{E},\\ \gamma &= \frac{E}{\sqrt{E^2 - c^2\boldsymbol{P}^2}}.\end{aligned} \tag{B.33}$$

The momentum vectors of the two particles in the center of mass frame are therefore

$$\boldsymbol{p}'_{1,\perp} = \boldsymbol{p}_{1,\perp} = -\boldsymbol{p}_{2,\perp} = -\boldsymbol{p}'_{2,\perp},$$

and

$$\boldsymbol{p}'_{1,\parallel} = \frac{E}{\sqrt{E^2 - c^2\boldsymbol{P}^2}}\left(\boldsymbol{p}_{1,\parallel} - \frac{\boldsymbol{P}}{E}E_1\right) = \frac{E_2\boldsymbol{p}_{1,\parallel} - E_1\boldsymbol{p}_{2,\parallel}}{\sqrt{E^2 - c^2\boldsymbol{P}^2}} = -\boldsymbol{p}'_{2,\parallel}.$$

The corresponding energies in the center of mass frame are

$$\begin{aligned}E'_1 &= \frac{E}{\sqrt{E^2 - c^2\boldsymbol{P}^2}}\left(E_1 - c^2\frac{\boldsymbol{P}}{E}\cdot\boldsymbol{p}_{1,\parallel}\right) = \frac{E_1^2 - c^2\boldsymbol{p}_1^2 + E_1E_2 - c^2\boldsymbol{p}_1\cdot\boldsymbol{p}_2}{\sqrt{E^2 - c^2\boldsymbol{P}^2}},\\ E'_2 &= \frac{E_2^2 - c^2\boldsymbol{p}_2^2 + E_1E_2 - c^2\boldsymbol{p}_1\cdot\boldsymbol{p}_2}{\sqrt{E^2 - c^2\boldsymbol{P}^2}}.\end{aligned}$$

For consistency we notice

$$E' = E'_1 + E'_2 = \sqrt{E^2 - c^2\boldsymbol{P}^2},$$

as also implied by $P'^2 = P^2$ with $\boldsymbol{P}' = 0$.

We also note that if we define the center of energy of the two-particle system,

$$\boldsymbol{R} = \frac{\boldsymbol{x}_1 E_1 + \boldsymbol{x}_2 E_2}{E_1 + E_2},$$

and if the particles do not interact, then the center of energy velocity is exactly the velocity (B.33) of the center of mass frame,

$$\dot{\boldsymbol{R}} = c^2 \frac{\boldsymbol{p}_1 + \boldsymbol{p}_2}{E_1 + E_2} = c\boldsymbol{\beta}.$$

If the particles interact, we need to also take into account the contributions from the fields which mediate the interactions to the center of energy and to the energy and momentum balances of the system, see (18.128, 18.129).

Appendix C: Completeness of Sturm-Liouville Eigenfunctions

Completeness of eigenfunctions of self-adjoint operators is very important in quantum mechanics. Formulating exact theorems and proofs in general situations is a demanding mathematical problem. However, the setting of Sturm-Liouville problems with homogeneous boundary conditions in one dimension is sufficiently simple to be treated in a single appendix.

Sturm-Liouville problems

Sturm-Liouville problems are linear boundary value problems consisting of a second order differential equation

$$\frac{d}{dx}\left(g(x)\frac{d\psi(x)}{dx}\right) - V(x)\psi(x) + E\varrho(x)\psi(x) = 0 \tag{C.1}$$

in an interval $a \leq x \leq b$ and homogeneous boundary conditions[1] (Sturm 1836, Liouville 1837)

$$\psi(a) = 0, \quad \psi(b) = 0. \tag{C.2}$$

The functions $g(x)$, $V(x)$ and $\varrho(x)$ are real and continuous in $a \leq x \leq b$, and we also assume that the functions $g(x)$ and $\varrho(x)$ are positive in $a \leq x \leq b$.

[1]General Sturm-Liouville boundary conditions would only require linear combinations of $\psi(x)$ and $\psi'(x)$ to vanish at the boundaries, but for our purposes it is sufficient to impose the special conditions $\psi(a) = 0$, $\psi(b) = 0$.

R. Dick, *Advanced Quantum Mechanics*, Graduate Texts in Physics,
DOI 10.1007/978-3-319-25675-7

In ket notation without reference to a particular representation, we would write equation (C.1) as

$$E\varrho(\mathrm{x})|\psi(E)\rangle = \frac{1}{\hbar^2}\mathrm{p}g(\mathrm{x})\mathrm{p}|\psi(E)\rangle + V(\mathrm{x})|\psi(E)\rangle.$$

We can assume $\psi(x) \equiv \langle x|\psi(E)\rangle$ to be a real function, and in this appendix we always assume that a and b are finite. We also require first order differentiability of $g(x)$ and continuity of $V(x)$ and $\varrho(x)$.

We can also assume $\psi'(a) > 0$. We know that $\psi'(a) \neq 0$ because $\psi'(a) = 0$ together with $\psi(a) = 0$ and the Sturm-Liouville equation (C.1) would imply $\psi(x) = 0$. Furthermore, linearity of the Sturm-Liouville equation implies that we can always change the sign of $\psi(x)$ to ensure $\psi'(a) > 0$.

Multiplication of equation (C.1) with $\psi(x)$ and integration yields

$$H[\psi] \equiv \int_a^b dx\left(g(x)\psi'^2(x) + V(x)\psi^2(x)\right) = E\int_a^b dx\,\varrho(x)\psi^2(x) \equiv E\langle\psi|\psi\rangle$$

where the last equation defines the scalar product

$$\langle\phi|\psi\rangle = \int_a^b dx\,\varrho(x)\phi(x)\psi(x). \tag{C.3}$$

It is easy to prove that (C.3) defines a scalar product since $\langle\psi|\psi\rangle \geq 0$ and $\langle\psi|\psi\rangle = 0 \Leftrightarrow \psi(x) = 0$, and

$$0 \leq \langle\psi + \lambda\phi|\psi + \lambda\phi\rangle = \langle\psi|\psi\rangle + 2\lambda\langle\psi|\phi\rangle + \lambda^2\langle\phi|\phi\rangle \tag{C.4}$$

has a minimum for

$$\lambda = -\frac{\langle\psi|\phi\rangle}{\langle\phi|\phi\rangle},$$

which after substitution in (C.4) yields the Schwarz inequality

$$\langle\psi|\phi\rangle^2 \leq \langle\psi|\psi\rangle\langle\phi|\phi\rangle.$$

The Sturm-Liouville equation (C.1) arises as an Euler-Lagrange equation from variation of the action

$$\begin{aligned} S[\psi] &= E\langle\psi|\psi\rangle - H[\psi] \\ &= \int_a^b dx\left(E\varrho(x)\psi^2(x) - g(x)\psi'^2(x) - V(x)\psi^2(x)\right) \end{aligned} \tag{C.5}$$

with fixed endpoints $\psi(a)$ and $\psi(b)$.

The stationary values of $S[\psi]$ for arbitrary fixed endpoints $\psi(a)$ and $\psi(b)$ are

$$S[\psi]\Big|_{\text{on-shell}} = g(a)\psi(a)\psi'(a) - g(b)\psi(b)\psi'(b),$$

where the designation "on-shell" means that $\psi(x)$ satisfies the Euler-Lagrange equation (C.1) of $S[\psi]$.

If we think of the Sturm-Liouville problem as a one-dimensional scalar field theory, $G(x) = 1/4g^2(x)$ would play the role of a metric in $a \le x \le b$ and $H[\psi]$ would be the energy of the field $\psi(x)$ if $\psi(x)$ is normalized, $\langle\psi|\psi\rangle = 1$.

Suppose $\psi_i(x)$ and $\psi_j(x)$ are solutions of the Sturm-Liouville problem (C.1, C.2) with eigenvalues E_i and E_j, respectively. Use of the Sturm-Liouville equation (C.1) and partial integration yields

$$E_i \int_a^x d\xi\, \varrho(\xi)\psi_j(\xi)\psi_i(\xi) = \int_a^x d\xi\, \psi_j(\xi)\left[V(\xi)\psi_i(\xi) - \frac{d}{d\xi}\left(g(\xi)\frac{d}{d\xi}\psi_i(\xi)\right)\right]$$
$$= \int_a^x d\xi\, \big(V(\xi)\psi_j(\xi)\psi_i(\xi) + g(\xi)\psi_j'(\xi)\psi_i'(\xi)\big) - g(x)\psi_j(x)\frac{d}{dx}\psi_i(x),$$

and after another integration by parts we find

$$(E_i - E_j)\int_a^x d\xi\, \varrho(\xi)\psi_i(\xi)\psi_j(\xi) = g(x)\left(\psi_i(x)\frac{d}{dx}\psi_j(x) - \psi_j(x)\frac{d}{dx}\psi_i(x)\right). \quad \text{(C.6)}$$

This equation implies for $E_i = E_j$

$$\frac{d}{dx}\ln\psi_i(x) = \frac{d}{dx}\ln\psi_j(x),$$

i.e. $\psi_i(x)$ has to be proportional to $\psi_j(x)$: There is no degeneracy of eigenvalues in the one-dimensional Sturm-Liouville problem.

For $x = b$, equation (C.6) implies the orthogonality property

$$(E_i - E_j)\langle\psi_i|\psi_j\rangle = 0$$

and taking into account the absence of degeneracy yields

$$\langle\psi_i|\psi_j\rangle \propto \delta_{ij}.$$

Liouville's normal form of Sturm's equation

We can gauge the functions $g(x)$ and $\varrho(x)$ away through a transformation of variables

$$x \to X = \int_a^x d\xi \sqrt{\frac{\varrho(\xi)}{g(\xi)}}, \quad \psi(x) \to \Psi(X) = (\varrho(x)g(x))^{1/4}\,\psi(x).$$

This yields

$$0 \leq X \leq B = \int_a^b dx \sqrt{\frac{\varrho(x)}{g(x)}}, \quad \Psi(0) = 0, \quad \Psi(B) = 0,$$

and the Sturm-Liouville equation (C.1) assumes the form of a one-dimensional Schrödinger equation,

$$\frac{d^2}{dX^2}\Psi(X) - V(X)\Psi(X) + E\Psi(X) = 0 \tag{C.7}$$

with

$$V(X) = \frac{V(x)}{\varrho(x)} + \frac{g(x)\varrho''(x) + \varrho(x)g''(x)}{4\varrho^2(x)} - \frac{5g(x)\varrho'^2(x)}{16\varrho^3(x)} - \frac{g'^2(x)}{16g(x)\varrho(x)}$$
$$+ \frac{g'(x)\varrho'(x)}{8\varrho^2(x)}.$$

Second order differentiability of $\varrho(x)$ and $g(x)$ is usually assumed. However, we only have to require continuity of the positive functions $\varrho(x)$ and $g(x)$ since we can deal with δ-function singularities in one-dimensional potentials,.

Equation (C.7) is *Liouville's normal form* of the Sturm-Liouville equation.

Nodes of Sturm-Liouville eigenfunctions

For the following reasoning we assume that we have smoothly continued the functions $V(x)$, $\varrho(x) > 0$ and $g(x) > 0$ for all values of $x \in \mathbb{R}$. It does not matter how we do that.

To learn more about the nodes of the eigenfunctions $\psi_i(x)$ of the Sturm-Liouville boundary value problem, let us now assume that $\psi(x, \lambda)$ and $\psi(x, \mu)$ are solutions of the incomplete initial value problems

$$\lambda\varrho(x)\psi(x, \lambda) = V(x)\psi(x, \lambda) - \frac{d}{dx}\left(g(x)\frac{d\psi(x, \lambda)}{dx}\right), \quad \psi(a, \lambda) = 0, \tag{C.8}$$

$$\mu\varrho(x)\psi(x, \mu) = V(x)\psi(x, \mu) - \frac{d}{dx}\left(g(x)\frac{d\psi(x, \mu)}{dx}\right), \quad \psi(a, \mu) = 0, \tag{C.9}$$

with $\lambda > \mu$, *but contrary to the boundary value problem (C.1, C.2) we do not impose any conditions at* $x = b$. In that case there exist solutions to the Sturm-Liouville equations for arbitrary values of the parameters λ, μ, and we can again require

$$\left.\frac{d\psi(x,\lambda)}{dx}\right|_{x=a} > 0, \quad \left.\frac{d\psi(x,\mu)}{dx}\right|_{x=a} > 0.$$

We recall the following facts from the theory of differential equations: The solution $\psi(x, \lambda)$ to the initial value problem (C.8) is unique up to a multiplicative constant, and $\psi(x, \lambda)$ depends continuously on the parameter λ.

The last fact is important, because it implies that the nodes $y(\lambda)$ of $\psi(x, \lambda)$, $\psi(y(\lambda), \lambda) = 0$, depend continuously on λ. Continuity of $y(\lambda)$ is used in the demonstration below that the boundary value problem (C.1, C.2) has a solution for every value of b.

Multiplication of equation (C.8) with $\psi(x, \mu)$ and equation (C.9) with $\psi(x, \lambda)$, integration from a to $x > a$, and subtraction of the equations yields

$$\begin{aligned}
&(\lambda-\mu)\int_a^x d\xi\, \varrho(\xi)\psi(\xi,\lambda)\psi(\xi,\mu) \\
&= \int_a^x d\xi \left[\psi(\xi,\lambda)\frac{d}{d\xi}\left(g(\xi)\frac{d\psi(\xi,\mu)}{d\xi}\right) - \psi(\xi,\mu)\frac{d}{d\xi}\left(g(\xi)\frac{d\psi(\xi,\lambda)}{d\xi}\right)\right] \\
&= g(x)\left(\psi(x,\lambda)\frac{d\psi(x,\mu)}{dx} - \psi(x,\mu)\frac{d\psi(x,\lambda)}{dx}\right). \qquad \text{(C.10)}
\end{aligned}$$

Now assume that $y(\mu)$ is the first node of $\psi(x, \mu)$ larger than a:

$$\psi(y(\mu), \mu) = 0, \quad y(\mu) > a.$$

Substituting $x = y(\mu)$ in (C.10) yields

$$(\lambda-\mu)\int_a^{y(\mu)} dx\, \varrho(x)\psi(x,\lambda)\psi(x,\mu) = g(y(\mu))\psi(y(\mu),\lambda)\left.\frac{d\psi(x,\mu)}{dx}\right|_{x=y(\mu)}. \qquad \text{(C.11)}$$

We know that

$$(\lambda-\mu)\varrho(x)\psi(x,\mu) > 0$$

for $a < x < y(\mu)$ and that

$$g(y(\mu))\left.\frac{d\psi(x,\mu)}{dx}\right|_{x=y(\mu)} < 0.$$

This implies $\psi(x, \lambda)$ must change its sign at least once for $a < x < y(\mu)$, and in particular $y(\lambda) < y(\mu)$:

The location of the leftmost node $y(\lambda) > a$ of the function $\psi(x, \lambda)$ moves closer to a if λ increases.

We are not really concerned with differentiability properties of the leftmost node $y(\lambda)$, but we can express the previous observation also as

$$y(\lambda) > a, \quad \frac{dy(\lambda)}{d\lambda} < 0.$$

Now assume that λ is small enough[2] so that even $y(\lambda) > b$. Then we can increase the parameter λ until we reach a value $\lambda = E_1$ such that $y(E_1) = b$. This is then the lowest eigenvalue of our original Sturm-Liouville boundary value problem (C.1), and the corresponding eigenfunction is

$$\psi_1(x) = \psi(x, \lambda = E_1). \tag{C.12}$$

The eigenfunction $\psi_1(x)$ for the lowest eigenvalue E_1 has no nodes in $a < x < b$.

Now we consider the first and the second node of $\psi(x, \mu)$ for $x > a$,

$$a < y(\mu) \equiv y_1(\mu) < y_2(\mu), \quad \psi(y_1(\mu), \mu) = 0, \quad \psi(y_2(\mu), \mu) = 0,$$

and we integrate from $y_1(\mu)$ to $y_2(\mu)$,

$$\begin{aligned}
&(\lambda - \mu)\int_{y_1(\mu)}^{y_2(\mu)} dx\, \varrho(x)\psi(x, \lambda)\psi(x, \mu) \\
&= \int_{y_1(\mu)}^{y_2(\mu)} dx \left[\psi(x, \lambda)\frac{d}{dx}\left(g(x)\frac{d\psi(x, \mu)}{dx}\right) - \psi(x, \mu)\frac{d}{dx}\left(g(x)\frac{d\psi(x, \lambda)}{dx}\right)\right] \\
&= g(y_2(\mu))\, \psi(y_2(\mu), \lambda)\frac{d\psi(x, \mu)}{dx}\bigg|_{x=y_2(\mu)} \\
&\quad - g(y_1(\mu))\, \psi(y_1(\mu), \lambda)\frac{d\psi(x, \mu)}{dx}\bigg|_{x=y_1(\mu)}.
\end{aligned}$$

We know

$$(\lambda - \mu)\varrho(x)\psi(x, \mu) < 0$$

[2]The alert reader might worry that all $y(\lambda)$ might be smaller than b, so that there is no finite small value λ with $y(\lambda) > b$, or otherwise that all $y(\lambda)$ might be larger than b, so that no finite value E_1 with $y(E_1) = b$ would exist. These cases can be excluded through Sturm's comparison theorem, to be discussed later.

for $y_1(\mu) < x < y_2(\mu)$, and

$$g(y_1(\mu))\frac{d\psi(x,\mu)}{dx}\bigg|_{x=y_1(\mu)} < 0, \quad g(y_2(\mu))\frac{d\psi(x,\mu)}{dx}\bigg|_{x=y_2(\mu)} > 0.$$

This tells us that $\psi(x,\lambda)$ has to change sign in the interval $y_1(\mu) < x < y_2(\mu)$, i.e. it must have at least one node there. We know that the first node $y_1(\lambda) < y_1(\mu)$ is outside of this interval. Therefore we can infer that at least the second node $y_2(\lambda)$ of $\psi(x,\lambda)$ must be smaller than $y_2(\mu)$: $y_2(\lambda) < y_2(\mu)$. We can repeat this reasoning for the pair of adjacent nodes $y_{n-1}(\mu)$, $y_n(\mu)$ of $\psi(x,\mu)$, and we always find for $\lambda > \mu$ that $y_n(\lambda) < y_n(\mu)$,

$$a < y_n(\lambda), \quad \psi(y_n(\lambda),\lambda) = 0, \quad \frac{dy_n(\lambda)}{d\lambda} < 0.$$

All nodes of the function $\psi(x,\lambda)$ on the right hand side of $x = a$ move closer to a if λ increases.

Therefore we can repeat the reasoning above which had let us to the first solution $\psi_1(x)$ with eigenvalue E_1 of our Sturm-Liouville problem. To find the second eigenfunction, we increase $\lambda > E_1$ until we hit a value $\lambda = E_2$ such that $y_2(E_2) = b$, and the corresponding eigenfunction

$$\psi_2(x) = \psi(x, E_2)$$

will have exactly one node $y_1(E_2)$ in the interval, $a < y_1(E_2) < b$.

The corresponding result for $y_n(\lambda)$ tells us that in the n-th step we will find a parameter $\lambda = E_n$ with $y_n(E_n) = b$ and eigenfunction

$$\psi_n(x) = \psi(x, E_n),$$

and this function will have $n-1$ nodes $a < y_1(E_n) < y_2(E_n) < \ldots < y_{n-1}(E_n) < y_n(E_n) = b$ inside the interval.

Sturm's comparison theorem and estimates for the locations of the nodes $y_n(\lambda)$

Sturm's comparison theorem makes a statement about the change of the nodes $y_n > a$ of the solution $\psi(x,\lambda)$ of

$$\frac{d}{dx}\left(g(x)\frac{d\psi(x,\lambda)}{dx}\right) + (\lambda\varrho(x) - V(x))\,\psi(x,\lambda) = 0, \quad \psi(a,\lambda) = 0, \qquad \text{(C.13)}$$

if the functions $g(x)$, $\varrho(x)$ and $V(x)$ change. To prove the comparison theorem, we do *not* use Liouville's normal form, but perform the following simple transformation of variables,

$$X = \int_a^x \frac{dx'}{g(x')}, \quad \Psi(X, \lambda) = \psi(x, \lambda).$$

This transforms (C.13) into the following form,

$$\frac{d^2\Psi(X, \lambda)}{dX^2} + (\lambda R(X) - V(X))\, \Psi(X, \lambda) = 0, \quad \Psi(0, \lambda) = 0, \tag{C.14}$$

$$R(X) = g(x)\varrho(x) > 0, \quad V(X) = g(x)V(x),$$

and the nodes $Y_n > 0$ of $\Psi(X, \lambda)$ are related to the nodes $y_n > a$ of $\psi(x, \lambda)$ through

$$Y_n = \int_a^{y_n} \frac{dx}{g(x)}. \tag{C.15}$$

Now we consider another Sturm-Liouville problem of the form (C.14), but with different functions

$$\lambda S(X) - W(X) > \lambda R(X) - V(X),$$

$$\frac{d^2\Phi(X, \lambda)}{dX^2} + (\lambda S(X) - W(X))\, \Phi(X, \lambda) = 0, \quad \Phi(0, \lambda) = 0, \tag{C.16}$$

and we denote the positive nodes of $\Phi(X, \lambda)$ with Z_n. We also require again $\Psi'(0) > 0$, $\Phi'(0) > 0$. Equations (C.14, C.16) imply

$$\int_{Y_{n-1}}^{Y_n} dX\, [V(X) - W(X) - \lambda\, (R(X) - S(X))]\, \Psi(X, \lambda)\Phi(X, \lambda)$$
$$= \Phi(Y_n, \lambda) \left.\frac{d\Psi(X, \lambda)}{dX}\right|_{X=Y_n} - \Phi(Y_{n-1}, \lambda) \left.\frac{d\Psi(X, \lambda)}{dX}\right|_{X=Y_{n-1}}. \tag{C.17}$$

The following terms in (C.17) have all the same sign,

$$[V(X) - W(X) - \lambda\, (R(X) - S(X))]\, \Psi(X, \lambda)\Big|_{Y_{n-1}<X<Y_n},$$

$$\left.\frac{d\Psi(X, \lambda)}{dX}\right|_{X=Y_{n-1}}, \quad -\left.\frac{d\Psi(X, \lambda)}{dX}\right|_{X=Y_n}.$$

This implies that $\Phi(X, \lambda)$ must change its sign in $Y_{n-1} < X < Y_n$, and since this must hold for every $n \geq 1$ we find

$$Z_n < Y_n.$$

Increasing $\lambda R(X) - V(X)$ moves the nodes $Y_n > 0$ of the function $\Psi(X,\lambda)$ to the left. From this we can first derive bounds for the nodes $Y_n > 0$ which arise from the nodes of the solutions of

$$\begin{aligned}
&\Psi''_{min}(X,\lambda) + (\lambda R_{max} - V_{min})\,\Psi_{min}(X,\lambda)\\
&= \Psi''_{min}(X,\lambda) + g_{max}\,(\lambda\varrho_{max} - U_{min})\,\Psi_{min}(X,\lambda) = 0, \qquad \text{(C.18)}\\
&\Psi_{min}(0,\lambda) = 0,
\end{aligned}$$

and

$$\begin{aligned}
&\Psi''_{max}(X,\lambda) + (\lambda R_{min} - V_{max})\,\Psi_{max}(X,\lambda)\\
&= \Psi''_{max}(X,\lambda) + g_{min}\,(\lambda\varrho_{min} - U_{max})\,\Psi_{max}(X,\lambda) = 0, \qquad \text{(C.19)}\\
&\Psi_{max}(0,\lambda) = 0.
\end{aligned}$$

Here we use the bounds of the continuous functions $g(x)$, $V(x)$, $\varrho(x)$ on $a \le x \le b$,

$$\begin{aligned}
&0 < g_{min} \le g(x) \le g_{max}, \quad U_{min} \le V(x) \le U_{max},\\
&0 < \varrho_{min} \le \varrho(x) \le \varrho_{max}.
\end{aligned}$$

The solutions of both equations (C.18) and (C.19) have nodes if (recall that both $g(x) > 0$ and $\varrho(x) > 0$)

$$\lambda > U_{min}/\varrho_{max},$$

and the two solutions are

$$\begin{aligned}
\Psi_{min}(X,\lambda) &\propto \sin\!\left(\sqrt{g_{max}\,(\lambda\varrho_{max} - U_{min})}X\right),\\
\Psi_{max}(X,\lambda) &\propto \sin\!\left(\sqrt{g_{min}\,(\lambda\varrho_{min} - U_{max})}X\right).
\end{aligned}$$

This yields bounds for the nodes $Y_n > 0$ of $\Psi(X,\lambda)$,

$$\frac{n\pi}{\sqrt{g_{max}\,(\lambda\varrho_{max} - U_{min})}} \le Y_n \le \frac{n\pi}{\sqrt{g_{min}\,(\lambda\varrho_{min} - U_{max})}}. \qquad \text{(C.20)}$$

However, we also know from equation (C.15) that $g_{min}Y_n \le y_n - a \le g_{max}Y_n$, and therefore[3]

[3]These bounds can be strengthened by a longer proof, but the present result is completely sufficient for our purposes.

$$a + \frac{g_{min} n\pi}{\sqrt{g_{max}\left(\lambda \varrho_{max} - U_{min}\right)}} \leq y_n \leq a + \frac{g_{max} n\pi}{\sqrt{g_{min}\left(\lambda \varrho_{min} - U_{max}\right)}}. \tag{C.21}$$

This implies in particular that *there is no accumulation point for the nodes* y_n *of* $\psi(x, \lambda)$*, and* y_n *must grow like* n *for large* n.

For our previous proof that $\psi_1(x)$ (C.12) has its first node at $y_1 = b$, we needed the assumption that there are small enough values of λ such that the first node $y_1(\lambda)$ of $\psi(x, \lambda)$ satisfies $y_1(\lambda) > b$. We can now confirm that from the lower bound in (C.21). It will suffice to choose

$$\frac{U_{min}}{\varrho_{max}} < \lambda < \frac{U_{min}}{\varrho_{max}} + \frac{g_{min}^2 \pi^2}{\varrho_{max} g_{max} (b-a)^2}. \tag{C.22}$$

We also needed the assumption that for large enough λ the first node $y_1(\lambda) > a$ would be smaller than b. This is easily confirmed from the upper bound in (C.21). It is sufficient to choose

$$\lambda > \frac{U_{max}}{\varrho_{min}} + \frac{g_{max}^2 \pi^2}{\varrho_{min} g_{min} (b-a)^2}. \tag{C.23}$$

Eigenvalue estimates for the Sturm-Liouville problem

We have found that the Sturm-Liouville boundary value problem (C.1, C.2) has an increasing, non-degenerate set of eigenvalues

$$E_1 < E_2 < \ldots$$

and arises as an Euler-Lagrange equation for the action

$$\begin{aligned} S[\psi] &= E\langle\psi|\psi\rangle - H[\psi] \\ &= \int_a^b dx \left(E\varrho(x)\psi^2(x) - g(x)\psi'^2(x) - V(x)\psi^2(x)\right). \end{aligned} \tag{C.24}$$

For every continuous function $\psi(x)$ in $a \leq x \leq b$ we define the normalized function

$$\hat{\psi}(x) = \frac{\psi(x)}{\sqrt{\langle\psi|\psi\rangle}}.$$

Since $S[\psi]$ is homogeneous in ψ, $\psi(x)$ is a stationary point of $S[\psi]$ if and only if $\hat{\psi}(x)$ is a stationary point of

$$S[\hat{\psi}] = E - H[\hat{\psi}],$$

which implies also that $\hat{\psi}(x)$ is a stationary point of the functional

$$H[\hat{\psi}] = \frac{H[\psi]}{\langle\psi|\psi\rangle} = \frac{\int_a^b dx\left[g(x)\psi'^2(x) + V(x)\psi^2(x)\right]}{\int_a^b dx\,\varrho(x)\psi^2(x)}. \tag{C.25}$$

We have already found that there is a discrete subset $\hat{\psi}_n(x)$, $n \in \mathbb{N}$, of stationary points of $H[\hat{\psi}]$ which satisfy the boundary conditions $\hat{\psi}_n(a) = 0$, $\hat{\psi}_n(b) = 0$, and are mutually orthogonal,

$$\langle\hat{\psi}_m|\hat{\psi}_n\rangle = \delta_{mn}.$$

Use of the Sturm-Liouville equation and the boundary conditions yields the values of the functional $H[\hat{\psi}]$ at the stationary points $\hat{\psi}_n(x)$,

$$H[\hat{\psi}_n] = E_n.$$

We already know $E_1 < E_2 < \ldots$, and therefore we have found that the functional $H[\hat{\psi}]$ has a minimum

$$H[\hat{\psi}_1] = E_1$$

on the space of functions

$$\mathcal{F}_{a,b} = \{\psi(x), a \leq x \leq b|\psi(a) = 0, \psi(b) = 0, \langle\psi|\psi\rangle = 1\},$$

and in general we have a minimum

$$H[\hat{\psi}_n] = E_n$$

on the space of functions

$$\mathcal{F}_{a,b}^{(n)} = \{\psi(x), a \leq x \leq b|\psi(a) = 0, \psi(b) = 0, \langle\psi|\psi\rangle = 1, \langle\psi_i|\psi\rangle = 0,\\ 1 \leq i \leq n-1\}.$$

The explicit form of $H[\hat{\psi}]$ in equation (C.25) shows that all the eigenvalues E_n increase if $g(x)$ increases or $V(x)$ increases or $\varrho(x)$ decreases.

However, those continuous functions must be bounded on the finite interval $a \leq x \leq b$,

$$0 < g_{min} \leq g(x) \leq g_{max}, \quad U_{min} \leq V(x) \leq U_{max},$$
$$0 < \varrho_{min} \leq \varrho(x) \leq \varrho_{max}.$$

Therefore we can replace those functions with their extremal values to derive estimates for the eigenvalues E_n.

The Sturm-Liouville problems for the extremal values are

$$g_{min/max}\psi_n''(x) + \left(E_{n,min/max}\varrho_{max/min} - U_{min/max}\right)\psi_n(x) = 0,$$
$$\psi_n(a) = 0, \quad \psi_n(b) = 0,$$

with solutions

$$\psi_n(x) \propto \sin\left(n\pi\frac{x-a}{b-a}\right)$$

and corresponding eigenvalues

$$E_{n,min/max} = \frac{1}{\varrho_{max/min}}\left(U_{min/max} + g_{min/max}\frac{n^2\pi^2}{(b-a)^2}\right).$$

This implies the bounds

$$\frac{1}{\varrho_{max}}\left(U_{min} + g_{min}\frac{n^2\pi^2}{(b-a)^2}\right) \leq E_n \leq \frac{1}{\varrho_{min}}\left(U_{max} + g_{max}\frac{n^2\pi^2}{(b-a)^2}\right). \qquad \text{(C.26)}$$

In particular, *at most a finite number of the lowest eigenvalues E_n can be negative, and the eigenvalues for large n must grow like n^2.*

Both of these observations are crucial for the proof that the set $\psi_n(x)$ of eigenfunctions of the Sturm-Liouville problem (C.1, C.2) provide a complete basis for the expansion of piecewise continuous functions in $a \leq x \leq b$.

Completeness of Sturm-Liouville eigenstates

We now assume that the Sturm-Liouville eigenstates are normalized,

$$\langle\psi_i|\psi_j\rangle = \delta_{ij}.$$

Let $\phi(x)$ be an arbitrary smooth function on $a \leq x \leq b$ with $\phi(a) = 0$ and $\phi(b) = 0$, and define

$$\varphi_n(x) = \phi(x) - \sum_{i=1}^{n}\psi_i(x)\langle\psi_i|\phi\rangle.$$

Then we have

$$0 \leq \langle\varphi_n|\varphi_n\rangle = \langle\phi|\phi\rangle - \sum_{i=1}^{n}\langle\psi_i|\phi\rangle^2,$$

i.e. for all n we have a *Bessel inequality*

$$\langle\phi|\phi\rangle \geq \sum_{i=1}^{n}\langle\psi_i|\phi\rangle^2.$$

We also have $\langle\varphi_n|\psi_i\rangle = 0$, $1 \leq i \leq n$, and $\varphi_n(a) = 0$, $\varphi_n(b) = 0$, i.e.

$$\varphi_n(x) \in \mathcal{F}_{a,b}^{(n+1)}.$$

Therefore the minimum property of the eigenvalue E_{n+1} implies

$$E_{n+1} \leq \frac{H[\varphi_n]}{\langle\varphi_n|\varphi_n\rangle}. \tag{C.27}$$

We have

$$\begin{aligned} H[\varphi_n] &= H[\phi] - 2\sum_{i=1}^{n}\langle\psi_i|\phi\rangle\int_a^b dx\left(g(x)\phi'(x)\psi_i'(x) + V(x)\phi(x)\psi_i(x)\right) \\ &\quad + \sum_{i,j=1}^{n}\langle\psi_i|\phi\rangle\langle\psi_j|\phi\rangle\int_a^b dx\left(g(x)\psi_i'(x)\psi_j'(x) + V(x)\psi_i\psi_j(x)\right). \end{aligned}$$

In the first sum, partial integration and use of the Sturm-Liouville equation yields

$$\begin{aligned} \int_a^b dx\left(g(x)\phi'(x)\psi_i'(x) + V(x)\phi(x)\psi_i(x)\right) &= E_i\int_a^b dx\,\varrho(x)\phi(x)\psi_i(x) \\ &= E_i\langle\psi_i|\phi\rangle. \end{aligned}$$

In the double sum, partial integration and use of the Sturm-Liouville equation yields

$$\begin{aligned} \int_a^b dx\left(g(x)\psi_i'(x)\psi_j'(x) + V(x)\psi_i\psi_j(x)\right) &= E_i\int_a^b dx\,\varrho(x)\psi_i(x)\psi_j(x) \\ &= E_i\delta_{ij}. \end{aligned}$$

This implies

$$H[\varphi_n] = H[\phi] - \sum_{i=1}^{n}E_i\langle\psi_i|\phi\rangle^2. \tag{C.28}$$

Since at most finitely many of the eigenvalues E_i can be negative, equation (C.28) tells us that the functional $H[\varphi_n]$ must remain bounded from above for $n \to \infty$, e.g. for

$$E_1 < E_2 < \cdots < E_N < 0 \leq E_{N+1} < \ldots$$

we have the bound

$$H[\varphi_n] \le H[\phi] + \sum_{i=1}^{N} |E_i| \langle \psi_i | \phi \rangle^2.$$

On the other hand, equation (C.27) yields for $n > N$ (to ensure $E_{n+1} > 0$),

$$\langle \varphi_n | \varphi_n \rangle = \langle \phi | \phi \rangle - \sum_{i=1}^{n} \langle \psi_i | \phi \rangle^2 \le \frac{H[\varphi_n]}{E_{n+1}}$$

and since E_{n+1} grows like n^2 for large n while $H[\varphi_n]$ must remain bounded, we find the completeness relation

$$\lim_{n\to\infty} \langle \varphi_n | \varphi_n \rangle = \lim_{n\to\infty} \int_a^b dx\, \varrho(x) \left(\phi(x) - \sum_{i=1}^{n} \psi_i(x) \langle \psi_i | \phi \rangle \right)^2 = 0 \tag{C.29}$$

or equivalently,

$$\langle \phi | \phi \rangle = \lim_{n\to\infty} \sum_{i=1}^{n} \langle \phi | \psi_i \rangle \langle \psi_i | \phi \rangle.$$

Completeness of the series

$$\sum_{i=1}^{\infty} \psi_i(x) \langle \psi_i | \phi \rangle \sim \phi(x)$$

in the sense of equation (C.29) is denoted as *completeness in the mean*, and is sometimes also expressed as

$$\text{l.i.m.}_{n\to\infty} \sum_{i=1}^{n} \psi_i(x) \langle \psi_i | \phi \rangle = \phi(x),$$

where l.i.m. stands for "limit in the mean". Completeness in the mean says that the series $\sum_{i=1}^{\infty} \psi_i(x) \langle \psi_i | \phi \rangle$ approximates $\phi(x)$ in the least squares sense.

Completeness in the mean also implies for the two piecewise continuous functions f and g

$$f(x) \pm g(x) \sim \sum_{i=1}^{\infty} \psi_i(x) \langle \psi_i | f \rangle \pm \sum_{i=1}^{\infty} \psi_i(x) \langle \psi_i | g \rangle$$

and therefore

$$\langle f | g \rangle = \frac{1}{4} \left(\langle f + g | f + g \rangle - \langle f - g | f - g \rangle \right) = \lim_{n\to\infty} \sum_{i=1}^{n} \langle f | \psi_i \rangle \langle \psi_i | g \rangle. \tag{C.30}$$

Completeness in the sense of (C.30) is enough for quantum mechanics, because it says that we can use the completeness relation

$$\underline{1} = \lim_{n\to\infty} \sum_{i=1}^{n} |\psi_i\rangle\langle\psi_i|$$

in the calculation of matrix elements between sufficiently smooth functions (where "sufficiently smooth = continuously differentiable to a required order" depends on the operators we use). This is all that is really needed in quantum mechanics. However, for piecewise smooth functions, the relation also holds pointwise almost everywhere (see Remark 3 below).

I would like to add a few remarks:

1. The completeness property (C.29) also applies to piecewise continuous functions in $a \le x \le b$ and functions which do not vanish at the boundary points, because every piecewise continuous function can be approximated in the mean by a smooth function which vanishes at the boundaries.
2. If $\phi(x)$ is a smooth function satisfying the Sturm-Liouville boundary conditions, as we have assumed in the derivation of (C.29), the series under the integral sign will even converge uniformly to $\phi(x)$,

$$\lim_{n\to\infty} \sum_{i=1}^{n} \psi_i(x)\langle\psi_i|\phi\rangle = \phi(x),$$

i.e. for all $a \le x \le b$ and all values $\epsilon > 0$, there exists an $n(\epsilon)$ such that

$$\left|\phi(x) - \sum_{i=1}^{n} \psi_i(x)\langle\psi_i|\phi\rangle\right| < \epsilon \quad \text{if} \quad n \ge n(\epsilon). \tag{C.31}$$

Uniformity of the convergence refers to the fact that the same $n(\epsilon)$ ensures (C.31) for all $a \le x \le b$.
3. If $\phi(x)$ is piecewise smooth in $a \le x \le b$, it can still be expanded pointwise in Sturm-Liouville eigenstates. Except for points of discontinuity of $\phi(x)$, and except for the boundary points if $\phi(x)$ does not satisfy the same boundary conditions as the eigenfunctions $\psi_i(x)$, the expansion

$$\phi(x) = \lim_{n\to\infty} \sum_{i=1}^{n} \psi_i(x)\langle\psi_i|\phi\rangle$$

holds pointwise, and the series converges uniformly to $\phi(x)$ in every closed interval which excludes discontinuities of $\phi(x)$ (and the series converges to the arithmetic mean in the points of discontinuity). The boundary points must also be excluded if $\phi(x)$ does not satisfy the Sturm-Liouville boundary conditions.

Appendix D: Properties of Hermite Polynomials

We use the following equation as a definition of Hermite polynomials,

$$H_n(x) = \exp\!\left(\frac{1}{2}x^2\right)\left(x - \frac{d}{dx}\right)^n \exp\!\left(-\frac{1}{2}x^2\right), \tag{D.1}$$

because we initially encountered them in this form in the solution of the harmonic oscillator in Chapter 6. We can use the identity

$$\left(x + \frac{d}{dx}\right) f(x) = \exp\!\left(-\frac{1}{2}x^2\right)\frac{d}{dx}\left[\exp\!\left(\frac{1}{2}x^2\right) f(x)\right]$$

to rewrite equation (D.1) in the form

$$\begin{aligned} H_n(x) &= \exp\!\left(\frac{1}{2}x^2\right)\left[2x - \exp\!\left(-\frac{1}{2}x^2\right)\frac{d}{dx}\exp\!\left(\frac{1}{2}x^2\right)\right]^n \exp\!\left(-\frac{1}{2}x^2\right) \\ &= \left[\exp\!\left(\frac{1}{2}x^2\right)\left[2x - \exp\!\left(-\frac{1}{2}x^2\right)\frac{d}{dx}\exp\!\left(\frac{1}{2}x^2\right)\right]\exp\!\left(-\frac{1}{2}x^2\right)\right]^n \\ &= \left(2x - \frac{d}{dx}\right)^n 1, \end{aligned} \tag{D.2}$$

or we can use the identity

$$\left(x - \frac{d}{dx}\right) f(x) = -\exp\!\left(\frac{1}{2}x^2\right)\frac{d}{dx}\left[\exp\!\left(-\frac{1}{2}x^2\right) f(x)\right]$$

to rewrite equation (D.1) in the Rodrigues form

$$H_n(x) = \exp(x^2)\left(-\frac{d}{dx}\right)^n \exp(-x^2). \tag{D.3}$$

R. Dick, *Advanced Quantum Mechanics*, Graduate Texts in Physics,
DOI 10.1007/978-3-319-25675-7

The Rodrigues formula implies

$$\sum_{n=0}^{\infty} H_n(x)\frac{z^n}{n!} = \sum_{n=0}^{\infty}\left[\exp(x^2)\,\frac{\partial^n}{\partial z^n}\exp(-(x-z)^2)\right]_{z=0}\frac{z^n}{n!}$$
$$= \exp(x^2)\exp(-(x-z)^2) = \exp(2xz - z^2)\,. \tag{D.4}$$

The residue theorem then also yields the representation

$$H_n(x) = \frac{n!}{2\pi\mathrm{i}}\oint dz\,\frac{\exp(2xz - z^2)}{z^{n+1}}, \tag{D.5}$$

where the integration contour encloses $z = 0$ in the positive sense of direction, i.e. counter clockwise.

Another useful integral representation for the Hermite polynomials follows from (D.2) and the equation

$$\int_{-\infty}^{\infty} du\,(2u)^n \exp(-(u+v)^2) = \int_{-\infty}^{\infty} du\left(-2v - \frac{\partial}{\partial v}\right)^n \exp(-(u+v)^2)$$
$$= \left(-2v - \frac{\partial}{\partial v}\right)^n \sqrt{\pi}.$$

This yields in particular for $v = -\mathrm{i}x$,

$$\int_{-\infty}^{\infty} du\,(2u)^n \exp(-(u-\mathrm{i}x)^2) = \mathrm{i}^n\sqrt{\pi}H_n(x). \tag{D.6}$$

Combination of equations (D.4) and (D.6) yields Mehler's formula[1],

$$\sum_{n=0}^{\infty} H_n(x)H_n(x')\frac{z^n}{n!} = \sum_{n=0}^{\infty} H_n(x)\frac{1}{\sqrt{\pi}n!}\int_{-\infty}^{\infty} du\,(-2\mathrm{i}uz)^n \exp(-(u-\mathrm{i}x')^2)$$
$$= \frac{1}{\sqrt{\pi}}\int_{-\infty}^{\infty} du\,\exp(-4\mathrm{i}xuz + 4u^2z^2)\exp(-(u-\mathrm{i}x')^2)$$
$$= \frac{1}{\sqrt{1-4z^2}}\exp\left(-4z\frac{z\,(x^2+x'^2) - xx'}{1-4z^2}\right). \tag{D.7}$$

This requires $|z| < 1/2$ for convergence. In Sections 6.3 and 13.1 we need this in the form for $|z| < 1$,

[1] F.G. Mehler, J. Math. 66, 161 (1866).

$$\sum_{n=0}^{\infty} H_n(x)H_n(x')\frac{z^n}{2^n n!}\exp\left(-\frac{x^2+x'^2}{2}\right)$$
$$=\frac{1}{\sqrt{1-z^2}}\exp\left(-\frac{\left(1+z^2\right)\left(x^2+x'^2\right)-4zxx'}{2\left(1-z^2\right)}\right). \tag{D.8}$$

Indeed, applications of this equation for the harmonic oscillator are usually in the framework of distributions and require the limit $|z| \to 1$. In principle we should therefore replace the corresponding phase factors z in Sections 6.3 and 13.1 with $z\exp(-\epsilon)$, and take the limit $\epsilon \to +0$ after applying any distributions which are derived from (D.8).

Appendix E: The Baker-Campbell-Hausdorff Formula

The Baker-Campbell-Hausdorff formula explains how to combine the product of operator exponentials $\exp(A)\cdot\exp(B)$ into a single operator exponential $\exp[\Phi(A,B)]$, if the series expansion for $\Phi(A,B)$ provided by the Baker-Campbell-Hausdorff formula converges.

We try to determine $\Phi(A,B)$ as a power series in a parameter λ,

$$\exp[\lambda A]\cdot\exp[\lambda B]=\exp[\Phi(\lambda A,\lambda B)],\quad \Phi(\lambda A,\lambda B)=\sum_{n=1}^{\infty}\lambda^n c_n(A,B).$$

We also use the notation of the adjoint action of an operator A on an operator B,

$$A^{(ad)}\circ B=-[A,B].$$

We start with

$$\exp[\alpha A]\cdot\exp[\beta B]=\exp[\Phi(\alpha A,\beta B)].$$

This implies with Lemma (6.22) the equations

$$\begin{aligned}
B&=\exp[-\Phi(\alpha A,\beta B)]\frac{\partial}{\partial\beta}\exp[\Phi(\alpha A,\beta B)]=\sum_{n=1}^{\infty}\frac{(-)^n}{n!}\overset{n}{[}\Phi(\alpha A,\beta B),\partial_\beta]\\
&=-\sum_{n=1}^{\infty}\frac{(-)^n}{n!}\overset{n-1}{[}\,\Phi(\alpha A,\beta B),\partial_\beta\Phi(\alpha A,\beta B)]\\
&=\sum_{n=1}^{\infty}\frac{1}{n!}\left(\Phi(\alpha A,\beta B)^{(ad)}\right)^{n-1}\circ\partial_\beta\Phi(\alpha A,\beta B)\\
&=\frac{\exp[\Phi(\alpha A,\beta B)^{(ad)}]-1}{\Phi(\alpha A,\beta B)^{(ad)}}\circ\partial_\beta\Phi(\alpha A,\beta B)
\end{aligned}$$

R. Dick, *Advanced Quantum Mechanics*, Graduate Texts in Physics,
DOI 10.1007/978-3-319-25675-7

and

$$A = -\exp[\Phi(\alpha A, \beta B)]\frac{\partial}{\partial\alpha}\exp[-\Phi(\alpha A, \beta B)] = -\sum_{n=1}^{\infty}\frac{1}{n!}{}^{n}[\Phi(\alpha A, \beta B), \partial_\alpha]$$

$$= \sum_{n=1}^{\infty}\frac{1}{n!}{}^{n-1}[\ \Phi(\alpha A, \beta B), \partial_\alpha\Phi(\alpha A, \beta B)]$$

$$= \sum_{n=1}^{\infty}\frac{1}{n!}\left(-\Phi(\alpha A, \beta B)^{(ad)}\right)^{n-1} \circ \partial_\alpha\Phi(\alpha A, \beta B)$$

$$= \frac{1-\exp[-\Phi(\alpha A, \beta B)^{(ad)}]}{\Phi(\alpha A, \beta B)^{(ad)}} \circ \partial_\alpha\Phi(\alpha A, \beta B).$$

For the inversion of these equations, we notice

$$\begin{aligned}\left(\frac{\exp(z)-1}{z}\right)^{-1} &= \frac{z}{\exp(z)-1} = z\frac{\exp(-z/2)}{\exp(z/2)-\exp(-z/2)}\\ &= \frac{z}{2}\frac{\exp(z/2)+\exp(-z/2)}{\exp(z/2)-\exp(-z/2)} - \frac{z}{2}\\ &= \frac{z}{2}\coth\frac{z}{2} - \frac{z}{2} = 1 + \sum_{n=1}^{\infty}\frac{(-)^{n+1}}{(2n)!}B_n z^{2n} - \frac{z}{2},\end{aligned}$$

$$\begin{aligned}\left(\frac{1-\exp(-z)}{z}\right)^{-1} &= \frac{z}{1-\exp(-z)} = z\frac{\exp(z/2)}{\exp(z/2)-\exp(-z/2)}\\ &= \frac{z}{2}\frac{\exp(z/2)+\exp(-z/2)}{\exp(z/2)-\exp(-z/2)} + \frac{z}{2}\\ &= \frac{z}{2}\coth\frac{z}{2} + \frac{z}{2} = 1 + \sum_{n=1}^{\infty}\frac{(-)^{n+1}}{(2n)!}B_n z^{2n} + \frac{z}{2},\end{aligned}$$

where the coefficients B_n are Bernoulli numbers.

The previous equations yield (with $\Phi(\alpha A, \beta B)^{(ad)} \circ A = -[\Phi(\alpha A, \beta B), A]$)

$$\begin{aligned}\partial_\alpha\Phi(\alpha A, \beta B) &= \frac{\Phi(\alpha A, \beta B)^{(ad)}}{2}\coth\frac{\Phi(\alpha A, \beta B)^{(ad)}}{2} \circ A\\ &\quad - \frac{1}{2}[\Phi(\alpha A, \beta B), A],\end{aligned}$$

$$\begin{aligned}\partial_\beta\Phi(\alpha A, \beta B) &= \frac{\Phi(\alpha A, \beta B)^{(ad)}}{2}\coth\frac{\Phi(\alpha A, \beta B)^{(ad)}}{2} \circ B\\ &\quad + \frac{1}{2}[\Phi(\alpha A, \beta B), B],\end{aligned}$$

$$\begin{aligned}\partial_\lambda \Phi(\lambda A, \lambda B) &= \Big[\partial_\alpha \Phi(\alpha A, \beta B) + \partial_\beta \Phi(\alpha A, \beta B)\Big]_{\alpha=\beta=\lambda} \\ &= \frac{\Phi(\lambda A, \lambda B)^{(ad)}}{2} \coth \frac{\Phi(\lambda A, \lambda B)^{(ad)}}{2} \circ (A+B) \\ &\quad + \frac{1}{2}[A-B, \Phi(\lambda A, \lambda B)],\end{aligned}$$

i.e.

$$\begin{aligned}\partial_\lambda \Phi(\lambda A, \lambda B) &= A + B + \sum_{n=1}^{\infty} \frac{(-)^{n+1}}{(2n)!} B_n \cdot [\Phi(\lambda A, \lambda B)^{(ad)}]^{2n} \circ (A+B) \\ &\quad + \frac{1}{2}[A-B, \Phi(\lambda A, \lambda B)].\end{aligned} \tag{E.1}$$

Equation (E.1) provides us with a recursion relation for the n-th order coefficient functions $c_n(A,B)$,

$$\begin{aligned}(n+1)c_{n+1}(A,B) &= \frac{1}{2}[A-B, c_n(A,B)] + \sum_{m=1}^{\lfloor n/2 \rfloor} \frac{(-)^{m+1}}{(2m)!} B_m \\ &\times \sum_{\substack{1 \le k_1, k_2, \ldots k_{2m} \\ k_1 + \ldots + k_{2m} = n}} [c_{k_{2m}}(A,B), [\ldots, [c_{k_2}(A,B), [c_{k_1}(A,B), A+B]] \ldots]],\end{aligned} \tag{E.2}$$

with

$$c_0(A,B) = 0, \quad c_1(A,B) = A + B.$$

The floor function $\lfloor x \rfloor$ maps to the next lowest integer smaller or equal to x, i.e. $\lfloor n/2 \rfloor = n/2$ if n is even, $\lfloor n/2 \rfloor = (n-1)/2$ if n is odd.

The result (E.2) yields for the next three terms

$$\begin{aligned}c_2(A,B) &= \frac{1}{2}[A,B], \\ c_3(A,B) &= \frac{1}{12}[A-B, [A,B]] + \frac{1}{6} B_1 [A+B, [A+B, A+B]] \\ &= \frac{1}{12}[A, [A,B]] + \frac{1}{12}[B, [B,A]], \\ c_4(A,B) &= \frac{1}{96}[A-B, [A, [A,B]] + [B, [B,A]]] \\ &\quad + \frac{1}{16} B_1 [A+B, [[A,B], A+B]]\end{aligned}$$

$$
\begin{aligned}
&= \frac{1}{96}\Big([A,[A,[A,B]]] - [B,[B,[B,A]]] + [A,[B,[B,A]]] \\
&\quad -[B,[A,[A,B]]] - [A,[A,[A,B]]] + [B,[B,[B,A]]] \\
&\quad +[A,[B,[B,A]]] - [B,[A,[A,B]]]\Big) \\
&= \frac{1}{48}[A,[B,[B,A]]] - \frac{1}{48}[B,[A,[A,B]]] = \frac{1}{24}[A,[B,[B,A]]].
\end{aligned}
$$

The Jacobi identity

$$[A,[B,C]] + [B,[C,A]] + [C,[A,B]] = 0$$

was used in the last step for c_4.

Appendix F: The Logarithm of a Matrix

Exponentials of square matrices $\underline{G}$, $\underline{M} = \exp \underline{G} = \sum_{n=0}^{\infty} \underline{G}^n/n!$, are frequently used for the representation of symmetry transformations. Indeed, the properties of continuous symmetry transformations are often discussed in terms of their first order approximations $\underline{1} + \underline{G}$, where it is assumed that continuity of the symmetries allows for parameter choices such that $\max |G_{ij}| \ll 1$. It is therefore of interest that the logarithm $\underline{G} = \ln \underline{M}$ of invertible square matrices can also be defined, although the existence of $\underline{G}$ does not imply that it can be chosen to satisfy $\max |G_{ij}| \ll 1$ for $\underline{M}$ close to the unit matrix, see below.

Suppose $\underline{M}$ is a complex invertible square matrix which is related to its Jordan canonical form through

$$\underline{M} = \underline{T} \cdot \oplus_n \underline{J}_n \cdot \underline{T}^{-1}.$$

Each of the smaller square matrices $\underline{J}_n$ has the form

$$\underline{J} = \lambda \underline{1} \tag{F.1}$$

or the form

$$\underline{J} = \begin{pmatrix} \lambda & 1 & 0 & 0 & \ldots & 0 & 0 & 0 \\ 0 & \lambda & 1 & 0 & \ldots & 0 & 0 & 0 \\ \vdots & \vdots & \vdots & \vdots & \vdots & \vdots & \vdots & \vdots \\ 0 & 0 & 0 & 0 & \ldots & \lambda & 1 & 0 \\ 0 & 0 & 0 & 0 & \ldots & 0 & \lambda & 1 \\ 0 & 0 & 0 & 0 & \ldots & 0 & 0 & \lambda \end{pmatrix}, \tag{F.2}$$

and $\det(\underline{M}) \neq 0$ implies that none of the eigenvalues λ can vanish.

R. Dick, *Advanced Quantum Mechanics*, Graduate Texts in Physics,
DOI 10.1007/978-3-319-25675-7

In the case (F.1) we have

$$\underline{J} = \exp(\ln \lambda \underline{1}), \quad \ln \underline{J} = \ln \lambda \underline{1}.$$

However, it is also possible to construct the logarithm of a Jordan block matrix (F.2). The direct sum of the logarithms of all the matrices $\underline{J}_n$ then yields the logarithm of the matrix $\underline{M}$,

$$\underline{M} = \exp(\underline{T} \cdot \oplus_n \ln \underline{J}_n \cdot \underline{T}^{-1}), \quad \ln \underline{M} = \underline{T} \cdot \oplus_n \ln \underline{J}_n \cdot \underline{T}^{-1}.$$

Suppose the Jordan matrix (F.2) is a $(\nu+1) \times (\nu+1)$ matrix. We define $(\nu+1) \times (\nu+1)$ matrices $\underline{N}_n$, $0 \le n \le \nu$, according to $(\underline{N}_n)_{ij} = \delta_{i+n,j}$, i.e. $\underline{N}_0$ is the $(\nu+1) \times (\nu+1)$ unit matrix and $\underline{N}_{1 \le n \le \nu}$ has non-vanishing entries 1 only in the n-th diagonal above the main diagonal. These matrices satisfy the multiplication law $\underline{N}_m \cdot \underline{N}_n = \Theta(\nu - m - n + \epsilon)\underline{N}_{m+n}$, which also implies $\underline{N}_n = (\underline{N}_1)^n$.

Each $(\nu+1) \times (\nu+1)$ Jordan block can be written as $\underline{J} = \lambda \underline{N}_0 + \underline{N}_1$, and its logarithm can be defined through

$$\underline{X} = \ln \underline{J} = \begin{pmatrix} \ln\lambda & \lambda^{-1} & -\lambda^{-2}/2 & \lambda^{-3}/3 & \ldots & (-)^{\nu-1}\lambda^{-\nu}/\nu \\ 0 & \ln\lambda & \lambda^{-1} & -\lambda^{-2}/2 & \ldots & (-)^{\nu-2}\lambda^{-(\nu-1)}/(\nu-1) \\ \vdots & \vdots & \vdots & \vdots & \vdots & \vdots \\ 0 & 0 & 0 & 0 & \ldots & -\lambda^{-2}/2 \\ 0 & 0 & 0 & 0 & \ldots & \lambda^{-1} \\ 0 & 0 & 0 & 0 & \ldots & \ln\lambda \end{pmatrix}$$

$$= \underline{N}_0 \ln \lambda - \sum_{n=1}^{\nu} \frac{(-\lambda)^{-n}}{n} \underline{N}_n. \tag{F.3}$$

We can prove $\exp(\underline{X}) = \underline{J}$ in the following way. The N-th power of $\underline{X}$ is (here $0 < \epsilon < 1$ is introduced to avoid the ambiguity of the Θ function at 0)

$$\begin{aligned} \underline{X}^N = {} & \underline{N}_0 (\ln \lambda)^N + (-)^N \sum_{1 \le n_1, n_2 \ldots n_N}^{\nu+1-N} \Theta\left(\nu + \epsilon - \sum_{i=1}^{N} n_i\right) \\ & \times \frac{(-\lambda)^{-n_1 - n_2 - \ldots - n_N}}{n_1 \cdot n_2 \cdot \ldots \cdot n_N} \underline{N}_{n_1 + n_2 + \ldots + n_N} \cdot \\ & - (-)^N N \ln \lambda \sum_{1 \le n_1, n_2 \ldots n_{N-1}}^{\nu+2-N} \Theta\left(\nu + \epsilon - \sum_{i=1}^{N-1} n_i\right) \\ & \times \frac{(-\lambda)^{-n_1 - n_2 - \ldots - n_{N-1}}}{n_1 \cdot n_2 \cdot \ldots \cdot n_{N-1}} \underline{N}_{n_1 + n_2 + \ldots + n_{N-1}} \end{aligned}$$

$$+(-)^N \binom{N}{2} (\ln \lambda)^2$$

$$\times \sum_{1 \le n_1, n_2 \dots n_{N-2}}^{\nu+3-N} \Theta\left(\nu + \epsilon - \sum_{i=1}^{N-2} n_i\right)$$

$$\times \frac{(-\lambda)^{-n_1-n_2-\dots-n_{N-2}}}{n_1 \cdot n_2 \cdot \ \dots \ \cdot n_{N-2}} \underline{N}_{n_1+n_2+\dots+n_{N-2}} + \dots$$

$$-N(\ln \lambda)^{N-1} \sum_{n=1}^{\nu} \frac{(-\lambda)^{-n}}{n} \underline{N}_n.$$

We can combine terms in the form

$$\underline{X}^N = \underline{N}_0 (\ln \lambda)^N + \sum_{m=1}^{N} (-)^m \binom{N}{m} (\ln \lambda)^{N-m}$$

$$\times \sum_{1 \le n_1, n_2 \dots n_m}^{\nu+1-m} \Theta\left(\nu + \epsilon - \sum_{i=1}^{m} n_i\right) \frac{(-\lambda)^{-n_1-n_2-\dots-n_m}}{n_1 \cdot n_2 \cdot \ \dots \ \cdot n_m} \underline{N}_{n_1+n_2+\dots+n_m}$$

$$= \underline{N}_0 (\ln \lambda)^N + \sum_{M=1}^{\nu} (-\lambda)^{-M} \underline{N}_M \sum_{m=1}^{\min(N,M)} (-)^m \binom{N}{m} (\ln \lambda)^{N-m}$$

$$\times \sum_{\substack{1 \le n_1, n_2 \dots n_m \\ n_1+n_2+\dots+n_m = M}}^{M+1-m} \frac{1}{n_1 \cdot n_2 \cdot \ \dots \ \cdot n_m}.$$

This is after isolation of the term with $M = 1$ in the sum,

$$\underline{X}^N = \underline{N}_0 (\ln \lambda)^N + N \frac{(\ln \lambda)^{N-1}}{\lambda} \underline{N}_1 + \Theta(\nu - 2 + \epsilon)$$

$$\times \sum_{M=2}^{\nu} (-\lambda)^{-M} \underline{N}_M \left(\sum_{m=1}^{\min(N,M)} (-)^m \binom{N}{m} (\ln \lambda)^{N-m} \right.$$

$$\left. \times \sum_{\substack{1 \le n_1, n_2 \dots n_m \\ n_1+n_2+\dots+n_m = M}}^{M+1-m} \frac{1}{n_1 \cdot n_2 \cdot \ \dots \ \cdot n_m} \right).$$

Only the first two terms survive in

$$\exp(\underline{X}) = \underline{1} + \sum_{N=1}^{\infty} \frac{\underline{X}^N}{N!} = \lambda \underline{N}_0 + \underline{N}_1 = \underline{J}$$

because the sum over N in the term of order M reduces to

$$\sum_{N=m}^{\infty}\frac{(\ln\lambda)^{N-m}}{(N-m)!}=\lambda,$$

and the remaining sums yield for $M\geq 1$

$$\sum_{m=1}^{M}\frac{(-)^m}{m!}\sum_{\substack{1\leq n_1,n_2\ldots n_m\\ n_1+n_2+\ldots+n_m=M}}^{M+1-m}\frac{1}{n_1\cdot n_2\cdot\ldots\cdot n_m}$$

$$=\frac{1}{2\pi\mathrm{i}}\oint_{|z|<1}dz\sum_{m=1}^{M}\frac{(-)^m}{m!}\sum_{n_1,n_2\ldots n_m=1}^{\infty}\frac{z^{n_1+n_2+\ldots+n_m-M-1}}{n_1\cdot n_2\cdot\ldots\cdot n_m}$$

$$=\frac{1}{2\pi\mathrm{i}}\oint_{|z|<1}dz\sum_{m=1}^{\infty}\frac{(-)^m}{m!}\sum_{n_1,n_2\ldots n_m=1}^{\infty}\frac{z^{n_1+n_2+\ldots+n_m-M-1}}{n_1\cdot n_2\cdot\ldots\cdot n_m}$$

$$=\frac{1}{2\pi\mathrm{i}}\oint_{|z|<1}dz\sum_{m=1}^{\infty}\frac{(-)^m}{m!}\left(\sum_{n=1}^{\infty}\frac{z^n}{n}\right)^m z^{-M-1}$$

$$=\frac{1}{2\pi\mathrm{i}}\oint_{|z|<1}dz\sum_{m=1}^{\infty}\frac{[\ln(1-z)]^m}{m!}z^{-M-1}$$

$$=\frac{1}{2\pi\mathrm{i}}\oint_{|z|<1}dz\left(z^{-M-1}-z^{-M}\right)=-\delta_{M,1}.$$

Equation (F.3) is a special case of a general procedure to define functions $\underline{M}\to f(\underline{M})$ of square matrices [16], and for every $n\in\mathbb{Z}$, the matrix $\underline{X}+2\pi\mathrm{i}n\underline{N}_0$ is also a logarithm of $\underline{J}$.

A glance at (F.3) tells us that we should avoid matrices with Jordan blocks in their eigenvalue decomposition if we want to find logarithms with the property $\max|(\ln\underline{M})_{ij}|\ll 1$ for $\max|M_{ij}-\delta_{ij}|\ll 1$. This can be achieved if we use hermitian and unitary matrices, and if $\underline{M}$ does not satisfy this condition, we can use its polar decomposition

$$\underline{M}=\underline{H}\cdot\underline{U}=(\underline{M}\cdot\underline{M}^{+})^{1/2}\cdot[(\underline{M}\cdot\underline{M}^{+})^{-1/2}\cdot\underline{M}] \quad \text{(F.4)}$$

in terms of a hermitian and a unitary factor, or a symmetric and an orthogonal factor if $\underline{M}$ is real. The factors will then have logarithms with small matrix elements if $\underline{M}$ is close to the unit matrix, i.e. the analysis of continuous symmetries in finite-dimensional vector spaces eventually requires the analysis of up to two first order transformations $\underline{1}+\ln\underline{H}$ and $\underline{1}+\ln\underline{U}$. This is the case e.g. for Lorentz transformations, where $\underline{H}$ is the pure boost part and $\underline{U}$ is the rotation.

Appendix G: Dirac γ matrices

It is useful for the understanding and explicit construction of γ matrices to discuss their properties in a general number d of spacetime dimensions. γ matrices in more than four spacetime dimensions are regularly used in theories which hypothesize the existence of extra spacetime dimensions. On the other hand, variants of the Dirac equation in two space dimensions or three spacetime dimensions have also become relevant in materials science for the description of electrons in Graphene and other two-dimensional materials.

γ-matrices in d dimensions

The condition (21.35), $\{\gamma_\mu, \gamma_\nu\} = -2\eta_{\mu\nu}$, implies that any product of n gamma coefficients $\gamma_\alpha \cdot \gamma_\beta \cdot \ldots \cdot \gamma_\omega$ can be reduced to a product of $n-2$ coefficients if two indices have the same value. We can also re-order any product such that the indices have increasing values. These observations imply that the d coefficients γ_μ can produce at most 2^d linearly independent combinations

$$\mathbf{1},\, \gamma_0,\, \gamma_1,\, \ldots,\, \gamma_{d-1},\, \gamma_0 \cdot \gamma_1,\, \gamma_0 \cdot \gamma_2,\, \ldots,\, \gamma_0 \cdot \gamma_1 \cdot \ldots \cdot \gamma_{d-1}. \tag{G.1}$$

We are actually interested in matrix representations of the algebra generated by (21.35), and consider the coefficients γ^μ and the objects in (G.1) as matrices in the following. We first discuss the case that d is an *even number of spacetime dimensions*, and we define multi-indices J through

$$\Gamma_J = \gamma_{\mu_1} \cdot \gamma_{\mu_2} \cdot \ldots \cdot \gamma_{\mu_n}, \quad \mu_1 < \mu_2 < \ldots < \mu_n, \quad n(J) = n. \tag{G.2}$$

It is easy to prove that

$$\mathrm{tr}(\Gamma_J) = 0. \tag{G.3}$$

R. Dick, *Advanced Quantum Mechanics*, Graduate Texts in Physics,
DOI 10.1007/978-3-319-25675-7

For even $n(J)$ this follows from the anti-commutativity of the γ-matrices and the cyclic invariance of the trace. For odd $n(J)$ this follows from the fact that there is at least one γ-matrix not contained in Γ_J, e.g. γ_1, and therefore

$$\mathrm{tr}(\Gamma_J) = -\mathrm{tr}(\gamma_1^2 \cdot \Gamma_J) = -\mathrm{tr}(\gamma_1 \cdot \Gamma_J \cdot \gamma_1) = \mathrm{tr}(\gamma_1^2 \cdot \Gamma_J) = -\mathrm{tr}(\Gamma_J) = 0.$$

The product $\Gamma_I \cdot \Gamma_J$ reduces either to a Γ-matrix Γ_K if $I \neq J$, or otherwise

$$\Gamma_I^2 = \pm\mathbf{1},$$

and this implies orthogonality of all the Γ-matrices and $\mathbf{1}$,

$$\mathrm{tr}(\Gamma_I \cdot \Gamma_J) \propto \delta_{IJ}.$$

For *even number of spacetime dimensions d* this implies that all the 2^d matrices in (G.1) are indeed linearly independent, and therefore a minimal matrix representation of (21.35) requires at least $(2^{d/2} \times 2^{d/2})$-matrices. We will see by explicit construction that such a representation exists, and because $2^{d/2}$ is the minimal dimension, the representation must be irreducible, i.e. cannot split into smaller matrices acting in spaces of lower dimensions. The representation also turns out to be unique up to similarity transformations

$$\gamma_\mu \to \gamma'_\mu = A \cdot \gamma_\mu \cdot A^{-1}.$$

For *odd number of spacetime dimensions d*, we also define the matrices Γ_J according to (G.1), but now the previous proof of $\mathrm{tr}(\Gamma_J) = 0$ only goes through for all the matrices Γ_J *except for the last matrix in the list*,

$$\Gamma_{0,1,\ldots d-1} = \gamma_0 \cdot \gamma_1 \cdot \ldots \cdot \gamma_{d-1}.$$

For odd d, this matrix contains an odd number of γ-matrices, and it contains all γ-matrices, such that the previous proof of vanishing trace for odd $n(J)$ does not go through for this particular matrix. Furthermore, this matrix has the properties

$$[\Gamma_{0,1,\ldots d-1}, \Gamma_J] = 0, \tag{G.4}$$

$$\Gamma^2_{0,1,\ldots d-1} = (-)^{(d+2)(d-1)/2}\mathbf{1} = (-)^{(d-1)/2}\mathbf{1}. \tag{G.5}$$

Commutativity with all other matrices implies that in every irreducible representation

$$\Gamma_{0,1,\ldots d-1} = \pm(-)^{(d-1)/4}\mathbf{1}, \tag{G.6}$$

see the following subsection for the proof.

This also implies that every product Γ_J of $n(J) \geq (d+1)/2$ γ matrices is up to a numerical factor a product Γ_I of $n(I) = d - n(J) \leq (d-1)/2$ γ matrices,

$$\left[\Gamma_J\right]_{n(J)\geq(d+1)/2} = \left[\mathbf{1}\cdot\Gamma_J\right]_{n(J)\geq(d+1)/2} \propto \left[\Gamma_{0,1,\ldots d-1}\cdot\Gamma_J\right]_{n(J)\geq(d+1)/2}$$
$$\propto \left[\Gamma_I\right]_{n(I)\leq(d-1)/2}.$$

Therefore there are only 2^{d-1} linearly independent matrices in (G.1) for odd d, and the minimal possible dimension of the representation is only $2^{(d-1)/2}$. The explicit construction later on confirms that the minimal dimension also works for odd number of spacetime dimensions. There are two different equivalence classes of matrix representations with dimension $2^{(d-1)/2}$.

We can summarize the results on the dimensions of γ matrices in d space(-time) dimensions by the statement the irreducible representations of the Dirac algebra are provided by $2^{\lfloor d/2\rfloor} \times 2^{\lfloor d/2\rfloor}$ matrices, where the floor function in the exponents rounds to the next lowest integer and is also often written in terms of Gauss brackets: $\lfloor d/2\rfloor \equiv [d/2]_G = d/2$ if d is even, $\lfloor d/2\rfloor \equiv [d/2]_G = (d-1)/2$ if d is odd.

Proof that in irreducible representations $\Gamma_{0,1,\ldots d-1} \propto \mathbf{1}$ for odd spacetime dimension d

$\Gamma_{0,1,\ldots d-1}$ commutes with all Γ_J. Suppose that we have an irreducible matrix representation of (G.1) in a vector space V of dimension $\dim V$. If λ is an eigenvalue of $\Gamma_{0,1,\ldots d-1} \propto \mathbf{1}$,

$$\det(\Gamma_{0,1,\ldots d-1} - \lambda\mathbf{1}) = 0,$$

we have

$$\dim\Big((\Gamma_{0,1,\ldots d-1} - \lambda\mathbf{1})\cdot V\Big) \leq \dim V - 1,$$

and

$$\Gamma_J\cdot(\Gamma_{0,1,\ldots d-1} - \lambda\mathbf{1})\cdot V = (\Gamma_{0,1,\ldots d-1} - \lambda\mathbf{1})\cdot\Gamma_J\cdot V.$$

The last equation would imply that $(\Gamma_{0,1,\ldots d-1} - \lambda\mathbf{1})\cdot V$, if non-empty, would be an invariant subspace under the action of the γ-matrices, in contradiction to the irreducibility of V. Therefore we must have

$$(\Gamma_{0,1,\ldots d-1} - \lambda\mathbf{1})\cdot V = \emptyset, \quad \Gamma_{0,1,\ldots d-1} = \lambda\mathbf{1}$$

in every irreducible representation. Equation (G.5) tells us that

$$\lambda = \pm(-)^{(d-1)/4}.$$

The proof is simply an adaptation of the proof of Schur's lemma from group theory.

Recursive construction of γ-matrices in different dimensions

We will use the following conventions for the explicit construction of γ-matrices: Up to similarity transformations, the γ-matrices in *two spacetime dimensions* are

$$\gamma_0 = \begin{pmatrix} 0 & 1 \\ 1 & 0 \end{pmatrix}, \quad \gamma_1 = \begin{pmatrix} 0 & 1 \\ -1 & 0 \end{pmatrix}. \tag{G.7}$$

For the recursive construction in higher dimensions $d \geq 3$ we now assume that γ_μ, $0 \leq \mu \leq d-2$, are γ-matrices in $d-1$ dimensions.

For the construction of γ-matrices in an *odd number d of spacetime dimensions* there are two inequivalent choices,

$$\Gamma_0 = \pm \mathrm{i}^{(d-3)/2}\gamma_0\gamma_1 \ldots \gamma_{d-2} = \pm\begin{pmatrix} -\underline{1} & \underline{0} \\ \underline{0} & \underline{1} \end{pmatrix}, \quad \Gamma_i = \gamma_i, \quad 1 \leq i \leq d-2,$$
$$\Gamma_{d-1} = -\mathrm{i}\gamma_0. \tag{G.8}$$

For the construction of γ-matrices in an *even number $d \geq 4$ of spacetime dimensions* there is only one equivalence class of representations,

$$\Gamma_0 = \begin{pmatrix} \underline{0} & \underline{1} \\ \underline{1} & \underline{0} \end{pmatrix}, \quad \Gamma_i = \begin{pmatrix} \underline{0} & -\gamma_0\gamma_i \\ \gamma_0\gamma_i & \underline{0} \end{pmatrix}, \quad 1 \leq i \leq d-2,$$
$$\Gamma_{d-1} = \begin{pmatrix} \underline{0} & -\gamma_0 \\ \gamma_0 & \underline{0} \end{pmatrix}. \tag{G.9}$$

Note that it does not matter from which of the two possible representations $\pm\gamma_0$ in the odd number $d-1$ of lower dimensions we start since Γ_0 intertwines the two possibilities,

$$\Gamma_0\Gamma_i\Gamma_0 = -\Gamma_i, \quad 1 \leq i \leq d-1.$$

The possibility of similarity transformations implies that there are infinitely many equivalent possibilities to construct these bases of γ-matrices. The construction described here was motivated from the desire to have Weyl bases (i.e. all γ^μ have only off-diagonal non-vanishing ($2^{(d/2)-1} \times 2^{(d/2)-1}$) blocks) in even dimensions, and to have the next best solution, *viz.* Dirac bases (i.e. $\gamma^0 = \pm\text{diag}(\underline{1}, -\underline{1})$, all γ^i like in a Weyl basis), in odd dimensions. Note that all the representations (G.8) and (G.9) of the γ-matrices in odd or even dimensions fulfill

$$\gamma_0^+ = \gamma_0, \quad \gamma_i^+ = -\gamma_i,$$

or equivalently

$$\gamma_\mu^+ = \gamma_0\gamma_\mu\gamma_0. \tag{G.10}$$

Every set of γ-matrices is equivalent to a set satisfying equation (G.10). We will prove this in the following subsection.

Proof that every set of γ-matrices is equivalent to a set which satisfies equation (G.10)

In this section we do not use summation convention but spell out all summations explicitly.

We define $2^{\lfloor d/2\rfloor} \times 2^{\lfloor d/2\rfloor}$ matrices $X_0 = \gamma_0$, $X_i = \mathrm{i}\gamma_i$,

$$\{X_\mu, X_\nu\} = 2\delta_{\mu\nu}$$

and prove that the matrices X_μ are equivalent to a set of unitary matrices Y_μ. Since the matrices Y_μ also satisfy $Y_\mu^{-1} = Y_\mu$, unitarity also implies hermiticity of Y_μ. We use the abbreviation $N = 2^{\lfloor d/2\rfloor}$, and consider the set $\mathbb{S}$ of $N \times N$ matrices

$$\mathbf{1}, \quad X_I = X_{\mu_1} \cdot \ldots \cdot X_{\mu_n}, \quad n \le \hat{n} = \begin{cases} d, & d \text{ even} \\ \frac{d-1}{2}, & d \text{ odd} \end{cases}$$

This set does not form a group, but only a group modulo $\mathbb{Z}_2$. But this is sufficient for the standard argument for equivalence to a set of unitary matrices.

The $N \times N$ matrix

$$H = \mathbf{1} + \sum_I X_I^+ \cdot X_I = H^+$$

is invariant under right translations in the set $\mathbb{S}$ (i.e. right multiplication of all elements by some fixed element Z), because that just permutes the elements, up to possible additional minus signs which cancel in H,

$$H = Z^+ \cdot Z + \sum_I (X_I \cdot Z)^+ \cdot (X_I \cdot Z).$$

H also has N positive eigenvalues, because

$$H\cdot\psi_\alpha = h_\alpha\psi_\alpha, \quad \psi_\alpha^+\cdot\psi_\beta = \delta_{\alpha\beta}, \tag{G.11}$$

implies

$$h_\alpha = \psi_\alpha^+\cdot H\cdot\psi_\alpha = 1 + \sum_I |X_I\cdot\psi_\alpha|^2 > 0. \tag{G.12}$$

If we define the matrix Ψ with columns ψ_α, equations (G.11) and (G.12) imply

$$\mathrm{diag}(h_1,\ldots h_N) = \Psi^+\cdot H\cdot\Psi, \quad H = \Psi\cdot\mathrm{diag}(h_1,\ldots h_N)\cdot\Psi^+.$$

Now define

$$Y_\mu = \Psi\cdot\mathrm{diag}(\sqrt{h_1},\ldots\sqrt{h_N})\cdot\Psi^+\cdot X_\mu\cdot(\Psi\cdot\mathrm{diag}(\sqrt{h_1},\ldots\sqrt{h_N})\cdot\Psi^+)^{-1}.$$

These matrices are indeed unitary,

$$\begin{aligned}
Y_\mu^+\cdot Y_\mu &= \left(\Psi\cdot\mathrm{diag}\left(\sqrt{h_1},\ldots\sqrt{h_N}\right)\cdot\Psi^+\right)^{-1}\cdot X_\mu^+ \\
&\quad\times\left(\Psi\cdot\mathrm{diag}\left(\sqrt{h_1},\ldots\sqrt{h_N}\right)\cdot\Psi^+\right)^{2}\cdot X_\mu \\
&\quad\times\left(\Psi\cdot\mathrm{diag}\left(\sqrt{h_1},\ldots\sqrt{h_N}\right)\cdot\Psi^+\right)^{-1} \\
&= \left(\Psi\cdot\mathrm{diag}\left(\sqrt{h_1},\ldots\sqrt{h_N}\right)\cdot\Psi^+\right)^{-1}\cdot X_\mu^+\cdot H\cdot X_\mu \\
&\quad\times\left(\Psi\cdot\mathrm{diag}\left(\sqrt{h_1},\ldots\sqrt{h_N}\right)\cdot\Psi^+\right)^{-1} \\
&= \left(\Psi\cdot\mathrm{diag}\left(\sqrt{h_1},\ldots\sqrt{h_N}\right)\cdot\Psi^+\right)^{-1} \\
&\quad\times[X_\mu^+\cdot X_\mu + \textstyle\sum_I(Z_I\cdot X_\mu)^+\cdot(Z_I\cdot X_\mu)] \\
&\quad\times\left(\Psi\cdot\mathrm{diag}\left(\sqrt{h_1},\ldots\sqrt{h_N}\right)\cdot\Psi^+\right)^{-1} \\
&= \left(\Psi\cdot\mathrm{diag}\left(\sqrt{h_1},\ldots\sqrt{h_N}\right)\cdot\Psi^+\right)^{-1}\cdot H \\
&\quad\times\left(\Psi\cdot\mathrm{diag}\left(\sqrt{h_1},\ldots\sqrt{h_N}\right)\cdot\Psi^+\right)^{-1} = \mathbf{1},
\end{aligned}$$

which concludes the proof of equivalence of the matrices X_μ to a set of matrices Y_μ which are both unitary and hermitian.

Equivalence of the γ matrices to hermitian or anti-hermitian matrices also implies that every reducible representation of γ matrices is fully reducible.

Uniqueness theorem for γ matrices

Every irreducible matrix representation of the algebra generated by (21.35) is equivalent to one of the representations constructed in the previous section.

We first consider the case of *even number of dimensions d*. The theorem says that in this case every irreducible matrix representation of (21.35) is equivalent to the representation in terms of $2^{d/2} \times 2^{d/2}$ constructed in equation (G.9).

Proof. Suppose the $N_1 \times N_1$ matrices $\gamma_{1,\mu}$ and the $N_2 \times N_2$ matrices $\gamma_{2,\mu}$, $0 \leq \mu \leq d-1$, are two sets of matrices which satisfy the conditions (21.35). V_1 is the N_1-dimensional vector space in which the matrices $\gamma_{1,\mu}$ act. We use the representations from the previous section, equation (G.9), for the matrices $\gamma_{2,\mu}$. This implies $N_1 \geq N_2 = 2^{d/2}$.

We denote the components of the matrices $\gamma_{1,\mu}$ and $\gamma_{2,\mu}$ with $\gamma_{1,\mu}{}^a{}_b$ and $\gamma_{2,\mu}{}^\alpha{}_\beta$, respectively, and define again multi-indices J for the two sets of γ matrices (cf. equation (G.2)),

$$\Gamma_{r,J} = \gamma_{r,\mu_1} \cdot \gamma_{r,\mu_2} \cdot \ldots \cdot \gamma_{r,\mu_n}, \; 1 \leq r \leq 2, \; \mu_1 < \mu_2 < \ldots < \mu_n, \; n(J) = n.$$

The squares of these matrices satisfy

$$\Gamma_{r,I}{}^2 = \pm\mathbf{1} = s_I \mathbf{1}, \tag{G.13}$$

where the sign factor

$$s_I = (-)^{n(I)[n(I)+1]/2} \eta_{\mu_1\mu_1} \tag{G.14}$$

arises as the product of the factor $(-)^{n(I)[n(I)-1]/2}$ from the permutations of γ matrices times a factor $(-)^{n(I)}$ from the sign on the right hand side of (21.35). Only $\eta_{\mu_1\mu_1}$ appears on the right hand side of (G.14) because we have only one timelike direction. For the case of general spacetime signature one could simply include the product $\eta_{\mu_1\mu_1}\eta_{\mu_2\mu_2} \ldots \eta_{\mu_n\mu_n}$.

The results from the previous section for even d tell us that a set $\Gamma_{r,I}$ with fixed r, after augmentation with the $N_r \times N_r$ unit matrix $\Gamma_{r,0} = \mathbf{1}$, contains 2^d linearly independent matrices.

We define the $N_1 \cdot N_2$ different $N_1 \times N_2$ matrices $E_a{}^\alpha$ with components

$$(E_a{}^\alpha)^b{}_\beta = \delta_a{}^b \delta^\alpha{}_\beta.$$

We use these matrices to form the $N_1 \times N_2$ matrices

$$\Omega_a{}^\alpha = E_a{}^\alpha + \sum_J s_J \Gamma_{1,J} \cdot E_a{}^\alpha \cdot \Gamma_{2,J},$$

i.e. in components,

$$(\Omega_a{}^\alpha)^b{}_\beta = \delta_a{}^b \delta^\alpha{}_\beta + \sum_J s_J (\Gamma_{1,J})^b{}_a (\Gamma_{2,J})^\alpha{}_\beta. \tag{G.15}$$

Suppose $I \neq J$. The conditions (21.35) imply that there is always a multi-index $K \neq I$ such that

$$\Gamma_{r,I} \cdot \Gamma_{r,J} = \pm \Gamma_{r,K},$$

and inversion of this equation yields

$$s_I s_J \Gamma_{r,J} \cdot \Gamma_{r,I} = \pm s_K \Gamma_{r,K}.$$

This implies

$$\begin{aligned}
\Gamma_{1,I} \cdot \Omega_a{}^\alpha &= \Gamma_{1,I} \cdot E_a{}^\alpha + E_a{}^\alpha \cdot \Gamma_{2,I} + \sum_{J \neq I} s_J \Gamma_{1,I} \cdot \Gamma_{1,J} \cdot E_a{}^\alpha \cdot \Gamma_{2,J} \\
&= \Big(s_I \Gamma_{1,I} \cdot E_a{}^\alpha \cdot \Gamma_{2,I} + E_a{}^\alpha \\
&\quad + \sum_{J \neq I} s_I s_J \Gamma_{1,I} \cdot \Gamma_{1,J} \cdot E_a{}^\alpha \cdot \Gamma_{2,J} \cdot \Gamma_{2,I} \Big) \cdot \Gamma_{2,I} \\
&= \Big(s_I \Gamma_{1,I} \cdot E_a{}^\alpha \cdot \Gamma_{2,I} + E_a{}^\alpha + \sum_{K \neq I} s_K \Gamma_{1,K} \cdot E_a{}^\alpha \cdot \Gamma_{2,K} \Big) \cdot \Gamma_{2,I} \\
&= \Omega_a{}^\alpha \cdot \Gamma_{2,I}.
\end{aligned} \tag{G.16}$$

We know that the matrices (G.15) are not null matrices, $\Omega_a{}^\alpha \neq \mathbf{0}$, because we know that the 2^d matrices $\{\mathbf{1}, \Gamma_{2,J}\}$ are linearly independent,

$$\delta_a{}^b \mathbf{1} + \sum_J s_J (\Gamma_{1,J})^b{}_a \Gamma_{2,J} \neq 0.$$

This implies that the N_1-dimensional vector space V_1 with basis vectors $\boldsymbol{e}_{1,b}$, $1 \leq b \leq N_1$, contains non-vanishing sets of $N_2 = 2^{d/2} \leq N_1$ basis vectors

$$\boldsymbol{e}_{1,\beta} = \boldsymbol{e}_{1,b} (\Omega_a{}^\alpha)^b{}_\beta, \quad 1 \leq \beta \leq 2^{d/2} \leq N_1,$$

which are invariant under the action of the γ matrices,

$$\boldsymbol{e}_{1,b}(\gamma_{1,\mu})^b{}_c(\Omega_a{}^\alpha)^c{}_\delta = \boldsymbol{e}_{1,b}(\Omega_a{}^\alpha)^b{}_\beta(\gamma_{2,\mu})^\beta{}_\delta.$$

Therefore the representation of γ matrices in V_1 is either reducible into invariant subspaces of dimension $2^{d/2}$, or we have $N_1 = 2^{d/2}$. In the latter case we must have

$$\det(\Omega_a{}^\alpha) \neq 0,$$

because representations spaces of dimension $2^{d/2}$ are irreducible, and therefore

$$\gamma_{1,\mu} = \Omega_a{}^\alpha \cdot \gamma_{2,\mu} \cdot (\Omega_a{}^\alpha)^{-1}$$

is equivalent to the representation from the previous section for even d. Thus concludes the proof for even d.

For *odd d* we observe that the matrices γ_μ, $0 \leq \mu \leq d-2$, form a set of γ matrices for a $(d-1)$-dimensional Minkowski space, which according to the previous result is either reducible or equivalent to the corresponding representation (G.9) from the previous section. However, using those matrices, the missing matrix γ_{d-1} can easily be constructed according to the prescription

$$\gamma_{d-1} = \pm(-)^{(d-1)(d-2)/4}\gamma_0 \cdot \gamma_1 \cdot \ldots \cdot \gamma_{d-2}. \tag{G.17}$$

Now assume that the matrices γ_μ, $0 \leq \mu \leq d-2$, are $2^{(d-1)/2} \times 2^{(d-1)/2}$ matrices, i.e. they form an *irreducible* representation of γ matrices for a $(d-1)$-dimensional Minkowski space. In that case completeness of the set

$$\Gamma_J = \gamma_{\mu_1} \cdot \gamma_{\mu_2} \cdot \ldots \cdot \gamma_{\mu_n}, \quad 0 \leq \mu_1 < \mu_2 < \ldots < \mu_n \leq d-2 \tag{G.18}$$

in $GL(2^{(d-1)/2})$ implies that (G.17) are the only options for the construction of γ_{d-1}. Completeness of the set (G.18) also implies that the two options for the sign in (G.17) correspond to two inequivalent representations.

On the other hand, if the matrices γ_μ, $0 \leq \mu \leq d-2$, form a *reducible* representation of γ matrices for a $(d-1)$-dimensional Minkowski space, they must be equivalent to matrices with irreducible $2^{(d-1)/2} \times 2^{(d-1)/2}$ matrices $\hat{\gamma}_\mu$, $0 \leq \mu \leq d-2$, in diagonal blocks. Then one can easily prove from the anti-commutation relations and the completeness of the set

$$\hat{\Gamma}_J = \hat{\gamma}_{\mu_1} \cdot \hat{\gamma}_{\mu_2} \cdot \ldots \cdot \hat{\gamma}_{\mu_n}, \quad 0 \leq \mu_1 < \mu_2 < \ldots < \mu_n \leq d-2$$

in $2^{(d-1)/2}$-dimensional subspaces that the matrix γ_{d-1} must consist of $2^{(d-1)/2} \times 2^{(d-1)/2}$ blocks which are proportional to $\hat{\gamma}_0 \cdot \hat{\gamma}_1 \cdot \ldots \cdot \hat{\gamma}_{d-2}$. The property $\gamma_{d-1}^2 = -\mathbf{1}$ can then be used to demonstrate that γ_{d-1} must be equivalent to a matrix which only has matrices

$$\hat{\gamma}_{d-1} = \pm(-)^{(d-1)(d-2)/4}\hat{\gamma}_0 \cdot \hat{\gamma}_1 \cdot \ldots \cdot \hat{\gamma}_{d-2}$$

in diagonal $2^{(d-1)/2} \times 2^{(d-1)/2}$ blocks, i.e. a representation of γ matrices for odd number d of dimensions is either equivalent to one of the two irreducible $2^{(d-1)/2}$-dimensional representations distinguished by the sign in (G.17), or it is a reducible representation.

In the recursive construction of γ matrices described above, I separated the two equivalence classes of irreducible representations through the sign of γ_0 instead of γ_{d-1}, cf. (G.8). We can cast the sign from γ_{d-1} to γ_0 through the similarity transformation

$$\gamma_0 \to \gamma_0 \cdot \gamma_{d-1} \cdot \gamma_0 \cdot \gamma_0 \cdot \gamma_{d-1} = -\gamma_0,$$

$$\gamma_{d-1} \to \gamma_0 \cdot \gamma_{d-1} \cdot \gamma_{d-1} \cdot \gamma_0 \cdot \gamma_{d-1} = -\gamma_{d-1},$$

$$\gamma_i \to \gamma_0 \cdot \gamma_{d-1} \cdot \gamma_i \cdot \gamma_0 \cdot \gamma_{d-1} = \gamma_i, \quad 1 \le i \le d-2.$$

Contraction and trace theorems for γ matrices

Here we explicitly refer to four spacetime dimensions again. The generalizations to any number of spacetime dimensions are trivial.

Equation (21.35) implies

$$\gamma^\sigma \gamma_\sigma = -4.$$

The higher order contraction theorems then follow from (21.35) and application of the next lower order contraction theorem, e.g.

$$\gamma^\sigma \gamma^\mu \gamma_\sigma = \{\gamma^\sigma, \gamma^\mu\}\gamma_\sigma - \gamma^\mu \gamma^\sigma \gamma_\sigma = 2\gamma^\mu,$$
$$\gamma^\sigma \gamma^\mu \gamma^\nu \gamma_\sigma = 4\eta^{\mu\nu}, \quad \gamma^\sigma \gamma^\mu \gamma^\nu \gamma^\rho \gamma_\sigma = 2\gamma^\rho \gamma^\nu \gamma^\mu.$$

The trace of a product of an odd number of γ matrices vanishes. The trace of a product of two γ matrices is determined by their basic anti-commutation property,

$$\mathrm{tr}(\gamma_\mu \gamma_\nu) = -4\eta_{\mu\nu}. \tag{G.19}$$

The trace of a product of four γ matrices is easily evaluated using their anti-commutation properties and cyclic invariance of the trace

$$\begin{aligned}\mathrm{tr}(\gamma_\kappa \gamma_\lambda \gamma_\mu \gamma_\nu) &= 8\eta_{\kappa\lambda}\eta_{\mu\nu} - \mathrm{tr}(\gamma_\kappa \gamma_\lambda \gamma_\nu \gamma_\mu) = 8\eta_{\kappa\lambda}\eta_{\mu\nu} - \mathrm{tr}(\gamma_\mu \gamma_\kappa \gamma_\lambda \gamma_\nu) \\ &= 8\eta_{\kappa\lambda}\eta_{\mu\nu} - 8\eta_{\mu\kappa}\eta_{\lambda\nu} + \mathrm{tr}(\gamma_\kappa \gamma_\mu \gamma_\lambda \gamma_\nu) \\ &= 8\eta_{\kappa\lambda}\eta_{\mu\nu} - 8\eta_{\mu\kappa}\eta_{\lambda\nu} + 8\eta_{\mu\lambda}\eta_{\kappa\nu} - \mathrm{tr}(\gamma_\kappa \gamma_\lambda \gamma_\mu \gamma_\nu),\end{aligned}$$

i.e.

$$\mathrm{tr}\big(\gamma_\kappa\gamma_\lambda\gamma_\mu\gamma_\nu\big) = 4\eta_{\kappa\lambda}\eta_{\mu\nu} - 4\eta_{\mu\kappa}\eta_{\lambda\nu} + 4\eta_{\mu\lambda}\eta_{\kappa\nu}. \tag{G.20}$$

For yet higher orders we observe

$$\begin{aligned}\mathrm{tr}\big(\gamma_{\alpha_1}\ldots\gamma_{\alpha_{2n}}\gamma_\mu\gamma_\nu\big) &= -2\eta_{\mu\nu}\mathrm{tr}(\gamma_{\alpha_1}\ldots\gamma_{\alpha_{2n}}) - \mathrm{tr}\big(\gamma_\mu\gamma_{\alpha_1}\ldots\gamma_{\alpha_{2n}}\gamma_\nu\big)\\ &= -2\eta_{\mu\nu}\mathrm{tr}(\gamma_{\alpha_1}\ldots\gamma_{\alpha_{2n}}) + \mathrm{tr}\big(\gamma_{\alpha_1}\gamma_\mu\gamma_{\alpha_2}\ldots\gamma_{\alpha_{2n}}\gamma_\nu\big)\\ &\quad + 2\eta_{\mu\alpha_1}\mathrm{tr}(\gamma_{\alpha_2}\ldots\gamma_{\alpha_{2n}}\gamma_\nu)\\ &= -2\eta_{\mu\nu}\mathrm{tr}(\gamma_{\alpha_1}\ldots\gamma_{\alpha_{2n}}) - \mathrm{tr}\big(\gamma_{\alpha_1}\ldots\gamma_{\alpha_{2n}}\gamma_\mu\gamma_\nu\big)\\ &\quad -2\sum_{i=1}^{2n}(-)^i\eta_{\mu\alpha_i}\mathrm{tr}\big(\gamma_{\alpha_1}\ldots\gamma_{\alpha_{i-1}}\gamma_{\alpha_{i+1}}\ldots\gamma_{\alpha_{2n}}\gamma_\nu\big),\end{aligned}$$

i.e. we have a recursion relation

$$\begin{aligned}\mathrm{tr}\big(\gamma_{\alpha_1}\ldots\gamma_{\alpha_{2n}}\gamma_\mu\gamma_\nu\big) &= -\eta_{\mu\nu}\mathrm{tr}(\gamma_{\alpha_1}\ldots\gamma_{\alpha_{2n}})\\ &\quad -\sum_{i=1}^{2n}(-)^i\eta_{\mu\alpha_i}\mathrm{tr}\big(\gamma_{\alpha_1}\ldots\gamma_{\alpha_{i-1}}\gamma_{\alpha_{i+1}}\ldots\gamma_{\alpha_{2n}}\gamma_\nu\big).\end{aligned}$$

This yields for products of six γ matrices

$$\begin{aligned}\mathrm{tr}\big(\gamma_\rho\gamma_\sigma\gamma_\kappa\gamma_\lambda\gamma_\mu\gamma_\nu\big) &= -4\eta_{\rho\sigma}\eta_{\kappa\lambda}\eta_{\mu\nu} + 4\eta_{\rho\kappa}\eta_{\sigma\lambda}\eta_{\mu\nu} - 4\eta_{\rho\lambda}\eta_{\kappa\sigma}\eta_{\mu\nu} + 4\eta_{\sigma\kappa}\eta_{\lambda\nu}\eta_{\mu\rho}\\ &\quad -4\eta_{\sigma\lambda}\eta_{\kappa\nu}\eta_{\mu\rho} + 4\eta_{\sigma\nu}\eta_{\lambda\kappa}\eta_{\mu\rho} - 4\eta_{\rho\kappa}\eta_{\lambda\nu}\eta_{\mu\sigma} + 4\eta_{\rho\lambda}\eta_{\kappa\nu}\eta_{\mu\sigma}\\ &\quad -4\eta_{\rho\nu}\eta_{\lambda\kappa}\eta_{\mu\sigma} + 4\eta_{\rho\sigma}\eta_{\lambda\nu}\eta_{\mu\kappa} - 4\eta_{\rho\lambda}\eta_{\sigma\nu}\eta_{\mu\kappa} + 4\eta_{\rho\nu}\eta_{\lambda\sigma}\eta_{\mu\kappa}\\ &\quad -4\eta_{\rho\sigma}\eta_{\kappa\nu}\eta_{\mu\lambda} + 4\eta_{\rho\kappa}\eta_{\sigma\nu}\eta_{\mu\lambda} - 4\eta_{\rho\nu}\eta_{\kappa\sigma}\eta_{\mu\lambda},\end{aligned} \tag{G.21}$$

and the trace of the product of eight γ matrices contains 105 terms,

$$\begin{aligned}\mathrm{tr}\big(\gamma_\alpha\gamma_\beta\gamma_\rho\gamma_\sigma\gamma_\kappa\gamma_\lambda\gamma_\mu\gamma_\nu\big) &= 4\eta_{\rho\sigma}\eta_{\kappa\lambda}\eta_{\mu\nu}\eta_{\alpha\beta} - 4\eta_{\rho\kappa}\eta_{\sigma\lambda}\eta_{\mu\nu}\eta_{\alpha\beta} + 4\eta_{\rho\lambda}\eta_{\kappa\sigma}\eta_{\mu\nu}\eta_{\alpha\beta}\\ &- 4\eta_{\sigma\kappa}\eta_{\lambda\nu}\eta_{\mu\rho}\eta_{\alpha\beta} + 4\eta_{\sigma\lambda}\eta_{\kappa\nu}\eta_{\mu\rho}\eta_{\alpha\beta} - 4\eta_{\sigma\nu}\eta_{\lambda\kappa}\eta_{\mu\rho}\eta_{\alpha\beta} + 4\eta_{\rho\kappa}\eta_{\lambda\nu}\eta_{\mu\sigma}\eta_{\alpha\beta}\\ &- 4\eta_{\rho\lambda}\eta_{\kappa\nu}\eta_{\mu\sigma}\eta_{\alpha\beta} + 4\eta_{\rho\nu}\eta_{\lambda\kappa}\eta_{\mu\sigma}\eta_{\alpha\beta} - 4\eta_{\rho\sigma}\eta_{\lambda\nu}\eta_{\mu\kappa}\eta_{\alpha\beta} + 4\eta_{\rho\lambda}\eta_{\sigma\nu}\eta_{\mu\kappa}\eta_{\alpha\beta}\\ &- 4\eta_{\rho\nu}\eta_{\lambda\sigma}\eta_{\mu\kappa}\eta_{\alpha\beta} + 4\eta_{\rho\sigma}\eta_{\kappa\nu}\eta_{\mu\lambda}\eta_{\alpha\beta} - 4\eta_{\rho\kappa}\eta_{\sigma\nu}\eta_{\mu\lambda}\eta_{\alpha\beta} + 4\eta_{\rho\nu}\eta_{\kappa\sigma}\eta_{\mu\lambda}\eta_{\alpha\beta}\\ &- 4\eta_{\sigma\kappa}\eta_{\lambda\mu}\eta_{\nu\beta}\eta_{\alpha\rho} + 4\eta_{\sigma\lambda}\eta_{\kappa\mu}\eta_{\nu\beta}\eta_{\alpha\rho} - 4\eta_{\sigma\mu}\eta_{\lambda\kappa}\eta_{\nu\beta}\eta_{\alpha\rho} + 4\eta_{\kappa\lambda}\eta_{\mu\beta}\eta_{\nu\sigma}\eta_{\alpha\rho}\\ &- 4\eta_{\kappa\mu}\eta_{\lambda\beta}\eta_{\nu\sigma}\eta_{\alpha\rho} + 4\eta_{\kappa\beta}\eta_{\mu\lambda}\eta_{\nu\sigma}\eta_{\alpha\rho} - 4\eta_{\sigma\lambda}\eta_{\mu\beta}\eta_{\nu\kappa}\eta_{\alpha\rho} + 4\eta_{\sigma\mu}\eta_{\lambda\beta}\eta_{\nu\kappa}\eta_{\alpha\rho}\\ &- 4\eta_{\sigma\beta}\eta_{\mu\lambda}\eta_{\nu\kappa}\eta_{\alpha\rho} + 4\eta_{\sigma\kappa}\eta_{\mu\beta}\eta_{\nu\lambda}\eta_{\alpha\rho} - 4\eta_{\sigma\mu}\eta_{\kappa\beta}\eta_{\nu\lambda}\eta_{\alpha\rho} + 4\eta_{\sigma\beta}\eta_{\mu\kappa}\eta_{\nu\lambda}\eta_{\alpha\rho}\\ &- 4\eta_{\sigma\kappa}\eta_{\lambda\beta}\eta_{\nu\mu}\eta_{\alpha\rho} + 4\eta_{\sigma\lambda}\eta_{\kappa\beta}\eta_{\nu\mu}\eta_{\alpha\rho} - 4\eta_{\sigma\beta}\eta_{\lambda\kappa}\eta_{\nu\mu}\eta_{\alpha\rho} + 4\eta_{\rho\kappa}\eta_{\lambda\mu}\eta_{\nu\beta}\eta_{\alpha\sigma}\end{aligned}$$

$$
\begin{aligned}
&-4\eta_{\rho\lambda}\eta_{\kappa\mu}\eta_{\nu\beta}\eta_{\alpha\sigma}+4\eta_{\rho\mu}\eta_{\lambda\kappa}\eta_{\nu\beta}\eta_{\alpha\sigma}-4\eta_{\kappa\lambda}\eta_{\mu\beta}\eta_{\nu\rho}\eta_{\alpha\sigma}+4\eta_{\kappa\mu}\eta_{\lambda\beta}\eta_{\nu\rho}\eta_{\alpha\sigma}\\
&-4\eta_{\kappa\beta}\eta_{\mu\lambda}\eta_{\nu\rho}\eta_{\alpha\sigma}+4\eta_{\rho\lambda}\eta_{\mu\beta}\eta_{\nu\kappa}\eta_{\alpha\sigma}-4\eta_{\rho\mu}\eta_{\lambda\beta}\eta_{\nu\kappa}\eta_{\alpha\sigma}+4\eta_{\rho\beta}\eta_{\mu\lambda}\eta_{\nu\kappa}\eta_{\alpha\sigma}\\
&-4\eta_{\rho\kappa}\eta_{\mu\beta}\eta_{\nu\lambda}\eta_{\alpha\sigma}+4\eta_{\rho\mu}\eta_{\kappa\beta}\eta_{\nu\lambda}\eta_{\alpha\sigma}-4\eta_{\rho\beta}\eta_{\mu\kappa}\eta_{\nu\lambda}\eta_{\alpha\sigma}+4\eta_{\rho\kappa}\eta_{\lambda\beta}\eta_{\nu\mu}\eta_{\alpha\sigma}\\
&-4\eta_{\rho\lambda}\eta_{\kappa\beta}\eta_{\nu\mu}\eta_{\alpha\sigma}+4\eta_{\rho\beta}\eta_{\lambda\kappa}\eta_{\nu\mu}\eta_{\alpha\sigma}-4\eta_{\rho\sigma}\eta_{\lambda\mu}\eta_{\nu\beta}\eta_{\alpha\kappa}+4\eta_{\rho\lambda}\eta_{\sigma\mu}\eta_{\nu\beta}\eta_{\alpha\kappa}\\
&-4\eta_{\rho\mu}\eta_{\lambda\sigma}\eta_{\nu\beta}\eta_{\alpha\kappa}+4\eta_{\sigma\lambda}\eta_{\mu\beta}\eta_{\nu\rho}\eta_{\alpha\kappa}-4\eta_{\sigma\mu}\eta_{\lambda\beta}\eta_{\nu\rho}\eta_{\alpha\kappa}+4\eta_{\sigma\beta}\eta_{\mu\lambda}\eta_{\nu\rho}\eta_{\alpha\kappa}\\
&-4\eta_{\rho\lambda}\eta_{\mu\beta}\eta_{\nu\sigma}\eta_{\alpha\kappa}+4\eta_{\rho\mu}\eta_{\lambda\beta}\eta_{\nu\sigma}\eta_{\alpha\kappa}-4\eta_{\rho\beta}\eta_{\mu\lambda}\eta_{\nu\sigma}\eta_{\alpha\kappa}+4\eta_{\rho\sigma}\eta_{\mu\beta}\eta_{\nu\lambda}\eta_{\alpha\kappa}\\
&-4\eta_{\rho\mu}\eta_{\sigma\beta}\eta_{\nu\lambda}\eta_{\alpha\kappa}+4\eta_{\rho\beta}\eta_{\mu\sigma}\eta_{\nu\lambda}\eta_{\alpha\kappa}-4\eta_{\rho\sigma}\eta_{\lambda\beta}\eta_{\nu\mu}\eta_{\alpha\kappa}+4\eta_{\rho\lambda}\eta_{\sigma\beta}\eta_{\nu\mu}\eta_{\alpha\kappa}\\
&-4\eta_{\rho\beta}\eta_{\lambda\sigma}\eta_{\nu\mu}\eta_{\alpha\kappa}+4\eta_{\rho\sigma}\eta_{\kappa\mu}\eta_{\nu\beta}\eta_{\alpha\lambda}-4\eta_{\rho\kappa}\eta_{\sigma\mu}\eta_{\nu\beta}\eta_{\alpha\lambda}+4\eta_{\rho\mu}\eta_{\kappa\sigma}\eta_{\nu\beta}\eta_{\alpha\lambda}\\
&-4\eta_{\sigma\kappa}\eta_{\mu\beta}\eta_{\nu\rho}\eta_{\alpha\lambda}+4\eta_{\sigma\mu}\eta_{\kappa\beta}\eta_{\nu\rho}\eta_{\alpha\lambda}-4\eta_{\sigma\beta}\eta_{\mu\kappa}\eta_{\nu\rho}\eta_{\alpha\lambda}+4\eta_{\rho\kappa}\eta_{\mu\beta}\eta_{\nu\sigma}\eta_{\alpha\lambda}\\
&-4\eta_{\rho\mu}\eta_{\kappa\beta}\eta_{\nu\sigma}\eta_{\alpha\lambda}+4\eta_{\rho\beta}\eta_{\mu\kappa}\eta_{\nu\sigma}\eta_{\alpha\lambda}-4\eta_{\rho\sigma}\eta_{\mu\beta}\eta_{\nu\kappa}\eta_{\alpha\lambda}+4\eta_{\rho\mu}\eta_{\sigma\beta}\eta_{\nu\kappa}\eta_{\alpha\lambda}\\
&-4\eta_{\rho\beta}\eta_{\mu\sigma}\eta_{\nu\kappa}\eta_{\alpha\lambda}+4\eta_{\rho\sigma}\eta_{\kappa\beta}\eta_{\nu\mu}\eta_{\alpha\lambda}-4\eta_{\rho\kappa}\eta_{\sigma\beta}\eta_{\nu\mu}\eta_{\alpha\lambda}+4\eta_{\rho\beta}\eta_{\kappa\sigma}\eta_{\nu\mu}\eta_{\alpha\lambda}\\
&-4\eta_{\rho\sigma}\eta_{\kappa\lambda}\eta_{\nu\beta}\eta_{\alpha\mu}+4\eta_{\rho\kappa}\eta_{\sigma\lambda}\eta_{\nu\beta}\eta_{\alpha\mu}-4\eta_{\rho\lambda}\eta_{\kappa\sigma}\eta_{\nu\beta}\eta_{\alpha\mu}+4\eta_{\sigma\kappa}\eta_{\lambda\beta}\eta_{\nu\rho}\eta_{\alpha\mu}\\
&-4\eta_{\sigma\lambda}\eta_{\kappa\beta}\eta_{\nu\rho}\eta_{\alpha\mu}+4\eta_{\sigma\beta}\eta_{\lambda\kappa}\eta_{\nu\rho}\eta_{\alpha\mu}-4\eta_{\rho\kappa}\eta_{\lambda\beta}\eta_{\nu\sigma}\eta_{\alpha\mu}+4\eta_{\rho\lambda}\eta_{\kappa\beta}\eta_{\nu\sigma}\eta_{\alpha\mu}\\
&-4\eta_{\rho\beta}\eta_{\lambda\kappa}\eta_{\nu\sigma}\eta_{\alpha\mu}+4\eta_{\rho\sigma}\eta_{\lambda\beta}\eta_{\nu\kappa}\eta_{\alpha\mu}-4\eta_{\rho\lambda}\eta_{\sigma\beta}\eta_{\nu\kappa}\eta_{\alpha\mu}+4\eta_{\rho\beta}\eta_{\lambda\sigma}\eta_{\nu\kappa}\eta_{\alpha\mu}\\
&-4\eta_{\rho\sigma}\eta_{\kappa\beta}\eta_{\nu\lambda}\eta_{\alpha\mu}+4\eta_{\rho\kappa}\eta_{\sigma\beta}\eta_{\nu\lambda}\eta_{\alpha\mu}-4\eta_{\rho\beta}\eta_{\kappa\sigma}\eta_{\nu\lambda}\eta_{\alpha\mu}+4\eta_{\rho\sigma}\eta_{\kappa\lambda}\eta_{\mu\beta}\eta_{\alpha\nu}\\
&-4\eta_{\rho\kappa}\eta_{\sigma\lambda}\eta_{\mu\beta}\eta_{\alpha\nu}+4\eta_{\rho\lambda}\eta_{\kappa\sigma}\eta_{\mu\beta}\eta_{\alpha\nu}-4\eta_{\sigma\kappa}\eta_{\lambda\beta}\eta_{\mu\rho}\eta_{\alpha\nu}+4\eta_{\sigma\lambda}\eta_{\kappa\beta}\eta_{\mu\rho}\eta_{\alpha\nu}\\
&-4\eta_{\sigma\beta}\eta_{\lambda\kappa}\eta_{\mu\rho}\eta_{\alpha\nu}+4\eta_{\rho\kappa}\eta_{\lambda\beta}\eta_{\mu\sigma}\eta_{\alpha\nu}-4\eta_{\rho\lambda}\eta_{\kappa\beta}\eta_{\mu\sigma}\eta_{\alpha\nu}+4\eta_{\rho\beta}\eta_{\lambda\kappa}\eta_{\mu\sigma}\eta_{\alpha\nu}\\
&-4\eta_{\rho\sigma}\eta_{\lambda\beta}\eta_{\mu\kappa}\eta_{\alpha\nu}+4\eta_{\rho\lambda}\eta_{\sigma\beta}\eta_{\mu\kappa}\eta_{\alpha\nu}-4\eta_{\rho\beta}\eta_{\lambda\sigma}\eta_{\mu\kappa}\eta_{\alpha\nu}+4\eta_{\rho\sigma}\eta_{\kappa\beta}\eta_{\mu\lambda}\eta_{\alpha\nu}\\
&-4\eta_{\rho\kappa}\eta_{\sigma\beta}\eta_{\mu\lambda}\eta_{\alpha\nu}+4\eta_{\rho\beta}\eta_{\kappa\sigma}\eta_{\mu\lambda}\eta_{\alpha\nu}.
\end{aligned}
\tag{G.22}
$$

Appendix H: Spinor representations of the Lorentz group

The explicit form of the Lagrange density (21.74) for the Dirac field and the appearance of the factor $\overline{\Psi} = \Psi^+ \cdot \gamma^0$ are determined by the requirement of Lorentz invariance of $\mathcal{L}$ and the transformation properties of spinors under Lorentz transformations. However, before we can elaborate on these points, we have to revisit the Lorentz transformation (B.8), which is also denoted as the vector representation because it acts on spacetime vectors. We can discuss this in general numbers n of spatial dimensions and $d = n + 1$ of spacetime dimensions.

Generators of proper orthochronous Lorentz transformations in the vector and spinor representations

We can write the two factors of a proper orthochronous Lorentz transformation (B.8) as exponentials of Lie algebra elements,

$$\Lambda(\boldsymbol{u}, \underline{\epsilon}) = \Lambda(\boldsymbol{u}) \cdot \Lambda(\underline{\epsilon}) = \exp(\boldsymbol{u} \cdot \boldsymbol{K}) \cdot \exp\!\left(\frac{1}{2}\epsilon^{ij} L_{ij}\right). \tag{H.1}$$

For the boost part we use explicit construction to prove that every proper Lorentz boost can be written in the form $\exp(\boldsymbol{u} \cdot \boldsymbol{K})$.

For the rotation part we can use the general result that every element of a compact Lie group can be written as a single exponential of a corresponding Lie algebra element, or we can use the fact that a general $n \times n$ rotation matrix consists of n orthonormal row vectors, which fixes the general form in terms of $n(n-1)/2$ parameters, and then demonstrate that the $n(n-1)/2$ parameters ϵ^{ij} of $\exp(\epsilon^{ij} L_{ij}/2)$ provide a general parametrization of n orthonormal row vectors.

R. Dick, *Advanced Quantum Mechanics*, Graduate Texts in Physics,
DOI 10.1007/978-3-319-25675-7

Alternatively, we can consider (H.1) as an example for the polar decomposition (F.4) and infer the representation in terms of matrix exponentials from the results on matrix logarithms in Appendix F.

The boost part is

$$\Lambda(\boldsymbol{u}) = \exp(\boldsymbol{u}\cdot\boldsymbol{K}) = \exp\left(\epsilon^{i0}L_{i0}\right) = \exp\left(\mathrm{i}\epsilon^{i0}M_{i0}\right) \tag{H.2}$$

and the spatial rotation is

$$\Lambda(\underline{\epsilon}) = \exp\left(\frac{1}{2}\epsilon^{ij}L_{ij}\right) = \exp\left(\frac{\mathrm{i}}{2}\epsilon^{ij}M_{ij}\right), \tag{H.3}$$

where ϵ_{ij} is the rotation angle in the ij plane. The generators are (in the vector representation),

$$\left(L_{\mu\nu}\right)^{\rho}{}_{\sigma} = \mathrm{i}\left(M_{\mu\nu}\right)^{\rho}{}_{\sigma} = \left(\eta^{\rho}{}_{\mu}\eta_{\nu\sigma} - \eta^{\rho}{}_{\nu}\eta_{\mu\sigma}\right). \tag{H.4}$$

These matrices generate the Lie algebra so$(1, d-1)$,

$$\begin{aligned}[L_{\mu\nu}, L_{\rho\sigma}] &= \eta_{\nu\rho}L_{\mu\sigma} + \eta_{\mu\sigma}L_{\nu\rho} - \eta_{\mu\rho}L_{\nu\sigma} - \eta_{\nu\sigma}L_{\mu\rho} \\ &= -\left(L_{\mu\nu}\right)_{\rho}{}^{\lambda}L_{\lambda\sigma} - \left(L_{\mu\nu}\right)_{\sigma}{}^{\lambda}L_{\rho\lambda}.\end{aligned} \tag{H.5}$$

In 4-dimensional Minkowski space, the angles ϵ_{ij} are related to the rotation angles φ_i around the x^i-axis according to

$$\varphi_i = \frac{1}{2}\epsilon_{ijk}\epsilon_{jk}, \quad \epsilon_{ij} = \epsilon_{ijk}\varphi_k. \tag{H.6}$$

To see how the boost vector $\boldsymbol{u}$ is related to the velocity $\boldsymbol{v} = c\boldsymbol{\beta}$, we will explicitly calculate the boost matrix $\Lambda(\boldsymbol{u})$. We have with a contravariant row index and a covariant column index, as in (B.8),

$$\boldsymbol{u}\cdot\boldsymbol{K} = u^iL_{i0} = \mathrm{i}u^iM_{i0} = \begin{pmatrix} 0 & -u_1 \ldots & -u_{d-1} \\ -u_1 & 0 \ \ldots & 0 \\ \vdots & \vdots & \vdots \\ -u_{d-1} & 0 \ \ldots & 0 \end{pmatrix} = \begin{pmatrix} 0 & -\boldsymbol{u}^T \\ -\boldsymbol{u} & \underline{0} \end{pmatrix},$$

$$(\boldsymbol{u}\cdot\boldsymbol{K})^2 = \begin{pmatrix} \boldsymbol{u}^2 & \boldsymbol{0}^T \\ \boldsymbol{0} & \boldsymbol{u}\otimes\boldsymbol{u}^T \end{pmatrix}, \quad (\boldsymbol{u}\cdot\boldsymbol{K})^{2n} = u^{2n}\begin{pmatrix} 1 & \boldsymbol{0}^T \\ \boldsymbol{0} & \hat{\boldsymbol{u}}\otimes\hat{\boldsymbol{u}}^T \end{pmatrix}, \tag{H.7}$$

$$(\boldsymbol{u}\cdot\boldsymbol{K})^{2n+1} = u^{2n+1}\begin{pmatrix} 0 & -\hat{\boldsymbol{u}}^T \\ -\hat{\boldsymbol{u}} & \underline{0} \end{pmatrix}. \tag{H.8}$$

For the interpretation of the $(d-1)\times(d-1)$ matrices $\hat{\boldsymbol{u}}\otimes\hat{\boldsymbol{u}}^T$ and $\underline{1}-\hat{\boldsymbol{u}}\otimes\hat{\boldsymbol{u}}^T$, note that for every $(d-1)$-dimensional spatial vector $\boldsymbol{r}$

$$\boldsymbol{r}_{\parallel} = \hat{\boldsymbol{u}}(\hat{\boldsymbol{u}}^T\cdot\boldsymbol{r}) = (\hat{\boldsymbol{u}}\otimes\hat{\boldsymbol{u}}^T)\cdot\boldsymbol{r}$$

is the part $\boldsymbol{r}_{\parallel}$ of the vector which is parallel to $\boldsymbol{u}$, and

$$\boldsymbol{r}_{\perp} = \boldsymbol{r}-\boldsymbol{r}_{\parallel} = (\underline{1}-\hat{\boldsymbol{u}}\otimes\hat{\boldsymbol{u}}^T)\cdot\boldsymbol{r}$$

is the part of the vector which is orthogonal to $\boldsymbol{u}$.

Substitution of the results (H.7, H.8) into (H.2) yields for the boost in the direction $\hat{\boldsymbol{u}} = \hat{\boldsymbol{\beta}}$

$$\begin{aligned}\Lambda(\boldsymbol{u}) &= \begin{pmatrix}\cosh(u) & \boldsymbol{0}^T\\ \boldsymbol{0} & \underline{1}+\hat{\boldsymbol{u}}\otimes\hat{\boldsymbol{u}}^T(\cosh(u)-1)\end{pmatrix} + \sinh(u)\begin{pmatrix}0 & -\hat{\boldsymbol{u}}^T\\ -\hat{\boldsymbol{u}} & \underline{0}\end{pmatrix}\\ &= \begin{pmatrix}\gamma & -\gamma\boldsymbol{\beta}^T\\ -\gamma\boldsymbol{\beta} & \underline{1}-\hat{\boldsymbol{u}}\otimes\hat{\boldsymbol{u}}^T+\gamma\hat{\boldsymbol{u}}\otimes\hat{\boldsymbol{u}}^T\end{pmatrix},\end{aligned}$$

i.e.

$$\gamma = \cosh(u),\quad \beta = \tanh(u),\quad u = \operatorname{artanh}(\beta) = \frac{1}{2}\ln\left(\frac{1+\beta}{1-\beta}\right).$$

The parameter u is usually denoted as the *boost parameter* or *rapidity* of the Lorentz transformation.

It may also be worthwhile to write down the corresponding rotation matrix in 4-dimensional Minkowski space. If we use the 3×3 matrices from Section 7.4 for the spatial subsections of the rotation matrices[1] L_{mn},

$$(L_i)_{jk} = \frac{1}{2}\epsilon_{imn}(L_{mn})_{jk} = \epsilon_{ijk}, \tag{H.9}$$

the rotation matrices take the following form,

$$\Lambda(\underline{\epsilon}) = \exp\begin{pmatrix}1 & \boldsymbol{0}^T\\ \boldsymbol{0} & \boldsymbol{\varphi}\cdot\boldsymbol{L}\end{pmatrix} = \begin{pmatrix}1 & \boldsymbol{0}^T\\ \boldsymbol{0} & \exp(\boldsymbol{\varphi}\cdot\boldsymbol{L})\end{pmatrix}, \tag{H.10}$$

with the 3×3 rotation matrix

$$\exp(\boldsymbol{\varphi}\cdot\boldsymbol{L}) = \hat{\boldsymbol{\varphi}}\otimes\hat{\boldsymbol{\varphi}}^T + \left(\underline{1}-\hat{\boldsymbol{\varphi}}\otimes\hat{\boldsymbol{\varphi}}^T\right)\cos\varphi + \hat{\boldsymbol{\varphi}}\cdot\boldsymbol{L}\sin\varphi. \tag{H.11}$$

Application of the matrix $\hat{\boldsymbol{\varphi}}\cdot\boldsymbol{L}$ generates a vector product,

$$(\hat{\boldsymbol{\varphi}}\cdot\boldsymbol{L})\cdot\boldsymbol{r} = -\,\hat{\boldsymbol{\varphi}}\times\boldsymbol{r}.$$

[1] In this Appendix we use underscore only for 2×2 matrices.

The anticommutation relations (21.35) imply that the properly normalized commutators of γ-matrices,

$$S_{\mu\nu} = \frac{i}{4}[\gamma_\mu, \gamma_\nu] \tag{H.12}$$

also provide a representation of the Lie algebra so(1,d-1) (H.5),

$$\begin{aligned}[S_{\mu\nu}, S_{\rho\sigma}] &= i\left(\eta_{\mu\rho}S_{\nu\sigma} + \eta_{\nu\sigma}S_{\mu\rho} - \eta_{\nu\rho}S_{\mu\sigma} - \eta_{\mu\sigma}S_{\nu\rho}\right) \\ &= i(L_{\mu\nu})_\rho{}^\lambda S_{\lambda\sigma} + i(L_{\mu\nu})_\sigma{}^\lambda S_{\rho\lambda}.\end{aligned} \tag{H.13}$$

See equations (H.16–H.18) for the proof.

This representation of the Lorentz group is realized in the transformation of Dirac spinors $\psi(x)$ under Lorentz transformations

$$x' = \Lambda(\epsilon)\cdot x = \exp\left(\frac{1}{2}\epsilon^{\mu\nu}L_{\mu\nu}\right)\cdot x,$$

$$\psi'(x') = U(\Lambda)\cdot\psi(x) = \exp\left(\frac{i}{2}\epsilon^{\mu\nu}S_{\mu\nu}\right)\cdot\psi(x). \tag{H.14}$$

The anticommutation relations (21.35) also imply invariance of the γ-matrices under Lorentz transformations $x' = \Lambda(\epsilon)\cdot x$,

$$\gamma'^\mu = \Lambda^\mu{}_\nu(\epsilon)\exp\left(\frac{i}{2}\epsilon^{\kappa\lambda}S_{\kappa\lambda}\right)\cdot\gamma^\nu\cdot\exp\left(-\frac{i}{2}\epsilon^{\rho\sigma}S_{\rho\sigma}\right) = \gamma^\mu, \tag{H.15}$$

see equation (H.19). This invariance property of the γ-matrices also implies form invariance of the Dirac equation under Lorentz transformations,

$$i\hbar\gamma^\mu\partial'_\mu\psi'(x') - mc\psi'(x') = \exp\left(\frac{i}{2}\epsilon^{\kappa\lambda}S_{\kappa\lambda}\right)\cdot\left(i\hbar\gamma^\mu\partial_\mu\psi(x) - mc\psi(x)\right),$$

i.e. *all inertial observers can use the same set of γ-matrices, and the Dirac equation has the same form for all of them.*

Verification of the Lorentz commutation relations for the spinor representations

The anti-commutation relations (21.35) imply

$$\begin{aligned}[\gamma_\mu\gamma_\nu, \gamma_\rho] &= \gamma_\mu\{\gamma_\nu, \gamma_\rho\} - \{\gamma_\mu, \gamma_\rho\}\gamma_\nu = 2\eta_{\mu\rho}\gamma_\nu - 2\eta_{\nu\rho}\gamma_\mu \\ &= 2(L_{\mu\nu})_\rho{}^\sigma\gamma_\sigma,\end{aligned} \tag{H.16}$$

where the matrices $L_{\mu\nu}$ were given in (H.4). Equation (H.16) also implies

$$[S_{\mu\nu}, \gamma_\rho] = \mathrm{i}(L_{\mu\nu})_\rho{}^\sigma \gamma_\sigma \tag{H.17}$$

and

$$\begin{aligned}
[S_{\mu\nu}, S_{\rho\sigma}] &= \frac{\mathrm{i}}{4}[S_{\mu\nu}, [\gamma_\rho, \gamma_\sigma]] = \frac{\mathrm{i}}{4}[[S_{\mu\nu}, \gamma_\rho], \gamma_\sigma] - \frac{\mathrm{i}}{4}[[S_{\mu\nu}, \gamma_\sigma], \gamma_\rho] \\
&= -\frac{1}{4}(L_{\mu\nu})_\rho{}^\lambda[\gamma_\lambda, \gamma_\sigma] + \frac{1}{4}(L_{\mu\nu})_\sigma{}^\lambda[\gamma_\lambda, \gamma_\rho] \\
&= \mathrm{i}(L_{\mu\nu})_\rho{}^\lambda S_{\lambda\sigma} + \mathrm{i}(L_{\mu\nu})_\sigma{}^\lambda S_{\rho\lambda}.
\end{aligned} \tag{H.18}$$

Equation (H.17) implies the Lorentz invariance of the γ-matrices,

$$\begin{aligned}
\exp\!\left(\frac{\mathrm{i}}{2}\epsilon^{\mu\nu}S_{\mu\nu}\right)\gamma_\rho \exp\!\left(-\frac{\mathrm{i}}{2}\epsilon^{\kappa\lambda}S_{\kappa\lambda}\right) &= \left[\exp\!\left(-\frac{1}{2}\epsilon^{\mu\nu}L_{\mu\nu}\right)\right]_\rho{}^\sigma \gamma_\sigma \\
&= \Lambda^{-1}(\epsilon)_\rho{}^\sigma \gamma_\sigma.
\end{aligned} \tag{H.19}$$

Scalar products of spinors and the Lagrangian for the Dirac equation

The hermiticity relation (G.10) implies the following hermiticity property of the Lorentz generators,

$$S^+_{\mu\nu} = \gamma^0 S_{\mu\nu}\gamma^0,$$

and therefore

$$\psi'^+(x') = \psi^+(x)\cdot\gamma^0 \exp\!\left(-\frac{\mathrm{i}}{2}\epsilon^{\mu\nu}S_{\mu\nu}\right)\gamma^0.$$

The adjoint spinor

$$\overline{\psi}(x) = \psi^+(x)\cdot\gamma^0$$

therefore transforms inversely to the spinor $\psi(x)$,

$$\overline{\psi}'(x) = \overline{\psi}(x)\cdot\exp\!\left(-\frac{\mathrm{i}}{2}\epsilon^{\mu\nu}S_{\mu\nu}\right),$$

and the product of spinors

$$\overline{\psi}(x)\cdot\phi(x) = \psi^+(x)\cdot\gamma^0\cdot\phi(x)$$

is Lorentz invariant. This yields a Lorentz invariant Lagrangian for the Dirac equation,

$$\mathcal{L} = \frac{\mathrm{i}\hbar c}{2}\left(\overline{\psi}(x)\cdot\gamma^{\mu}\cdot\partial_{\mu}\psi(x) - \partial_{\mu}\overline{\psi}(x)\cdot\gamma^{\mu}\cdot\psi(x)\right) - mc^2\overline{\psi}(x)\cdot\psi(x). \tag{H.20}$$

The spinor representation in the Weyl and Dirac bases of γ-matrices

In *even dimensions*, the construction (G.9) yields γ-matrices of the form

$$\gamma_0 = \begin{pmatrix} 0 & \underline{1} \\ \underline{1} & 0 \end{pmatrix}, \quad \gamma_i = \begin{pmatrix} 0 & \underline{\sigma}_i \\ -\underline{\sigma}_i & 0 \end{pmatrix}, \tag{H.21}$$

with hermitian $(2^{(d/2)-1} \times 2^{(d/2)-1})$ matrices $\underline{\sigma}_i$, which satisfy

$$\{\underline{\sigma}_i, \underline{\sigma}_j\} = 2\delta_{ij}. \tag{H.22}$$

The spinor representation of the Lorentz generators in this Weyl basis is

$$S_{0i} = \frac{\mathrm{i}}{2}\gamma_0\gamma_i = \frac{\mathrm{i}}{2}\begin{pmatrix} -\underline{\sigma}_i & 0 \\ 0 & \underline{\sigma}_i \end{pmatrix}, \tag{H.23}$$

$$S_{ij} = \frac{\mathrm{i}}{4}[\gamma_i, \gamma_j] = -\frac{\mathrm{i}}{4}\begin{pmatrix} [\underline{\sigma}_i, \underline{\sigma}_j] & 0 \\ 0 & [\underline{\sigma}_i, \underline{\sigma}_j] \end{pmatrix}. \tag{H.24}$$

This is the advantage of a Weyl basis: The $2^{d/2}$ components of a spinor explicitly split into two Weyl spinors with $2^{(d/2)-1}$ components. The two Weyl spinors transform separately under proper orthochronous Lorentz transformations. A Dirac spinor representation in even dimensions is therefore reducible under the group of proper orthochronous Lorentz transformations. However, the form of S_{0i} tells us that the two Weyl spinors are transformed into each other under time or space inversions. Therefore the representation of the full Lorentz group really requires the full $2^{d/2}$-dimensional Dirac spinor.

The rotation generators in the Dirac representation in even dimensions are the same as in the Weyl basis, but the boost generators become

$$S_{0i} = -\frac{\mathrm{i}}{2}\begin{pmatrix} 0 & \underline{\sigma}_i \\ \underline{\sigma}_i & 0 \end{pmatrix}. \tag{H.25}$$

For an *odd number of spacetime dimensions* our construction provides γ-matrices of the form,

$$\gamma_0 = \pm\begin{pmatrix} -\underline{1} & 0 \\ 0 & \underline{1} \end{pmatrix}, \quad \gamma_i = \begin{pmatrix} 0 & \underline{\sigma}_i \\ -\underline{\sigma}_i & 0 \end{pmatrix}, \quad 1 \leq i \leq d-2,$$

$$\gamma_{d-1} = -\mathrm{i}\begin{pmatrix} 0 & \underline{1} \\ \underline{1} & 0 \end{pmatrix}.$$

The rotation generators S_{ij}, $1 \leq i,j \leq d-2$, are the same as in $d-1$ dimensions, but rotations of the $(i, d-1)$ plane are generated by

$$S_{i,d-1} = \frac{1}{2}\begin{pmatrix} \underline{\sigma}_i & 0 \\ 0 & -\underline{\sigma}_i \end{pmatrix}, \tag{H.26}$$

and the boost generators are off-diagonal,

$$S_{0i} = \mp\frac{\mathrm{i}}{2}\begin{pmatrix} 0 & \underline{\sigma}_i \\ \underline{\sigma}_i & 0 \end{pmatrix}, \quad S_{0,d-1} = \pm\frac{1}{2}\begin{pmatrix} 0 & -\underline{1} \\ \underline{1} & 0 \end{pmatrix}. \tag{H.27}$$

The proper orthochronous Lorentz group therefore mixes all the $2^{(d-1)/2}$ components of a Dirac spinor in odd dimensions.

Construction of the vector representation from the spinor representation

Equation (21.35) implies

$$\mathrm{tr}(\gamma_\mu \gamma_\nu) = -2^{\lfloor d/2 \rfloor}\eta_{\mu\nu}. \tag{H.28}$$

This and the invariance of the γ-matrices (H.15) can be used to reconstruct the vector representation of a proper orthochronous Lorentz transformation from the corresponding spinor representation,

$$\Lambda^\mu{}_\nu(\epsilon) = -2^{-\lfloor d/2 \rfloor}\mathrm{tr}\left[\exp\left(-\frac{\mathrm{i}}{2}\epsilon^{\kappa\lambda}S_{\kappa\lambda}\right)\cdot\gamma^\mu\cdot\exp\left(\frac{\mathrm{i}}{2}\epsilon^{\rho\sigma}S_{\rho\sigma}\right)\cdot\gamma_\nu\right]. \tag{H.29}$$

We can also use equation (H.28) to transform every vector into a spinor of order 2 (or every tensor of order n into a spinor of order $2n$),

$$x(\gamma) = x^\mu\gamma_\mu, \quad x^\mu = -2^{-\lfloor d/2 \rfloor}\mathrm{tr}[\gamma^\mu \cdot x(\gamma)],$$

and the invariance of the γ-matrices implies

$$x'^\mu = \Lambda^\mu{}_\nu(\epsilon)x^\nu \quad \Leftrightarrow \quad x'(\gamma) = \exp\left(\frac{\mathrm{i}}{2}\epsilon^{\kappa\lambda}S_{\kappa\lambda}\right)\cdot x(\gamma)\cdot\exp\left(-\frac{\mathrm{i}}{2}\epsilon^{\rho\sigma}S_{\rho\sigma}\right).$$

Construction of the free Dirac spinors from Dirac spinors at rest

We use $c = 1$ and $d = 4$ in this section. The Dirac equation in momentum space (21.41) is for a Dirac spinor $\psi(E,\mathbf{0})$ at rest

$$(m - E\gamma^0)\psi(E,\mathbf{0}) = 0. \tag{H.30}$$

The hermitian 4×4 matrix γ^0 can only have eigenvalues ± 1, which each must be two-fold degenerate because γ^0 is traceless. Therefore Dirac spinors at rest must correspond to energy eigenvalues $E = \pm m$. To construct the free Dirac spinors for arbitrary on-shell momentum 4-vector we can then use a boost into a frame where the fermion has on-shell momentum 4-vector $\pm p$,

$$\begin{pmatrix} \pm E \\ \mathbf{0} \end{pmatrix} \to \begin{pmatrix} \pm\sqrt{\mathbf{p}^2+m^2} \\ \pm\mathbf{p} \end{pmatrix} = \Lambda\cdot\begin{pmatrix} \pm m \\ \mathbf{0} \end{pmatrix}, \tag{H.31}$$

and equation (H.14) then implies

$$\psi(\pm\sqrt{\mathbf{p}^2+m^2},\pm\mathbf{p}) = U(\Lambda)\cdot\psi(\pm m,\mathbf{0}).$$

The Lorentz boost which takes us from the rest frame of the fermion into a frame where the fermion has on-shell momentum 4-vector $\pm p$ is

$$\begin{aligned}\Lambda(\boldsymbol{u}) = \{\Lambda^\mu{}_\nu(\boldsymbol{u})\} &= \begin{pmatrix} \gamma & -\gamma\boldsymbol{\beta}^T \\ -\gamma\boldsymbol{\beta} & \underline{1} - \hat{\boldsymbol{u}}\otimes\hat{\boldsymbol{u}}^T + \gamma\hat{\boldsymbol{u}}\otimes\hat{\boldsymbol{u}}^T \end{pmatrix} \\ &= \frac{1}{m}\begin{pmatrix} \sqrt{\mathbf{p}^2+m^2} & \mathbf{p}^T \\ \mathbf{p} & m\underline{1} - m\hat{\mathbf{p}}\otimes\hat{\mathbf{p}}^T + \sqrt{\mathbf{p}^2+m^2}\,\hat{\mathbf{p}}\otimes\hat{\mathbf{p}}^T \end{pmatrix},\end{aligned}$$

i.e. with $E(\mathbf{p}) \equiv \sqrt{\mathbf{p}^2+m^2}$,

$$\begin{aligned}\gamma &= \cosh(u) = \frac{1}{m}\sqrt{\mathbf{p}^2+m^2} = \frac{E(\mathbf{p})}{m}, \quad \gamma\boldsymbol{\beta} = \hat{\boldsymbol{u}}\sinh(u) = -\frac{\mathbf{p}}{m}, \\ \boldsymbol{v} &= \boldsymbol{\beta} = -\frac{\mathbf{p}}{\sqrt{\mathbf{p}^2+m^2}}.\end{aligned} \tag{H.32}$$

The minus sign makes perfect sense: We have to transform from the particle's rest frame into a frame which moves with speed $\boldsymbol{v} = -\boldsymbol{v}_{particle}$ relative to the particle to observe the particle with speed $\boldsymbol{v}_{particle} = \mathbf{p}/E(\mathbf{p})$.

The rapidity parameter of the particle is

$$u = \mathrm{artanh}(\beta) = \frac{1}{2}\ln\left(\frac{1+\beta}{1-\beta}\right) = \frac{1}{2}\ln\left(\frac{\sqrt{\boldsymbol{p}^2+m^2}+|\boldsymbol{p}|}{\sqrt{\boldsymbol{p}^2+m^2}-|\boldsymbol{p}|}\right)$$
$$= \ln\left(\frac{\sqrt{\boldsymbol{p}^2+m^2}+|\boldsymbol{p}|}{m}\right).$$

The general boost matrix acting on the spinors is

$$U(\boldsymbol{u}) = \exp\left(\mathrm{i}u^i S_{i0}\right) = \exp\left(\frac{1}{2}u^i\gamma_0\gamma_i\right) = \cosh\left(\frac{u}{2}\right) + \hat{\boldsymbol{u}}\cdot\gamma_0\boldsymbol{\gamma}\sinh\left(\frac{u}{2}\right),$$
$$U^2(\boldsymbol{u}) = \exp\left(2\mathrm{i}u^i S_{i0}\right) = \exp\left(u^i\gamma_0\gamma_i\right) = \cosh(u) + \hat{\boldsymbol{u}}\cdot\gamma_0\boldsymbol{\gamma}\sinh(u).$$

In the present case we have

$$U^2(\boldsymbol{u}) = \frac{1}{m}\left(\sqrt{\boldsymbol{p}^2+m^2} - \boldsymbol{p}\cdot\gamma_0\boldsymbol{\gamma}\right),$$

i.e. we can also write

$$U(\boldsymbol{u}) = \frac{1}{\sqrt{m}}\sqrt{\sqrt{\boldsymbol{p}^2+m^2} - \boldsymbol{p}\cdot\gamma_0\boldsymbol{\gamma}}. \tag{H.33}$$

The corresponding boost matrices in the Dirac representation (21.36) are

$$\gamma_0\gamma_i = \begin{pmatrix} 0 & -\underline{\sigma}_i \\ -\underline{\sigma}_i & 0 \end{pmatrix},$$
$$U(\boldsymbol{u}) = \begin{pmatrix} \cosh\left(\frac{u}{2}\right) & -\hat{\boldsymbol{u}}^T\cdot\underline{\boldsymbol{\sigma}}\sinh\left(\frac{u}{2}\right) \\ -\hat{\boldsymbol{u}}\cdot\underline{\boldsymbol{\sigma}}\sinh\left(\frac{u}{2}\right) & \cosh\left(\frac{u}{2}\right) \end{pmatrix} = \frac{1}{\sqrt{m}}\begin{pmatrix} E(\boldsymbol{p}) & \boldsymbol{p}\cdot\underline{\boldsymbol{\sigma}} \\ \boldsymbol{p}\cdot\underline{\boldsymbol{\sigma}} & E(\boldsymbol{p}) \end{pmatrix}^{1/2}.$$

For the evaluation of the hyperbolic functions, we note

$$\cosh\left(\frac{u}{2}\right) = \sqrt{\frac{\cosh(u)+1}{2}} = \sqrt{\frac{E(\boldsymbol{p})+m}{2m}},$$
$$\sinh\left(\frac{u}{2}\right) = \sqrt{\frac{\cosh(u)-1}{2}} = \sqrt{\frac{E(\boldsymbol{p})-m}{2m}} = \frac{|\boldsymbol{p}|}{\sqrt{2m(E(\boldsymbol{p})+m)}}.$$

This yields

$$U(\boldsymbol{u}) = \frac{1}{\sqrt{2m(E(\boldsymbol{p})+m)}}\begin{pmatrix} E(\boldsymbol{p})+m & \boldsymbol{p}\cdot\underline{\boldsymbol{\sigma}} \\ \boldsymbol{p}\cdot\underline{\boldsymbol{\sigma}} & E(\boldsymbol{p})+m \end{pmatrix}. \tag{H.34}$$

The rest frame spinors satisfying equation (H.30) in the Dirac representation are

$$u(\mathbf{0},\tfrac{1}{2}) = \begin{pmatrix} \sqrt{2m} \\ 0 \\ 0 \\ 0 \end{pmatrix}, \quad u(\mathbf{0},-\tfrac{1}{2}) = \begin{pmatrix} 0 \\ \sqrt{2m} \\ 0 \\ 0 \end{pmatrix},$$

$$v(\mathbf{0},-\tfrac{1}{2}) = \begin{pmatrix} 0 \\ 0 \\ \sqrt{2m} \\ 0 \end{pmatrix}, \quad v(\mathbf{0},\tfrac{1}{2}) = \begin{pmatrix} 0 \\ 0 \\ 0 \\ \sqrt{2m} \end{pmatrix},$$

and application of the spinor boost matrix (H.34) yields the spinors $u(\boldsymbol{p},\pm\frac{1}{2})$ and $v(\boldsymbol{p},\pm\frac{1}{2})$ in agreement with equations (21.45–21.48). The initial construction there from $m - \gamma \cdot p$ gave us the negative energy solutions $v(-\boldsymbol{p},\pm\frac{1}{2})$ for momentum 4-vector $(-E(\boldsymbol{p}),\boldsymbol{p})$, whereas the construction from equation (H.31) gave us directly the negative energy solutions $v(\boldsymbol{p},\pm\frac{1}{2})$ for momentum 4-vector $-p = (-E(\boldsymbol{p}),-\boldsymbol{p})$, which in either derivation are finally used in the general free solution (21.49).

Appendix I: Transformation of fields under reflections

In this Appendix we will assume $d = 4$ for the number of spacetime dimensions. The proper orthochronous Lorentz transformations were introduced in Appendix B and we have discussed exponential representations of boosts and rotations in equations (H.1–H.27). However, the relativistic line element $ds^2 = -\eta_{\mu\nu}dx^\mu dx^\nu$ is also invariant under reflections[1]

$$\mathrm{P}_\mu : dx^\mu \to -dx^\mu, \quad dx^\nu \to dx^\nu (\nu \neq \mu).$$

The product of any two spatial reflections is a rotation of the corresponding spatial plane by π, cf. (H.10, H.11),

$$\mathrm{P}_i\mathrm{P}_j = \exp(\mathrm{i}\pi M_{ij}),$$

and this implies that we can write any particular spatial reflection as a combination of the reflection $\mathrm{P} = \mathrm{P}_1\mathrm{P}_2\mathrm{P}_3$ of all spatial directions with a rotation by π,

$$\mathrm{P}_i = \mathrm{P}\exp\left(\frac{\mathrm{i}}{2}\pi\epsilon_{ijk}M_{jk}\right).$$

Therefore it is sufficient to discuss the two discrete Lorentz transformations $\mathrm{T} = \mathrm{P}_0$ (reversal of time direction) and P. The spatial inversion P is also denoted as a *parity transformation*.

[1]The reflections $dt \to -dt$ or $dx^i \to -dx^i$ (or up to constant shifts, $t \to -t$, $x^i \to -x^i$) are usually denoted as time or space *inversions*. This convention likely originated from the fact that in algebraic fields (here "field" refers to the mathematical definition of a set which allows for addition, subtraction, multiplication, and division where possible) $x \to -x$ is the inversion operation with respect to addition. However, the operations P_μ are reversals of time or spatial directions which arise from reflections at 3-dimensional hyperplanes located at some coordinate value X^μ: $x^\mu \to 2X^\mu - x^\mu$. Therefore we prefer the designation *reflections* for these transformations.

R. Dick, *Advanced Quantum Mechanics*, Graduate Texts in Physics,
DOI 10.1007/978-3-319-25675-7

We can determine the transformation properties of fields under P and T from the requirement that electrodynamics should be invariant under these transformations, i.e. we postulate that the equations

$$[\hbar\partial - \mathrm{i}QA]^2\phi - m^2c^2\phi = 0, \quad \gamma^\mu[\mathrm{i}\hbar\partial_\mu + qA_\mu]\Psi - mc\Psi = 0, \tag{I.1}$$

$$-\frac{1}{\mu_0 c}\partial_\mu F^{\mu\nu} = q\overline{\Psi}\gamma^\nu\Psi + \mathrm{i}Qc\left[\partial^\nu\phi^+\cdot\phi - \phi^+\cdot\partial^\nu\phi + \mathrm{i}\frac{Q}{\hbar}\phi^+ A^\nu\phi\right] \tag{I.2}$$

hold in this form also for an observer that uses reflected spatial axes or uses decreasing values of t to label the future.

We know already from classical electrodynamics how electromagnetic fields and charge distributions transform under P and T, see e.g. [19],

$$\mathrm{T}:\ t' = -t, \quad \boldsymbol{x}' = \boldsymbol{x}, \quad j_0'(\boldsymbol{x},t) = j_0(\boldsymbol{x},-t), \quad \boldsymbol{j}'(\boldsymbol{x},t) = -\boldsymbol{j}(\boldsymbol{x},-t),$$
$$\boldsymbol{E}'(\boldsymbol{x},t) = \boldsymbol{E}(\boldsymbol{x},-t), \quad \boldsymbol{B}'(\boldsymbol{x},t) = -\boldsymbol{B}(\boldsymbol{x},-t),$$

$$\mathrm{P}:\ t' = t, \quad \boldsymbol{x}' = -\boldsymbol{x}, \quad j_0'(\boldsymbol{x},t) = j_0(-\boldsymbol{x},t), \quad \boldsymbol{j}'(\boldsymbol{x},t) = -\boldsymbol{j}(-\boldsymbol{x},t),$$
$$\boldsymbol{E}'(\boldsymbol{x},t) = -\boldsymbol{E}(-\boldsymbol{x},t), \quad \boldsymbol{B}'(\boldsymbol{x},t) = \boldsymbol{B}(-\boldsymbol{x},t).$$

The transformation properties of the electromagnetic fields imply that (up to gauge transformations) the vector potentials transform according to

$$\mathrm{T}: A_0'(\boldsymbol{x},t) = A_0(\boldsymbol{x},-t), \quad \boldsymbol{A}'(\boldsymbol{x},t) = -\boldsymbol{A}(\boldsymbol{x},-t),$$
$$\mathrm{P}: A_0'(\boldsymbol{x},t) = A_0(-\boldsymbol{x},t), \quad \boldsymbol{A}'(\boldsymbol{x},t) = -\boldsymbol{A}(-\boldsymbol{x},t).$$

The components of $A_\mu(x)$ transform under P like the derivative operators ∂_μ, such that the covariant derivatives transform like $D_0' = D_0$, $D_i' = -D_i$. We can therefore get the correct transformation behavior of the currents on the right hand side of Maxwell's equations (I.2) and preserve the matter equations (I.1) if we transform the matter fields (up to gauge transformations) according to

$$\mathrm{P}: \phi'(\boldsymbol{x},t) = \phi(-\boldsymbol{x},t), \quad \Psi'(\boldsymbol{x},t) = \gamma^0\Psi(-\boldsymbol{x},t).$$

On the other hand, the partial derivatives and vector potentials pick up relative minus signs under time reversal T:

$$\hbar\partial_0 - \mathrm{i}qA_0(\boldsymbol{x},t) = -\hbar\partial_0' - \mathrm{i}qA_0'(\boldsymbol{x},t'), \quad \hbar\nabla - \mathrm{i}q\boldsymbol{A}(\boldsymbol{x},t) = \hbar\nabla + \mathrm{i}q\boldsymbol{A}(\boldsymbol{x},t').$$

The transformation properties of scalar and spinor fields under time reversal therefore need to invoke complex conjugations to preserve the matter equations of motion (I.1), and they need to reverse the signs of some of the derivatives of the Dirac field after complex conjugation while leaving the other derivative terms unchanged. In a Dirac of Weyl basis of γ matrices (21.36, 21.37), this can be achieved (up to gauge transformations) through the transformation laws[2]

$$\mathrm{T}: \phi'(\boldsymbol{x},t) = \phi^*(\boldsymbol{x},-t), \quad \Psi'(\boldsymbol{x},t) = \gamma_1\gamma_3\Psi^*(\boldsymbol{x},-t).$$

Relativistic electrodynamics is invariant under P and T and also under charge conjugation (21.3, 21.79). However, as a general property relativistic field theories only need to be invariant under the combination CPT, see e.g. Vol. I of [41], which also provides original references for the CPT theorem. In our conventions, CPT acts on scalar and spinor fields and real vector potentials according to

$$\mathrm{CPT}: \phi'(x) = \phi(-x), \quad \Psi'(x) = \gamma_5\Psi(-x), \quad A'_\mu(x) = A_\mu(-x).$$

Here the γ_5 matrix is defined as

$$\gamma_5 = \mathrm{i}\gamma^0\gamma^1\gamma^2\gamma^3.$$

It takes the following explicit forms in the Dirac or Weyl representations:

$$\gamma_5^{(D)} = \begin{pmatrix} \underline{0} & \underline{1} \\ \underline{1} & \underline{0} \end{pmatrix}, \quad \gamma_5^{(W)} = \begin{pmatrix} \underline{1} & \underline{0} \\ \underline{0} & -\underline{1} \end{pmatrix}.$$

[2]Time reversal and charge conjugation also require a transposition of operator products if the fermionic field Ψ is not a c number field but an operator.

Appendix J: Green's functions in d dimensions

We denote the number of spatial dimensions with d in this appendix, and we *suspend* the use of summation convention until we reach (J.62).

Green's functions are solutions of linear differential equations with δ function source terms. Basic one-dimensional examples are provided by the conditions

$$\frac{d}{dx}S(x) - \kappa S(x) = -\delta(x), \quad \frac{d^2}{dx^2}G(x) - \kappa^2 G(x) = -\delta(x), \tag{J.1}$$

with solutions

$$G(x) = \frac{a}{2\kappa}\exp(-\kappa|x|) + \frac{a-1}{2\kappa}\exp(\kappa|x|) + A\exp(\kappa x) + B\exp(-\kappa x), \tag{J.2}$$

and

$$\begin{aligned} S(x) &= \frac{d}{dx}G(x) + \kappa G(x) \\ &= a\Theta(-x)\exp(\kappa x) + (a-1)\Theta(x)\exp(\kappa x) + 2\kappa A\exp(\kappa x) \\ &= C\exp(\kappa x) + \Theta(-x)\exp(\kappa x) = C'\exp(\kappa x) - \Theta(x)\exp(\kappa x), \\ C' &= C + 1 = 2\kappa A + a. \end{aligned} \tag{J.3}$$

That the functions (J.2, J.3) satisfy the conditions (J.1) is easily confirmed by using

$$\frac{d}{dx}|x| = \Theta(x) - \Theta(-x), \quad \frac{d}{dx}\Theta(\pm x) = \pm\delta(x).$$

The solutions of the conditions in the limit $\kappa \to 0$ are

$$G(x) = \alpha x + \beta - \frac{|x|}{2}, \quad S(x) = \frac{d}{dx}G(x) = \alpha + \frac{\Theta(-x) - \Theta(x)}{2}. \tag{J.4}$$

R. Dick, *Advanced Quantum Mechanics*, Graduate Texts in Physics,
DOI 10.1007/978-3-319-25675-7

The appearance of integration constants signals that we can impose boundary conditions on the Green's functions. An important example for this is the requirement of vanishing Green's functions at spatial infinity, which can be imposed if the real part of κ does not vanish. For positive real κ this implies the one-dimensional Green's functions

$$G(x) = \frac{1}{2\kappa}\exp(-\kappa|x|), \quad S(x) = \Theta(-x)\exp(\kappa x).$$

However, in one dimension we cannot satisfy the boundary condition of vanishing Green's functions at infinity if $\kappa = 0$, and we will find the same result for the scalar Green's function $G(\boldsymbol{x})$ in two dimensions. We can satisfy conditions that the Green's functions (J.4) should vanish on a half-axis $x < 0$ or $x > 0$ for $\kappa = 0$ by choosing $\alpha = \mp 1/2$, $\beta = 0$. On the other hand, if $\kappa = \mathrm{i}k$ is imaginary with $k > 0$, the Green's function

$$G(x) = \frac{\mathrm{i}}{2k}\exp(\mathrm{i}k|x|)$$

describes the spatial factor of outgoing waves $\exp[\mathrm{i}(k|x| - \omega t)]$, i.e. the one-dimensional version of outgoing spherical waves.

Green's functions for Schrödinger's equation

We are mostly concerned with Green's functions associated with time-independent Hamilton operators

$$H = \frac{\mathbf{p}^2}{2m} + V(\mathbf{x}) = \int d^d\boldsymbol{x}\,|\boldsymbol{x}\rangle\left(-\frac{\hbar^2}{2m}\Delta + V(\boldsymbol{x})\right)\langle\boldsymbol{x}|.$$

Note that the number of spatial dimensions d is left as a discrete variable.

The inversion condition for the energy-dependent Schrödinger operator,

$$(E - H)\,\mathcal{G}_{d,V}(E) = 1 \tag{J.5}$$

is in $\boldsymbol{x}$ representation the condition

$$\left(E + \frac{\hbar^2}{2m}\Delta - V(\boldsymbol{x})\right)\langle\boldsymbol{x}|\mathcal{G}_{d,V}(E)|\boldsymbol{x}'\rangle = \delta(\boldsymbol{x} - \boldsymbol{x}'). \tag{J.6}$$

The equations (J.5) and (J.6) show that we should rather talk about a Green's *operator* $\mathcal{G}_{d,V}(E)$ (or a *resolvent* in mathematical terms), with matrix elements $\langle\boldsymbol{x}|\mathcal{G}_{d,V}(E)|\boldsymbol{x}'\rangle$. We will instead continue to use the designation Green's function both for $\mathcal{G}_{d,V}(E)$ and the Fourier transformed operator $\mathcal{G}_{d,V}(t)$ and for all their representations in $\boldsymbol{x}$ or $\boldsymbol{k}$ space variables (or their matrix elements with respect to any other quantum states). The designation Green's function originated from the matrix elements $\mathcal{G}_{d,V}(\boldsymbol{x},\boldsymbol{x}';E) \equiv \langle\boldsymbol{x}|\mathcal{G}_{d,V}(E)|\boldsymbol{x}'\rangle$. These functions preceded the resolvent $\mathcal{G}_{d,V}(E)$ because the inception of differential equations preceded the discovery of abstract operator concepts and bra-ket notation.

The Green's function $\mathcal{G}_{d,V}(E)$ can eventually be calculated perturbatively in terms of the free Green's function $\mathcal{G}_d(E) \equiv \mathcal{G}_{d,V=0}(E)$. The equations

$$(E-H_0)\,\mathcal{G}_{d,V}(E) = 1 + V\mathcal{G}_{d,V}(E), \quad (E-H_0)\,\mathcal{G}_d(E) = 1, \tag{J.7}$$

yield

$$\begin{aligned}\mathcal{G}_{d,V}(E) &= \mathcal{G}_d(E) + \mathcal{G}_d(E)V\mathcal{G}_{d,V}(E)\\ &= \mathcal{G}_d(E) + \mathcal{G}_d(E)V\mathcal{G}_d(E) + \mathcal{G}_d(E)V\mathcal{G}_d(E)V\mathcal{G}_{d,V}(E)\\ &= \sum_{n=0}^{\infty}\mathcal{G}_d(E)\,(V\mathcal{G}_d(E))^n = \sum_{n=0}^{\infty}(\mathcal{G}_d(E)V)^n\,\mathcal{G}_d(E).\end{aligned} \tag{J.8}$$

From the geometric series appearing in (J.8) we can also find the representations

$$\mathcal{G}_{d,V}(E) = \mathcal{G}_d(E)\frac{1}{1-V\mathcal{G}_d(E)} = \frac{1}{1-\mathcal{G}_d(E)V}\mathcal{G}_d(E)$$

which are of course equivalent to the original condition $(E-H)\mathcal{G}_{d,V}(E) = 1$ through

$$\begin{aligned}(E-H_0-V)^{-1} &= \left((E-H_0)\left[1-(E-H_0)^{-1}V\right]\right)^{-1}\\ &= \left(1-(E-H_0)^{-1}V\right)^{-1}(E-H_0)^{-1}\end{aligned}$$

and the corresponding relation with $E-H_0$ extracted on the right hand side of V.

Whether the formal iteration (J.8) yields a sensible numerical approximation depends on the potential V, the energy E, and on the states for which we wish to calculate the corresponding matrix element of $\mathcal{G}_{d,V}(E)$. We defined $H_0 = \mathbf{p}^2/2m$ as the free Hamiltonian, and we have used the first two terms of (J.8) in potential scattering theory in the Born approximation. Other applications of series like (J.8) in perturbation theory would include a solvable part V_0 of the potential in H_0 and use only a perturbation $V' = V - V_0$ for the iterative solution. However, our main concern in the following will be the free Green's function $\mathcal{G}_d(E)$.

The variable E in (J.5) can be complex, but $\mathcal{G}_{d,V}(E)$ will become singular for values of E in the spectrum of H. It is therefore useful to explicitly add a small imaginary part if E is constrained to be real, which is the most relevant case for us. To discuss the implications of a small imaginary addition to E, consider Fourier transformation of (J.5) into the time domain. Substitution of

$$\mathcal{G}_{d,V}(E) = \int_{-\infty}^{\infty} dt\, \mathcal{G}_{d,V}(t) \exp(iEt/\hbar), \tag{J.9}$$

$$\mathcal{G}_{d,V}(t) = \frac{1}{2\pi\hbar} \int_{-\infty}^{\infty} dE\, \mathcal{G}_{d,V}(E) \exp(-iEt/\hbar), \tag{J.10}$$

yields

$$\left(i\hbar \frac{d}{dt} - H \right) \mathcal{G}_{d,V}(t) = \delta(t). \tag{J.11}$$

We can solve this equation in the form

$$\mathcal{G}_{d,V}(t) = \frac{a}{i\hbar}\Theta(t)\mathcal{K}_{d,V}(t) + \frac{a-1}{i\hbar}\Theta(-t)\mathcal{K}_{d,V}(t) = \frac{a-\Theta(-t)}{i\hbar}\mathcal{K}_{d,V}(t), \tag{J.12}$$

if $\mathcal{K}_{d,V}(t)$ is the solution of the time-dependent Schrödinger equation

$$\left(i\hbar \frac{d}{dt} - H \right) \mathcal{K}_{d,V}(t) = 0$$

with initial condition $\mathcal{K}_{d,V}(0) = 1$. Indeed, we have found this solution and used it extensively in Chapter 13. It is the time evolution operator

$$\mathcal{K}_{d,V}(t) = U(t) = \exp\left(-\frac{i}{\hbar} Ht \right). \tag{J.13}$$

Equations (J.12) and (J.13) imply that the Green's function in the energy representation is

$$\begin{aligned} \mathcal{G}_{d,V}(E) &= \frac{a}{i\hbar} \int_{0}^{\infty} dt\, \exp[i(E - H + i\epsilon)t/\hbar] \\ &\quad - \frac{1-a}{i\hbar} \int_{-\infty}^{0} dt\, \exp[i(E - H - i\epsilon)t/\hbar] \\ &= \frac{a}{E - H + i\epsilon} + \frac{1-a}{E - H - i\epsilon}, \end{aligned} \tag{J.14}$$

with a small shift $\epsilon > 0$.

The time-dependent Green's function (J.12) solves the inhomogeneous equation

$$\left(i\hbar \frac{d}{dt} - H \right) F(t) = J(t)$$

in the form

$$\begin{aligned} F(t) &= F_0(t) + \int_{-\infty}^{\infty} dt'\, \mathcal{G}_{d,V}(t-t')J(t') \\ &= F_0(t) + \frac{a}{\mathrm{i}\hbar}\int_{-\infty}^{t} dt'\, \exp\!\left(-\frac{\mathrm{i}}{\hbar}H(t-t')\right)J(t') \\ &\quad + \frac{a-1}{\mathrm{i}\hbar}\int_{t}^{\infty} dt'\, \exp\!\left(-\frac{\mathrm{i}}{\hbar}H(t-t')\right)J(t'), \end{aligned} \tag{J.15}$$

where $F_0(t)$ is an arbitrary solution of the Schrödinger equation

$$\left(\mathrm{i}\hbar\frac{d}{dt} - H\right)F_0(t) = 0.$$

The Green's function (J.12, J.14) with $a = 1$ is the *retarded Green's function*, because the solution (J.15) receives only contributions from $J(t')$ at times $t' < t$ for $a = 1$. The Green's function with $a = 0$ is denoted as an *advanced Green's function*, because it determines $F(t)$ from back evolution of future values of $J(t)$.

We will now specialize to the retarded free Green's function. So far we have found the following representations for this Green's function,

$$\mathcal{G}_d(t) = \frac{\Theta(t)}{\mathrm{i}\hbar}\exp\!\left(-\frac{\mathrm{i}t}{2m\hbar}\mathbf{p}^2\right), \tag{J.16}$$

$$\mathcal{G}_d(E) = -\frac{2m}{\hbar^2}G_d(E) = \frac{1}{E + \mathrm{i}\epsilon - (\mathbf{p}^2/2m)}. \tag{J.17}$$

The rescaled Green's function $G_d(E)$ is an inverse Poincaré operator

$$\left(\Delta + \frac{2mE}{\hbar^2}\right)\langle \boldsymbol{x}|G_d(E)|\boldsymbol{x}'\rangle = -\delta(\boldsymbol{x}-\boldsymbol{x}'), \tag{J.18}$$

and has been introduced to make the connection with electromagnetic Green's functions and potentials more visible.

The equations (J.16, J.17) do not generate any spectacular dependence on the number d of spatial dimensions in the $\boldsymbol{k}$-space representation of the retarded free Green's functions,

$$\begin{aligned} \langle \boldsymbol{k}|\mathcal{G}_d(t)|\boldsymbol{k}'\rangle &= \frac{\Theta(t)}{\mathrm{i}\hbar}\exp\!\left(-\mathrm{i}\frac{\hbar t}{2m}\boldsymbol{k}^2\right)\delta(\boldsymbol{k}-\boldsymbol{k}') \equiv \mathcal{G}_d(\boldsymbol{k},t)\delta(\boldsymbol{k}-\boldsymbol{k}'), \\ \langle \boldsymbol{k}|G_d(E)|\boldsymbol{k}'\rangle &= \frac{\delta(\boldsymbol{k}-\boldsymbol{k}')}{\boldsymbol{k}^2 - (2mE/\hbar^2) - \mathrm{i}\epsilon} \equiv G_d(\boldsymbol{k},E)\delta(\boldsymbol{k}-\boldsymbol{k}'), \end{aligned} \tag{J.19}$$

and also the d-dependence of the mixed representations is not particularly noteworthy, e.g.

$$\langle \boldsymbol{x}|\mathcal{G}_d(t)|\boldsymbol{k}\rangle = \langle \boldsymbol{x}|\boldsymbol{k}\rangle\mathcal{G}_d(\boldsymbol{k},t) = \frac{\Theta(t)}{\mathrm{i}\hbar\sqrt{2\pi}^d}\exp\!\left(\mathrm{i}\boldsymbol{k}\cdot\boldsymbol{x} - \mathrm{i}\frac{\hbar t}{2m}\boldsymbol{k}^2\right),$$

$$\langle \boldsymbol{x}|G_d(E)|\boldsymbol{k}\rangle = \langle \boldsymbol{x}|\boldsymbol{k}\rangle G_d(\boldsymbol{k},E) = \frac{1}{\sqrt{2\pi}^d}\frac{\exp(\mathrm{i}\boldsymbol{k}\cdot\boldsymbol{x})}{\boldsymbol{k}^2 - (2mE/\hbar^2) - \mathrm{i}\epsilon}.$$

The $\boldsymbol{x}$-representation of the time-dependent Green's function,

$$\langle \boldsymbol{x}|\mathcal{G}_d(t)|\boldsymbol{x}'\rangle = \frac{1}{(2\pi)^d}\int d^d\boldsymbol{k}\,\mathcal{G}_d(\boldsymbol{k},t)\exp[\mathrm{i}\boldsymbol{k}\cdot(\boldsymbol{x}-\boldsymbol{x}')] \equiv \mathcal{G}_d(\boldsymbol{x}-\boldsymbol{x}',t),$$

is

$$\begin{aligned}\mathcal{G}_d(\boldsymbol{x},t) &= \frac{\Theta(t)}{\mathrm{i}\hbar(2\pi)^d}\int d^d\boldsymbol{k}\,\exp\!\left[\mathrm{i}\left(\boldsymbol{k}\cdot\boldsymbol{x} - \frac{\hbar t}{2m}\boldsymbol{k}^2\right)\right]\\ &= \frac{\Theta(t)}{\mathrm{i}\hbar}\sqrt{\frac{m}{2\pi\mathrm{i}\hbar t}}^{\,d}\exp\!\left(\mathrm{i}\frac{m\boldsymbol{x}^2}{2\hbar t}\right).\end{aligned} \tag{J.20}$$

This equation holds in the sense that $\mathcal{G}_d(\boldsymbol{x}-\boldsymbol{x}',t-t')$ has to be integrated with an absolutely or square integrable function $J(\boldsymbol{x}',t')$ to yield a solution $F(\boldsymbol{x},t)$ (J.15) of an inhomogeneous Schrödinger equation.

The representation of the retarded free Green's function in the time-domain is interesting in its own right, but in terms of dependence on the number d of dimensions, the operator $\mathrm{i}\hbar\mathcal{G}_d(t)$ and its representations $\mathrm{i}\hbar\mathcal{G}_d(\boldsymbol{k},t)$ and $\mathrm{i}\hbar\mathcal{G}_d(\boldsymbol{x},t)$ are simply products of d copies of the corresponding one-dimensional Green's function $\mathrm{i}\hbar\mathcal{G}_1(t)$ and its representations. Free propagation in time separates completely in spatial dimensions[1].

The interesting dimensional aspects of the Green's function appear if we represent it in the energy domain and in $\boldsymbol{x}$-space,

$$\begin{aligned}\langle \boldsymbol{x}|G_d(E)|\boldsymbol{x}'\rangle &= \frac{1}{(2\pi)^d}\int d^d\boldsymbol{k}\,G_d(\boldsymbol{k},E)\exp[\mathrm{i}\boldsymbol{k}\cdot(\boldsymbol{x}-\boldsymbol{x}')]\\ &\equiv G_d(\boldsymbol{x}-\boldsymbol{x}',E).\end{aligned} \tag{J.21}$$

This requires a little extra preparation.

[1]This is a consequence of the separation of the free non-relativistic Hamiltonian $H_0 = \mathbf{p}^2/2m$. However, this property does not hold in relativistic quantum mechanics, and therefore the free time-dependent Green's function in the relativistic case is not a product of one-dimensional Green's functions, see (J.44).

Polar coordinates in d dimensions

Evaluation of the d-dimensional Fourier transformation in (J.21) involves polar coordinates in d-dimensional $\boldsymbol{k}$ space. Furthermore, it is also instructive to derive the zero energy Green's function $G_d(0)$ directly in $\boldsymbol{x}$ space, which is also conveniently done in polar coordinates. Therefore we use $\boldsymbol{x}$ space as a paradigm for the discussion of polar coordinates in d dimensions with the understanding that in $\boldsymbol{k}$ space, $r = \sqrt{\boldsymbol{x}^2}$ is replaced with $k = \sqrt{\boldsymbol{k}^2}$.

We define polar coordinates $r, \theta_1, \dots \theta_{d-1}$ in d dimensions through

$$\begin{aligned}
x_1 &= r\sin\theta_1 \cdot \sin\theta_2 \cdot \ldots \cdot \sin\theta_{d-2} \cdot \sin\theta_{d-1}, \quad \varphi = \frac{\pi}{2} - \theta_{d-1} \\
x_2 &= r\sin\theta_1 \cdot \sin\theta_2 \cdot \ldots \cdot \sin\theta_{d-2} \cdot \cos\theta_{d-1}, \\
&\vdots \\
x_{d-1} &= r\sin\theta_1 \cdot \cos\theta_2, \\
x_d &= r\cos\theta_1.
\end{aligned}$$

This yields corresponding tangent vectors along the radial coordinate lines, cf. (5.18),

$$\boldsymbol{a}_r = \frac{\partial \boldsymbol{x}}{\partial r} = \boldsymbol{e}_r,$$

and along the θ_i coordinate lines

$$\boldsymbol{a}_i = \frac{\partial \boldsymbol{x}}{\partial \theta_i} = r\sin\theta_1 \cdot \sin\theta_2 \cdot \ldots \sin\theta_{i-1}\boldsymbol{e}_i.$$

Here we defined the unit tangent vector along the θ_i coordinate line

$$\boldsymbol{e}_i = \frac{\boldsymbol{a}_i}{|\boldsymbol{a}_i|}.$$

This should not be confused with Cartesian basis vectors since we do not use any Cartesian basis vector in this section.

The induced metric is $g_{\mu\nu} = \boldsymbol{a}_\mu \cdot \boldsymbol{a}_\nu$, see Section 5.4. This yields in the present case

$$g_{\mu\nu}\Big|_{\mu\neq\nu} = 0, \quad g_{rr} = 1,$$

and

$$g_{ii} = r^2\sin^2\theta_1 \cdot \sin^2\theta_2 \cdot \ldots \cdot \sin^2\theta_{i-1}, \quad 1 \le i \le d-1.$$

The Jacobian determinant (5.28) of the transformation from polar to Cartesian coordinates and the related volume measure (5.27) are then

$$\sqrt{g} = r^{d-1}\sin^{d-2}\theta_1 \cdot \sin^{d-3}\theta_2 \cdot \ldots \cdot \sin\theta_{d-2}$$

and

$$d^d\boldsymbol{x} = drd\theta_1 \ldots d\theta_{d-1}\, r^{d-1}\sin^{d-2}\theta_1 \cdot \sin^{d-3}\theta_2 \cdot \ldots \cdot \sin\theta_{d-2}. \tag{J.22}$$

In particular, the hypersurface area of the $(d-1)$-dimensional unit sphere is

$$S_{d-1} = 2\pi \prod_{n=1}^{d-2} \int_0^\pi d\theta\, \sin^n\theta = \frac{2\sqrt{\pi}^d}{\Gamma(d/2)}. \tag{J.23}$$

The gradient operator $\boldsymbol{\nabla} = \sum_\mu \boldsymbol{a}^\mu \partial_\mu$ is

$$\boldsymbol{\nabla} = \boldsymbol{e}_r \frac{\partial}{\partial r} + \sum_{i=1}^{d-1} \frac{\boldsymbol{e}_i}{r\sin\theta_1 \cdot \sin\theta_2 \cdot \ldots \cdot \sin\theta_{i-1}} \frac{\partial}{\partial\theta_i}.$$

For the calculation of the Laplace operator, we need the derivatives (recall that we do not use summation convention in this appendix)

$$\boldsymbol{e}_j \cdot \frac{\partial \boldsymbol{e}_r}{\partial\theta_j} = \sin\theta_1 \cdot \sin\theta_2 \cdot \ldots \sin\theta_{j-1}$$

and

$$\boldsymbol{e}_j \cdot \frac{\partial \boldsymbol{e}_i}{\partial\theta_j} = \delta_{j,i+1}\cos\theta_i + \Theta(j > i+1)\cos\theta_i \cdot \sin\theta_{i+1} \cdot \ldots \cdot \sin\theta_{j-1}.$$

This yields

$$\Delta = \frac{\partial^2}{\partial r^2} + \frac{d-1}{r}\frac{\partial}{\partial r} + \frac{1}{r^2}\sum_{i=1}^{d-1} \frac{1}{\sin^2\theta_1 \cdot \sin^2\theta_2 \cdot \ldots \cdot \sin^2\theta_{i-1}} \frac{\partial^2}{\partial\theta_i^2}$$
$$+\frac{1}{r^2}\sum_{i=1}^{d-2}\sum_{j=i+1}^{d-1} \frac{\cot\theta_i}{\sin^2\theta_1 \cdot \sin^2\theta_2 \cdot \ldots \cdot \sin^2\theta_{i-1}} \frac{\partial}{\partial\theta_i}.$$

We only need the radial part of the Laplace operator for the direct calculation of the zero energy Green's function $G_d(\boldsymbol{x}, E=0) \equiv G_d(r)$. The condition

$$\Delta G_d(r) = \frac{1}{r^{d-1}}\frac{d}{dr} r^{d-1} \frac{d}{dr} G_d(r) = -\delta(\boldsymbol{x})$$

implies after integration over a spherical volume with radius r,

$$S_{d-1}r^{d-1}\frac{d}{dr}G_d(r) = \frac{2\sqrt{\pi}^d}{\Gamma(d/2)}r^{d-1}\frac{d}{dr}G_d(r) = -1.$$

This yields

$$G_d(r) = \begin{cases} (a-r)/2, & d=1, \\ -(2\pi)^{-1}\ln(r/a), & d=2, \\ \Gamma\left(\frac{d-2}{2}\right)\left(4\sqrt{\pi}^d r^{d-2}\right)^{-1}, & d\geq 3. \end{cases} \tag{J.24}$$

The integration constant determines for $d=1$ and $d=2$ at which distance a the Green's function vanishes. For $d\geq 3$ the vanishing integration constant $\propto a^{2-d}$ is imposed by the usual boundary condition $\lim_{r\to\infty}G_{d\geq 3}(r)=0$.

The free Green's function in the $\boldsymbol{x}$-representation with full energy dependence is still translation invariant and isotropic, $\langle \boldsymbol{x}|G_d(E)|\boldsymbol{x}'\rangle \equiv G_d(\boldsymbol{x}-\boldsymbol{x}',E) = G_d(|\boldsymbol{x}-\boldsymbol{x}'|,E)$, and can be gotten from integration of the condition

$$\Delta G_d(\boldsymbol{x},E) + \frac{2m}{\hbar^2}EG_d(\boldsymbol{x},E) = -\delta(\boldsymbol{x}). \tag{J.25}$$

The result (J.24) motivates an *ansatz*

$$G_d(\boldsymbol{x},E) = f_d(r,E)G_d(r). \tag{J.26}$$

This will solve (J.25) if the factor $f_d(r,E)$ satisfies

$$\frac{d^2}{dr^2}f_d(r,E) + \frac{3-d}{r}\frac{d}{dr}f_d(r,E) + \frac{2m}{\hbar^2}Ef_d(r,E) = 0, \quad f_d(0,E)=1.$$

This yields together with the requirement $G_d(\boldsymbol{x},E)|_{E<0}\in\mathbb{R}$ and analyticity in E (and the convention $\sqrt{-E}|_{E>0} = -\mathrm{i}\sqrt{E}$),

$$\begin{aligned} G_d(\boldsymbol{x},E) = {} & \frac{\Theta(-E)}{\sqrt{2\pi}^d}\left(\frac{\sqrt{-2mE}}{\hbar r}\right)^{\frac{d-2}{2}} K_{\frac{d-2}{2}}\left(\sqrt{-2mE}\frac{r}{\hbar}\right) \\ & + \mathrm{i}\frac{\pi}{2}\frac{\Theta(E)}{\sqrt{2\pi}^d}\left(\frac{\sqrt{2mE}}{\hbar r}\right)^{\frac{d-2}{2}} H^{(1)}_{\frac{d-2}{2}}\left(\sqrt{2mE}\frac{r}{\hbar}\right), \end{aligned} \tag{J.27}$$

where the conventions and definitions from [1] were used for the modified Bessel and Hankel functions.

The result (J.27) tells us that outgoing spherical waves of energy $E > 0$ in d dimensions are given by Hankel functions,

$$G_d(\boldsymbol{x}, E > 0) = \frac{\mathrm{i}\pi}{2\sqrt{2\pi}^d}\left(\frac{\sqrt{2mE}}{\hbar r}\right)^{\frac{d-2}{2}} H^{(1)}_{\frac{d-2}{2}}\left(\sqrt{2mE}\frac{r}{\hbar}\right),$$

with asymptotic form

$$G_d(\boldsymbol{x}, E > 0)\Big|_{kr\gg 1} \simeq \frac{1}{2k}\left(\frac{k}{2\pi r}\right)^{\frac{d-1}{2}} \exp\left(\mathrm{i}kr - \mathrm{i}\frac{d-3}{4}\pi\right),$$

while d-dimensional Yukawa potentials of range a are described by modified Bessel functions,

$$V_d(r) = \frac{1}{\sqrt{2\pi}^d r^{d-2}}\left(\frac{r}{a}\right)^{\frac{d-2}{2}} K_{\frac{d-2}{2}}\left(\frac{r}{a}\right),$$

with asymptotic form

$$V_d(r \gg a) \simeq \frac{\exp(-r/a)}{2\sqrt{a}^{d-3}\sqrt{2\pi r}^{d-1}}.$$

The result (J.27) can also be derived through Fourier transformation (J.21) from the energy-dependent retarded Green's function in $\boldsymbol{k}$ space,

$$G_d(\boldsymbol{x}, E) = \frac{1}{(2\pi)^d}\int d^d\boldsymbol{k}\, G_d(\boldsymbol{k}, E)\exp(\mathrm{i}\boldsymbol{k}\cdot\boldsymbol{x}),$$

$$G_d(\boldsymbol{k}, E) = \frac{1}{\boldsymbol{k}^2 - (2mE/\hbar^2) - \mathrm{i}\epsilon},$$

or in terms of poles in the complex k plane (where $k \equiv |\boldsymbol{k}|$ for $d > 1$),

$$G_d(\boldsymbol{k}, E) = \frac{\Theta(E)}{(k - \sqrt{2mE/\hbar^2} - \mathrm{i}\epsilon)(k + \sqrt{2mE/\hbar^2} + \mathrm{i}\epsilon)} + \frac{\Theta(-E)}{(k - \mathrm{i}\sqrt{-2mE/\hbar^2})(k + \mathrm{i}\sqrt{-2mE/\hbar^2})}. \tag{J.28}$$

This yields for $d > 1$ (for the ϑ integral see [32], p. 457, no. 6)

$$
\begin{aligned}
G_d(\boldsymbol{x},E) &= \frac{S_{d-2}}{(2\pi)^d}\int_0^\infty dk\int_0^\pi d\vartheta\,\frac{k^{d-1}\exp(\mathrm{i}kr\cos\vartheta)}{k^2-(2mE/\hbar^2)-\mathrm{i}\epsilon}\sin^{d-2}\vartheta\\
&= \frac{1}{2^{d-1}\sqrt{\pi}^{d+1}\Gamma\left(\frac{d-1}{2}\right)}\int_0^\infty dk\int_0^\pi d\vartheta\,\frac{k^{d-1}\exp(\mathrm{i}kr\cos\vartheta)}{k^2-(2mE/\hbar^2)-\mathrm{i}\epsilon}\sin^{d-2}\vartheta\\
&= \frac{1}{\sqrt{2\pi}^d\sqrt{r}^{d-2}}\int_0^\infty dk\,\frac{\sqrt{k}^d}{k^2-(2mE/\hbar^2)-\mathrm{i}\epsilon}J_{\frac{d-2}{2}}(kr)\\
&= \frac{\Theta(-E)}{\sqrt{2\pi}^d}\left(\frac{\sqrt{-2mE}}{\hbar r}\right)^{\frac{d-2}{2}}K_{\frac{d-2}{2}}\left(\sqrt{-2mE}\frac{r}{\hbar}\right)\\
&\quad+\mathrm{i}\frac{\pi}{2}\frac{\Theta(E)}{\sqrt{2\pi}^d}\left(\frac{\sqrt{2mE}}{\hbar r}\right)^{\frac{d-2}{2}}H^{(1)}_{\frac{d-2}{2}}\left(\sqrt{2mE}\frac{r}{\hbar}\right).
\end{aligned}
\tag{J.29}
$$

For the k integral for $E > 0$ see [33], p. 179, no. 28. The real part of the integral for $E < 0$ is given on p. 179, no. 35. The integrals can also be performed with symbolic computation programs, of course. The k integral actually diverges for $d \geq 5$, but recall that we have found the same solution for arbitrary d from the *ansatz* (J.26). Fourier transformation of (J.28) for $d = 1$ directly yields the result (20.6), which also coincides with (J.29) for $d = 1$.

The time evolution operator in various representations

We have seen that the Green's function $\mathcal{G}_{d,V}(t)$ in the time domain is intimately connected to the time evolution operator

$$U(t) = \exp(-\mathrm{i}Ht/\hbar)$$

through equations (J.12, J.13). We can also define an energy representation for the time evolution operator in analogy to equations (J.9, J.10),

$$U(E) = \int_{-\infty}^{\infty} dt\,U(t)\exp(\mathrm{i}Et/\hbar) = 2\pi\hbar\delta(E-H). \tag{J.30}$$

Indeed, we have encountered this representation of the time evolution operator already in the frequency decomposition (5.14) of states,

$$|\psi(\omega)\rangle = \sqrt{2\pi}\,U(\hbar\omega)|\psi(t=0)\rangle.$$

The free d-dimensional evolution operator in the time domain is simply the product of d one-dimensional evolution operators,

$$U_0(t) = \exp\left(-\frac{\mathrm{i}t}{2m\hbar}\mathbf{p}^2\right),$$

$$\langle \boldsymbol{k}|U_0(t)|\boldsymbol{k}'\rangle = U_0(\boldsymbol{k},t)\delta(\boldsymbol{k}-\boldsymbol{k}'), \quad U_0(\boldsymbol{k},t) = \exp\left(-\mathrm{i}\frac{\hbar t}{2m}\boldsymbol{k}^2\right),$$

$$\langle \boldsymbol{x}|U_0(t)|\boldsymbol{k}\rangle = \frac{1}{\sqrt{2\pi}^d}\exp\left(\mathrm{i}\boldsymbol{k}\cdot\boldsymbol{x} - \mathrm{i}\frac{\hbar t}{2m}\boldsymbol{k}^2\right),$$

and

$$\langle \boldsymbol{x}|U_0(t)|\boldsymbol{x}'\rangle = \frac{1}{(2\pi)^d}\int d^3\boldsymbol{k}\,\exp\!\left[\mathrm{i}\boldsymbol{k}\cdot(\boldsymbol{x}-\boldsymbol{x}')\right]U_0(\boldsymbol{k},t) = U_0(\boldsymbol{x}-\boldsymbol{x}',t),$$

$$U_0(\boldsymbol{x},t) = \sqrt{\frac{m}{2\pi\mathrm{i}\hbar t}}^{\,d}\exp\left(\mathrm{i}\frac{m\boldsymbol{x}^2}{2\hbar t}\right).$$

Just like for the free Green's functions, the dependence on d becomes more interesting in the energy domain. The equation

$$U_0(\boldsymbol{k},E) = 2\pi\hbar\delta\left(E - \frac{\hbar^2\boldsymbol{k}^2}{2m}\right) = \pi\sqrt{\frac{2m}{E}}\delta\left(|\boldsymbol{k}| - \frac{\sqrt{2mE}}{\hbar}\right)$$

yields

$$\begin{aligned}
U_0(\boldsymbol{x},E) &= \frac{1}{(2\pi)^d}\int d^d\boldsymbol{k}\,\exp(\mathrm{i}\boldsymbol{k}\cdot\boldsymbol{x})U_0(\boldsymbol{k},E)\\
&= \frac{\Theta(E)S_{d-2}}{2^d(\pi\hbar)^{d-1}}\sqrt{2m}^{\,d}\sqrt{E}^{\,d-2}\int_0^\pi d\vartheta\,\exp\left(\mathrm{i}\sqrt{2mE}\frac{r}{\hbar}\cos\vartheta\right)\sin^{d-2}\vartheta\\
&= \Theta(E)\frac{m}{\hbar}\left(\frac{1}{\pi\hbar r}\sqrt{\frac{mE}{2}}\right)^{\frac{d-2}{2}} J_{\frac{d-2}{2}}\left(\sqrt{2mE}\frac{r}{\hbar}\right).
\end{aligned} \tag{J.31}$$

We have encountered several incarnations of the time evolution equation

$$|\psi(t)\rangle = U_0(t-t')|\psi(t')\rangle,$$

e.g. with $\langle\boldsymbol{k}|\psi\rangle \equiv \langle\boldsymbol{k}|\psi(0)\rangle$,

$$\begin{aligned}
\langle\boldsymbol{x}|\psi(t)\rangle &= \int d^d\boldsymbol{k}\,\langle\boldsymbol{x}|U_0(t)|\boldsymbol{k}\rangle\langle\boldsymbol{k}|\psi\rangle\\
&= \frac{1}{\sqrt{2\pi}^d}\int d^d\boldsymbol{k}\,\exp\left[\mathrm{i}\left(\boldsymbol{k}\cdot\boldsymbol{x} - \frac{\hbar t}{2m}\boldsymbol{k}^2\right)\right]\psi(\boldsymbol{k})\\
&= \int d^d\boldsymbol{x}'\,U_0(\boldsymbol{x}-\boldsymbol{x}',t-t')\langle\boldsymbol{x}'|\psi(t')\rangle.
\end{aligned}$$

Equation (J.31) implies with $|\psi\rangle \equiv |\psi(t=0)\rangle$ the $(\boldsymbol{x}, \omega)$ representation for free states in terms of their $(\boldsymbol{x}, t=0)$ representations,

$$\begin{aligned}
\langle \boldsymbol{x}|\psi(\omega)\rangle &= \frac{1}{\sqrt{2\pi}}\int dt\, \exp(\mathrm{i}\omega t)\,\langle \boldsymbol{x}|\psi(t)\rangle \\
&= \frac{1}{\sqrt{2\pi}}\int d^d x' \int dt\, \exp(\mathrm{i}\omega t)\,\langle \boldsymbol{x}|U_0(t)|\boldsymbol{x}'\rangle\langle \boldsymbol{x}'|\psi\rangle \\
&= \frac{1}{\sqrt{2\pi}}\int d^d x'\, \langle \boldsymbol{x}|U_0(\hbar\omega)|\boldsymbol{x}'\rangle\langle \boldsymbol{x}'|\psi\rangle \\
&= \Theta(\omega)\frac{m}{\sqrt{2\pi}\hbar}\left(\frac{1}{\pi}\sqrt{\frac{m\omega}{2\hbar}}\right)^{\frac{d-2}{2}} \\
&\quad \times \int d^d x'\, J_{\frac{d-2}{2}}\left(\sqrt{\frac{2m\omega}{\hbar}}|\boldsymbol{x}-\boldsymbol{x}'|\right)\frac{\langle \boldsymbol{x}'|\psi\rangle}{\sqrt{|\boldsymbol{x}-\boldsymbol{x}'|}^{d-2}}.
\end{aligned} \tag{J.32}$$

In turn, equation (5.12) implies for the initial state

$$|\psi\rangle = \frac{1}{\sqrt{2\pi}}\int d\omega\, |\psi(\omega)\rangle,$$

and therefore the kernel in (J.32) must yield a d-dimensional δ function,

$$\frac{m}{\sqrt{2\pi}^d\hbar}\int_0^\infty d\omega \left(\sqrt{\frac{2m\omega}{\hbar}}\frac{1}{|\boldsymbol{x}|}\right)^{\frac{d-2}{2}} J_{\frac{d-2}{2}}\left(\sqrt{\frac{2m\omega}{\hbar}}|\boldsymbol{x}|\right) = \delta(\boldsymbol{x}),$$

or in terms of magnitude of wave number,

$$\frac{1}{\sqrt{2\pi}^d}\int_0^\infty dk\, k\left(\frac{k}{r}\right)^{\frac{d-2}{2}} J_{\frac{d-2}{2}}(kr) = \delta(\boldsymbol{x}). \tag{J.33}$$

However, this is just the familiar relation

$$\frac{1}{(2\pi)^d}\int d^d k\, \exp(\mathrm{i}\boldsymbol{k}\cdot\boldsymbol{x}) = \delta(\boldsymbol{x})$$

after evaluation of the angular integrals in polar coordinates in $\boldsymbol{k}$ space.

In particular, for $d=1$ we find

$$\langle x|\psi(\omega)\rangle = \Theta(\omega)\sqrt{\frac{m}{\pi\hbar\omega}}\int dx'\, \cos\left(\sqrt{\frac{2m\omega}{\hbar}}(x-x')\right)\langle x'|\psi\rangle. \tag{J.34}$$

In two dimensions we find

$$\langle \boldsymbol{x}|\psi(\omega)\rangle = \Theta(\omega)\frac{m}{\sqrt{2\pi}\hbar}\int d^2x'\, J_0\!\left(\sqrt{\frac{2m\omega}{\hbar}}|\boldsymbol{x}-\boldsymbol{x}'|\right)\langle \boldsymbol{x}'|\psi\rangle, \tag{J.35}$$

and in three dimensions we find

$$\langle \boldsymbol{x}|\psi(\omega)\rangle = \Theta(\omega)\frac{m}{\sqrt{2\pi^3}\hbar}\int d^3x'\, \sin\!\left(\sqrt{\frac{2m\omega}{\hbar}}|\boldsymbol{x}-\boldsymbol{x}'|\right)\frac{\langle \boldsymbol{x}'|\psi\rangle}{|\boldsymbol{x}-\boldsymbol{x}'|}. \tag{J.36}$$

The free states in the $(\boldsymbol{x},\omega)$ representation are a superposition of stationary radial waves, where each point contributes with weight $\langle \boldsymbol{x}'|\psi\rangle$.

Relativistic Green's functions in d spatial dimensions

Although the theory in this subsection is relativistic, we will not use manifestly Lorentz covariant 4-vector notation like $x = (ct, \boldsymbol{x})$, $k = (\omega/c, \boldsymbol{k})$ for the space-time and momentum variables because we are also interested in mixed representations of the Green's functions like $G(\boldsymbol{x},\omega)$. For the manifestly covariant notation see the note at the end of this Appendix.

The relativistic free scalar Green's function in the time domain must satisfy

$$\left(\Delta - \frac{1}{c^2}\frac{\partial^2}{\partial t^2} - \frac{m^2c^2}{\hbar^2}\right)G_d(\boldsymbol{x},t;\boldsymbol{x}',t') = -\delta(\boldsymbol{x}-\boldsymbol{x}')\delta(t-t'). \tag{J.37}$$

This yields after transformation into $(\boldsymbol{k},\omega)$ space

$$G_d(\boldsymbol{k},\omega;\boldsymbol{k}',\omega') = G_d(\boldsymbol{k},\omega)\delta(\boldsymbol{k}-\boldsymbol{k}')\delta(\omega-\omega'), \tag{J.38}$$

where the factor $G_d(\boldsymbol{k},\omega)$ is

$$G_d(\boldsymbol{k},\omega) = \frac{1}{\boldsymbol{k}^2 - \frac{\omega^2}{c^2} + \frac{m^2c^2}{\hbar^2} - \mathrm{i}\epsilon}. \tag{J.39}$$

The shift $-\mathrm{i}\epsilon$, $\epsilon > 0$, into the complex plane is such that this reproduces the retarded non-relativistic Green's function (J.19) in the non-relativistic limit

$$\omega \Rightarrow \frac{mc^2 + E}{\hbar},$$

when terms of order $\mathcal{O}(E^2)$ are neglected. However, in the relativistic case this yields both retarded and advanced contributions in the time domain. This convention for the poles in the relativistic theory was introduced by Richard Feynman[2] and yields the Green's functions of Stückelberg and Feynman.

The solution in $x = (ct, \boldsymbol{x})$ space is then

$$G_d(\boldsymbol{x}, t; \boldsymbol{x}', t') = G_d(\boldsymbol{x} - \boldsymbol{x}', t - t'),$$

$$G_d(\boldsymbol{x}, t) = \frac{1}{2\pi}\int d\omega\, G_d(\boldsymbol{x}, \omega)\exp(-\mathrm{i}\omega t), \tag{J.40}$$

$$G_d(\boldsymbol{x}, \omega) = \frac{1}{(2\pi)^d}\int d^d\boldsymbol{k}\, G_d(\boldsymbol{k}, \omega)\exp(\mathrm{i}\boldsymbol{k}\cdot\boldsymbol{x}). \tag{J.41}$$

The integral is the same as in (J.29) with the substitution

$$\frac{2m}{\hbar^2}E \to \frac{\omega^2}{c^2} - \frac{m^2c^2}{\hbar^2},$$

i.e.

$$\begin{aligned} G_d(\boldsymbol{x}, \omega) = {} & \frac{\Theta(mc^2 - \hbar|\omega|)}{\sqrt{2\pi}^d}\left(\frac{\sqrt{m^2c^4 - \hbar^2\omega^2}}{\hbar c r}\right)^{\frac{d-2}{2}} \\ & \times K_{\frac{d-2}{2}}\left(\sqrt{m^2c^4 - \hbar^2\omega^2}\frac{r}{\hbar c}\right) \\ & + \mathrm{i}\frac{\pi}{2}\frac{\Theta(\hbar|\omega| - mc^2)}{\sqrt{2\pi}^d}\left(\frac{\sqrt{\hbar^2\omega^2 - m^2c^4}}{\hbar c r}\right)^{\frac{d-2}{2}} \\ & \times H^{(1)}_{\frac{d-2}{2}}\left(\sqrt{\hbar^2\omega^2 - m^2c^4}\frac{r}{\hbar c}\right). \end{aligned} \tag{J.42}$$

The $\omega = 0$ Green's functions

$$\left(\Delta - \frac{m^2c^2}{\hbar^2}\right)G_d(\boldsymbol{x}) = -\delta(\boldsymbol{x}), \quad G_d(\boldsymbol{x}) = \frac{1}{\sqrt{2\pi}^d}\left(\frac{mc}{\hbar r}\right)^{\frac{d-2}{2}} K_{\frac{d-2}{2}}\left(\frac{mc}{\hbar}r\right),$$

yield again the results (J.24) in the limit $m \to 0$, albeit with diverging integration constants in low dimensions,

$$a_{d=1} = \frac{\hbar}{mc}, \quad a_{d=2} = \frac{2\hbar}{mc}\exp(-\gamma).$$

In terms of poles in the complex k plane, the complex shift in (J.39) implies

[2] R.P. Feynman, Phys. Rev. 76, 749 (1949).

$$G_d(\boldsymbol{k},\omega) = \frac{c^2\Theta(\hbar\omega - mc^2)}{\left(ck - \sqrt{\omega^2 - (mc^2/\hbar)^2} - \mathrm{i}\epsilon\right)\left(ck + \sqrt{\omega^2 - (mc^2/\hbar)^2} + \mathrm{i}\epsilon\right)}$$
$$+ \frac{c^2\Theta(mc^2 - \hbar\omega)}{\left(ck - \mathrm{i}\sqrt{(mc^2/\hbar)^2 - \omega^2}\right)\left(ck - \mathrm{i}\sqrt{(mc^2/\hbar)^2 - \omega^2}\right)}. \qquad (J.43)$$

However, in terms of poles in the complex ω plane, equation (J.39) implies

$$G_d(\boldsymbol{k},\omega) = -\frac{c^2}{\left(\omega - c\sqrt{k^2 + (mc/\hbar)^2} + \mathrm{i}\epsilon\right)\left(\omega + c\sqrt{k^2 + (mc/\hbar)^2} - \mathrm{i}\epsilon\right)}.$$

Fourier transformation to the time domain therefore yields a representation of the relativistic free Green's function which explicitly shows the combination of retarded positive frequency and advanced negative frequency components,

$$G_d(\boldsymbol{k},t) = \frac{1}{2\pi}\int d\omega\, G_d(\boldsymbol{k},\omega)\exp(-\mathrm{i}\omega t)$$
$$= \mathrm{i}c\Theta(t)\frac{\exp\left(-\mathrm{i}\sqrt{\boldsymbol{k}^2 + (mc/\hbar)^2}\,ct\right)}{2\sqrt{\boldsymbol{k}^2 + (mc/\hbar)^2}}$$
$$+ \mathrm{i}c\Theta(-t)\frac{\exp\left(\mathrm{i}\sqrt{\boldsymbol{k}^2 + (mc/\hbar)^2}\,ct\right)}{2\sqrt{\boldsymbol{k}^2 + (mc/\hbar)^2}}. \qquad (J.44)$$

On the other hand, shifting both poles into the lower complex ω plane,

$$G_d^{(r)}(\boldsymbol{k},\omega) = -\frac{c^2}{\left(\omega - c\sqrt{k^2 + (mc/\hbar)^2} + \mathrm{i}\epsilon\right)\left(\omega + c\sqrt{k^2 + (mc/\hbar)^2} + \mathrm{i}\epsilon\right)},$$

yields the retarded relativistic Green's function

$$G_d^{(r)}(\boldsymbol{k},t) = \frac{1}{2\pi}\int d\omega\, G_d^{(r)}(\boldsymbol{k},\omega)\exp(-\mathrm{i}\omega t)$$
$$= c\Theta(t)\frac{\sin\left(\sqrt{\boldsymbol{k}^2 + (mc/\hbar)^2}\,ct\right)}{\sqrt{\boldsymbol{k}^2 + (mc/\hbar)^2}} = c^2\Theta(t)\mathcal{K}_d(\boldsymbol{k},t), \qquad (J.45)$$

cf. equation (21.9). If $\mathcal{K}_d(\boldsymbol{x}, t)$ exists, then one can easily verify that the properties

$$\left(\Delta - \frac{1}{c^2}\frac{\partial^2}{\partial t^2} - \frac{m^2c^2}{\hbar^2}\right)\mathcal{K}_d(\boldsymbol{x}, t) = 0,$$

$$\mathcal{K}_d(\boldsymbol{x}, 0) = 0, \quad \frac{\partial}{\partial t}\mathcal{K}_d(\boldsymbol{x}, t)\bigg|_{t=0} = \delta(\boldsymbol{x})$$

imply that $G_d^{(r)}(\boldsymbol{x}, t) = c^2\Theta(t)\mathcal{K}_d(\boldsymbol{x}, t)$ is a retarded Green's function.

On the other hand, shifting both poles into the upper complex ω plane,

$$G_d^{(a)}(\boldsymbol{k}, \omega) = -\frac{c^2}{\left(\omega - c\sqrt{k^2 + (mc/\hbar)^2} - \mathrm{i}\epsilon\right)\left(\omega + c\sqrt{k^2 + (mc/\hbar)^2} - \mathrm{i}\epsilon\right)},$$

yields the advanced relativistic free Green's function

$$\begin{aligned} G_d^{(a)}(\boldsymbol{k}, t) &= \frac{1}{2\pi}\int d\omega\, G_d^{(a)}(\boldsymbol{k}, \omega)\exp(-\mathrm{i}\omega t) \\ &= -c\Theta(-t)\frac{\sin\left(\sqrt{\boldsymbol{k}^2 + (mc/\hbar)^2}\, ct\right)}{\sqrt{\boldsymbol{k}^2 + (mc/\hbar)^2}} \\ &= -c^2\Theta(-t)\mathcal{K}_d(\boldsymbol{k}, t) = G_d^{(r)}(\boldsymbol{k}, -t). \end{aligned} \quad \text{(J.46)}$$

Retarded relativistic Green's functions in $(\boldsymbol{x}, t)$ representation

Evaluation of the Green's functions $G_d^{(r)}(\boldsymbol{x}, t)$ and $G_d(\boldsymbol{x}, t)$ for the massive Klein-Gordon equation is very cumbersome if one uses standard Fourier transformation between time and frequency. It is much more convenient to use Fourier transformation with imaginary frequency, which is known as Laplace transformation. We will demonstrate this for the retarded Green's function. We try a Laplace transform of $G_d^{(r)}(\boldsymbol{x}, t)$ in the form[3]

[3]Assuming only $\Re w \geq 0$ assumes that the retarded Green's functions are integrable along the time axis. This makes physical sense since the impact of a perturbation which occurred at time $t' = 0$ at the point $\boldsymbol{x}' = 0$ that is felt at the point $\boldsymbol{x}$ should decrease with time. The assumption can also be justified *a posteriori* from the explicit results (J.58–J.60), which show that the Green's functions for $d \leq 3$ oscillate and decay for $t \to \infty$. For Laplace transforms of less well behaved functions $G(\boldsymbol{x}, t)$ one can require $\Re w > \nu$ if $\exp(-\nu t)G(\boldsymbol{x}, t)$ is bounded for $t \to \infty$. The vertical integration contour for the inverse transformation (J.48) must then be to the right of $\nu - \mathrm{i}\infty \to \nu + \mathrm{i}\infty$.

$$g_d(\boldsymbol{x}, w) = \int_0^\infty dt\, \exp(-wt) G_d^{(r)}(\boldsymbol{x}, t), \quad \Re w \geq 0. \tag{J.47}$$

The completeness relation for Fourier monomials,

$$\delta(t) = \frac{1}{2\pi}\int_{-\infty}^{\infty} d\omega\, \exp(-\mathrm{i}\omega t) = \frac{1}{2\pi\mathrm{i}}\int_{-\mathrm{i}\infty}^{\mathrm{i}\infty} dw\, \exp(wt)$$

then yields the inversion of (J.47),

$$G_d^{(r)}(\boldsymbol{x}, t) = \frac{1}{2\pi\mathrm{i}}\int_{-\mathrm{i}\infty}^{\mathrm{i}\infty} dw\, \exp(wt) g_d(\boldsymbol{x}, w). \tag{J.48}$$

The condition (J.37) on the d-dimensional scalar Green's functions then implies

$$\left(\Delta - \frac{w^2}{c^2} - \frac{m^2c^2}{\hbar^2}\right) g_d(\boldsymbol{x}, w) = -\delta(\boldsymbol{x}) \tag{J.49}$$

with solution

$$g_d(\boldsymbol{x}, w) = \frac{1}{(2\pi)^d}\int d^d\boldsymbol{k}\, \frac{\exp(\mathrm{i}\boldsymbol{k}\cdot\boldsymbol{x})}{\boldsymbol{k}^2 + (w/c)^2 + (mc/\hbar)^2}.$$

In one dimension this yields

$$g_1(x, w) = \frac{c\exp\!\left(-\sqrt{w^2 + (mc^2/\hbar)^2}\,|x|/c\right)}{2\sqrt{w^2 + (mc^2/\hbar)^2}}. \tag{J.50}$$

In higher dimensions, we need to calculate

$$\begin{aligned} g_d(\boldsymbol{x}, w) &= \frac{S_{d-2}}{(2\pi)^d}\int_0^\infty dk \int_0^\pi d\vartheta\, k^{d-1}\sin^{d-2}\vartheta\, \frac{\exp(\mathrm{i}kr\cos\vartheta)}{k^2 + (w/c)^2 + (mc/\hbar)^2} \\ &= \frac{1}{\sqrt{2\pi}^d}\int_0^\infty dk\, \frac{k^{d-1}}{k^2 + (w/c)^2 + (mc/\hbar)^2}\, \frac{1}{\sqrt{kr}^{d-2}} J_{\frac{d-2}{2}}(kr). \end{aligned} \tag{J.51}$$

We can formally reduce (J.51) for $d \geq 3$ to the corresponding integrals in lower dimensions by using the relation

$$\left(-\frac{1}{x}\frac{d}{dx}\right)^n \frac{J_\nu(x)}{x^\nu} = \frac{J_{\nu+n}(x)}{x^{\nu+n}},$$

However, this would not save the day for the non-existent functions $G_{d\geq 4}^{(r)}(\boldsymbol{x}, t)$, although we can find functions $g_d(\boldsymbol{x}, w)$ (J.54, J.55) for every number d of dimensions.

see number 9.1.30, p. 361 in [1]. This yields for $d = 2n + 1$

$$\begin{aligned}\frac{1}{\sqrt{kr}^{d-2}}J_{\frac{d-2}{2}}(kr) &= k^{-2n}\left(-\frac{1}{r}\frac{\partial}{\partial r}\right)^n \sqrt{kr}J_{-\frac{1}{2}}(kr)\\ &= \sqrt{\frac{2}{\pi}}k^{-2n}\left(-\frac{1}{r}\frac{\partial}{\partial r}\right)^n \cos(kr),\end{aligned} \tag{J.52}$$

and for $d = 2n + 2$,

$$\frac{1}{\sqrt{kr}^{d-2}}J_{\frac{d-2}{2}}(kr) = k^{-2n}\left(-\frac{1}{r}\frac{\partial}{\partial r}\right)^n J_0(kr). \tag{J.53}$$

The resulting relations for the Green's functions in the $(\boldsymbol{x}, w)$ representations are then

$$\begin{aligned}g_{2n+1}(\boldsymbol{x}, w) &= \left(-\frac{1}{2\pi r}\frac{\partial}{\partial r}\right)^n \frac{1}{\pi}\int_0^\infty dk\,\frac{\cos(kr)}{k^2 + (w/c)^2 + (mc/\hbar)^2}\\ &= \left(-\frac{1}{2\pi r}\frac{\partial}{\partial r}\right)^n \frac{c\exp\!\left(-\sqrt{w^2 + (mc^2/\hbar)^2}\,r/c\right)}{2\sqrt{w^2 + (mc^2/\hbar)^2}},\end{aligned} \tag{J.54}$$

and

$$\begin{aligned}g_{2n+2}(\boldsymbol{x}, w) &= \left(-\frac{1}{2\pi r}\frac{\partial}{\partial r}\right)^n \frac{1}{\pi}\int_0^\infty dk\,\frac{kJ_0(kr)}{k^2 + (w/c)^2 + (mc/\hbar)^2}\\ &= \left(-\frac{1}{2\pi r}\frac{\partial}{\partial r}\right)^n \frac{1}{2\pi}K_0\!\left(\sqrt{w^2 + (mc^2/\hbar)^2}\,\frac{r}{c}\right).\end{aligned} \tag{J.55}$$

Inverse Laplace transformation yields the retarded Green's functions $G_d^{(r)}(\boldsymbol{x}, t)$,

$$G_{2n+1}^{(r)}(\boldsymbol{x}, t) = \left(-\frac{1}{2\pi r}\frac{\partial}{\partial r}\right)^n \frac{c}{2}\Theta(ct - r)J_0\!\left(mc\sqrt{c^2t^2 - r^2}/\hbar\right), \tag{J.56}$$

$$G_{2n+2}^{(r)}(\boldsymbol{x}, t) = \left(-\frac{1}{2\pi r}\frac{\partial}{\partial r}\right)^n \frac{c}{2\pi}\frac{\Theta(ct - r)}{\sqrt{c^2t^2 - r^2}}\cos\!\left(mc\sqrt{c^2t^2 - r^2}/\hbar\right). \tag{J.57}$$

We note that the retarded Green's functions for fixed r decrease like $t^{-(n+1)} = t^{-\lfloor (d+1)/2\rfloor}$ for $t \to \infty$, except the δ function singularities become unacceptable for $d \geq 4$.

The retarded relativistic Green's functions in one, two and three dimensions are therefore

$$G_1^{(r)}(x, t) = \frac{c}{2}\Theta(ct - |x|)J_0\!\left(mc\sqrt{c^2t^2 - x^2}/\hbar\right), \tag{J.58}$$

$$G_2^{(r)}(\boldsymbol{x},t) = \frac{c}{2\pi}\frac{\Theta(ct-r)}{\sqrt{c^2t^2-r^2}}\cos\!\left(mc\sqrt{c^2t^2-r^2}/\hbar\right), \tag{J.59}$$

and

$$G_3^{(r)}(\boldsymbol{x},t) = \frac{c}{4\pi r}\delta(r-ct) - \frac{mc^2}{4\pi\hbar}\frac{\Theta(ct-r)}{\sqrt{c^2t^2-r^2}}J_1\!\left(mc\sqrt{c^2t^2-r^2}/\hbar\right). \tag{J.60}$$

The $(\boldsymbol{x},t)$ representations of the corresponding advanced Green's functions then follow from (J.46) as

$$G_d^{(a)}(\boldsymbol{x},t) = G_d^{(r)}(\boldsymbol{x},-t).$$

The propagator function $\mathcal{K}_d(\boldsymbol{x},t)$ for the free Klein-Gordon fields follows from (J.45, J.46) as

$$c^2\mathcal{K}_d(\boldsymbol{x},t) = G_d^{(r)}(\boldsymbol{x},t) - G_d^{(a)}(\boldsymbol{x},t). \tag{J.61}$$

The functions $G_{d\geq 4}^{(r)}(\boldsymbol{x},t)$ and $\mathcal{K}_{d\geq 4}(\boldsymbol{x},t)$ do not exist, but the corresponding functions $G_d^{(r)}(\boldsymbol{k},t) = c^2\Theta(t)\mathcal{K}_d(\boldsymbol{k},t)$ (J.45) and $G_d^{(r)}(\boldsymbol{k},\omega)$ exist in any number of dimensions.

Green's functions for Dirac operators in d dimensions

We now restore summation convention. The Green's functions for the free Dirac operator must satisfy

$$\left(\mathrm{i}\gamma^\mu\partial_\mu - \frac{mc}{\hbar}\right)S_d(\boldsymbol{x},t) = -\,\delta(\boldsymbol{x})\delta(t). \tag{J.62}$$

Since the Dirac operator is a factor of the Klein-Gordon operator, the solutions of the equations (J.62) and (J.37) are related by

$$S_d(\boldsymbol{x},t) = \left(\mathrm{i}\gamma^\mu\partial_\mu + \frac{mc}{\hbar}\right)G_d(\boldsymbol{x},t) \tag{J.63}$$

and

$$\begin{aligned} G_d(\boldsymbol{x},t) &= \int d^d\boldsymbol{x}'\int dt'\, S_d(\boldsymbol{x}'-\boldsymbol{x},t'-t)\cdot S_d(\boldsymbol{x}',t') \\ &= \int d^d\boldsymbol{x}'\int dt'\, S_d(\boldsymbol{x}',t')\cdot S_d(\boldsymbol{x}'+\boldsymbol{x},t'+t). \end{aligned} \tag{J.64}$$

The free Dirac Green's function in wave number representation is (here $k^2 \equiv k^\mu k_\mu$)

$$S_d(k) = \hbar \frac{mc - \hbar\gamma^\mu k_\mu}{\hbar^2 k^2 + m^2c^2 - \mathrm{i}\epsilon}, \tag{J.65}$$

where the pole shifts again correspond to the Feynman propagator with retarded and advanced components.

Green's functions in covariant notation

The relativistic free scalar Green's function satisfies

$$\frac{p^2 + m^2c^2}{\hbar^2} G_d = 1, \tag{J.66}$$

i.e. in the $k = (\omega/c, \boldsymbol{k})$ domain

$$\langle k|G_d|k'\rangle = G_d(k)\delta(k - k'), \tag{J.67}$$

where the factor $G_d(k)$ is

$$G_d(k) = \frac{1}{k^2 + (m^2c^2/\hbar^2) - \mathrm{i}\epsilon}. \tag{J.68}$$

This yields after transformation into $x = (ct, \boldsymbol{x})$ space ($D = d + 1$),

$$\begin{aligned}\langle x|G_d|x'\rangle &= \frac{1}{(2\pi)^D}\int d^Dk \int d^Dk' \,\langle k|G_d|k'\rangle \exp[\mathrm{i}(k\cdot x - k'\cdot x')],\\ &= \frac{1}{(2\pi)^D}\int d^Dk\, G_d(k) \exp[\mathrm{i}k\cdot(x - x')] = G_d(x - x'),\end{aligned} \tag{J.69}$$

which satisfies

$$\left(\partial^2 - \frac{m^2c^2}{\hbar^2}\right)\langle x|G_d|x'\rangle = \left(\partial'^2 - \frac{m^2c^2}{\hbar^2}\right)\langle x|G_d|x'\rangle = -\delta(x - x'). \tag{J.70}$$

The relation to (J.37–J.60) is

$$\langle x|G_d|x'\rangle = \frac{1}{c}G_d(\boldsymbol{x}, t; \boldsymbol{x}', t'), \quad \langle k|G_d|k'\rangle = cG_d(\boldsymbol{k}, \omega; \boldsymbol{k}', \omega'),$$

$$G_d(k) = G_d(\boldsymbol{k}, \omega).$$

Translation invariance (J.67, J.69) implies that the Green's function in mixed representation is proportional to plane waves, $\langle x|G_d|k\rangle = G_d(k)\langle x|k\rangle$.

The fermionic Green's function satisfies

$$\frac{\gamma\cdot p + mc}{\hbar}S_d = 1, \quad S_d = \frac{mc - \gamma\cdot p}{\hbar}G_d, \tag{J.71}$$

or in various representations,

$$\left(\mathrm{i}\gamma\cdot\partial - \frac{mc}{\hbar}\right)\langle x|S_d|x'\rangle = -\delta(x-x'), \tag{J.72}$$

$$\langle k|S_d|k'\rangle = S_d(k)\delta(k-k'), \quad \langle x|S_d|k\rangle = S_d(k)\langle x|k\rangle, \tag{J.73}$$

$$S_d(k) = \left(\frac{mc}{\hbar} - \gamma\cdot k\right)G_d(k) = \frac{(mc/\hbar) - \gamma\cdot k}{k^2 + (m^2c^2/\hbar^2) - \mathrm{i}\epsilon}, \tag{J.74}$$

$$\begin{aligned}\langle x|S_d|x'\rangle &= S_d(x-x') = \frac{1}{c}S_d(\mathbf{x}-\mathbf{x}', t-t')\\ &= \frac{1}{(2\pi)^D}\int d^Dk\, S_d(k)\exp[\mathrm{i}k\cdot(x-x')]\\ &= \left(\mathrm{i}\gamma\cdot\partial + \frac{mc}{\hbar}\right)G_d(x-x').\end{aligned} \tag{J.75}$$

The pole shifts in (J.68, J.74) correspond to the Feynman conventions. For the retarded Green's functions $G_d^{(r)}$ and $S_d^{(r)}$ both poles have to be shifted into the lower k^0 plane. Note that as a consequence of (J.75) the fermionic Green's function also satisfies

$$\mathrm{i}\partial'_\mu S_d(x-x')\gamma^\mu + \frac{mc}{\hbar}S_d(x-x') = \delta(x-x'). \tag{J.76}$$

Green's functions as reproducing kernels

Suppose that $\mathcal{V}$ is a $(d+1)$-dimensional spacetime volume with boundary $\partial\mathcal{V}$. The equation (J.70) and the Klein-Gordon equation imply for a free field $\phi(x)$ and for x in $\mathcal{V}$ the relation

$$\begin{aligned}\phi(x) &= \int_{\mathcal{V}} d^Dx'\,\phi(x')\left(\frac{m^2c^2}{\hbar^2} - \partial'^2\right)G_d(x-x')\\ &= \int_{\mathcal{V}} d^Dx'\,\partial'^\mu\left[G_d(x-x')\,\overleftrightarrow{\partial'}_\mu\,\phi(x')\right].\end{aligned}$$

The Gauss theorem in D spacetime dimensions then yields a representation for all values of $\phi(x)$ inside of $\mathcal{V}$ in terms of the values of the Klein-Gordon field on the boundary $\partial\mathcal{V}$,

$$\phi(x) = \int_{\partial\mathcal{V}} d^d x'\, n^\mu \left[G_d(x-x')\, \overleftrightarrow{\partial'}_\mu\, \phi(x') \right], \tag{J.77}$$

where n^μ is an outward bound normal vector with $n_0 = 1$ on spacelike boundaries $t' =$ constant, $t' > t$, and $n_0 = -1$ on $t' =$ constant, $t' < t$. If $G_d(x-x')$ is in particular the retarded Green's function,

$$G_d^{(r)}(x-x') = c\Theta(t-t')\mathcal{K}_d(\boldsymbol{x}-\boldsymbol{x}', t-t'), \tag{J.78}$$

or the advanced Green's function,

$$G_d^{(a)}(x-x') = c\Theta(t'-t)\mathcal{K}_d(\boldsymbol{x}-\boldsymbol{x}', t'-t), \tag{J.79}$$

and $\partial\mathcal{V}$ contains only the spacelike surface $t' < t$, or only the spacelike surface $t' > t$, then (J.77) is the solution (21.8) of the initial value problem ($t' < t$) or future value problem[4] ($t' > t$) for the Klein-Gordon field.

For free Dirac fields the Dirac equation and (J.76) implies for $x \in \mathcal{V}$ the equation

$$\psi(x) = \mathrm{i} \int_{\partial\mathcal{V}} d^d x'\, n_\mu S_d(x-x')\gamma^\mu \psi(x'). \tag{J.80}$$

This yields again the initial/final value solution (21.64) if $\partial\mathcal{V}$ contains only the spacelike surface $t' < t$ or only the spacelike surface $t' > t$, since the retarded and advanced Green's functions are related to the time evolution kernel (21.65) through

$$S_d^{(r)}(x-x') = \mathrm{i}\Theta(t-t')\mathcal{W}_d(\boldsymbol{x}-\boldsymbol{x}', t-t')\gamma^0,$$
$$S_d^{(a)}(x-x') = -\mathrm{i}\Theta(t'-t)\mathcal{W}_d(\boldsymbol{x}-\boldsymbol{x}', t-t')\gamma^0,$$

see also equations (J.75), (J.78), (J.79) and (21.67).

Liénard-Wiechert potentials in low dimensions

The massless retarded Green's functions solve the basic electromagnetic wave equation for the electromagnetic potentials in Lorentz gauge,

[4]The future value problem or backwards evolution problem asks the question: Which field configuration $\phi(x)$ at time t yields the prescribed field configuration $\phi(x')$ at time $t' > t$ through time evolution with the equations of motion?

$$\left(\partial_\mu\partial^\mu - \frac{m^2c^2}{\hbar^2}\right)A^\nu(x) = -\mu_0 j^\nu(x), \quad \partial_\mu A^\mu(x) = 0,$$

$$A^\mu(x) = \mu_0 \int d^{d+1}x'\, G_d^{(r)}(x - x')j^\mu(x').$$

In three dimensions this yields the familiar Liénard-Wiechert potentials from the contributions of the currents on the backward light cone of the spacetime point x,

$$A^\mu_{d=3}(\boldsymbol{x}, t) = \frac{\mu_0}{4\pi}\int d^3\boldsymbol{x}'\,\frac{1}{|\boldsymbol{x}-\boldsymbol{x}'|}j^\mu\left(\boldsymbol{x}', t - \frac{1}{c}|\boldsymbol{x}-\boldsymbol{x}'|\right). \tag{J.81}$$

However, in one and two dimensions, the Liénard-Wiechert potentials sample charges and currents from the complete region inside the backward light cone,

$$A^\mu_{d=1}(x, t) = \frac{\mu_0 c}{2}\int_{-\infty}^{\infty} dx' \int_{-\infty}^{t-(|x-x'|/c)} dt'\, j^\mu(x', t'), \tag{J.82}$$

$$A^\mu_{d=2}(\boldsymbol{x}, t) = \frac{\mu_0 c}{2\pi}\int d^2\boldsymbol{x}' \int_{-\infty}^{t-(|\boldsymbol{x}-\boldsymbol{x}'|/c)} dt'\,\frac{j^\mu(\boldsymbol{x}', t')}{\sqrt{c^2(t-t')^2 - (\boldsymbol{x}-\boldsymbol{x}')^2}}. \tag{J.83}$$

Stated differently, a δ function type charge-current fluctuation in the spacetime point x' generates an outwards traveling spherical electromagnetic perturbation on the forward light cone starting in x' if we are in three spatial dimensions. However, in one dimension the same kind of perturbation fills the whole forward light cone uniformly with electromagnetic fields, and in two dimensions the forward light cone is filled with a weight factor $[c^2(t - t')^2 - (\boldsymbol{x} - \boldsymbol{x}')^2]^{-1/2}$. How can that be? The electrostatic potentials (J.24) for $d = 1$ and $d = 2$ hold the answer to this. Those potentials indicate linear or logarithmic confinement of electric charges in low dimensions. Therefore a positive charge fluctuation in a point x' must be compensated by a corresponding negative charge fluctuation nearby. Both fluctuations fill their overlapping forward light cones with opposite electromagnetic fields, but those fields will exactly compensate in the overlapping parts in one dimension, and largely compensate in two dimensions. The net effect of these opposite charge fluctuations at a distance a is then electromagnetic fields along a forward light cone of thickness a, i.e. electromagnetic confinement in low dimensions effectively ensures again that electromagnetic fields propagate along light cones. This is illustrated in Figure J.1.

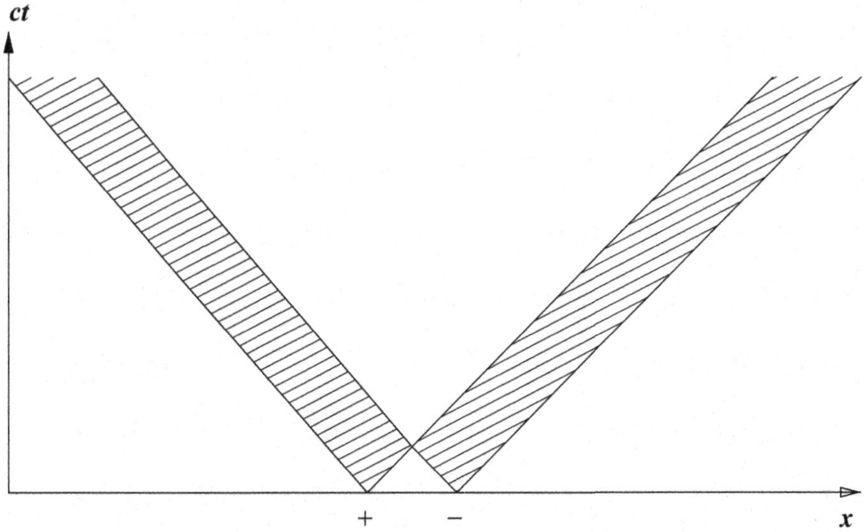

Fig. J.1 The contributions of nearby opposite charge fluctuations at time $t = 0$ in one spatial dimension generate net electromagnetic fields in the hatched "thick" light cone region

Bibliography

1. M. Abramowiz, I.A. Stegun (eds.), *Handbook of Mathematical Functions*, 10th printing (Wiley, New York, 1972)
2. A.O. Barut, R. Raczka, *Theory of Group Representations and Applications* (World Scientific, Singapore, 1986)
3. H.A. Bethe, E.E. Salpeter, *Quantum Mechanics of One- and Two-Electron Atoms* (Springer, Berlin, 1957)
4. B. Bhushan (ed.), *Springer Handbook of Nanotechnology*, 2nd edn. (Springer, New York, 2007)
5. J. Callaway, *Quantum Theory of the Solid State* (Academic Press, Boston, 1991)
6. J.F. Cornwell, *Group Theory in Physics*, vols. I & II (Academic Press, London, 1984)
7. R. Courant, D. Hilbert, *Methods of Mathematical Physics*, vols. 1 & 2 (Interscience Publ., New York, 1953, 1962)
8. G.W.F. Drake (ed.), *Springer Handbook of Atomic, Molecular, and Optical Physics* (Springer, New York, 2006)
9. A.R. Edmonds, *Angular Momentum in Quantum Mechanics*, 2nd edn. (Princeton University Press, Princeton, 1960)
10. R.P. Feynman, A.R. Hibbs, *Quantum Mechanics and Path Integrals* (McGraw-Hill, New York, 1965)
11. P. Fulde, *Electron Correlations in Molecules and Solids*, 2nd edn. (Springer, Berlin, 1993)
12. M. Getzlaff, *Fundamentals of Magnetism* (Springer, Berlin, 2008)
13. C. Grosche, F. Steiner, *Handbook of Feynman Path Integrals* (Springer, Berlin, 1998)
14. W. Heitler, *The Quantum Theory of Radiation* (Oxford University Press, Oxford, 1957)
15. H. Hellmann, *Einführung in die Quantenchemie* (Deuticke, Leipzig, 1937)
16. N.J. Higham, *Functions of Matrices – Theory and Computation* (Society of Industrial and Applied Mathematics, Philadelphia, 2008)
17. H. Ibach, H. Lüth, *Solid State Physics – An Introduction to Principles of Materials Science*, 3rd edn. (Springer, Berlin, 2003)
18. C. Itzykson, J.-B. Zuber, *Quantum Field Theory* (McGraw-Hill, New York, 1980)
19. J.D. Jackson, *Classical Electrodynamics*, 3rd edn. (Wiley, New York, 1999)
20. S. Kasap, P. Capper (eds.), *Springer Handbook of Electronic and Photonic Materials* (Springer, New York, 2006)
21. T. Kato, *Perturbation Theory for Linear Operators* (Springer, Berlin, 1966)
22. C. Kittel, *Quantum Theory of Solids*, 2nd edn. (Wiley, New York, 1987)
23. H. Kleinert, *Path Integrals in Quantum Mechanics, Statistics, Polymer Physics, and Financial Markets*, 5th edn. (World Scientific, Singapore, 2009)

R. Dick, *Advanced Quantum Mechanics*, Graduate Texts in Physics,
DOI 10.1007/978-3-319-25675-7

24. L.D. Landau, E.M. Lifshitz, *Quantum Mechanics: Non-relativistic Theory* (Pergamon, Oxford, 1977)
25. O. Madelung, *Introduction to Solid-State Theory* (Springer, Berlin, 1978)
26. L. Marchildon, *Quantum Mechanics: From Basic Principles to Numerical Methods and Applications* (Springer, New York, 2002)
27. E. Merzbacher, *Quantum Mechanics*, 3rd edn. (Wiley, New York, 1998)
28. A. Messiah, *Quantum Mechanics*, vols. 1 & 2 (North-Holland, Amsterdam, 1961, 1962)
29. P.M. Morse, H. Feshbach, *Methods of Theoretical Physics*, vol. 2 (McGraw-Hill, New York, 1953)
30. J. Orear, A.H. Rosenfeld, R.A. Schluter, *Nuclear Physics: A Course Given by Enrico Fermi at the University of Chicago* (University of Chicago Press, Chicago, 1950)
31. M. Peskin, D.V. Schroeder, *An Introduction to Quantum Field Theory* (Addison-Wesley, Reading, 1995)
32. A.P. Prudnikov, Yu.A. Brychkov, O.I. Marichev, *Integrals and Series*, vol. 1 (Gordon and Breach Science Publ., New York, 1986)
33. A.P. Prudnikov, Yu.A. Brychkov, O.I. Marichev, *Integrals and Series*, vol. 2 (Gordon and Breach Science Publ., New York, 1986)
34. M.E. Rose, *Elementary Theory of Angular Momentum* (Wiley, New York, 1957)
35. F. Schwabl, *Quantum Mechanics*, 4th edn. (Springer, Berlin, 2007)
36. F. Schwabl, *Advanced Quantum Mechanics*, 4th edn. (Springer, Berlin, 2008)
37. R.U. Sexl, H.K. Urbantke, *Relativity, Groups, Particles* (Springer, New York, 2001)
38. J.C. Slater, *Quantum Theory of Molecules and Solids*, vol. 1 (McGraw-Hill, New York, 1963)
39. A. Sommerfeld, *Atombau und Spektrallinien*, 3rd edn. (Vieweg, Braunschweig, 1922). English translation *Atomic Structure and Spectral Lines* (Methuen, London, 1923)
40. J. Stöhr, H.C. Siegmann, *Magnetism – From Fundamentals to Nanoscale Dynamics* (Springer, New York, 2006)
41. S. Weinberg, *The Quantum Theory of Fields*, vols. 1 & 2 (Cambridge University Press, Cambridge, 1995, 1996)
42. E.P. Wigner, *Gruppentheorie und ihre Anwendungen auf die Quantenmechanik der Atomspektren* (Vieweg, Braunschweig, 1931). English translation *Group Theory and its Application to the Quantum Mechanics of Atomic Spectra* (Academic Press, New York, 1959)

Index

R. Dick, *Advanced Quantum Mechanics*, Graduate Texts in Physics,
DOI 10.1007/978-3-319-25675-7

Printed in the United States
By Bookmasters

Printed in the United States
By Bookmasters